WileyPLUS

ALL THE HELP, RESOURCES, AND PERSONAL SUPPORT YOU AND YOUR STUDENTS NEED!

www.wileyplus.com/resources

1st DAY OF CLASS ...AND BEYOND!

2-Minute Tutorials and all of the resources you and your students need to get started

WileyPLUS
Student Partner Program

Student support from an experienced student user

Wiley Faculty Network

Collaborate with your colleagues, find a mentor, attend virtual and live events, and view resources
www.WhereFacultyConnect.com

WileyPLUS

Quick Start

Pre-loaded, ready-to-use assignments and presentations created by subject matter experts

Technical Support 24/7
FAQs, online chat, and phone support
www.wileyplus.com/support

© Courtney Keating/ iStockphoto

Your *WileyPLUS* Account Manager, providing personal training and support

FOURTH EDITION

The Heart of Mathematics

An Invitation to Effective Thinking

EDWARD B. BURGER
Williams College

MICHAEL STARBIRD
The University Of Texas At Austin

WILEY

Vice President and Publisher	Laurie Rosatone
Project Editor	Jennifer Brady
Developmental Editor	Anne Scanlan-Rohrer
Marketing Manager	Kimberly Kanakes
Marketing Assistant	Patrick Flatley
Media Assistant	Courtney Welsh
Senior Production Editor	Kenneth Santor
Designer	Maureen Eide
Photo Editor	Sheena Goldstein
Photo Department Manager	Hilary Newman
Cover Photo	SSPL via Getty Images, Inc.

This book was typeset in 10/12 Times Ten Roman at MPS Limited and printed and bound by Quad Graphics/Versailles. The cover was printed by Quad Graphics.

This book is printed on acid-free paper.

Founded in 1807, John Wiley & Sons, Inc. has been a valued source of knowledge and understanding for more than 200 years, helping people around the world meet their needs and fulfill their aspirations. Our company is built on a foundation of principles that include responsibility to the communities we serve and where we live and work. In 2008, we launched a Corporate Citizenship Initiative, a global effort to address the environmental, social, economic, and ethical challenges we face in our business. Among the issues we are addressing are carbon impact, paper specifications and procurement, ethical conduct within our business and among our vendors, and community and charitable support. For more information, please visit our website: www.wiley.com/go/citizenship.

ISBN 13   978-1-118-15659-9
ISBN 13   BRV   978-1-118-23570-6

Printed in the United States of America.

10 9 8 7 6 5 4 3 2 1

About the Authors

Edward B. Burger,
Michael Starbird,
and some fuzzy trees.

Edward B. Burger is the Francis Christopher Oakley Third Century Professor of Mathematics at Williams College and most recently served as Vice Provost for Strategic Educational Initiatives at Baylor University. He is the author of over 60 research articles, books, and video series (starring in over 3,000 on-line videos). The Mathematical Association of America honored him on several occasions: He received the 2001 Deborah and Franklin Tepper Haimo National Award for Distinguished Teaching of Mathematics; he was named the 2001-2003 Pólya Lecturer; in 2004 he was awarded the Chauvenet Prize and in 2006 he was a recipient of the Lester R. Ford Prize. In 2006, *Reader's Digest* listed Burger in their annual "100 Best of America" as America's Best Math Teacher. In 2010 he was named the winner of the 2010 Robert Foster Cherry Award for Great Teaching – the largest and most prestigious prize in higher education teaching across all disciplines in the English-speaking world. Also in 2010 he won a Telly Award for his appearance in "Mathletes" for NBC-TV as part of their Winter Olympic coverage. *The Huffington Post* named him one of their 2010 *Game Changers*; "HuffPost's *Game Changers* salutes 100 innovators, visionaries, mavericks, and leaders who are reshaping their fields and changing the world." Most recently in 2012, Microsoft Worldwide Education selected him as one of their "Global Heroes in Education." In 2013 Burger was inducted as a Fellow of the American Mathematical Society.

Michael Starbird is a University Distinguished Teaching Professor of Mathematics at The University of Texas at Austin. He has been at UT his whole career except for leaves, including as a Visiting Member of the Institute for Advanced Study in Princeton, New Jersey and a member of the technical staff of the Jet Propulsion Laboratory in Pasadena, California. He has received more than a dozen teaching awards including the Mathematical Association of America's 2007 Deborah and Franklin Tepper Haimo

National Award for Distinguished Teaching of Mathematics, the Texas statewide Minnie Stevens Piper Professor award, the UT Regents' Outstanding Teaching Award (in the inaugural year of the award), and most of the UT-wide teaching awards including the Jean Holloway Award for Excellence in Teaching, the Friar Centennial Teaching Fellowship, the Chad Oliver Plan II Teaching Award, the President's Associates Teaching Excellence Award, the Dad's Association Centennial Teaching Fellowship, the Eyes of Texas Excellence Award (twice), and others. He is a member of UT's Academy of Distinguished Teachers. He has produced DVD courses for The Teaching Company in the Great Courses Series on calculus, statistics, probability, geometry, and the joy of thinking and has given hundreds of lectures and workshops. In 2013 Starbird was inducted as a Fellow of the American Mathematical Society.

Burger and Starbird have co-authored two general audience books: *Coincidences, Chaos, and All That Math Jazz: Making Light of Weighty Ideas,* and *The 5 Elements of Effective Thinking.*

About the Cover

The dodecahedron (a 12-sided regular solid) represents mathematics. But to make mathematics come to life, as with any discipline, you must bring yourself and engage in thinking. Thus we view the cover as a blank slate – a fresh white canvas ready for you to create and engage. So this unusual "white album"-style cover attempts to provoke thought. Bring yourself to mathematics and make it your own. Make the cover your own – transform the white canvas into a work of original art. It really is a metaphor for the creative elements of mathematical thinking and a celebration of original thought. If you do put your own mark on the cover, send us an image via *www.heartofmath.com*. We'd love to see your creativity in action and may even offer a prize for the most thought-provoking and attractive design.

Contents

Welcome!

We wrote this book to be read. We designed many attractions—a kit for grasping concepts hands-on, jokes (some aren't too lame), 3D pictures and glasses, and a style of presentation that we hope invites you to discover new ideas. Most of all, this book contains intriguing lessons for thinking that can change your life.

> *Of course, no one actually reads their math textbooks.*
>
> **AN ANONYMOUS MATH STUDENT**

A World of Ideas

Most people do not have an accurate picture of mathematics. For many, mathematics is the torture of tests, homework, and problems, problems, problems. The very word *problems* suggests unpleasantness and anxiety. But mathematics is not "problems."

Some people view mathematics as a set of formulas to be applied to a list of problems at the ends of textbook chapters. Toss that idea into the trash. Formulas in algebra, trigonometry, and calculus are incredibly useful. But, in this book, you will see that mathematics is a network of intriguing ideas—not a dry, formal list of techniques.

We want you to discover what mathematics really is and to become a fan. However, if you are not intrigued by the romance of the subject, that's fine too, because at least you will have a firmer understanding of what it is you are judging. Mathematics is a living, breathing, changing organism with many facets to its personality. It is creative, powerful, and even artistic.

Wanderer Overlooking the Sea of Fog by Caspar David Friedrich (1817).

> *. . . mathematicians are really seeking to behold the things themselves, which can be seen only with the eye of the mind.*
>
> **PLATO**

Mathematics uses penetrating techniques of thought that we can all use to solve problems, analyze situations, and sharpen the way we look at our world. This book emphasizes basic strategies of thought and analysis. These strategies have their greatest value to us in dealing with real-life decisions and situations that are completely outside mathematics. These "life lessons," inspired by mathematical thinking, empower us to better grapple with and conquer the problems and issues that we all face in our lives—from love to business, from art to politics. If you can conquer infinity and the fourth dimension, then what can't you do?

As you read this book, we hope you discover the beauty and fascination of mathematics, admire its goals for this book. We want you to look at your life, your habits of thought, and your perception of the world in a new way. And we hope you enjoy the view.

Part of the power of mathematics lies in its inexorable quest for elegance, symmetry, order, and grace. Seeking pattern, order, and understanding is a transforming process that mathematics can help us develop.

FALLING OFF THE MATH CLIFF

① A boy begins his wondrous journey.

② He perseveres.

③ Math cannot become any more difficult than this, can it?

④ Yes.

⑤ Yet he does not give up!

⑥ For a brief moment, there is a glimmer of comprehension.

⑦ Actual midair pedalling.

⑧ The plummet.

A Mathematical Journey

The realm of mathematics contains some of the greatest ideas of humankind—ideas comparable to the works of Shakespeare, Plato, and Michelangelo. These mathematical ideas helped shape history, and they can add texture, beauty, and wonder to our lives.

To make our mathematical excursion as pleasant as possible, we have tried to make it all fun—fun to read, fun to do, and fun to think about. We hope you explore some, learn some, think some, enjoy some, and add a new aspect to your view of everything. We hope you laugh at our bad jokes and silly remarks, forgive our sometimes unbridled enthusiasm, but also embrace the profound issues at hand.

The road through this book is not free from perils, bumps, and jolts. Sometimes you will confront issues that start beyond your comprehension, but they won't stay beyond your comprehension. The journey to true understanding can be difficult and frustrating, but stay the course and be patient. There is light at the end of the tunnel—and throughout the journey, too.

What's the point of it all? Well, the bottom line is that mathematics involves profound ideas. Making these ideas our own empowers us with the strength, the techniques, and the confidence to accomplish wonders.

> *It may well be doubted whether, in all the range of science, there is any field so fascinating to the explorer—so rich with hidden treasures—so fruitful in delightful surprises—as Pure Mathematics.*
>
> **LEWIS CARROLL**

Travel Tips—Read the Book

We have some suggestions about how to use this book:

> *In mathematics I can report no deficiency, except it be that men do not sufficiently understand the excellent use of Pure Mathematics.*
>
> **ROGER BACON**

Answer our questions ➤ We often pose questions in the middle of a section and invite you to give an answer or a guess before continuing. Please attempt to answer these questions. If you don't know an answer for sure, guess. Don't be afraid to make lots of mistakes—that is the only way to learn. It is much better to guess wrong than not to think about the question at all.

Think ➤ This is our main goal. We want you to contemplate some of the greatest and most intriguing creations of human thought. Constantly stop and think.

Be active, not passive ➤ Our wish is for you to be an active participant. Take the concepts and make them your own. Look beyond the mathematical ideas, and don't be satisfied with mere knowledge. Challenge yourself to attain the power to figure things out on your own.

Play ➤ This technology icon indicates that relevant interactive software is available at www.heartofmath.com. Check it out and enjoy it!

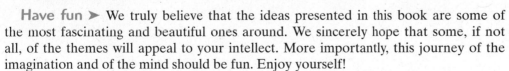

Have fun ➤ We truly believe that the ideas presented in this book are some of the most fascinating and beautiful ones around. We sincerely hope that some, if not all, of the themes will appeal to your intellect. More importantly, this journey of the imagination and of the mind should be fun. Enjoy yourself!

Finally, reading mathematics is much different from reading about many other subjects. Here's how we read mathematics. We read a sentence or two, stop reading, think about what we've read, and then realize we're completely and utterly confused. Usually we discover that we didn't really understand the previous paragraph. But, we don't get frustrated . . . it's the nature of the beast. Instead, we either reread some previous sections or just reread the previous sentence. The fuzziness slowly begins to fade ever so slightly, and the concept begins to come into focus. Then we attempt to think about the issue and work

> *If we do not expect the unexpected, we will never find it.*
>
> **HERACLITUS**

> *Shall any gazer see with mortal eyes
> Or any searcher know by mortal
> mind— Veil after veil will lift—but
> there must be Veil upon veil behind.*
>
> SIR EDWIN ARNOLD

with it on our own or with friends. It is at this point that we begin to appreciate and understand the ideas presented. One of the great features of mathematics is that once we do understand an idea, our grasp of it is completely solid. There is no vagueness or uncertainty. So, adopt high standards for what you view as "understanding." Be actively engaged as you read. Draw pictures, explain ideas to friends. Put yourself in the position of the discoverer of each idea. Ask questions, search for answers, and let those answers guide you to still more questions.

With all good wishes,

Edward B. Burger
Michael Starbird

Surfing the Book

I t's too early to get caught up in details. Instead, let's just surf the book and get a quick overview of what's ahead. The whole book revolves around just two basic themes:

- EFFECTIVE THINKING
- SOME TRULY GREAT IDEAS

What mathematical sites lie ahead? Instead of just starting in with a hot and spicy math topic, we thought it would be more fun to surf the entire book and get a quick, big-picture overview of what is on the horizon. We hope these "home pages" will pique your curiosity and tantalize your intellect. Keep an open mind; forgo any previous biases and prejudices toward mathematics; and do not censor any inventive thoughts or sparks of interest you may develop toward the subject. Let's surf.

© Shannon Stent / iStockphoto.

Heartofmath.com

1 Welcome   2 Games   3 Number   4 Infinity   5 Gems   6 Space   7 Graphs   8 Chaos   9 Chance   10 Meaning   Deciding   Farewell

Fun and Games
An Introduction to Rigorous Thought

Can this book help you think more effectively, more inventively, solve life problems more creatively, and analyze issues more logically?

The short answer is "Yes."

Is there a better way to meet the powerful world of logical thought than through Fun and Games?

The short answer is "No."

Is this book strange and sometimes over the edge?

The short answer is "Absolutely."

Just hang with us and see how far we'll go.

This site is an invitation to think and have fun with genies, damsels, and Dodgeball and, in the process, develop a system of logical inquiry that we will use throughout the book and throughout our lives.

Who can better develop your thinking skills than you? As you resolve the many dilemmas in these crazy stories, you will automatically discover your own path to logical and strategic thought. Don't feel like going at it alone? Get a friend or a roommate to try some with you . . . it's all fun and games.

GO TO PAGE 3

GO TO PAGE 26

GO TO PAGE 7

. . . the primary question was not what do we know, but how do we know it.

ARISTOTLE

...are we having fun yet?

Heartofmath.com

Number
Contemplation

Wherever there is a number, there is beauty.

PROCLUS

Worried about balding? How about this one: Are there two hairy people on Earth with exactly the same number of hairs on their bodies? Does Rogaine change the answer?

What do the reproductive habits of 13th-century rabbits have in common with the Parthenon?

More than you think.

Arc art and music branches of mathematics?

You betcha Bach!

Don't give up! Think working on really challenging questions that others have tried to solve is fruitless? Ask Andrew Wiles. In 1994 he answered a 350-year-old question—it only took him seven years. Hey, intellectual triumphs happen—it just takes tenacity!

Can you tell time? If so, then you might have a promising career at decoding the numbers at the bottom of Universal Product Codes. Want to know how?

XQE TPS LPBE AX TZ?

So numbers are no biggie? In ancient Greece you were thrown from a ship and drowned if you told people about certain numbers. Sound irrational?

How close is 1 to 0.99999 . . . ? Closer than you might think.

Ancient questions about numbers still remain unanswered. Act now . . . mathematicians are standing by.

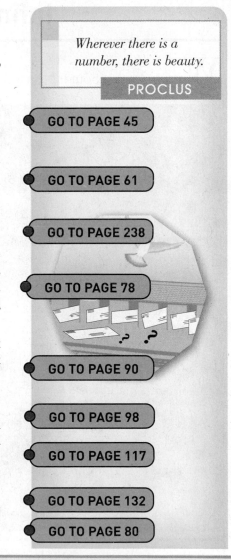

GO TO PAGE 45

GO TO PAGE 61

GO TO PAGE 238

GO TO PAGE 78

GO TO PAGE 90

GO TO PAGE 98

GO TO PAGE 117

GO TO PAGE 132

GO TO PAGE 80

If 2 can be 1, who is 1 to become??

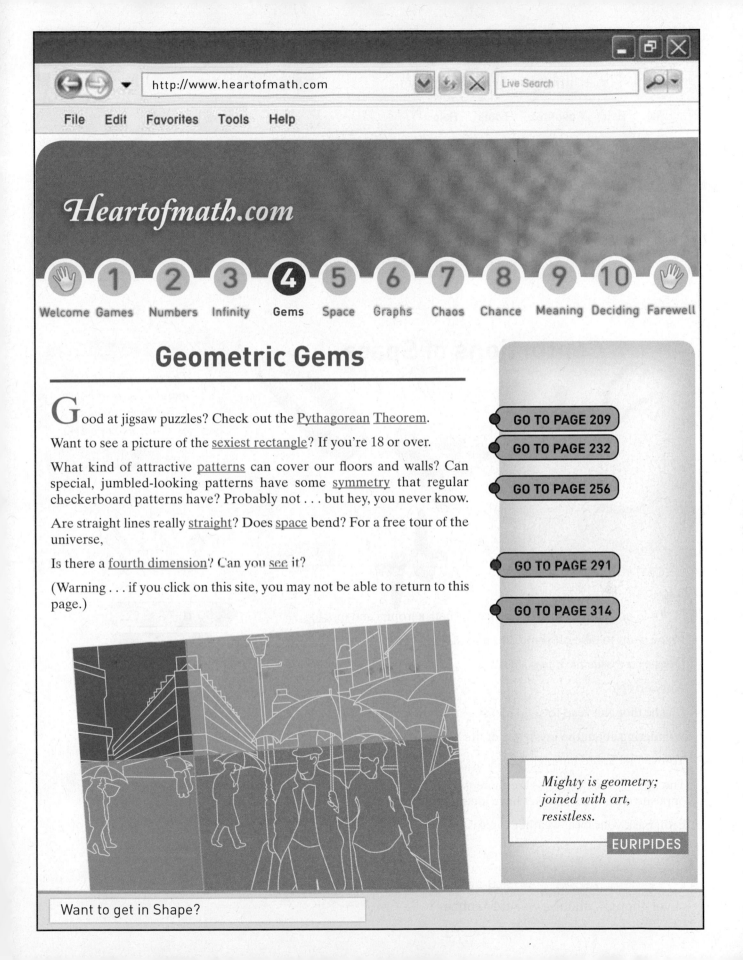

Geometric Gems

Good at jigsaw puzzles? Check out the Pythagorean Theorem.

Want to see a picture of the sexiest rectangle? If you're 18 or over.

What kind of attractive patterns can cover our floors and walls? Can special, jumbled-looking patterns have some symmetry that regular checkerboard patterns have? Probably not . . . but hey, you never know.

Are straight lines really straight? Does space bend? For a free tour of the universe,

Is there a fourth dimension? Can you see it?

(Warning . . . if you click on this site, you may not be able to return to this page.)

GO TO PAGE 209

GO TO PAGE 232

GO TO PAGE 256

GO TO PAGE 291

GO TO PAGE 314

Mighty is geometry; joined with art, resistless.

EURIPIDES

Want to get in Shape?

Heartofmath.com

Contortions of Space

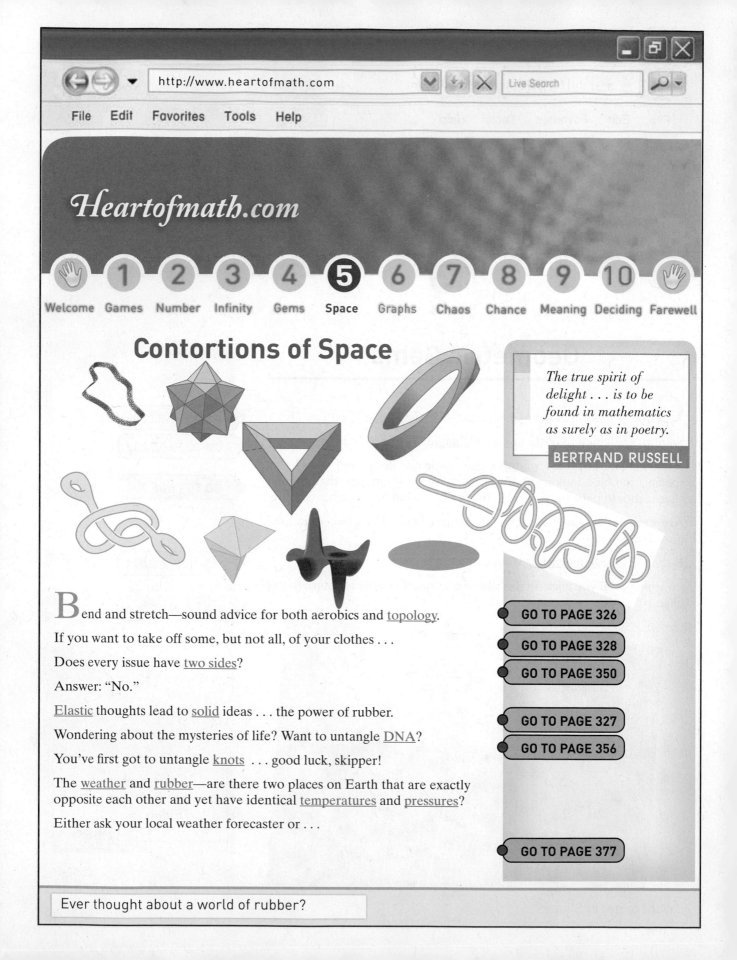

The true spirit of delight . . . is to be found in mathematics as surely as in poetry.

BERTRAND RUSSELL

Bend and stretch—sound advice for both aerobics and <u>topology</u>.

If you want to take off some, but not all, of your clothes . . .

Does every issue have <u>two sides</u>?

Answer: "No."

<u>Elastic</u> thoughts lead to <u>solid</u> ideas . . . the power of rubber.

Wondering about the mysteries of life? Want to untangle <u>DNA</u>?

You've first got to untangle <u>knots</u> . . . good luck, skipper!

The <u>weather</u> and <u>rubber</u>—are there two places on Earth that are exactly opposite each other and yet have identical <u>temperatures</u> and <u>pressures</u>?

Either ask your local weather forecaster or . . .

GO TO PAGE 326

GO TO PAGE 328

GO TO PAGE 350

GO TO PAGE 327

GO TO PAGE 356

GO TO PAGE 377

Ever thought about a world of rubber?

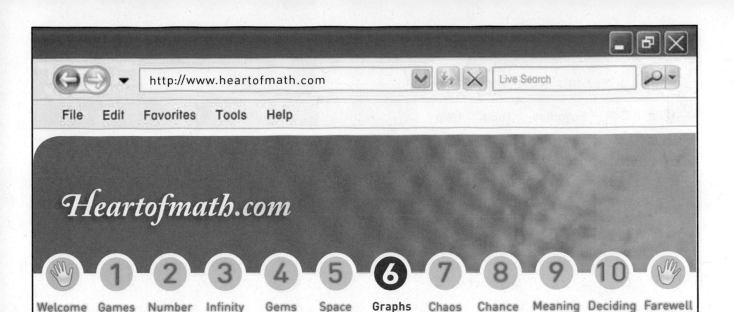

Heartofmath.com

Welcome | 1 Games | 2 Number | 3 Infinity | 4 Gems | 5 Space | 6 Graphs | 7 Chaos | 8 Chance | 9 Meaning | 10 Deciding | Farewell

Modeling Our World through Graphs

I found I could say things with color and shapes that I couldn't say any other way— things I had no words for..

GEORGIA O'KEEFE

GO TO PAGE 386

GO TO PAGE 401

GO TO PAGE 414

GO TO PAGE 434

Can you always cover all bases without revisiting any base? Depends on the number of bases and your path.

Counting corners reveals more than merely cutting corners. Don't believe us?

Can you always keep it on the down low?

How edgy is your social network?

You are being connected...

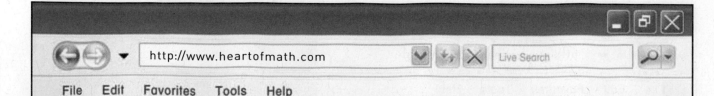

Heartofmath.com

Welcome   Games   Number   Infinity   Gems   Space   Graphs   **Chaos**   Chance   Meaning   Deciding   Farewell

1   2   3   4   5   6   **7**   8   9   10

Fractals and Chaos

Can <u>pictures</u> or ideas be <u>infinitely intricate</u>?

<u>Fractals</u>—is there anything that is not one? Probably not . . . but what is one?

Can objects <u>straddle</u> two dimensions?

Can we <u>predict</u> the <u>population</u>, the <u>weather</u>, or even the positions of the planets in the future?

Answer: "No."

A <u>butterfly</u> flaps its wings in Brazil. Two weeks later there is a tornado in Kansas—kiss <u>Dorothy</u> good-bye. Are these events related?

Nature is sheer and utter <u>chaos</u>. Why bother cleaning your room?

GO TO PAGE 462

GO TO PAGE 480

GO TO PAGE 493

GO TO PAGE 527

GO TO PAGE 553

GO TO PAGE 559

> *God has put a secret art into the forces of Nature so as to enable it to fashion itself out of chaos into a perfect world system.*
>
> **IMMANUEL KANT**

. . . details . . . details . . . details . . .

http://www.heartofmath.com

Heartofmath.com

Welcome 1 Games 2 Number 3 Infinity 4 Gems 5 Space 6 Graphs 7 Chaos **8 Chance** 9 Meaning 10 Deciding Farewell

Taming Uncertainty

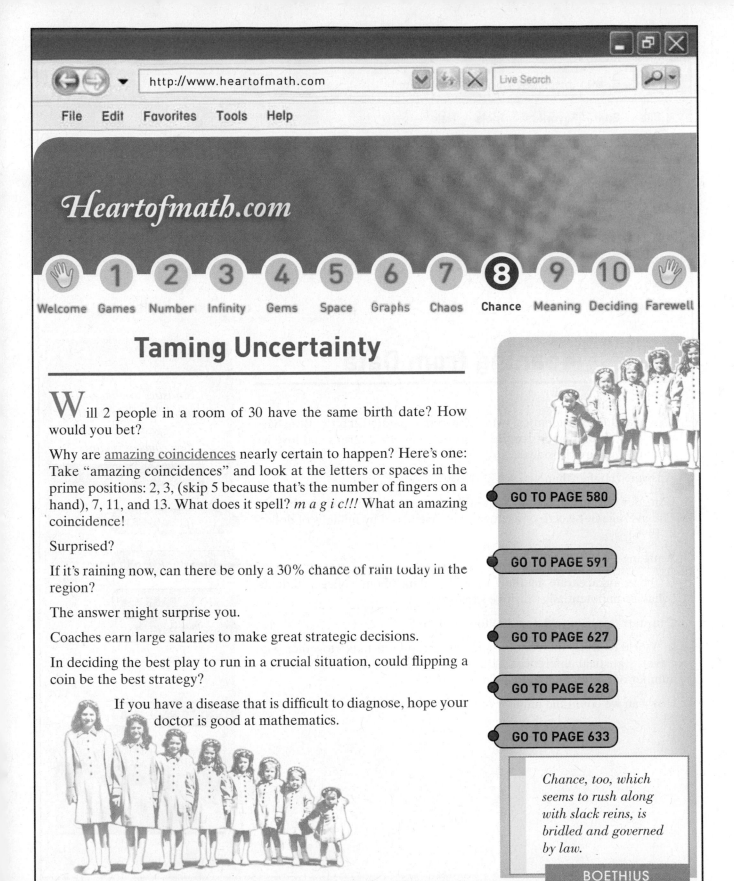

Will 2 people in a room of 30 have the same birth date? How would you bet?

Why are amazing coincidences nearly certain to happen? Here's one: Take "amazing coincidences" and look at the letters or spaces in the prime positions: 2, 3, (skip 5 because that's the number of fingers on a hand), 7, 11, and 13. What does it spell? *m a g i c!!!* What an amazing coincidence!

Surprised?

If it's raining now, can there be only a 30% chance of rain today in the region?

The answer might surprise you.

Coaches earn large salaries to make great strategic decisions.

In deciding the best play to run in a crucial situation, could flipping a coin be the best strategy?

If you have a disease that is difficult to diagnose, hope your doctor is good at mathematics.

GO TO PAGE 580

GO TO PAGE 591

GO TO PAGE 627

GO TO PAGE 628

GO TO PAGE 633

> *Chance, too, which seems to rush along with slack reins, is bridled and governed by law.*
>
> BOETHIUS

Chances are . . .

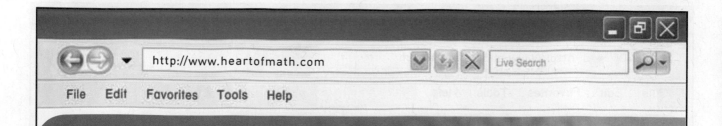

Heartofmath.com

Welcome	Games	Number	Infinity	Gems	Space	Graphs	Chaos	Chance	**Meaning**	Deciding	Farewell

Meaning from Data

John and Jim are identical twins who were separated at birth. Both have married women named Jennifer who watch *Seinfeld* reruns and love ice cream. What are the odds?

Answer: "Higher than you might think."

Consider all graduates of Lakeside School over its whole history. The average net worth of each graduate increased by millions of dollars in 1996.

Amazing . . . or not?

Is there an accurate method to gather data about college students' behavior in potentially sensitive situations?

Can statistics help to defeat the forces of evil?

In World War II, the serial numbers on captured tanks together with clever statistical inferences helped the Allied forces to estimate the number of German tanks available to the Nazis.

How can we count the number of tigers in the wild?

> *Statistics are like bikinis. What they reveal is suggestive, but what they conceal is vital.*
>
> **AARON LEVENSTEINS**

GO TO PAGE 738

GO TO PAGE 650

GO TO PAGE 651

GO TO PAGE 749

GO TO PAGE 748

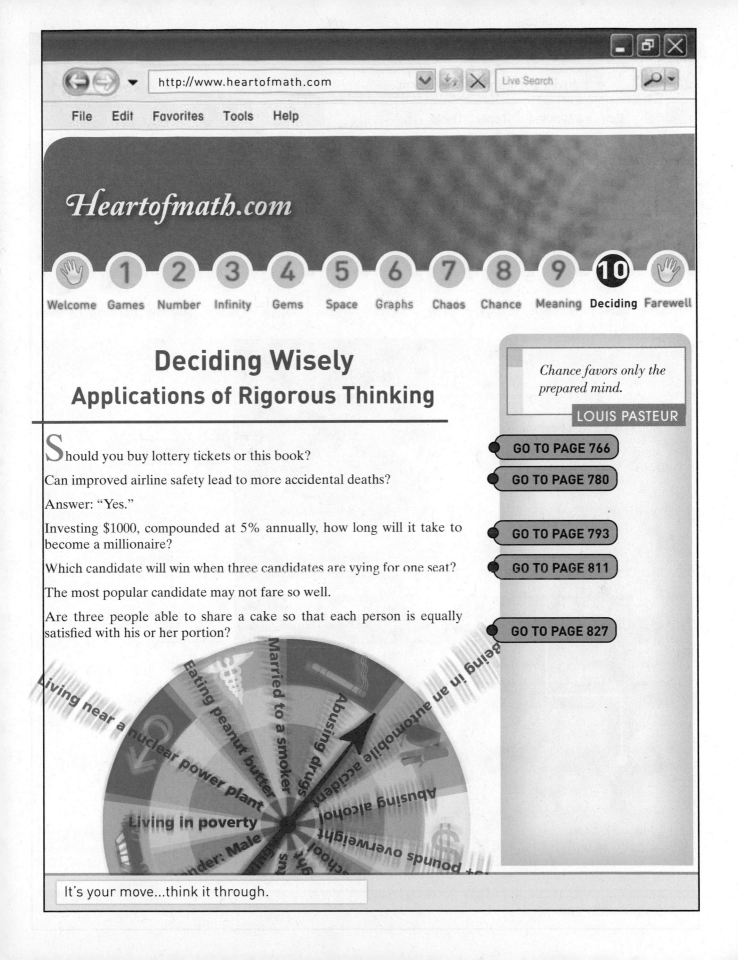

Heartofmath.com

Welcome   1 Games   2 Number   3 Infinity   4 Gems   5 Space   6 Graphs   7 Chaos   8 Chance   9 Meaning   10 Deciding   Farewell

Deciding Wisely
Applications of Rigorous Thinking

Chance favors only the prepared mind.

LOUIS PASTEUR

Should you buy lottery tickets or this book?

Can improved airline safety lead to more accidental deaths?

Answer: "Yes."

Investing $1000, compounded at 5% annually, how long will it take to become a millionaire?

Which candidate will win when three candidates are vying for one seat?

The most popular candidate may not fare so well.

Are three people able to share a cake so that each person is equally satisfied with his or her portion?

GO TO PAGE 766

GO TO PAGE 780

GO TO PAGE 793

GO TO PAGE 811

GO TO PAGE 827

It's your move...think it through.

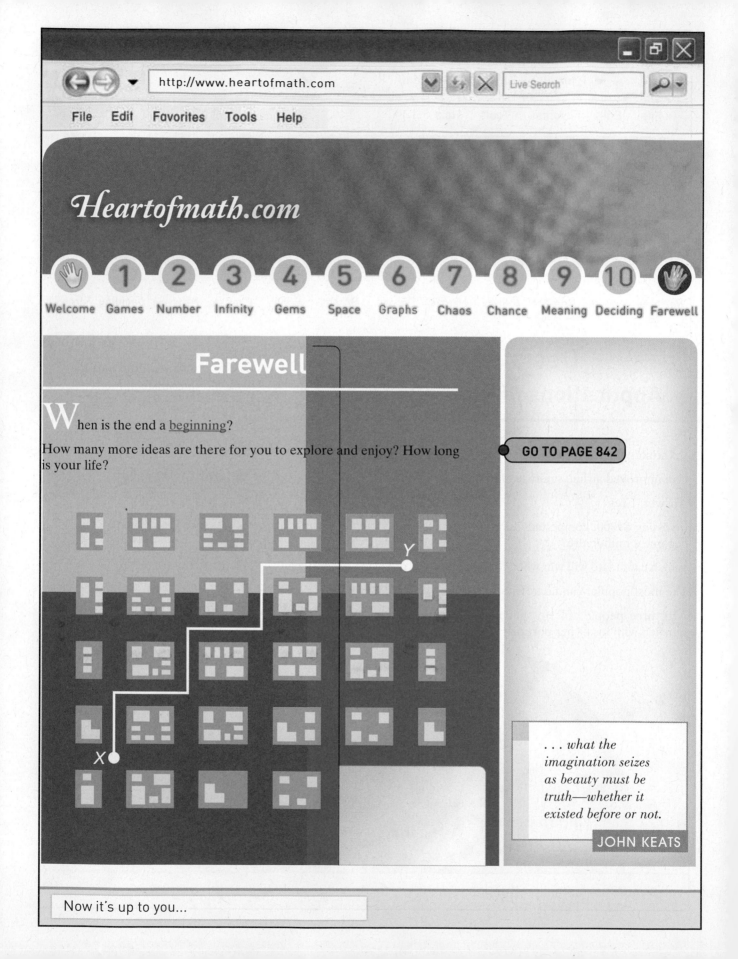

Farewell

When is the end a beginning?

How many more ideas are there for you to explore and enjoy? How long is your life?

GO TO PAGE 842

. . . what the imagination seizes as beauty must be truth—whether it existed before or not.

JOHN KEATS

Now it's up to you...

Now that we have a sense of what's ahead, let's dig in . . .

CHAPTER ONE

Fun and Games

An Introduction to Rigorous Thought

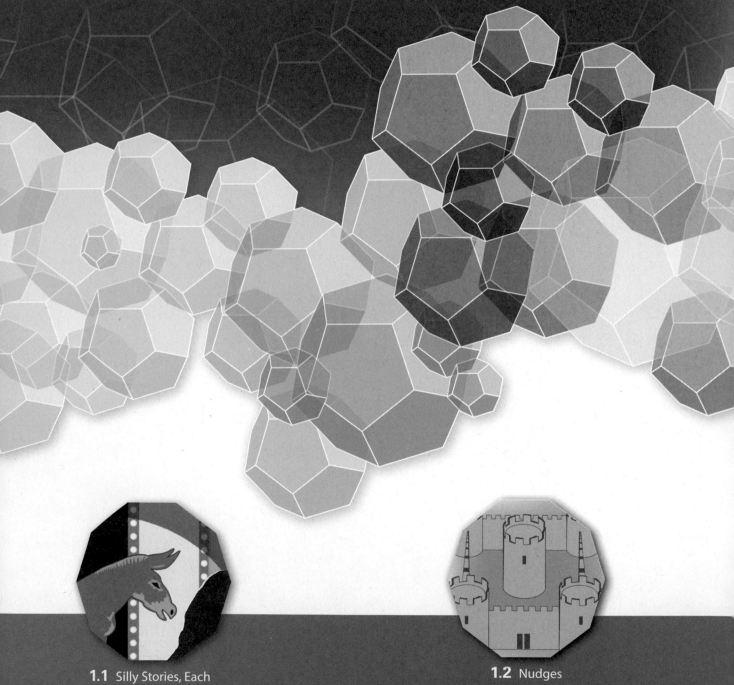

1.1 Silly Stories, Each
with a Moral

1.2 Nudges

Fun and games, rigorous thought: Here we will discover that in mathematics these go hand in hand. Who says that profound ideas and important insights come only from hard work? Sure, we can consider the discipline of mathematics broadly, from a philosophical perspective, and we can be intrigued by its mysterious wonders. But when we get right down to it, we think mathematics is just plain fun. We hope that one day you will, too.

We start with two fundamental observations:

1. Mathematics involves logical and creative thinking.
2. Thinking can be fun.

By grappling with conundrums serious or otherwise, we can discover significant concepts. As we grope for solutions to silly stories, we begin to develop effective strategies for serious thinking.

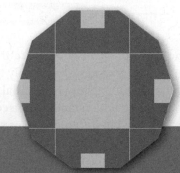

1.3 The Punch Lines

1.4 From Play to Power

SILLY STORIES, EACH WITH A MORAL

Conundrums that Evoke Techniques of Effective Thinking

Henri de Toulouse-Lautrec, *La Vache Enragée,* 1896. Silly, but still art!

> *Beware gentle knight, there is no greater monster than reason.*
>
> MIGUEL DE CERVANTES

You now have a mission. Your mission, should you decide to accept it, is to read the following stories and attempt to answer the questions they raise. The only rules are as follows:

1. Make an earnest attempt to solve each puzzle.

2. Be creative.

3. Don't give up: If you get stuck, look at the story in a different way.

4. If you become frustrated, stop working, move on, and then return to the story later.

5. Share these stories with your family and friends.

6. HAVE FUN!

> *A journey of one thousand miles begins with a single step.*
>
> CHINESE PROVERB

How do we approach issues in life that need to be resolved? A critical step is simply to begin. Think a bit and then move forward. Taking that first step, though essential, is often scary; more often than not, we do not possess a clear understanding of a complete solution or even see how a solution will eventually fall into place. This situation is like being asked to walk through a forest in the dark. Without knowing the terrain, the natural tendency is to freeze like a deer in headlights. However, we must learn not to let this understandable fear paralyze us intellectually; we must take a step. It is only by stumbling through many small intellectual steps that we are eventually able to make any progress at all.

For example, imagine we're soccer players with the ball at midfield. In this position we can't possibly know how a goal will be achieved, and we can't stop to envision the entire progression of the future before kicking the ball. Instead, we move with the understanding that the specific goal strategy will become clear as opportunities arise.

Just try out ideas with these stories—loosen up, try to kick the ball, and don't worry if you miss. Remember:

> *Truth comes out of error more easily than out of confusion.*
>
> FRANCIS BACON

After you have given considerable thought to a story, move to the corresponding part in Section 1.2, "Nudges," where leading questions and suggestions provide a gentle push in the right direction, in case you need a hint. There we also identify some strategies for tackling both mathematical questions and, more importantly, questions that will arise in your life.

Section 1.3, "The Punch Lines," provides solutions and commentary about how the questions and their resolutions fit into the mathematical landscape. As you think about these stories, you will discover some profound ideas that capture the essence of some deep and beautiful mathematical concepts.

As you proceed, remember the rules on page 4, especially rule 6.

Story 1. That's a Meanie Genie

On an archeological dig near the highlands of Tibet, Alley discovered an ancient oil lamp. Just for laughs she rubbed the lamp. She quickly stopped laughing when a huge puff of magenta smoke spouted from the lamp, and an ornery genie named Murray appeared. Murray, looking at the stunned Alley, exclaimed, "Well, what are you staring at? Okay, okay, you've found me; you get your three wishes. So, what will they be?" Alley, although in shock, realized she had an incredible opportunity. Thinking quickly, she said, "I'd like to find the Rama Nujan, the jewel that was first discovered by Hardy the High Lama." "You got it," replied Murray, and instantly nine identical-looking stones appeared. Alley looked at the stones and was unable to differentiate any one from the others.

Finally she said to Murray, "So where is the Rama Nujan?" Murray explained, "It is embedded in one of these stones. You said you wished to find it. So now you get to find it. Oh, by the way, you may take only one of the stones with you, so choose wisely!" "But they look identical to me. How will I know which one has the Rama Nujan in it?" Alley questioned. "Well, eight of the stones weigh the same, but the stone containing the jewel weighs slightly more than the others," Murray responded with a devilish grin.

*Visit www.heartofmath.com to see these Web pages come to life and to discover more.

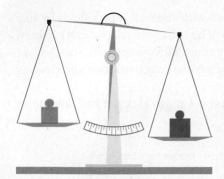

Alley, becoming annoyed, whispered under her breath, "Gee, I wish I had a balance scale." Suddenly a balance scale appeared. "That was wish two!" declared Murray. "Hey, that's not fair!" Alley cried. "You want to talk fair? You think it's fair to be locked in a lamp for 1729 years? You know you can't get cable TV in there, and there's no room for a satellite dish! So don't talk to me about fair," Murray exclaimed. Realizing he had gone a bit overboard, Murray proclaimed, "Hey, I want to help you out, so let me give you a tip: That balance scale may be used only once." "What? Only once?" she said, thinking out loud. "I wish I had another balance scale." ZAP! Another scale appeared. "Okay, kiddo, that was wish three," Murray snickered. "Hey, just one minute," Alley said, now regretting not having asked for one million dollars or something more standard. "Well at least this new scale works correctly, right?" "Sure, just like the other one. You may use it only once." "Why?" Alley inquired. "Because it is a 'wished' balance scale," he said, "so the rule is 'one scale, one balancing'; it's just like the rule against using one wish to wish for a hundred more wishes." "You are a very obnoxious genie." "Hey, I don't make up the rules, lady, I just follow them," he said.

So, Alley may use each of the two balance scales exactly once. Is it possible for Alley to select the slightly heavier stone containing the Rama Nujan from among the nine identical-looking stones? Explain why or why not.

Story 2. Damsel in Distress

Long ago, knights in shining armor battled dragons and rescued damsels in distress on a daily basis. Although it is not often stressed in many stories of chivalry, the rescue often involved logical thinking and creative problem solving by the damsel. Here then is a typical knightly encounter.

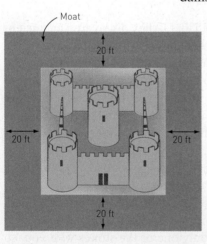

Once upon a time, a notorious knight captured a damsel and imprisoned her in a castle surrounded by a square moat that was infested with extraordinarily hungry alligators. The moat was 20 feet across, and no drawbridge existed because after depositing the damsel in the castle, the evil knight had taken it with him (giving his horse one major hernia).

After a time, a good knight rode up and said, "Hail, sweet damsel, for I am here, and thou art there. Now what are we going to do?"

The knight, though good, was not too bright and consequently paced back and forth along the moat looking anxiously at the alligators and trying feebly to think of a plan. While doing so, he stumbled upon two sturdy beams of wood suitable for walking across but lacking sufficient length. Alas, the moat was 20 feet across, but the beams were each only 19 feet long and 8 inches wide. He tried to stretch them and then tried to think. Neither effort proved successful. He had no nails, screws, saws, Superglue, or any other method of joining the two beams to extend their length.

What to do? What to do? Fortunately, the damsel, after a suitable time to allow the good knight to attempt to solve the puzzle on his own, called to the knight and gave him a few hints that enabled him to rescue her. What was the maiden's suggestion?

This story from medieval times foreshadows our journey into the geometric and the visual.

Story 3. The Fountain of Knowledge

During an incredibly elaborate hazing stunt during pledge week, Trey Sheik suddenly found himself alone in the Sahara Desert. His desire to become a fraternity brother was now overshadowed by his desire to find something to drink (these desires, of course, are not unrelated). As he wandered aimlessly through the desert sands, he began to regret his involvement in the whole frat scene. Both hours and miles had passed and Trey was near dehydration. Only now did Trey appreciate the advantages of sobriety. Suddenly, he came upon an oasis.

There, sitting in a shaded kiosk beside a small pool of mango nectar, was an old man named Al Donte. Big Al not only ran the mango bar but was also a travel agent and could book Trey on a two-humped camel back to Michigan. At the moment, however, Trey desired nothing but a large drink of that beautifully translucent and refreshing mangoade. Al informed Trey that he sold the juice only in 8-ounce servings and the cost for one serving was $3.50. Trey frantically searched his pockets, and though he found much sand, he also discovered that he had exactly $3.50.

Trey's jubilation at the thought of liquid coating his parched throat was quickly shattered when Al casually announced that he did not have an 8-ounce glass; all he had was a 6-ounce glass and a 10-ounce glass—neither of which had any markings on it. Al, being a man of his word, would not hear of selling any more or any less than an 8-ounce serving of his libation. Trey, in desperation, wondered whether it was possible to use only the unmarked 6- and 10-ounce glasses to produce exactly 8 ounces in the 10-ounce glass. Do you think it's possible? If so, explain how, and if not, explain why. This pledge-week prank does whet our appetites for a world of numbers.

6-oz glass   10-oz glass

Story 4. Dropping Trou

Before reading on, remember that truth is sometimes stranger than fiction. The highlight of Professor Burger's April 1993 talk to more than 300 Williams College students and their parents occurred when, after removing his shoes, he tied his feet together with a stout rope, leaped onto the table, dramatically removed his belt, unzipped his zipper, and dropped his pants. The purple cows (Williams mascots) mooing about on his baggy boxer shorts completed an image not soon forgotten in the annals of mathematical talks. The more conservative parents in the audience were contemplating transferring their sons and daughters to a less "progressive" school.

But then, at the moment of maximum shock and bewilderment, Professor Burger performed the seemingly impossible feat of rehabilitating his fast-sinking reputation. Without removing the rope attached to his feet, he turned his pants inside out and pulled his trousers back to their accustomed position (though now inside out). Thus he simultaneously restored his modesty *and* his credibility by demonstrating the mathematical triumph of reversing his pants without removing the rope that was tying his feet together.

Please attempt to duplicate Professor Burger's amazing feat—in the privacy of your room, of course. You will need a rope or cord about 5 feet long. One end of

David Bowman ©Williams Record

Edward Burger: exposed on
April 24, 1993.

*Visit www.heartofmath.com to see these Web pages come to life and to discover more.

the rope should be tied snugly around one ankle and the other end tied equally snugly about the other ankle. Now, without removing the rope, try to take your pants off, turn them inside out, and put them back on so that you, the rope, and your pants are all exactly as they were at the start, with the exception of your pants being inside out. While some may find this experiment intriguing, others may find it in poor taste. Everyone will agree, however, that surprising outcomes arise when we bend and contort objects and space.

Story 5. Dodgeball

Dodgeball is a game for two players—Player One and Player Two (although any two people can play it, even if they are not named "Player One" and "Player Two"). Each player has a special game board (shown below) and is given six turns.

Player One begins by filling in the first horizontal row of his game board with a run of X's and O's. That is, on the first line of his board, he will write either an X or an O in each box. Then Player Two places either an X or an O in the first box of her board. So at this point, Player One has filled in the first row of his board with six letters, and Player Two has filled in the first box of her board with one letter.

The game continues with Player One writing down either an X or an O in each box of the second horizontal row of his board. Then Player Two writes one letter (an X or an O) in the second box of her board. The game proceeds in this fashion until all of Player One's boxes are filled with X's and O's; thus, Player One has produced six rows of six marks each, and Player Two has produced one row of six marks. All marks are visible to both players at all times. Player One wins if any of his rows exactly matches Player Two's row (Player One matches Player Two).

*

Player One's game board

1						
2						
3						
4						
5						
6						

Player Two's game board

1	2	3	4	5	6

*Visit www.heartofmath.com to see these Web pages come to life and to discover more.

Player Two wins if her row does not match any of Player One's rows (Player Two dodges Player One).

Would you rather be Player One or Player Two? Who has the advantage? Can you devise a strategy for either side that will always result in victory? This little game holds within it the key to understanding the sizes of infinity.

Story 6. A Tight Weave

Sir Pinsky, a famous name in carpets, has a worldwide reputation for pushing the limits of the art of floor covering. The fashion world stands agog at the clean lines and uncanny coherence of his purple and gold creations. Some call him square because his designs so richly employ that quaint quadrilateral. But squares in the hands of a master can create textures beyond the weavers' world, although not beyond human imagination.

One day Sir Pinsky began a creation with, as always, a perfect, purple square. However, one square seemed too plain, so in the exact center of it he added a gold square. He saw that the central square implicitly defined eight purple squares surrounding it. As he pondered, he realized that those eight purple squares were identical to his original large square except for two things: (1) each was one-third the size of the whole square; and (2) none of them had a gold square in its center.

He wondered whether he could further modify his design so that each of the eight small squares would replicate the entire design except for being one-third its size. After much thought, he solved this puzzle and created a design with which his name is associated. Can you sketch and describe his design? Create this design in stages, adding more gold squares at each stage.

Suppose the original square rug is 1 yard by 1 yard. How much gold material would be needed for the second stage? How much for the third stage? Continue computing the area of the gold squares at various stages of the process, and then guess how much gold material will be needed to create the final floor covering. The answer is surprising.

Though our carpet designer is thoroughly modern in all ways, the source of his inspiration is ancient. In Chapter 6, "Fractals and Chaos," we will see an example of this style in the 19th-century Buddhist tapestry, *Vaishravana Mandala*.

Story 7. Let's Make a Deal

"Let's make a deal!" Monty Hall enthuses to the gentleman dressed as a giant singing raisin. The gleeful raisin, whose name is Warren Piece, is ready to wheel and deal as Monty Hall explains the game. "Behind one of these three doors is the Cadillac of your dreams. It is as long as a train and comes complete with a Jacuzzi. Of course,

*Visit www.heartofmath.com to see these Web pages come to life and to discover more.

if you spend too much time in the Jacuzzi, your skin will wrinkle, but hey, you're a raisin, your skin's already wrinkled." Monty Hall continues by warning that, "Behind the other doors, however, are two other modes of transportation: two old pack mules. They don't come with Jacuzzis, although given their exotic odor, you may want to give them a bath." Of course, the crowd is laughing and applauding, just as the studio sign instructs.

Monty sums it up: "So, there are three closed doors. Behind one is a luxurious car, and behind the other two are mules. Now comes the moment of truth. What door do you pick?" The audience erupts, "Take Door Number 1, take Door Number 1!!" "Door Number 2, Door Number 2!!" "Door Number 3's the one. Choose 3." Poor Warren Piece looks around at the crowd, confused and nervous. He considers Door Number 1, then 2, then 3. Finally Monty prompts, "Okay, Warren, which do you want?"

The raisin-clad Warren shouts, "Okay, okay, I'll take Door Number 3, Door Number 3." As Monty Hall quiets the overly excited audience, he tells Warren, "I'll tell you what I'm going to do. I'm going to show you what's behind one of the doors you didn't pick. Let's take a look at what's behind Door Number 2." With that, Monty Hall turns to the Vanna White of the 1960s and says, "Please show us what is behind Door Number 2." The door dramatically swings open, the audience erupts, and Warren breathes once more—behind Door Number 2 is a mule! Monty, knowing where the mules are, always opens one of the mule doors first.

Monty continues, "We now see that the Cadillac is *not* behind Door Number 2. You guessed Door Number 3. I'll tell you what I'm going to do. If you want, I'll let you change your mind and choose Door Number 1 instead. It's up to you. Do you want to stick to your original choice, or do you want to switch?" The audience goes nuts. "Stick, stick," yell half. "Switch, switch," advise the others. What to do, what to do?

We now invite you to add your voice to the cacophony—although you need not shout. What should Warren Piece do? Should he switch choices, stick to his original guess, or does it not matter? Here a classic TV game show raises the question: How can we accurately measure the uncertain?

Story 8. Rolling Around in Vegas

Recently the swaggering burly billionaire, Mr. Bones, introduced an exciting new dice game at his glitzy High-Rollin' Bones' Hotel and Casino. An oversized gold bowl containing four dice is presented to the player. The player inspects each die, removes which-

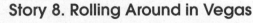

*Visit www.heartofmath.com to see these Web pages come to life and to discover more.

ever die seems the luckiest, and throws a $100 chip in the bowl. Then Mr. Bones chooses one of the three remaining dice, takes a $100 chip (showing his likeness) from his personal collection, and modestly places it into the bowl. Next the player and Mr. Bones roll their respective dice. Whoever rolls the higher number wins the two chips. Simple.

To make the game interesting, the four dice are not the run-of-the-mill dice we remember from the gambling-free days of our youth. While each die *does* have six sides as usual, their faces are marked in unusual ways. The kit that accompanies each new copy of *The Heart of Mathematics: An invitation to effective thinking* contains these four special dice. Roll 'em on out of your kit.

One die has two 6's and four 2's. Another has three 5's and three 1's. The third has four 4's and two blank faces. The last die has 3's on each face. The dice are not weighted—that is, any face is just as likely to land face-up as any other.

Deep Pockets Drew strides up to the bowl to choose the winning die. Which die should Drew draw? Drew considers the die that has all 3's. Which die could Mr. Bones select that will beat the all-3's die two-thirds of the time? After finding that die, we know that the all-3's die would not be a particularly wise choice.

Next Deep Pockets Drew considers the die with four 4's and two blank faces. Why will the die with three 5's and three 1's beat it two-thirds of the time? After verifying this dicey dominance, we know that selecting the die with four 4's and two 0's would not be a smart move.

Drew next considers the die with three 5's and three 1's. Why will the die with two 6's and four 2's beat it two-thirds of the time? After confirming this superiority, we know that the die with three 5's and three 1's would not be the best die.

Only one possibility remains: the die with two 6's and four 2's. Is there a die that will beat it two-thirds of the time? Your surprising discovery will show that none of the four dice is the "best" one to select, because each one can be beaten by one of the other three dice two-thirds of the time. Amazing.

So now Drew can put the dice in a circular order where each one beats its clockwise neighbor two-thirds of the time. What is that order? After doing the math, Deep Pockets Drew chooses not to play, and as a result his pockets become deeper.

This intriguing dice game surprisingly leads to the seemingly unrelated insight that the idea of a fair and democratic voting system is impossible—so much for "a government of the people, by the people, and for the people."

Story 9. Watsamattawith U?

Watsamattawith University (WU) is a fine institution, but a paradoxical place. They have comfortable dorm rooms, yet all the students sleep in class; their track team streaks from place to place, yet their cheeks are red with embarrassment as they lose every meet; every student is vegetarian, yet their dining facility is named Holstein Hall; and their student senate is called the House of Representatives. Go figure!

And go figure, indeed—for that is exactly what the registrar of WU did in computing the average GPA of the current graduating class. Every year she computes the average GPA of the male students, the female students, and then all the students from that graduating class.

This year she noticed something most peculiar. The average GPA of the male students in the current graduating class was higher than the average GPA of the male students from last year's graduating class, and the average GPA of the female students in the current graduating class was higher than the average GPA of the female students from last year's graduating class. Sounds great for the current graduates. Unfortunately, she discovered that the average GPA of the entire graduating class was actually *lower* than the average GPA of last year's class. What!?

Congratulations, graduating class . . . you collectively have a remarkable GPA.

Given that there were no errors in the registrar's computations, is it possible that such a phenomenon could occur or is this scenario so ridiculously impossible that merely asking the question deserves the response: *Watsamattawith U?!* If such a scenario is possible, explain how by describing an example where the GPAs of the males goes up, the GPA of the females goes up, but the GPA of all the students goes down. Otherwise respond with . . . well, you know.

Story 10. Dot of Fortune

One day three college students were selected at random from the studio audience to play the ever-popular TV game show, "Dot of Fortune." One of the students had already discovered the power and beauty of mathematical thinking, while the other two were not nearly so fortunate. The stage contained no mirrors, reflective surfaces, or television monitors. The three students were seated around a small round table and blindfolded. As Pat, the host, explained the rules of the game, Vanna affixed a conspicuous but small colored dot to each student's forehead.

"So, contestants," Pat explained, "at the sound of the bell you will remove your blindfolds. You will see your two companions sitting quietly at the table, each with a dot on his or her forehead. Each dot is either red or white. You cannot, of course, see the dot on your own forehead. After you have observed the dots on your companions' foreheads, you will raise your hand if you see at least one red dot. If you do not see a red dot, you will keep your hands on the table. The object of the game is to deduce the color of your own dot. As soon as you know the color of your dot, hit the buzzer in front of you. Do you understand the rules of the game?" All the students understood the rules, although the math fan understood them better.

"Are you ready?" asked Vanna after affixing a red dot to each student's forehead. After the contestants nodded, Vanna rang the bell and they removed their blindfolds. The studio audience quivered with anticipation. The students looked at one another's dots, and all raised their hands. After some time, the math fan hit her buzzer, knowing what color dot she had. Explain how she knew this. Why did the other students not know? This game requires creative logical reasoning—a powerful means to make discoveries whether they are in math, in life, or even (although rarely) on prime-time TV.

CAUTION!
Proceed to the next section only after you have given considerable thought to each of the stories.

1.2 NUDGES

Leading Questions and Hints for Resolving the Stories

© SuperStock/SuperStock

Honoré Daumier *Crispin and Scapin,* ca. 1858–1860. Shh . . . Don't tell!

Life Lesson

Often we discover a solution only after we move beyond what appears to be the obvious or straightforward approach.

> *When we cannot use the compass of mathematics or the torch of experience . . . it is certain we cannot take a single step forward.*
>
> **VOLTAIRE**

Story 1. That's a Meanie Genie

Initially, we might think that finding the jewel is impossible because Alley is allowed to make only two comparisons. Instead of comparing stones individually, perhaps she should compare one *collection* of stones with another *collection* of stones. Now suppose Alley compares one group with another using the first scale. What can she conclude? What should she do next?

Story 2. Damsel in Distress

Thinking about variations on a situation can shed light on which features are essential and which are not. In this case we might consider a variation in which the damsel in distress is on the other side of a 20-foot river rather than surrounded by a square moat. Unfortunately for the maiden, if she were separated from bliss by a river, she would go blissless, because the two 19-foot beams, in the absence of tools, would still not enable the knight to rescue her. Could the square shape of the moat come into play in the solution?

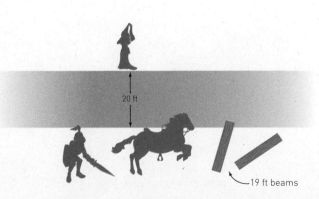

20 ft

19 ft beams

14

Looking at extremes is a potent technique of analysis in many situations and may be helpful here. The extremes, either geometrical ones as in this situation or conceptual ones in other situations, frequently reveal features we might have otherwise overlooked.

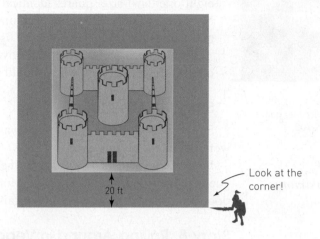

20 ft

Look at the corner!

Story 3. The Fountain of Knowledge

To solve this puzzle, combine trial and error with careful observation. As we observe the outcomes of various attempts, we can teach ourselves what may be possible. Try filling up the 10-ounce glass, and then use it to fill the 6-ounce glass. What do you have now—anything new?

Story 4. Dropping Trou

We hope that you physically attempt this exercise. By actually trying a task on your own, it's often possible to discover insights that otherwise may have been hidden from view (particularly in this case).

You will notice that the rope does restrict the amount of movement of your pants. Your mission is to discover means to work around such constraints. For example, try moving parts of the pants through other parts. You may first want to try this task wearing shorts rather than long pants.

Story 5. Dodgeball

Play this game a few times with a friend. Switch roles so that each of you has the opportunity to be Player One and Player Two. Remember, if you are Player One, your goal is to match one of your rows with your opponent's row. If you are Player Two, you want to dodge all six of your opponent's rows; that is, you want your row to differ in at least one spot from each of the six rows of your opponent. Who would you rather be: Player One or Player Two?

Story 6. A Tight Weave

Consider a purple square that has a smaller gold square in its center. How do each of the eight surrounding squares differ from the whole picture? They are much the same except that the whole picture has a gold square in the middle, and each

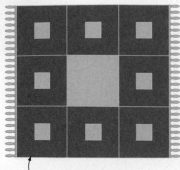

We want each square to look like a smaller copy of the entire rug.

---

of the eight surrounding squares is solid purple. How could you modify those eight one-third-size surrounding squares to make them look like smaller copies of the entire picture?

Now let's ask the question again: "In the picture you now have, is each of the eight one-third-size squares identical to smaller copies of the whole picture?" No. How would you modify each one-third-size square-with-a-gold-center to make it identical to the whole new figure? Are you done?

Draw several steps of this repetitive process. At each stage, add up the areas of all the gold squares. When should you stop this process?

Story 7. Let's Make a Deal

Suppose the raisin's initial guess was wrong. What would be the result if he were to change his answer?

Suppose instead of three doors, there were ten doors. After Warren Piece guessed Door Number 3, suppose Monty Hall opened eight of the remaining doors and all had mules. Should our raisin switch in that game? Why?

Story 8. Rolling Around in Vegas

To compare two dice, consider making a chart with the columns labeled by the six numbers on one die (some of them duplicates) and the rows labeled by the six numbers of the other die (some of them duplicates). Then put the corresponding numbers in each square and see in how many squares one die beats the other. If one die wins 24 of the 36 times, then that die will win two-thirds of the time. A chart for the die with three 5's and three 1's versus the die with four 4's and two blanks has been started for you.

	1	1	1	5	5	5
0	(0,1)			(0,5)		
0	(0,1)			(0,5)		
4	(4,1)			(4,5)		
4	(4,1)			(4,5)		
4	(4,1)			(4,5)		
4	(4,1)			(4,5)		

Story 9. Watsamattawith U?

The natural initial reaction to the question is *Watsamattawith U?!* However, statistical issues often can be both subtle and counterintuitive. Suppose there are 300 students in each of the graduating classes. In last year's class, half were male and half were female. Suppose that the average GPA of the men was 2.0 while the average GPA of the women was 3.5. Then the average GPA for last year's graduating class was 2.75—which is halfway between 2.0 and 3.5. Can you create a graduating class of 300 students for which the average GPA of the men is 2.1, the average GPA of the women is 3.6, and yet the average GPA of the entire graduating class descended to 2.65? Bonus "nudge": Answer—*Yes, you can!*

Story 10. Dot of Fortune

Sometimes no action is action enough. Put yourself in the position of one of the three contestants. You know that the dot on your forehead is either red or white. The trick to figuring out this conundrum is to imagine what would happen if you were wearing a white dot. You are sitting at the table looking at two red dots. What would your two companions be seeing? What could they deduce? What would they do? What did they do? What can you conclude?

STOP!
Do not proceed to the next section until you have thought about the stories, read the previous section, and tried to come up with answers. No peeking!

1.3 THE PUNCH LINES

Solutions and Further Commentary

© Tribalium/iStockphoto

> *Mathematics seems to endow one with something like a new sense.*
>
> **CHARLES DARWIN**

Story 1. That's a Meanie Genie

Alley identifies which stone contains the jewel with no problem because she has read *The Heart of Mathematics*. She arranges the stones into three groups of three and places one group on one side of the first balance scale and another group on the other side. What can she conclude? If both sides weigh the same, then she knows that the (heavier) jewel must be in the third group of three. If, however, one side is heavier than the other, then she knows that the jewel is one of the three that weighed more. In either case, after only one weighing, Alley is able to identify a group of only three stones among which is the Rama Nujan.

She then takes two of these three stones and places one on each side of the second scale. If one weighs more than the other, then she knows that this stone is the one containing the jewel. If they both weigh the same, then she knows that the third stone must contain the jewel. Thus, by weighing the stones only twice, Alley is able to find the jewel.

Take partial steps whenever possible. Notice that, instead of trying to identify the jewel immediately, Alley first reduces the pool of choices from nine to three. Thus she first makes the problem easier. "Divide and conquer" is an important and useful technique in both mathematics and life.

Life Lesson
Break a hard problem into easier ones.

*
WWW.

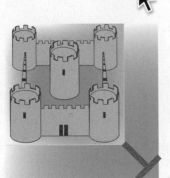

Story 2. Damsel in Distress

Focusing attention on the corner of the moat suggests using one of the beams to span the corner. Of course, we need to check that the two 19-foot beams are long enough to make the configuration in the picture.

There are at least two ways to verify that this picture is correct. One way is to construct a physical model. The picture shown here is a physical model scaled

*Visit www.heartofmath.com to see these Web pages come to life and to discover more.

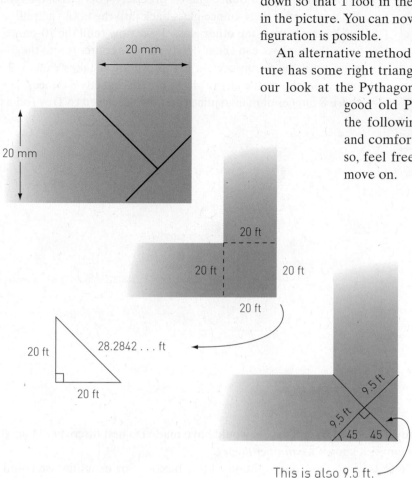

down so that 1 foot in the story corresponds to 1 millimeter in the picture. You can now measure and ensure that this configuration is possible.

An alternative method would be to observe that the picture has some right triangles. This observation foreshadows our look at the Pythagorean Theorem. After we examine good old Pythagoras's theorem (Chapter 4), the following paragraphs will seem soothing and comforting. If for now you find them less so, feel free to glance through them and just move on.

Notice that the corner of the moat forms a 20-foot-by-20-foot square. By the Pythagorean Theorem, the distance from the outer corner of the shore to the inner corner of the castle island is equal to the square root of $20^2 + 20^2$. Using a calculator, we see that the distance is 28.2842 . . . feet.

Placing the 19-foot beam diagonally across the corner of the moat as far out as it can go creates a triangle that cuts off the corner. If we draw a line from the center of the beam to the outer corner of the moat, we create two identical 45-degree right triangles, as shown. Since the length of half the beam is 9.5 feet, we learn that the center of the beam is also 9.5 feet from the outer corner of the moat.

Since the total diagonal distance from the outer corner of the moat to the corner of the castle island is 28.2842 . . . feet, the distance to the center of the beam is (28.2842 . . . feet − 9.5 feet) = 18.7842 . . . feet. Since that distance is just less than 19 feet, the other beam will just barely span the distance between the beam and the island. In gratitude for her rescue, the damsel provided the good knight with a romantic lesson in *geometry*.

Story 3. The Fountain of Knowledge

Suppose we fill up the 10-ounce glass with mango juice and slowly pour it into the 6-ounce glass, stopping at the moment the 6-ounce glass is full. Notice that

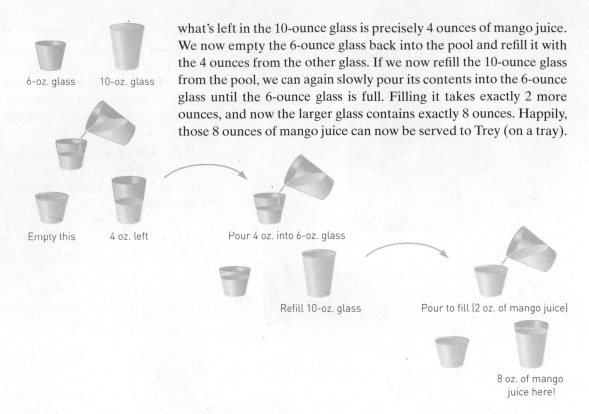

6-oz. glass   10-oz. glass

Empty this   4 oz. left

Pour 4 oz. into 6-oz. glass

Refill 10-oz. glass

Pour to fill (2 oz. of mango juice)

8 oz. of mango juice here!

what's left in the 10-ounce glass is precisely 4 ounces of mango juice. We now empty the 6-ounce glass back into the pool and refill it with the 4 ounces from the other glass. If we now refill the 10-ounce glass from the pool, we can again slowly pour its contents into the 6-ounce glass until the 6-ounce glass is full. Filling it takes exactly 2 more ounces, and now the larger glass contains exactly 8 ounces. Happily, those 8 ounces of mango juice can now be served to Trey (on a tray).

If Trey had found a solution, he would have made his first discovery in an area of mathematics known as *number theory*.

There is more than one solution to this puzzle. For example, we could have begun by filling the 6-ounce glass and pouring its entire contents into the 10-ounce glass. See if you can use this starting point to find an alternative solution.

Story 4. Dropping Trou

The sequence of diagrams shown illustrates a solution to this knotty puzzle. Notice that by bending, contorting, and twisting your pants around, you can produce different configurations. Questions involving bending, contorting, and twisting lead to interesting and surprising discoveries. The notion of bending space is the fundamental notion in an area of mathematics called *topology*.

Often, thinking only in the abstract does not reveal new insights. Make the issue concrete and physical whenever possible.

Many people believe mathematical issues exist outside the realm of our life experience. In truth, many surprising and even counterintuitive mathematical discoveries can be made by freeing ourselves from old, unsubstantiated biases and experimenting with new ways of thinking and seeing.

Life Lesson

> **By doing we often discover valuable insights.**

*Visit www.heartofmath.com to see these Web pages come to life and to discover more.

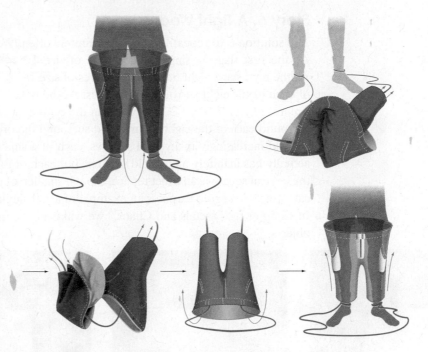

Method: With pants on rope, bring one of the ends (cuff) of the right leg through the inside of the left leg; pull all the way through. When done, pants will be right side out (still) but the rope will now go through the pants. Now reach each hand into the inside of each pants leg and grab the cuffs. Simultaneously, pull the cuffs up through the pants. The pants will be inside out and the rope will no longer be around the pants.

*

Story 5. Dodgeball

We want to be Player Two. Here is a strategy that will guarantee victory. Player One fills in the first row of six boxes in his table. As Player Two, we look at the first letter and ignore the last five. If his first letter is an X, we write an O; if it's an O, we write an X. Notice that, no matter what happens later, after this point, we are certain that the row we will create will definitely not be the same as Player One's first row. The two rows will differ in at least the first box. Player One now writes down his second row of six letters. We examine only the second letter in this new row. If that letter is an X, we write an O; if that letter is an O, we write an X. Now we are sure that no matter what follows, our row will not be the same as Player One's second row because the rows definitely differ in the second letter. If we repeat this process, we will have created a row of X's and O's that is different from the six rows created by Player One.

Creating a row that does not match any of our opponent's rows has a powerful application in the study of *infinity*. Although this modest little game has only six steps, the concept behind it has tremendous ramifications, as we shall see in Chapter 3, "Infinity."

As a final note, we pose the following question: Suppose that we are Player One, and our opponent—who is trying to follow the strategy described above to win—makes a mistake by placing the wrong letter in the first box. Can you now describe a strategy for us, as Player One, to ensure a win? Give this new challenge a try.

Life Lesson

Often simple observations can have deep consequences.

*Visit www.heartofmath.com to see these Web pages come to life and to discover more.

Story 6. A Tight Weave

The solution is to repeat the process infinitely often. We start with a purple square. At the first stage, a single gold square of size 1/3 × 1/3 is placed in the center. At the next stage eight more gold squares of size 1/9 × 1/9 are placed in the centers of each of the eight surrounding squares. At the next stage, 8 × 8, or 64, more gold squares of size 1/27 × 1/27 are placed in the centers of each of the eight squares that surround each of the eight squares that surround the original square. At each stage, we add increasingly many gold squares, each of a smaller size. So the final picture actually has infinitely many gold squares, but each of the eight squares surrounding the central square is an exact replica, though smaller, of the whole picture. This intricate purple and gold carpet is an example of a self-similar object known as a *fractal*. In Chapter 6, "Fractals and Chaos," we will examine many such infinitely intricate objects.

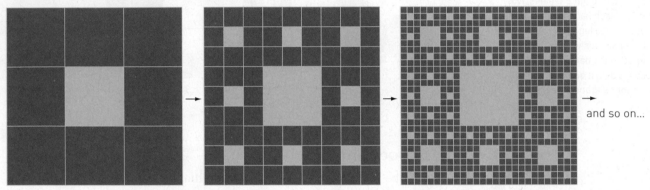

and so on...

What is the area of all the (infinitely many) gold squares? Since all those gold squares lie within the rug that is 1 yard square, we know the area cannot be more than 1. At the first stage, we have one gold square of size 1/3 × 1/3, so its area is 1/9. At the next stage, we add eight more gold squares, each of size 1/9 × 1/9, so their areas total $8 \times (1/9)^2$, making the total area of gold squares at stage two equal to $1/9 + 8 \times (1/9)^2 = 0.2098\ldots$. At the third stage, we add 8^2 more squares, each of area $(1/9)^3$. Thus, the total area of gold squares at stage three equals $1/9 + 8 \times (1/9)^2 + 8^2 \times (1/9)^3 = 0.2976\ldots$. Repeating, we begin to see a pattern. The fourth stage, for example, would have a gold area equal to $1/9 + 8 \times (1/9)^2 + 8^2 \times (1/9)^3 + 8^3 \times (1/9)^4 = 0.3757\ldots$. Thus, the total area of gold squares in the final pattern would be the infinite sum:

$$\frac{1}{9} + 8 \times \left(\frac{1}{9}\right)^2 + 8^2 \times \left(\frac{1}{9}\right)^3 + 8^3 \times \left(\frac{1}{9}\right)^4 + 8^4 \times \left(\frac{1}{9}\right)^5 + 8^5 \times \left(\frac{1}{9}\right)^6 + \ldots .$$

What does it equal? Even though there are infinitely many terms, we know that the whole area must be a number not greater than 1. What number is it?

The gold area at the 5th stage is 0.4450 . . . ;

at the 10th stage it is 0.6920 . . . ;

at the 15th stage it is 0.8291 . . . ;

at the 25th stage it is 0.9474 . . . ;

at the 50th stage it is 0.9972 . . . ;

at the 100th stage it is 0.999992 . . . .

Life Lesson
Data can help uncover surprising observations and help build intuition and understanding.

From this pattern of numbers, it becomes clear that the gold area becomes increasingly close to 1—and that is a great guess for the area.

A clever way to calculate the total area is to add up all the infinitely many terms. We start by giving a name to the total; let's call that number SUM. On the next page you see the infinite sum that SUM represents. Directly under that, you see what (8/9)SUM equals. Notice that multiplying each term of SUM by 8/9 just shifts that term to the right. For example, $(8/9)(1/9) = 8 \times (1/9)^2$.

$$\text{sum} = \frac{1}{9} + 8 \times \left(\frac{1}{9}\right)^2 + 8^2 \times \left(\frac{1}{9}\right)^3 + 8^3 \times \left(\frac{1}{9}\right)^4 + 8^4 \times \left(\frac{1}{9}\right)^5 + 8^5 \times \left(\frac{1}{9}\right)^6 + \ldots$$

$$\left(\frac{8}{9}\right)\text{sum} = 8 \times \left(\frac{1}{9}\right)^2 + 8^2 \times \left(\frac{1}{9}\right)^3 + 8^3 \times \left(\frac{1}{9}\right)^4 + 8^4 \times \left(\frac{1}{9}\right)^5 + 8^5 \times \left(\frac{1}{9}\right)^6 + \ldots$$

Since all the terms of (8/9)SUM are directly under an identical term of SUM, it is easy to subtract (8/9)SUM from SUM, because all the terms drop out except the first term:

$$\text{SUM} - \left(\frac{8}{9}\right)\text{SUM} = \frac{1}{9} \quad \text{and so:}$$
$$\left(\frac{1}{9}\right)\text{SUM} = \frac{1}{9}$$

Since (1/9)SUM = 1/9, what is SUM? It must equal 1! In other words, the area of the gold squares is equal to the area of the entire rug. Thus, even though there are many purple threads remaining in the final pattern, as we begin to see in the illustration on page 22, the purple contributes no area to the rug. Surprise! We will see many more counterintuitive mysteries of infinity in our studies of numbers, fractals, and, of course, infinity itself.

Story 7. Let's Make a Deal

Fortunately, Warren Piece enjoys mathematics as a hobby, so he believes he can solve this conundrum. He thinks carefully, assesses the chances each way, and confidently proclaims (while still jumping up and down, of course), "I switch my guess to Door Number 1, Monty."

Monty Hall turns and says, "Okay. Let's see what deal you've made. What is behind Door Number 1?" The door swings slowly open, and the crowd gasps as they see behind Door Number 1 the most beautiful finned chassis that General Motors ever painted pink. Bedlam reigns. "How did you know?" asks Monty Hall over the din. Warren Piece explains.

"When I originally guessed Door Number 3, I had a one-third chance of being right and a two-thirds chance of being wrong. Thus it's more likely I was wrong than right. When you opened Door Number 2 and revealed no car, I hoped I was wrong originally—which, recall, was more likely than not. If I was wrong originally, then the car must be behind the remaining door, Door Number 1. So I switched, knowing that the probability of my winning after switching was 2 out of 3, whereas the chance of my having picked it correctly the first time was only 1 out of 3." Monty Hall was compelled to ask, "How did you figure that out?"—to which our hero raisin sagely replied, "By studying *The Heart of Mathematics: An invitation to effective thinking*." (Not a bad coast-to-coast TV plug for our book, even coming from a guy named Warren Piece in a raisin suit.)

Some people might think that it doesn't matter if he switches or not. However, the chances of finding the car are indeed greater by switching. One way to demonstrate this is to list all the possible ways the Cadillac and the mules can be placed behind the doors:

	Door Number 1	Door Number 2	Door Number 3
Case 1	Cadillac	mule	mule
Case 2	mule	Cadillac	mule
Case 3	mule	mule	Cadillac

Warren Piece first picked Door Number 3, and the likelihood of his finding the car there was 1 out of 3, that is, one-third, which is not too likely. Next, Monty opened a door showing a mule. Let's see what happens if Warren were to switch in each of the three possible scenarios. In Case 1, Monty opens Door Number 2. If Warren switches in this case, he wins the Cadillac. In Case 2, Monty opens Door Number 1. If Warren switches in this case, he would win. In Case 3, Monty could open either Door Number 1 or 2. If Warren switches in this case, sadly, he would be the owner of a mule. Therefore, overall, the likelihood of his winning the car by switching is 2 out of 3, or two-thirds, which is twice as likely as the one-third chance of selecting the car if he sticks to his original guess. This brief encounter with probability illustrates that counterintuitive outcomes can occur during attempts to measure the unknown.

Story 8. Rolling Around in Vegas

The die with four 4's and two blanks beats the die with all 3's two-thirds of the time. There are four 4's on the 4-0 die so it comes up 4 two-thirds of the time it is rolled. Whenever it comes up 4 it beats the 3-die, which always comes up 3. So the die with four 4's and two 0's will win two-thirds of the time.

The die with three 5's and three 1's will beat the die with four 4's and two blanks, because one-third of the time it is rolled, the 4-0 die will come up with a 0, in which case it loses, no matter what the 5-1 die comes up. In two-thirds of the cases where the 4-0 die comes up with a 4, half those times, or one-third of the time that the 4-0 die is rolled, the 5-1 die will come out with a 5, in which case the 5-1 die wins anyway. So altogether, two-thirds of the time the 5-1 die wins over the 4-0 die. We can also see this fact by making a chart of all 36 outcomes of rolling the two dice against each other and noticing that in 24 of those outcomes the 5-1 die wins. Those 24 winning squares are colored in gold in the chart below.

	1	1	1	5	5	5
0	(0,1)	(0,1)	(0,1)	(0,5)	(0,5)	(0,5)
0	(0,1)	(0,1)	(0,1)	(0,5)	(0,5)	(0,5)
4	(4,1)	(4,1)	(4,1)	(4,5)	(4,5)	(4,5)
4	(4,1)	(4,1)	(4,1)	(4,5)	(4,5)	(4,5)
4	(4,1)	(4,1)	(4,1)	(4,5)	(4,5)	(4,5)
4	(4,1)	(4,1)	(4,1)	(4,5)	(4,5)	(4,5)

The die with the two 6's and four 2's will beat the die with three 5's and three 1's, because one-third of the time it is rolled, the 6-2 die will come up with a 6, in which case it wins no matter what the 5-1 die comes up. In two-thirds of the cases where the 6-2 die comes up with a 2, half those times, or one-third of the times the 6-2 die is rolled, the 5-1 die will come out with a 1, in which case the 6-2 die wins anyway. So altogether, two-thirds of the time the 6-2 die wins over the 5-1 die. Again, we can also see this fact by making a chart of all 36 outcomes of rolling the two dice against each other and noticing that in 24 of those outcomes the 6-2 die wins. Those 24 winning squares are colored in gold in the chart on the next page.

	1	1	1	5	5	5
2	(2,1)	(2,1)	(2,1)	(2,5)	(2,5)	(2,5)
2	(2,1)	(2,1)	(2,1)	(2,5)	(2,5)	(2,5)
2	(2,1)	(2,1)	(2,1)	(2,5)	(2,5)	(2,5)
2	(2,1)	(2,1)	(2,1)	(2,5)	(2,5)	(2,5)
6	(6,1)	(6,1)	(6,1)	(6,5)	(6,5)	(6,5)
6	(6,1)	(6,1)	(6,1)	(6,5)	(6,5)	(6,5)

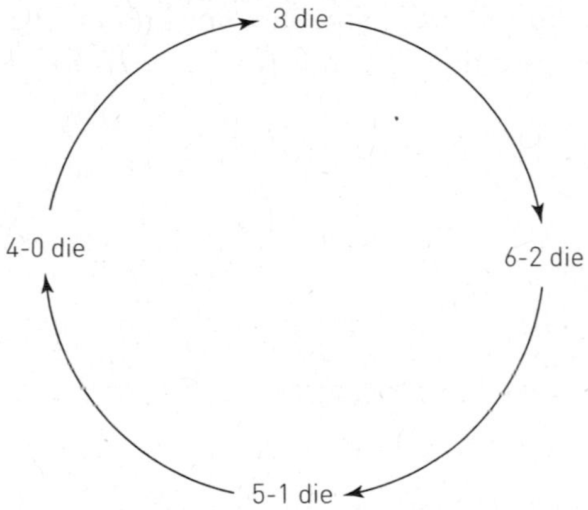

Finally, the die with all 3's can beat the die with two 6's and four 2's two-thirds of the time. To see this, just note that a 2 will appear two-thirds of the times that the 6-2 die is rolled, and each time a 2 appears, the 3-die will win.

So in the ordering shown here, each die beat the one next to it in the clockwise direction two-thirds of the time.

Story 9. Watsamattawith U?

Suppose there are 300 students in each of the graduating classes. In last year's class, 150 were male and 150 were female. Suppose that the average GPA of last year's men was 2.0 and the average GPA of last year's women was 3.5. Then the average GPA for last year's graduating class was 2.75. One way to arrive at that answer is to replace all the men's GPAs with the male average and all the women's GPAs with the female average and then average all 300 GPAs as follows:

$$(2.0 \times 150 + 3.5 \times 150)/300 = 825/300 = 2.75.$$

Now assume that this year's graduating class is comprised of 200 men and only 100 women. If the average GPA of the male students rose to 2.1 (higher than last year's average) and the average GPA of the female students became 3.6 (again higher than last year's average), then the GPA for this year's class can be found by computing: $(2.1 \times 200 + 3.6 \times 100)/300 = (420 + 360)/300 = 780/300 = 2.6!$ Amazing . . . until we realized we had the ability to change the proportion of men to women. By having a higher proportion of the poorer male students, even though the GPA of the males increased from 2.0 to 2.1, since there were more males this year, they dragged down the GPA of the student body. The number of male students compared to female students was a quantity that could be changed,

3 die

4-0 die

6-2 die

5-1 die

Life Lesson

Whenever possible, move from qualitative thinking to quantitative thinking.

Life Lesson

Be on the lookout for opportunities in which you have the ability to vary a parameter or circumstance.

but we might not have thought about that possibility. Hidden features of a statistical situation like this are sometimes called "lurking variables."

Story 10. Dot of Fortune

The math fan sees a red dot on the forehead of each of the other two players. She knows she has either a white dot or a red dot on her own forehead. Let's see what happens if we suppose her dot is white.

What would her two companions at the table see? Each would see one red dot and one white dot, and each would see two arms raised. Each would be thinking, "Do I have a red dot or a white dot on my forehead? If I have a white dot, then the red-dotted person would not have her hand up. Therefore, I must have a red dot." After making this easy deduction, this person would hit the buzzer.

But what did these two people actually do? Or, more to the point, what did they *not* do? They did not hit their buzzers! If either of them had seen a white dot and a red dot and two raised hands, he or she would have been able to deduce that his or her own dot was red. Since neither person buzzed right away, neither must have seen a white dot on the math fan's forehead. Therefore, the math fan waited just long enough to know that the other two players could not deduce their own dot colors, and then she buzzed, confident that her dot was red.

A final question of the story is, Why did the other students not know? The answer to that question is, of course, because they had not read *The Heart of Mathematics*.

Life Lesson

There is great power to be found in logical and creative thinking.

1.4 FROM PLAY TO POWER

Discovering Strategies of Thought for Life

Grandmaster Maurice Ashley plays Cameron Clarke, age 10, during the "It's your move" chess exhibition.

> *Imagination is more important than knowledge.*
>
> ALBERT EINSTEIN

Our stories illustrate strategies of thinking. Even in such a lighthearted setting, certain techniques of thought emerge as powerful means to illuminate the unknown—techniques applicable to any situation we may face in life. We'll encounter more "life lessons" elsewhere in *The Heart of Mathematics;* here we've summarized a few. Although some may seem obvious or trivial, don't take them lightly—they can be surprisingly useful for analyzing and enjoying life's adventure.

Lessons for Life

1. *Just do it.*
2. *Make mistakes and fail, but never give up.*
3. *Keep an open mind.*
4. *Explore the consequences of new ideas.*
5. *Seek the essential.*
6. *Understand the issue.*
7. *Understand simple things deeply.*
8. *Break a difficult problem into easier ones.*
9. *Examine issues from several points of view.*
10. *Look for patterns and similarities.*

MINDSCAPES · Invitations to Further Thought

We now provide some additional stories for further amusement and enlightenment. We call them "Mindscapes" because they are vistas for the mind that encourage you to expand your way of thinking.

For each of the following situations, contemplate, analyze, and resolve the puzzle. Also, guess which branch of mathematics each situation represents: Logic,

27

Number Theory, Infinity, Geometry, Topology, Chaos, or Probability. Of course, we haven't discussed any of these areas in depth yet, but just take a guess—being wrong is fine.

Finally, we invite you to provide an aesthetic critique of each question and your solutions. In other words, did you find either the question or your solution interesting? Which questions were the most challenging? Do you like one of your solutions better than the others? At the end of this section we provide some hints for some of the questions. Use them sparingly.

> *"Contrariwise,"* continued *Tweedle-dee, "if it was so, it might be; and if it were so, it would be; but as it isn't, ain't. That's logic."*
>
> LEWIS CARROLL

1. **Late-night cash.** Suppose that David Letterman and Paul Shaffer have the same amount of money in their pockets. How much must Dave give to Paul so that Paul will have $10 more than Dave?

2. **Politicians on parade.** There were 100 politicians at a certain convention. Each politician was either crooked or honest. We are given the following two facts:

   a. At least one of the politicians was honest.

   b. Given any two of the politicians, at least one of the two was crooked.

   Can it be determined from these facts how many of the politicians were honest and how many were crooked? If so, how many? If not, why not?

3. **The profit.** A dealer bought an item for $7, sold it for $8, bought it back for $9, and sold it for $10. How much profit did she make?

4. **The truth about . . .** Fifty-six biscuits are to be fed to 10 pets; each pet is either a cat or a dog. Each dog is to get six biscuits, and each cat is to get five. How many dogs are there? (Try to find a solution without performing any algebra.)

5. **It's in the box.** There are two boxes: one marked A and one marked B. Each box contains either $1 million or a deadly snake that will kill you instantly. You must open one box. On box A there is a sign that reads: "At least one of these boxes contains $1 million." On box B there is a sign that reads: "A deadly snake that will kill you instantly is in box A." You are told that either both signs are true or both signs are false. Which box do you open? Be careful! The wrong answer is fatal!

6. **Lights out.** Two rooms are connected by a hallway that has a bend in it so that it is impossible to see one room while standing in the other. One of the rooms has three light switches. You are told that exactly one of the switches turns on a light in the other room, and the other two are not connected to any lights. What is the fewest number of times you would have to walk to the

other room to figure out which switch turns on the light? And the follow-up question is: Why is the answer to the preceding question "one"? (Look out: This question uses properties of real lights as well as logic.)

7. **Out of sight but not out of mind.** The infamous band Slippery Even When Dry ended their concert and checked into the Fuzzy Fig Motel. The guys in the band (Spike, Slip, and Milly) decided to share a room. They were told by Chip, the night clerk who was taking a home-study course on animal husbandry, that the room cost $25 for the night.

   Milly, who took care of the finances, collected $10 from each band member and gave Chip $30. Chip handed Milly the change, $5 in singles. Milly, knowing how bad Slip and Spike were at arithmetic, pocketed two of the dollars, turned to the others, and said, "Well guys, we got $3 change, so we each get a buck back." He then gave each of the other two members a dollar and pocketed the last one for himself.

   Once the band members left the office, Chip, who witnessed this little piece of deception, suddenly realized that something strange had just happened. Each of the three band members first put in $10, so there was a total of $30 at the start. Then Milly gave each guy and himself $1 back. That means that each person put in only $9, which is a total of $27 ($9 from each of the three). But Milly had skimmed off $2, so that gives a total of $29. But there was $30 to start with. Chip wondered what happened to that extra dollar and who had it. Can you please resolve and explain the issue to Chip?

8. **Comedy Central.** A good comedian armed with strong material can really kill an audience. Bad comedians, on the other hand, are caught and brought to Comedy Central Correctional (CCC), where they do their timing. But moving C- and D-list comedians to lockdown is not as easy as it may appear. One day, in the outback of Los Angeles, three Comedy Correctional officers were escorting three bad comedians, who had just bombed at the Laff Stop, to CCC. Suddenly, along their march to CCC, they came upon the banks of the mighty Los Angeles River and needed to cross its turbulent and deep waters. None of the six could swim, but all could row. Fortunately, on the river's shore was a small rowboat available for use.

   Since the boat was small and the officers and comedians were all on the portly side, it was clear that only two persons could cross at one time. The problem was: While each comedian was not particularly great, if ever there were a moment when the comedians outnumbered the officers, then the comedians—together—were funny enough to kill their audience (i.e., the officers). Given this reality, the officers decided that being prudent was better than being dead, so they agreed that at no time would they allow any group of officers to be outnumbered by comedians during the crossing. For their part, the comedians did not fear being outnumbered by the officers because they realized that an excess of officers would result only in more discussion among the officers, thus relieving the comedians of the burden of being "on."

   How do the officers and comedians all cross the river using only the one boat yet at no time letting the comedians outnumber the officers on either side of the river?

9. **Whom do you trust?** Congresswoman Smith opened the *Post* and saw that a bean-counting scandal had been leaked to the press. Outraged, Smith

immediately called an emergency meeting with the five other members of the Special Congressional Scandal Committee, the busiest committee on Capitol Hill.

Once they were all assembled in Smith's office, Smith declared, "As incredible as it sounds, I know that three of you always tell the truth. So now I'm asking all of you, Who spilled the beans to the press?"

Congressman Schlock spoke up, "It was either Wind or Pocket."

Congressman Wind, outraged, shouted, "Neither Slie nor I leaked the scandal."

Congressman Pocket then chimed in, "Well, both of you are lying!"

This provoked Congressman Greede to say, "Actually, I know that one of them is lying and the other is telling the truth."

Finally, Congressman Slie, with steadfast eyes, stated, "No, Greede, that is not true."

Assuming that Congresswoman Smith's first declaration is true, can you determine who spilled the beans?

10. **A commuter fly.** A passenger train left Austin, Texas, at 12:00 p.m. bound for Dallas, exactly 210 miles away; it traveled at a steady 50 miles per hour. At the same instant, a freight train left Dallas headed for Austin on the same track, traveling at 20 miles per hour. At this same high noon, a fly leaped from the nose of the passenger train and flew along the track at 100 miles per hour. When the fly touched the nose of the oncoming freight train, she turned and flew back along the track at 100 miles per hour toward the passenger train. When she reached the nose of the passenger train, she instantly turned and flew back toward the freight train. She continued turning and flying until, you guessed it, she was squashed as the trains collided head on.

How far had the fly flown before her untimely demise?

11. **A fair fare.** Three strangers, Bob, Mary, and Ivan, meet at a taxi stand and decide to share a cab to cut down the cost. Each has a different destination, but all the destinations are on the highway leading from the airport, so no circuitous driving is required. Bob's destination is 10 miles away, Mary's is 20 miles, and Ivan's is 30 miles. The taxi costs $1.50 per mile including the tip, regardless of the number of passengers. How much should each person pay? (*Caution:* There is more than one way of looking at this situation.)

12. **Getting a pole on a bus.** For his 13th birthday, Adam was allowed to travel down to Sarah's Sporting Goods store to purchase a brand new fishing pole. With great excitement and anticipation, Adam boarded the bus on his own and arrived at Sarah's store. Although the collection of fishing poles was tremendous, there was only one pole for Adam and he bought it: a 5-foot, one-piece fiberglass "Trout Troller 570" fishing pole.

When Adam's return bus arrived, the driver reported that Adam could not board the bus with the fishing pole. Objects longer than 4 feet were not allowed on the bus. Adam remained at the bus stop holding his beautiful 5-foot Trout Troller. Sarah, who had observed the whole ordeal, rushed out and said, "We'll get your fishing pole on the bus!" Sure enough, when the same bus and the same driver returned, Adam boarded the bus with his fishing pole, and the driver welcomed him aboard with a smile. How was Sarah able to have Adam board the bus with his 5-foot fishing pole without breaking or bending the bus-line rules or the pole?

13. **Tea time.** Carmilla Snobnosey lifted the delicate Spode teapot and poured exactly 3 ounces of the aromatic brew into the flowered, shell china teacup. She placed the cream pitcher, also containing exactly 3 ounces, on the Revere silver tray and carried the offering to Podmarsh Hogslopper.

"Would you like some tea and cream, Mr. Hogslopper?" she asked.

"Yup. Thanks. Ow doggie, sure looks hot. I'd better cool it down with this here milk," he responded politely and carefully poured exactly 1 ounce of cream into his steaming tea and stirred. "That oughta do it," he said when the steam stopped rising from the tea. "Here, I'll just give you back that there cream." Whereupon he carefully spooned exactly 1 ounce from his teacup back into the creamer. Podmarsh blushed as he looked at a tea leaf or two floating in the cream and realized his faux pas. Caught at an awkward pass, he decided to smooth things over with an intriguing puzzle.

"Ya know, Mrs. Snobnosey, I wonder if the tea is more diluted than the cream, or if the cream is more diluted than the tea?"

Resolve the dilution problem.

14. **A shaky story.** Stacy and Sam Smyth were known for throwing a heck of a good party. At one of their wild gatherings, five couples were present (this included the Smyths, of course). The attendees were cordial, and some even shook hands with other guests.

Although we have no idea who shook hands with whom, we do know that no one shook hands with themselves and no one shook hands with his or her own spouse. Given these facts, a guest might not shake anyone's hand or might shake as many as eight other people's hands. At midnight, Sam Smyth gathered the crowd and asked the nine other people how many hands each of them had shaken.

Much to Sam's amazement, each person gave a different answer. That is, someone didn't shake any hands, someone else shook one hand, someone else shook two hands, someone else shook three hands, and so forth, down to the last person, who shook eight hands. Given this outcome, determine the exact number of hands that Stacy Smyth shook.

*

15. **Murray's brother.** On another archeological dig, Alley discovered another ancient oil lamp. Again she rubbed the lamp, and a different genie named Curray appeared. After Alley explained her run-in with Murray, Curray responded, "Well, since you know my brother Murray, it's like we're almost family. I'm going to give you four wishes instead of three. What do you say?" Since things had worked out so well the last time, she said, "I already found the Rama Nujan, so now I'd like to find the Dormant Diamond." "You got it," replied Curray. And instantly 12 identical-looking stones appeared. She then used her last three wishes to acquire three balance scales. Each scale was clearly labeled, "One Use Only." Alley looked at the stones and was unable to differentiate any one from the others. Curray explained, "The diamond is embedded in one of the stones. Eleven of the stones weigh the same,

but the stone containing the jewel weighs either slightly more or slightly less than the others. I am not telling you which—you must find the right stone and tell me whether it is heavier or lighter."

Alley could use each of the three balance scales exactly once. She was able to select the stone containing the Dormant Diamond from among the 12 identical-looking stones and determine whether it was heavier or lighter than each of the 11 other stones. This puzzle is a challenge. Try to figure out how Alley might have accomplished this feat.

Further Challenges

16. **Cutting (chess) boards.** Suppose we are given a standard 8 × 8 checkerboard and an immense supply of dominoes. Each domino can cover exactly two adjacent squares on the checkerboard (first checkerboard below). As a warm-up, verify that the checkerboard can be covered completely by dominoes where each domino covers exactly two squares and the dominoes do not overlap one another. Assume next that two squares of the checkerboard have been cut off as shown (second checkerboard). Your challenge now is to determine if you can cover this cut checkerboard with nonoverlapping dominoes so that again, each domino covers exactly two squares. Finally, your last challenge is to consider the same question for the truncated checkerboard (last checkerboard). Does your answer change? Justify your answers.

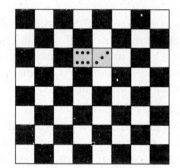
One domino covers two adjacent squares

Cut checkerboard

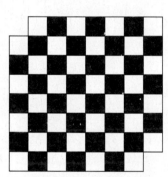
Trucated checkerboard

17. **Siegfried & You.** Consider the following mathematical illusion. A regular deck of 52 playing cards is shuffled several times by an audience member until everyone agrees that the cards are completely shuffled. Then, without looking at the cards themselves, the magician divides the deck into two equal piles of 26 cards. The magician taps both piles of face-down cards three times. Then, one by one, the cards of both piles are revealed. Magically, the magician was able to have the cards arrange themselves so that the number of cards showing black suits in the first pile is identical to the number of cards showing red suits in the second pile. Your challenge is to figure out the secret to this illusion and thén perform it for your friends.

18. **Penny for your thoughts.** Some number of pennies are spread out on a table. They lie either heads up or tails up (see the figure on the next page). Unfortunately, you are blindfolded and thus both the coins and the table upon which they sit are hidden from view. You can feel your way across the table and thus can count the total number of pennies on the table's surface,

but you cannot determine if any individual penny rests heads up or down (perhaps you're wearing gloves). You are informed of one fact (beyond the total number of pennies on the table): Someone tells you the number of pennies that are lying heads up. While remaining blindfolded, you may now rearrange the coins, turn any of them over, and move them in any way you wish as long as the final configuration has all the pennies resting (heads or tails up) on the table. Your challenge is to arrange the pennies into two collections and then turn over whatever pennies you wish so that both collections have the same number of heads-up pennies.

19. When will the world end? The Towers of Hanoi is a puzzle consisting of three pegs and a collection of punctured disks of different diameters that can be placed around any of the pegs. The puzzle begins with all the disks on a single peg in descending order of diameter, with the largest disk on the bottom (first figure below). The object is to transfer all the disks to another peg so that they end up residing on this new peg in the original descending order given the following two rules: Every move consists of removing the top disk on one peg

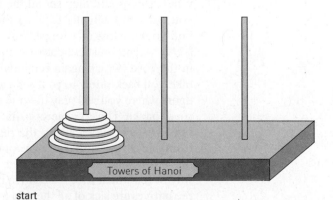

start

*Visit www.heartofmath.com to see these Web pages come to life and to discover more.

and placing it on top of the pile on another peg; and at no time can a larger disk be placed on top of a smaller disk (final figure). Describe a solution to the puzzle if there are four disks, then again if there are five disks, and again if there are six disks. Can you discover a pattern to the minimum number of moves required to solve the puzzle, given how many disks there are?

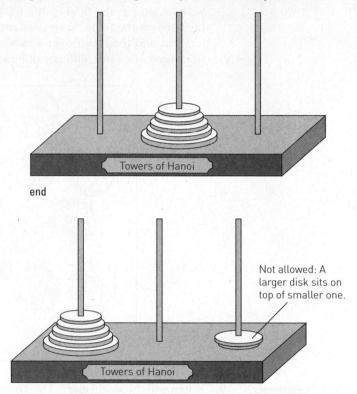

end

Not allowed: A larger disk sits on top of smaller one.

There is a legend that certain monks had a magnificent edition of this puzzle consisting of 64 gold disks and three diamond pins. They were able to move one disk per second. The legend is that the world would end once the monks completed their mission. Use the pattern you found to predict when the world will end—a useful piece of information as you plan your future.

20. **The fork in the road.** You are vacationing on a mythical island resort in which it never rains and they get all the cable stations you know and love. One day you decide to actually turn off the TV and go outside for a hike. Soon you find yourself lost in a forest. You yearn for the main ingredient of the Food Network and must get back to your resort pronto. You finally arrive at a fork in the road (which again reminds you of food). You know that one path will take you back safely to your villa and the other one will lead you into a den of tigers where you will play the role of the main course. You have no idea which road to take. The good news is that you see a native by the fork in the road who knows which road leads to the resort; the bad news is that the natives come from one of two tribes: the Liars or the Truth-Tellers. The Liars always lie, while the Truth-Tellers always tell the truth. The other bad news is that there is no way to tell which tribe this native belongs to. No matter which tribe, however, the natives are sick of all the silly tourists (like you) and all their TV talk. They can only stand one question—that is, you can only ask one question in order to find your way back. What question do you ask the native?

Hints and Problem-Solving Techniques

Here are some suggestions for tackling these puzzles. Hints and solutions to selected Mindscapes in later chapters appear at the end of the book.

1. **Late-night cash.** Try it! After you act it out, explain what happened.

2. **Politicians on parade.** What if more than one politician is honest? Read fact (b) carefully.

Life Lesson
Carefully understand and analyze the facts at hand.

3. **The profit.** Different people will get different answers, and each person will argue that his or hers is correct. Act out the transactions and see what happens. After you try this, go back and figure out why other answers are incorrect.

Life Lesson
Experimentation is an effective means of resolving difficult issues.

4. **The truth about . . .** What if all the animals were cats? How many extra biscuits would you have? Consider turning some of those cats into dogs. This transformation leads to an algebra-free solution.

Life Lesson
Often a clever idea can be more potent than conventional wisdom.

5. **It's in the box.** Consider the two possibilities carefully. You don't want to slip up on this one.

Life Lesson
Carefully consider the outcomes of various scenarios.

6. **Lights out.** Suppose you turn on a switch, wait a half hour, and then turn the switch off. If you were then to walk into the other room, could you tell if the light had been on for half an hour? Ponder this question, and use it to resolve the original puzzle.

Life Lesson
Don't overlook or dismiss facts that seem insignificant or irrelevant.

7. **Out of sight but not out of mind.** Don't be fooled by all the numbers. Force yourself to figure out what was paid out and what was given back.

Life Lesson

Don't believe unsubstantiated claims, even if they sound scientific. Until you understand the issue for yourself, be skeptical!

If that doesn't help, get 30 $1 bills and act out the entire episode. Once you discover the truth, go back and find out where the problem is in the story.

Life Lesson

Experimentation is a powerful means for discovering patterns and developing insights.

See how many different ways you can devise to understand and explain what actually happened.

Life Lesson

Once you find an argument that resolves an issue, it is a great challenge to find a different argument. However, in attempting to find other arguments, we often gain further insight into and understanding of the situation. Also, the first argument we come up with may not be the best one.

8. **Comedy Central.** Professor Starbird shares his grandmother's solution.

When my grandmother was 92, I gave this challenge to her along with three nickels to represent the comedians and three Life Savers candies to represent the officers. We set up a line on her table to represent the river so that she could slide the comedians and the officers (the nickels and the Life Savers) back and forth singly or in pairs, thereby solving the puzzle. When I arrived the next day for my visit, she was delighted to tell me that she had solved the puzzle.

"How did you do it?" I asked.

She replied triumphantly, "I ate the officers."

We give her half credit. A useful aspect of her method was to model the puzzle using a concrete representation. Making a written table with two columns would also be a good way to represent the setting. One column would be one bank of the river, and the other column would be the other bank. Each row would represent the situation after a crossing. So the first row would have three C's, three O's, and a B for the boat in the left-hand column and nothing in the right. The next row might have two C's and two O's in the left column and one C, one O, and the B in the right column. Going from row to row must be possible by moving one or two C's or O's along with the B to the other column.

Life Lesson

> *Once you have an effective representation of this question, a little experimentation will lead to an answer.*

Life Lesson

> *Devising a good representation of a problem is frequently the biggest step toward finding a solution.*

9. **Whom do you trust?** To find the person who leaked the story, you must determine who is telling the truth. Ask yourself whether you can determine the truthfulness or deceit of any one person.

   If Pocket is telling the truth, then Schlock and Wind are liars, and the remaining three—Pocket, Greede, and Slie—are telling the truth. Could those three all be telling the truth? If not, then you know for certain that Pocket is lying.

   Since Slie contradicts Greede, you know that one of them is lying. Which one?

Life Lesson

> *A rock of certainty can be the foundation of a tower of truth.*

10. **A commuter fly.** On close inspection, notice that the fly changes directions an infinite number of times during her travels. It is possible to compute how far the fly has flown before she encounters the freight train for the first time. Once you know this, it's possible to compute the distance she travels before encountering the passenger train on the return trip. You could compute those distances and find a pattern and then solve the problem by adding up the infinite list of distances. However, there is a much easier way to solve this puzzle.

    How much time will pass before the trains collide? How far will the fly fly in that length of time? Case closed.

    This story is not complete without our telling an anecdote about the famous mathematician John von Neumann. Von Neumann was notorious for being extremely fast and accurate at calculating numbers in his head—oddly enough, not a skill that all mathematicians possess. One day he was walking with a friend who asked him the question about the fly between the trains. Instantly, von Neumann stated the answer. The questioner said, "Oh, you saw the trick." To which von Neumann replied, "Yes, it was an easy infinite series."

    If you are not von Neumann, the fly-between-the-trains story provides a good life lesson.

Life Lesson

> *Look at problems from different perspectives.*

Go out of your way to think about different ways to view a problem. In this case, if you know how long the fly flies, you can compute the distance the fly travels. You have now reduced the original problem to a different, though related, problem. In this case, the different problem is much simpler to solve than the original one.

Life Lesson

Look at related situations.

11. **A fair fare.** This question does not have one definitive answer. However, a look at a related problem may persuade you that one possibility is best. What if, instead of staying in one taxi the whole time, the three travelers traveled the first 10 miles together and then all got out and paid the first cabby. The first traveler then left, and the remaining two got another cab, rode 10 miles, and again got out and paid the second cabby. Then the last traveler took a cab alone for the remaining 10 miles. This rephrasing of the original problem makes the division of payment seem more obvious.

12. **Getting a pole on a bus.** It seems impossible to get the 5-foot pole on the bus, given that the largest length of an item allowed on the bus is 4 feet. Sarah gave Adam a large box to put the pole in. Now give the dimensions of the box and explain why it does the trick.

Life Lesson

Often an inventive solution arises from looking at a situation in an unusual way.

13. **Tea time.** This question contains much unnecessary and distracting information. A close look at the story reveals that the description of the dinnerware and the names of the people are extraneous details. But what may not be quite so obvious is that the number of ounces in the teacup and the creamer and the amounts poured and spooned are also irrelevant.

    Don't be distracted by this extraneous information. Suppose the problem did not contain those facts at all and instead was stated as follows:

    A creamer and a teacup each have exactly the same amount of cream and tea, respectively. An undisclosed amount of mixing of the cream and tea goes on, but after the mixing, each of the two containers still contains the same amount of liquid as the other. Is the tea more diluted than the cream, or is the cream more diluted than the tea?

    Having less information might force you to look at the situation differently and, consequently, to understand and solve it.

Life Lesson

Look at problems from different perspectives.

14. **A shaky story.** Exactly one person at the party said to Sam that he or she shook eight hands. Note the obvious fact that each person with whom that

person shook hands must have shaken hands with at least one person. Now determine how many hands that person's spouse shook. See if this approach leads to any insights. If it still does not, consider an easier problem: Suppose that there were just three couples, or even two couples. Search for a pattern.

Life Lesson

If you have a hard problem, first work on a simpler, related problem to develop insight.

15. **Murray's brother.** This genie has posed a difficult challenge. It can be fun to work on, but do not work on it too long if you get frustrated. In this puzzle, we must squeeze every ounce (or even gram) of information from every weighing.

Life Lesson

Don't ignore information.

Each weighing must be designed to give us maximum information. After a weighing, we learn many things. Let's begin by putting four stones on each side of a scale and recording what we observe. If the scale balances, we know that all eight stones weigh the same, and the diamond is not among those eight. So the mystery stone is among the remaining four, but we still do not know whether it is heavier or lighter than the others. Can you now find the Dormant Diamond and determine whether it is heavy or light?

Suppose that the four-against-four weighing does not balance. This imbalance gives us much information. We know that the unweighed four stones all weigh the same. We know that each of the four stones on the light side of the scale are potentially light, but none of them is potentially heavier than the 11 other stones. We know similar things about the four stones on the other side of the scale. We will have to keep track of the stones and consider putting potentially light stones with potentially heavy ones to help sort things out. For example, suppose we weigh a potentially light stone with a potentially heavy stone on one side of the scale and two stones that are known to be normal on the other side. Then, depending on which way the scale tips, we can conclude which of the two stones is the Dormant Diamond.

You might think about the last step to help you find an intermediate solution. That is, you might specify what collections of stones and knowledge would allow you to find the diamond in one more weighing. For example, suppose you figure out that the diamond is among three stones that are potentially heavier than the others. Could you find the diamond in one more weighing? Or suppose you had narrowed the field to three stones, one potentially heavier than normal and two potentially lighter than normal. Could you find the diamond in one more weighing? This technique of working backward is often useful.

This balance-scale conundrum is tricky and difficult. Everyone, including experienced mathematicians, would have to think hard to solve it. It can be fun to work on if you enjoy this type of puzzle. Play with it; think carefully about what you know; carefully keep track of all the information you gather. But if you're not enjoying yourself, then just move on.

16. **Cutting (chess) boards.** In the first truncated chessboard, only the top row has been changed. Can you cover this new row with dominoes? Doesn't seem too hard, right? The second truncated board is a different story. How would your covering method change? If you're having trouble finding a successful covering, can you think of a clear explanation as to why such a covering is impossible? Look closely at what lies underneath a domino that covers two squares. Then look at the deleted squares of the second truncated board.

Life Lesson

Don't underestimate the role of visual patterns or properties.
Seek the essential.

17. **Siegfried & You.** It's very important to recall that a deck of cards has 26 black cards and 26 red cards. Consider a particular example of this trick. Suppose it happens that the first pile created by the magician has 10 black cards and 16 red cards. What's in the second pile? Try another example. What happens in general?

Life Lesson

Carefully understand and analyze the facts at hand. If you are faced with a difficult challenge, work on a simple case to develop insight.

18. **Penny for your thoughts.** First try this challenge with two pennies having one head showing. (It's OK to have your eyes open while you experiment, but remember that you want a method that works with your eyes closed.) How can you get two collections? Once you've created your two collections, can you flip one or two coins so that both collections have the same number of heads? (Remember that zero is a number!)

Now think about a slightly bigger case: three pennies with one head showing. What sizes will your two collections be? Consider the cases for which collection has the single head. In each case, can you do some flipping to equalize the heads?

Do you have an idea that applies easily to larger examples? Think about the previous challenge: The deck had 26 black cards, so in *any* pile of 26 cards, each "missing" black card was replaced by a red card, and the missing black cards were all in the other pile. This is what made the trick work.

Suppose you had 10 pennies with six heads showing. Think about what an arbitrary collection of six coins would look like. If it doesn't contain all the heads, where are the missing heads? How many pennies show tails in your collection of six? Now remember that you are allowed to flip as many coins as you like!

Life Lesson

Try an example and look for ways to relate to previous experiences.
Consider extreme cases, such as the smallest one.

19. **When will the world end?** Before trying the puzzle with four or five disks, try it with two and then three. With only two disks the puzzle is easy, taking

only three moves. As you approach the puzzle with three disks, think about what you have to do before you can move the bottom disk. After you move the bottom disk, what do you have to do to finish? For the puzzle with four disks, what do you have to accomplish before you can move the bottom disk? What about after you move the bottom disk? Do you see the pattern? Fill in the table below and see if you can guess a formula for the number of moves required for a puzzle with *n* disks. Use your formula to help predict when the world will end.

Number of Disks	Number of Moves
2	3
3	
4	
5	
.	.
.	.
.	.
64	

Life Lesson

Look at simpler cases and build up; uncover patterns as you go.

20. **The fork in the road.** Because the goal is to choose left or right, suppose any question you ask will have one of those two words as an answer. The challenge is to create a single question with two properties: Both Liars and Truth-Tellers will give the same answer AND you know what the answer means. That is, you know whether the answer is telling you the road to take or the road not to take.

   For now, let's suppose that the left road returns to the villa and illustrate the situation with a simple table. When we ask the most straightforward question, we get two different answers, so we would not get definitive information. We want to modify our question so that only one of the responses changes. Think about how a native from one tribe would read the answer given by a member of the other tribe. Any ideas?

	"Which road leads to the villa?"	Better question:
Answer from Liar	RIGHT	
Answer from Truth-Teller	LEFT	

Life Lesson

Consider different perspectives within the challenge at hand.

CHAPTER TWO

Number Contemplation

2.1 Counting

2.2 Numerical Patterns in Nature

2.3 Prime Cuts of Numbers

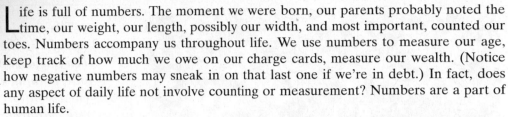

> *Arithmetic has a very great and elevating effect, compelling the soul to reason about abstract number . . .*
>
> PLATO

Life is full of numbers. The moment we were born, our parents probably noted the time, our weight, our length, possibly our width, and most important, counted our toes. Numbers accompany us throughout life. We use numbers to measure our age, keep track of how much we owe on our charge cards, measure our wealth. (Notice how negative numbers may sneak in on that last one if we're in debt.) In fact, does any aspect of daily life not involve counting or measurement? Numbers are a part of human life.

In this chapter we explore the notion of number. Just as numbers play a fundamental role in our daily lives, they also play a fundamental role within the realm of mathematics. We will come to see the richness of numbers and delve into their surprising traits. Some collections of numbers fit so well together that they actually lead to notions of aesthetics and beauty, whereas other numbers are so important that they may be viewed as basic building blocks. Relationships among numbers turn out to have powerful implications in our modern world, for example, within the context of secret codes and the Internet. Exploring the numbers we know leads us to discover whole new worlds of numbers beyond our everyday understanding. Within this expanded universe of number, many simple questions are still unanswered—mystery remains.

One of the main goals of this book is to illustrate methods of investigating the unknown, wherever in life it occurs. In this chapter we highlight some guiding principles and strategies of inquiry by using them to develop ideas about numbers.

Remember that an intellectual journey does not always begin with clear definitions and a list of facts. It often involves stumbling, experimenting, and searching for patterns. By investigating the world of numbers we encounter powerful modes of thought and analysis that can profoundly influence our daily lives.

2.4 Crazy Clocks and Checking Out Bars

2.5 Public Secret Codes and How to Become a Spy

2.6 The Irrational Side of Numbers

2.7 Get Real

How the Pigeonhole Principle Leads to Precision Through Estimation

© tuja66/iStockphoto

> *The simple modes of number*
> *are of all other the most distinct;*
> *even the least variation, which is a*
> *unit, making each combination as*
> *clearly different from that which*
> *approacheth nearest*
> *to it, as the most remote; two being*
> *as distinct from one,*
> *as two hundred; and the idea of two*
> *as distinct from the idea*
> *of three, as the magnitude of the*
> *whole earth is from*
> *that of a mite.*
>
> **JOHN LOCKE**

We begin with the numbers we first learned as children: 1, 2, 3, 4, . . . (The " . . . " indicate that there are more, but we don't have enough room to list them.) These numbers are so natural to us they are actually called *natural numbers*. These numbers are familiar, but often familiar ideas lead to surprising outcomes, as we will soon see.

The most basic use of numbers is counting, and we will begin by just counting approximately. That is, we'll consider the power and the limitations of making rough estimates. In a way, this is the weakest possible use we can make of numbers, and yet we will still find some interesting outcomes. So let's just have some fun with plain old counting.

Quantitative Estimation

One powerful technique for increasing our understanding of the world is to move from qualitative thinking to quantitative thinking whenever possible. Some people still count: "1, 2, 3, many." Counting in that fashion is effective for a simple existence but does not cut it in a world of trillion-dollar debts and gigabytes of hard-drive storage. In our modern world there are practical differences between thousands, millions, billions, and trillions. Some collections are easy to count exactly because there are so few things in them: the schools in the Big Ten Conference, the collection of letters you've written home in the past month, and the clean underwear in your dorm room. Other collections are more difficult to count exactly—such as the grains of sand in the Sahara Desert, the stars in the sky, and the hairs on your roommate's body. Let's look more closely at this last example.

It would be difficult, awkward, and frankly just plain weird to count the number of hairs on your roommate's body. Without undertaking that perverse task, we nevertheless pose the following.

Question Do there exist two nonbald people on the planet who have exactly the same number of hairs on their bodies?

It appears that we cannot answer this question since we don't know (and don't intend to find out) the body-hair counts for anyone. But can we estimate body-hair counts well enough to get some idea of what those numbers might be? In particular, can we at least figure out a number that we could state with confidence is larger than the number of hairs on the body of any person on Earth?

How Hairy Are We?

Let's take the direct approach to this body-hair business. One of the authors counted the number of hairs on a 1/4-inch × 1/4-inch square area on his scalp and counted about 100 hairs—that's roughly 1600 hairs per square inch. From this

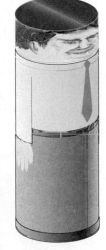

modest follicle count, we can confidently say that no person on Earth has as many as 16,000 hairs in any square inch anywhere on his or her body. The author is about 72 inches tall and 32 inches around. If the author were a perfect cylinder, he would have 72-inch × 32-inch or about 2300 square inches of skin on the sides and about another 200 square inches for the top of his head and soles of his feet, for a total of 2500 square inches of skin. Since the author is not actually a perfect cylinder (he has, for example, a neck), 2500 square inches is an overestimate of his skin area. There are people who are taller and bigger than this author, but certainly there is no one on this planet who has 10 times as much skin as this author. Therefore, no body on Earth will have more than 25,000 square inches of skin. We already agreed that each square inch can have no more than 16,000 hairs on it. Thus we deduce that no person on this planet can have more than 400 million (400,000,000) hairs on his or her body.

A 1″ square containing 2000 hairs—not too physically likely.

How Many Are We?

An almanac or a Web site would tell us that there are more than 7 billion (7,000,000,000) people on this planet. Given this information, can we answer our question: Do there exist two nonbald people on the planet who have *exactly the same* number of hairs on their bodies? We urge you to think about this question and try to answer it before reading on.

Why Many People Are Equally Hairy

There are more than 6.8 billion people on Earth, but each person has many fewer than 400 million hairs on his or her body. Could it be that no two people have the same number of body hairs? What would that mean? It would mean that each of the 6.8 billion people would have a different number of body hairs. But we know that the number of body hairs on each person is less than 400 million. So, there are less than 400 million different possible body-hair numbers. Therefore, not all 6.8 billion people can have different body-hair counts.

Suppose we have 400 million rooms—each numbered in order. Suppose each person did know his or her body-hair count, and we asked each person in the world to go into the room whose number is equal to his or her body-hair number. Could everyone go into a different room? Of course not! We have more than 6 billion people and only 400 million room choices—some room or rooms must have more than one person. In other words, there definitely exist two people, in fact many people, who have the same number of body hairs.

By using some simple estimates, we have been able to answer a question that first appeared unanswerable. The surprising twist is that in this case a rough estimate led to a conclusion about an exact equality. However, there are limitations to our analysis. For example, we are unable to name two *specific* people who have the same body-hair counts even though we know they are out there.

The Power of Reasoning

In spite of the silliness of our hair-raising question we see the power of reasoned analysis. We were faced with a question that on first inspection appeared unanswerable, but through creative thought we were able to crack it. When we are first faced with a new question or problem, the ultimate path of logical reasoning is often hidden from sight. When we try, think, fail, think some more, and try some more, we finally discover a path.

We solved the hairy-body question, but that question in itself is not of great value. However, once we have succeeded in resolving an issue, it is worthwhile to isolate the approach we used, because the method of thought may turn out to be far more important than the problem it solved. In this case, the key to answering our question was the realization that there are more people on the planet than there are body hairs on any individual's body. This type of reasoning is known as the *Pigeonhole principle*. If we have an antique desk with slots for envelopes (known as *pigeonholes*) like the one shown, and we have more envelopes than slots, then certainly some slot must contain at least two envelopes. This Pigeonhole principle is a simple idea, but it is a useful tool for drawing conclusions when the size of a collection exceeds the number of possible variations of some distinguishing trait.

Once we understand the Pigeonhole principle, we become conscious of something that has always been around us—we see it everywhere. For example, in a large swim meet, some pairs of swimmers will get exactly the same times to the tenths of a second. Some days more than 100 people will die in car wrecks. With each breath, we breathe an atom that Einstein breathed before us. Each person will arrive at work during the exact same minute many times during his or her life. Many trees have the same number of leaves. Many people get the same SAT score.

Number Personalities

The natural numbers 1, 2, 3, . . . , besides being useful in counting, have captured the imagination of people around the world from different cultures and different eras. The study of natural numbers began several thousand years ago and continues to this day. Mathematicians who are intrigued by numbers come to know them individually. In the eye of the mathematician, individual numbers have their own personalities—unique characteristics and distinctions from other numbers. In subsequent sections of this chapter, we will discover some intriguing properties of numbers and uncover their nuances. For now, however, we wish to share a story that captures the human side of mathematicians. Of course, mathematicians, like people in other professions, display a large range of personalities, but this true story of Ramanujan and Hardy depicts almost a caricature of the "pure" mathematician. It illustrates part of the mythology of mathematics and provides insight into the personality of an extraordinary mathematician.

This interaction of two mathematicians on such an abstract plane even during serious illness is poignant. They clearly thought each number was worthy of special consideration. To affirm their special regard for each number, we now demonstrate conclusively that every natural number is interesting by means of a whimsical, though ironclad, proof.

Srinivasa Ramanujan

G. H. Hardy

Ramanujan and Hardy

One of the most romantic tales in the history of the human exploration of numbers involves the life and work of the Indian mathematician Srinivasa Ramanujan. Practically isolated from the world of academics, libraries, and mathematicians, Ramanujan made amazing discoveries about natural numbers.

In 1913, Ramanujan wrote to the great English mathematician G. H. Hardy at Cambridge University, describing his work. Hardy immediately recognized that Ramanujan was a unique jewel in the world of mathematics, because Ramanujan had not been taught the standard ways to think about numbers and thus was not biased by the rigid structure of a traditional education; yet he was clearly a mathematical genius. Since the pure nature of mathematics transcends languages, customs, and even formal training, Ramanujan's imaginative explorations have since given mathematicians everywhere an exciting and truly unique perspective on numbers.

Ramanujan loved numbers as his friends, and found each to be a distinct wonder. A famous illustration of Ramanujan's deep connection with numbers is the story of Hardy's visit to Ramanujan in a hospital. Hardy later recounted the incident: "I remember once going to see him when he was lying ill at Putney. I had ridden in taxi cab number 1729 and remarked that the number seemed to me rather a dull one and that I hoped it was not an unfavorable omen. 'No,' he replied, 'it is a very interesting number; it is the smallest number expressible as the sum of two cubes in two different ways.'" Notice that, indeed, $1729 = 12^3 + 1^3$, and also $1729 = 10^3 + 9^3$.

The Intrigue of Numbers.

Every natural number is interesting.

Proof that Natural Numbers Are Interesting

Let's first consider the number 1. Certainly 1 is interesting, because it is the first natural number and it is the only number with this property: If we pick any number and then multiply it by 1, the answer is the original number we picked. So, we agree that the first natural number is interesting.

Let us now consider the number 2. Well, 2 is the first even number, and that is certainly interesting—and, if that weren't enough, remember that 2 is the smallest number of people required to make a baby. Thus, we know that 2 is genuinely interesting.

We now consider the number 3. Is 3 interesting? Well, there are only two possibilities: Either 3 *is* interesting, or 3 *is not* interesting. Let us suppose that 3 *is not* interesting. Then notice that 3 has a spectacular property: It is the *smallest* natural number that is not interesting—which is certainly an interesting property! Thus we see that 3 is, after all, quite interesting.

Knowing now that 1, 2, and 3 are all interesting, we can make an analogous argument for 4 or any other number. In fact, suppose now that k is a certain natural number with the property that the first k natural numbers are all interesting. That is, 1, 2, 3, . . . , k are all interesting. We know this fact is true if k is 1, and, in fact, it is true for larger values of k as well (2, 3, and 4, for example).

We now consider the very next natural number: $k + 1$. Is $k + 1$ interesting? Suppose it were not interesting. Then it would be the smallest natural number that is not interesting (all the smaller natural numbers would be known to be interesting). Well, that's certainly interesting! Thus, $k + 1$ must be interesting, too. Since we have shown that there can be no smallest uninteresting number, we must conclude that every natural number is interesting.

Our proof employs a logical "domino effect" to establish the validity of the theorem. This proof technique is known as *mathematical induction* and is used to prove many important mathematical results (most of which are not nearly as silly as the one we just established here).

A Look BACK

NATURAL NUMBERS are the natural place to begin our journey. This deceptively simple collection of numbers plays a significant role in our lives. We can understand our world more deeply by moving from qualitative to quantitative understanding. Counting and estimating, together with the Pigeonhole principle, lead to surprising insights about everyday events. Natural numbers help us understand our world, but they also constitute a world of their own. The whimsical assertion that every natural number is interesting foreshadows our quest to discover the variety and individual essence of these numbers.

Our strategy for understanding the richness of numbers was to start with the most basic and familiar use for numbers—counting. Looking carefully at the simple and the familiar is a powerful technique for creating and discovering new ideas.

> *To speak algebraically, Mr. M.*
> *is execrable, but Mr. C. is*
> *(x 1 1)-ecrable.*
>
> EDGAR ALLAN POE

Life Lessons
Understand simple things deeply.

MINDSCAPES   Invitations to Further Thought

In this section, Mindscapes marked **(H)** *have hints for solutions at the back of the book. Mindscapes marked* **(ExH)** *have expanded hints at the back of the book. Mindscapes marked* **(S)** *have solutions.*

Developing Ideas

1. **Muchos mangos.** You inherit a large crate of mangos. The top layer has 18 mangos. Peering through the cracks in the side of the crate, you estimate there are five layers of mangos inside. About how many mangos did you inherit?

2. **Packing balls.** Your best friend is about to turn 21 and you want to send him a box full of Ping-Pong balls. You have a square box measuring 12 inches on each side and you wonder how many Ping-Pong balls would fit inside. Suppose you have just enough balls to cover the bottom of the box in a single layer. How could you estimate the number that would fill the box?

3. **Alternative rock.** You have an empty CD rack consisting of five shelves and you just bought five totally kickin' CDs. Can each CD go on a different shelf? What if you had six new CDs?

4. **The Byrds.** You have 16 new CDs to put on your empty five-shelf CD rack. Can you place the CDs so that each shelf contains three or fewer CDs? Can you arrange them so that each shelf contains exactly three?

5. **For the birds.** Explain the Pigeonhole principle.

Solidifying Ideas

6. **Treasure chest (ExH).** Someone offers to give you a million dollars ($1,000,000) in one-dollar ($1) bills. To receive the money, you must lie down; the million one-dollar bills will be placed on your stomach.

   If you keep them on your stomach for 10 minutes, the money is yours! Do you accept the offer? Carefully explain your answer using quantitative reasoning.

7. **Order please.** Order the following numbers from smallest to largest: number of telephones on the planet; number of honest members of Congress;

number of people; number of grains of sand; number of states in the United States; number of cars.

8. **Penny for your thoughts (H).** Two thousand years ago, a noble Arabian king wished to reward his minister of science. Although the modest minister resisted any reward from the king, the king finally forced him to state a desired reward. Impishly the minister said that he would be content with the following token: "Let us take a checkerboard. On the first square I would be most grateful if you would place one piece of gold. Then on the next square twice as much as before, thus placing two pieces, and on each subsequent square, placing twice as many pieces of gold as in the previous square. I would be most content with all the gold that is on the board once your majesty has finished." This sounded extremely reasonable, and the king agreed. Given that there are 64 squares on a checkerboard, roughly how many pieces of gold did the king have to give to our "modest" minister of science? Why did the king have him executed?

9. **Twenty-nine is fine.** Find the most interesting property you can, unrelated to size, that the number 29 has and that the number 27 does not have.

10. **Perfect numbers.** The only natural numbers that divide evenly into 6, other than 6 itself, are 1, 2, and 3. Notice that the sum of all those numbers equals the original number 6 $(1 + 2 + 3 = 6)$. What is the next number that has the property of equaling the sum of all the natural numbers other than itself that divide evenly into it? Such numbers are called *perfect numbers*. No one knows whether or not there are infinitely many perfect numbers. In fact, no one knows whether there are *any* odd perfect numbers. These two unsolved mysteries are examples of long-standing open questions in the theory of numbers.

11. **Many fold (S).** Suppose you were able to take a large piece of paper of ordinary thickness and fold it in half 50 times. What would the height of the folded paper be? Would it be less than a foot? About one yard? As long as a street block? As tall as the Empire State Building? Taller than Mount Everest?

12. **Only one cake.** Suppose we had a room filled with 370 people. Will there be at least two people who have the same birthday?

13. **For the birds.** Years ago, before overnight delivery services and e-mail, people would send messages by carrier pigeon and would keep an ample supply of pigeons in pigeonholes on their rooftops. Suppose you have a certain number of pigeons, let's say P of them, but you have only $P - 1$ pigeonholes. If every pigeon must be kept in a hole, what can you conclude? How does the principle we discussed in this section relate to this question?

14. **Sock hop (ExH).** You have 10 pairs of socks, five black and five blue, but they are not paired up. Instead, they are all mixed up in a drawer. It's early in the morning, and you don't want to turn on the lights in your dark room. How many socks must you pull out to guarantee that you have a pair of one color? How many must you pull out to have two good pairs (each pair is the same color)? How many must you pull out to be certain you have a pair of black socks?

15. **The last one.** Here is a game to be played with natural numbers. You start with any number. If the number is even, you divide it by 2. If the number is odd, you triple it (multiply it by 3), and then add 1. Now you repeat the

process with this new number. Keep going. You win (and stop) if you get to 1. For example, if we start with 17, we would have:

$$17, 52, 26, 13, 40, 20, 10, 5, 16, 8, 4, 2, 1 \rightarrow \text{we see a 1, so we win!}$$

Play four rounds of this game starting with the numbers 19, 11, 22, and 30. Do you think you will always win no matter what number you start with? No one knows the answer!

Creating New Ideas

16. **See the three.** What proportion of the first 1000 natural numbers have a 3 somewhere in them? For example, 135, 403, and 339 all contain a 3, whereas 402, 677, and 8 do not.

17. **See the three II (H).** What proportion of the first 10,000 natural numbers contains a 3?

18. **See the three III.** Explain why almost all million-digit numbers contain a 3.

19. **Commuting.** One hundred people in your neighborhood always drive to work between 7:30 and 8:00 A.M. and arrive 30 minutes later. Why must two people always arrive at work at the same time, within a minute?

20. **RIP (S).** The Earth has more than 6.8 billion people and almost no one lives 100 years. Suppose this longevity fact remains true. How do you know that some year soon, more than 50 million people will die?

Further Challenges

21. **Say the sequence.** The following are the first few terms in a sequence. Can you figure out the next few terms and describe how to find all the terms in the sequence?

    1

    11

    21

    1211

    111221

    312211

    . . .

22. **Lemonade.** You want to buy a new car, and you know the model you want. The model has three options, each one of which you can either take or not take, and you have a choice of four colors. So far 100,000 cars of this model have been sold. What is the largest number of cars that you can guarantee to have the same color and the same options as each other?

In Your Own Words

23. **With a group of folks.** In a small group, discuss and work through the reasoning for why there are two people on Earth having the same number of hairs on their bodies. After your discussion, write a brief narrative describing your analysis and conclusion in your own words.

For the Algebra Lover

Here we celebrate the power of algebra as a powerful way of finding unknown quantities by naming them, of expressing infinitely many relationships and connections clearly and succinctly, and of uncovering pattern and structure.

24. **Ramanujan noodles (H).** Ramanujan tells you that his Social Security number ends with the four digits 2261 and that 2261 can be expressed as the sum of two cubes. He also says that 4 is one of those two numbers to be cubed. If x represents the other number to be cubed, then write an equation that relates 4 and x to the number 2261. Now find the value for x and verify that your solution is correct.

25. **Bird count.** You want to know how many pigeons you have altogether in your seven pigeonholes. Your smartest pigeon (who's been studying *The Heart of Mathematics*) left you a note listing the number of pigeons in each hole as x, x, $3x$, $5x$, x, $2x$, and $2x$. The note also says that the total number of pigeons is $x^2 - 100$. Write an equation relating the information your smart bird has given you. Then solve for x and determine the number of birds you have.

26. **Many pennies.** Suppose you have a 3×3 checkerboard (so it has 9 squares). You put one penny on the first square, then 3 pennies on the next square, then 3^2 pennies on the next square, and so on until there are pennies on all nine squares. Write an expression that gives the total number of pennies on the board. Find the sum.

29. **Park clean-up.** You run a volunteer organization to help clean up local parks. You have 108 volunteers to assign to jobs in 6 different parks. The two medium-sized parks require half as many volunteers as the one largest park. The two small parks require half as many as the medium-sized parks. And one very small park requires only one-fifth the volunteers that the largest park requires. How many volunteers should be sent to each park?

30. **Where's the birdie?** One of your pigeons decides to fly off at noon one day in search of the meaning of life. She flies for a total of five hours. The distance, $D(t)$, she is from her pigeonhole is given by the function $D(t) = 5t - t^2$, where t measures the number of hours since she flew the coop. How far away is she at 3:00 PM? How far away is she at 5:00 PM? What does this tell you about where she finds what she's looking for?

2.2 NUMERICAL PATTERNS IN NATURE

Discovering the Beauty of the Fibonacci Numbers

> *There is no inquiry which is not finally reducible to a question of Numbers; for there is none which may not be conceived of as consisting in the determination of quantities by each other, according to certain relations.*
>
> **AUGUSTE COMTE**

© George Peters/iStockphoto

We can discover patterns by looking closely at our world.

Often when we see beauty in nature, we are subconsciously sensing hidden order—order that itself has an independent richness. Thus we stop and smell the roses—or, more accurately, count the daisies. In the previous section, we contented ourselves with estimation, whereas here we move to exact counting. The example of counting daisies is an illustration of discovering numerical patterns in nature through direct observation. The pattern we find in the daisy appears elsewhere in nature and also gives rise to issues of aesthetics that touch such diverse fields as architecture and painting. We begin our investigation, however, firmly rooted in nature.

Have you ever examined a daisy? Sure, you've picked off the white petals one at a time while thinking: "Loves me . . . loves me not," but have you ever taken a good hard look at what's left once you've finished plucking? A close inspection of the yellow in the middle of the daisy reveals unexpected structure and intrigue. Specifically, the yellow area contains clusters of spirals coiling out from the center. If we examine the flower closely, we see that there are, in fact, two sets of spirals—a clockwise set and a counterclockwise set. These two sets of spirals interlock to produce a hypnotic interplay of helical form.

Interlocking spirals abound in nature. The cone flower and the sunflower both display nature's signature of dual, locking spirals. Flowers are not the only place in nature where spirals occur. A pinecone's exterior is composed of two sets of interlocking spirals. The rough and prickly facade of a pineapple also contains two collections of spirals.

© Victor Burnside/iStockphoto

© Imad Birkholz/iStockphoto

© George Peters/iStockphoto

© Emilie Duchesne/iStockphoto

© Wendy Townrow/iStockphoto

Be Specific: Count

In our observations we should not be content with general impressions. Instead, we move toward the specific. In this case we ponder the quantitative quandary: How many spirals are there? An approximate count is: lots. Is the number of clockwise spirals the same as the number of counterclockwise spirals? You can physically verify that the pinecone has 5 spirals in one direction and 8 in the other. The pineapple has 8 and 13. The daisy and cone flower both have 21 and 34. The sunflower has a staggering 55 and 89. In each case, we observe that the number of spirals in one direction is nearly twice as great as the number of spirals in the opposite direction. Listing all those numbers in order we see

5, 8, 13, 21, 34, 55, 89.

Is there any pattern or structure to these numbers?

Suppose we were given just the first two numbers, 5 and 8, on that list of spiral counts. How could we use these two numbers to build the next number? How can we always generate the next number on our list?

We note that 13 is simply 5 plus 8, whereas 21, in turn, is 8 plus 13. Notice that this pattern continues. What number would come after 89? Given this pattern, what number should come before 5? How about before that? How about before that? And before that?

Leonardo's Legacy: the Fibonacci Sequence

The rule for generating successive numbers in the sequence is to add up the previous two terms. So the next number on the list would be $55 + 89 = 144$.

Through spiral counts, nature appears to be generating a sequence of numbers with a definite pattern that begins

1 1 2 3 5 8 13 21 34 55 89 144 . . . .

This sequence is called the *Fibonacci sequence*, named after the mathematician Leonardo of Pisa (better known as Fibonacci—a shortened form of Filius Bonacci, *son of Bonacci*), who studied it in the 13th century. After seeing this surprising pattern, we hope you feel compelled to count for yourself the spirals in the previous pictures of flowers. In fact, you may now be compelled to count the spirals on a pineapple every time you go to the grocery store.

Why do the numbers of spirals always seem to be consecutive terms in this list of numbers? The answer involves issues of growth and packing. The yellow florets in the daisy begin as small buds in the center of the plant. As the plant grows, the young buds move away from the center toward a location where they have the most room to grow—that is, in the direction that is least populated by older buds. If one simulates this tendency of the buds to find the largest open area as a model of growth on a computer, then the spiral counts in the geometrical pattern so constructed will appear in our list of numbers. The Fibonacci numbers are an illustration of surprising and beautiful patterns in nature. The fact that nature and number patterns reflect each other is indeed a fascinating concept.

A powerful method for finding new patterns is to take the abstract patterns that we directly observe and look at them by themselves. In this case, let's move beyond the vegetable origins of the Fibonacci numbers and just think about the Fibonacci sequence as an interesting entity in its own right. We conduct this investigation with the expectation that interesting relationships that we find among Fibonacci numbers may also be represented in our lives.

Fibonacci Neighbors

We observed that flowers, pinecones, and pineapples all display consecutive pairs of Fibonacci numbers. These observations point to some natural bond between adjacent Fibonacci numbers. In each case, the number of spirals in one direction

The Granger Collection

Leonardo of Pisa, or Fibonacci

Life Lesson

Unexpected patterns are often a sign of hidden, underlying structure.

Fraction of Adjacent Fibonacci Numbers	Decimal Equivalent
$\frac{1}{1}$	1.0
$\frac{2}{1}$	2.0
$\frac{3}{2}$	1.5
$\frac{5}{3}$	1.666...
$\frac{8}{5}$	1.6
$\frac{13}{8}$	1.625
$\frac{21}{13}$	1.6153...
$\frac{34}{21}$	1.6190...
$\frac{55}{34}$	1.6176...
$\frac{89}{55}$	
$\frac{144}{89}$	
$\frac{233}{144}$	

was not quite twice as great as the number of spirals in the other direction. Perhaps we can find richer structure and develop a deeper understanding of the Fibonacci numbers by moving from an estimate ("not quite twice") to a precise value. So, let's measure the relative size of each Fibonacci number in comparison to the next one. We measure the relative size of one number in comparison to another by considering their ratio—that is, by dividing one of the numbers into the other. Here we list the quotients of adjacent Fibonacci numbers. Use a calculator to compute the last three terms in the chart.

What do we notice about these answers? In the right-hand column of the display, notice that the pairs of Fibonacci numbers are getting larger and larger in size. But what about their *relative* sizes?

Converging Quotients

The relative sizes—that is, the quotients of consecutive Fibonacci numbers in the right column—seem to oscillate. They get bigger, then smaller, then bigger, then smaller, but they are apparently becoming increasingly close to one another and are converging toward some intermediate value. What is the exact value for the target number toward which these ratios are heading?

To find it, let's look again at those quotients of Fibonacci numbers, but this time let's write those fractions in a different way. Looking at the same information from a different vantage point often leads to insight. If we're careful with the arithmetic and remember the rule for building the Fibonacci numbers, we will uncover an unusual pattern of 1's. Notice how we use the pattern of 1's from one quotient to produce a pattern of 1's for the next quotient. Each step below involves the facts that $a/b = 1/(b/a)$ and that each Fibonacci number can be written as the sum of the previous two.

$$\frac{1}{1} = 1$$

$$\frac{2}{1} = \frac{1+1}{1} = 1 + \frac{1}{1}$$

$$\frac{3}{2} = \frac{2+1}{2} = \frac{2}{2} + \frac{1}{2} = 1 + \frac{1}{\frac{2}{1}} = 1 + \frac{1}{1+\frac{1}{1}}$$

$$\frac{5}{3} = \frac{3+2}{3} = \frac{3}{3} + \frac{2}{3} = 1 + \frac{1}{\frac{3}{2}} = 1 + \frac{1}{1+\frac{1}{1+\frac{1}{1}}}$$

$$\frac{8}{3} = \frac{5+3}{5} = \frac{5}{5} + \frac{3}{5} = 1 + \frac{1}{\frac{5}{3}} = 1 + \frac{1}{1+\frac{1}{1+\frac{1}{1+\frac{1}{1}}}}$$

Let's look at what we're doing. Replacing the top Fibonacci number in the numerator of our fraction by the sum of the previous two Fibonacci numbers allows us to see a pattern. For example,

$$\frac{233}{144} = \frac{144 + 89}{144} = 1 + \frac{89}{144} = 1 + \frac{1}{\frac{144}{89}}.$$

Now notice that 144/89 would be the previous fraction on our list. So 144/89 would have already been written as a long fraction of 1's.

If we continue this process we see that the ratio of any two adjacent Fibonacci numbers is a number that looks like this:

$$1 + \cfrac{1}{1 + \cfrac{1}{1 + \cfrac{1}{1 + \cfrac{\cdot}{\cdot \cdot + 1}}}}.$$

Unending 1's

As we compute the quotient of ever larger Fibonacci numbers, the ratios head toward the strange expression: $1 + 1/(1 + 1/(1 + \ldots))$ in which we mean by "..." that this fraction never ends, the quotient goes on forever. Only in mathematics can we create something that is truly unending. Let's give this unending number a name. We call this number φ (φ is the lowercase Greek letter *phi*, and in our journey through geometry a few chapters from now, we'll find out why it is called φ—stay tuned). So we see that

$$\varphi = 1 + \cfrac{1}{1 + \cfrac{1}{1 + \cfrac{1}{1 + \cdot \cdot \cdot}}}$$

where the dots mean this process goes on forever. Remember, our goal is to figure out what number the quotients of consecutive Fibonacci numbers approach, and we now see that they approach φ. But what exactly does φ equal? Since it's described in such an interesting form—as an infinitely long fraction containing only 1's—it seems impossible to know the precise value of φ, or is it possible? Right now, the answer is not clear. So let's look for some pattern within that exotic expression for φ.

Before attempting to answer the preceding question, we first ask a warm-up question: We are going to write φ out again; however, this time, notice that we have placed a frame around part of φ. Here is our question: What does the number inside the frame equal?

Life Lesson
New perspectives often reveal new insights.

$$\varphi = 1 + \cfrac{1}{1 + \cfrac{1}{1 + \cfrac{1}{1 + \cdots}}}$$

The answer is: The number in the frame is φ again. Why? Well, suppose we were just shown the number inside the frame without any of that other stuff around it. We'd look at that new number and realize that the 1's go on forever in the same pattern as before, and thus that number is just φ. Stay with this picture until you see the idea behind it. Therefore, we just discovered that

$$\varphi = 1 + \frac{1}{\varphi}.$$

Solving for φ

Now we have an equation involving just φ, and this will allow us to solve for the exact value of φ. First, we can subtract 1 from both sides to get

$$\varphi - 1 = \frac{1}{\varphi}.$$

Multiplying through by φ we get

$$\varphi^2 - \varphi = 1$$

or just

$$\varphi^2 - \varphi - 1 = 0.$$

This "quadratic equation" can be solved using the quadratic formula, which implies that

$$\varphi = \frac{1 \pm \sqrt{5}}{2}.$$

But since φ is bigger than 1, we must have

$$\varphi = \frac{1 + \sqrt{5}}{2}.$$

Using a calculator, express $(1 + \sqrt{5})/2$ as a decimal and compare it with the data from our previous calculator experimentation on the quotients of consecutive Fibonacci numbers. Well, there we have it—our goal was to find the exact value of φ, and through a process of observation and thought we succeeded.

The Golden Ratio

At the moment, the only interesting features of φ we have seen are that it is the fixed value the quotients of consecutive Fibonacci numbers approach and that it can be expressed in a remarkable way as an endless "fraction within a fraction within a fraction . . . " just using 1's. We started with simple observations of flowers and pinecones. We saw a numerical pattern among our observations. The pattern led us to the number $(1 + \sqrt{5})/2$.

The number $\varphi = (1 + \sqrt{5})/2$ is called the *Golden Ratio* and, besides its connection with nature's spirals, it captures the proportions of some especially pleasing shapes in art, architecture, and geometry. Just to foreshadow what is to come when we revisit the Golden Ratio in the geometry chapter, here is a question: What are the proportions of the most attractive rectangle? In other words, when someone says "rectangle" to you, and you think of a shape, what is it? Light some scented candles, put on a Yanni CD, close your eyes, and dream about the most attractive and pleasing rectangle you can imagine. Once that image is etched in your mind, open your eyes, put out the candles, and pick from the four choices below the rectangle that you think is most representative of that magical rectangle dancing in your mind.

Many people feel that the second rectangle from the left is the most aesthetically pleasing—the one that captures the notion of "rectangleness." That rectangle is called the *Golden Rectangle*, and we will examine it in detail in Chapter 4. The ratio of the dimensions of the sides of the Golden Rectangle is a number that we will see is rich with intrigue. If we divide the length of the longer side by the length of the shorter side, we again come upon φ: the Golden Ratio. A 3-inch × 5-inch index card is close to being a Golden Rectangle. Notice that its dimensions, 3 and 5, are consecutive Fibonacci numbers. In the geometry chapter we will consider the (at times controversial) aesthetic issues involving φ and make some interesting connections between the Fibonacci numbers and the Golden Rectangle in art.

To Be or Not to Be Fibonacci

After finding Fibonacci numbers hidden in the spirals of nature, it saddens us to realize that not all numbers are Fibonacci. However, we are delighted to announce that in fact every natural number is a neat *sum* of Fibonacci numbers. In particular, *every* natural number is either a Fibonacci number or it is expressible uniquely as a sum of nonconsecutive Fibonacci numbers. Here is one way to find the sum:

1. Write down a natural number.
2. Find the largest Fibonacci number that does not exceed your number. That Fibonacci number is the first term in your sum.

3. Subtract that Fibonacci number from your number and look at this new number.

4. Find the largest Fibonacci number that does not exceed this new number. That Fibonacci number is the second number in your sum.

5. Continue this process.

For example, consider the number 38. The largest Fibonacci number not exceeding 38 is 34. So consider $38 - 34 = 4$. The largest Fibonacci number not exceeding 4 is 3, and $4 - 3 = 1$, which is a Fibonacci number. Therefore, $38 = 34 + 3 + 1$. Similarly, we can build any natural number just by adding Fibonacci numbers in this manner. In one sense, Fibonacci numbers are building blocks for the natural numbers through addition.

Fun and Games with Fibonacci

Fibonacci numbers not only appear in nature; they can also be used to accumulate wealth. (Moral: Math pays.) We can see this moral for ourselves in a game called *Fibonacci nim*, which is played with two people. All we need is a pile of sticks (toothpicks, straws, or even pennies will do). Person One moves first by taking any number of sticks (at least one but not all) away from the pile. After Person One moves, it is Person Two's move, and the moves continue to alternate between them. Each person (after the first move) may take away as many sticks as he or she wishes; the only restriction is that he or she must take at least one stick but no more than two times the number of sticks the previous person took. The player who takes the last stick wins the game.

Suppose we start with ten sticks, and Person One removes three sticks, leaving seven. Now Person Two may take any number of the remaining sticks from one to six (six is two times the number Person One took). Suppose Person Two removes five, leaving two in the pile. Now Person One is permitted to take any number of sticks from one to ten ($10 = 2 \times 5$), but because there are only two sticks left, Person One takes the two sticks and wins. Play Fibonacci nim with various friends and with different numbers of starting sticks. Get a feel for the game and its rules—but don't wager quite yet.

If we are careful and use the Fibonacci numbers, we can always win. Here is how. First, we make sure that the initial number of sticks we start with is *not* a Fibonacci number. Now we must be Person One, and we find some poor soul to be Person Two. If we play it just right, we will always win. The secret is to write the number of sticks in the pile as a sum of nonconsecutive Fibonacci numbers. Figure out the *smallest* Fibonacci number occurring in the sum, and remove that many sticks from the pile on the first move. Now it is your luckless opponent's turn. No matter what he or she does, we will repeat the preceding procedure. That is, once he or she is done, we count the number of sticks in the pile, express the number as a sum of nonconsecutive Fibonacci numbers, and then remove the number of sticks that equals the smallest Fibonacci number in the sum. It is a fact that, no matter what our poor opponent does, we will always be able to remove that number of sticks without breaking the rules. Experiment with this game and try it. Wager at will—or not.

A Look BACK

WE DEFINE THE FIBONACCI numbers successively by starting with 1, 1, and then adding the previous two terms to get the next term. These numbers are rich with structure and appear in nature. The numbers of clockwise and counterclockwise spirals in flowers and other plants are consecutive Fibonacci numbers. The ratio of consecutive Fibonacci numbers approaches the Golden Ratio, a number with especially pleasing proportions. While not all numbers are Fibonacci, every natural number can be expressed as the sum of distinct, nonconsecutive Fibonacci numbers.

The story of Fibonacci numbers is a story of pattern. As we look at the world, we can often see order, structure, and pattern. The order we see provides a mental concept that we can then explore on its own. As we discover relationships in the pattern, we frequently find that those same relationships refer back to the world in some intriguing way.

Life Lessons

Understand simple things deeply.

MINDSCAPES   Invitations to Further Thought

In this section, Mindscapes marked (H) have hints for solutions at the back of the book. Mindscapes marked (ExH) have expanded hints at the back of the book. Mindscapes marked (S) have solutions.

Developing Ideas

1. **Fifteen Fibonaccis.** List the first 15 Fibonacci numbers.

2. **Born φ.** What is the precise number that the symbol φ represents? What sequence of numbers approaches φ?

3. **Tons of ones.** Verify that $1 + \dfrac{1}{1 + \dfrac{1}{1}}$ equals 3/2.

4. **Twos and threes.** Simplify the quantities $2 + \dfrac{2}{2 + \dfrac{2}{2}}$ and $3 + \dfrac{3}{3 + \dfrac{3}{3}}$.

5. **The family of φ.** Solve the following equations for x:

$$x = 2 + \frac{1}{x}, \ x = 3 + \frac{1}{x}.$$

Solidifying Ideas

6. **Baby bunnies.** This question gave the Fibonacci sequence its name. It was posed and answered by Leonardo of Pisa, better known as Fibonacci.

Suppose we have a pair of baby rabbits: one male and one female. Let us assume that rabbits cannot reproduce until they are one month old and that they have a one-month gestation period. Once they start reproducing, they produce a pair of bunnies each month (one of each sex). Assuming that no pair ever dies, how many pairs of rabbits will exist in a particular month?

During the first month, the bunnies grow into rabbits. After two months, they are the proud parents of a pair of bunnies. There will now be two pairs of rabbits: the original, mature pair and a new pair of bunnies. The next month, the original pair produces another pair of bunnies, but the new pair of bunnies is unable to reproduce until the following month. Thus we have:

Time in Months	Start	1	2	3	4	5	6	7
Number of Pairs	1	1	2					

Continue to fill in this chart and search for a pattern. Here is a suggestion: Draw a family tree to keep track of the offspring.

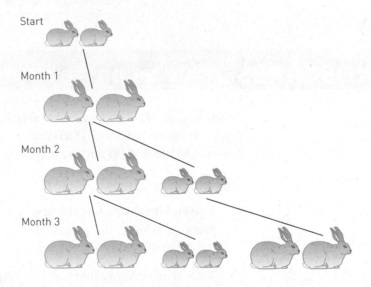

We'll use the symbol F_1 to stand for the first Fibonacci number, F_2 for the second Fibonacci number, F_3 for the third Fibonacci number, and so forth. So $F_1 = 1$, and $F_2 = 1$, and, therefore, $F_3 = F_2 + F_1 = 2$, and $F_4 = F_3 + F_2 = 3$, and so on. In other words, we write F_n for the nth Fibonacci number where n represents any natural number; for example, we denote the 10th Fibonacci number as F_{10} and hence we have $F_{10} = 55$. So, the rule for generating the next Fibonacci number by adding up the previous two can now be stated symbolically, in general, as:

$$F_n = F_{n-1} + F_{n-2}.$$

7. **Discovering Fibonacci relationships (S).** By experimenting with numerous examples in search of a pattern, determine a simple formula for $(F_{n+1})^2 + (F_n)^2$; that is, a formula for the sum of the squares of two consecutive Fibonacci numbers. (See Mindscape 6 for a description of the notation F_n.)

8. **Discovering more Fibonacci relationships.** By experimenting with numerous examples in search of a pattern, determine a simple formula for $(F_{n+1})^2 - (F_{n-1})^2$; that is, a formula for the difference of the squares of two Fibonacci numbers. (See Mindscape 6 for a description of the notation F_n.)

9. **Late bloomers (ExH).** Suppose we start with one pair of baby rabbits, and again they create a new pair every month, but this time let's suppose that it takes two months before a pair of bunnies is mature enough to reproduce. Make a table for the first 10 months, indicating how many pairs there would be at the end of each month. Do you see a pattern? Describe a general formula for generating the sequence of rabbit-pair counts.

10. **A new start.** Suppose we build a sequence of numbers using the method of adding the previous two numbers to build the next one. This time, however, suppose our first two numbers are 2 and 1. Generate the first 15 terms. This sequence is called the *Lucas sequence* and is written as $L_1, L_2, L_3, \ldots$. Compute the quotients of consecutive terms of the Lucas sequence as we did with the Fibonacci numbers. What number do these quotients approach? What role do the initial values play in determining what number the quotients approach? Try two other first terms and generate a sequence. What do the quotients approach?

11. **Discovering Lucas relationships.** By experimenting with numerous examples in search of a pattern, determine a formula for $L_{n-1} + L_{n+1}$; that is, a formula for the sum of a Lucas number and the Lucas number that comes after the next one. (*Hint:* The answer will not be a Lucas number. See Mindscape 10.)

12. **Still more Fibonacci relationships.** By experimenting with numerous examples in search of a pattern, determine a formula for $F_{n-1} + F_{n+1}$; that is, a formula for the sum of a Fibonacci number and the Fibonacci number that comes after the next one. (See Mindscape 6 for a description of the notation F_n.) (*Hint:* The answer will not be a Fibonacci number. Try Mindscape 10 first.)

13. **Even more Fibonacci relationships.** By experimenting with numerous examples in search of a pattern, determine a formula for $F_{n+2} - F_{n-2}$; that is, a formula for the difference between a Fibonacci number and the fourth Fibonacci number before it. (See Mindscape 6 for a description of the notation F_n.) (*Hint:* The answer will not be a Fibonacci number. Try Mindscape 10 first.)

14. **Discovering Fibonacci and Lucas relationships.** By experimenting with numerous examples in search of a pattern, determine a formula for $F_n + L_n$; that is, a formula for the sum of a Fibonacci number and the corresponding Lucas number. (See Mindscape 6 for a description of the notation F_n.)

15. **The enlarging area paradox (S).** The square shown here has sides equal to 8 (a Fibonacci number) and thus has area equal to $8 \times 8 = 64$. The square can be cut up into four pieces whereby the short sides have lengths 3 and 5, as illustrated. (You will find these pieces in the kit that accompanied your book.) Now use those pieces to construct a rectangle having base 13 and height 5. The area of this rectangle is $13 \times 5 = 65$. So by moving around the four pieces, we increased the area by 1 unit! Using the puzzle pieces from your kit, do the experiment and then explain this impossible feat. Show this puzzle to your friends and record their reactions.

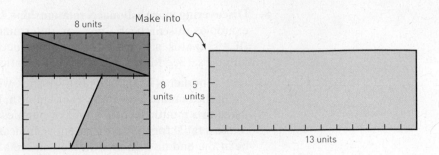

8 units

Make into

8 units 5 units

13 units

16. **Sum of Fibonacci (H).** Express each of the following natural numbers as a sum of distinct, nonconsecutive Fibonacci numbers: 52, 143, 13, 88.

17. **Some more sums.** Express each of the following natural numbers as a sum of distinct, nonconsecutive Fibonacci numbers: 43, 90, 2000, 609.

18. **Fibonacci nim: The first move.** Suppose you are about to begin a game of Fibonacci nim. You start with 50 sticks. What is your first move?

19. **Fibonacci nim: The first move II.** Suppose you are about to begin a game of Fibonacci nim. You start with 100 sticks. What is your first move?

20. **Fibonacci nim: The first move III.** Suppose you are about to begin a game of Fibonacci nim. You start with 500 sticks. What is your first move?

21. **Fibonacci nim: The next move.** Suppose you are playing a round of Fibonacci nim with a friend. The game begins with 15 sticks. You start by removing two sticks; your friend then takes four. How many sticks should you take next to win?

22. **Fibonacci nim: The next move II.** Suppose you are playing a round of Fibonacci nim with a friend. The game begins with 50 sticks. You start by removing three sticks; your friend then takes five; you then take eight; your friend then takes ten. How many sticks should you take next to win?

23. **Fibonacci nim: The next move III.** Suppose you are playing a round of Fibonacci nim with a friend. The game begins with 90 sticks. You start by removing one stick; your friend then takes two; you take three; your friend takes six; you take two; your friend takes one; you take two; your friend takes four; you take one, and then your friend takes two. How many sticks should you take next to win?

24. **Beat your friend.** Play Fibonacci nim with a friend: Explain the rules, but do not reveal the secret to winning. Use 20 sticks to start. Play carefully, and beat your friend. Play again with another (non-Fibonacci) number of sticks to start. Record the number of sticks removed at each stage of each game. Finally, reveal the secret strategy and record your friend's reaction.

25. **Beat another friend.** Play Fibonacci nim with another friend: Explain the rules, but do not reveal the secret to winning. Use 30 sticks to start. Play carefully and beat your friend. Play again with another (non-Fibonacci) number of sticks to start. Record the number of sticks removed at each stage of each game. Finally, reveal the secret strategy and record your friend's reaction.

Creating New Ideas

26. **Discovering still more Fibonacci relationships.** By experimenting with numerous examples in search of a pattern, determine a formula for $F_{n+1} \times$

$F_{n-1} - (F_n)^2$; that is, a formula for the product of a Fibonacci number and the Fibonacci number that comes after the next one, minus the square of the Fibonacci number in between them. (*Hint:* The answer will be different depending on whether n is even or odd. Consider examples of different cases separately.)

27. **Finding factors (S).** By experimenting with numerous examples, find a way to factor F_{2n} into the product of two natural numbers that are from famous sequences. That is, consider every other Fibonacci number starting with the second 1 in the sequence, and factor each in an interesting way. Discover a pattern. (*Hint:* Mindscape 10 may be relevant.)

28. **The rabbits rest.** Suppose we have a pair of baby rabbits—one male and one female. As before, the rabbits cannot reproduce until they are one month old. Once they start reproducing, they produce a pair of bunnies (one of each sex) each month. Now, however, let us assume that each pair dies after three months, immediately after giving birth. Create a chart showing how many pairs we have after each month from the start through month nine.

29. **Digging up Fibonacci roots.** Using the square root key on a calculator, evaluate each number in the top row and record the answer in the bottom row.

Number	$\sqrt{\left(\dfrac{F_3}{F_1}\right)}$	$\sqrt{\left(\dfrac{F_4}{F_2}\right)}$	$\sqrt{\left(\dfrac{F_5}{F_3}\right)}$	$\sqrt{\left(\dfrac{F_6}{F_4}\right)}$	$\sqrt{\left(\dfrac{F_7}{F_5}\right)}$	$\sqrt{\left(\dfrac{F_8}{F_6}\right)}$	$\sqrt{\left(\dfrac{F_9}{F_7}\right)}$
Computed Value							

Looking at the chart, make a guess as to what special number $\sqrt{F_{n+2}/F_n}$ approaches as n gets larger and larger.

30. **Tribonacci.** Let's start with the numbers 0, 0, 1, and generate future numbers in our sequence by adding up the previous three numbers. Write out the first 15 terms in this sequence, starting with the first 1. Use a calculator to evaluate the value of the quotients of consecutive terms (dividing the smaller term into the larger one). Do the quotients seem to be approaching a fixed number?

31. **Fibonacci follies.** Suppose you are playing a round of Fibonacci nim with a friend. You start with 15 sticks. You start by removing two sticks; your friend then takes one; you take two; your friend takes one. What should your next move be? Can you make it without breaking the rules of the game? Did you make a mistake at some point? If so, where?

32. **Fibonacci follies II.** Suppose you are playing a round of Fibonacci nim with a friend. You start with 35 sticks. You start by removing one stick; your friend then takes two; you take three; your friend takes six; you take three; your friend takes two. What should your next move be? Can you make it without breaking the rules of the game? Did you make a mistake at some point? If so, where?

33. **Fibonacci follies III.** Suppose you are playing a round of Fibonacci nim with a friend. You start with 21 sticks. You start by removing one stick; your friend then takes two. What should your next move be? Can you make it without breaking the rules of the game? What went wrong?

34. **A big fib (ExH).** Suppose we have a natural number that is not a Fibonacci number—let's call it N. Suppose that F is the largest Fibonacci number that

does not exceed N. Show that the number $N - F$ must be smaller than the Fibonacci number that comes right before F.

35. **Decomposing naturals (H).** Use the result of Mindscape 34, together with the notion of systematically reducing a problem to a smaller problem, to show that every natural number can be expressed as a sum of distinct, nonconsecutive Fibonacci numbers.

Further Challenges

36. **How big is it?** Is it possible for a Fibonacci number greater than 2 to be exactly twice as big as the Fibonacci number immediately preceding it? Explain why or why not. What would your answer be if we removed the phrase "greater than 2"?

37. **Too small.** Suppose we have a natural number that is not a Fibonacci number—let's call it N. Let's write F for the largest Fibonacci number that does not exceed N. Show that it is impossible to have a sum of two distinct Fibonacci numbers each less than F add up to N.

38. **Beyond Fibonacci.** Suppose we create a new sequence of natural numbers starting with 0 and 1. Only this time, instead of adding the two previous terms to get the next one, let's generate the next term by adding 2 *times* the previous term to the term before it. In other words: $G_{n+1} = 2G_n + G_{n-1}$. Such a sequence is called a *generalized Fibonacci sequence*. Write out the first 15 terms in this generalized Fibonacci sequence. Adapt the methods that were used in this section to figure out that the quotient of consecutive Fibonacci numbers approaches $(1 + \sqrt{5}) / 2$ to discover the exact number that G_{n+1}/G_n approaches as n gets large.

39. **Generalized sums.** Let G_n be the generalized Fibonacci sequence defined in Mindscape 38. Can every natural number be expressed as the sum of distinct, nonconsecutive generalized Fibonacci numbers? Show why, or give several counterexamples. What if you were allowed to use consecutive generalized Fibonacci numbers? Do you think you could do it then? Illustrate your hunch with four or five specific examples.

40. **It's hip to be square (H).** Adapt the methods of this section to prove that the numbers $\sqrt{F_{n+2} / F_n}$ approach φ as n gets larger and larger. (Here, F_n stands for the usual nth Fibonacci number. See Mindscape 6.)

In Your Own Words

41. **Personal perspectives.** Write a short essay describing the most interesting or surprising discovery you made in exploring the material in this section. If any material seemed puzzling or even unbelievable, address that as well. Explain why you chose the topics you did. Finally, comment on the aesthetics of the mathematics and ideas in this section.

42. **With a group of folks.** In a small group, discuss and work through the reasoning for how the quotients of consecutive Fibonacci numbers approach the Golden Ratio. After your discussion, write a brief narrative describing the rationale in your own words.

43. **Creative writing.** Write an imaginative story (it can be humorous, dramatic, whatever you like) that involves or evokes the ideas of this section.

44. **Power beyond the mathematics.** Provide several real-life issues—ideally, from your own experience—that some of the strategies of thought presented in this section would effectively approach and resolve.

For the Algebra Lover

Here we celebrate the power of algebra as a powerful way of finding unknown quantities by naming them, of expressing infinitely many relationships and connections clearly and succinctly, and of uncovering pattern and structure.

45. **Rabbit line.** A new species of rabbits has a linear breeding pattern. If you start with 2 rabbits, every year you get 5 more. Find an expression for the number of rabbits you have after t years. So, for example, at year $t = 0$ we have 2 rabbits; at year $t = 1$ we have 7 rabbits. How many years have passed if you have 102 rabbits? (These rabbits also have the special skill of never dying.)

46. **Finding x (H).** Solve for x: $x = 1 + \dfrac{6}{x}$.

47. **Appropriate address.** Fibonacci's house number is the product of the number 107 together with three consecutive odd integers. If x denotes the first of these three odd numbers, write an expression that represents all possible Fibonacci house numbers. Try a few small values of x. Which x do you think gives Fibonacci's actual house number? Why?

48. **Zen bunnies.** Your rabbits do yoga every morning in a special pen. When drawn in the xy-plane, it's the region consisting of all points that satisfy the inequalities $2 \le x \le 5$ and $1 \le y \le 3$. Sketch the region and label key points.

49. **The power of gold (H).** In 1843 Jacques Binet (not to be confused with the comedian Jack Benny), derived the following formula:

$$y = \frac{\left(\dfrac{1 + \sqrt{5}}{2}\right)^{n} - \left(\dfrac{1 - \sqrt{5}}{2}\right)^{n}}{\sqrt{5}}$$

Evaluate this formula for $n = 1, 2$, and 3; that is, find the values for y when $n = 1, 2$, and 3. What do you notice about your answers? Guess the value for y when $n = 10$. Why did we place Binet's formula in this section?

2.3 PRIME CUTS OF NUMBERS

How the Prime Numbers Are the Building Blocks of All Natural Numbers

Charles Demuth, I *Saw the Figure 5 in Gold,* 1928.

> *. . . number is merely the product of our mind.*
> — KARL FRIEDRICH GAUSS

The natural numbers, 1, 2, 3, . . . , help us describe and understand our world. They in turn form their own invisible world filled with abstract relationships, some of which can be revealed through simple addition and multiplication. These basic operations lead to subtle insights about our familiar numbers.

Our strategy for uncovering the structure of the natural numbers is to break down complex objects and ideas into their fundamental components. This simple yet powerful technique recurs frequently throughout this book, and throughout our lives. As we become accustomed to using this strategy, we will see that complicated situations are often best analyzed first by investigating the building blocks of an idea or an object and then by understanding how these building blocks combine to create a complex whole. The natural numbers are a good arena for observing this principle in action.

Building Blocks

What are the fundamental components of the natural numbers? How can we follow the suggestion of breaking down numbers into smaller components?

There are many ways to express large natural numbers in terms of smaller ones. For example, we might first think of addition: Every natural number can be constructed by just adding $1 + 1 + 1 + 1 + \cdots + 1$ enough times. This method demonstrates perhaps the most fundamental feature of natural numbers: They are simply a sequence of counting numbers, each successive one bigger than its predecessor. However, this feature provides only a narrow way of distinguishing one natural number from another.

Divide and Conquer

How can one natural number be expressed as the product of smaller natural numbers? This innocent-sounding question leads to a vast field of interconnections among the natural numbers that mathematicians have been exploring for thousands of years. Our adventure begins by recalling the arithmetic from our youth and looking at it afresh.

One method of writing a natural number as a product of smaller ones is first to divide and then to see if there is a remainder. We were introduced to division a long time ago in third or fourth grade—we weren't impressed. Somehow it paled in comparison to, say, recess. The basic reality of long division is that either it comes out even or there is a remainder. If the division comes out even, we then know that the smaller number divides evenly into the larger number and that our number can be factored. For example, 12 divided by 3 is 4, so $3 \times 4 = 12$ and 3 and 4 are factors of 12. More generally, suppose that n and m are any two natural numbers. We say that n divides evenly into m if there is a natural number q such that

$$nq = m.$$

The integers n and q are called *factors* of m.

If the division does not come out even, the remainder is less than the number we tried to divide. For example, 16 divided by 5 is 3 with a remainder of 1.

This whole collection of elementary school flashbacks can be summarized in a statement that sounds far more impressive than "long division," namely, the *Division Algorithm*.

> **The Division Algorithm.**
>
> *Suppose n and m are natural numbers. Then there exist unique numbers q (for quotient) and r (for remainder), that are either natural numbers or zero, such that*
>
> $$m = nq + r \quad and$$
>
> $0 \le r \le n - 1$ *(r is greater than or equal to 0 but less than or equal to n − 1).*

Prime Time

Factoring a big number into smaller ones gives us some insights into the larger number. This method of breaking up natural numbers into their basic components leads us to the important notion of prime numbers.

There certainly are natural numbers that cannot be factored as the product of two smaller natural numbers. For example, 7 cannot be factored into two smaller natural numbers, nor can these: 2, 3, 5, 11, 13, 17, which, including 7, are the first seven such numbers (let's ignore the number 1). These unfactorable numbers are called *prime numbers*. A prime number is a natural number greater than 1 that cannot be expressed as the product of two smaller natural numbers.

The prime numbers form the multiplicative building blocks for all natural numbers greater than 1. That is, every natural number greater than 1 is either prime or it can be expressed as a product of prime numbers.

The Prime Factorization of Natural Numbers.

Every natural number greater than 1 is either a prime number or it can be expressed as a product of prime numbers.

Let's first look at a specific example and then see why the Prime Factorization of Natural Numbers is true for all natural numbers greater than 1.

Is 1386 prime? No, 1386 can be factored as

$$1386 = 2 \times 693.$$

Is 693 prime? No. It can be factored as

$$693 = 3 \times 231.$$

The number 3 is prime, but

$$231 = 3 \times 77.$$

So far we have

$$1386 = 2 \times 3 \times 3 \times 77.$$

Is 77 prime? No,

$$77 = 7 \times 11$$

but 7 and 11 are both primes. Thus we have

$$1386 = 2 \times 3 \times 3 \times 7 \times 11.$$

So, we have factored 1386 into a product of primes. This simple "divide and conquer" technique works for any number!

Proof of the Prime Factorization of Natural Numbers

Let n be any natural number greater than 1 that is not a prime. Then there must be a factorization of n into two smaller natural numbers both greater than 1 (why?), say $n = a \times b$. We now look at a and b. If both are primes, we end our proof, since n is equal to a product of two primes. If either a or b is not prime, then it can be factored into two smaller natural numbers, each greater than 1. If we continue in this manner with each of the factors, since the factors are getting smaller, we will find in a finite number of steps a factorization of n where all the factors are prime numbers. So, n is a product of primes, and our proof is complete.

Really Big Numbers: We Can, but We Can't

Even though we know that every natural number can theoretically be factored, as a practical matter, factoring really large numbers is currently impossible. For example, suppose someone asks us to factor this long number:

> 22030198713510925601925734109347109238710248356189726582736458764058761984569183246501923847019237418726501826354109832740192384701923460198256109238476019823740192386745087236409823674019823601923860923640192834701298347012934871029386102398461039871023984710923871029387102398719238710938.

We might have some difficulty. In fact, the fastest computers in the world working full time would take centuries to factor this number. Our inability to factor really big numbers is the key to devising public key codes for sending private information over the Web and using e-mail and ATMs. We consider this modern application of prime numbers in Section 2.5, "Public Secret Codes and How to Become a Spy."

What's the Largest Prime?

The Prime Factorization of Natural Numbers reveals that prime numbers are the building blocks for the natural numbers; that is, to build any natural number, all we need to do is multiply the appropriate prime numbers together. Of course, whenever we are presented with building blocks, whether they are Lincoln Logs, Legos, elementary particles, or primes, the question arises: How many blocks do we have? In this case, we ask how many prime numbers are there? Are there more than a million primes? Are there more than a billion primes? Are there infinitely many primes?

Since there are infinitely many natural numbers, it seems reasonable to think that we would need bigger and bigger primes in order to build up the bigger and bigger natural numbers. Therefore, we might conclude that there are infinitely many primes. This argument, however, isn't valid. Do you see why? It's not valid because we can build natural numbers as large as we wish just by multiplying a couple of small prime numbers. For example, using just 2 and 3, we are able to make huge numbers:

$$45{,}349{,}632 = 2^8 \, 3^{11}.$$

The point is that, since we are allowed to use tons of 2's and 3's in our product, we can construct numbers as large as we wish. Therefore, at this moment it seems plausible that there may be only finitely many prime numbers, even though every large (and small) natural number is a product of primes.

Justus van Gent, *Euclid,* 1473–1475.

© Fine Art Images/SuperStock

Infinitely Many Primes

Although the previous argument was invalid, it turns out that there are infinitely many prime numbers, and thus there is no largest prime number. Euclid discovered the following famous valid proof more than 2000 years ago. It is beautiful in that the idea is clever and uncomplicated. In fact, on first inspection we may think there is some sleight of hand going on, as if some fast-talking salesman in a polyester plaid sports jacket is trying to sell us a used car—before we know it, we're the not-so-proud owner of a 1973 Gremlin. But this is not the case. Think along with us as we develop this argument. The ideas fit together beautifully, and, if you stay with it, the argument will suddenly "click." Let's now examine one of the great triumphs of human reasoning.

The Infinitude of Primes.

There are infinitely many prime numbers.

The Strategy Behind the Proof

The strategy for proving that there are infinitely many prime numbers is to show that, for each and every given natural number, we can always find a prime number that is larger than that natural number. Since we can consider larger and larger numbers as our natural number, this claim would imply that there are larger and larger prime numbers. Thus we would show that there must be infinitely many primes, because we could find primes as large as we want without bound.

Before moving forward with the general idea of the proof, let's illustrate the key ingredient with a specific example. Suppose we wanted to find a prime number that is greater than 4. How could we proceed? Well, we could just say "5" and be done with it, but that is not in the spirit of what we are trying to do. Our goal is to discover a method that can be generalized and used to find a prime number that exceeds an *arbitrary* natural number, not just the pathetically small number 4. So we seek a systematic means of finding a prime that exceeds, in this case, 4. What should we do? Our challenge is to:

1. find a number bigger than 4 that is not evenly divisible by 2;
2. find a number bigger than 4 that is not evenly divisible by 3; and
3. find a number bigger than 4 that is not evenly divisible by 4.

Each of these tasks individually is easy. To satisfy (1), we just pick an odd number. To satisfy (2), we just pick a number that has a remainder when we divide by 3. To satisfy (3), we just pick a number that has a remainder when we divide by 4. We now build a number that meets all those conditions simultaneously. Let's call this new number N for *new*. Here is an N that meets all three conditions simultaneously:

$$N = (1 \times 2 \times 3 \times 4) + 1.$$

So, N is really just the number 25, but, since we are trying to discover a general strategy, let's not think of N as merely 25 but instead think of N as the more impressive $(1 \times 2 \times 3 \times 4) + 1$.

We notice that N is definitely larger than 4. By the Prime Factorization of Natural Numbers, we know that N is either a prime number or a product of prime numbers. In the first possibility, if N is a prime number, then we have just found a prime number that is larger than 4—which was our goal. We now must consider the only other possibility: that N is not prime. If N is not prime, then N is a product of prime numbers. Let's call one of those prime factors of N "*OUR-PRIME*." So *OUR-PRIME* is a prime that divides evenly into N.

Now what can we say about *OUR-PRIME*? Is *OUR-PRIME* equal to 2? Well, if we divide 2 into N we see that, since

$$N = 2 \times (1 \times 3 \times 4) + 1,$$

we have a remainder of 1 when 2 is divided into N. Therefore, 2 does not divide evenly into N, and so *OUR-PRIME* is not 2. Is *OUR-PRIME* equal to 3? No, for the same reason:

$$N = 3 \times (1 \times 2 \times 4) + 1,$$

so we get a remainder of 1 when 3 is divided into N; therefore, 3 does not divide evenly into N and hence is not a factor of N. Likewise, we will get a remainder of 1 when 4 is divided into N, and, therefore, 4 is not a factor of N. So 2, 3, and 4 are not factors of N. Hence we conclude that all the factors of N must be larger than 4 (there are no factors of N that are 4 or smaller). But that means that, since *OUR-PRIME* is a prime that is a factor of N, *OUR-PRIME* must be a prime number greater than 4. Therefore, we have just found a prime number greater than 4. Mission accomplished!

Following this same strategy, show that there must be a prime greater than 5. Can you use the preceding method to show there must be a prime greater than 10,000,000? Try it now.

Finally, the Proof

Now let's use the method we developed in the specific example to prove our theorem in general. Remember that we wish to demonstrate that, for any particular natural number, there is a prime number that exceeds that particular number. Let m represent an arbitrary natural number. Our goal now is to show that there is a prime number that exceeds m. To accomplish this lofty quest, we will, just as before, construct a new number using all the numbers from 1 to m. We'll call this new number N (for *new* number) and define it to be 1 plus the product of all the natural numbers from 1 to m—in other words (or more accurately in other symbols):

$$N = (1 \times 2 \times 3 \times 4 \times \cdots \times m) + 1.$$

It is fairly easy to see that N is larger than m. By the Prime Factorization of Natural Numbers we know that there are only two possibilities for N: Either N is prime or N is a product of primes. If the first possibility is true, then we have found what we wanted since N is larger than m. But if it is not true, then we must consider the more challenging possibility: that N is a product of prime numbers.

If N is a product of primes, then we can choose one of those prime factors and call it *BIG-PRIME*. So, *BIG-PRIME* is a prime factor of N. Let's now try to pin down the value of *BIG-PRIME*. We'll start off small.

Does *BIG-PRIME* equal 2? Well, if we divide 2 into N we see that, since

$$N = 2 \times (1 \times 3 \times 4 \times \cdots \times m) + 1,$$

we have a remainder of 1 when 2 is divided into N. Therefore, 2 does not divide evenly into N, and so *BIG-PRIME* cannot equal 2.

Does *BIG-PRIME* equal 3? No, for the same reason:

$$N = 3 \times (1 \times 2 \times 4 \times \cdots \times m) + 1,$$

so we get a remainder of 1 when 3 is divided into N. In fact, what is the remainder when any number from 2 to m is divided into N? The remainder is always 1, by the same reasoning that we used for 2 and 3.

Okay, so we see that none of the numbers from 2 through m divides evenly into N. That fact means that none of the numbers from 2 through m is a factor of N. Therefore, what can be said about the size of the factor *BIG-PRIME*? Answer: It must be BIG since we know that N has no factors between 2 and m. Hence any factor of N must be larger than m. Therefore, *BIG-PRIME* is a prime number that is larger than m.

Well, we did it! We just showed that there is a prime that exceeds m. Since this procedure works for any value of m, this argument shows that there are arbitrarily large prime numbers. Therefore, we must have infinitely many prime numbers, and we have completed the proof.

The Clever Part of the Proof

In the proof, each step by itself isn't too hard, but the entire argument, taken as a whole, is subtle. What is the most clever part of the proof? In other words, where is the most imagination required? Which step in the argument would have been hardest to think up on your own?

We believe that the most ingenious part of the proof is the idea of constructing the auxiliary number N (one more than the product of all the numbers from 1 to m). Once we have the idea of considering that N, we can finish the proof. But it took creativity and contemplation to arrive at that choice of N. We might well say to ourselves, "Gee, I wouldn't have thought of making up that N." Generally, slick proofs such as this are arrived at only after many attempts and false starts—just as Euclid no doubt experienced before he thought of this proof. Very few people can understand arguments of this type on first inspection, but once we can hold the whole proof in our minds, we will regard it as straightforward and persuasive and appreciate its aesthetic beauty. Ingenuity is at the heart of creative mathematical reasoning, and therein lies the power of mathematical thought.

Prime Demographics

Now that we know for sure that there are infinitely many prime numbers, we wonder how the primes are distributed among the natural numbers. Is there some pattern to their distribution? There are infinitely many primes, but how rare are they among the numbers? What proportion of the natural numbers are prime numbers? Half? A third? To explore these questions, let's start by looking at the natural numbers and the primes among them. Here are the first few with the primes printed in bold:

1, **2**, **3**, 4, **5**, 6, **7**, 8, 9, 10, **11**, 12, **13**, 14, 15, 16, **17**, 18, **19**, 20, 21, 22, **23**, 24, 25, 26, 27, 28, **29**, . . .

Out of the first 24 natural numbers, nine are primes. We see that 9/24 = 0.375 of the first 24 natural numbers are primes—that's just a little over one-third. Extrapolating from this observation, would we guess that just over one-third of *all* natural numbers are prime numbers? We could try an experiment; namely, we could continue to list the natural numbers and find the proportions of primes and see whether that proportion remains about one-third of the total number. If we do this experiment, we will learn an important lesson in life: Don't be too hasty to generalize based on a small amount of evidence.

Before high-speed computers were available, calculating (or just estimating) the proportion of prime numbers in the natural numbers was a difficult task. In fact, years ago "computers" were people who did computations. Such people were amazingly accurate, but they required a great deal of time and dedication to accomplish what today's electronic computers can do in seconds. An 18th-century Austrian arithmetician, J. P. Kulik, spent 20 years of his life creating, by hand, a table of the first 100 million primes. His table was never published, and sadly the volume containing the primes between 12,642,600 and 22,852,800 has disappeared.

Today, software computes the number of primes less than *n* for increasingly large values of *n*, and the computer prints out the proportion: (number of primes less than *n*)/*n*. Computers have no difficulty producing such a table for values of *n* up to the billions, trillions, and beyond. If we examine the results, we notice that the proportion of primes slowly goes downward. That is, the percentage of numbers less than a million that are prime is smaller than the percentage of numbers less than a thousand that are prime. The primes, in some sense, get sparser and sparser as the numbers get bigger and bigger.

A Conjecture About Patterns

In the early 1800s, Karl Friedrich Gauss (*below left*), one of the greatest mathematicians ever—known by many as the Prince of Mathematics—and A. M. Legendre (*below right*), another world-class mathematician, made an insightful observation about primes. They noticed that, even though primes do not appear to occur in any predictable pattern, the proportion of primes is related to the so-called natural logarithm—a function relating to exponents that we may or may not remember from our school daze.

Years ago, one needed to interpolate huge tables to find the logarithm. Today, we have scientific calculators that compute logarithms instantly and painlessly. Get out a scientific calculator and look for the LN key. Hit the LN key

Karl Friedrich Gauss   A. M. Legendre

and then type "3." You should see 1.09861. . . . We encourage you to try some natural-logarithm experiments on your calculator. How does the size of the natural logarithm of a number compare with the size of the number itself?

Gauss and Legendre conjectured that the proportion of the number of primes among the first n natural numbers is approximately $1/\mathrm{Ln}(n)$. The following chart, constructed with the aid of a computer (over which Gauss and Legendre would drool), shows the number of primes up to n, the proportions of primes, and a comparison with $1/\mathrm{Ln}(n)$.

n	Number of Primes up to n	Proportion of Primes up to n (Number of Primes $\leq n$)/n	$1/\mathrm{Ln}(n)$	Proportion $-$ $1/\mathrm{Ln}(n)$
10	4	0.4	0.43429...	−0.03429...
100	25	0.25	0.21714...	0.03285...
1000	168	0.168	0.14476...	0.02323...
10,000	1229	0.1229	0.10857...	0.01432...
100,000	9592	0.09592	0.08685...	0.00906...
1,000,000	78,498	0.078498	0.07238...	0.00611...
10,000,000	664,579	0.0664579	0.06204...	0.00441...
100,000,000	5,761,455	0.05761455	0.05428...	0.00332...
1,000,000,000	50,847,534	0.050847534	0.04825...	0.00259...

Notice how the last column seems to be getting closer and closer to zero—that is, the proportion of primes in the first n natural numbers is approximately $1/\mathrm{Ln}(n)$, and the fraction (number of primes less than n)/n is becoming increasingly closer to $1/\mathrm{Ln}(n)$ the bigger n gets. Our observations show that

(number of primes up to n)/n is roughly $1/\mathrm{Ln}(n)$.

Multiplying both of these quantities by n, it appears that

(number of primes up to n) is roughly $n/\mathrm{Ln}(n)$.

These observations culminate in what is called the *Prime Number Theorem*, which gives the approximate number of primes less than or equal to n as n gets really big.

The Prime Number Theorem.

As n gets larger and larger, the number of prime numbers less than or equal to n approaches n/Ln(n).

The Proof Had to Wait

Having sufficient insight into the nature of numbers to make such a conjecture is a tremendous testament to the intuition of Gauss and Legendre. But, even though they conjectured that the Prime Number Theorem was true, they were unable to prove it. This famous conjecture remained unproved for nearly a century until, in 1896, two mathematicians, Hadamard and de la Vallée Poussin, simultaneously but independently constructed proofs of the Prime Number Theorem. That is, working on their own, they each found, in the same year, a proof of this extremely important and difficult result. The proof of this theorem is extremely long and uses the machinery of many branches of mathematics (including, of all things, imaginary numbers). This mathematical episode illustrates a hopeful and powerful fact about human thought: The human mind is capable of solving extremely difficult problems and answering truly hard questions. This observation should give us all hope for finding solutions to some of the many problems we face in our world today.

Some problems are solved in relatively short time periods; others take longer. In some cases long-standing mysteries concerning numbers are resolved only after centuries of vain attempts. When such stubborn questions are finally answered, the solutions represent a triumph for humanity. In the mid-1990s, with the diligence and drive of several generations of mathematicians, a 350-year-old problem was finally solved. Such a solution is a celebration of the human spirit and justly made the front page of *The New York Times* and the *Wall Street Journal*. This is the story of Fermat's Last Theorem.

The (Mostly True) Story of Fermat's Last Theorem

It was a dark and stormy night. Rain pelted the wavy window pane. Wind seeped in through cracks in the window frame. A candle flickered uncertainly in the cold breeze. The year was 1637.

Pierre de Fermat sat in the great chair of his library poring over the leather-bound Latin translation of the ancient tome of Diophantus's *Arithmetica.* He concentrated intensely on the ideas that Diophantus had written 2000 years before. During those two millennia whole civilizations had risen and fallen, but the study of numbers bridged Fermat and Diophantus, spanning centuries and even transcending death itself.

As the night wore on, inchoate ideas, at first jumbled and ill-defined, began to weave a pattern in Fermat's mind. He gasped as insight and understanding began to take shape. He was seeing how the parts fit together, how the numbers played on one another. Finally, it was clear to him.

On that fateful night in 1637 he wrote in the margin of Diophantus's *Arithmetica* the statement that would make him famous forever. He wrote in Latin:

> *It is impossible to write a cube as a sum of two cubes, a fourth power as a sum of two fourth powers, and, in general, any power beyond the second as a sum of two similar powers. For this, I have discovered a truly wondrous proof, but the margin is too small to contain it.*

He saw that, for any two natural numbers, if we take the first number raised to the third power, then add it to the second number raised to the third power, that sum will *never* equal a natural number raised to the third power. For example, suppose we consider the natural numbers 2 and 14. We notice that

$$2^3 = 2 \times 2 \times 2 = 8, \quad 14^3 = 14 \times 14 \times 14 = 2744,$$

Pierre de Fermat

Life Lesson

Truly difficult problems can be solved. We must have tenacity and patience.

so

$$2^3 + 14^3 = 2752,$$

and there is no natural number that, when raised to the third power, equals 2752 ($14^3 = 2744$ is too small, and $15^3 = 3375$ is too large).

What Fermat saw was that for *any* natural number greater than or equal to 3, let's call that number n, it is impossible to find three natural numbers, let's call them x, y, and z, such that x raised to the nth power (n x's multiplied together) added to y raised to the nth power equals z raised to the nth power. In other words, in the language of mathematical symbols, for any natural number n greater than or equal to 3, there do not exist any natural numbers x, y, and z that satisfy the equation

$$x^n + y^n = z^n.$$

Fermat claimed that he saw a "wondrous proof," but did he? Did he discover a correct proof, or had he deceived himself? No one knows for certain, because Fermat never wrote his "proof" down. In fact, Fermat was notorious for making such statements with usually little or no justification or proof. By the 1800s, all of Fermat's statements had been resolved, all but the preceding one—his "last" one. This unproved assertion became known as *Fermat's Last Theorem.*

We will never know whether Fermat had discovered a correct proof of his Last Theorem, but we do know one thing. He did not discover the proof that Andrew Wiles of Princeton University produced in 1994, 357 years after Fermat wrote his tantalizing marginal note. If Fermat had somehow conceived of Wiles's deep and complicated proof, he would not have written, "The margin of this book is not large enough to contain it." He would have written, "The proof would require a moving van to carry it." Wiles's proof drew on entirely new branches of mathematics and incorporated ideas undreamed of in the 17th century.

Some of the greatest minds in mathematics have worked on Fermat's Last Theorem. The statement of Fermat's Last Theorem does not strike us as intrinsically important or interesting—it just states that a certain type of equation never can be solved with natural numbers. What *is* interesting and important are all the deep mathematical ideas that arose during attempts to prove Fermat's Last Theorem, many of which led to new branches of mathematics. Although Fermat's Last Theorem has not yet been used for practical purposes, the new theories developed to attack it turned out to be valuable in many practical technological advances. From an aesthetic perspective, it is difficult to determine which questions are more important than others. When problems resist attempts by the best mathematical minds over many years, the problems gain prestige. Fermat's Last Theorem resisted all attacks for 357 years, but it finally succumbed.

Andrew Wiles's complete proof of Fermat's Last Theorem is over 130 pages long, and it relies on many important and difficult theorems, including some new theorems from geometry (although it appears surprising that geometry should play a role in solving this problem involving natural numbers). When mathematicians expose connections between seemingly disparate areas of mathematics, they feel an electric excitement and pleasure. In mathematics, as in nature,

Denise Applewhite/Princeton University

Andrew Wiles

elements fit together and interrelate. As we will begin to discover, deep and rich connections weave their way through the various mathematical topics, forming the very fabric of truth.

The Vast Unknown

Many people think mathematics is a static, ancient body of facts, formulas, and techniques. In reality, much of it is a wondrous mystery with many questions unanswered and more still yet to be asked. Many people think we will soon know all there is to know. But this impression is not the case, even though such thinking has persisted throughout history.

> *Everything that can be invented has been invented.*
>
> CHARLES H. DUELL, COMMISSIONER
> U.S. PATENT OFFICE, 1899

> *We don't know a millionth of one percent about anything.*
>
> THOMAS EDISON

Human thought is an ever-expanding universe—especially in mathematics. We know a small amount, and our knowledge allows us to glimpse a small part of what we do not know. Vastly larger is our *ignorance* of what we do not know. An important shift in perspective on mathematics and other areas of human knowledge occurs when we move from the sense that we know most of the answers to the more accurate and comforting realization that we will not run out of mysteries.

So, after celebrating, as we have, some of the great mathematical achievements that were solved only after many decades of human creativity and thought, we close this section by gazing forward to questions that remain unsolved—*open* questions. From among the thousands of questions on which mathematicians are currently working, here are two famous ones about prime numbers that were posed hundreds of years ago and are still unsolved. Fermat's Last Theorem has been conquered; but somehow the mathematical force is not ready to let go its hold and let these two fall.

The Twin Prime Question.

Are there infinitely many pairs of prime numbers that differ from one another by two? (11 and 13, 29 and 31, and 41 and 43 are examples of some such pairs.)

The Goldbach Question.

Can every positive, even number greater than 2 be written as the sum of two primes? (Pick some even numbers at random, and see whether you can write them each as a sum of two primes.)

Computer analysis allows us to investigate a tremendous number of cases, but the results of such analyses do not provide ironclad proof for all cases.

We have seen how to decompose natural numbers into their fundamental building blocks, and we have discovered further mysteries and structures in this realm. Can we use these antique, abstract results about numbers in our modern lives? The amazing and perhaps surprising answer is a resounding *yes!*

A Look BACK

THE PRIME NUMBERS are the basic multiplicative building blocks for natural numbers, since every natural number greater than 1 can be factored into primes. We can prove that there are infinitely many primes by showing that we can always find a prime number larger than any specified number. The strategy is to take the product of all numbers up to the specified number and then add 1. This new large integer must have all its prime factors greater than the original specified number.

The study of primes goes back to ancient times. Some questions remained unanswered for a long time before being resolved. And others remain unanswered.

We discovered proofs of the Prime Factorization of Natural Numbers and the Infinitude of Primes by carefully exploring specific examples and searching for patterns. Considering specific examples while thinking about the general case guided us to new discoveries.

Life Lessons

Understanding a specific case well is a major step toward discovering a general principle.

MINDSCAPES Invitations to Further Thought

In this section, Mindscapes marked (H) have hints for solutions at the back of the book. Mindscapes marked (ExH) have expanded hints at the back of the book. Mindscapes marked (S) have solutions.

Developing Ideas

1. **Primal instincts.** List the first 15 prime numbers.

2. **Fear factor.** Express each of the following numbers as a product of primes: 6, 24, 27, 35, 120.

3. **Odd couple.** If n is an odd number greater than or equal to 3, can $n + 1$ ever be prime? What if n equals 1?

4. **Tower of power.** The first four powers of 3 are $3^1 = 3$, $3^2 = 9$, $3^3 = 27$, and $3^4 = 81$. Find the first 10 powers of 2. Find the first five powers of 5.

5. **Compose a list.** Give an infinite list of natural numbers that are not prime.

Solidifying Ideas

6. **A silly start.** What is the smallest number that looks prime but really isn't?

7. **Waiting for a nonprime.** What is the smallest natural number n, greater than 1, for which $(1 \times 2 \times 3 \times \ldots \times n) + 1$ is *not* prime?

8. **Always, sometimes, never.** Does a prime multiplied by a prime ever result in a prime? Does a nonprime multiplied by a nonprime ever result in a prime? Always? Sometimes? Never? Explain your answers.

9. **The dividing line.** Does a nonprime divided by a nonprime ever result in a prime? Does it ever result in a nonprime? Always? Sometimes? Never? Explain your answers.

10. **Prime power.** Is it possible for an extremely large prime to be expressed as a large integer raised to a very large power? Explain.

11. **Nonprimes (ExH).** Are there infinitely many natural numbers that are not prime? If so, prove it.

12. **Prime test.** Suppose you are given a number n and are told that 1 and the number n divide into n. Does that mean n is prime? Explain.

13. **Twin primes.** Find the first 15 pairs of twin primes.

14. **Goldbach.** Express the first 15 even numbers greater than 2 as the sum of two prime numbers.

15. **Odd Goldbach (H).** Can every odd number greater than 3 be written as the sum of two prime numbers? If so, prove it; if not, find the smallest counter-example and show that the number given is definitely not the sum of two primes.

16. **Still the 1 (S).** Consider the following sequence of natural numbers: 1111, 11111, 111111, 1111111, 11111111, ... Are all these numbers prime? If not, can you describe infinitely many of these numbers that are definitely not prime?

17. **Zeros and ones.** Consider the following sequence of natural numbers made up of 0's and 1's: 11, 101, 1001, 10001, 100001, 1000001, 10000001, ... Are all these numbers prime? If not, find the first such number that is not prime and express it as a product of prime numbers.

18. **Zeros, ones, and threes.** Consider the following sequence of natural numbers made up of 0's, 1's, and 3's: 13, 103, 1003, 10003, 100003, 1000003, 10000003, ... Are all these numbers prime? If not, find the first such number that is not prime and express it as a product of prime numbers.

19. **A rough count.** Using results discussed in this section, estimate the number of prime numbers that are less than 10^{10}.

20. **Generating primes (H).** Consider the list of numbers: $n^2 + n + 17$, where n first equals 1, then 2, 3, 4, 5, 6, ... What is the smallest value of n for which $n^2 + n + 17$ is not a prime number? (*Bonus:* Try this for $n^2 - n + 41$. You'll see an amazingly long string of primes!)

21. **Generating primes II.** Consider the list of numbers: $2^n - 1$, where n first equals 2, then 3, 4, 5, 6, ... What is the smallest value of n for which $2^n - 1$ is not a prime number?

22. **Floating in factors.** What is the smallest natural number that has three distinct prime factors in its factorization?

23. **Lucky 13 factor.** Suppose a certain number when divided by 13 yields a remainder of 7. What is the smallest number we would have to subtract from our original number to have a number with a factor of 13?

24. **Remainder reminder (S).** Suppose a certain number when divided by 13 yields a remainder of 7. If we add 22 to our original number, what is the remainder when this new number is divided by 13?

25. **Remainder roundup.** Suppose a certain number when divided by 91 yields a remainder of 52. If we add 103 to our original number, what is the remainder when this new number is divided by 7?

Creating New Ideas

26. **Related remainders (H).** Suppose we have two numbers that both have the same remainder when divided by 57. If we subtract the two numbers, are there any numbers that we know will definitely divide evenly into this difference? What is the largest number that we are certain will divide into the difference? Use this observation to state a general principle about two numbers that have the same remainder when divided by another number.

27. **Prime differences.** Write out the first 15 primes all on one line. On the next line, underneath each pair, write the difference between the larger number and the smaller number in the pair. Under this line, below each pair on the previous line, write the difference between the larger number and the smaller number. Continue in this manner. Your "triangular" table should begin with:

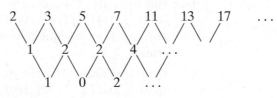

and so on . . .

Once your chart is made, imagine that all the primes were listed on that first line. What would you guess is the pattern for the sequence of numbers appearing in the first entry of each line? The actual answer is not known. It remains an open question! What do you think?

28. **Minus two.** Suppose we take a prime number greater than 3 and then subtract 2. Will this new number always be a prime? Explain. Are there infinitely many primes for which the answer to the question is yes? How does this last question relate to a famous open question?

29. **Prime neighbors.** Does there exist a number n such that both n and $n + 1$ are prime numbers? If so, find such an n; if not, show why not.

30. **Perfect squares.** A perfect square is a number that can be written as a natural number squared. The first few perfect squares are 1, 4, 9, 16, 25, 36. How many perfect squares are less than or equal to 36? How many are less than or equal to 144? In general, how many perfect squares are less than or equal to n^2? Using all these answers, estimate the number of perfect squares less than or equal to N. (*Hint:* Your estimate may involve square roots and should be the exact answer whenever N is itself a perfect square.)

31. **Perfect squares versus primes.** Using a calculator or a computer, fill in the last two columns of the following chart.

n	Number of Primes up to n	Approximate Number of Perfect Squares up to n	Number of Primes/ Number of Perfect Squares
10	4		
100	25		
1000	168		
10,000	1229		
100,000	9592		
1,000,000	78,498		
10,000,000	664,579		
100,000,000	5,761,455		
1,000,000,000	50,847,534		

Given the information found in your chart, what do you conclude about the proportion of prime numbers to perfect squares? Are prime numbers more or less common than perfect squares? Using the Prime Number Theorem, estimate the quotient of the number of primes up to n divided by the number of perfect squares up to n. Use a computer or a graphing calculator to graph your answer. Does the graph confirm your original conjecture?

32. **Prime pairs.** Suppose that p is a prime number greater than or equal to 3. Show that $p + 1$ cannot be a prime number.

33. **Remainder addition.** Let A and B be two natural numbers. Suppose that, when A is divided by n, the remainder is a, and, when B is divided by n, the remainder is b. How does the remainder when $A + B$ is divided by n compare with the remainder when $a + b$ is divided by n? Try some specific examples first. Can you prove your answer?

34. **Remainder multiplication.** Let A and B be two natural numbers. Suppose that the remainder when A is divided by n is a, and the remainder when B is divided by n is b. How does the remainder when $A \times B$ is divided by n compare with the remainder when $a \times b$ is divided by n? Try some specific examples first. Can you prove your answer?

35. **A prime-free gap (S).** Find a run of six consecutive natural numbers, none of which is a prime number. (*Hint:* Prove that you can start with $[1 \times 2 \times 3 \times 4 \times 5 \times 6 \times 7] + 2$.)

Further Challenges

36. **Prime-free gaps.** Using Mindscape 35, show that, for a given number, there exists a run of that many consecutive natural numbers, none of which is a prime number.

37. **Three primes (ExH).** Prove that it is impossible to have three consecutive integers, all of which are prime.

38. **Prime plus three.** Prove that if you take any prime number greater than 11 and add 3 to it the sum is not prime.

39. **A small factor.** Prove that if a number greater than 1 is not a prime number, then it must have a prime factor less than or equal to its square root. (*Hint:* Suppose that all the factors are greater than its square root, and show that the product of any two of them must be larger than the original number.) This Mindscape shows that, to determine if a number is a prime, we have only to check divisibility of the primes up to the square root of the number.

40. **Prime products (H).** Suppose we make a number by taking a product of prime numbers and then adding the number 1—for example, $(2 \times 5 \times 17) + 1$. Compute the remainder when any of the primes used is divided into the number. Show that none of the primes used can divide evenly into the number. What can you conclude about the primes that divide evenly into the number? Can you use this line of reasoning to give another proof that there are infinitely many prime numbers?

In Your Own Words

41. **Personal perspectives.** Write a short essay describing the most interesting or surprising discovery you made in exploring the material in this section. If any material seemed puzzling or even unbelievable, address that as well. Explain why you chose the topics you did. Finally, comment on the aesthetics of the mathematics and ideas in this section.

42. **With a group of folks.** In a small group, discuss and work through the proof that there are infinitely many primes. After your discussion, write a brief narrative describing the proof in your own words.

43. **Creative writing.** Write an imaginative story (it can be humorous, dramatic, whatever you like) that involves or evokes the ideas of this section.

44. **Power beyond the mathematics.** Provide several real-life issues—ideally, from your own experience—that some of the strategies of thought presented in this section would effectively approach and resolve.

For the Algebra Lover

Here we celebrate the power of algebra as a powerful way of finding unknown quantities by naming them, of expressing infinitely many relationships and connections clearly and succinctly, and of uncovering pattern and structure.

45. **Seldom prime.** Suppose that x is a natural number and consider the associated number y given by $y = x^2 - 1$. Show that y is prime if $x = 2$. Now prove that for any natural number x not equal to 2, the associated number y is not a prime number.

46. **A special pair of twins.** A composite number x is the product of two twin primes p and q, in which $p < q$. These special twin primes also satisfy the following "cool" equation having only prime coefficients: $3q - 2p = 17$. The first sentence of this Mindscape describes two equations, one explicitly and one implicitly. The former is $x = pq$. Find the other equation. Solve this

equation for q and solve the "cool" equation for q. Using your work, find the prime p and then find q. Given your answers, solve for x.

47. **Special K p.** A prime p satisfies the equation $p^3 - 4p = 105$. Factor the left side to help you discover the value of p. Are there any other primes p that satisfy an equation of the form $p^3 - 4p = K$, for other numbers K besides 105?

48. **Prime estimate.** The Prime Number Theorem states that as n gets very large, the number of primes less than or equal to n gets close to $n/Ln(n)$. Use a graphing calculator (hand-held or on the Web) to plot the graph of $y = n/Ln(n)$. Estimate how many primes there are less than or equal to 500.

49. **One real root (H).** Find one value of x for which $x^3 + 1^3 = 2^3$. Give your answer in its exact form, not as a decimal approximation from your calculator. How did you know in advance that your answer would not be a natural number?

CRAZY CLOCKS AND CHECKING OUT BARS

Cyclical Clock Arithmetic and Bar Codes

Use your *Heart of Mathematics* 3D glasses to view this picture.

> *A rule to trick th' arithmetic.*
>
> **RUDYARD KIPLING**

Cycles are familiar parts of life. The seasons, phases of the moon, day and night, birth and death—all are among the most powerful natural forces that define our lives, and all are cycles. Whole cultural traditions revolve around this cyclic reality of life; consider, for example, the notion of reincarnation, unless you already considered it in a previous life. We can use these cycles as models to develop analogous constructs in the realm of numbers. Such explorations create yet another kind of cycle, because the abstract mathematical insights refer back to the world, and we find applications of these abstractions in our daily lives.

Our strategy for examining cycles in the world of numbers is to find a phenomenon in nature (in this case cyclicity) and to develop a mathematical model that captures some features of the natural process. This method of reasoning by analogy is a powerful way to develop new ideas, because we use existing ideas, events, and phenomena to guide us in creating new insights.

Most people would not believe that there is deep and powerful number theory going on when they glance at their watch or check out at the grocery store: "Sure, numbers are involved. The time of day is expressed in numbers, as is the price of an item—but these numbers are neither deep nor powerful anything!" Most people, however, are sadly mistaken. The fabulous world of exotic number theory lies hidden in everyday objects.

Time

What time is it? Suppose your watch says it is 9:00 (9 o'clock), and you are to meet the love of your life in 37 hours. What time will your watch read when you fall into the arms of your soul mate? Careful—the answer is not 46 o'clock. The answer is 10:00. So some type of strange arithmetic must be required, since 9 plus

37 equals 10—wacky! How does one perform arithmetic in the context of telling time by a clock? Unlike the natural numbers, which get larger and larger when we add them together, a clock cycles around, and in 12 hours the clock returns to its original position (assuming we're using a 12-hour clock). Counting with a clock in some sense is easier than standard counting, because the numbers never get too large. For example, to add 37 to 9 we could count as follows:

$$9, \quad 10, \quad 11, \quad 12,$$
$$1, \quad 2, \quad 3, \quad 4, \quad 5, \quad 6, \quad 7, \quad 8, \quad 9, \quad 10, \quad 11, \quad 12,$$
$$1, \quad 2, \quad 3, \quad 4, \quad 5, \quad 6, \quad 7, \quad 8, \quad 9, \quad 10, \quad 11, \quad 12,$$
$$1, \quad 2, \quad 3, \quad 4, \quad 5, \quad 6, \quad 7, \quad 8, \quad 9, \quad 10.$$

Notice how, once we get to 12, we start all over again and cycle back to 1. This procedure involves a kind of arithmetic different from standard arithmetic.

Clock Arithmetic

For the moment, let's refer to this arithmetic—where we return to 1 after 12—as *clock arithmetic*. Let's look at a few examples of clock arithmetic. We have already seen that $9 + 37 = 10$. What is $6 + 12$? The answer is 6, since adding 12 just spins us back around to where we started. So 12 is just like 0 in this clock arithmetic. This observation allows us to perform this new arithmetic in a different way. For example, with $9 + 37$, we could notice that 37 is equal to $12 + 12 + 12 + 1$. But remember that adding 12 is just like adding 0, so really 37 is equivalent to 1. Therefore, $9 + 37 = 9 + 1 = 10$.

Let's consider a different kind of question: What does $(4 \times 7) + 20$ equal in clock arithmetic? Well, $4 \times 7 = 28$, but 28 is equivalent to 4, since $28 = 12 + 12 + 4$. Now 20 is equivalent to 8, since $20 = 12 + 8$. Therefore, $(4 \times 7) + 20 = 4 + 8 = 12$. So the answer is 12.

What would happen to our arithmetic if we had a crazy clock that looked like this?

Notice now that adding 7 spins us back to where we started, and thus adding 7 is the same as adding 0. So now $6 + 4 = 10$ is equivalent to 3, since $10 = 7 + 3$. In other words, with this crazy clock, 4 hours after 6 o'crazy clock is actually 3 o'crazy clock. Why would anyone ever bother with such a crazy clock? Well, actually we use this crazy clock, not for telling time, but for telling days in a week. Once again we see that the notion of cycles is natural and important. In fact, as we will now discover, this kind of crazy-clock arithmetic helps us find errors in grocery prices, our checking accounts, UPS package deliveries, airline tickets, and driver's license numbers; it even helps us check out Shakespeare—*read on, MacDuff.*

Equivalence

As we look at cycles, we are developing an idea of *equivalence*. The notion of equivalence occurs in clock arithmetic, for example, when we note that 37 is equivalent to 1. As we develop the idea of cyclical arithmetic, this concept of equivalence will become a central theme.

Let's carefully define a type of arithmetic that will capture the spirit of our previous observations and generalize the notion of clock arithmetic. First we'll

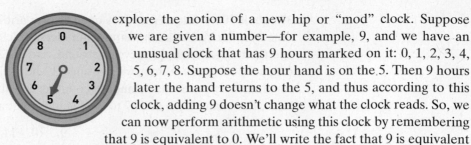

explore the notion of a new hip or "mod" clock. Suppose we are given a number—for example, 9, and we have an unusual clock that has 9 hours marked on it: 0, 1, 2, 3, 4, 5, 6, 7, 8. Suppose the hour hand is on the 5. Then 9 hours later the hand returns to the 5, and thus according to this clock, adding 9 doesn't change what the clock reads. So, we can now perform arithmetic using this clock by remembering that 9 is equivalent to 0. We'll write the fact that 9 is equivalent to 0 as $9 \equiv 0$, where the symbol "$\equiv$" means "equivalent." Of course, 9 is not *equal* to 0, but using this clock we see that 9 is *equivalent* to 0. Let's call this arithmetic *mod 9 clock arithmetic*. The key is that we can perform arithmetic as usual with the understanding that a 9 may be replaced by a 0. For example:

$$13 + 25 = (9 + 4) + (9 + 9 + 7) \equiv (0 + 4) + (0 + 0 + 7) =$$
$$4 + 7 = 11 = 9 + 2 \equiv 0 + 2 = 2 \text{ mod } 9,$$

so $13 + 25 \equiv 2 \text{ mod } 9$. We write the phrase "mod 9" at the end to remind us and indicate to others what kind of mod clock we are using to perform the arithmetic. In terms of the clock itself, we could have computed the preceding sum by placing the hour hand on the 0 and then moving the hand around 13 hours (which brings us to 4) and then moving from that 4 another 25 hours, which brings us to 2.

Once we see how mod 9 clock arithmetic works, we can abstract the idea by dispensing with the visual aid of the clock and calling it *mod 9 arithmetic*. This notation is convenient, since in mathematical jargon "mod" actually stands for "modulo" or "modular," which means just doing arithmetic using the equivalence of $9 \equiv 0 \text{ mod } 9$ (translation: "9 is equivalent to 0 modulo 9" or "9 is equivalent to 0 in mod 9 arithmetic").

In this next example, notice how we are able to replace large numbers with smaller equivalent numbers by just writing them in terms of 9's (also notice how remainders are making an appearance in our work):

$$(3 \times 5) + (7 \times 100) = (9 + 6) + (7 \times ((9 \times 11) + 1)) \equiv$$
$$(0 + 6) + (7 \times ((0 \times 11) + 1)) = 6 + 7 = 13 = 9 + 4 \equiv$$
$$0 + 4 = 4 \text{ mod } 9,$$

so, $(3 \times 5) + (7 \times 100) \equiv 4 \text{ mod } 9$. Once we get the hang of it, this arithmetic is pretty easy. We just pull out multiples of 9 and replace them by 0's. Of course, we can now think about other mod clocks. We can do this modular arithmetic with any clock, as long as we know how many hours it has on it. For any particular natural number n, we write "mod n" to mean that we are thinking about arithmetic on a mod clock that has n hours on it (marked 0, 1, 2, . . . , $n - 1$), and so adding n to any number just brings us back to where we started—thus n is equivalent to 0.

Practice Makes Perfect

Let's really make this new arithmetic our own. Here we ask a few questions. Some present true equivalences, and others do not. Check each one, and determine which are correct and which are wrong. For the ones that are incorrect, figure out a correct answer. Notice that in each case we are using a different mod number, so first have a look at the "mod n" part to see what kind of arithmetic to use.

The Hot Questions

As a warm-up, we'll answer the first question.

1. Is $26 + 31^5 \equiv 0 \bmod 29$?

This statement is true since we are considering a mod 29 clock, so 29 is equivalent to 0. Therefore, $26 + 31^5 = 26 + (29 + 2)^5 \equiv 26 + 2^5 = 26 + 32 = 26 + (29 + 3) \equiv 26 + 3 = 29 \equiv 0 \bmod 29$. (Notice how we did not need to figure out that $31^5 = 28{,}629{,}151$.) Now it's your turn.

2. Is $7^2 + (5 \times 57) \equiv 40 \bmod 48$?

3. Is $2^4 + 5^{301} + (6 \times 31) \equiv 3 \bmod 5$?

4. Is $9^{2000} \equiv 1 \bmod 80$? (*Hint:* Write 2000 as 2×1000.)

The Cool Answers

2. is incorrect: $7^2 + (5 \times 57) \equiv 49 + (5 \times 9) \equiv 1 + 45 = \mathbf{46} \bmod 48$.

3. is also incorrect: $2^4 + 5^{301} + (6 \times 31) \equiv 16 + 0^{301} + (1 \times 1)$
$$\equiv 1 + 1 = \mathbf{2} \bmod 5.$$

4. is correct: $9^{2000} = 9^{2 \times 1000} = (9^2)^{1000} = (81)^{1000} \equiv 1^{1000} = \mathbf{1} \bmod 80$.

Notice how we can work with enormous powers of numbers without even breaking a sweat by just carefully reducing the numbers in clever ways to smaller equivalent numbers in the modular arithmetic. Now let's see how we use modular arithmetic in our daily lives without our even realizing it.

The Mod World of Modular Arithmetic

The Clairvoyant Kleenex Consultant

You have the flu and feel awful. You moan, you groan, you sneeze, you wheeze—let's face it, you're sick. As you sit up in bed, you feel lightheaded—not because you have a fever, but because you have been watching too many hours of mind-numbing daytime TV. You notice in your boredom that on the bottom of your Kleenex box there is a toll-free number for consumer service—1-800-KLEENEX, which amuses you (because of your lightheadedness). Although you are extremely satisfied with the tissues, you decide to dial the number and talk to somebody because you are feeling lonely. The perky Kleenex representative on the other end of the telephone line asks you to read the 12-digit bar code appearing on the bottom of the box. You look and see those thin, fat, and medium lines that make up the Universal Product Code (UPC), which is now tattooed on nearly every product. As all those lines dance in your head you read:

0 3 6 0 0 0 2 8 1 5 0 9

The chipper voice immediately responds by saying, "I think you made a mistake. Could you please read them again?" You glance back at the bar code, still

EAN

< 036000 285109 >

bleary-eyed, and realize that you indeed made a mistake. In fact, the numbers appearing under the bar code are 0 3 6 0 0 0 2 8 5 1 0 9; you reversed the 5 and the 1. But how did your telephone partner immediately know you made a mistake? Perhaps Kleenex reps are clairvoyant, but they definitely use modular arithmetic.

Check Digits

A bar code and its associated numbers (often 12 or 13 digits) make up the UPC. The first six digits encode information about the manufacturer, and the next five digits encode information about the product. That leaves us with the last digit, which is called the *check digit*. The check digit provides a means of detecting if a UPC number is incorrect. Here is how the check digit works with 12 digits: We line up the first 11 digits of the UPC—let's call them: $d_1, d_2, d_3, d_4, d_5, d_6, d_7, d_8, d_9, d_{10}, d_{11}$. We now combine them in an unusual way. We multiply every other number by 3 and then add up all the numbers. That is, we compute

$$3d_1 + d_2 + 3d_3 + d_4 + 3d_5 + d_6 + 3d_7 + d_8 + 3d_9 + d_{10} + 3d_{11}.$$

We now select the check digit (we'll call it c) to be the number from 0 to 9 such that when we add the preceding sum to c, we get an answer that is equivalent to 0 mod 10 (in other words, that sum is evenly divided by 10). In our Kleenex example, we would want to find the number c so that:

$$(3 \times 0) + 3 + (3 \times 6) + 0 + (3 \times 0) + 0 + (3 \times 2) + 8 +$$
$$(3 \times 5) + 1 + (3 \times 0) + c \equiv 0 \bmod 10,$$

or, in other words (numbers),

$$51 + c \equiv 0 \bmod 10, \text{ or equivalently (mod 10)}: 1 + c \equiv 0 \bmod 10.$$

Notice that $1 + 9 = 10$, which is equivalent to 0 in mod 10 arithmetic. So the check digit c should equal 9. Note that the last digit of the UPC is indeed a 9. The Kleenex customer service rep was able to take the UPC you read and check it by multiplying every other number by 3, adding all those numbers (including the check digit) and determining if that sum is equivalent to 0 mod 10. Let's compute this sum for the number you originally read off:

$$0 \quad 3 \quad 6 \quad 0 \quad 0 \quad 0 \quad 2 \quad 8 \quad 1 \quad 5 \quad 0 \quad 9$$
$$(3 \times 0) + 3 + (3 \times 6) + 0 + (3 \times 0) + 0 + (3 \times 2) + 8 +$$
$$(3 \times 1) + 5 + (3 \times 0) + 9 = 52 \equiv 2 \bmod 10.$$

Since that sum is not equivalent to 0 mod 10, the service rep knew that there must have been an error. This check digit system is actually used on *most* 12-digit UPCs. Find some UPCs on various products and compute the sum (multiplying every other number by 3). Your final answer should be equivalent to 0 mod 10. If a single error is made (such as a number read incorrectly), the check digit will always detect the error. This knowledge of UPCs is great for impressing friends, family, and people at parties. With practice, you can have someone give you all the digits except a particular one (say the first one), and, as long as you know which digit is left out, you can predict what the number should be. We provide some practice in the Mindscapes.

Be careful, however. This system is able to detect certain errors, but not all possible errors. For example, if you had switched the 2 and the 5 while reading the UPC to the Kleenex representative, this system would not have detected the error. This simple system is always able to detect an error if exactly one digit is wrong and, in most cases, if two adjacent numbers are switched. The point, however, is that mod 10 arithmetic is the key to this error-checking code. There are many different types of error-detection systems—many of which use modular arithmetic. We'll just point out one more here.

Checking Checks

Take a look at a check from any U.S. bank. In the lower left-hand corner notice nine digits written in that funky 1960s computer font. That is the bank identification number, and the last digit is the check digit. To check the check's check digit, we compute a slightly more complicated sum. Let's call the nine digits $n_1, n_2, n_3, n_4, n_5, n_6, n_7, n_8, n_9$ (that last digit is the check digit). With this error-checking system we compute:

$$7n_1 + 3n_2 + 9n_3 + 7n_4 + 3n_5 + 9n_6 + 7n_7 + 3n_8 + 9n_9$$

and again consider this number mod 10. If the sum is not equivalent to 0 mod 10, then an error has occurred. For example, suppose the nine numbers are:

$$1\ 2\ 0\ 0\ 1\ 0\ 1\ 4\ 3.$$

We would compute:

$$(7 \times 1) + (3 \times 2) + (9 \times 0) + (7 \times 0) + (3 \times 1) + (9 \times 0) +$$
$$(7 \times 1) + (3 \times 4) + (9 \times 3) = 62 \equiv 2 \mod 10.$$

Since we do not get 0, we conclude that there must be an error somewhere. We know for certain those numbers are definitely not a correct bank identification number. Awesome, huh? The Williamstown Savings Bank identification number begins with:

$$2\ 1\ 1\ 8\ 7\ 2\ 9\ 4\ \blacksquare$$

What must its last (check) digit be?

The answer is 6. This error-detection system is able to catch any one error (when one digit is read incorrectly) and is able to catch most switches of two consecutive digits and even most switches of two digits one apart from each other (for example, if you read 5 3 9 when the number was actually 9 3 5). In the Mindscapes, we provide other illustrations, including airline tickets, so the next time you are waiting around in an airport for your connecting flight, look at your boarding pass and do some modular arithmetic.

Josey Customer
5650 Southworth Street
Williamstown, MA 01267

001
83~174
1211

PAY TO THE
ORDER OF_____ $_____

_____ DOLLARS

Williamstown Savings Bank
1234 Main Street, Williamstown, MA USA

FOR _____

⑆211872946⑆ 344100 27 ⑈0001

In the next section we will discover that modular arithmetic is at the core of keeping all our secrets secure on the Information Superhighway.

A Look BACK

THE NOTION OF CLOCK ARITHMETIC or modular arithmetic mathematically models the idea of cycles and the cyclic nature of our lives. This form of arithmetic enables us not only to figure out the day of the week 210 days hence but also to find errors in data using check digits. As we will see, modular arithmetic also leads to the notion of public key secret codes. These last two advances are extremely valuable in our technological world.

Our world contains many examples of cycles. One way to develop mathematical ideas is to look at natural phenomena, model them using mathematics, and then explore the abstract ideas contained in the model. We can take general notions and refine them to develop ideas. We can then explore our new world of the mind without referring back to nature. In the process, however, we often find that our thought experiments are useful in the real world.

Life Lessons

Create abstract ideas by modeling nature.

•

Explore ideas systematically, investigate consequences, and formulate general principles.

MINDSCAPES ⟩ Invitations to Further Thought

In this section, Mindscapes marked (H) *have hints for solutions at the back of the book. Mindscapes marked* (ExH) *have expanded hints at the back of the book. Mindscapes marked* (S) *have solutions.*

Developing Ideas

1. **A flashy timepiece.** You own a very expensive watch that is currently flashing "3:00." What time will it read in 12 hours? In 14 hours? In 25 hours? In 240 hours? What time is it when an elephant sits on it?

2. **Living in the past.** Your watch currently reads "8:00." What time did it read 24 hours earlier? Ten hours earlier? Twenty-five hours earlier? What time did it read 2400 hours earlier?

3. **Mod prods.** Which number from 0 to 6 is equivalent to 16 mod 7? Which number from 0 to 6 is equivalent to 24 mod 7? Which number from 0 to 6 is equivalent to 16×24 mod 7? What number is equivalent to (16 mod 7) $\times$ (24 mod 7) mod 7? What do you notice about the last two quantities you computed?

4. **Mod power (ExH).** Reduce 7 mod 3. Reduce 7^2 mod 3. Reduce [7 mod 3]^2 mod 3. Would you rather find 7 mod 3 first and then square it, or square 7 and then find 7^2 mod 3? What if you had to reduce 7^{1000} mod 3? Okay, you guessed it, now go ahead and reduce 7^{1000} mod 3.

5. **A tower of mod power.** Reduce 13 mod 11. Reduce 13^2 mod 11. Compare $(13 \text{ mod } 11)^2$ with 13^2 mod 11. Now reduce 13^3 mod 11 and 13^4 mod 11 without raising 13 to the power 3 or 4.

Solidifying Ideas

6. **Hours and hours.** The clock now reads 10:45. What time will the clock read in 96 hours? What time will the clock read in 1063 hours? Suppose the clock reads 7:10. What did the clock read 23 hours earlier? What did the clock read 108 hours earlier?

7. **Days and days.** Today is Saturday. What day of the week will it be in 3724 days? What day of the week will it be in 365 days?

8. **Months and months (H).** It is now July. What month will it be in 219 months? What month will it be in 120,963 months? What month was it 89 months ago?

9. **Celestial seasonings (S).** Which of the following is the correct UPC for Celestial Seasonings Ginseng Plus Herb Tea? Show why the other numbers are not valid UPCs.

0	7 1 7 3 4	0 0 0 2 1	8
0	7 0 7 3 4	0 0 0 2 1	8
0	7 0 7 4 3	0 0 0 2 1	8

10. **SpaghettiOs.** Which of the following is the correct UPC for Campbell's SpaghettiOs? Show why the other numbers are not valid UPCs.

0	5 1 0 0 0	0 2 5 6 2	4
0	5 1 0 0 0	0 2 5 2 6	4
0	5 1 0 0 0	0 2 5 2 6	5

11. **Progresso.** Which of the following is the correct UPC for Progresso minestrone soup? Show why the other numbers are not valid UPCs.

0	4 1 1 9 6	0 1 0 1 2	1
0	5 2 0 1 0	0 0 1 2 1	2
0	0 5 0 5 5	0 0 5 0 5	3

12. **Tonic water.** Which of the following is the correct UPC for Canada Dry tonic water? Show why the other numbers are not valid UPCs.

0	1 6 9 0 0	0 0 3 0 3	4
0	2 4 0 0 1	1 0 6 9 1	3
0	1 0 0 1 0	2 0 1 1 0	5

13. **Real mayo (H).** The following is the UPC for Hellmann's 8-oz. Real Mayonnaise. Find the missing digit.

0	4 8 0 0 1	2 6 ■ 0 4	2

14. **Applesauce.** The following is the UPC for Lucky Leaf Applesauce. Find the missing digit.

0	2 8 5 0 0	1 1 0 7 0	■

15. **Grand Cru.** The following is the UPC for Celis Ale Grand Cru. Find the missing digit.

    ■   3 5 8 8 8   4 1 2 0 1   9

16. **Mixed nuts.** The following is the UPC for Planter's 6.5-oz. Mixed Nuts. Find the missing digit.

    0   2 9 ■ 0 0   0 7 3 6 7   8

17. **Blue chips.** The following is the UPC for Garden of Eatin' 10-oz. Blue Corn Chips. Find the missing digit.

    0   1 5 8 3 9   ■ 0 0 0 1   5

18. **Lemon.** The following is the UPC for RealLemon Lemon Juice. Find the missing digit.

    0   5 3 0 0 0   1 5 1 0 8   ■

19. **Decoding (S).** A friend with lousy handwriting writes down a UPC. Unfortunately, you can't tell his 4's from his 9's or his 1's from his 7's. If the code looks like 9 0 3 0 6 8 8 2 3 5 1 7, is there any way to deal with the ambiguity? If so, what is the actual UPC? If it is not possible to determine the correct UPC, explain why.

20. **Check your check.** Find the bank code on your check. Verify that it is a valid bank code.

21. **Bank checks.** Determine the check digits for the following bank codes:

    3 1 0 6 1 4 8 3 ■   0 2 5 7 1 1 0 8 ■

22. **More bank checks (ExH).** Determine the check digits for the following bank codes:

    6 2 9 1 0 0 2 7 ■   5 5 0 3 1 0 1 1 ■

23. **UPC your friends.** Have a friend find a product that has a 12-digit UPC. Ask your friend to carefully read aloud the digits but to skip one digit and say "blank" in its place. Figure out the missing digit. Do this with several different products if you wish. Explain to your friend how you did it. Record the UPCs, the missing digit, and your friend's reactions.

24. **Whoops.** A UPC for a product is

    0   5 1 0 0 0   0 2 5 2 6   5

    Explain why the errors in the following misread versions of this UPC would not be detected as errors:

    0   5 1 0 0 0   0 2 6 2 5   5
    0   5 0 0 0 0   0 5 5 2 6   5

25. **Whoops again.** A bank code is

    0 1 1 7 0 1 3 9 8

    Explain why the errors in the following misread versions of this bank code would not be detected:

    7 1 1 0 0 1 3 9 8
    0 1 1 7 0 8 3 9 1

Creating New Ideas

26. **Mod remainders (S).** Where would 129 be on a mod 13 clock (which goes from 0 to 12)? What is the remainder when 129 is divided by 13?

27. **More mod remainders.** Where would 2015 be on a mod 7 clock? What is the remainder when 2015 is divided by 7? Generalize your observations and state a connection between mod clocks and remainders.

28. **Money orders.** U.S. Postal Money Orders have a 10-digit serial number and a check digit. The check digit is the number between 0 and 6 that represents what the 10-digit serial number is equivalent to using a mod 7 clock. This check digit is the same as the remainder when the serial number is divided by 7. What is the check digit for a money order with serial number 6830910275?

29. **Airline tickets.** An airline ticket identification number is a 14-digit number. The check digit is the number between 0 and 6 that represents what the identification number is equivalent to using a mod 7 clock. Thus, the check digit is just the remainder when the identification number is divided by 7. What is the check digit for the airline ticket identification number 1 006 1559129884?

30. **UPS.** United Parcel Service uses the same check digit method used on U.S. Postal Money Orders and airline tickets for its package-tracking numbers. What would be the check digit for UPS tracking number 84200912?

31. **Check a code.** U.S. Postal Money Order serial numbers, airline ticket identification numbers, UPS tracking numbers, and Avis and National rental car identification numbers all use the mod 7 check digit procedure. Find an example and check the check digit. For instance, get a copy of an airline ticket and check the identification number.

32. **ISBN-13.** The 13-digit book identification number, called the International Standard Book Number (ISBN-13), has its last digit as the check digit. The check digit works on a mod 10 clock. If the ISBN-13 has digits $d_1, d_2, d_3, d_4, d_5, d_6, d_7, d_8, d_9, d_{10}, d_{11}, d_{12}, d_{13}$, then to check if this number is valid, we compute the following number:

$$1d_1 + 3d_2 + 1d_3 + 3d_4 + 1d_5 + 3d_6 + 1d_7 + 3d_8 + 1d_9 + 3d_{10} + 1d_{11} + 3d_{12} + 1d_{13}$$

If the ISBN-13 is correct, this new calculated number should be equivalent to 0 mod 10, that is, the sum should have its last digit being 0. For example, consider the ISBN 978-0-691-15666-8. To check this ISBN, we compute (mod 10):

$$(1 \times 9) + (3 \times 7) + (1 \times 8) + (3 \times 0) + (1 \times 6) +$$
$$(3 \times 9) + (1 \times 1) + (3 \times 1) + (1 \times 5) + (3 \times 6) + (1 \times 6) + (3 \times 6) + (1 \times 8) =$$
$$9 + 21 + 8 + 0 + 6 + 27 + 1 + 3 + 5 + 18 + 6 + 18 + 8 \equiv$$
$$9 + 1 + 8 + 0 + 6 + 7 + 1 + 3 + 5 + 8 + 6 + 8 + 8 \equiv$$
$$70 \equiv 0 \text{ mod } 10.$$

Therefore, this number is a valid ISBN-13. Verify this check method for the ISBN-13 on the copyright page of this book.

33. **ISBN-13 check (H).** Find the check digits for the following ISBN-13s:

    978-0470-42476- ■;   978-0-375-50487- ■.

34. **ISBN-13 error.** The ISBN-13 978-4-1165-9105-4 is incorrect. Two adjacent digits have been transposed. The check digit is not part of the pair of reversed digits. What is the correct ISBN?

35. **Brush up your Shakespeare.** Find a book containing a play by Shakespeare and check its ISBN-13.

Further Challenges

36. **Mods and remainders.** Use the Division Algorithm (see Section 2.3) to show that the remainder when a number n is divided by m is equal to the position n would be on a mod m clock (a mod m clock goes from 0 to $m - 1$).

37. **Catching errors (H).** Give some examples in which the UPC check digit does not detect an error of two switched adjacent digits. Try to determine a general condition whereby a switching error in those digits would not be detected. (*Hint:* Consider the difference of the digits.)

38. **Why three?** In the UPC, why is 3 the number every other digit is multiplied by rather than 6? (*Hint:* Multiply every digit from 0 to 9 by 3 and look at the answers mod 10. Do the same with 6 and compare your results.) Are there other numbers besides 3 that would function effectively? What number might you try?

39. **A mod surprise.** For each number n from 1 to 4, compute n^2 mod 5. Then for each n, compute n^3 mod 5 and finally n^4 mod 5. Do you notice anything surprising?

40. **A prime magic trick.** Pick a prime number and call it p. Now pick any natural number smaller than p and call it a. Compute a^{p-1} mod p. What do you notice? You can use this observation as the basis for a magic trick. Have a friend think of a natural number less than p (but keep it to him- or herself). Tell that person that you will predict and write what the remainder will be when a^{p-1} is divided by p. Write your answer and seal it in an envelope, and then ask what the person's number was. Now, to your friend's amazement, compute the remainder when a^{p-1} is divided by p and reveal the hidden prediction. Record your friend's reaction. The next section uses this observation in a powerful way. Check it out.

In Your Own Words

41. **Personal perspectives.** Write a short essay describing the most interesting or surprising discovery you made in exploring the material in this section. If any material seemed puzzling or even unbelievable, address that as well. Explain why you chose the topics you did. Finally, comment on the aesthetics of the mathematics and ideas in this section.

42. **With a group of folks.** In a small group, discuss and work through the four mod questions posed in the section and the Clairvoyant Kleenex Consultant story. After your discussion, write a brief narrative describing the methods in your own words.

43. **Creative writing.** Write an imaginative story (it can be humorous, dramatic, whatever you like) that involves or evokes the ideas of this section.

44. **Power beyond the mathematics.** Provide several real-life issues—ideally, from your own experience—that some of the strategies of thought presented in this section would effectively approach and resolve.

For the Algebra Lover

Here we celebrate the power of algebra as a powerful way of finding unknown quantities by naming them, of expressing infinitely many relationships and connections clearly and succinctly, and of uncovering pattern and structure.

45. **One congruence, two solutions.** Find two different natural numbers that are each a solution to $3x \equiv 5 \bmod 7$.

46. **Chinese remainder.** Find one natural number x that simultaneously satisfies both $x \equiv 0 \bmod 2$ and $x \equiv 1 \bmod 3$.

47. **More remainders.** Find one natural number z that simultaneously satisfies $z \equiv 1 \bmod 2$, $z \equiv 1 \bmod 3$, and $z \equiv 1 \bmod 5$. (*Hint*: What do you know about the number $z - 1$?)

48. **Quotient coincidence.** Suppose x is a natural number. When you divide x by 7 you get a quotient of q and a remainder of 6. When you divide x by 11 you get the same quotient but a remainder of 2. Find x.

49. **Mod function (H).** The value $y = 10$ satisfies the congruence $y \equiv 3 \bmod 7$. So does $y \equiv 17$. Find a formula for y in terms of the variable x so that no matter what natural number you plug in for x, the result is a value of y that satisfies $y \equiv 3 \bmod 7$.

2.5 PUBLIC SECRET CODES AND HOW TO BECOME A SPY

Encrypting Information Using Modular Arithmetic and Primes

> *No more fiction: we calculate; but that we may calculate, we had to make fiction first.*
>
> FRIEDRICH NIETZSCHE

Did you ever notice that 007 is a prime? Coincidence?

With the thawing of the Cold War, have spies fallen on hard times? Might we expect to see people in trench coats standing on street corners with signs that read, "Ex-spy. Will break codes for food"? Actually, no, because although the spy business is a bit slack now, in this age of the Information Superhighway secret codes have become indispensable. Whether you're withdrawing money from an ATM, sending your VISA card number over the Web to make a purchase, or making a stock transaction, stolen data could mean stolen money. Information on the Internet needs to be secured through the use of codes. Modular arithmetic is at the heart of modern coding, and thus coding is possibly the most powerful example of the unforeseen applicability of abstract mathematical ideas—in this case to the digital world. Who would have thought that *cryptography*—the study of secret codes—would become an important part of daily life and that the exploration of numbers would be central to coding?

This section is difficult. To master every part of the mathematics involved requires a significant effort. Luckily for us, the idea of public key cryptography is interesting even without fully delving into the mathematical details that make it work. The first part of this section offers an overview of the delicate ideas behind public key cryptography, while the second part outlines the technical details of the method in practice and the mathematical reasoning for why the scheme works. It is perfectly fine (if not advisable) to focus on the first part and either skim or skip the second part. This challenging section is evidence that as the world changes, ideas that seem marginal today may become central tomorrow. Good luck.

Coding and Decoding

How can we code and decode messages? One possibility is to replace one letter by another letter. For example, suppose we created the following coding scheme:

Message	A	B	C	D	E	F	G	H	I	J	K	L	M	N	O	P	Q	R	S	T	U	V	W	X	Y	Z
Coded As	T	H	E	Q	U	I	C	K	B	R	O	W	N	F	X	J	M	P	D	V	L	A	Z	Y	G	S

If you wanted to send the message:

YOUR JOKES ARE LAME,

then you could send the coded message:

GXLP RXOUD TPU WTNU.

The major problem with this code is that breaking it is easy even without knowing the key. That is, any enemy who captured a sufficiently long encrypted message could figure out the original message. More elaborate coding methods are harder to break but still can be deciphered if the codes are shared. That is, suppose you are receiving messages from both Bill and Hillary, and each encodes his or her message using the same scheme. If Hillary captures Bill's encrypted message, couldn't she decode it by simply reversing the encoding procedure? It seems that we must trust our friends—a grave drawback to any shared coding scheme. Ideally, we would prefer a code by which people are able to encode messages to us but are at the same time unable to decode other messages that have been encoded by the same process. Is such a coding scheme possible? If someone could encode a message, then all he or she would have to do to decode messages in the same type of coding scheme is reverse the coding procedure. However, this plausible statement turns out to be false, and therein lies the core of modern coding methods.

In this section we will look at a coding technique invented during the last few decades that uses a 350-year-old theorem about modular arithmetic to encrypt and decode secret messages.

Life Lesson

Attractive ideas in one realm often have unexpected uses elsewhere.

Public Key Codes

The method uses an encrypting and decoding scheme that is fundamentally new in the coding business. The new wrinkle is the invention of the public key code. *Public key codes* are codes that allow us to encode any message but prevent us from decoding other messages encrypted by the same technique. Such codes are called *public key codes* because we can tell the entire world how to encode messages to us. We can even tell our enemies. We can take out an ad in the newspaper telling everyone how to encode messages to us; it's no secret; it's public. The key is that we and only we can *decode* an encrypted message. Isn't this notion counterintuitive? How can such a coding scheme work? We'll take a look at one such scheme known as the RSA public key code.

Before jumping into the technical details of this coding scheme, let's try to make the basic idea of the encoding and decoding aspects of the RSA code plausible. For this purpose, we journey to Carson City.

The Carson City Kid and the Perfect Shuffle Code

The Carson City Kid was the master of cards (actually he was no kid, but it sounds better than "the Carson City Yuppie"). His hands were quicker than the eye, and his morals were just as fast. One thing the Kid could do without fail was what is known in the trade as a *perfect shuffle*. That is, he would cut the deck of 52 cards precisely in half and then shuffle them perfectly—one card from the top half, then one from the bottom half, and so on, intermixing the cards exactly—one from one side, one from the other. For the larcenous among our readers, the advantage of a perfect shuffle is that, contrary to typical random shuffles, perfect shuffles only *appear* to bring disorder to the deck. The original ordering is restored after exactly 52 perfect shuffles.

The Kid was an enterprising soul who did not want to spend his life in casinos, rolling in money and surrounded by glamorous and attractive people. Instead, he decided to go into the secret message biz and be surrounded by glabrous and atrocious people. His method was simple. He knew that most people could not execute perfect shuffles. They could do only five or six shuffles before messing up. The Kid's method was straightforward: The code sender would take a deck of 52 blank playing cards and write the message using one letter per card. Then the sender spy would carefully do five perfect shuffles—leaving the deck of cards in an apparently random order. The spy receiving the shuffled deck would then hand the coded message (the shuffled deck) to the Kid.

The Kid knew exactly what to do. He quickly shuffled the deck with 47 more perfect shuffles and voilà! The cards had rearranged themselves exactly into their original order, and so the message could be read.

The Basic Theme of the "Public" Aspect of the RSA Coding Scheme in 10 Sentences

Let's select two enormous prime numbers—and we mean enormous—say each having about 300 digits—and multiply those numbers together. How can we multiply them together? Computers are whizzes at *multiplying* natural numbers—even obscenely long ones. *Factoring* large numbers, however, is hard—even for computers. Computers are smart but not infinitely smart—there are limits to the size of natural numbers that they can factor. In fact, our product is much too large for even the best computers to factor. So if we announce that huge product to the world, even though it can be factored in theory, in practice it cannot. Thus we are able to announce the gigantic number to everyone, and yet no one but we would know its two factors. This huge product is the *public* part of the RSA public key code. Somehow, the fact that only we know how to factor that number allows us to decode messages while others cannot. It's not obvious why this factoring fact is helpful in making secret codes, but we'll see that it's really at the heart of the mattter.

Of course, the Kid's technique is too simple to use in practice. With determination, a person who captured the five-shuffled deck could do the reverse of those perfect shuffles. However, the Kid's technique demonstrates a mathematical fact that revolutionized the coding business.

The RSA Coding Scheme

In 1977, Ronald Rivest, Adi Shamir, and Leonard Adleman discovered a public key coding scheme that uses modular arithmetic. This public key coding method is referred to as the *RSA Coding Scheme* and is now used millions of times each day. Kid Carson's 47 perfect shuffles that return the deck to its original order capture the spirit of this RSA public key coding scheme. A shuffling procedure encodes a message, and only the receiver knows how to continue to shuffle the message further in a way that unshuffles the message—no one besides the receiver can perform that additional shuffling. So, now there are two basic questions we hope you are wondering: (1) What are we shuffling? and (2) How do we keep shuffling to get back to where we started?

Shamir, Rivest, Adleman. (not in RSA order!)

How the Scheme Works: Shuffling Numbers

We will shuffle numbers. That is, we will first convert our message to numbers and then shuffle those numbers. How do we shuffle the numbers? Let's take a prime number, say 5. Pick any number that does not have 5 as a factor, for example, 8. To shuffle 8, let's just raise 8 to higher powers and look at the remainders of those powers of 8 when we divide by 5. In other words, raise 8 to higher powers and look at those numbers mod 5.

Powers of 8	Powers of 8 mod 5
$8^1 = \quad\quad 8$	3
$8^2 = \quad\quad 64$	4
$8^3 = \quad\quad 512$	2
$8^4 = \quad 4096$	1
$8^5 = 32{,}768$	3

Life Lesson

Look for patterns.

The second column represents a type of shuffling of a four-card deck where the "shuffling" is accomplished by multiplying by 8. Notice that after five shuffles of multiplication by 8 mod 5 we get back to 3.

Let's try this again with a different-size deck. Suppose we pick the prime 7 and choose a number that does not have 7 as a factor—say 10. Let's shuffle 10 by raising it to higher powers and considering those powers mod 7.

Powers of 10		Powers of 10 mod 7
$10^1 =$	10	3
$10^2 =$	100	2
$10^3 =$	1000	6
$10^4 =$	10,000	4
$10^5 =$	100,000	5
$10^6 =$	1,000,000	1
$10^7 =$	10,000,000	3

Here we notice that after seven shuffles of powers mod 7 we get back to 3. Now it's your turn. Try this shuffling yourself. Let's set the prime number to be 5. Now pick some numbers that have no factor of 5 and shuffle them by raising them to powers mod 5. Try this shuffling with at least two different numbers. What do you notice? How many shuffles get us back to where we started mod 5? Let's look for patterns.

By experimenting, we discover that, if we shuffle 5 times mod 5, we get back to where we started. We also notice that, after we shuffle 4 times mod 5, we always get 1 as the answer. This observation turns out to be a mathematical fact—known as *Fermat's Little Theorem*.

Fermat's Little Theorem.

If p is a prime number and n is any integer that does not have p as a factor, then n^{p-1} is equal to 1 mod p. In other words, n^{p-1} will always have a remainder of 1 when divided by p.

It is Fermat's Little Theorem, proved more than 350 years ago, that is the basis of our shuffling procedure. Now let's tackle the RSA public key code scheme.

An Illustrative, Cryptic Example

We introduce the RSA public key code method by considering a specific example. Using a diabolically clever idea that will be explained later, we construct and publicize a pair of numbers to the world—in this example the numbers 7 and 143. In real life the numbers would be much larger, perhaps having several hundred digits each. At the same time we construct the public numbers, we also construct and keep secret a decoding number, in this case 103. The public part of the public key code does not contain the key to unlock the code; instead, the key is the

Life Lesson

Ground your under-standing in the specific.

secret decoding number that is kept only by the receiver of encrypted messages. It never needs to be transmitted to anyone else. We'll explain later how all these numbers were created.

We not only publicize the numbers 7 and 143 but also explain exactly how to use them to encrypt a message. Here are the instructions, which could be published in the newspaper.

Encoding Messages

Suppose that W is a secret Swiss bank account number (less than 143) that the sender wants to encrypt and send to someone. The sender computes W^7 (remember that 7 is the first public number) and then computes the remainder when W^7 is divided by 143, the second public number. That is,

$$W^7 = 143q + C,$$

where the remainder C is an integer between 0 and 142. Or, expressed in modular arithmetic,

$$W^7 \equiv C \bmod 143.$$

The number C is now the coded version of W.

Decoding Messages

The receiver receives the coded message C and now must decode it. This decoding process requires the receiver to compute C^{103} (recall that 103 is the secret number that no one but the receiver knows) and then to compute the remainder when C^{103} is divided by 143. That is,

$$C^{103} = 143q + D,$$

where the remainder D is an integer between 0 and 142. Or, expressed in modular arithmetic,

$$C^{103} = D \bmod 143.$$

The amazing fact is that D (the decoded message) will always be identical to W (the original, uncoded message). Thus the receiver decoded the coded message C to produce the original message W.

Suppose someone sets up the public key code described above, announces the public numbers (7 and 143) and the coding method, and keeps the number 103 secret. Now let's further suppose that a friend wishes to secretly send her Swiss bank account number, 71. The table here shows the sequence of events to code the message 71.

To encode the number 71, the sender computes 71^7, which equals 9,095,120,158,391, and then computes the remainder when this number is divided by 143. The remainder turns out to be 124. So 124 is the encoded version of 71, and that is what the sender sends to the receiver. Now the receiver has to decode 124.

Remember that 103 is the secret decoding key known only to the receiver. To decode 124, the receiver first computes 124 to the 103rd power and then finds the remainder when that number is divided by 143.

Receiver	Announces to the world the numbers (7 and 143) and coding instructions. Tells no one secret decoding number 103.		Receiver receives the coded message "124." Using the secret number 103, receiver computes 124^{103} mod 143, which is 71 — the original message!
Sender		Wants to send "71" to her friend. Knows the coding numbers (7 and 143). Sender computes 71^7 mod 143, which is 124. 124 is the coded version of 71. Sender sends 124 to receiver.	

The number 124^{103} equals

41921187047849896446113000569294530888483668997732045634627122565220914671133939555703940592675185212029512808239919702590414929088043093696556512787027350058759384015077439569484127475589434834019120344958849410 6624,

which is a pretty big number; however, the remainder when it is divided by 143 is (drum roll, please) 71.

Of course, if we wanted to send or receive words (rather than just numbers) we could first convert the letters into numbers in some straightforward way (for example, A = 01, B = 02, ... Z = 26, space = 27) and then encode the number, just as we did with 71. Raising the encrypted number to the 103rd power and taking the remainder when that result is divided by 143 results in the original message. The incredible fact is that this method of retrieving the original number always works.

Take any number n from 0 to 142, compute the remainder when the number n^7 is divided by 143, and call that remainder C. Then compute the remainder when C^{103} is divided by 143, and we will miraculously have the original number n with which we started! Of course, there's a reason behind such seemingly amazing coincidences.

Questions the Coding Scheme Generates

The preceding example leads to many questions. Here are a few:

1. We said that, in real life, the numbers we use would have perhaps several hundred digits each. Is it practical, even for a computer, to raise numbers with hundreds of digits to powers of several hundred digits?

2. Where did the numbers 7, 143, and 103 come from?

3. Why in the world does this process work? Is this coding and decoding process just an amazing fluke?

How Do We Raise Huge Numbers to Large Powers?

As we illustrated, with the aid of a computer we can raise 124 to the 103rd power and write out the whole 200-plus digit answer.

Question If you take a 100-digit number and raise it to a 100-digit power, approximately how many digits would the answer have? Would it be physically possible to write out such a number?

Answer No. A 100-digit number raised to a 100-digit power would have about 10^{102} **digits**. If we wanted to write out such a number, we would have to write pretty small, since physicists' best current estimate for the number of atoms in the universe is less than 10^{80}, a tiny fraction of 10^{102}. Thus a huge number of digits would have to be written on each atom in the universe. In other words, no computer could possibly do that calculation.

Well, then, aren't we stuck? We have just seen that no computer can actually raise a 100-digit number to a 100-digit power. We must be clear about what we really want.

Actually, we are not stuck. All we are really after is the *remainder* that we get after dividing that huge power by some number. Finding this remainder can be accomplished by modular arithmetic without ever computing the big power. Such a procedure does not strain even a laptop computer. We and our computers are able to compute remainders efficiently without needing more atoms than our universe possesses.

Where Do Those Numbers Come From?

The numbers 143, 7, and 103 stem from an ingenious combination of just a few ideas from number theory. Here is the procedure for choosing the numbers. We give the method in flowchart form and describe it in prose. As the flowchart on the next page shows, the process begins by our picking two different prime numbers. In this case, we pick $p = 11$ and $q = 13$. Their product is 143, and that is one of the public numbers. The first public number in this coding scheme is always the product of two different primes. In real life, these primes would each have a couple hundred digits.

To select the numbers 7 and 103, we follow some intermediate steps. The first step is to compute $(p - 1)(q - 1)$ or, in this case, $10 \times 12 = 120$. Now we select any number e that shares no common factors with 120 (e stands for *encoding*). It can be any such number. In our example, we chose 7. That number is the other public number.

Okay, in all honesty, what we are about to do next appears to be coming out of thin air, but its value will soon become clear. We find integers d and y such that $7d - 120y = 1$. In this case, we find that we can take d to equal 103 and y to be 6, since those values satisfy the equation

$$(7 \times 103) - (120 \times 6) = 1.$$

The number d (d for *decoding*), which in this case equals 103, is the secret decoding number that we keep to ourselves. That's it! Whenever we select the numbers p, q, e, and d in the manner described previously, the coding scheme will always work.

Receiver

Selects two different prime numbers, in this case 11 and 13, but tells no one what they are.

Multiplies them together: 11 × 13 = 143. This becomes one of the public numbers. The public will know the product but will not be able to factor it since the number, in practice, would be too large.

Subtracts 1 from each of the two primes, 11 − 1 = 10 and 13 − 1 = 12, and then multiplies these answers together to get 120. The receiver then selects a number at random that has no common factor with 120. In this case the receiver selects 7, which becomes the other number publicly announced.

Using the numbers 120 and 7, the receiver finds integers d and y so that they satisfy the equation: $7d - 120y = 1$. One such solution is $(7 \times 103) - (120 \times 6) = 1$. The value of d—in this case 103—is the secret decoding number that only the receiver knows or can figure out, since figuring out a solution to the equation required the factorization of 143—which no one knows but the receiver.

Why Do Those Numbers Work?

Let's see why this coding and decoding scheme always works. Before we give an overview of why the RSA coding scheme works, we have to confess that what follows is difficult. What makes it difficult? The answer is that there are many steps. Although each step on its own is no great intellectual feat, when we string them together, one after another, the logic and modular arithmetic can get out of hand. These details are more interesting to some than to others. So readers who decide to invest the energy to learn what follows must expect to struggle and to reread the information several times. Other readers may decide to limit their investment in this topic and move on—remember Vietnam.

We begin our explanation by a quick recap: First we picked two primes,

$$p = 11 \quad \text{and} \quad q = 13.$$

Their product is 143, and that is one of the public numbers. We next computed

$$(p - 1)(q - 1)$$

or, in this case,

$$10 \times 12 = 120$$

and selected a number, e, that shares no common factors with 120—we chose 7. That number is the other public number. We then found integers 103 and 6 that satisfy

$$(7 \times 103) - (120 \times 6) = 1.$$

We keep the number 103 secret, because it's the secret decoding number, and we announce the pair (7 and 143) to the entire world. We're now ready to see how coding and decoding work with the message 71. Remember that, to encode the message, we computed the remainder when 71^7 is divided by 143 and got 124. So, in modular arithmetic terms,

$$71^7 \equiv 124 \mod 143.$$

Then, to decode, we raised 124 to the 103rd power and found the remainder when we divided the answer by 143 (that is, we figured out $124^{103} \mod 143$). We now want to determine why this remainder must always equal the original message (in this case 71) that our sender sent. Specifically, we wish to understand why

$$124^{103} \equiv 71 \mod 143.$$

The Numerical Details

Recall that 143 is 11×13. We consider the two primes 11 and 13 separately and start with 11. Recall that 124 is the remainder when 71^7 is divided by 143. Therefore, by the Division Algorithm,

$$71^7 = 143k + 124,$$

for some integer k. We now write 143 as 11×13 and notice that

$$71^7 = (11 \times 13 \times k) + 124 \equiv 0 + 124 \mod 11.$$

So we can conclude that

$$(71^7)^{103} \equiv (124)^{103} \mod 11.$$

Remember now how we selected 103:

$$(7 \times 103) = 1 + (120 \times 6).$$

Using this fact we see that

$$(71^7)^{103} = 71^{(7 \times 103)} = 71^{1+(120 \times 6)} = 71 \times (71)^{(120 \times 6)} = 71 \times (71^{72})^{10}.$$

Since the prime 11 is not a factor of 71, we know by Fermat's Little Theorem—our shuffling fact—that

$$(71^{72})^{10} \equiv 1 \mod 11 \text{ (notice that the power 10 is just } 11 - 1).$$

So, putting all these observations together, we see that

$$124^{103} \equiv (71^7)^{103} = 71 \times (71^{72})^{10} \equiv 71 \times 1 = 71 \mod 11.$$

Therefore, by subtracting 71 from both sides, we see that

$$124^{103} - 71 \equiv 0 \mod 11.$$

In other words, we now see that 11 divides evenly into $124^{103} - 71$.

If we repeat with the prime 13 every step we did with 11, we see that 13 also divides evenly into $124^{103} - 71$.

We have just shown that both 11 and 13 divide evenly into $124^{103} - 71$. Since 11 and 13 are different primes, we know that the product of 11 and 13, namely 143, must divide evenly into $124^{103} - 71$. In other words, when $124^{103} - 71$ is divided by 143 the remainder is 0. Given this fact, what is the remainder when just 124^{103} alone is divided by 143? It has to equal 71, since, as we have already seen, once we subtract off the 71, its remainder is zero. Thus the remainder when 124^{103} is divided by 143 is 71. We decoded and got the original message back. This numerical illustration includes all the essential ideas why this procedure works in general.

The RSA Code in General

To build an RSA public key code, the receiver first selects two different prime numbers—let's call them p and q. In practice the receiver chooses primes that have perhaps 200 digits each. Now we define the integers n and m as follows: $n = pq$, and $m = (p - 1)(q - 1)$.

Next the receiver selects any positive integer e (*encoding* power) such that m and e share no common factors. Since e and m share no common factors, it turns out we can always find positive integers d (*decoding* power) and y such that $ed = 1 + my$.

The receiver announces to the entire world the pair of numbers (e and n). The important thing, however, is to tell no one what d is. All the other numbers are no longer needed. The receiver probably should destroy any documents with the numbers p or q on them. Although we know that n can be factored into the product of p and q, the number n is so large that nobody besides the receiver can find those factors. The receiver no longer needs the factors p and q, and, if they get into the wrong hands, the security of our code could be violated.

If a sender wishes to send the receiver the message W, the sender computes the remainder when W^e is divided by n—let's call the remainder C (*coded* message). The integer C is the coded version of the message W. Since the message W has now been encoded to the integer C, the sender could take out an ad in the paper and tell everyone the integer C. Only the receiver, however, will be able to decode it by finding the remainder when C^d is divided by n—let's call it D (*decoded* message). It turns out that D will be exactly W. That is, D is the original message, so the receiver just decoded the message. The proof of why it always works involves the same logic used in our previous example.

The RSA scheme is both interesting and extremely practical. In fact, RSA has become widely accepted commercially. Think we're just kidding? In July of 1996, RSA Data Security, the company formed to promote and sell RSA solutions, was sold for about $400 million to Security Dynamics. This astronomical figure gives some indication of how seriously this coding scheme is being taken, and it all comes from divisibility, remainders, and crazy clocks. More proof that math pays! Innovative ideas are valuable in our society, regardless of their source.

Breaking the Code

Before closing this section, we address a natural question, namely, "How could someone break this code?" Remember that the world knows both n and e. If someone could factor n into its prime factorization $p \times q$, then knowing p, q, and e, he or she could figure out $m = (p - 1)(q - 1)$ and then figure out d (do you see how?). Once *d* is known, the code can be broken. So how could we avoid having someone break the code? Well, certainly we don't want anyone to be able to factor our number *n*. How could we arrange that? We take a look at current computer technology. We determine what the largest integer is that can be factored by the best computers. We then select our primes *p* and *q* so that each is larger than the largest integer that can be factored by machine.

Might some clever insight into how numbers work allow computers to factor even extremely large numbers and thus factor our *n*? If someone came up with such a scheme, the RSA public key code scheme would fail.

Is there another way of breaking the code without factoring *n*? This is an unsolved question in number theory. In other words, no one knows if the only way to break the code is to actually factor *n*. Perhaps there is a devilishly sneaky alternative method of breaking the code without ever factoring *n*. This possibility is a big mystery. Since no one currently knows how to break the code, the RSA system provides a safe coding scheme, at least for now. It's interesting to see where the borders of knowledge are.

A Look BACK

A SECRET NUMBER can be encoded by raising it to a power mod *n*, where *n* is the product of two large primes. It can be decoded and the original secret number retrieved by raising the result to yet another power mod *n*. The secrecy of this scheme is dependent on our inability to factor numbers that are the products of two primes, each having hundreds of digits. It seems plausible that, if we can encrypt a message using some process, then we could reverse the process and decode other messages that were encrypted by that method. However, such is not the case. Public key encryption exists in the realm between the plausible and the actual.

Understanding the public key encryption and decoding scheme is best attempted using a specific example. All the ideas for the general scheme are clearly present in the special case. Often we learn generally applicable principles by concentrating on specific cases.

Life Lessons

Ground your understanding in the specific.

MINDSCAPES ⟩ Invitations to Further Thought

In this section, Mindscapes marked **(H)** *have hints for solutions at the back of the book. Mindscapes marked* **(ExH)** *have expanded hints at the back of the book. Mindscapes marked* **(S)** *have solutions.*

Developing Ideas

1. **What did you say?** The message below was encoded using the scheme on page 99. Decode the original message.

    VKBD BD VKU EXPPUEV TFDZUP

2. **Secret admirer.** Use the scheme on page 99 to encode the message "I LOVE YOU."

3. **Setting up secrets.** Let $p = 7$ and $q = 17$. Are p and q both prime numbers? Find $(p - 1)(q - 1)$, the number we call m. Now let e equal 5. Does e have any factors in common with m? Finally, verify that $77e - 4m = 1$.

4. **Second secret setup.** Let $p = 5$ and $q = 19$. Are p and q both prime numbers? Find $(p - 1)(q - 1)$, the number we call m. Now select e to equal 11. Does e have any factors in common with m? Now suppose $d = 59$ and $y = 9$. Verify that $e \times d - m \times y = 1$.

5. **Secret squares.** Reduce the following quantities: $1^2 \bmod 3$; $2^2 \bmod 3$; $3^2 \bmod 3$; $4^2 \bmod 3$; $5^2 \bmod 3$; $6^2 \bmod 3$. Do you notice a pattern?

Solidifying Ideas

6. **Petit Fermat 5.** Compute 2^4 (mod 5). Compute 4^4 (mod 5). Compute 3^4 (mod 5). Oh, what the heck, compute n^4 (mod 5) for all numbers n from 1 through 4.

7. **Petit Fermat 7.** Compute 4^6 (mod 7). Compute 5^6 (mod 7). Compute 2^6 (mod 7). Oh, why not, compute n^6 (mod 7) for all numbers n from 1 through 6.

8. **Top secret (ExH).** In our discussion, the two public numbers 7 and 143 were given. How would you encode the message "4"? The secret decoding number is 103. Without performing the calculation (unless you have a computer that can do modular arithmetic for you), how would you decode the encrypted message you just made if you were the receiver?

9. **Middle secret (H).** In our discussion, the two public numbers 7 and 143 were given. How would you encode the message "3"? The secret decoding number is 103. Without performing the calculation (unless you have a computer that can do modular arithmetic for you), how would you decode the encrypted message you just made if you were the receiver?

10. **Bottom secret.** In our discussion, the two public numbers 7 and 143 were given. How would you encode the message "11"? The secret decoding number is 103. Without performing the calculation (unless you have a computer that can do modular arithmetic for you), how would you decode the encrypted message you just made if you were the receiver?

11. **Creating your code (S).** Suppose you wish to devise an RSA coding scheme for yourself. You select $p = 3$ and $q = 5$. Compute m, and then find (by trial and error if necessary) possible values for e and d.

12. **Using your code.** Given the coding scheme you devised in Mindscape 11, show how a friend would encode "HI" (as in our discussion, use 01 for A, 02 for B, . . . , 26 for Z to convert the letters to numbers). Now decode the coded message. Did you return to your original HI?

The following list of random information may be useful in the subsequent three Mindscapes:

$73^7 \equiv 83 \bmod 143$	$83^{143} \equiv 58 \bmod 103$	$8^{103} \equiv 83 \bmod 143$
$74^7 \equiv 35 \bmod 143$	$74^{143} \equiv 51 \bmod 103$	$74^{103} \equiv 61 \bmod 143$
$61^7 \equiv 74 \bmod 143$	$38^{143} \equiv 29 \bmod 103$	$83^{103} \equiv 73 \bmod 143$
$83^7 \equiv 8 \bmod 143$	$35^{143} \equiv 5 \bmod 103$	$38^{103} \equiv 103 \bmod 143$
$38^7 \equiv 25 \bmod 143$	$8^{143} \equiv 72 \bmod 103$	$35^{103} \equiv 74 \bmod 143$

13. **Public secrecy.** Using the list in Mindscape 12, with the public numbers 7 and 143, how would you encode "83"? How would you decode the message using the decoding number 103? What numbers in the list must you refer to for the encoding and decoding operations?

14. **Going public.** Using the list in Mindscape 12, with the public numbers 7 and 143, how would you encode "61"? How would you decode the message using the decoding number 103? What numbers in the list must you refer to for the encoding and decoding operations?

15. **Secret says (H).** Using the list in Mindscape 12, with the public numbers 7 and 143, and decoding number 103, how would you decode "38"? What numbers in the list must you refer to for this decoding operation?

Creating New Ideas

16. **Big Fermat (S).** Compute 5^{600} (mod 7). (*Hint*: Recall your answer to Mindscape 7, "Petit Fermat 7.") Compute $8^{1,000,000}$ (mod 11).

17. **Big and powerful Fermat (ExH).** Recall how exponents work, for example, $7^{15} = 7^{(12+3)} = 7^{12} \times 7^3$. Now, using exponent antics, compute 5^{668} (mod 7).

18. **The value of information.** How large should the primes p and q be in the RSA coding scheme? Of course, if you pick ridiculously large ones, then the product of the two would be impossible to factor from a practical point of view. Do you really need the primes to be that large? What if you're just sending a little love message to a special friend? Do you think the CIA will want to break your code? What determines the size of the primes you need?

19. **Something in common.** Suppose that p is a prime number and n is a number that has p as a factor. What is n^{p-1} mod p in this case?

20. **Faux pas Fermat.** Compute 1^5 mod 6, 2^5 mod 6, 3^5 mod 6, 4^5 mod 6, and 5^5 mod 6. What if you raise the numbers to the power 2? Compute n^6 mod 9 for numbers n from 1 to 8. What is the answer when n and 9 have no common factors? Do you think there is a way to extend Fermat's Little Theorem when the mod number is not prime? Make a guess (yes or no).

Further Challenges

21. **Breaking the code.** If you could factor a large public number into its two prime factors, how could you break the code? Outline a procedure.

22. **Signing your name.** Suppose you get a message that claims to be coming from your friend Joseph Shlock. It says that you should invest all your savings in pork kidneys. How do you know if the message really came from your friend Shlock and not your arch-enemy Irving Satan? Ideally, it would be great if each message were "signed" by the sender in such a manner that no one could "forge" the signature. Using the RSA scheme (in reverse), devise a method for verifying a message's origin. (*Hint:* Could the encoding procedure be used to reveal the signature?)

In Your Own Words

23. **Personal perspectives.** Write a short essay describing the most interesting or surprising discovery you made in exploring the material in this section. If any material seemed puzzling or even unbelievable, address that as well. Explain

why you chose the topics you did. Finally, comment on the aesthetics of the mathematics and ideas in this section.

24. **With a group of folks.** In a small group, discuss and work through the RSA public key coding scheme. After your discussion, write a brief narrative describing the method in your own words.

25. **Creative writing.** Write an imaginative story (it can be humorous, dramatic, whatever you like) that involves or evokes the ideas of this section.

26. **Power beyond the mathematics.** Provide several real-life issues—ideally, from your own experience—that some of the strategies of thought presented in this section would effectively approach and resolve.

For the Algebra Lover

Here we celebrate the power of algebra as a powerful way of finding unknown quantities by naming them, of expressing infinitely many relationships and connections clearly and succinctly, and of uncovering pattern and structure.

27. **Powers of 2.** Compute 2^2 mod 3, 2^4 mod 5, and 2^6 mod 7. What do you notice? Does Fermat predict your answers?

28. **FOILed!** FOIL the expression $(a-1)(b-1)$. Suppose you know that $ab = 323$ and that $(a-1)(b-1) = 288$. Use these two pieces of information to find the value of $a+b$. Now suppose you also know that a and b are prime. Find a and b.

29. **FOILed again!** FOIL the expression $(x-1)(y-1)$. Suppose you know that $xy = 91$ and that $(x-1)(y-1) = 72$. Use these two pieces of information to find the value of $x+y$. Now suppose you also know that x and y are prime. Find x and y.

30. **Secret primes.** You know that p and q are primes and that $(p-1)(q-1) = 24$. Suppose you also know that $q-p = 2$. Use the second equation to substitute into the first equation and solve for q. Then find p.

31. **More secrets.** You know that p and q are primes and that $(p-1)(q-1) = 60$. Suppose you also know that $q-p = 4$. Find p and q.

THE IRRATIONAL SIDE OF NUMBERS

Are There Numbers Beyond Fractions?

Auguste Rodin, *The Thinker* (*Le Penseur*), 1880–1882. Sometimes rational thought leads to insights into a surprisingly irrational reality.

© JOE CICAK/iStockphoto

> *God made the integers, all else is the work of humankind.*
>
> LEOPOLD KRONECKER

The natural numbers are the first and most natural measures of quantity; however, suppose we have more than one of something but less than two? Clearly we need fractional quantities to make such measurements. Fractions let us measure any quantity to any desired degree of precision. In principle, we could measure a length to within one-millionth of an inch. But are there lengths that even the most precise fraction cannot measure exactly? Specifically, is every number a fraction? That question of measuring lengths challenged the ancient Greeks and eventually forced them to a totally counterintuitive realization about lengths that later led to an entirely new notion of "number." This discovery of the Greeks, which we will soon discover for ourselves, is another powerful illustration of our major theme: By asking clear questions and examining the familiar in a careful and logical manner, we uncover hidden richness.

Is every number a fraction? To answer this question we make an assumption and follow the consequences of doing so. Letting logic lead, we suppose that every length could be measured exactly as a fraction, and then we see what other results we would be compelled, by logic, to accept. Exploring the logical consequences of an assumption is a valuable way to determine whether the assumption is reasonable.

A Rational Mindset

The ancient Greeks, and probably people before them, devised a reasonable method of measuring parts of things. If we take an object and break it into 10 equal pieces and take 9 of them, then we have measured nine-tenths (9/10). Those who sell gasoline at the corner store have learned this lesson well—every gas price ends in 9/10 of a cent. This clever ploy allows the neighborhood convenience store to milk us for a smidgen extra on each gallon without our really noticing.

Theoretically we could take an object, break it into a million pieces, and take 375,687 of them and get 375,687 millionths (375,687/1,000,000). So by taking a large enough number of pieces, we can measure parts of things to any degree of accuracy we want; unfortunately, even then we may not be correct. But we are getting a bit ahead of ourselves. The Greeks thought that the natural numbers were natural gifts from the gods. Ratios of those number essences together with their negatives and zero produce the rational numbers. A *rational number,* therefore, is a number that can be written as a fraction a/b or $-a/b$ in which a and b are natural numbers or $a = 0$. Some examples of rational numbers are 1/2, 22/7, 109/51, –35/219, 15 (15/1), and 0.

To get accustomed to this idea, find a rational number between 1 and 2. Now find a rational number between 1001/1003 and 1002/1003. Why is there always a rational number between any two other rational numbers?

Notice that, even if two different rational numbers are very close together, we can always find many (in fact, infinitely many) rational numbers between them. Since we can cut things up into as many equal-size pieces as we wish, it seems reasonable to conjecture that every number is rational, which parallels the beliefs of the early Greeks.

The ancient Greeks viewed the natural numbers as "god-given" numbers. A ratio of natural numbers (that is, a fraction or rational number) was, in the ancient Greeks' eyes, not a pure number, but instead a combination of two genuine numbers. They believed that every length has a measure that is either a natural number or a ratio of natural numbers.

Given common observations and life experiences, this idea seems both natural and rational (excuse the pun). In fact, the atomic theory of matter and quantum mechanics suggest that matter has a limit to its divisibility, and, hence, for physical objects, there may be a specific number of indivisible units that make up the object. So if we break an object arbitrarily in two and wish to measure how big each piece is, we would count the number of particles in one piece and divide that number by the total number of particles in the original object to see what fraction of the object we have.

Of course, mathematics is not constrained by mere physical reality. Physical reality is just the starting point for mathematics.

The Pythagoreans' Secret Society and the Square Root of 2

Life Lesson

Explore the consequences of assumptions.

Let's now assume, as the Greeks did, the reasonable hypothesis that all length measures (what we now view as all positive numbers) are rational numbers and see where that assumption leads us. Unfortunately, it leads us into some deep trouble. Along the way, however, we will learn that an effective method for discovering new ideas and truths is to explore the consequences of assumptions.

Pythagoras (580 BCE–500 BCE) and his followers formed a school devoted to discovering great ideas, many of which were mathematical. This school, sometimes referred to as the "brotherhood," was actually a community of families and some believe the inspiration for Plato's Republic. The Pythagorean School had many unusual rules (for example, they were not allowed to eat beans), but devoted themselves to studying numbers as a means of attaining a better understanding of the gods. The Pythagorean School was a secret society. Its members developed important mathematical concepts and kept them to themselves.

The Pythagorean Theorem, which we will physically see and touch in Chapter 4, "Geometric Gems," tells us that in a right triangle the square of the hypotenuse equals the sum of the squares of the lengths of the two shorter sides; that is, $a^2 + b^2 = c^2$ in right triangles such as this one.

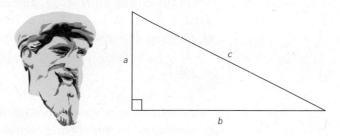

After the Pythagoreans discovered this relationship, they considered a triangle with $a = b = 1$:

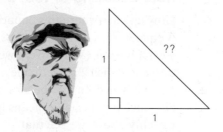

and wondered what the length of the hypotenuse was. If they called the length H, they knew that $1^2 + 1^2 = H^2$. So $H^2 = 2$, which is to say that H is the square root of 2, denoted today as $H = \sqrt{2}$.

Assume It's Rational

The Pythagoreans believed that all lengths are rational, so the number H, the square root of 2, would have to be equal to some fraction, say a/b, in which both a and b are natural numbers. Let's see where that assumption led the Pythagoreans and where it leads us.

The first thing we can do with our assumption that H equals a/b is to cancel any common factors in a and b until we are left with an equivalent rational number with the numerator and denominator having no common factors. For example,

$$\frac{60}{45} = \frac{2^2 \times 3 \times 5}{3^2 \times 5} = \frac{2^2}{3} = \frac{4}{3}.$$

Notice how we factored each number into primes and then canceled primes that are common to both numerator and denominator. So the rational 60/45 is equal to 4/3. Notice that 3 and 4 have no common factors.

No Common Factors

Let's return to H, the square root of 2. We are assuming that H is a rational number a/b, and by following the cancellation process, we find that we can write H

as c/d, in which c and d share no common factor other than 1. In particular (and this observation is important), if 2 divides evenly into either of the numbers c or d, then 2 will *not* divide evenly into the other one. That is, $H = c/d$ and **not both** c and d are even numbers (since otherwise they would have a common factor of 2).

On the one hand, $H = \sqrt{2}$, and on the other hand, $H = c/d$. So putting both hands together, we have $\sqrt{2} = c/d$. To simplify that equation, let's square both sides of $\sqrt{2} = c/d$. Doing so would produce

$$2 = \frac{c^2}{d^2}.$$

Since natural numbers are easier to visualize than fractions, let's "clear the denominator" by multiplying each side by d^2:

$$2d^2 = c^2.$$

Here's Looking at You, c

Now let's see where this equation leads. What kind of number is c^2? Well, we see it equals $2d^2$, so 2 is a factor of c^2 and therefore c^2 must be even. But if c^2 is even, c itself is even. (Why? Well, we know 2 is a prime number, and if 2 divides evenly into $c \times c$ then 2 must divide evenly into just c alone.) So c is an EVEN number. Since c is an even number, then it must equal 2 times some other number, say $c = 2n$ (since that's what it means to be even, after all). If we substitute $2n$ for c in the equality above, we see that

$$2d^2 = c^2 = (2n)^2 = (2n)(2n) = 4n^2.$$

Looking at this equation, we have this unstoppable desire to divide both sides by 2, which leads to $d^2 = 2n^2$, which, in turn, leads to trouble.

A Troubling "d" Tour

What kind of number is d^2? It must be even. Thus d itself must also be EVEN. Remember that we started by assuming that the square root of 2 was equal to a rational number a/b. We made some legal deductions from that assumption, namely, after canceling, $a/b = c/d$ where c and d are not both even. Then we deduced that c and d must both be even—directly contradicting ourselves. So this situation is impossible. Therefore, whatever we assumed must be false.

But what did we assume? A rereading of our argument shows that we assumed only one thing, namely that $\sqrt{2}$ is a rational number. Since our assumption led us to a contradiction, our assumption must be false. So, $\sqrt{2}$ is not a rational number; that is, $\sqrt{2}$ is not equal to the ratio of two integers. A number that is not rational is called *irrational*. The observation about the square root of 2 is so important that we highlight it here:

Square Root of 2 Is Irrational.
The square root of 2 is an irrational number.

Making It Your Own

You can make this argument your own by working through the ideas and explaining them to a friend. The argument is beautiful and elegant, because a powerful and surprising result arises from some basic ideas creatively strung together. This proof appears in Euclid's book entitled *Elements* and is usually attributed to Euclid (ca. 300 BCE). However, evidence exists that Aristotle (384 BCE–322 BCE) also knew about this argument. Of course, we'll never know for sure who first discovered this counterintuitive revelation together with this elegant line of reasoning. Among most mathematicians, the proofs that there are infinitely many prime numbers and that the square root of 2 is irrational are considered to be among the most beautiful arguments in the field.

Accepting Reality

We may not like the idea that an entirely new kind of number must exist— numbers that are not rational; however, there is no use fighting it. No matter how much our previous worldview led us to believe that all numbers are expressible as the ratio of two integers, it just isn't so. We must accept the proven truth, embrace a new worldview, and explore the reality of numbers as they are. Once we prove something, we must add it to our list of truths and move on.

Of course, we are modern people, and we are unlikely to find the existence of irrational numbers a challenge to our philosophical biases about how the world is organized. But the ideas of proportion and ratios of natural numbers played a more central role in the ancient Greeks' understanding of reality. For them, the notion of irrational numbers significantly challenged their ideas about reality.

The Pythagoreans reacted strongly to the disturbing discovery of irrational lengths (numbers), and they kept the idea secret. It is said that when one of their members was caught revealing the secret, he was taken on a boat ride and thrown overboard. Think we're joking? Proclus around the 5th century CE gave a brief account: "It is well known that the man who first made public the theory of irrationals perished in a shipwreck in order that the inexpressible and unimaginable should ever remain veiled. And so the guilty man, who fortuitously touched on and revealed this aspect of living things, was taken to the place where he began and there is forever beaten by the waves." In contrast, students today who divulge the mysteries of the irrational numbers and other scientific phenomena find themselves scooped up by major corporations offering impressive salaries. Nice!

Beyond the Square Root of 2

Mathematicians are extremely frugal when it comes to ideas: Once they have one, they try to recycle and reuse it. By pushing an idea to its limits, we often uncover more than we first expected.

To illustrate this lesson, let's see if we can adapt the ideas used in the proof that the square root of 2 is irrational to show that the square root of 3 is irrational. If you try to work through the ideas on your own before reading on, you will gain a much deeper understanding of the ideas at work.

Life Lesson

Once we discover an important idea, we should use it to deduce new or more general consequences.

Square Root of 3

Let's assume that $\sqrt{3}$ is a rational number a/b, where a and b are natural numbers. By following the cancellation process, we find that we can write $\sqrt{3}$ as c/d, where c and d share no common factor other than 1. In particular (and this observation is important), if 3 divides evenly into either of the numbers c or d, then 3 will not divide evenly into the other one. That is, $\sqrt{3} = c/d$ and **not both** c and d have the factor of 3 (since otherwise we could cancel). If we now square both sides of the equation $\sqrt{3} = c/d$, we see that

$$3 = \frac{c^2}{d^2}.$$

If we multiply each side by d^2 to get integers on both sides of the equation, we have

$$3d^2 = c^2.$$

Now let's see what this equation means. We see that c^2 equals $3d^2$, so 3 is a factor of c^2. But if c^2 has a factor of 3, then c itself must have a factor of 3. (Why? Well, we know 3 is a prime number, and if 3 divides evenly into $c \times c$, then 3 must divide evenly into just c alone.) Since c has a factor of 3, it must equal 3 times some other integer, say $c = 3n$. If we substitute $3n$ for c in the preceding equality, we see that

$$3d^2 = (3n)^2 = (3n)(3n) = 9n^2.$$

We can now divide both sides by 3 and get $d^2 = 3n^2$. This last equation shows that d^2 must have a factor of 3 in it, and therefore d must have a factor of 3 as well. Thus we see that c and d share a common factor of 3. But we selected c and d so that they have no common factor greater than 1. This is a contradiction, and, therefore, this situation is impossible—hence our assumption must be false. So, $\sqrt{3}$ is not a rational number—it is irrational. Notice how this argument parallels our first one.

Other Irrationals

In fact, we can use this method to show that $\sqrt{6}$ and many other examples are irrational. We invite you to try these in the Mindscapes.

With a bit of care, we can extend this idea even further and show that $\sqrt{2} + \sqrt{3}$ is irrational. Again, let's assume that $\sqrt{2} + \sqrt{3}$ is actually a rational number, say a/b. If we square both sides, $(\sqrt{2} + \sqrt{3})^2 = (a/b)^2$, we have to be a bit careful expanding the left side. We do it here:

$$\begin{aligned}
(\sqrt{2} + \sqrt{3})^2 &= (\sqrt{2} + \sqrt{3})(\sqrt{2} + \sqrt{3}) \\
&= (\sqrt{2} + \sqrt{3})\sqrt{2} + (\sqrt{2} + \sqrt{3})\sqrt{3} \\
&= \sqrt{2}\sqrt{2} + \sqrt{3}\sqrt{2} + \sqrt{2}\sqrt{3} + \sqrt{3}\sqrt{3} \\
&= 2 + \sqrt{6} + \sqrt{6} + 3 \\
&= 5 + 2\sqrt{6}.
\end{aligned}$$

So we see that $5 + 2\sqrt{6} = a^2/b^2$, which means that $\sqrt{6} = (a^2 - 5b^2)/2b^2$. But the number on the right side is a fraction, so that means that $\sqrt{6}$ is a rational number. However, you will show in the Mindscapes that $\sqrt{6}$ is irrational. Therefore, we have reached another contradiction. Our first assumption must have been false, and we conclude that $\sqrt{2} + \sqrt{3}$ is irrational.

Irrational Power

If we whittle the idea we are using down to its core, we can use it to prove that other more exotic numbers are irrational. Suppose $3^A = 9$. What would A equal? If $3^C = 27$, what would C equal? These questions are not too hard: $A = 2$, and $C = 3$. But suppose B is the number such that $3^B = 10$. What is B? We know from the previous two questions that B is bigger than 2 and smaller than 3, but there is no use trying to figure out exactly what decimal number it equals, because we can't: It's an irrational number. Why? Well, suppose that B were a rational number, say u/v, where both u and v are natural numbers. Then $3^{u/v} = 10$. If we raise both sides to the vth power, then the v's cancel out in the power on the left side:

$$(3^{u/v})^v = 3^u = 10^v.$$

Since u and v are each at least 1, then 3 must divide evenly into 10^v, which is absurd: The only prime numbers that divide evenly into 10^v are 2's and 5's. This contradiction means that our assumption was false, so B must be an irrational number.

The number B is called a *logarithm*. If $3^B = 10$, then we would say that B is the logarithm of 10 in base 3. Using a calculator we can estimate B and see that $B = 2.09590327428938460. \ldots$

π

Our method to show certain numbers are irrational does not work for more exotic numbers. The circumference of a circle having diameter 1 is equal to the famous number π (pi). Although the rational number 22/7 is almost equal to π, it is not exactly equal to π: 22/7 = 3.142857142857. . . , while π = 3.141592. . . .

The Greeks and subsequent mathematicians studied π intensely: In 1650 BCE, Egyptians estimated that π ≈ 256/81, and roughly 500 years later, mathematicians in India approximated π as 62,832/20,000, which is incredibly close to π. It was not until 1761, however, that someone proved that, in fact, π is irrational. The first person to prove this important fact was Johann Lambert, and he used techniques from calculus. As an amusing postscript to Lambert's result and to the earlier works of Greek, Egyptian, and Indian mathematicians, we note that some ground was lost in 1897: The Indiana State Legislature considered a bill to declare π equal to 4, which was "offered as a contribution to education to be used only by the State of Indiana free of cost. . . ." Fortunately, the legislature did not pass the bill and math instructors in Indiana are not breaking any laws when they describe π correctly.

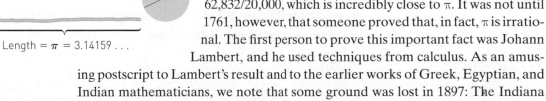

Length = π = 3.14159 . . .

Still Unknown

In general, it is difficult to determine if numbers are rational or irrational. As a modest illustration, nobody knows if any of the numbers on the following list are irrational—it is possible (but not likely) that some are actually rational: 2^π, π^π, π^{π}. Don't they all "look" irrational? Yes, but no one knows how to prove it for sure.

We now see that numbers come in two flavors: rational and irrational. The collection of all these numbers—rational and irrational—forms the real numbers, which leads us to our final journey through the notion of number.

A Look BACK

BEYOND THE WORLD of natural numbers are the rational numbers, fractions. But some numbers are irrational—not rational. We can show that $\sqrt{2}$ is irrational by assuming the contrary. If $\sqrt{2}$ were rational, then it would be equal to a fraction written in lowest terms. That assumption implies that both the numerator and the denominator would have to have a factor of 2, which would contradict the fact that the fraction was in lowest terms. Thus $\sqrt{2}$ is not a fraction. This strategy can be used to demonstrate that other numbers are irrational. This reasoning allowed us to move from the comfortable world of natural numbers and their ratios to the real world of irrationality.

An effective strategy for analyzing life is to make an assumption and see what consequences follow logically. If a logical consequence is a contradiction, then the assumption must be wrong.

Life Lessons

Follow assumptions to their logical conclusions.

MINDSCAPES   Invitations to Further Thought

In this section, Mindscapes marked **(H)** *have hints for solutions at the back of the book. Mindscapes marked* **(ExH)** *have expanded hints at the back of the book. Mindscapes marked* **(S)** *have solutions.*

Developing Ideas

1. **A rational being.** What is the definition of a rational number?
2. **Fattened fractions.** Reduce these overweight fractions to lowest terms:

$$\frac{6}{24}, \quad \frac{15}{9}, \quad -\frac{14}{42}, \quad \frac{125}{10}, \quad -\frac{121}{11}.$$

3. **Rational arithmetic.** The numbers 1/2, 2/3, 5/2, and 6/5 are rational. Show that the numbers below are also rational by expressing each as a ratio of two integers.

$$\frac{1}{2} + \frac{5}{2}, \quad \frac{1}{2} - \frac{2}{3}, \quad \left(\frac{1}{2}\right) \times \left(\frac{6}{5}\right), \quad \frac{\frac{1}{2}}{\frac{2}{3}}, \quad \frac{\left(\frac{5}{2}\right) \times \left(\frac{6}{5}\right)}{\frac{2}{3}}$$

4. **Decoding decimals.** Show that each of the decimal numbers below is actually a rational number by expressing it as a ratio of two integers.

$$0.02, \ 6.23, \ 2.71828, \ -168.5, \ -0.00005$$

5. **Odds and ends.** Square the numbers from 1 to 12. Do the even numbers have even squares? Do the odd numbers have odd squares? Make a conjecture based on your observations.

Solidifying Ideas

6. **Irrational rationalization.** We know that $\sqrt{2}$ is irrational. Therefore $3\sqrt{2} \ / \ 5\sqrt{2}$ must also be irrational. Is this conclusion correct? Why or why not?

7. **Rational rationalization.** We know 2/5 and 7/3 are rational. Therefore (2/5)/(7/3) is also rational. Is this conclusion correct? Why or why not?

8. **Rational or not (ExH).** For each of the following numbers, determine if the number is rational or irrational. Give brief reasons justifying your answers.

$$\frac{4}{9}, \quad 1.75, \quad \frac{\sqrt{20}}{3\sqrt{5}}, \quad \frac{\sqrt{2}}{14}, \quad 3.14159$$

9. **Irrational or not.** Determine if each of the following numbers is rational or irrational. Give brief reasons justifying your answers.

$$\sqrt{\frac{16}{20}}, \quad \sqrt{\frac{12}{7.5}}, \quad -147, \quad 0, \quad \frac{\sqrt{3}}{3}$$

In Mindscapes 10–16, show that the value given is irrational.

10. $\sqrt{5}$ **(H).**

11. $2\sqrt{3}$.

12. $\sqrt{7}$.

13. $\sqrt{3} + \sqrt{5}$.

14. $\sqrt{2} + \sqrt{7}$.

15. $\sqrt{10}$ **(S).**

16. $1 + \sqrt{10}$.

17. **An irrational exponent (H).** Suppose that E is the number such that $12^E = 7$. Show that E is an irrational number.

18. **Another irrational exponent.** Suppose that E is the number such that $13^E = 8$. Show that E is an irrational number.

19. **Still another exponent (ExH).** Suppose that E is the number such that $14^E = 9$. Show that E is an irrational number.

20. **Another rational exponent.** Suppose that E is the number such that $8^E = 4$. Show that E is a rational number. In the previous two Mindscapes, you developed an argument that showed that an exponent was irrational. Where does that argument break down in this case?

21. **Rational exponent.** Suppose that E is the number such that $(\sqrt{2})^E = 2\sqrt{2}$. Show that E is a rational number. In Mindscapes 18 and 19, you developed an argument that showed that an exponent was irrational. Where does that argument break down in this case?

22. **Rational sums.** Show that the sum of any two rational numbers is another rational number. (*Hint:* Let a/b be one rational number and c/d be the other. Now show that $a/b + c/d$ is another rational number.)

23. **Rational products.** Show that the product of any two rational numbers is another rational number. (*Hint:* Adapt the previous hint.)

24. **Root of a rational.** Show that $\sqrt{(1/2)}$ is irrational.

25. **Root of a rational (S).** Show that $\sqrt{(2/3)}$ is irrational.

Creating New Ideas

26. π. Using the fact that π is irrational, show that $\pi + 3$ is also irrational.

27. **2π.** Using the fact that π is irrational, show that 2π is also irrational.

28. **π^2.** Suppose that we know only that π^2 is irrational. Use that fact to show that π is irrational.

29. **A rational in disguise.** Show that the number $\left(\sqrt{2}^{\sqrt{2}}\right)^{\sqrt{2}}$ is a rational number even though it might look irrational. What familiar number does it equal?

30. **Cube roots (H).** The cube root of 2, denoted as $\sqrt[3]{2}$, is the number such that if it were cubed (raised to the third power), it would equal 2. That is, $(\sqrt[3]{2})^3 = 2$. Show that $\sqrt[3]{2}$ is irrational.

31. **More cube roots.** Show that $\sqrt[3]{3}$ is irrational.

32. **One-fourth root.** Show that the fourth root of 5, $\sqrt[4]{5}$, is irrational.

33. **Irrational sums (S).** Does an irrational number plus an irrational number equal an irrational number? If so, show why. If not, give some counterexamples.

34. **Irrational products (H).** Does an irrational number multiplied by an irrational number equal an irrational number? If so, show why. If not, give some counterexamples.

35. **Irrational plus rational.** Does an irrational number plus a rational number equal an irrational number? If so, show why. If not, give some counterexamples.

Further Challenges

36. $\sqrt{p}$. Show that for any prime number p, $\sqrt{p}$ is an irrational number.

37. $\sqrt{pq}$. Show that, for any two different prime numbers p and q, $\sqrt{pq}$ is an irrational number.

38. $\sqrt{p} + \sqrt{q}$. Show that, for any prime numbers p and q, $\sqrt{p} + \sqrt{q}$ is an irrational number.

39. $\sqrt{4}$. The square root of 4 is equal to 2, which is a rational number. Carefully modify the argument for showing that $\sqrt{2}$ is irrational to try to show that $\sqrt{4}$ is irrational. Where and why does the argument break down?

40. **Sum or difference (H).** Let a and b be any two irrational numbers. Show that either $a + b$ or $a - b$ must be irrational.

In Your Own Words

41. **Personal perspectives.** Write a short essay describing the most interesting or surprising discovery you made in exploring the material in this section. If any material seemed puzzling or even unbelievable, address that as well. Explain why you chose the topics you did. Finally, comment on the aesthetics of the mathematics and ideas in this section.

42. **With a group of folks.** In a small group, discuss and work through the arguments that the square root of 2 and the square root of 3 are irrational. After your discussion, write a brief narrative describing the arguments in your own words.

43. **Creative writing.** Write an imaginative story (it can be humorous, dramatic, whatever you like) that involves or evokes the ideas of this section.

44. **Power beyond the mathematics.** Provide several real-life issues—ideally, from your own experience—that some of the strategies of thought presented in this section would effectively approach and resolve.

For the Algebra Lover

Here we celebrate the power of algebra as a powerful way of finding unknown quantities by naming them, of expressing infinitely many relationships and connections clearly and succinctly, and of uncovering pattern and structure.

45. **Rational x.** Simplify the following expressions to show that x is rational in each case.

$$x = \frac{\frac{3}{5} + \frac{3}{5}}{\frac{17}{5}} \qquad x = \frac{\frac{5}{3}}{1 + \frac{11}{4}} \qquad x = \frac{4x^2 \quad 100}{(3x + 15)(x \quad 5)}$$

46. **High 5.** Suppose that x is a positive number satisfying the equation $x^2 = 5$. Is x a rational or irrational number? Justify your answer.

47. **Don't be scared (H).** Consider the scary equation. $7x^3 - 19x^2 + 10x - \sqrt{2} = 5$. Without solving the equation for x, determine if the solution for x is a rational or irrational number.

48. **A hunt for irrationals.** Find all solutions to the equation $x^3 - 3x = 0$. How many of your solutions are rational? How many are irrational?

49. **A hunt for rationals.** For any number x, the equation $2x^2 - x - 3 = (2x - 3)(x + 1)$ is always valid. (Such equations are called *identities*). Use that identity to help you find all values for x that satisfy $2x^2 - x - 3 = 0$. Are the solutions you found rational or irrational? More generally, a quadratic equation always has two solutions. Suppose you are given two rational numbers. Can you always find an equation $ax^2 + bx + c = 0$ such that the solutions for x are the two given rational numbers and the numbers a, b, and c are rational?

The Point of Decimals and Pinpointing Numbers on the Real Line

Illustration from a 17th-century letter by Felipe Guáman Poma, showing an Incan treasurer holding a quipu. During the 15th and 16th centuries, the Inca used quipus, a system of knotted cords, to record numerical information, such as population and trade with other tribes.

Why are wise few, fools numerous in the excesse? 'Cause, wanting number, they are numberlesse.

AUGUSTA LOVELACE

Our development of the notion of "number" took us from the familiar natural numbers and rationals to the more mysterious realm of irrational numbers. While these collections of numbers are distinctive, they all fit together in a basic way: Given any two different numbers, one is bigger than the other. The numbers are all ordered. That orderly hierarchy of numbers by size allows us to represent all numbers on one line. In this final section on number, we explore the connections between the number line and the notion of number.

The guiding principle for this part of the exploration of number is to bring global coherence to separate ideas. By examining the totality of numbers as one entity, we will discover new surprises and develop a better understanding of both the rational and the irrational. Initially some of these discoveries may contradict our intuition. Our exploration involves looking at the types of numbers we know and deducing how those numbers must be interconnected on the number line. This point of view leads to the representation of numbers in decimal form. We must be open-minded and accept logical consequences that we deduce. Once we accept correct conclusions, we will understand the collection of all numbers on the number line—the real numbers—as a coherent idea aptly called the *continuum*.

Lining Up

The real number line has appeared in elementary school textbooks as long as school cafeterias have been serving students sloppy joes. Here we start from scratch but soon make unexpected discoveries—just as we did with our sloppy joes—about the familiar idea of the number line.

We begin with the number line itself:

The integer points are labeled, but we would like to be able to label or describe every point on this line. To make progress in this direction, let's consider the points halfway between each consecutive pair of integers. For example, the number 5/2 is the point that sits exactly midway between 2 and 3. In fact, any rational number corresponds to a specific point on this line. For example, we can locate the point to which 37/23 corresponds by dividing each interval between consecutive integers into 23 equal pieces. Then we start at 0 and jump from mark to mark: 1/23, then 2/23, then 3/23, and so on. When we get to 23/23, we see that is the point that is also labeled 1. Jumping 14 more times gets us to 37/23. A similar procedure allows us to find a point on the line corresponding to any rational number.

Rationals Everywhere

The points associated with rational numbers are all over the line: No matter where we are standing on the line, we can always find a rational number point as close as we wish. Suppose we want to find a rational number point that is within a distance of 1/10,000 of where we are standing. We just divide each segment between every two consecutive integers into 10,000 equal pieces and make those 10,000 marks, then mark off all the points that correspond to rational numbers having 10,000 in their denominators (5876/10,000, for example). Therefore, no matter where we are, we will be within 1/10,000 of one of these rational points.

Now that we see that the rationals are essentially everywhere, we may ask: Are there any unlabeled or undescribed points left on our line? The previous section provides us with the answer to this question. Let's construct a point on the number line that definitely does not correspond to a rational number.

An Irrational Point

Here is a way of finding a number on the number line that is not a rational number: Build a square whose base is the interval from 0 to 1. Next draw the diagonal from 0 to the upper-right corner of the square. Using a compass, copy the length of that diagonal onto the number line and make a mark there. What number did we just mark? The square root of 2.

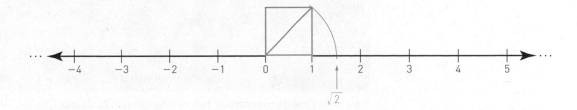

Life Lesson

Look for new ways of expressing an idea.

In the previous section we showed that the square root of 2 is irrational. Thus there are points on the line that cannot be labeled with rational numbers, and we are faced with the question: Is there a uniform method to label every single point on the line—rational and irrational?

The Decimal Point

Let's label each point by describing ever more precisely where it sits on our line. This process is familiar, because it is the idea that generates the decimal expansion of numbers. The *decimal expansion* of a number provides us with a road map that allows us to home in and locate the number on our line. For example, let's consider the decimal expansion of the square root of 2:

$$\sqrt{2} = 1.414213562\ldots.$$

The number to the left of the decimal point tells us that our number will be somewhere between 1 and 2. Where between? We cut the interval from 1 to 2 into 10 equal pieces (10 since we are looking at the *deci*[10]mal expansion). The next digit, in this case 4, tells us in which small interval our number is located. We then take that small interval and cut it up into 10 (very small) equal pieces. The next digit (in our example, 1) tells us in which very small interval our number resides. Notice that, as we continue this process, we whittle away and create smaller and smaller intervals. This process allows us to home in on the number we seek, in this case $\sqrt{2}$.

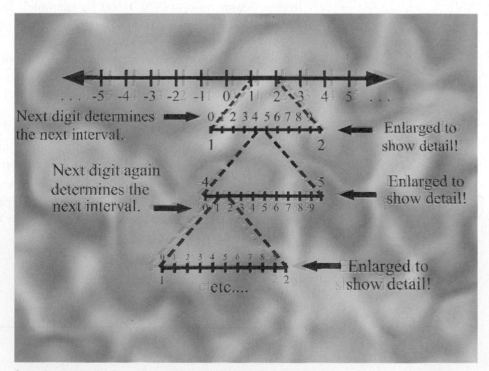

Stare deeply into the line with your *Heart of Mathematics* 3D glasses.

We are getting closer and closer to $\sqrt{2}$. Each little interval, once it is cut up into 10 equal pieces, looks like the larger parent interval that it came from. For $\sqrt{2}$, this homing process never ends. We keep localizing our point in smaller and smaller nested intervals, and we get closer and closer, but we have to do this process infinitely many times to actually hit the $\sqrt{2}$ point exactly.

A Home Base

To name the point $\sqrt{2}$ on the real line, we divided intervals into 10 equal parts, because we are accustomed to using 10 as the basic number for counting. However, we could well have located the number $\sqrt{2}$ by dividing each interval into any other number of equal parts. For example, we could always divide each subinterval into two parts. In that case, since the point $\sqrt{2}$ lies in the first half of the interval, we put a 0 after the point. Dividing that half interval into half again, we would note that $\sqrt{2}$ is in the second half, so we would put a 1 as the next digit. Dividing into halves at each time gives a way to locate $\sqrt{2}$ using just 0's and 1's. This representation is known as the *base 2 expansion*. In base 2, $\sqrt{2}$ would have a representation of the form:

$$\sqrt{2} = 1.01101010000010011110\ldots._2.$$

The key point to remember is that each preceding digit just tells us which of the two subintervals—the left = 0, the right = 1—we fall into as we head toward the square root of 2.

Base 2 (sometimes referred to as *binary*) representation of numbers is quite useful since each point on the number line can be located using only 0's and 1's. This economy of symbols is convenient for computers, which store information by sequences of ons and offs, represented by 1's and 0's. Other bases are useful for other purposes, but the strategy for finding a representation in any base remains the same.

Rational vs. Irrational Decimals

We have seen that some of the points on the real line, like $\sqrt{2}$, have decimal expansions that require infinitely many digits before they are completely specified. The number 1/2, however, has a decimal expansion of 0.5. This distinction between the decimal representation of the irrational $\sqrt{2}$ and the rational 0.5 leads us to consider the relationship between the decimal expansions of rational numbers and the decimal expansions of irrational numbers.

Question Given a decimal expansion of a number, can we tell if the number is rational or irrational?

A reasonable guess is that a decimal expansion represents a rational number precisely when its decimal expansion terminates (such as 0.5 or 12.76), and it is an irrational number precisely when it has a decimal expansion that goes on forever (such as the decimal expansion of $\sqrt{2}$). What do you think about this guess? Is it correct? Partly correct? Completely incorrect? None of the above?

Our guess is *partly correct*. If the decimal expansion of a number terminates, then the number must be rational. To see why, we notice that, if the decimal expansion terminates, then we can just shift the decimal point all the way over to the right and then divide by 10 raised to the power equal to the number of places we moved the decimal point. For example,

$$12.76 = \frac{1276}{100}, \quad 6.3709 = \frac{63,709}{10,000}, \quad 14.35670381 = \frac{1,435,670,381}{10^8}.$$

However, just because a decimal expansion goes on forever does not imply that the number is irrational. Consider the rational number

$$\frac{1}{3} = 0.3333333333\ldots = 0.\overline{3}.$$

In this case we just see 3 repeating forever. We call a decimal expansion *periodic* if from some point on and then forever onward the pattern of digits repeats. For example,

$$3.5959595959\ldots = 3.\overline{59} \quad \text{and} \quad 9.345276994994994994994\ldots = 9.345276\overline{994}$$

are examples of periodic decimal expansions. Periodic decimal expansions go on forever and do not terminate. The interesting fact is that they are all rational numbers.

Periodic Decimals

To see why periodic decimal numbers are rational, consider the number $3.\overline{59} = 3.5959595959\ldots$ We now illustrate a method that will allow us to figure out what rational number this decimal represents. The key idea is to multiply the decimal by a power of 10 (that just shifts the decimal point to the right) so that we can align the periodic part with the periodic part of another copy of the decimal expansion. Then if we subtract these numbers, the infinitely long periodic part cancels away. In our example (we'll call the number W for "What number is it?"), we multiply the decimal expansion by 100 and thus shift the decimal two places to the right. We then subtract the original decimal expansion. Notice how the periodic parts line up perfectly and how they all drop out when we subtract.

$$
\begin{aligned}
100W &= 359.5959595959\ldots \\
- \quad W &= 3.5959595959\ldots \\
\hline
99W &= 356.0000000000\ldots
\end{aligned}
$$

So, we see that $99W = 356$, so $W = 356/99$, a rational number. Using a calculator, we can check that 356/99 has a decimal expansion of $3.5959595959\ldots$ This method always works. We just have to make sure that the periodic parts line up. So numbers whose decimal representation is periodic are always rational.

Divide and Conquer: Repeat

To find the decimal expansion for a rational number, we use long division. Notice that, as we perform the long division, the intermediate differences we get (through the subtraction) are always between 0 and the number by which we are dividing. Since there are only finitely many natural numbers between 0 and the number we are dividing by, at some point we will see a difference that we have already seen before. This is just another example of the Pigeonhole principle. Once this repetition begins, it will go on forever. The following example illustrates this observation more clearly.

Let's find the decimal expansion of 1141/990 using long division:

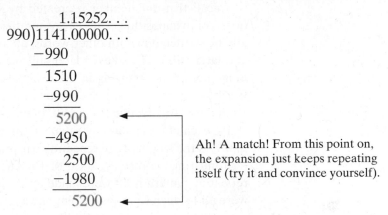

We've just discovered exactly when a decimal expansion is a rational number. We record this insight formally as:

Decimal Expansion of Rationals and Irrationals.

A decimal expansion represents a rational number precisely when either the decimal expansion terminates or the decimal expansion is eventually periodic. Thus, a decimal expansion is an irrational number precisely when it does not terminate and is not periodic.

We can now use this theorem to prove some interesting results. For example:

0.123456789. . . Is Irrational.

The real number

0.1234567891011121314151617181920212223242526. . .

(formed by writing the natural numbers in order, in juxtaposition) is an irrational number.

Proof that "0.123456789. . ." Is Irrational

There are only two possibilities: 0.123. . . is either a rational number or an irrational number. Let us assume it's a rational number. If we could show that this possibility is in fact impossible, that would prove that the other possibility (that the number is irrational) must be true.

If this number were a rational number, its decimal expansion would have to be eventually periodic. In other words, there would have to be a finite string of numbers in the decimal expansion that, from some point on, repeats forever. What would be the length of this finite string that repeats? We don't know. All we are sure of is that the length is finite—let's call that finite length of the repeating pattern L. For example, in the number 45.3219811981198119811... the repeating pattern "1981" has length 4, so here L would be 4.

Recall that our number is created by writing down the natural numbers in order, in juxtaposition. Thus every number that is made just from 1's will eventually be written down. In other words, at some point we will see 11 and later 111 and later still 1111 and even later still 11111, etc. Thus we see that there will be long runs of 1's occurring in the decimal expansion. In fact, we will have arbitrarily long runs of 1's. For example, the natural number consisting of a billion 1's in a row will eventually appear. In particular, we can find infinitely many runs of 1's, each of which has more than 10 times L number of 1's in a row (remember that L is the length of the repeating pattern). So at some point the repeating pattern must march through a sea of 1's. The only way that could happen is if the repeating pattern itself were made up exclusively of 1's. If the repeating pattern were all 1's, then from some point onward, all we would see in the decimal expansion of this number would be 1's, contrary to how the number was constructed. This contradiction tells us that our assumption that the number is rational must be false. Therefore, the decimal number 0.1234... is not rational and must be irrational, which completes our proof.

No Holes, No Neighbors

Characterizing the decimal expansions of rational numbers allows us to journey deep into the dense jungle of the real number line. *Gaze closely at the number line. Look really closely. Put your nose right up to the page. Stare deeply into the hypnotic line.* Perhaps you are getting sleepy . . . well, snap out of it and wake up! Notice that there are no holes in the number line—instead the line flows smoothly and produces a continuous and unending stream of real numbers—*the continuum*. This "unholey" image leads to a question that has a strange answer.

Suppose we are a particular real number—to ground our thinking, let's suppose we are 0. Now who are our immediate neighbors? In particular, what is the next real number after 0? Can we name it? Suppose someone guessed 1/2. We could easily show that 1/2 is not the next real number after 0; after all, 1/4 is closer to 0 than 1/2. Suppose that someone else guessed 3/702 (= 0.004273...). We could take half of that number and find the number 3/1404 (= 0.002136...), which is even closer to 0. In fact, if anyone gave us any number greater than 0, we could just divide that number in half and find another even smaller number that is also greater than 0. What can we conclude? The answer is that there is no next real number immediately following 0. The moment we specify a number bigger than 0, we could find another number (in fact infinitely many) that is between 0 and the specified number. The real number line flows continuously without breaks and between any two points on the line we can always find lots of points in between them. Hence, there is no next real number after 0.

Our reasoning could be used to show that there is no next real number after 1 or even after $\sqrt{2}$. We have therefore verified the following:

> **There Is No Next Real Number.**
> *Given any particular real number, there is no next real number immediately following it.*

Redundancy in Representation of Reals

We now have a sense of the connections among points on the number line, their decimal expansions, and the notions of rational and irrational. Every point on the real number line can be represented as a decimal number, but we have not considered the following question:

Is there only one way to write a number in its decimal expansion?

Although it would be convenient for each real number to have only one decimal expansion, unfortunately there are some real numbers that have more than one decimal expansion. We illustrate this fact with an example.

What rational number is 0.999999. . . ? We'll call this number N (for Nines). Let's answer this question using our method of multiplying by a power of 10, lining up the repeating period, and then subtracting.

$$
\begin{array}{r}
10\,N = 9.9999\ldots \\
-\ \ N = 0.9999\ldots \\
\hline
9N = 9.0000\ldots
\end{array}
$$

Life Lesson

Don't let personal biases get in the way of new discoveries.

So $9N = 9$; and what must N be? $N = 1$. We just proved that $1 = 0.99999\ldots$. Does this equation look strange? Sometimes mathematical results, even when proven rigorously, are so counterintuitive that we remain skeptical of their validity. The fact that $1 = 0.9999\ldots$ is a great example of this phenomenon. Even though we have given a rigorous mathematical proof of this amazing fact, we now give an intuitive argument that may be more convincing.

Suppose we believe that 1 is not equal to 0.99999. . . (remember that those 9's go on forever without ever stopping). Then one of these numbers would be bigger than the other. Which one would be larger? Certainly 1 would be larger than 0.9999. . . . If these numbers were not equal, then there must be some numbers between 0.99999. . . and 1 on the number line. For example, the average of those two numbers would have to be between them. That average must be a number that is larger than 0.99999. . . and at the same time smaller than 1. What could that number be? It would have to start off with a 0 (otherwise it would not be less than 1). What would the next digit be? It must be a 9 since anything else would make the number smaller than 0.9999. . . . What about the next digit? It would have to be 9 as well. If we continue in this manner we see that we are building 0.99999. . . . But that is not bigger than 0.99999. . . . So there are no numbers between 0.9999. . . and 1, and hence they must be equal. Did you like this argument? It's amusing to think about.

Here is another way of looking at all those 9's. For some strange reason, people feel comfortable with the fact that $1/3 = 0.33333$. . . and $2/3 = 0.6666$. . . . Well, if we add those together we see $1 = 0.9999$. . . ! We have proved the following:

> **0.9999. . . = 1.**
> *1 = 0.999999. . . .*

What is another decimal expansion for the number 0.499999 . . . ? Use the ideas given above and give it a try.

Random Reals

Finally, before closing this section and this chapter we pose an intriguing question:

> *If we randomly pick a real number—that is, we take a pin, close our eyes, and place the pin on some point on the real number line—what is the likelihood that the number we picked is a rational number? Is it a 50-50 chance?*

A reasonable answer would be 50-50 since a real number is either rational or irrational. Unfortunately, this reasonable-sounding answer is far from correct. Although we will not be able to give a rigorous answer to this question until we journey through the world of the infinite in the next chapter, we are able to give a plausible argument that answers the question.

How could we randomly pick a real number besides closing our eyes and dropping a pin on a number line? Well, one way is to randomly choose digits among 0, 1, 2, 3, . . . , 8, 9 and write them down to build a decimal number. We could get a 10-sided die with the sides numbered from 0 to 9. Let's suppose that we always start with 0 so our random number will be between 0 and 1. Now we roll a 10-sided die or have a random number generator spit out digits and we record them:

0.79356284565748388365300483628304726118394573378. . . .

We don't stop! We do this forever and thus create a real number. What is the likelihood that this random number is a rational number? Well, for it to be rational, from some point on, the number must have a pattern that repeats forever. But what does that mean? It means that from some point on, we keep repeating the identical pattern without ever deviating. How likely is it that we will repeat a finite pattern forever given that we are generating the digits randomly? The answer is: not likely at all—in fact, it should "never" happen. There would have to be an amazing and even unheard-of conspiracy in the random digits to have them all, from some point on, follow a periodic pattern. Thus the probability that we randomly generate a rational number is actually zero. So if we just randomly pick a real number, it is "certain" to be an irrational number.

Irrationals Abound

This huge preponderance of irrational numbers might be a tough pill to swallow since we are so accustomed and comfortable with rational numbers and since we noticed earlier that the rationals seem to be everywhere on the number line. But, mathematically, rational numbers are actually hard to find. We will see exactly what "probability zero" and "certain" mean in the probability chapter. For now, if we accept the preceding informal analysis, we are faced with an extremely interesting question: If a real number selected at random is "never" a rational number and "always" an irrational number (whatever the notions of "never" and "always" mean), then does that mean that there are, somehow, more irrational numbers than rational numbers? Certainly there are infinitely many rational numbers and infinitely many irrational numbers. Could one of these infinite sets actually be greater than the other? Perhaps what first appears familiar and natural (the rational numbers) will in fact be the exotic and strange, whereas what appeared to be foreign and strange (irrational numbers) will actually turn out to be more the norm! These curious questions set the stage for our next adventure: the world of the infinite.

A Look BACK

THE RATIONAL AND IRRATIONAL numbers taken together form the real numbers—the collection of points on a line. We are able to use the decimal expansion of a real number to locate any real number on the number line. The decimal expansion also allows us to distinguish rational numbers from irrational numbers. A number is rational precisely if its decimal expansion eventually repeats; otherwise it is irrational. Using these ideas, we are able to devise means of converting repeating decimals to fractions and also to prove that certain real numbers, such as $0.123456789101112\ldots$, are irrational, whereas, surprisingly, $0.9999\ldots = 1$. The real line presents a picture of numbers orderly arranged. No number has an immediate neighbor, a number just above it or just below it.

Our strategy for exposing this view of the real numbers was to seek a unified view of all the types of numbers we had developed before. We looked for a relationship that encompassed all the ideas we had generated, in this case, the ideas of rational and irrational numbers. The simple ordering of numbers suggested that we could effectively represent all numbers as points on a line and that we could name each point on the line or number, rational or irrational, using a decimal representation. Some discoveries required us to give up biases and accept logical conclusions. Being open-minded about new ideas is a difficult and important lesson in every arena of life.

> . . . an irrational number . . .
> lies hidden in a kind of cloud of
> infinity.
>
> MICHAEL STIFEL

Life Lessons

Seek unifying ideas.

•

Keep an open mind.

MINDSCAPES Invitations to Further Thought

In this section, Mindscapes marked (H) *have hints for solutions at the back of the book. Mindscapes marked* (ExH) *have expanded hints at the back of the book. Mindscapes marked* (S) *have solutions.*

Developing Ideas

1. **X marks the "X-act" spot.**

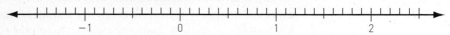

On the number line above, place an *X* on the approximate location for each of the following numbers:

$$\frac{3}{2}, \quad -\frac{1}{3}, \quad 2.3, \quad -1.1, \quad 0.9, \quad 1.05, \quad -0.55.$$

2. **Moving the point.** Simplify each of the expressions.

$$10 \times (3.14), \quad 1000 \times (0.123123\ldots), \quad 10 \times (0.4999\ldots), \quad \frac{98.6}{100}, \quad \frac{0.333\ldots}{10}$$

3. **Watch out for ones!** Express 1/9 in decimal form. (*Hint:* Use long division to divide 9 into 1.)

4. **Real redundancy.** Suppose $M = 0.4999\ldots$. Then what does $10M$ equal? Find two expressions for the quantity $10M - M$ and set those two expressions equal to each other. (*Hint:* One expression is simply $9M$.) Can you solve your equation to discover something marvelous about M?

5. **Being irrational.** Explain what it means for a number to be irrational.

Solidifying Ideas

6. **Always, sometimes, never.** A number with an unending decimal expansion is (choose one: always, sometimes, never) irrational. Explain and illustrate your answer with examples.

7. **Square root of 5.** *The $\sqrt{5}$ has an unending decimal expansion, but it might eventually repeat.* Is this statement true or false? Explain.

8. **A rational search (ExH).** Find a rational number that is bigger than 12.0345691 but smaller than 12.0345692.

9. **Another rational search.** Find a rational number that is bigger than 3.14159 but smaller than 3.14159001.

10. **An irrational search (H).** Describe an irrational number that is bigger than 5.7 but smaller than 5.72.

11. **Another irrational search.** Describe an irrational number that is bigger than 0.0001 but smaller than 0.00010001.

12. **Your neighborhood.** Suppose we tell you that we are thinking of a number that begins with 10.0398XXXXX, where "XXXXX" are digits that we have hidden from view. What is the smallest our number could possibly be? What is the largest our number could possibly be?

13. **Another neighborhood.** Suppose we tell you that we are thinking of a number that begins with 5.5501XXXXX..., where "XXXXX..." are digits that we have hidden from view. What is the smallest our number could possibly be? What is the largest our number could possibly be?

In Mindscapes 14–16, express each fraction in its decimal expansion.

14. $\frac{6}{7}$ (S).

15. $\frac{17}{20}$.

16. $\frac{21.5}{15}$.

In Mindscapes 17–25, express each number as a fraction.

17. 1.28901.
18. 20.4545.
19. 12.999.
20. 2.222222....
21. 43.12121212...(S).
22. 5.6312121212....
23. 0.0101010101....
24. 71.239999999....(ExH).
25. **Just not rational (H).** Show that the number

    0.0100100010000100000100000010000000100...

    is irrational.

Creating New Ideas

26. **Farey fractions.** Let F_n be the collection of all rational numbers between 0 and 1 (we write 0 as 0/1 and 1 as 1/1) whose numerators and denominators do not exceed n. So, for example,

    $$F_1 = \left\{\frac{0}{1}, \frac{1}{1}\right\}, \quad F_2 = \left\{\frac{0}{1}, \frac{1}{2}, \frac{1}{1}\right\}, \quad F_3 = \left\{\frac{0}{1}, \frac{1}{3}, \frac{1}{2}, \frac{2}{3}, \frac{1}{1}\right\},$$

    F_n is called the *nth Farey fraction*. List F_4, F_5, F_6, F_7, and F_8. Make a large number line segment between 0 and 1 and write in the Farey fractions. How can you generate F_8 using F_7? Generalize your observations and describe how to generate F_n. (*Hint:* Try adding fractions a wrong way.)

27. **Even irrational.** Show that the number

    0.2468101214161820222426283032343638 40...

    is irrational.

28. **Odd irrational (H).** Show that the number

    1.357911131517192123252729313335373941...

    is irrational.

29. **A proof for** π. Suppose we look at the first one billion decimal digits of π. Those digits do not repeat. Does that prove that π is irrational? Why or why not? What if we examined the first trillion digits?

30. **Irrationals and zero.** Is there an irrational number that is closer to zero than any other irrational? If so, describe it. If not, give a sequence of irrational numbers that get closer and closer to zero. (*Hint:* Start by considering $\sqrt{2}/2$ and $\sqrt{2}/3$.)

31. **Irrational with 1's and 2's (S).** Is it possible to build an irrational number whose decimal digits are just 1's and 2's? If so, describe such a number and show why it's irrational. If not, explain why.

32. **Irrational with 1's and some 2's.** Is it possible to build an irrational number whose decimal digits are just 1's and 2's and only finitely many 2's appear? If so, describe such a number and show why it's irrational. If not, explain why.

33. **Half steps.** Suppose you are just a point and are standing on the number line at 1 but are dreaming of 0. You take a step to the point 1/2, the midpoint between 0 and 1. You proceed to move closer to 0 by taking a step that is half of the previous one. You continue this process again and again. Will you ever land on 0? Explain. Is this observation hard to accept?

34. **Half steps again (ExH).** Suppose now that you are a very, very, very short line segment (your length is less than 1/100,000,000,000). You are standing on the number line so that your center is right on 1, but, again, you are dreaming of 0. You shift your segment so that your center is at 1/2, midway between 0 and 1. You proceed to move closer to 0 by taking a step that is half of the previous one. You continue this process again and again. Will your segment ever contain 0? Explain. Is this observation less puzzling than the previous Mindscape? Why?

35. **Cutting** π. Is it possible to cut up the interval between 3 and 4 into pieces of exactly the same size such that one of the pieces has the point π on its right edge? Why or why not?

Further Challenges

36. **From infinite to finite.** Find a real number that has an unending and nonrepeating decimal expansion, with the property that if you square the number, the decimal expansion of the squared number terminates.

37. **Rationals (H).** Show that, between any two different real numbers, there is always a rational number.

38. **Irrationals.** Show that, between any two different real numbers, there is always an irrational number.

39. **Terminator.** Show that if a rational number has a decimal expansion that terminates (or alternatively, has a tail of zeros that goes on forever), then the rational number can be written as a fraction where the only prime numbers dividing the denominator are 2 and 5.

40. **Terminator II.** Show that if the denominator of a fraction has only factors of 2 and 5, then the decimal expansion for that number must terminate in a tail of zeros.

In Your Own Words

41. **Personal perspectives.** Write a short essay describing the most interesting or surprising discovery you made in exploring the material in this section.

If any material seemed puzzling or even unbelievable, address that as well. Explain why you chose the topics you did. Finally, comment on the aesthetics of the mathematics and ideas in this section.

42. **With a group of folks.** In a small group, discuss and work through the arguments that the number $0.12345678910\ldots$ is irrational and that $0.99999\ldots = 1$. After your discussion, write a brief narrative describing the arguments in your own words.

43. **Creative writing.** Write an imaginative story (it can be humorous, dramatic, whatever you like) that involves or evokes the ideas of this section.

44. **Power beyond the mathematics.** Provide several real-life issues—ideally, from your own experience—that some of the strategies of thought presented in this section would effectively approach and resolve.

For the Algebra Lover

Here we celebrate the power of algebra as a powerful way of finding unknown quantities by naming them, of expressing infinitely many relationships and connections clearly and succinctly, and of uncovering pattern and structure.

45. **An unknown digit.** Let x be a digit satisfying the decimal equation: $10 - x = 0.xxxxx\ldots$ Solve for the unknown digit x.

46. **Is x rational?** Suppose that I is a fixed but unknown irrational number. Consider the equation $4x - I = 2/3$. Is it possible to determine if the value for x that satisfies the equation is rational or irrational? Explain your answer.

47. **Is y irrational?** You decide to create the digits of a decimal number y between 0 and 1 using the function $f(n) = 3n + 1$. Here's your system. Compute $f(1)$ and put the result as the first digit of y to the right of the decimal point. Compute $f(2)$ and put the result as the second digit of y. Compute $f(3)$ and put the result as the third and fourth digits of y. Compute $f(4)$ and put the result as the fifth and sixth digits of y. And so on. What's the tenth digit of y? Do you think y is irrational? Why or why not?

48. **Is z irrational?** Follow the same construction as described in the previous Mindscape to create a decimal number z. This time use the function $g(n) = n^2 + n + 1$, and answer the given questions about z.

49. **Triple digits (H).** Suppose a, b, and c are digits that satisfy the three simultaneous equations

$4a - b - c = 3$

$2a - b + c = 11$

$a + b + 3c = 26$

Find the number with decimal expansion $a.bc0000\ldots$.

CHAPTER THREE

Infinity

201…

3.1 Beyond Numbers

3.2 Comparing the Infinite

3.3 The Missing Member

> *To see the world in a grain of sand, And a heaven in a wild-flower; Hold infinity in the palm of your hand, And eternity in an hour.*
>
> WILLIAM BLAKE

Infinity—where all superlatives meet . . . bigger than the biggest, more than all . . . infinitely beyond what could be conceived by the mind . . . greater than any quantity . . . more vast than can be counted . . . beyond all numbers. The notion of infinity triggers a sense of mystical wonder and boundless incomprehensibility. Among all ideas of human thought, infinity is one of the most mysterious and romantic. It often resides within the realm of the spiritual where it captures the essence of ultimate grandeur and vastness without end.

Using the power of mathematical reasoning, we will conquer this seemingly incomprehensible idea of infinity and claim it for our own. Infinity will continue to evoke images of vastness and power; however, after we understand it mathematically, our initial sense of mystery will be replaced by a whole new world of richness. We will discover that there is not just one superlatively vast quantity called *infinity;* instead, even infinity itself has peers and superiors. Instead of one infinity, we will see a constellation of infinities, each different from the others, each with unique features and qualities that can be explored. Our single image of one luminous but ill-defined infinity will burst into infinitely many clear infinities—richly related and intertwined.

We approach infinity by keeping our feet firmly on the ground and recalling ideas we have known since childhood. As children, we shared M&M's with our best friend like this: "One for me, one for you; one for me, one for you. . . ." We take this familiar, everyday idea, specify its meaning with great precision, explore its consequences, and discover that it is the key to understanding infinity. In our everyday lives, we are often confronted with vagueness and mystery. Our journey toward the infinite will illustrate a powerful method of moving from the fuzzy to the focused. The heart of the strategy is to make familiar ideas precise. Such thinking lets us journey from the familiar to the mysterious and empowers us to discover within the mysterious, the familiar.

3.4 Travels Toward the Stratosphere of Infinities

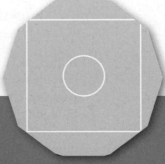

3.5 Straightening Up the Circle

BEYOND NUMBERS

What Does Infinity Mean?

The Andromeda galaxy. Does this galaxy contain infinitely many stars?

© Paul LeFevre/iStockphoto

> *The known is finite, the unknown infinite; intellectually we stand on an island in the midst of an illimitable ocean of inexplicability. Our business in every generation is to reclaim a little more land.*
>
> THOMAS H. HUXLEY

We seek to put the study of infinity on a firm and logical footing. Since infinity is such a vast and intimidating topic, we prefer to begin by closely looking at the familiar ideas of numbers and counting. We want to count infinite collections, but we don't know how to count that high. Thus we seek different ways to count collections of ordinary objects in the hope that one method might work for infinite collections as well.

Where would we begin to search for the infinite? It's well beyond 1, 2, and 3; certainly past

41359724545124715156120963071048.

Infinity is even beyond

9142452345245001282106162966380902777871210598012670019872665
2198733061118542109827629991176653276533279073847295892835729
8399428310342167161237402372034017487102381084793047109237109
2471098097109230192740192380198231092309182741092730239845235
0293740198230192317240918230471093750193401984019734019824091
7304109740192401984019734091834017236817634651234917864817943.

In fact, as large and incomprehensible as this number is, it is sobering to realize that almost all numbers are far larger still. On the road to the infinite, we would pass this enormous number almost instantly and soon view it as a tiny jot. Even though this number appears early in the infinite list of all numbers, to our minds it seems vast. In reality, we do not even have an intuitive feel for such a large number. The number lacks a sensible name—magnitudes such as millions, billions, or trillions do not make a dent in it. Just reading the digits aloud without error would be a challenge, and finding any of its prime divisors might be nearly impossible.

Even though we are comfortable with large numbers in the abstract, in truth we have little real understanding of such enormous quantities. Given our inability to grasp even these—relatively speaking—modestly sized numbers, is it possible for our minds to fathom the notion of the infinite? Let's first ponder a basic question.

What Does "2" Mean?

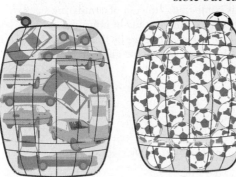

Life Lesson
Understand simple things deeply.

Where should we begin our voyage toward the infinite? Our journey to infinity begins with 2. We are all familiar with 2, and we will use that intuitive intimacy to develop a concept of size that will take us well beyond what we know.

If we have two apples and two hands, we could put one apple in each hand, and no hand would be un-appled, or any apple un-handed. If we have two socks and we put one sock on each hand, then our socks and our hands correspond exactly. This observation also implies that we can put one apple in each sock to demonstrate that the socks and apples also correspond; however, we do not recommend this last experiment unless the socks are clean.

Grappling with infinity requires us to imagine scenarios that are not quite possible but can be clearly conceived in the mind. Suppose we have a huge barrel full of Volvos and a barrel full of soccer balls, and we want to know if there are more Volvos than soccer balls, more soccer balls than Volvos, or the same number of each. How would we decide?

Probably we would just count the number of Volvos and soccer balls in each of the barrels and compare the two numbers. Certainly that method works, but an alternative method allows us to grapple with magnitudes beyond what we can count. Let's devise a method for comparing the barrels without counting Volvos or soccer balls.

We could take one Volvo from the first barrel and one soccer ball from the second, pair them up and put them aside (probably putting the ball in the car's trunk). Then we could pair another Volvo from the first barrel with another soccer ball from the second. If we continue pairing in this fashion, we could

. . .

tell whether we have the same number of Volvos and soccer balls without ever knowing how many we actually have.

This simple idea is important. We have just described a method for determining when two collections contain the same number of objects without actually counting them. Two collections whose objects can be paired evenly—one from one collection with one from the other collection—are said to have a *one-to-one correspondence*.

Correspondences

As we explore the notion of one-to-one correspondences, we must do something that is extremely difficult: We must forget. We must forget that 8 means something; we must forget that 37 means something. We must strip from our minds the names of numbers, leaving behind only the idea that *two collections of objects are equally numerous, precisely if there is a one-to-one correspondence between the elements of the two collections*. In taking this step, we are moving away from thinking of a number as an attribute of a collection ("I have five fingers on my right hand") to thinking instead of an idea of comparison ("I have just as many fingers on my right hand as on my left, because I can touch all the fingertips on one hand with the fingertips of the other hand").

This focus on comparison, on one-to-one correspondence, allows us to examine the infinite and turn our vague sense of awe into concrete understanding.

THE FUNDAMENTAL IDEA in the study of infinity is that two collections have the same size if there is a one-to-one correspondence between the members of one collection and the members of the other collection. This compelling concept of comparing sizes via one-to-one correspondence is the rock on which the whole study of infinity is built.

We distilled this important idea of one-to-one correspondence by thinking hard about something we know well—counting. Generally, the most reliable guide to the unknown is a deep understanding of the simple and familiar.

Life Lessons

Understand simple things deeply.

MINDSCAPES Invitations to Further Thought

In this section, Mindscapes marked **(H)** *have hints for solutions at the back of the book. Mindscapes marked* **(ExH)** *have expanded hints at the back of the book. Mindscapes marked* **(S)** *have solutions.*

Developing Ideas

1. **Still the one.** What is a one-to-one correspondence?

2. **I get around.** Consider the following pairing:

   Honda . . . Deb
   Saab . . . Ed
   Lexus . . . Mike
   Trail-a-Bike . . . Julia

   Who corresponds to the Saab? What mode of transportation corresponds to Julia?

3. **Numerical nephew.** At a family gathering, your four-year-old nephew approaches you and proudly proclaims he has found the biggest number. How would you gently refute his naïve notion?

4. **Pile of packs.** You walk into class late and notice a bunch of backpacks lying against one wall. How could you check to see if there's a one-to-one correspondence between the backpacks and the students in the room? Is there a way to pair up each backpack with a student?

5. **Bunch of balls.** Your first job every morning at tennis camp is to get the ball machine ready for action. You open up some new cans of tennis balls and empty them into a large hopper. Is there a one-to-one correspondence between the balls and the cans?

Solidifying Ideas

6. **The same, but unsure how much (H).** We have used a method of checking whether two sets of objects have the same number of things by pairing and removing one object from each set until no objects remain. If we run out of objects from both sets at the same time, then we know that the sets contain the same number of things. Otherwise, we know that one set is larger than the other. Describe several scenarios in which we can compare the sizes of two collections without computing individual sizes—for example, people filling all seats in an auditorium.

7. **Taking stock (S).** It turns out that there is a one-to-one correspondence between the New York Stock Exchange symbols for companies and the companies themselves (for example, PE is Philadelphia Electric Company). Explain why this correspondence must be one-to-one. What would happen if it were not? Describe potential problems.

8. **Don't count on it.** The following are two collections of the symbols @ and ©:

@ @

© ©.

Are there more @'s than ©'s? Describe how you can quickly answer the question without counting, and explain the connection with the notion of a one-to-one correspondence.

9. **Here's looking @ ®.** The following collections contain the symbols @ and ®:

@ @

® ®.

Are there more @'s than ®'s? Describe how you can quickly answer the question without actually counting, and explain the connection with the notion of a one-to-one correspondence.

10. **Enough underwear.** When Deb packs for a trip, she doesn't count the number of days she will be away and then count out that many pairs of underwear. Instead, she places underwear into her suitcase, one at a time, and says the name of each day she will be away: "Monday" (and places one in), "Tuesday" (and places another in), "Wednesday," etc. Using this method, does Deb know the number of underwear she placed in her bag? Does she have enough underwear for her trip? Discuss the connection this true story has with the notion of a one-to-one correspondence.

11. **791ZWV.** Suppose a stranger tells you that the license plate number on his car is 791ZWV. If you had a listing of all automobiles in the United States together with their license plate numbers, would you be able to precisely identify the stranger's vehicle? If so, explain why. If not, explain why, and identify what additional information you would require to identify it exactly. Discuss the connection between this situation and the notion of a one-to-one correspondence.

12. **245-2345.** Suppose a stranger tells you that her telephone number is 245-2345. Would you be able to dial her number and be certain that you reach her? If so, explain why. If not, explain why not, identify what additional information you would require to be certain to reach her home. Discuss the connection between this situation and the notion of a one-to-one correspondence.

13. **Social security (H).** Is there a one-to-one correspondence between U.S. residents and their Social Security numbers? Explain why or why not.

14. **Testing one two three.** A professor wishes to distribute one examination to each student in the class. What is the most efficient way for her to determine whether she has more students than exams: Pass out the exams or count?

15. **Laundry day (ExH).** Suppose you are given a bag of quarters. The laundry machine requires $1.75 worth of quarters. One way to count how many washes you can do is to take out one quarter and say "25¢," then take out another and say "50¢," and so on. In practice, however, you might well use a different method that uses a notion of one-to-one correspondence. Explain such a method.

Creating New Ideas

16. **Hair counts.** Do there exist two nonbald people on Earth such that there is a one-to-one correspondence between the collection of hairs on one person's body and the collection of hairs on the other person's body? Feel free to use facts from previous chapters, but explain how they provide an answer to the one-to-one correspondence question.

17. **Social number (S).** Social Security numbers contain nine digits. Suppose that all nine-digit numbers are allowable Social Security numbers. Is there a one-to-one correspondence between allowable Social Security numbers and U.S. residents? You may assume that the U.S. population is about 300 million. Explain your answer.

18. **Musical chairs.** Musical chairs is a fun game in which a group of people march around a row of chairs while music is played. There is one more person than there are chairs. The moment the music stops, everyone scrambles for a chair. The person left chairless loses and moves to the sidelines. Then everyone in a chair gets up, one chair is removed, and the music and marching begin again. At what points in this game do we have a one-to-one correspondence between chairs and people, and at what points do we not have such a correspondence? Explain the correspondences as the game is played until there is a winner.

19. **Dining hall blues.** One day in Ralph P. Uke Dining Hall, the students in line discovered that no more forks were available (neither clean nor dirty). While there was much jubilation in the line, what can you conclude about the set of all students in the dining hall and the set of all forks? (Assume that every

student with a fork has only one.) Do we know how many forks there are? How many students? Discuss how the idea of one-to-one correspondence is relevant in addressing these questions.

20. **Dorm life (H).** Every student at a certain college is assigned to a dorm room. Does this imply that there is a one-to-one correspondence between dorm rooms and students? Explain your answer.

Further Challenges

21. **Pigeonhole principle.** Recall the Pigeonhole principle from the first section in Chapter 2. Restate this principle in terms of a correspondence. Suppose you try a method of assigning pigeons to holes and, after filling all the holes, some pigeons remain. If you remove the pigeons and try again, is there any hope of placing each pigeon in an individual hole the second time? Suppose you have an infinite number of pigeons and pigeonholes. Is it possible that a first attempt to give each pigeon an individual hole failed but a second attempt succeeded?

22. **Mother and child.** Every child has one and only one birth mother. Does this imply that a one-to-one correspondence exists between the set of all children and the set of all birth mothers? Explain your answer.

In Your Own Words

23. **With a group of folks.** In a small group, discuss the notion of a one-to-one correspondence and why two collections that have a one-to-one correspondence should be considered the same size. After your discussion, write a brief narrative describing your conclusions in your own words.

For the Algebra Lover

Here we celebrate the power of algebra as a powerful way of finding unknown quantities by naming them, of expressing infinitely many relationships and connections clearly and succinctly, and of uncovering pattern and structure.

24. **Coast to coast.** Jessica is working part-time from 7:00 AM to 11:00 AM in New York and needs to know the current co rresponding time in California. There is a one-to-one correspondence between those Eastern times (expressed in hours as x) and Pacific times (expressed in hours as y). Explain why the function $y = x - 3$ represents the one-to-one correspondence between Eastern and Pacific time zones that Jessica needs. What Pacific time corresponds to $x = 10$? What Eastern time corresponds to $y = 5$?

25. **An interesting correspondence.** Suppose you invest some amount of money (we will call that unknown amount of dollars d) into a savings account that pays 1.25% simple interest annually. We let I represent the amount of interest you will have after one year. Thus there is a one-to-one correspondence between the amount you invest d and the amount of interest I you will earn after one year. That correspondence is given by the function $I = 0.0125d$. Use this function to find the value of I corresponding to $d = \$10,000$. Express in words what your answer represents.

26. **Chicken Little.** With increased attention to eating healthier, locally grown food, your hometown now has many families wanting to raise chickens.

Local ordinances require the use of chicken coops, so your little sister starts a summer business building luxury coops. It takes her one week to design and build a coop that meets chicken coop building codes as well as the individual preferences of each customer, so there is a one-to-one correspondence between the weeks she works in the summer and the coops she builds. She starts work on Monday, June 4 and earns a profit of $100 on each coop. How much money will she have made by the end of summer on August 31?

27. **Table for four.** The table below shows a one-to-one correspondence between the four values of x and four values of y. Find a formula for y in terms of x that, when evaluated at x, produces the corresponding y value.

x	–1	0	1	2
y	2	5	8	11

28. **Square table.** The table below shows a one-to-one correspondence between the natural numbers 1, 2, 3, … and the perfect squares 1, 4, 9, … Let the perfect squares be denoted by y and the natural numbers be denoted by n. Find a formula for y in terms of n that reflects the correspondence given in the table.

n	1	2	3	4	5	6	7	8	…
y	1	4	9	16	25	36	49	64	…

COMPARING THE INFINITE

Pairing Up Collections via a One-to-One Correspondence

I saw . . . a quantity passing through infinity and changing its sign from plus to minus. I saw exactly how it happened . . . but it was after dinner and I let it go.

SIR WINSTON CHURCHILL

Michelangelo Buonarroti, *The Creation of Adam* (1508–1512). An early one-to-one correspondence.

We now enter the world of infinity armed with one idea—a criterion for comparison: the one-to-one correspondence. We will test the consequences of this idea by comparing some infinite collections of familiar objects. As usual, our most productive strategy is to examine the familiar before we journey toward the unknown.

Since numbers are really the only infinite collections we know, we turn to them for our first examples to help us become accustomed to the idea of one-to-one correspondence. We start with the most basic collection of numbers we can think of and then consider related but different collections. Our goal is to determine whether various collections can be put into one-to-one correspondence, since one-to-one correspondence is the fundamental principle on which our investigation of infinity is built.

Familiar but Infinite

What is familiar and concrete to one person may be foreign and abstract to another, but as far as numbers are concerned, we probably all agree on which are the most familiar. In 1886, Leopold Kronecker, a number theorist, made a statement about what is basic in the world of mathematics: "God created the positive integers; all the rest is human creation."

One, two, three, . . . these are the positive integers. For every positive integer, there is a next bigger one. Although we may think of these positive integers successively, we may also think of all of them at once—that is, think about the collection of all positive integers. The collection (also referred to as the *set*) of positive integers is so basic and natural to our way of thinking that it is called the set of *natural numbers*.

1   2   3   4   5   6   7   8   . . .   10,023

. . .   32,376,201   . . .

The set of natural numbers. (They're all there, but we didn't have time to write them all out.)

147

The set of all natural numbers is our first infinite set, and it has a comfortable feel about it. Among infinite sets, the natural numbers seem the most natural. By examining this and related collections of numbers, we will begin to develop a better and more precise idea of infinity.

Natural Numbers with 1 Removed

Suppose we are given another copy of the set of natural numbers—in a different font:

1, 2, 3, ...

Unfortunately, we absentmindedly lost the number *1*. Thus, this new set is the collection of natural numbers with the number *1* removed. Specifically, our set consists of

2, 3, 4, ...

Now we have a new infinite set. But it has one less element than the set of natural numbers . . . or has it?

On the one hand, we can observe through life experience that, if we have a barrel of Volvos and one is removed, we then have fewer Volvos left. It seems reasonable to conclude that we have fewer natural numbers if we remove the number *1*. On the other hand, we may think, "Hey, infinity is infinity is infinity, so the new set doesn't contain fewer elements."

Our intuition is pulled in two directions. One direction is the "infinity is infinity" camp; the other is the "take one away, you have one less" school. We will soon discover that both these arguments will lead us astray. What's wrong with our intuition? Nothing, except that our insights and life experiences involving collections of everyday objects will not always apply to infinite collections.

If our intuition leads to two opposite conclusions and both sound reasonable, we are compelled to investigate until we understand the consequences of both ideas. If we believe that the set of natural numbers with the number 1 removed is the same size as the set of natural numbers, then we need a rigorous and logical reason. Vague thoughts of "infinity is infinity" will not suffice in a quantitative court of law. However, we must remember to avoid little distractions, such as our entire life history, that tell us when we remove an object, we are left with fewer things. Let's keep an open mind and remember that our criterion for determining the equivalence of two collections is not a vague, undefined feeling developed through years of experience but, instead, is a clear criterion stated crisply and explicitly as the existence of a one-to-one correspondence. Since we have formulated an explicit definition, let's rely on it in preference to general impressions. Infinity is a large, wild beast; but, if we remain focused on our principle of comparison, we will have infinity tamely eating out of the palm of our hand.

A Search for a One-to-One Correspondence

To determine whether there are as many natural numbers as there are integers starting with *2, 3, 4, . . .* , we must ask if there is a one-to-one correspondence between the elements of the set

2, 3, 4, . . .

and those of the set of natural numbers

$$1, 2, 3, \ldots.$$

If we are from the school of "take one away, we have one less," it may appear that there cannot be a one-to-one correspondence between our two sets, since, if we paired the numbers in the two sets, we'd see the following:

Natural Numbers	1	2	3	4	5	6	7	8	9	10	11	12	13	14	15	16	17	18	19	...
A Pairing	↓	↓	↓	↓	↓	↓	↓	↓	↓	↓	↓	↓	↓	↓	↓	↓	↓	↓	↓	
Our New Set	2	3	4	5	6	7	8	9	10	11	12	13	14	15	16	17	18	19		...

Notice how the "1" in the natural numbers is alone and is not paired with any number in our new set. Hence this pairing is not a one-to-one correspondence. Does this failure imply that no one-to-one correspondence exists? The answer is a resounding no! A one-to-one correspondence may still exist. We merely conclude that the *particular* pairing we just created isn't a one-to-one correspondence.

Life Lesson

Just because a specific attempt failed does not mean that the task at hand is impossible.

The Naturals Minus 1 Equals the Naturals

It turns out that, in fact, a one-to-one correspondence between these two sets does exist, which will not surprise members of the "infinity is infinity" camp. We illustrate such a one-to-one correspondence below. But members of the "infinity is infinity" camp should not be too smug just yet.

Natural Numbers	1	2	3	4	5	6	7	8	9	10	11	12	13	14	15	16	17	18	19	...
A New Pairing																				
Our New Set	2	3	4	5	6	7	8	9	10	11	12	13	14	15	16	17	18	19	20	...

Suppose we dump all the natural numbers into the Natural Number barrel and the natural numbers with *1* removed in the New Set barrel. In the new pairing, we grab 1 from the Natural Number barrel and *2* from the New Set barrel, pair them, and toss them. Next, we pair 2 with *3*, then 3 with *4*, 4 with *5*, and so on.

Notice that every element from the Natural Number barrel is paired with exactly one of the elements from our New Set barrel, and each number from our New Set barrel is associated with exactly one natural number. The moment we mention a particular number from one list, we know who it pairs with from the other list. After completing this infinite process, no numbers remain; both barrels are empty.

We can use the following symbol to express this one-to-one correspondence: ↔. The symbol compactly illustrates which numbers are paired. Thus, $n \leftrightarrow n + 1$ means that the number n from the Natural Number barrel is paired with the number $n + 1$ from the New Set barrel. Once we know which number n represents, we immediately know its mate is $n + 1$.

For example, if n is 4, then we see its mate is **4 + 1,** which is **5**. So 4 from the Natural Number barrel is paired with **5** from the New Set barrel. Notice that this pairing is exactly the one we described in the picture.

Cardinality

The existence of this one-to-one correspondence means that these two sets have the same number of elements. We must be careful here. We really should not say that these sets have the same "number" of elements, since infinity is not actually a number. How can we get around this thorn? We create new terminology. We use the phrase *cardinality of a set* to mean the "number" of things in the set, with the understanding that the set may contain infinitely many things. If a set contains only finitely many things, then its cardinality is just the number of things in the set.

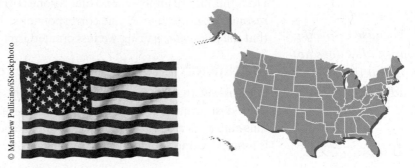

© Matthew Pullicino/iStockphoto

How would you show a one-to-one correspondence between the set of stars and the set of states?

Two sets have the *same cardinality* if there is a one-to-one correspondence between the elements of one set and the elements of the other set—such as the set of stars on the U.S. flag and the set of states in the United States. In this case, both sets have finite cardinality, and that cardinality equals 50.

As an example involving infinite collections, we saw that the set of natural numbers and the set of natural numbers greater than 1 have the same cardinality.

The Ping-Pong Ball Conundrum

Let's now consider a story that frees us from the confines of our physical reality and allows us to use our imagination. At the end of the story, we pose several questions, and we encourage you to take a moment to guess an answer.

The scene opens with a very large barrel center stage with a raised, angled trough. In the trough, lined up like little round soldiers, we see an unending row of white, numbered Ping-Pong balls ordered sequentially starting with 1. There is one Ping-Pong ball for each natural number.

Now we embark on a mental experiment. This experiment will last for exactly one minute, and in that time we will perform an infinite number of tasks. But 60 seconds after we start, our stopwatch's alarm will beep, and we will stop—period.

Task 1

We begin with 60 seconds on the clock. In half that time, 30 seconds, we must pour the first 10 Ping-Pong balls (numbered 1 through 10) into the large barrel and then reach into the barrel, find the Ping-Pong ball numbered 1, and remove it (just throw it away). We are left with nine balls in the barrel and one thrown away.

Task 2

Now we have 30 seconds remaining, and we begin to pick up the pace. In half the time remaining, 15 seconds, we must dump into the barrel the next 10 Ping-Pong balls (numbered 11 through 20) and then reach in and scoop out the ball numbered 2. We are now left with 18 balls in the barrel and two thrown away. But we don't stop yet.

Tasks 3, 4, and 5

The third task must be accomplished in half the time remaining, 7.5 seconds. We must quickly drop in the next 10 balls (numbered 21 through 30) and then find and remove the ball numbered 3. We are then left with 27 balls in the barrel—and a few beads of perspiration on our foreheads.

We're still a long way from finished, but perhaps the pattern is emerging. We continue to cut the time remaining in half and, in each individual half period, drop the next 10 balls into the barrel and fish out the ball with the lowest number.

Task 6

At the sixth stage, we drop the balls numbered 51 through 60 into the barrel and remove the ball numbered 6, leaving 54 balls in the barrel. We have to work fast, because this entire stage must be completed in 0.9375 of a second (that is, half of half of half of half of half of 60 seconds).

Remaining Tasks

Clearly, we must pick up the pace. In fact, we see that soon we will have to move faster than the speed of sound and even faster than the speed of light—now, that's fast. Physically impossible? Why yes, of course, but it's also physically impossible to have infinitely many Ping-Pong balls—if we did, they would take up all the space in our universe, and their weight would squash us like small bugs. Happily, this is an exercise of the mind; thus, we use our power of imagination to save us from these potentially dangerous balls and also to allow us to dump in and pull out balls at incredible speeds.

Our experiment of the imagination is hurried and frantic; mercifully, however, it lasts only one minute. Sixty seconds after we begin, the stopwatch beeps and we stop and attempt to catch our breath. As we catch our breath, we glance into the barrel. What do we see? Is it empty? Does it contain some Ping-Pong balls? Does it contain infinitely many Ping-Pong balls? We invite you to consider these questions.

On the Ball

Do Ping-Pong balls remain in the barrel after the minute has expired? If so, then we have an annoying request: Name one! Any ball remaining has a number stamped on its surface. What is that number? Recall that the balls are numbered 1, 2, 3, . . . forever—there is no last ball. What is the number of a ball in the barrel? Could it be 5? No, because we know exactly when the ball numbered 5 was removed—namely, at the fifth stage (remember, we dumped balls 41–50 into the barrel and removed the ball numbered 5). Could ball 45,671,803 remain? Well no, because we know exactly when we removed it from the barrel—specifically, at the 45,671,803rd stage of the experiment. Thus, which balls are left? The answer is amazing and surprising: "NONE!" Since all the balls were numbered and we systematically removed them, we can state with exact precision the time when any particular ball was removed. Thus, at the end of the minute, the barrel is empty. This counterintuitive and perhaps unbelievable answer is puzzling— especially because the number of Ping-Pong balls in the barrel increased by nine at each stage.

This experiment, the question posed, and particularly the answer presented require serious thought, but we can convince ourselves that the barrel is empty. This Ping-Pong ball conundrum is a dynamic illustration of the dramatic difference between the finite and the infinite. In actuality, this activity produced a one-to-one pairing between the intervals of half times (stages) and the numbered balls. We were able to pair them evenly. Again, our intuition, based on finite collections, is not always accurate when applied to infinity. Nevertheless, through reason, the experience we are gaining will help us develop an understanding of infinity. To further develop this understanding, let's explore other infinite collections.

Looking for Giants

When we saw that removing the number *1* from the set of natural numbers did not decrease the size or the cardinality of the set, we surprised those whose intuition previously dictated that removing an element should make a set smaller. We will also decimate the intuition of those who remain in the "infinity is infinity" camp by demonstrating that some infinite sets are even more infinite than the natural numbers! How can we even begin this seemingly impossible quest for giant sets?

We need to think of some infinite sets that are likely to be even larger than the set of natural numbers. A possibility is the set of *all* integers: positive, negative, and zero. This set contains infinitely many more numbers than the set of natural numbers—namely, all the negative integers and zero—so it appears to be a good candidate. Intuitively, there appear to be *twice* as many integers as natural numbers.

Integers Equal Naturals

Unfortunately, merely adding infinitely many negative numbers and zero is not enough to increase the size of the set of natural numbers. To prove this statement, as always, we need to exhibit an explicit one-to-one correspondence between the set of natural numbers and the set of all integers. You can find that correspondence yourself. To describe such a correspondence, draw a table that lists the natural numbers, 1, 2, 3, . . . , down the left side of the paper. Next to each natural number, write one integer (positive, negative, or zero) in such a systematic fashion that all integers will eventually appear on the list. Wait to look at our answer until you've given it a try.

Natural Numbers		All Integers
1	⟷	0
2	⟷	1
3	⟷	−1
4	⟷	2
5	⟷	−2
6	⟷	3
7	⟷	−3
⋮	⋮	⋮
$2n$	⟷	n
(The even numbers)		
$2n + 1$	⟷	$-n$
(The odd numbers)		
⋮	⋮	⋮

A one-to-one correspondence between the natural numbers (that is, the positive integers) and all integers (including negative integers and zero).

Notice that in this pairing, every natural number appears exactly once in the left-hand column, and every integer (positive, negative, or zero) appears exactly once at some time in the right-hand column. Thus, we have an even pairing, a one-to-one correspondence. What number is paired with −9 in the right column? What number is paired with 9 in the left column? We can choose random elements from either set and figure out under this pairing with what those random elements get paired—this experimentation is the best way to get a feel for the correspondence. Is any number not paired? No.

So, we have proven that the set of natural numbers has the same cardinality as the set of all integers. Notice how the two symbolic expressions toward the bottom of our list capture the spirit of the pairing; namely, the even natural numbers are paired with the positive integers, whereas, the odd natural numbers are paired with the negative integers or zero. For example, the natural number 32 can be written as 2×16, so the symbolic expression $2n \leftrightarrow n$ shows us that 32 is paired with 16. However, 33 can be written as $(2 \times 16) + 1$; thus, the symbolic expression $2n + 1 \leftrightarrow -n$ shows us that 33 is paired with −16. So, if we looked down our list, at some point we would see the following pairings:

32	⟷	16
33	⟷	−16

Rational Numbers

We might feel more or less stymied in our quest for an infinite set bigger than the set of natural numbers. But let's not forget that lots of numbers are not integers. For example, how about the rational numbers? Recall that the set of rational numbers is the set of all ratios of integers (fractions). Between any two integers there are infinitely many rationals.

Within the set of rational numbers, we actually have infinitely many different infinite sets. The set of rational numbers must be huge.

Unfortunately, in our search for different sizes of infinity, the rationals are still not numerous enough. We rely once again on the definition of same cardinality; namely, two sets have the same cardinality if there is a one-to-one correspondence between the elements of one set and the elements of the other. The trick here is to write down the rational numbers in a convenient and systematic way so that we know we have listed them all. The idea is to put all the rational numbers with numerator 1 in one column, all those with numerator 2 in another column, and so on, as shown in the following diagram.

⋮	⋮	⋮	⋮	⋮
1/5	2/5	3/5	4/5	5/5 …
1/4	2/4	3/4	4/4	5/4 …
1/3	2/3	3/3	4/3	5/3 …
1/2	2/2	3/2	4/2	5/2 …
1/1	2/1	3/1	4/1	5/1 …

0

… −5/1	−4/1	−3/1	−2/1	−1/1
… −5/2	−4/2	−3/2	−2/2	−1/2
… −5/3	−4/3	−3/3	−2/3	−1/3
… −5/4	−4/4	−3/4	−2/4	−1/4
… −5/5	−4/5	−3/5	−2/5	−1/5
⋮	⋮	⋮	⋮	⋮

Notice that all the rationals in the same row have the same denominator. For example, to find 37/112 in the diagram, we just go 37 spaces to the right of 0 and 112 spaces up. So, we can see that all the rational numbers appear somewhere in the pattern. In fact, each rational number appears many times, because the fractions are not all reduced to lowest terms. For example, 1/2 appears and so does 2/4, 3/6, and so on, but that redundancy is okay, because, at this point, we merely want a systematic method of recording every rational number without omitting any. Notice that all the positive rational numbers appear in the upper-right part of the diagram, all the negative rationals appear in the lower left, and 0 is in the middle. So far, we have described a way of writing down all the rational numbers.

Rationals Equal Naturals

To show the one-to-one correspondence between the rational numbers and the natural numbers, we will thread a single rectangular spiral through all the rationals, starting in the middle at 0 and moving counterclockwise outward. To see the one-to-one correspondence with the natural numbers, we will count the rational numbers as we encounter them along the spiral and make them boldface to remind us that we have paired that rational with a natural number. We start with the rational **0** corresponding to the natural number 1; then, moving to the right and up, the rational **1/1** = 1 corresponds to the natural number 2; the rational −**1/1** = −1 corresponds to 3; the rational **2/1** = 2 corresponds to 4. We next come to 2/2, which has already been counted, so we skip it and move to **1/2,** which corresponds to 5; then −**2/1** = −2 corresponds to 6. We skip −2/2, since it equals −1, which already corresponds to 3, and move to −**1/2,** which corresponds to 7; and so on. Notice that every rational number will eventually be reached and put in correspondence with some natural number. This one-to-one correspondence shows that the set of all rational numbers has the same cardinality as the set of the natural numbers.

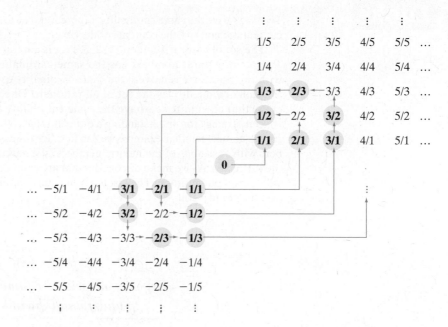

We may now straighten out the spiral and produce the following list showing the one-to-one correspondence. Notice that if someone examined just the list, he or she would have difficulty detecting the pattern, which arises from the spiral previously illustrated.

Natural Numbers		Rational Numbers
1	⟷	0
2	⟷	1/1
3	⟷	−1/1
4	⟷	2/1
5	⟷	1/2
6	⟷	−2/1
7	⟷	−1/2
8	⟷	3/1
9	⟷	3/2
10	⟷	2/3
11	⟷	1/3
12	⟷	−3/1
…	…	…

Threading and counting along the spiral provide an important insight into sets having the same cardinality as the set of natural numbers. If we can write a set as an infinite list, we can make a one-to-one correspondence with the natural numbers.

We now see that the rational numbers did not provide us with an infinity larger than that of the natural numbers. Our quest for an even grander infinity has thus far failed. But perhaps in chasing sets larger than an infinite set, we should expect to have to go a long, long way.

Two SETS have the same cardinality if there is a one-to-one correspondence between the contents of one and the contents of the other.

The set of natural numbers, 1, 2, 3, . . . , is a natural first infinite set to investigate. The set of natural numbers has the same cardinality as the set of natural numbers with the number *1* removed; the same cardinality as the set of all integers; and even the same cardinality as the set of all rationals. One might naturally, but mistakenly, guess that all infinite sets have the same cardinality as the natural numbers.

Our strategy for understanding a difficult topic is to explore and experiment with the simplest, most familiar example we can find. In this case, we compared the natural numbers with variations of the natural numbers (the natural numbers minus 1, the integers, the rationals). As we make discoveries that are counterintuitive, we gain experience that will retrain our intuition. The discovery of counterintuitive truths liberates us to think of notions grander still.

Life Lessons

> *Experiments with the familiar help us*
> *to understand the unknown.*

MINDSCAPES Invitations to Further Thought

In this section, Mindscapes marked **(H)** *have hints for solutions at the back of the book. Mindscapes marked* **(ExH)** *have expanded hints at the back of the book. Mindscapes marked* **(S)** *have solutions.*

Developing Ideas

1. *Au natural.* Describe the set of natural numbers.

2. *Au not-so-natural.* Describe each of the sets given below in words.

$$\{3, 6, 9, 12, 15, 18, \ldots\}$$
$$\{1, 2, 3, 4, 5\}$$
$$\left\{\frac{1}{2}, \frac{1}{4}, \frac{1}{8}, \frac{1}{16}, \ldots\right\}$$
$$\{-1, -2, -3, -4, \ldots\}$$
$$\{1, 4, 9, 16, 25, 36, 49, 64, 81, 100\}$$

3. **Set setup.** We can denote the natural numbers symbolically as $\{1, 2, 3, 4, \ldots\}$. Use this notation to express each of the sets described below.

- The set of natural numbers less than 10.
- The set of all even natural numbers.
- The set of solutions to the equation $x^2 - 4 = 0$.
- The set of all reciprocals of the natural numbers.

4. **Little or large.** Which of the sets in Mindscape 3 are infinite sets? Which are finite?

5. **A word you can count on.** Define the *cardinality* of a set.

Solidifying Ideas

6. **Even odds.** Let E stand for the set of all even natural numbers (so $E = \{2, 4, 6, 8, \ldots\}$) and O stand for the set of all odd natural numbers (so $O = \{1, 3, 5, 7, \ldots\}$). Show that the sets E and O have the same cardinality by describing an explicit one-to-one correspondence between the two sets.

7. **Naturally even.** Let E stand for the set of all even natural numbers (so $E = \{2, 4, 6, 8, \ldots\}$). Show that the set E and the set of all natural numbers have the same cardinality by describing an explicit one-to-one correspondence between the two sets.

8. **Fives take over.** Let EIF be the set of all natural numbers ending in 5 (EIF stands for "ends in five"). That is,

$$EIF = \{5, 15, 25, 35, 45, 55, 65, 75, \ldots\}.$$

Describe a one-to-one correspondence between the set of natural numbers and the set EIF.

9. **Six times as much (ExH).** If we let $\mathbb{N}$ stand for the set of all natural numbers, then we write $6\mathbb{N}$ for the set of natural numbers all multiplied by 6 (so $6\mathbb{N} = \{6, 12, 18, 24, \ldots\}$). Show that the sets $\mathbb{N}$ and $6\mathbb{N}$ have the same cardinality by describing an explicit one-to-one correspondence between the two sets.

10. **Any times as much.** If we let $\mathbb{N}$ stand for the set of all natural numbers, and a stand for any particular natural number, then we write $a\mathbb{N}$ for the set of natural numbers all multiplied by a. Do the sets $\mathbb{N}$ and $a\mathbb{N}$ have the same cardinality? If so, describe an explicit one-to-one correspondence between the two sets.

11. **Missing 3 (H).** Let TIM be the set of all natural numbers except the number 3 (TIM stands for "three is missing"), so $TIM = \{1, 2, 4, 5, 6, 7, 8, 9, \ldots\}$. Show that the set TIM and the set of all natural numbers have the same cardinality by describing an explicit one-to-one correspondence between the two sets.

12. **One weird set.** Let OWS (you figure it out) be the set defined by

$$OWS = \{1, 3, 5, 7, 8, 10, 11, 12, 13, 14, 15, 16, \ldots\};$$

that is, after 8, the set contains all the natural numbers from 10 on. Show that the set OWS and the set of all natural numbers have the same cardinality by describing an explicit one-to-one correspondence between the two sets.

13. **Squaring off.** Let S stand for the set of all natural numbers that are perfect squares, so $S = \{1, 4, 9, 16, 25, 36, 49, 64, \ldots\}$. Show that the set S and the set

of all natural numbers have the same cardinality by describing an explicit one-to-one correspondence between the two sets.

14. **Counting Cubes (formerly Crows).** Let C stand for the set of all natural numbers that are perfect cubes, so

$$C = \{1, 8, 27, 64, 125, 216, 343, 512, \dots\}.$$

Show that the set C and the set of all natural numbers have the same cardinality by describing an explicit one-to-one correspondence between the two sets.

15. **Reciprocals.** Suppose R is the set defined by

$$R = \left\{ \frac{1}{1}, \frac{1}{2}, \frac{1}{3}, \frac{1}{4}, \frac{1}{5}, \frac{1}{6}, \dots \right\}$$

Describe the set R in words. Show that it has the same cardinality as the set of natural numbers.

16. **Hotel Cardinality (formerly California) (H).** It is the stranded traveler's fantasy. The Hotel Cardinality is a full-service luxury hotel with bar and restaurant. It has as many rooms as there are natural numbers. The room numbers are 1, 2, 3, 4, 5, ... You can see why stranded travelers love the Hotel Cardinality. There appears to be no need for the sad sign: No Vacancy. Suppose, however, that every room is occupied. Now it appears that the night manager must flash the No Vacancy sign. What if a weary traveler were to arrive late at night looking for a place to stay? Could the night manager figure out a way to provide the traveler with a private room (no sharing) without evicting another guest? The answer is yes. Describe how this accommodation can be made; of course, some guests will have to move to other rooms.

17. **Hotel Cardinality continued.** Given the scenario in Mindscape 16, suppose now that two more travelers arrive, each wanting his or her own private room. Is it possible for the night manager to make room for these folks without pushing anyone onto the streets?

18. **More Hotel C (ExH).** Given the scenario in Mindscape 16—that is, the hotel starts full—suppose now that the Infinite Life Insurance Company, which has lots of employees—in fact, there are as many employees as there are natural numbers—decides to provide each of its employees with a private room. Is it possible for the night manager to give each of the infinitely many employees his or her own room without kicking any of the other guests onto the streets? By the way, as you may have guessed, after this busy evening, the night manager quit and got a job as the night manager of a nearby Motel 6 (it only had 56 rooms).

19. **So much sand.** Prove that there cannot be an infinite number of grains of sand on Earth.

20. **Half way.** Suppose you take the line below, cut it in half, and then take the left piece and cut it in half, and then take the leftmost piece and cut it in half, and so on, without ever stopping. How many different pieces of the line would you have? Does the set of all pieces have the same cardinality as the set of all natural numbers? Justify your answer.

21. **Pruning sets.** Suppose you have a set. If you remove some of the things from the set, will the collection of remaining things always contain fewer items in it than the original set? If so, demonstrate why. If not, illustrate with an example.

22. **A natural prune.** Describe a collection of numbers you could remove from the set of natural numbers so that the set of remaining numbers contains fewer numbers than the set of all natural numbers. (*Bonus:* How many times did the word *numbers* appear in the previous sentence?)

23. **Prune growth.** Is it possible to remove things from a set so that the collection of remaining things is larger than the original set? Explain why or why not. Illustrate with an example.

24. **Same cardinality?** Suppose we have two sets and we are able to pair elements from one set with elements from the other set in such a way that all the elements from the first set are paired with elements from the second set; but there are elements from the second set that were never paired. Does this pairing imply that the two sets do not have the same cardinality? Justify your answer.

25. **Still the same? (S).** Suppose we have two sets, and every pairing of elements from one set with elements from the other set results in having elements from the second set never paired. Do these pairings imply that the two sets do not have the same cardinality? Justify your answer.

Creating New Ideas

26. **Modest rationals (H).** Devise and then describe a method to systematically list all rational numbers between 0 and 1.

27. **A window of rationals.** Using your answer to Mindscape 26, show that the set of rationals between 0 and 1 has the same cardinality as the set of natural numbers.

28. **Bowling ball barrel.** Suppose you have infinitely many bowling balls and two huge barrels. You take the bowling balls and put each ball in one of the two barrels. What can you conclude about the cardinality of at least one of the barrels? Prove your answer.

29. **Not a total loss.** Take the set of natural numbers and remove a finite number of numbers from it. Prove that this new set has the same cardinality as the set of natural numbers.

30. **Mounds of mounds.** A Peter-Paul Mounds package contains two delicious chocolate-covered coconut candy bars. Suppose you had infinitely many Mounds packages: one for each natural number. Does the collection of individual candy bars have the same cardinality as the set of natural numbers? If not, explain why. If so, provide a one-to-one correspondence.

31. **Piles of peanuts (ExH).** You have infinitely many piles of peanuts. In the first pile you have one peanut; in the second pile you have two; in the third you have three; and so on. How many nuts do you have? Does the set of all these nuts have the same cardinality as the set of natural numbers? If not, explain why. If so, provide a one-to-one correspondence.

32. **The big city (S).** Not-Finite City (also known as The Really Big Apple) is made up of infinitely many avenues running north and south (one avenue for each natural number) and infinitely many streets running east and west (one

street for each natural number). A traffic light is placed at every intersection of a street with an avenue. How many traffic lights are there? Does the set of all traffic lights have the same cardinality as the set of natural numbers? If not, explain why. If so, provide a one-to-one correspondence.

33. **Don't lose your marbles.** Suppose you have infinitely many large boxes (one for each natural number). In each box, you have infinitely many marbles (one for each natural number)—so the boxes are really big! Does the set of all marbles have the same cardinality as the set of natural numbers? Just make a guess and explain why you guessed what you guessed (you need not justify your answer rigorously).

34. **Make a guess.** Guess an infinite set that does not have the same cardinality as the set of natural numbers.

35. **Coloring.** Consider the infinite collection of circles below:

Suppose you have two markers, one red and one blue, and you color each circle one of the two colors. How many different ways can you color the circles? Do you think that the set of all possible circle colorings has the same cardinality as the set of all natural numbers? This question is tricky; just make a guess and explain why you guessed what you guessed (you need not rigorously justify your answer). You may first want to answer the question for just these four circles as a warm-up. We'll return to this Mindscape later.

Further Challenges

36. **Ping-Pong balls on parade (H).** This Mindscape is based on the experiment described in the section on adding and removing infinitely many Ping-Pong balls from a barrel. This time, suppose you dump into the barrel 10 Ping-Pong balls numbered 1–10 as before and remove number 1. But next you put in 100 Ping-Pong balls, numbered 11–110, and remove number 2. Then you put in 1000 Ping-Pong balls, numbered 111–1110, and remove number 3, and so on. The question is: How many Ping-Pong balls remain in the barrel after the stopwatch beeps? Infinitely many? Finitely many? Can you name one?

37. **Naked Ping-Pong balls.** This Mindscape is based on the experiment described in the section on adding and removing infinitely many Ping-Pong balls from a barrel. This time, the Ping-Pong balls are not numbered! You play the same game as in our experiment; but, now at each stage you just reach in and remove one Ping-Pong ball (you cannot fish around for a particular one since they now all look the same). How many Ping-Pong balls might remain in the barrel? This question is an interesting conundrum.

38. **Primes.** Show that the set of all prime numbers has the same cardinality as the set of all natural numbers.

39. **A grand union.** Suppose you have two sets, and each set has the same cardinality as the set of natural numbers. Take the elements of both sets and put them together to make one huge set. Prove that this new huge set has the same cardinality as the set of natural numbers.

40. **Unnoticeable pruning.** Suppose you have any infinite set. Is it always possible to remove some things from that set such that the collection of remaining things has the same cardinality as the original set? Explain why or why not, and illustrate your answer with an example.

In Your Own Words

41. **Personal perspectives.** Write a short essay describing the most interesting or surprising discovery you made in exploring the material in this section. If any material seemed puzzling or even unbelievable, address that as well. Explain why you chose the topics you did. Finally, comment on the aesthetics of the mathematics and ideas in this section.

42. **With a group of folks.** In a small group, discuss and work through the argument showing that the set of rational numbers has the same cardinality as the set of natural numbers. After your discussion, write a brief narrative describing the argument in your own words.

43. **Creative writing.** Write an imaginative story (it can be humorous, dramatic, whatever you like) that involves, or evokes, the ideas of this section.

44. **Power beyond the mathematics.** Provide several real-life issues—ideally, from your own experience—that some of the strategies of thought presented in this section would effectively approach and resolve.

For the Algebra Lover

Here we celebrate the power of algebra as a powerful way of finding unknown quantities by naming them, of expressing infinitely many relationships and connections clearly and succinctly, and of uncovering pattern and structure.

45. **How many mp3s?** In the Lost and Found Office of your school there are two boxes. One box contains a bunch of mp3 players and the other box contains a bunch of earbud headphones. You are told that these collections have the same cardinality. A deranged algebra instructor noticed that if $x^2 - x - 71$ represents the cardinality of the mp3 players, then $3x - 26$ represents the cardinality of the headphones. Untangle this cryptic observation to determine how many mp3 players there are. Suppose your lost mp3 player ended up in that first box. How easy would it be to locate yours?

46. **Pink ping pong possibilities.** You have a box containing 50 traditional, white ping pong balls, each with a natural number printed on it starting with 1 and going up to 50. Your friend has a box of pink ping pong balls, each with a distinct natural number printed on it starting with 8 and going up by 5's. If there's a one-to-one correspondence between the white balls and the pink balls, what's the largest label occurring on a pink ping pong ball?

47. **Plot the dots (H).** The table below gives a one-to-one correspondence between the x-values in the top row and the y-values in the bottom row. Plot the ordered pairs (x, y) in the xy-plane. Find a formula for y in terms of x that, when evaluated at x, gives the corresponding y value. Now find a formula for x in terms of y that, when evaluated at y, gives the corresponding x value.

x	0	1	2	3	4
y	6	4	2	0	−2

48. **1 to 1 or not 1 to 1?** Does the table below give a one-to-one correspondence between the x-values in the top row and the y-values in the bottom row? If yes, explain why. If not, explain why not. Either way, can you find a formula for y in terms of x? What about a formula for x in terms of y?

x	–2	–1	0	1	2	3
y	4	1	0	1	4	9

49. **Roommates.** Your school has 4000 students who want to live in dorms. There is a one-to-one correspondence between these students and available beds. If 80% of the beds are in double rooms, how many students will be living in double rooms? Suppose the remaining rooms are single rooms. How many dorm rooms are there altogether?

THE MISSING MEMBER

Georg Cantor Answers: Are Some Infinities Larger than Others?

Georg Cantor, the first person to tame infinity.

©Corbis

> *From the paradise created for us by Cantor, no one will drive us out.*
>
> DAVID HILBERT

Around 1872, the German mathematician Georg Cantor shook the foundations of infinity when he proved that the set of real numbers has more elements than the set of natural numbers. In other words, he proved that infinity is not one size but that some infinities are more infinite than others. At first such a notion might seem nonsensical. Once we have reached infinity, surely we cannot climb farther. But Cantor showed that there were yet higher mountains to scale.

To show that the real numbers are more numerous than the natural numbers, Cantor focused intently on what it would mean for the real numbers and the natural numbers to have the same cardinality. It would mean that the real and natural numbers could be put in one-to-one correspondence. Writing down what such a correspondence might look like gives us a visual clue as to why any attempted correspondence between the natural numbers and the real numbers could not include every real; some real is missing—*the missing member*.

From Bizarre to Intuitive

When Cantor conceived his idea near the end of the 19th century, many mathematicians resisted it strongly and attacked Cantor in a personal, abusive manner. These attacks, combined with other psychological issues, contributed to a bleak existence for Cantor, who spent much of the last part of his life in an insane asylum. People do not easily give up their intuition or beliefs, and the resistance of mathematicians to Cantor's "taming" of infinity was neither the first nor the last time that people resisted ideas that contradicted their preconceptions.

In 1637, Galileo was imprisoned for saying that the Earth moves around the Sun. The idea that the Earth moves is extremely counterintuitive. We accept the idea of a moving Earth principally because everyone else believes it. When we were children, we were told that the Earth moves, and so a moving Earth does not present a challenge to our beliefs. In fact, anyone who would today assert that the Earth is stationary

and the Sun revolves around the Earth would be regarded as a kook—and for good reason. However, historically speaking, a moving Earth was not easy to prove. Galileo was imprisoned because authorities of the time were not able to see how they could preserve their belief in the centrality of humanity and accept the radical idea that the Earth was just one of several planets revolving about the Sun.

Most people do not consider infinity to be of life-threatening importance. Cantor was not imprisoned for heretical thinking. Infinity is a little too "out there" for most people to get worked up about. But, for those mathematicians who were deeply immersed in such issues, the idea of having many different infinities was a tremendous blow. Out of vague and ill-defined notions of sizes of sets, Cantor distilled the fundamental idea of one-to-one correspondence. This idea is so basic that one feels compelled to accept its logical consequences. But these consequences contradict the intuition of most people—until their intuition is adjusted, after which the concept of more than one infinity becomes perfectly natural. Thanks to Cantor, who reached out and considered the counterintuitive, no mathematician today has a problem encompassing the idea of multiple infinities.

Sometimes understanding a fact requires us to change our minds in a dramatic way. However, that initially counterintuitive fact may at some future date attain the level of intuitive truth. For the Greeks, the existence of irrational numbers presented such a challenge. For a century now, mathematicians have come to understand the hierarchy of infinities. Time and again the bizarre and rare, after their discovery and assimilation, become the natural and familiar. These mental transitions are some of the great joys of thought.

Unequal Decimals

We now turn to the task of demonstrating that the set of real numbers has a strictly larger cardinality than the set of natural numbers. In other words, we now show that the set of real numbers is more infinite than the set of natural numbers.

Recall that each real number can be expressed as an infinitely long decimal expansion. For example,

$$243.4766668754468008876728758493457884445321\ldots$$

is a real number. Before moving forward, we must first make an easy observation about real numbers. Suppose we examine two decimal numbers, but we cover all the digits with question marks, except for the digit in the fifth place after the decimal point. So, we have two funny-looking numbers: ??.????2????... and ??.????4????.... We can't identify these numbers, because we can read only the fifth digit after the decimal point. But one thing we do know is that these two numbers are different. If they were the same, we could not have a 2 in the fifth place after the decimal point of one number and a 4 in the fifth place in the other. Likewise, if we have two numbers and one has a 2 and the other has a 4 in the 87th place after the decimal, then the two numbers must be different. This observation is not hard to understand, but it is a key to Cantor's reasoning.

Two Long Lists

Cantor proved that there are more real numbers than natural numbers through a clever, yet simple, idea. His basic strategy was to attempt an impossible task in order to understand why it couldn't be done. If the set of real numbers and the set of natural numbers have the same cardinality, then there would be a one-to-one correspondence between them. So his idea was to list the natural numbers down the left-hand side of a page, list reals in the right-hand column, and then show how to construct a real number that could not appear on the list. He showed that, once we commit ourselves to a list of reals in the right-hand column, one real number corresponding to each natural number, then we can describe a real decimal number that does not appear anywhere on that infinite list. So, we could not have listed all the real numbers in the right-hand column. Thus, the natural numbers and the real numbers could not be put in one-to-one correspondence, and so there are more real numbers than natural numbers.

Imagine a barrel containing all the natural numbers and another barrel containing all the real numbers. We will now reach in and grab the natural numbers one at a time and in order, pairing each with a real number we grabbed from the other barrel. We will then record the pairing, grab another two, and repeat. This procedure creates a list of all the natural numbers in one column and a list of real numbers in the other column. To illustrate this process, the pairing might begin like this:

Natural Numbers		Real Numbers
1	⟷	0.5562736349561738492134...
2	⟷	142.0273298163847273471873...
3	⟷	7.6123598736482351919723...
4	⟷	238.1852193647891209251902...
5	⟷	−0.00083738265191836548713...
6	⟷	31.847222356754445669033466...
7	⟷	658.3333333333543356708632...
8	⟷	−37.83958382139857446882891...
⋮	⋮	⋮
11	⟷	29.99907982742111199853769...
⋮	⋮	⋮

We can view this correspondence as two infinitely long columns: On the left we have a complete list of all natural numbers, and on the right we have a list of real numbers. We are now wondering whether every single real number will appear somewhere in the right-hand list. If the set of real numbers and the set of natural numbers have the same cardinality, then it would be possible to list all the reals in some order—one for each natural number. But, in fact, we will construct a real number in decimal form that does not appear anywhere in the right-hand column. That is, we will show that there are so many more real numbers than natural numbers that given *any* pairing between the natural numbers and the reals, a real number will always be left out—it is impossible to produce a one-to-one correspondence. Put another way, if we have the natural numbers in one barrel and the real numbers in another barrel, and we remove one natural number and one real number, pair them, and repeat, then after we run out of natural numbers, there will be real numbers left over! In fact, most of the real numbers would still be left in the barrel.

A Missing Real

We are going to write down a particular real number that we will call M, for "missing." We will write it in its decimal expansion. Our number M will be between 0 and 1, so its decimal expansion begins with 0.??? . . . . Now we must decide what the digits "??? . . . " are. Each digit will be one of two possibilities: a 2 or a 4. We will decide on the digits of our number M one at a time, successively, so we must be patient. We now describe the criterion by which we choose each digit of our number M.

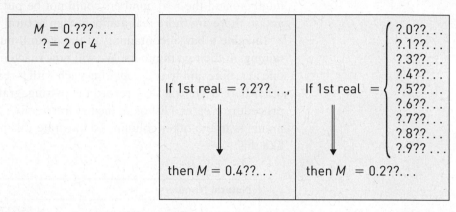

Determining the first digit of M by considering the first digit of the first real number on our list. Either way we know that M and the real number are not equal because the first digit after the decimal point is different.

We start with the first digit after the decimal point. Remember that we have a table that pairs one real number with each natural number. So some real number is paired with the natural number 1. We look at the first digit following the decimal point. Although this insight will not shake the very foundations of your universe, we boldly state that there are only two possibilities for that first digit: It is either 2 or it is not 2. We will use the first digit of that first real number to decide on the first digit of our number M—the real number we are building. If the first digit of that first real number is 2, then we will set the first digit after the decimal point of our number M to be 4. If, however, the first digit of that first real number is not 2, then we will set the first digit after the decimal point of our number M to be 2. Observe that, no matter what digits come next in M, we know for sure that the number M will not equal the real number paired with 1. Why? Because M and the real number paired with 1 have different first digits after the decimal point!

How will we define the digit that is in the second place after the decimal point of our number M? We take a look at the real number paired with the natural number 2, see what its second-place digit after the decimal point is, and ask if it equals 2. If that digit is 2, then we will set the second digit of our number M to be 4. If, however, that digit is not 2, we will set the second digit of our number M to be 2. Notice that we have defined M such that M's second digit after the decimal point is not the same as the second digit after the decimal point of the real number corresponding to 2. In particular, M cannot equal the second real number in the list—the real number corresponding to 2.

We continue to define the digits of M in this fashion. So, for example, to determine the 11th digit of M, we look at the 11th digit in the real number that is paired with the natural number 11. If that digit is 2, then we define the 11th digit of our number M to be 4; if that digit is not 2, then we define the 11th digit of our number M to be 2.

So Far, So Good?

To determine whether this process is clear, let's look at the lists of natural numbers and real numbers in the previous table and write down what M would be. The answer appears in the next paragraph, so write your answer first. Why is Cantor's argument referred to as Cantor's *diagonalization proof?*

If the one-to-one correspondence is the one given in the preceding section, then M would equal 0.24442424 . . . Incidentally, M's 11th digit after the decimal point is 4. Notice that, no matter what the given correspondence is, the number M will have only 2's and 4's in its decimal expansion.

Is *M* Really Missing?

At this point, we know how to construct M if we are given a table with natural numbers in one column, each corresponding to a real number. Of course, M is a real number. Could M appear anywhere on this list of reals? If so, where could it appear?

Is M the first real number on the list—the real number that corresponds to 1? No, because we selected the first digit of M so that it differs from the first digit of the real number paired with the natural number 1.

Is M the second number on the list? No, because we selected the second digit of M so that it differs from the second digit of the number associated with the natural number 2.

Is M the 1,582,987th number in the list? No, because we selected the 1,582,987th digit of M to differ from the 1,582,987th digit of the real number paired with the natural number 1,582,987.

What does all this mean? M cannot equal any real number in the table; therefore, M is not on the list! If we had entered different reals, then we would build a different M. But the point is that, for any particular attempt we make to list the natural numbers on the left and correspond a real number with each natural number, we can create a decimal number M that is not on that list. In other words, it is impossible for any correspondence of natural numbers and reals to contain all the real numbers. In particular, there is no one-to-one correspondence between the natural numbers and the reals!

~~Infinity~~ Infinities

We have just shown that, if we are given any correspondence from the natural numbers to the real numbers, then we can produce a real number that does not correspond to any one of the natural numbers. So, no matter how we try to pair numbers from the two sets, we will always have real numbers left over. Therefore, there are **different sizes of infinity!** We have found two different infinities: The cardinality of the natural numbers is **not the same** as the cardinality of the real numbers, even though both sets are infinite. There are more real numbers than there are natural numbers, or, phrased another way, the cardinality of the real numbers is larger than the cardinality of the natural numbers, even though the natural numbers are already infinite. Incredible!

Cantor's Theorem.
There are more real numbers than natural numbers.

The idea of infinities being larger than other infinities is not easy to digest. We must think about it, work through the preceding argument many times over, and explain it to someone. It is a great challenge to try to understand something that appears to contradict our personal hunches and intuition, especially for members of the now defunct "infinity is infinity" camp. We must master the logical arguments so solidly that they become irrefutable. Only then will our preexisting biases give way to a new idea of the infinite. This process requires much effort, but is well worth the investment.

CANTOR'S DIAGONALIZATION argument shows that, for any given correspondence from the natural numbers to the reals, we can always construct a new real number that does not appear on the list—that is, that does not correspond to any one of the natural numbers. Consequently, the cardinality of the real numbers is not the same as the cardinality of the natural numbers. There are more real numbers than natural numbers.

What strategy allowed us to discover a proof that there are more real numbers than natural numbers? We began by looking carefully at what must be true if the real numbers and the natural numbers had the same cardinality. We wrote down a possible correspondence—naturals on the left, reals on the right—and thought about the question: Are any real numbers missing? We noticed that real numbers have infinitely many digits past the decimal point, and for two real numbers to be different, they need to differ only in one of those places. Keeping this in mind while looking at the lists led us to the diagonalization argument.

When we feel that something is impossible, or if we just don't know whether it is true or false, a good way to find the truth is to explore carefully what would happen if the impossible or unknown were true. Systematically design what-if scenarios and play them to their conclusions.

> *I can see it, but don't believe it!.*
>
> GEORG CANTOR

Life Lessons

Explore consequences of possibilities.

MINDSCAPES   Invitations to Further Thought

In this section, Mindscapes marked **(H)** *have hints for solutions at the back of the book. Mindscapes marked* **(ExH)** *have expanded hints at the back of the book. Mindscapes marked* **(S)** *have solutions.*

Developing Ideas

1. **Shake 'em up.** What did Georg Cantor do that "shook the foundations of infinity"?

2. **Detecting digits.** Here's a list of three numbers between 0 and 1:

   0.12345

   0.24242

   0.98765

   What's the first digit of the first number? What's the second digit of the second number? What's the third digit of the third number?

3. **Delving into digits.** Consider the real number

   0.12345678910111213141516. . . .

   Describe in words how this number is constructed. What's its 14th digit? What's the 25th digit? What's the 31st digit?

4. **Undercover friend (ExH).** Your friend gives you a list of three five-digit numbers, but she only reveals one digit in each:

   3????

   ?8???

   ??2??

   Can you describe a five-digit number you know for certain will not be on her list? If so, give one; if not, explain why not.

5. **Underhanded friend.** Now your friend shows you a new list of three five-digit numbers, again with only a few digits revealed:

   6????

   ?5???

   ?????

   Can you describe a five-digit number you know for certain will not be on her list? If so, give one; if not, explain why not.

Solidifying Ideas

6. **Dodgeball.** Revisit the game of Dodgeball from Chapter 1, "Fun and Games." Play it several times with several people. Get the strategy down, and then explain to your opponents the underlying principle. Record the results of the games.

7. **Don't dodge the connection (S).** Explain the connection between the Dodgeball game and Cantor's proof that the cardinality of the reals is greater than the cardinality of the natural numbers.

8. **Cantor with 3's and 7's.** Rework Cantor's proof from the beginning. This time, however, if the digit under consideration is 3, then make the corresponding digit of M a 7; and if the digit is not 3, make the associated digit of M a 3.

9. **Cantor with 4's and 8's.** Rework Cantor's proof from the beginning. This time, however, if the digit under consideration is 4, then make the corresponding digit of M an 8; and if the digit is not 4, make the associated digit of M a 4.

10. **Think positive.** Prove that the cardinality of the positive real numbers is the same as the cardinality of the negative real numbers. (*Caution:* You need to describe a one-to-one correspondence; however, remember that you cannot list the elements in a table.)

11. **Diagonalization.** Cantor's proof is often referred to as "Cantor's diagonalization argument." Explain why this is a reasonable name.

12. **Digging through diagonals.** First, consider the following infinite collection of real numbers. Describe in your own words how these numbers are constructed (that is, describe the procedure for generating this list of numbers). Then, using Cantor's diagonalization argument, find a number not on the list. Justify your answer.

    0.12345678910111213141516171718...
    0.24681012141618202224262830302...
    0.36912151821242730333639424245...
    0.48121620242832364044485256660...
    0.51015202530354045505560657070...

13. **Coloring revisited (ExH).** In Mindscape 35 of the previous section we considered the following infinite collection of circles and all the different ways of coloring the circles with either red or blue markers. Show that the set of all possible circle colorings has a greater cardinality than the set of all natural numbers.

    ...

14. **A penny for their thoughts.** Suppose you had infinitely many people, each one wearing a uniquely numbered button: 1, 2, 3, 4, 5, ... (you can use all the people in the Hotel Cardinality if you don't know enough people yourself). You also have lots of pennies (infinitely many, so you're *really* rich; but don't try to carry them all around at once). Now you give each person a penny; then ask everyone to flip his or her penny at the same time. Then ask them to shout out in order what they flipped (H for heads and T for tails). So you might hear: HHTHHTTTHTTHTHTHTHTHHHHTH... or you might hear THTTTH THHTTHTHTTTTTHHHHTHTHTH..., and so forth. Consider the set of all possible outcomes of their flipping (all possible sequences of H's and T's). Does the set of possible outcomes have the same cardinality as the natural numbers? Justify your answer.

15. **The first digit (H).** Suppose that, in constructing the number M in the Cantor diagonalization argument, we declare that the first digit to the right of the decimal point of M will be 7, and then the other digits are selected as before (if the second digit of the second real number has a 2, we make the second

digit of M a 4; otherwise, we make the second digit a 2, and so on). Show by example that the number M may, in fact, be a real number on our list.

Creating New Ideas

16. **Ones and twos (H).** Show that the set of all real numbers between 0 and 1 just having 1's and 2's after the decimal point in their decimal expansions has a greater cardinality than the set of natural numbers. (So, the number $0.112111122212122211112\ldots$ is a number in this set, but $0.1161221212122122\ldots$ is not, because it contains digits other than just 1's and 2's.)

17. **Pairs (S).** In Cantor's argument, is it possible to consider pairs of digits rather than single digits? That is, suppose we look at the first two digits of the first real number on our list; and, if they are not 22, then we make the first two digits of M be 22. If the first two digits are 22, then we make the first two digits of M be 44. Similarly, let the next two digits of the next real number on our list determine the next two digits of M, and so on. If this procedure would still produce a number M not on our list, then provide the details for such a method. If this procedure does not work, explain or illustrate why it does not.

18. **Three missing.** Given a list of real numbers, as in the Cantor argument, explain how to construct three different real numbers that are not on the list.

19. **No Vacancy (H).** Recall the Hotel Cardinality, described in Mindscapes 16, 17, and 18 of the previous section. Create a collection of people so that it would be impossible for the night manager to give each person a room. Thus, for a really big group of people, a No Vacancy sign (or actually a Not Enough Room sign) might actually be necessary. Explain why it is not possible to give each person from your group a room.

20. **Just guess.** This is just a "guessing question." Do you think there are sets whose cardinality is actually larger than that of the set of real numbers? Or, do you think the infinity of reals is the largest infinity? Just make a guess and informally explain it.

Further Challenges

21. **Nines.** Would Cantor's argument work if we used 2 and 9 instead of 2 and 4 as the digits? That is, for each digit we ask whether the digit is a 2. If it is 2, we make the analogous digit of M a 9, and otherwise we make the digit a 2. Using this method, we are not guaranteed that the number M we construct is not on our list. Provide a scenario in which the constructed number is on the list! (*Hint:* Remember from Chapter 2 that $0.1999999\ldots = 0.2$. The number 9 is key here.)

22. **Missing irrational.** Could you modify the diagonalization procedure so that the missing real you produce is a rational number? How could you modify the diagonalization argument so that the missing real number you produce is an irrational number? (*Hint*: Using the construction in this section, each digit of M is a 2 or a 4. Modify the construction so the 10th place, the 100th place, the 1000th place, and so on, are either a 3 or a 5, whereas the rest are 2's and 4's. Why will such a number be irrational?)

In Your Own Words

23. **Personal perspectives.** Write a short essay describing the most interesting or surprising discovery you made in exploring the material in this section. If any

material seemed puzzling or even unbelievable, address that as well. Explain why you chose the topics you did. Finally, comment on the aesthetics of the mathematics and ideas in this section.

24. **With a group of folks.** In a small group, discuss and work through Cantor's argument showing that the set of real numbers has a greater cardinality than the set of natural numbers. After your discussion, write a brief narrative describing the argument in your own words.

For the Algebra Lover

Here we celebrate the power of algebra as a powerful way of finding unknown quantities by naming them, of expressing infinitely many relationships and connections clearly and succinctly, and of uncovering pattern and structure.

25. **Logging cardinality.** The function graphed here is the logarithmic function $y = \log x$. Notice how the y-axis is a vertical asymptote, that is, the graph gets closer and closer to the y-axis but never touches it. Use this graph to show that the collection of all positive real numbers has the same cardinality as the collection of all real numbers.

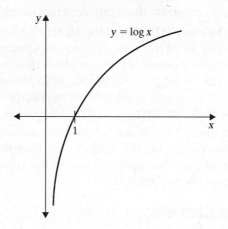

26. **U-graph it.** Using a graphic or on-line calculator, sketch the graph $y = \dfrac{x}{x^2 - 1}$

Use this graph to show that the collection of real numbers strictly between -1 and 1 (so all the numbers bigger than -1 but less than 1) has the same cardinality as the collection of all real numbers.

27. **Is a square a one-to-one correspondence? (H)** Sketch a graph of the function $y = x^2$. Using your graph, determine if this function provides a one-to-one correspondence between the real numbers on the x-axis and the real numbers on the y-axis. If it does give a one-to-one correspondence between the numbers on the two axes, justify your answer using your graph. If not, use your graph to explain why that function is not a one-to-one correspondence. In particular, does every number on the x-axis correspond to exactly one number on the y-axis? Does every number on the y-axis correspond to exactly one number on the x-axis?

28. **Is a cube a one-to-one correspondence?** Sketch a graph of the function $y = x^3$. Using your graph, determine if this function provides a one-to-one

correspondence between the real numbers on the x-axis and the real numbers on the y-axis. If it does give a one-to-one correspondence between the two axes, justify your answer using your graph. If not, use your graph to explain why that function is not a one-to-one correspondence. In particular, does every number on the x-axis correspond to exactly one number on the y-axis? Does every number on the y-axis correspond to exactly one number on the x-axis?

29. **Find the digit.** Your friend is thinking of a real number r between 0 and 1 in its decimal form. She tells you that she can write down the digits by evaluating the function $f(x) = 3x^2 + 5x - 3$ at $x = 1, 2, 3$, etc., and writing those answers in order. That is, $f(1) = 5$ and $f(2) = 19$, so her number r starts as 0.519. Notice that the third digit to the right of the decimal point is 9. Find the tenth digit to the right of the decimal point of r.

3.4 TRAVELS TOWARD THE STRATOSPHERE OF INFINITIES

The Power Set and the Question of an Infinite Galaxy of Infinities

William Blake, *The Ancient of Days*, 1794. People have always tried to measure the infinite.

> *There is no smallest among the small and no largest among the large; but always something still smaller and something still larger.*
>
> ANAXAGORAS

The theory of infinity described in the previous sections is truly beautiful. Why? Because we were able to analyze something logically that initially appeared unanalyzable, and our intellectual journey revealed new and surprising structure to ideas that were previously ill-formed and vague. Logical reasoning, interwoven with well-defined objects and a clever idea, completely disproved an intuitive feeling and replaced it with the solid foundation of a new understanding of the infinite. That's the power of mathematics, or perhaps more accurately, the power of mathematical thinking—the ability to identify and describe objects at hand, create a logical framework wherein these objects can exist, understand them deeply, analyze them carefully, and arrive at a valid conclusion. Of course, this way of thinking is not restricted to mathematical issues. We can use these techniques of logical reasoning every day to help resolve life issues and enrich our lives.

So, we've just discovered that there is an infinity larger than another infinity. Specifically, we have seen that the cardinality of the set of real numbers is greater than the cardinality of the set of natural numbers. This surprising discovery now leads to a vast number of questions—here are just a few.

1. Is there an infinity that is greater than the infinity of the set of natural numbers yet less than the larger infinity of the set of real numbers? (Is the cardinality of the reals the next bigger infinity after the cardinality of the natural numbers?)

2. Is there an infinity greater than the infinity of the set of real numbers?

3. Are there infinitely many different sizes of infinity?

4. Is there a largest infinity—one that encompasses all others?

By the end of this section, we will have answered three of these questions. The remaining question is much more complicated and has a surprising answer of sorts. Think about these questions, then try to guess which question has the unusual resolution. It's a great intellectual challenge in life to determine which questions are more difficult to answer than others, even if we are unable to answer them.

The strategy we employ here is a potent method for making progress in any situation. In other words, once we have discovered an idea, let's milk it for all it's worth. Recall that the diagonalization argument was discovered first in the Dodgeball game and then was used to show that there are more real numbers than natural numbers. Can we use the same idea to see whether we can find infinities even greater than the reals? If an idea provides an answer to one question, it is often powerful enough to answer many related questions as well.

Life Lesson

By attempting to order the difficulty of issues we are taking the first steps toward resolving them.

Endless Infinities

We shall soon see that no infinity is without superiors. An infinity's very existence implies another infinity vaster still. In fact, the collection of all infinities has no bound. There are infinitely many different sizes of infinity—so vast a collection that no infinity can even count the number of different sizes of infinity!

We all know that there is no integer larger than all the others. Each integer is bettered by its successor. With infinite sets, a similar pattern occurs. As we have done before, we use the phrase *cardinality of a set* to mean the size of the set. For any infinite set we are given, we will show how to construct a related but larger infinite set with cardinality strictly bigger than the cardinality of the set we were given. When we have absorbed and understood this construction, we will see that Cantor's diagonalization proof, which showed that there are more real numbers than natural numbers, provides a model for how to generate ever larger infinities. In fact, both arguments have at their root the exact same idea used in Dodgeball. Sometimes an answer to one question immediately provides the answers to more general questions.

Sets Within Sets

We want to start with a set and then create a larger set, but we begin by creating smaller sets. One method of generating a new set is to look at a few elements in the starting set. Let's solidify our thinking by looking at a specific example.

Suppose that **Suits** is the set containing ♣,♦,♥,♠. We could write this set as **Suits** = {♣,♦,♥,♠} (the fancy brackets indicate that **Suits** is a set containing all the symbols inside). We can make new sets by selecting just some of the elements from **Suits**. For example, we could define a set **Reds** to be the set containing the two elements ♦,♥ and a set **Blacks** to be the set containing ♣,♠. In other words, **Reds** = {♦,♥} and **Blacks** = {♣,♠}.

So, the elements of both **Reds** and **Blacks** are all from the original set **Suits**. Thus, the sets **Reds** and **Blacks** are each an example of a *subset* of **Suits**. In particular, given a set S, we say that another set T is a *subset* of S if every element of T is also in S. So, a subset of a set is just a collection of some of the elements from the original set. How many different subsets of the set of all card suits **Suits** are there? Here is the complete list of all **Suits'** subsets (there are a lot):

$$\{\ \},$$
$$\{\clubsuit\}, \{\diamondsuit\}, \{\heartsuit\}, \{\spadesuit\},$$
$$\{\clubsuit, \diamondsuit\}, \{\clubsuit, \heartsuit\}, \{\clubsuit, \spadesuit\}, \{\diamondsuit, \heartsuit\}, \{\diamondsuit, \spadesuit\}, \{\heartsuit, \spadesuit\},$$
$$\{\clubsuit, \diamondsuit, \heartsuit\}, \{\clubsuit, \diamondsuit, \heartsuit\}, \{\clubsuit, \diamondsuit, \spadesuit\}, \{\diamondsuit, \heartsuit, \spadesuit\},$$
$$\{\clubsuit, \diamondsuit, \heartsuit, \spadesuit\}.$$

We see all the collections of single elements, all possible pairs, and all triples. But looking at the preceding list, we may wonder about the first and the last subsets. The first looks strange: It has nothing in it. It is called the *empty set*—the set that contains nothing. The empty set is a subset of every set. The other subset that may look strange is the last one; it consists of the entire set itself. Well, according to our definition of *subset*, that last set is a subset of **Suits**. Those are all the subsets of **Suits**. By counting, we see that there are 16 subsets of **Suits**.

How Many Subsets?

Now we want you to experiment. Write down all the subsets of the set containing two coins: a penny and a dime. How many subsets are there?

Write down all the subsets of the set of musicians in a trio. How many subsets are there?

Do you see a pattern between the number of elements in a set and the number of subsets the set has?

© TriggerPhoto/iStockphoto

Guess (or better yet "conjecture") how many subsets there are of a set containing seven elements. Did you guess 14? Did you guess 49? Did you guess 32? Did you look at your previous examples and try to find the pattern? If you guessed any of these numbers or some other plausible answer, then you were doing the right thing. Making concrete, reasonable guesses is an effective strategy for figuring something out. Even if it's wrong, a good guess gives us ideas. We can analyze why it failed and then make appropriate adjustments. The repeated process of guessing, analyzing reasons for failure, and making adjustments often leads to insight and the correct answers.

Let's return to the question: How many subsets does a set of seven things have? If you guessed 128, you've probably discerned the pattern.

How many subsets does a set containing 583 elements have? Did you guess

31658291388557380359744322690514840324496812684955115509000007
11798908448136360789978004993358391097586685019425300658354369
74724391269341507853042325432566683503348940 8?

If so, then you really need to get out a bit more. Okay, seriously, notice that, in each case, the number of subsets of a set is a power of 2. That is, $2^4 = 2 \times 2 \times 2 \times 2 = 16$, $2^7 = 2 \times 2 \times 2 \times 2 \times 2 \times 2 \times 2 = 128$, and $2^{583} = 31{,}658{,}291$, blah, blah, blah, 489,408. Why is the number of subsets always a power of 2?

The Power of 2

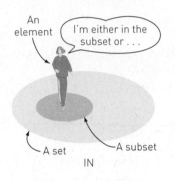

An element

I'm either in the subset or . . .

A set

A subset

IN

I'm not in the subset. There are only those two possibilities.

NOT IN

Let's think about that last question for a moment. Suppose we were an element of a set. For any particular subset we think of, we know that it was made by picking out some of the elements from the original set. Therefore, there are exactly two possibilities: Either we are in that subset (we got picked) or we are not in that subset (we didn't get picked). This basic observation is the key to unlocking the pattern.

Let's start with a set containing one element and work our way up to larger sets. Suppose we have a set consisting of one element—say the set is {♣}. Then there are exactly two subsets corresponding to the two possibilities—either ♣ is not in the subset or ♣ is in the subset. So the subsets of {♣} are

$$\{\ \},$$
$$\{♣\}.$$

Now let's suppose we consider the two-element set {♣, ♦}. Its subsets consist of all the subsets that do not contain ♦ together with all the subsets that do contain ♦ (remember that, for any subset, either ♦ is in that subset or it is not in that subset). Every subset of {♣, ♦} that does not contain ♦ must be a subset of just {♣}, so those subsets are { } and {♣}. We could create the subsets that do contain ♦ by throwing ♦ into each of those previous subsets to get the four subsets of {♣, ♦}:

{ }, {♦} (notice this is just { } where we throw in ♦),
{♣}, {♣, ♦} (notice this is just {♣} where we throw in ♦).

Now let's consider the three-element set {♣, ♦, ♥}. Its subsets consist of all the subsets that do not contain ♥ together with all the subsets that do contain ♥. But, as before, every subset that does not contain ♥ is a subset of {♣, ♦}. So, we have already listed all the subsets that do not contain ♥. We could create the remaining subsets by including ♥ in each of those subsets to get the eight subsets of {♣, ♦, ♥}:

$$
\begin{array}{ll}
\{\ \}, & \{♥\}, \\
\{♣\}, & \{♣, ♥\}, \\
\{♦\}, & \{♦, ♥\}, \\
\{♣, ♦\}, & \{♣, ♦, ♥\}.
\end{array}
$$

Following this pattern, we see that adding a new element doubles the number of subsets. So, if our set has one element, it has $2^1 = 2$ subsets. If we have a set with two elements, it has twice as many subsets—that is, $2^2 = 4$ subsets. A three-element set has $2^3 = 8$ subsets, and so on. We can record these observations as a theorem.

The Subset Count.

A set containing n elements has 2^n subsets.

The Power Set

Okay, suppose we are given a set (let's call the set S). We are now about to build a new set using our idea of subsets. This new set consists of all the subsets of S. That is, this new set is a set whose elements are actually subsets of S. This set seems strange at first, but let's absorb the idea. A set contains things that we call *elements;* however, the elements could be anything. They could be numbers, or suits of playing cards, or musicians. The elements could even be sets themselves. So, for example, if *Suits* = {♣, ♦, ♥, ♠}, then the set of all subsets of *Suits* is the set consisting of the following 16 things:

$$
\begin{array}{c}
\{\ \}, \\
\{♣\}, \{♦\}, \{♥\}, \{♠\}, \\
\{♣, ♦\}, \{♣, ♥\}, \{♣, ♠\}, \{♦, ♥\}, \{♦, ♠\}, \{♥, ♠\}, \\
\{♣, ♦, ♥\}, \{♣, ♦, ♠\}, \{♣, ♥, ♠\}, \{♦, ♥, ♠\}, \\
\{♣, ♦, ♥, ♠\}.
\end{array}
$$

Given a set S, we define the *power set of S*, written as $\mathscr{P}(S)$, to be the set consisting of all possible subsets of S. In other words, the power set of a set contains all the different possible subcollections we can create from the original set. So, the power set of {♣, ♦, ♥, ♠} is the set consisting of the 16 subsets of {♣, ♦, ♥, ♠} that are listed above. Notice that, in the preceding example, {♦, ♥} is an element of $\mathscr{P}$(*Suits*); {♠} is another element of $\mathscr{P}$(*Suits*); { } is an element of $\mathscr{P}$(*Suits*). Remember that the elements of $\mathscr{P}(S)$ are themselves sets. So $\mathscr{P}$(*Suits*) would be written within big brackets, and each of its elements would also have brackets. Here are some sets and their associated power sets.

Set S	Power set $\mathcal{P}(S)$
$\{1, 2\}$,	$\{\{\ \}, \{1\}, \{2\}, \{1, 2\}\}$
$\{a, b, c\}$,	$\{\{\ \}, \{a\}, \{b\}, \{c\}, \{a, b\}, \{a, c\}, \{b, c\}, \{a, b, c\}\}$
$\{♣, ♦, ♥, ♠\}$	$\{\{\ \}, \{♣\}, \{♦\}, \{♥\}, \{♠\}, \{♣, ♦\}, \{♣, ♥\}, \{♣, ♠\},$
	$\{♦, ♥\}, \{♦, ♠\}, \{♥, ♠\}, \{♣, ♦, ♥\}, \{♣, ♦, ♠\},$
	$\{♣, ♥, ♠\}, \{♦, ♥, ♠\}, \{♣, ♦, ♥, ♠\}\}$

Notice that we could rephrase the Subset Count theorem by saying that if a set has n elements, its power set has 2^n elements.

> ## Cantor's Power Set Theorem.
> *For any set S, (finite or infinite), the cardinality of the power set of S, $\mathcal{P}(S)$, is strictly greater than the cardinality of S.*

Power Sets Are Big

In all the examples we have seen, the number of elements in the power set of a set is much greater than the number of elements in the original set. For example, there are four elements in the set of all suits in a deck of cards; however, there are $2^4 = 16$ elements in the power set of this set. Let's face it, 16 is much bigger than 4. We know this observation is pretty obvious, but it's important—just keep reading.

It turns out that power sets are *always* larger than the original set *even if the original set is already infinite!* This theorem was one of Georg Cantor's important achievements.

Let's think about what this theorem means and its consequences. If we take an infinite set S and look at the set of all its subsets (that is, look at its power set $\mathcal{P}(S)$), the cardinality of the set $\mathcal{P}(S)$ will be greater than the cardinality of S. How can we use this fact to make more and more increasingly larger infinities? This is a challenging question! Please think about it and attempt to answer it before reading on.

An Infinity of Infinities

Suppose we take the set of natural numbers and call this set $\mathbb{N}$. Of course, $\mathbb{N}$ is an infinite set. Let's look at the power set $\mathcal{P}(\mathbb{N})$—that is, the set of all subsets of $\mathbb{N}$. Here are some examples of elements of $\mathcal{P}(\mathbb{N})$ (remember that each element of $\mathcal{P}(\mathbb{N})$ is a subset of $\mathbb{N}$): $\{1\}, \{2\}, \{3\}, \ldots, \{1, 2\}, \{1, 3\}, \{2, 3\}, \ldots$, each triple of numbers, each set of four numbers, each finite set of numbers. But also $\{2, 4, 6, 8, \ldots\}$, the subset consisting of all even numbers, is one element of $\mathcal{P}(\mathbb{N})$. The collection of prime numbers forms one element of $\mathcal{P}(\mathbb{N})$. The collection of Fibonacci numbers is one element of $\mathcal{P}(\mathbb{N})$. Each subset of $\mathbb{N}$ is an element of $\mathcal{P}(\mathbb{N})$.

Cantor's Power Set Theorem says that $\mathcal{P}(\mathbb{N})$ is a set that has more elements than $\mathbb{N}$ itself! Now $\mathcal{P}(\mathbb{N})$ is itself a set. So, suppose we consider its power set! That is, suppose we consider the set of all subsets of the set of all subsets of $\mathbb{N}$.

We would call this new set $\mathcal{P}(\mathcal{P}(\mathbb{N}))$. Naming any element of the set $\mathcal{P}(\mathcal{P}(\mathbb{N}))$ requires considerable care. But what do we know about the size of $\mathcal{P}(\mathcal{P}(\mathbb{N}))$? By Cantor's Power Set Theorem, we know that the size of $\mathcal{P}(\mathcal{P}(\mathbb{N}))$ is greater than the size of $\mathcal{P}(\mathbb{N})$. So, we have just found a third infinity that is greater than the previous two!

Of course, we could repeat this process indefinitely. For example, we could consider the set of all subsets of the set of all subsets of the set of all subsets of N, denoted by $\mathcal{P}(\mathcal{P}(\mathcal{P}(\mathbb{N})))$. The size of this set is greater than the size of $\mathcal{P}(\mathcal{P}(\mathbb{N}))$. So, we now see a way of producing infinitely many sets, each with a larger cardinality than the last. (Just for fun, try describing, in words, the set $\mathcal{P}(\mathcal{P}(\mathcal{P}(\mathcal{P}(\mathbb{N})))))$. It's a kind of mathematical tongue twister.)

Why the Power Set Theorem Is True

Cantor's Power Set Theorem opens the door to an infinity of infinities. Let's see why the theorem is true. First we'll restate the theorem.

Cantor's Power Set Theorem.

For any set S, (finite or infinite), the cardinality of the power set of S, $\mathcal{P}(S)$, is strictly greater than the cardinality of S.

A Plan of Attack

Remember that cardinality is determined by one-to-one correspondences. Suppose we look at a correspondence that has elements of S on one side and elements of the power set of S on the other. For example, let's look at the set $S = \{a, b, c, d, e, f, \ldots\}$ (we are just letting the letters a, b, c, d, and so on, stand for some of the elements of the set S). A pairing between the set S and its power set $\mathcal{P}(S)$ might look like the following (remember, elements of $\mathcal{P}(S)$ are subsets of S):

Elements of S	Elements of $\mathcal{P}(S)$ (subsets of S)
a ⟷	$\{c, f\}$
b ⟷	$\{a, b, d, g, j\}$
c ⟷	$\{a, g, r, s, t, u\}$
d ⟷	$\{\ \}$
e ⟷	S
$\vdots$	$\vdots$
x ⟷	some subset of S
$\vdots$	$\vdots$

A sample correspondence between elements of S and elements of $\mathcal{P}(S)$.

To see that the cardinality of $\mathcal{P}(S)$ is greater than the cardinality of S, we will show how to construct some subset of S (in other words, some element of $\mathcal{P}(S)$) that does not appear anywhere in the right-hand column. That is, given *any* pairing between the elements of S and the elements of $\mathcal{P}(S)$, we will show that there will always be elements of $\mathscr{P}(S)$ that are not paired with a partner from S. If this were a prom, all the elements of S would have dance partners from $\mathscr{P}(S)$, but there would be people from $\mathscr{P}(S)$ who would have no one from S with whom to dance. It's sad but true—there is no one-to-one pairing; we will always run out of people from S before running out of people from $\mathscr{P}(S)$ (even if there are infinitely many people in each set—one heck of a big prom!).

Since this method of constructing an unpartnered element of $\mathscr{P}(S)$ will work for any pairing we make, we will know that it is impossible to have a one-to-one correspondence between S and all of $\mathscr{P}(S)$. We will see that this method is really a recycled version of the idea that Cantor used to show that there are more real numbers than natural numbers. Thereby, Cantor not only proved two of the great theorems of mathematics, but he also started the recycling craze a hundred years before its time—terrific ideas deserve to be reused, expanded, exploited, modified, generalized, and developed. This strategy of mining ideas for all they are worth is a powerful take-home lesson from Cantor.

Life Lesson

Whenever we have an idea, we should push it as far as possible beyond its original context.

An Upset Subset—Building a Missing Subset

Remember that we're assuming we have a fixed correspondence that associates with each element of S a subset of S—that is, an element of $\mathscr{P}(S)$. Our goal is to construct an element of $\mathscr{P}(S)$ that is not paired with any element of S. To construct this missing subset of S, we consider each element of S and decide whether that element will be included in this mystery subset or not. Recall that in Cantor's proof there are more real numbers than natural numbers; our choice of the fifth digit, for example, of the number we were constructing guaranteed that the number we were constructing was not the one corresponding to the natural number 5. We did the same for each of the natural numbers and, thereby, concluded that the newly constructed number did not correspond to any natural number. Here we proceed in a similar manner.

Remember that we are given a pairing that associates each element of S with a subset of S, and that we now wish to construct a subset of S that doesn't correspond to any element of S in this pairing. Let's call this mystery subset **Mystery** and start constructing it by considering the element a from set S. We need to decide whether a should be in **Mystery** or excluded from **Mystery**. To answer this question, let's look at the set to which a corresponds in our given correspondence. As we see in the preceding table, a is paired with the subset $\{c, f\}$. Since a is not in $\{c, f\}$, we will put a in **Mystery**. Once we have made the decision to put a in **Mystery**, then we know for certain that whatever else **Mystery** may or may not contain, **Mystery** is not equal to $\{c, f\}$, because a is in **Mystery**, and a is not in $\{c, f\}$. Notice how similar this logic is to Cantor's previous argument.

Let's now move on to element b. How will we decide whether or not to put b in **Mystery**? Again let's look at the subset of S to which b corresponds. In the table we see that b corresponds to $\{a, b, d, g, j\}$. So should we put b in **Mystery**? No. Why? Because by making the decision to exclude b from **Mystery**, then whatever else may or may not be in **Mystery**, we know for sure that **Mystery** is not $\{a, b, d, g, j\}$, since we have decided that b is not in **Mystery**, whereas b is in $\{a, b, d, g, j\}$.

We're on a roll, so let's do a couple more. Should c be in **Mystery**? First we consult the table and see that c corresponds to $\{a, g, r, s, t, u\}$. So we will place c in

Mystery, since c is not in the subset to which c corresponds. Once we have decided to put c in **Mystery**, we know for certain that whatever **Mystery** turns out to be, it is not equal to the subset corresponding to c, since c is in **Mystery** and not in the subset corresponding to c.

Should d be in **Mystery**? Yes, because d is not in { } (the empty set). Notice that we're just using the same logic in each case.

Should e be in **Mystery**? No, because e is in S. Once again, the logic is the same.

Should x be in **Mystery**? Since we did not tell you the specific subset to which x corresponds, let's just write down our decision process. We will put x in **Mystery** if x is not in the subset corresponding to x, and we will not put x in **Mystery** if x is in the subset corresponding to x.

Finally, for the big finish, could **Mystery** appear anywhere in the right-hand column of our table? No, because if **Mystery** did appear, it would correspond to some specific element. That element would either be in **Mystery** or not be in **Mystery**. However, neither of those possibilities can be true. If that element were in **Mystery**, then it would be an element of the set it corresponds with and, consequently, *would not* be in **Mystery**. However, if that element were not in **Mystery**, then it would not be included in the set it corresponds to and, hence, *would* be in **Mystery**. Since that element can be neither in **Mystery** nor out of **Mystery**, it simply cannot exist. Thus, no element of S can correspond to **Mystery**; therefore, **Mystery** is a subset that cannot appear in the right-hand column of our table.

Since we have shown that no one-to-one correspondence can pair the elements of a set with its subsets, the cardinality of the power set cannot equal the cardinality of a set, and the proof of Cantor's theorem is complete.

The Whole Proof Shorter

In fact, the previous three paragraphs are the entire proof of Cantor's Power Set Theorem. If we had any set S at all, and any correspondence taking each element of S to a subset of S, then we could construct a **Mystery** subset that does not correspond to any element of S. How? For each element x of S, we will put x in **Mystery** if x is not in the subset corresponding to x, and we will not put x in **Mystery** if x is in the subset corresponding to x. By using that criterion, we know that **Mystery** cannot equal the subset corresponding to any element of S. (Notice that, in constructing the **Mystery** subset, we did not care whether S was finite or infinite.)

Since **Mystery** cannot correspond to any element of S, **Mystery** is not on the right side of our table, and the pairing cannot be a one-to-one correspondence. Therefore, there is no one-to-one correspondence, and we have proved that the power set of S must have a greater cardinality than S. So, we have proved Cantor's Power Set Theorem!

The Whole Proof Really Short

Just for fun, we will now write down the exact same proof using the extremely abbreviated, cryptic notation that makes mathematics succinct but difficult to read. To unravel and understand this one-line proof, a reader would have to produce the preceding explanation. We write it only for your amusement, so we will not bother to clearly define the terms or explain the notation. Feel free to use it to impress family and friends.

Complete Proof of Cantor's Power Set Theorem

Consider $f: S \to \mathcal{P}(S)$. Then $\{x \in S \mid x \notin f(x)\} \notin f(S)$.   *Q.E.D.*

Literal Translation

Consider a pairing from S to $\mathcal{P}(S)$. Then the particular subset consisting of all elements x from S, such that x is not an element of the set corresponding to x with the given pairing, is not paired with any element of S. *Quod erat demonstrandum* ("which was to be proved").

Just think: If we had thought of that first, we could have gotten a Ph.D. and become famous with less than a line of work. The importance and value of a new idea should not be measured by how much space is required to write it down. Length is not a good measure of depth.

The Answer Round

Remember that our entire discussion about power sets arose from our desire to answer some questions about infinity. Now we can conclude that there is indeed a set having greater cardinality than the cardinality of the set of real numbers—namely, the power set of the real numbers. This result answers the second question from the opening of this section. The set of all subsets of the real numbers is a collection even larger than the set of reals. By repeating this process (successively building the power set of the previous set), we see that there are infinitely many different sizes of infinity. This observation provides an answer to question 3 from the beginning of this section. What about questions 1 and 4?

Cardinality Comparisons

We saw that the cardinality of the set of natural numbers is smaller than the cardinality of the set of real numbers. The cardinality of the set of natural numbers is also smaller than the cardinality of the power set of the natural numbers. How do the cardinality of the real numbers and the cardinality of the power set of the natural numbers compare?

It turns out that the cardinality of the power set of the natural numbers, $\mathcal{P}(\mathbb{N})$, is exactly the same as the cardinality of the set of real numbers. Although the proof is not too hard, we suppress our desire to describe the details here. We mention that the cardinality of the reals is the same as the cardinality of $\mathcal{P}(\mathbb{N})$ because it allows us to state the following famous conjecture about the number of points on the real line.

The Continuum Hypothesis

Using Cantor's Power Set Theorem, we have seen how to create larger infinities. But how big are these infinities? Are we skipping a whole bunch of infinities in between the ones we are generating?

Question ▶ Is the cardinality of the set of real numbers the **next** larger infinity bigger than the cardinality of ℕ?

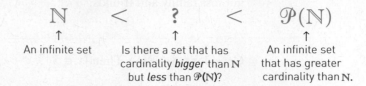

$$\mathbb{N} \quad < \quad ? \quad < \quad \mathscr{P}(\mathbb{N})$$

An infinite set — Is there a set that has cardinality *bigger* than ℕ but *less* than $\mathscr{P}(\mathbb{N})$? — An infinite set that has greater cardinality than ℕ.

We know that the cardinality of the real numbers is larger than the cardinality of ℕ, but perhaps there is an infinity between them. In other words, is there a collection of real numbers whose cardinality is strictly greater than the cardinality of the natural numbers yet strictly less than the cardinality of the real numbers? Is that possible? Is that impossible? The famous *Continuum Hypothesis* was a conjecture about this question.

Continuum Hypothesis.

There is no cardinality between the cardinality of the set of natural numbers and the cardinality of the set of real numbers (sometimes referred to as the continuum).

If you thought that the study of infinity was pretty wild so far, hold on to your hat. If you don't have a hat, get one quick and hold on. The amazing fact is that the Continuum Hypothesis is neither true nor false!

Please do not throw our book out the window; but if you do, please buy another copy. We have not taken leave of our senses. You read correctly, the Continuum Hypothesis is neither true nor false, and it can be *proven* that it is neither true nor false. Thus, question 1 from the beginning of this section was the one with the bizarre answer.

The Continuum Hypothesis pushes us, in some sense, right to the edge of mathematical truth. Kurt Gödel, in 1940, proved that it is impossible to disprove the Continuum Hypothesis; then in 1963, Paul Cohen proved that it is impossible to prove the Continuum Hypothesis. In other words, the Continuum Hypothesis can be neither proved nor disproved using the standard mathematical axioms and machinery! The Continuum Hypothesis is independent of our entire mathematical structure. This result is truly amazing. It brings us to the very edge of knowledge—the cliff of mathematics. Perhaps we should step back before we fall off.

Russell's Paradox

Before closing this section, given all the strange phenomena we have discussed, we should discuss a grand idea bigger than everything. Suppose we take every set of every size and simply imagine the set comprising every set—in other words, the ultimately largest set, since it would contain every set as an element, including itself. This set would be the grandmother of infinities, the all-embracing set of everything. But this idea, though satisfyingly grand, is inherently contradictory.

Cantor's Power Set Theorem implies that there is no greatest of all infinities—so the answer to question 4 from the beginning of this section is no. By Cantor's

Power Set Theorem, for any set, the set of all its subsets is strictly bigger. If we consider this hypothesized biggest set of all sets, we run into a contradiction, because it would be impossible for all its subsets to be elements of it—there are too many of them. So it cannot contain every set. The only way around this contradiction is to concede that this huge set is not a set! This statement is an insight of Bertrand Russell and is known as *Russell's Paradox*.

What went wrong? Why isn't the set of all sets a set? The answer is fundamental. If you look back through this chapter, you may notice that we never defined a particular term: *set*. Russell's Paradox shows us that even basic-sounding ideas have inherent limitations. The intuitive idea that a set is just a "bunch of things" turns out to be inadequate for truly enormous collections. Russell's Paradox forces us to confront the strange fact that some collections are just too unwieldy to call sets. Mathematicians were thus forced to move beyond their intuition, create an appropriate definition for the word set, and then carefully investigate its nuances and consequences. These studies led to an entire branch of mathematics: *Set Theory*. Instead of delving into the arcane essence of "setness," we thought it would be best to move on and explore infinity within the geometrical world. All set?

A Look BACK

THE POWER SET OF A SET S, denoted $\mathcal{P}(S)$, is the collection of all its subsets. A finite set with n elements has a power set with 2^n elements.

Cantor's Power Set Theorem shows that the cardinality of the power set of a set is always greater than the cardinality of the original set, even for infinite sets. The proof of this theorem uses the same diagonalization idea that Cantor had used to show that the real numbers are more numerous than the natural numbers.

Using Cantor's Power Set Theorem, we can start with the natural numbers $\mathbb{N}$ and build an infinite collection of sets, each with a greater cardinality than its predecessor—namely, $\mathbb{N}$, $\mathcal{P}(\mathbb{N})$, $\mathcal{P}(\mathcal{P}(\mathbb{N}))$, $\mathcal{P}(\mathcal{P}(\mathcal{P}(\mathbb{N})))$, and so forth. There is no largest infinite set, because we can always take the set of all its subsets to create a larger set.

This theory illustrates the power and value of exploiting the same idea as much as possible. A new idea is a new tool. Go beyond its original intended use and modify it to create yet more tools.

Life Lessons

Apply ideas widely.

MINDSCAPES   Invitations to Further Thought

In this section, Mindscapes marked (H) *have hints for solutions at the back of the book. Mindscapes marked* (ExH) *have expanded hints at the back of the book. Mindscapes marked* (S) *have solutions.*

Developing Ideas

1. **Which are which?** Look at the set $S = \{a, b, c, d\}$. Classify each of the five following items as either an *element* of S or a *subset* of S: $\{a\}$; a; $\{\ \}$; d; $\{a, b, c, d\}$.

2. **Power play.** Define the *power set* of a given set.

3. **Universal emptiness.** What set is a subset of every set?

4. **With three or without.** List all the subsets of {1, 2, 3} that do not contain 3. How many are there? Now list all the subsets that do contain 3. How many are there? How many subsets are there altogether?

5. **Solar power.** What is the cardinality of the power set of {Earth, Moon, Sun}? List the elements of this power set.

Solidifying Ideas

6. **All in the family (ExH).** A family of four tries to eat dinner together as much as possible. On Tuesday, some family members sit down for dinner. How many different possible groups of diners are there? (Their position at the table does not matter.) List all the possible groupings.

7. **Making an agenda (H).** There are eight members on the board of directors of the Couch Potato Corporation, not including the Chair-potato of the Board. Each board member e-mails the chairperson one agenda item that he or she would like to discuss at an upcoming meeting. All the agenda items sent to the chair are different. The chair chooses the agenda from the list received; however, the chair does not necessarily have to use all the items. How many different possible agendas (not counting order) could there be for their next meeting?

8. **The power of sets (S).** Let $S = \{!, @, \#, \$, \%, \&\}$. Below are two columns: on the left are names of certain sets, and on the right are confusing-looking thingies—each thingie is an element of one of the sets in the left-hand column. Your mission, if your instructor chooses to assign it, is as follows: For each thingie in the right-hand column, determine which set from the left-hand column contains that thingie as a member. As examples, we have done the first two for you: $\{!, @\}$ is a subset of S, so it is an element of the power set $\mathcal{P}(S)$, and @ is a member of S itself. Okay, now you're on your own! Good luck.

Set	Thingie
S	$\{!, @\}$
$\mathcal{P}(S)$	@
$\mathcal{P}(\mathcal{P}(S))$	$\{!, \#, \%\}$
$\mathcal{P}(\mathcal{P}(\mathcal{P}(S)))$	$\{ \{\{@, !\}\}, \{\{\$\}\} \}$
$\mathcal{P}(\mathcal{P}(\mathcal{P}(\mathcal{P}(S))))$	$\{ \{!\}, \{@\}, \{\#\}, \{\%\}, \{\&\} \}$
$\mathcal{P}(\mathcal{P}(\mathcal{P}(\mathcal{P}(\mathcal{P}(S)))))$	$\{ \{\{@\}\}, \{\{\#, \$\}\}, \{\{!\}, \{\%, \&\}\} \}$
	$\{\{\{!\}\}\}$
	$\{\#\}$
	$\{ \{@\}, \{\$, !\} \}$
	$!$

9. **Powerful words.** Suppose that S is a set. Describe the following sets, in words, as clearly as possible: $\mathcal{P}(\mathcal{P}(S))$; $\mathcal{P}(\mathcal{P}(\mathcal{P}(S)))$; $\mathcal{P}(\mathcal{P}(\mathcal{P}(\mathcal{P}(S))))$. (For example, $\mathcal{P}(S)$ might be described as *the set of all subsets of the set S.*)

10. **Identifying the power.** Let S be the set given by $S = \{m, a, t, h, f, u, n\}$. Which of the following are elements of $\mathcal{P}(S)$? $\{m, a, t, h\}$; a; $\{\ \}$; $\{\{m, a, t, h\}\}$; $\{\{m\}, \{a\}, \{t\}, \{h\}\}$; $\{m\}$.

11. **Two Cantor (ExH).** Suppose S is the set defined by $S = \{1, 2\}$. Consider the following pairing of elements of S with elements of $\mathcal{P}(S)$:

Elements of S	Elements of $\mathcal{P}(S)$
1	$\{2\}$
2	$\{1, 2\}$

    Using the idea of Cantor's proof, describe a particular subset of S that is not on this list.

12. **Another two.** Suppose S is the set defined by $S = \{1, 2\}$. Consider the following pairing of elements of S with elements of $\mathcal{P}(S)$:

Elements of S	Elements of $\mathcal{P}(S)$
1	$\{2\}$
2	$\{1\}$

    Using the idea of Cantor's proof, describe a particular subset of S that is not on this list.

13. **Cantor code.** Suppose that *Words* is the set defined by *Words* = {all, you, infinity, found, them, search, the, it}. Consider the following pairing of elements of *Words* with elements of $\mathcal{P}(Words)$:

Elements of *Words*	Elements of $\mathcal{P}(Words)$
all	{all, infinity, found}
you	{it, search, them}
infinity	{all, them, infinity}
found	{you, the, it}
them	{found, them}
search	{all, infinity, search}
the	{the, search}
it	{infinity, you, all}

    Using the idea of Cantor's proof, describe a particular element of $\mathcal{P}(Words)$ that is not on this list.

14. **Finite Cantor (H).** Suppose that S is the set defined by $S = \{@, \&, \%, \$, \#, !\}$. Consider the following pairing of elements of S with elements of $\mathcal{P}(S)$:

Elements of S	Elements of $\mathcal{P}(S)$
@	$\{@, !, \$\}$
&	$\{\ \}$
%	$\{\&, \$\}$
$	$\{\$\}$
#	$\{@, \%\}$
!	$\{@, \#, !\}$

    Using Cantor's proof, describe a particular subset of S that is not on this list.

15. **One real big set.** Describe (in words) a set whose cardinality is greater than the cardinality of the set of real numbers.

Creating New Ideas

16. **The Grand Real Hotel.** Suppose there was a really huge hotel called the Grand Real Hotel. This hotel is so large that there is an individual room associated with each individual real number (imagine the room numbers: "I'm staying in room number 63.7269711294…"). Does there exist a set that

would cause the Grand Real Hotel to put up a No Vacancy (or Not Enough Room) sign? If so, describe such a set. If not, explain why not.

17. **The Ultra Grand Hotel (S).** Could there be an infinite hotel so large that there is never a need to put up the Not Enough Room sign? Explain your answer.

18. **Powerful counting.** Let the set S be given by $S = \{@, \&, \%, \$, \#, !\}$. How many elements does the set $\mathcal{P}(\mathcal{P}(\mathcal{P}(\mathcal{P}(\mathcal{P}(\mathcal{P}(S))))))$ contain?

19. **Russell's barber's puzzle (H).** In a certain village there is one male barber who shaves all those men, and only those men, who do not shave themselves. Does the barber shave himself? Show that the answer cannot be yes or no and that this question is a paradox. This barber paradox is related to Russell's insight that the set of all sets cannot be a set. In particular, consider the set *NoWay*, whose elements are special sets—namely, all sets that do not contain themselves as elements. Is *NoWay* an element of itself?

20. **The number name paradox.** Let S be the set of all natural numbers that are describable in English words using no more than 50 characters (so, 240 is in S since we can describe it as "two hundred forty," which requires fewer than 50 characters). Assuming that we are allowed to use only the 27 standard characters (the 26 letters of the alphabet and the space character), show that there are only finitely many numbers contained in S. (In fact, perhaps you can show that there can be no more than 27^{50} elements in S.) Now, let the set T be all those natural numbers not in S. Show that there are infinitely many elements in T. Next, since T is a collection of natural numbers, show that it must contain a smallest number. Finally, consider the smallest number contained in T. Prove that this number must simultaneously be an element of S and not an element of S—a paradox!

Further Challenges

21. **Adding another.** Suppose that you have any infinite set (it could be really huge), and you wish to add one new element to the set. Prove that this new set (the old set with the new element thrown in) has the same cardinality as the original set. (*Hint:* The infinite set has a subset that is in one-to-one correspondence with the natural numbers. Use previous ideas to add the new element to this subset, and let the rest of the set correspond with itself.)

22. **Ones and twos.** Describe a one-to-one correspondence between the set of all decimal numbers between 0 and 1 that are made with only 1's and 2's (like 0.1211212212212. . .) and the set of all decimal numbers between 0 and 1 that are made with only 3's, 4's, 5's, and 6's (like 0.36543546354554. . .). (*Hint:* Decimals made of 1's and 2's only can have their first two places 0.11. . . , 0.12. . . , 0.21. . . , 0.22. . . .)

In Your Own Words

23. **Personal perspectives.** Write a short essay describing the most interesting or surprising discovery you made in exploring the material in this section. If any material seemed puzzling or even unbelievable, address that as well. Explain why you chose the topics you did. Finally, comment on the aesthetics of the mathematics and ideas in this section.

24. **With a group of folks.** In a small group, discuss and work through Cantor's proof that the cardinality of a set is always smaller than the cardinality of its

power set. After your discussion, write a brief narrative describing the proof in your own words.

25. **Creative writing.** Write an imaginative story (it can be humorous, dramatic, whatever you like) that involves or evokes the ideas of this section.

26. **Power beyond the mathematics.** Provide several real-life issues—ideally, from your own experience—that some of the strategies of thought presented in this section would effectively approach and resolve.

For the Algebra Lover

Here we celebrate the power of algebra as a powerful way of finding unknown quantities by naming them, of expressing infinitely many relationships and connections clearly and succinctly, and of uncovering pattern and structure.

27. **Enjoying the exponential function.** Consider the exponential function $y = 2^x$. If the variable x represents the cardinality of a given finite collection, then what does the corresponding value of y represent with respect to that given finite collection? Sketch a graph of this function. Find the value for y if $x = 5$.

28. **Tower of power (H).** Suppose a set S has x elements. Give a formula in terms of x for the cardinality of the power set of the power set of S. Use your formula to find the cardinality of the power set of the power set of S if S contains exactly three elements.

29. **Power play.** Simplify the following expressions:

$$2^5 2^{-3} 2^2 \qquad \frac{2^x 2^y}{2^z} \qquad \frac{6^x}{2^x 3^x} \qquad \frac{2^t 3^t 5^t}{10^t} \qquad \frac{5^3}{5^{-2} 25}$$

30. **Powerful products.** For each function given below, evaluate as indicated.

$f(x) = 2^x$: evaluate at $x = 5$, $x = 0$, $x = -1$, and $x = -3$.
$g(t) = 3^{-t}$: evaluate at $t = 1$, $t = 0$, $t = 2$, and $t = -2$.
$f(x) = (1/2)^x$: evaluate at $x = 3$, $x = 0$, $x = -2$, and $x = -3$.

31. **Generalizing "equality."** Throughout this chapter we have considered the notion of same cardinality. If two collections have the same cardinality, then they are "equally numerous" (with the understanding that their sizes might be infinite). In this chapter we discovered that there are infinitely many different sizes of infinity. It turns out that "same cardinality" is a generalization of what we mean when we say two things are "equal". In mathematics, we call such a generalization an *equivalence relation*. Specifically, an equivalence relation between two objects (say **A** and **B**) is any relation that is *reflexive* (that is, an object **A** must be equivalent to itself), *symmetric* (that is, if an object **A** is equivalent to an object **B**, then **B** must be equivalent to **A**), and *transitive* (that is, if an object **A** is equivalent to **B** and **B**, in turn, is equivalent to **C**, then **A** must be equivalent to **C**). Show that the relation of "same cardinality" is an equivalence relation between two collections.

STRAIGHTENING UP THE CIRCLE

Exploring the Infinite Within Geometrical Objects

© Fine Art Images/SuperStock

M. C. Escher's *Circle Limit IV*, 1956.

In our previous exploration of infinity, all the sets and correspondences have been rather abstract. Here we consider somewhat more concrete occurrences of infinity by looking at objects that we can draw. We will still look for one-to-one correspondences, but we can sometimes make connections by just drawing lines. So let's turn now to exploring infinity in the visual field.

Short Versus Long Lines

Consider the following two line segments S and B. (We just cut them out of a number line; that's why there are numbers written under them.)

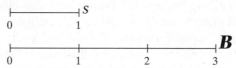

Remember that a line segment is just a bunch of points all lined up like soldiers but incredibly close together. Are there more points in line segment B than in line segment S? Think about this question and come up with a guess and a justification. We'll wait right here.

There are many reasons why one could argue that there are more points in line B than S. First, segment B is three times longer than segment S. Second, the letter B to the right of the second line segment is much bigger than the letter S to the right of the first line segment. Let's think about the first reason a bit more.

Wrong Reasoning

The following is a reasonable-sounding, though fallacious, argument:

> I've already learned that two sets have the same cardinality if there is
> a one-to-one correspondence between them. In other words, is there

a way of pairing the elements of the two sets so that each element of the first set is paired with exactly one element of the second set, and each element of the second set is paired with exactly one element of the first set? Here is why the cardinality of the set of points in line B is greater than the cardinality of the set of points in line S. Consider the following pairing:

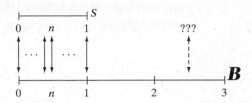

This pairing is easy to describe. Pair the point on S located at the number n with the point on B located at n. So, 0 in S pairs with 0 in B, 1/2 in S pairs with 1/2 in B, and 1 in S pairs with 1 in B. This pairing is a one-to-one correspondence between the points of S and the points of B from 0 to 1. But now notice that the point 2 1/2 in B is not paired with any point from S. In fact, no point in B after 1 is paired with any point in S. So, since we have a one-to-one correspondence between all the points of S and just some and not all of the points in B, the cardinality of points in B must not equal the cardinality of points in S. This must show that the cardinality of the set of points in B must be greater than the cardinality of the set of points in S.

Life Lesson

Examine arguments critically—don't just accept them.

Let's think about the preceding argument. It's pretty clever and sounds reasonable. The big question is, what's the problem? Let's examine this argument critically and objectively. We must not believe something just because it sounds reasonable.

Okay, the argument is not correct, but can we figure out where it went wrong? One sentence is false, and, from that point on, everything else is wrong. Can you find it? Give it a try before reading on.

Life Lesson

Making mistakes is a sign of creativity and strength; it is not a sign of weakness.

Before analyzing the situation, remember that an argument, even an incorrect one, is a good way to focus our thoughts on important issues. In this case, someone thought about an issue and put ideas together into a coherent explanation. This process is truly great. Every time someone does this, it is a time to celebrate.

Okay, enough celebrating—after all, the argument is wrong. Now let's think about the issue a bit more and see if we can discover the truth.

The Meaning of Not

Recall that, for two sets to have the same cardinality, there must exist a one-to-one correspondence between the two sets. So what does it mean for two sets *not* to have the same cardinality? It means it is impossible to find a one-to-one correspondence between the two sets. The point is that just finding a particular pairing that is not a one-to-one correspondence is not enough. Finding such a pairing just shows that that particular pairing didn't work. It does not show that a one-to-one correspondence doesn't exist. Maybe some other pairing is a one-to-one correspondence, even though the first one was not. This possibility is very important in the study of infinite sets, as the line segments S and B will illustrate.

It's a Tie

It turns out that the set of points of line S has the same cardinality as the set of points of line B. How can we prove this? All we need to do is construct a one-to-one correspondence between the two sets, and we can use geometry to help us by connecting the two segments as follows:

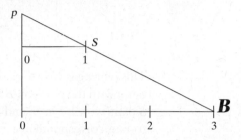

All we did is draw a line that passes through both of the 0's and a line that passes through the 1 on S and the 3 on B. Since S is shorter than B, these two lines must intersect. We call the point where they meet p (for "point," clever, huh?). We now describe the one-to-one correspondence. Suppose we have a point on the line S, let's call it w (for "wonder who this point will be paired with"). We will now pair up w with exactly one point from B. Which one? Draw a line from the point p through the point w and see where it intersects B. Let's call that point of intersection h (for "hit"). So, refer to the picture at left to see what it looks like.

Notice that the line through p and w hits B at *exactly* one point! Thus, every point of S is paired with exactly one point of B. Now, suppose we have a point from B—let's call it r (for "running out of clever names for points"). To see which point of S we will pair r with, consider the line that goes from p to r. Notice that it will have to cross S and will cross S at exactly one point—say l (for "last point we will mention"). We will pair r up with l as follows:

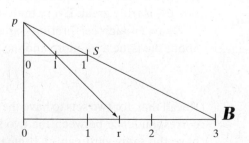

Therefore, in this manner, every point of B is paired with exactly one point of S. Thus, we have constructed a one-to-one correspondence between the set of points of S and the set of points of B, and so the sets have the same cardinality. In fact, this exact idea proves the following fact:

All Line Segments Are Created Equal.

For any two line segments, the cardinality of points of one segment equals the cardinality of points of the other.

Is this theorem surprising? It may seem a bit strange, especially since one line segment could be very long and the other could be very short. But now we see how to find a one-to-one correspondence. In some sense, there are so many points in a line segment that we can stretch the segment out like a rubber band and make a longer line segment without adding more points. This is the spirit of the one-to-one correspondence we created previously—a pairing through stretching.

Pick on Someone Your Own Size

Now we are experts about cardinalities of line segments. But what about an entire line that goes on forever in each direction? Let S (for "small segment") be the set of all points on a line segment strictly between 0 and 1 (so we will not include 0 or 1—only the points between them). And let's call the entire number line $\mathbb{R}$ (for "real numbers"). Here is a picture of the players:

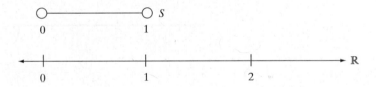

Notice the circles at the ends of S. Those circles are to remind us that neither 0 nor 1 is part of S. S is just the set of points in between 0 and 1.

Is the cardinality of the set of points in $\mathbb{R}$ greater than the cardinality of the set of points in S? We've seen that all line segments have the same cardinality, but now we are asking what happens if we consider a line that extends out infinitely in both directions. What do you think? Make a guess before reading on.

We are asking whether there is a one-to-one correspondence between the set of points in S and the set of real numbers, $\mathbb{R}$. In fact, a one-to-one correspondence does exist. That is, the cardinality of the set S equals the cardinality of the set $\mathbb{R}$. This claim seems incredible. How could there be just as many points between 0 and 1 as there are points on the entire real number line? After all, the number line goes on forever.

Round-Up

The one-to-one correspondence between these two sets turns out to be useful in several different settings besides this one. It is clever and tricky, so we'll proceed with caution. First, we notice that we will not change the cardinality of the line segment S if we bend it a little bit. We would be changing its shape, but not its cardinality.

Here are some different ways of bending *S*:

That last smiley face semicircle interests us the most. In fact, if we continued to bring 0 and 1 around, we would end up with the circle shown here.

In this shape, the circle at 0 is exactly on top of the circle at 1. We just took the segment and brought the ends together so that we created a big circle with one point missing: the north pole. Why is it missing? Because the segment *S* consists of all points between 0 and 1. So 0 and 1 are not part of the segment. Thus, after we place the missing point 0 on top of the missing point 1, we are left with just one hole.

Now we bring the number line $\mathbb{R}$ into the picture—actually, right below the picture—as follows:

Rolling Up the Real Line

The circle sits on the line at its south pole and touches the line at exactly one point on $\mathbb{R}$, namely, 0. We again draw lines to show the one-to-one correspondence. Let *a* (for "arbitrary point") be a point on the circle minus the north pole (so *a* cannot be the north pole). To see who gets paired with *a* on the number line, we draw a line from the north pole hole through the point *a* and continue until we hit the number line, as shown here.

The line hits the number line at one point, say, *t* (for "that point"). We will pair the point *a* on the circle with the point *t* on the number line. Similarly, if we had a point on the number line, we could find the point on the circle that it is paired with by connecting the point on the number line with the north pole hole. That line will intersect the circle at exactly one other point, and that is the point paired with the point on the number line. This process gives us a one-to-one correspondence between the points on the real line and the points on the circle minus the north pole. Hence the cardinality of the points on the circle minus the north pole equals the cardinality of the points on the real number line.

Recall that we already saw that the cardinality of the set of points of the circle minus the north pole is equal to the cardinality of the set of points of the interval S.

Putting all this together, we conclude that the cardinality of the points of S equals the cardinality of the set of all points on the number line. We just proved a remarkable result:

> **The Number Line Is Like a Line Segment.**
>
> *The cardinality of points between any two real numbers on a number line is equal to the cardinality of the set of all points on the real number line.*

Stereographic Projection

This cool one-to-one correspondence between the line segment curled up into a circle minus the north pole and the number line is so important that it has a fancy name: *stereographic projection*. The idea is that we are projecting all the points of the circle minus the north pole onto the number line. This pairing rule may be thought of in another way: In some sense we are wrapping the number line around the circle minus the north pole. Notice that, as we travel along the circle and approach the north pole from the left, those points are paired with points on the number line that are heading off to the left horizon. Similarly, as we approach the north pole from the right, those are the points paired with points on the number line that go off to the right horizon.

The Solid Square

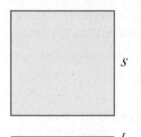

We have just discovered that not only do different-length line segments have the same cardinalities, but, even if we stretch out a line segment forever (making the real number line), we don't up the cardinality. In our quest to make geometric objects with really big cardinality, let's try, instead, to up the dimension and consider solid objects in the plane. Specifically, consider the objects shown.

S is a square and L is a line segment. Actually, it will be easier to view the line segment and the sides of the square as little pieces of the real number line so we can identify points in the square, using coordinates in the usual way:

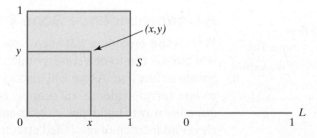

Recall that we can specify a particular point in the square by giving its horizontal distance away from 0 (called x) and its vertical distance away from 0 (called y) and denote the point as the pair (x, y). Is the cardinality of points inside the square S greater than the cardinality of the points of the line L?

Let's give this puzzler some thought. First of all, notice that the base of the square is exactly the same as line L. If we draw any horizontal line across the square, we would have another line that is exactly equal to L. Thus, we have lots of copies of the line L inside the square S. In fact, if we were to imagine inking up the line L (so it's all inky) and then dragging it up the paper exactly one unit, we would make a square ink blot exactly in the same shape as S. So we have a different copy of L in the square S for every single point on the vertical edge of S:

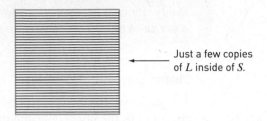

Just a few copies of L inside of S.

So the cardinality of L is clearly not bigger than the cardinality of S. Therefore, the cardinality of L must be less than or equal to the cardinality of S. In fact, our observation appears to indicate that in some sense S is much bigger than L, because it has a whole extra dimension (up!), which allows us to stack and store infinitely many L's, one for each real number in the vertical direction between 0 and 1. Thus, it is reasonable to conjecture that the cardinality of points of S is greater than the cardinality of points in L. However, so far all the examples we have looked at in this section have had the same cardinality, so perhaps the line and the square have the same cardinality as well.

Of course, since this is the last example of the chapter (the big finish), perhaps this is the one where the cardinalities will differ. But, enough mind-game distractions. In fact, the cardinality of the square is not greater than the cardinality of the line segment:

Squares Equal Lines.

The cardinality of points inside a square is the same as the cardinality of points on a line segment.

A Correspondence Gone Bad

Why is the preceding statement true? How would we prove it? As always, we will build a one-to-one correspondence and evenly pair points in the square with points on the line. As we will discover, the first attempt we make to craft a one-to-one correspondence will actually fail. (Remember that failing is good—it is an effective way to build new insights and discover the truth. We will then see that a clever adaptation of our failed attempt will actually lead us to success.)

To show that the cardinality of the square S is the same as the cardinality of the line segment L, we need to show that we can pair each point of S with a point of L and each point of L with a point of S. In other words, suppose that we had two barrels: In one barrel we put all the points from inside the square S, and in the other barrel we put all the points of the line segment L. If we could demonstrate a way of pairing elements of the S barrel with elements of the L barrel so that, once all the pairings are made, we have nothing left in either barrel, then that would demonstrate that the cardinality of the things in the S barrel is the same as the cardinality of things in the L barrel.

Okay, let's give it a try. Here's how each point of the square S can be paired with exactly one point of the line L. To describe this pairing, we first need to be able to name precisely the points in S and on L. We do this through decimal expansions. Recall from the last section in Chapter 2 that every real number can be expressed with an infinitely long decimal expansion:

$$\frac{7}{3} = 2.33333333333333333333333333333333333333333333\ldots$$
$$\pi = 3.14159265358979323846264338327950288419716933\ldots$$
$$\sqrt{2} = 1.41421356237309504880168872420969807856967187\ldots$$
$$\frac{13}{2} = 6.5000000000000000000000000000000\ldots$$

The last one could be written differently. Remember that we proved that $1 = 0.9999999999\ldots$. Similarly, we could write

$$\frac{13}{2} = 6.4999999999999999999999999999999\ldots$$

To make our naming of points unambiguous, we will always use the decimal representation ending in all nines whenever we have a choice. In other words, if a decimal ends with an infinite tail of zeros (like $0.50000\ldots$), we rewrite it so it has an infinite tail of nines ($0.49999\ldots$). This choice of using 9's, rather than 0's, is arbitrary, but we have now spoken and will accept no back talk (we can be pretty forceful when we want to be).

Shuffling Digits

If we take a point in the square, say (x, y), we can write both x and y as infinite decimals. For example, say,

$$x = 0.95780864030230509960000778732250684006584033\ldots$$

and

$$y = 0.44444487599098420000743009240006786357809389393725\ldots$$

To figure out which point on the line L we will pair (x, y) with, we do something interesting: We shuffle the digits of the decimal expansions of x and y together, like a deck of cards, to create a new decimal number. That is, we create a new decimal number by taking 0, and then tacking on the first digit in x and then the first digit in y and then the next digit in x followed by the next digit in y and then the third digit in x followed by the third digit in y, and so on. We produce one infinitely long decimal expansion. That new number is between 0 and 1, and it is the number on L that we will associate with the point (x, y) in the square S. So, for example, the preceding (x, y) point in S would be paired with the point

$$z = 0.9454748404846847053909203908540290906000007040370\ldots$$

We just interwove the digits of x and y, as if they had been shuffled. We wrote the digits of x in blue and the digits of y in red, so we can visually see the interwovenness of the digits in z. Notice that, using this method, two different points in the square get paired with two different points on the line. So, this correspondence pairs each point of S with a point in L in such a way that no two points of S are paired with the same point of L. This clever pairing seems to be a one-to-one correspondence; however, something is slightly wrong. What is wrong? Can you spot the subtle problem?

Whoops

On first inspection, it appears as though we have made a one-to-one correspondence between all the points of S and all the points of L. After all, if we took a decimal point on L, couldn't we unshuffle it to get a point in S? Surprisingly, some points of L were *not* hit in our shuffling correspondence. That is, there are some points in L that are never paired with points in S. We will not keep you in suspense for long. Let's build a point of L that is not hit.

Think of the point on L given by the decimal $0.3790909090909\ldots$. If we unshuffle, then the point in S to which we would expect this point to correspond would be $(0.399999\ldots, 0.700000\ldots)$. The problem is that we agreed that $0.700000\ldots$ would be represented as $0.699999\ldots$. So, since $(0.39999\ldots, 0.70000\ldots,)$ equals $(0.39999\ldots, 0.699999\ldots)$, that point in S goes to the point on L, $0.369999\ldots$ (which equals 0.37). So, the point $0.3790909090909\ldots$ in L is left without a partner from this shuffling pairing with S. In other words, with this pairing, even though we will have nothing left in the S barrel, there will be numbers remaining in the L barrel.

Success from Failure

We can try to modify good but failed attempts and ideas in order to achieve success. Let's look carefully at what went wrong and see whether we can fix it.

The problem is that alternating zeros got us in trouble. To avoid this defect, we modify the shuffling procedure so that this annoying glitch will disappear. Instead of shuffling the numbers perfectly (a digit from x followed by a digit from y, and so on), we shuffle groups of digits—we put down a few digits of x followed by a few digits of y, and so forth. How do we group the digits? We group all consecutive runs of zeros together and include the first nonzero digit. If there are no zeros, then the nonzero digits form their own individual groups. For example, for the numbers considered before, we would group their digits by

$$x = 0.9\ 5\ 7\ 8\ 08\ 6\ 4\ 03\ 02\ 3\ 05\ 09\ 9\ 6\ 00007\ 7\ 8\ 7\ 3\ 2\ 2\ 5\ 06\ 84\ldots$$

and

$$y = 0.4\ 4\ 4\ 4\ 4\ 4\ 87\ 5\ 9\ 9\ 09\ 8\ 4\ 2\ 00007\ 4\ 3\ 009\ 2\ 4\ 0006\ 7\ 8\ 6\ 3\ 5\ 7\ 8\ 09.\ldots$$

If we now shuffle the grouped digits together, we would pair the point (x, y) in S with the point in L given by

$$0.945474840846448037025390590909986400007270000784733009 22.\ldots$$

We can check that this new shuffling technique gives rise to a one-to-one correspondence. Certainly every point in the square is paired with one and only one element of the line. It is also the case that every element of the line is paired with one and only one element of the square. For example, the previous problem point on the line was 0.379090909. . . . Notice that this point gets paired with the point on the square at (0.390909090. . . , 0.709090909. . .). This shuffling provides us with a one-to-one correspondence, and so the points in a square and the points on the line have the same cardinality.

A Look BACK

MANY GEOMETRIC OBJECTS have the same cardinality, and that equivalence can sometimes be seen through creative visual correspondences. Every two intervals, no matter how long or short, have the same cardinality. The whole real line has the same cardinality as an interval. This last correspondence is produced using the stereographic projection.

A filled-in square has the same cardinality as an interval, but this one-to-one correspondence is not produced geometrically. This surprising equality of square and line is shown by demonstrating a subtle pairing between the two sets via shuffling grouped digits together.

Understanding issues of infinity is an intellectual triumph. These ideas and theories are challenging, but they are also beautiful, rich, and interesting, because they allow us to understand and analyze concepts that at first seemed beyond human comprehension. It is incredible how our minds are capable of grasping and taming notions that at first appear out of human reach. Therein lies part of the beauty and the power of mathematics. Surprising intellectual discoveries are a great tribute to the human spirit.

One important strategy illustrated in this section was the identification of failure as a step toward success. By making a mistake and then looking carefully at where the reasoning went wrong, we were able to modify an argument to make it work.

We also saw the value of taking familiar objects and ideas and viewing them from a different perspective. We can think of variations of common notions that result in uncommon insights.

Life Lessons

Look at ordinary things in extraordinary ways.

MINDSCAPES   Invitations to Further Thought

In this section, Mindscapes marked **(H)** *have hints for solutions at the back of the book. Mindscapes marked* **(ExH)** *have expanded hints at the back of the book. Mindscapes marked* **(S)** *have solutions.*

Developing Ideas

1. **Lining up.** Can you draw a line segment that has more points than the line segment L?

$$L \text{————}$$

2. **Reading between the lines.** Use the figure below to demonstrate that there is a one-to-one correspondence between the points on the line segment L and the points on the line segment M.

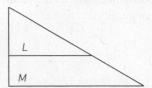

3. **De lines and Descartes.** Put line segments L and M into the xy-plane as shown. The red line is given by the equation $y = -x + 3$. Verify that the points $(0, 3)$ and $(3, 0)$ satisfy the equation.

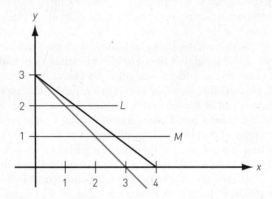

4. **Red line rendezvous (H).** Given the equation for the red line in Mindscape 3, find the point where L intersects the red line.

5. **Rendezvous two.** Given the equation for the red line in Mindscape 3, find the point where M intersects the red line.

Solidifying Ideas

6. **A circle is a circle (H).** Prove that a small circle has the same number of points as a large circle. Stated precisely, prove that the cardinality of points on a small circle is the same as the cardinality of points on a large circle. Describe a one-to-one correspondence between these two sets.

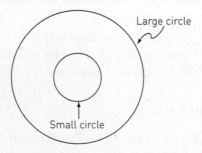

7. **A circle is a square.** Prove that a small circle has the same number of points as a large square. Stated precisely, prove that the cardinality of points on a

small circle is the same as the cardinality of points on a large square. Describe a one-to-one correspondence between these two sets.

8. **A circle is a triangle.** Prove that a small circle has the same number of points as a large triangle. Stated precisely, prove that the cardinality of points on a small circle is the same as the cardinality of points on a large triangle. Describe a one-to-one correspondence between these two sets.

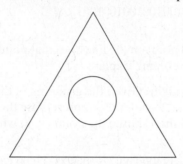

9. **Stereo connections (ExH).** Given the stereographic projection as a one-to-one correspondence between points on a line segment (rolled up) and points on the real line, identify which points on the line segment are paired with the points marked on the real line below.

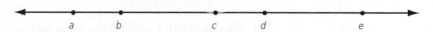

a   b   c   d   e

10. **More stereo connections.** Given the stereographic projection as a one-to-one correspondence between points on a line segment (rolled up) and points on the real line, identify which points on the real line are paired with the points marked on the line segment below.

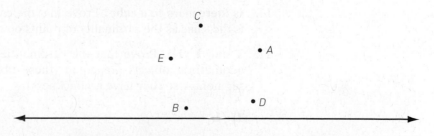

11. **Perfect shuffle problems (H).** Suppose we used our failed perfect shuffling of digits to mix the digits of the numbers (x, y) that describe a point on the square to get a one-to-one correspondence with the points on a line segment. What point in the square would be paired with the point

$$0.1200010001000100010001\ldots?$$

With what point on the line does that point in the square actually get paired? Is this a problem? Explain.

12. **More perfect shuffle problems.** Suppose we used our failed perfect shuffling of digits to mix the digits of the numbers (x, y) that describe a point on the square to get a one-to-one correspondence with the points on a line segment. What point on the square would be paired with the point

$$0.12001001001001001001\ldots?$$

With what point on the line does that point on the square actually get paired? Is this a problem? Explain.

13. **Grouping digits.** Given the grouping of digits method as a way to shuffle the digits of the two numbers representing the points of a square, how would you group and then shuffle the point (x, y) where

$$x = 0.64340004897211098000345003703030306\ldots$$
$$y = 0.49867400546080009707076002303303030112\ldots?$$

14. **Where it came from.** Given the grouping of digits method as the one-to-one correspondence between the square and the line segment, with what point on the square would the point

$$0.990040878000404820244000501101909090909090\ldots?$$

on the line get paired?

15. **Group fix (S).** Consider the point on the line from Mindscape 11. What point on the square would that point get paired with using the grouping-shuffling method?

Creating New Ideas

16. **Is there more to a cube?** Prove that the cardinality of points in a solid cube is the same as the cardinality of points on a line segment.

17. **T and L (H).** Prove that the cardinalities of points in the following two geometrical objects are equal (these objects are made up of little line segments—so they have no thickness).

T   L

18. **Infinitely long is long.** Must it be the case that every subset of the real line, even one that goes on forever, has a one-to-one correspondence with a set of points in the interval between 0 and 1?

19. **Plugging up the north pole (ExH).** What would happen if you filled in the north pole of the circle in the stereographic projection? Is there any point on the real line to which that extra point could correspond? Show that the cardinality of a circle is the same as the cardinality of $(0, 1]$ (an interval where 0 is not included, but 1 is included).

20. **3D stereo (S).** Let S' be the set of points on the surface of a ball (a hollow sphere) minus the north pole. How does the cardinality of points on the plane compare to the cardinality of points on S'? Justify your answer, and explain the title of this story.

Further Challenges

21. **Stereo images.** Given your answer to the preceding 3D stereo question, suppose we had a circle drawn on the sphere that passed through the north pole (which is missing) and the south pole (longitudinal lines). What object is this circle paired with on the plane?

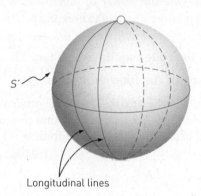

Longitudinal lines

What shapes would latitudinal lines get paired with?

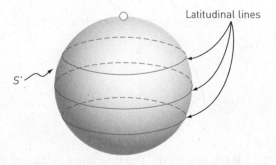

Latitudinal lines

22. **Grouped shuffle.** Carefully verify that the pairing described by shuffling groups of digits is, in fact, a one-to-one correspondence.

In Your Own Words

23. **Personal perspectives.** Write a short essay describing the most interesting or surprising discovery you made in exploring the material in this section. If any material seemed puzzling or even unbelievable, address that as well. Explain why you chose the topics you did. Finally, comment on the aesthetics of the mathematics and ideas in this section.

24. **With a group of folks.** In a small group, discuss and work through the proof that the set of points in a square has the same cardinality as the set of points on a line segment. After your discussion, write a brief narrative describing the proof in your own words.

25. **Creative writing.** Write an imaginative story (it can be humorous, dramatic, whatever you like) that involves or evokes the ideas of this section.

26. **Power beyond the mathematics.** Provide several real-life issues—ideally, from your own experience—that some of the strategies of thought presented in this section would effectively approach and resolve.

For the Algebra Lover

Here we celebrate the power of algebra as a powerful way of finding unknown quantities by naming them, of expressing infinitely many relationships and connections clearly and succinctly, and of uncovering pattern and structure.

27. **Giving the rolled-up interval a tan.** The graph depicted here is a piece of the graph of the function $y = \tan\left(\dfrac{\pi}{2}x\right)$. Notice that there are two vertical asymptotes: one at $x = -1$ and the other $x = 1$. Use the graph of this function to provide another justification that the cardinality of all the points between -1 and 1 (not including the endpoints -1 or 1) is the same as the cardinality of the collection of points on the y-axis.

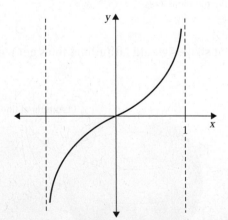

28. **Back and forth.** The function $y = 5x - 2$ gives a one-to-one correspondence between the points in the interval $[1, 4]$ on the x-axis and the points in the interval $[3, 18]$ on the y-axis. Find the value on the y-axis that corresponds to $x = 2$. Find the value on the x-axis that correspond to $y = 11$.

29. **Forth and back.** The function $y = -3x + 1$ gives a one-to-one correspondence between the points in the interval $[-2, 2]$ on the x-axis and the points in the interval $[7, -5]$ on the y-axis. Find the value on the y-axis that corresponds to $x = -1$. Find the value on the x-axis that corresponds to $y = 0$.

30. **Lining up (H).** Find a function that gives a one-to-one correspondence between the points in the interval $[0, 6]$ on the x-axis and the points in the interval $[1, 4]$ on the y-axis.

31. **Queuing up.** Find a function that gives a one-to-one correspondence between the points in the interval $[-3, 4]$ on the x-axis and the points in the interval $[-2, 12]$ on the y-axis.

CHAPTER FOUR

Geometric Gems

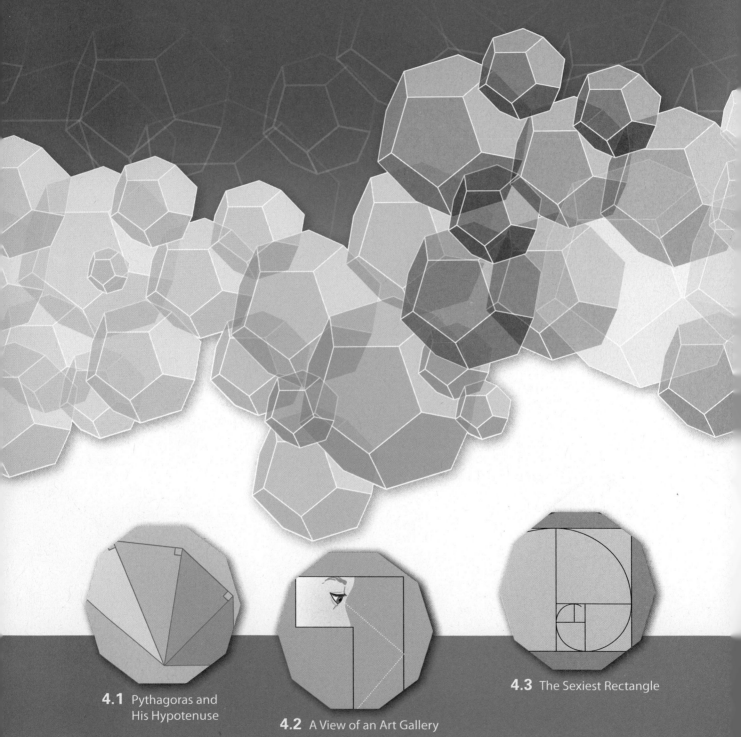

4.1 Pythagoras and
His Hypotenuse

4.2 A View of an Art Gallery

4.3 The Sexiest Rectangle

When we look at and touch the objects in our world, we encounter the physical form and spatial relationships of matter. We see shapes—some interesting, some ordinary, and some attractive. Geometry is the study of those shapes. Geometric relationships give us a basic sense of order, coherence, and beauty. The lines, circles, and spatial patterns of the physical world give depth to our intuition about shape and enable us to develop insights and recognize patterns in things geometric. Geometry captures the structure and nuances of our physical reality.

Geometry has been studied since ancient times. However, advances in geometry are still being made, and many questions remain unanswered. Here we will examine both ancient geometric discoveries and modern developments. The geometry of the plane includes not only the ancient Pythagorean Theorem but also such issues as guarding museums, investigating aesthetics, creating nonrepeating floor patterns, and searching for symmetry. Of course, geometry extends beyond the flat plane and describes objects in space. There too we see contributions ancient and modern, from the symmetric Platonic solids, which have been admired for millennia, to curved geometries that provide new tools for describing our universe. Finally, we journey into a world created by imagination—the fourth dimension.

Exploration of geometry allows us to experience ways of thinking that can be applied in many areas, both within mathematics and beyond it: keeping an open mind to all possibilities—even those that first appear impossible; understanding simple things deeply; exploiting our insights; searching for patterns; and breaking up difficult tasks into many easy ones. But perhaps the technique most powerfully illustrated by geometry is the idea of building understanding by actually doing and trying. Physically holding and manipulating objects or looking carefully at illustrations allows us to understand ideas more viscerally. We begin our travels through geometry by applying this strategy—physically doing and trying—to one of the most important and ancient geometrical gems, the Pythagorean Theorem.

4.4 Soothing Symmetry and Spinning Pinwheels

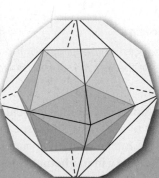

4.5 The Platonic Solids Turn Amorous

4.6 The Shape of Reality?

4.7 The Fourth Dimension

PYTHAGORAS AND HIS HYPOTENUSE

How a Puzzle Leads to the Proof of One of the Gems of Mathematics

© Yen Teoh/iStockphoto

It can't be wrong when it looks so *right!*

> *I'm very well acquainted, too, with matters mathematical,*
> *I understand equations, both the simple and quadratical,*
> *About Binomial Theorem I'm teeming with a lot of news,*
> *With many cheerful facts about the square of the hypotenuse.*
>
> WILLIAM S. GILBERT,
> *THE PIRATES OF PENZANCE*

Some geometric relationships are so profound that they have changed the shape of civilization—both literally and figuratively. Perhaps the best known and most fundamental theorem in all of mathematics is the *Pythagorean Theorem*. While we have seen this basic fact in high school, here we examine it more deeply.

The Pythagorean Theorem.

In any right triangle, the square of the length of the hypotenuse is equal to the sum of the squares of the lengths of the other two sides.

This theorem and its proof form one of the classical intellectual accomplishments of humankind, comparable to a work of Shakespeare. The proof presented here, which was discovered by the Indian mathematician Bhaskara in the 12th century, exemplifies aesthetics and beauty in mathematical arguments.

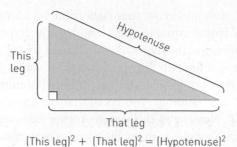

$(\text{This leg})^2 + (\text{That leg})^2 = (\text{Hypotenuse})^2$

A Puzzle Proof

Since the Pythagorean Theorem concerns the relationships among the sides of a right triangle, first we must become well acquainted with a right triangle—see it, feel it, handle it. In your kit you will find four copies of the same right triangle and one small square. Play with

these shapes, move them around, put them together, and search for interesting patterns: things that fit, relationships, familiarity. Just see what you find, and remember that no discovery is too small or too inconsequential.

As you play with these shapes, notice that the two nonright (which we jokingly refer to as "wrong") angles of one of the triangles fit together to form a right angle. Just move the triangles around to confirm this fact. You may remember from high school geometry that the sum of the angles of any triangle is 180° and that a right angle is 90°. Thus, the sum of the two remaining angles should be 90° as well—one of the rare occasions when two wrongs actually do make a right.

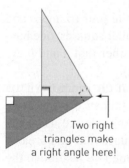

Two right triangles make a right angle here!

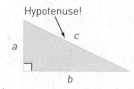

Life Lesson

Powerful and important discoveries can come from play.

Hypotenuse!

A Hypotenuse Square

Now we begin the proof of the Pythagorean Theorem in earnest. Let's call the length of the hypotenuse of the right triangle c and the lengths of the other two sides a and b. We may now state the Pythagorean Theorem as

$$a^2 + b^2 = c^2.$$

We first wish to use the five pieces from the kit to construct a large square with each side of length c, which is the length of the hypotenuse. You have four identical right triangles, and so you have four hypotenuses, one for each side of the square. Assemble your four right triangles and one little square into a square having side length equal to c, the length of the hypotenuse.

Remember that the hypotenuse of a right triangle is the side opposite the right angle. You may have to flip your triangles over to get them to fit together into a square. Make certain that you keep the hypotenuses toward the outside of the square you are building. Also recall that the angles at each corner of the square must be 90°. Now build this square having sides equal to the hypotenuse, using the four right triangles with the little square in the middle. Here we see in action the credo "doing and trying are powerful means of building understanding."

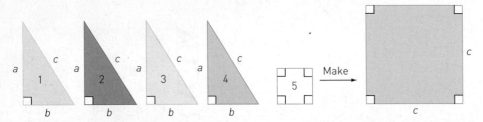

Using the pieces on the left, construct the square on the right.

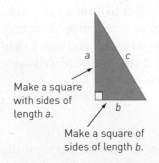

Make two squares.

Make a square with sides of length a.

Make a square of sides of length b.

You have now created a large square out of four right triangles and one small square. This big square's area is equal to the length of the hypotenuse of the original right triangle multiplied by itself or c^2. So we just discovered that the total area of all five pieces is equal to c^2.

The Two-Small-Squares Challenge

The goal now is to rearrange the five pieces we just used, the four triangles and the small square, into a new configuration to create two smaller squares: one having side length a and the other having side length b (remember that a and b are the lengths of the nonhypotenuse sides of the right triangle).

Once we have built these two squares, the area of one of the squares will be a^2, and the area of the other will be b^2. So, the area of this total new configuration will be the sum of the two areas: $a^2 + b^2$. But remember that we are using the exact same pieces we used previously to make the large square that had area c^2. Since the areas of these two configurations must be the same (we just move the pieces around, we don't change their size), we will therefore have shown that $a^2 + b^2 = c^2$. Try it now by moving the pieces around and making shapes.

A Hint

The two squares having areas a^2 and b^2 will not be freestanding squares. They will be parts of a shape that looks like a thickened L. *Don't read on until you have really tried to solve this puzzle. You can do it!*

The Answer

We can make the two squares by just moving two triangles from the original large square we constructed.

By drawing the line that divides that figure into two squares, one having area a^2 and the other having area b^2, we see that the total area is $a^2 + b^2$.

Life Lesson
Keep an open mind to all possibilities.

We see that the same four triangles and small square that previously made a large square of area c^2 now fit together to make a figure having area $a^2 + b^2$. Thus, for this right triangle anyway, the Pythagorean Theorem holds. Of course, this same proof could have been done starting with any right triangle and a small square with side $a - b$. So, this geometric demonstration completes a proof of the Pythagorean Theorem.

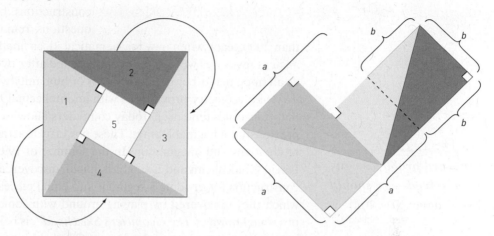

Although our proof is now complete, it will not be real until we have made it our own. Work through the proof yourself. Attempt not only to understand the logic of each and every step but also to capture the essence of the entire argument. Say, "Voilà!" or "Eureka!" or "Cool!" when you have drawn the line separating the two squares a^2 and b^2. Once you have digested the ideas, explain the proof to a roommate, a relative, or a friend. Use puzzle pieces and be patient if they do not immediately see it. It is only after we have explained the proof to several people, and witnessed their moment of understanding, that it is ours for life.

You may also experiment with the Pythagorean Theorem using the *Heart of Mathematics* Web site, www.heartofmath.com.

> *. . . a theorem as "the square of the hypote nuse of a right angled triangle is equal to the sum of the squares of the sides" is as dazzlingly beautiful now as it was in the day when Pythagoras discovered it.*
>
> **LEWIS CARROLL**

A Fascination with Geometrical Ideas

Geometrical issues have fascinated people since the dawn of thought and continue to inspire and tantalize us. The Greeks of antiquity viewed the line and the circle to be the most elegant and fundamental of all geometric forms. Consequently,

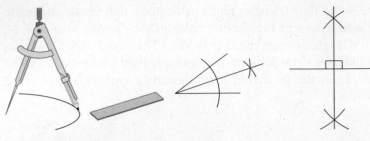

the Greeks were captivated with the study of geometrical constructions—identifying those shapes and lengths that can or cannot be drawn using just a compass (which makes circles) and a straight, unmarked edge (which makes lines). The Greeks devised many constructions, but a handful of construction questions remained unanswered for more than 2000 years, when new mathematical ideas finally settled them. The fact that ancient mysteries were definitively resolved after two millennia of effort can give us all hope that other great problems of humanity will also be solved.

While geometry provides us with an intellectual tie to ancient Greek thinkers that transcends time itself, today computers allow us to see geometrical objects in previously unimaginable ways. These powerful tools are spawning a new generation of creativity and imagination. In the summer of 1995, two eighth-grade students, David Goldenheim and Daniel Litchfield, discovered a completely new geometric construction for dividing a segment into equal pieces. Their original construction, which they discovered by playing around with some clever ideas on a computer program known as *The Geometer's Sketchpad*™, is believed to be the first such construction since the time of the ancient Greeks! Indeed, this interweaving of ancient ideas with modern thought forms the detailed fabric we now view as geometry.

Life Lesson

Hard problems can be solved—we should never give up.

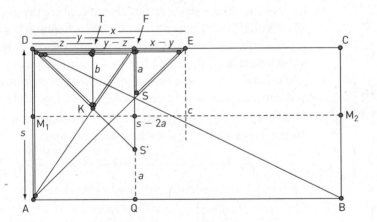

The eighth-grade construction

A Look BACK

THE PYTHAGOREAN THEOREM can be proved by physically building with puzzle pieces a square whose sides are the hypotenuse of a right triangle and showing how those pieces can be reassembled into a thickened L shape with area equal to the sum of the squares of the lengths of the other two sides.

The Pythagorean Theorem is an idea that may at first seem abstract, but when we hold and manipulate physical triangles, that abstract idea can become concrete and real. Using several ways of learning about an idea helps us to understand it.

Life Lessons

Experience ideas in as many ways as possible.

MINDSCAPES   Invitations to Further Thought

*In this section, Mindscapes marked (**H**) have hints for solutions at the back of the book. Mindscapes marked (**ExH**) have expanded hints at the back of the book. Mindscapes marked (**S**) have solutions.*

Developing Ideas

1. **The main event.** State the Pythagorean Theorem.

2. **Two out of three.** If a right triangle has legs of length 1 and 2, what is the length of the hypotenuse? If it has one leg of length 1 and a hypotenuse of length 3, what is the length of the other leg?

3. **Hypotenuse hype.** If a right triangle has legs of length 1 and x, what is the length of the hypotenuse?

4. **Assessing area.** Suppose you know the base of a rectangle has a length of 4 inches and a diagonal has a length of 5 inches. Find the area of the rectangle.

5. **Squares all around.** How does the figure below relate to the Pythagorean Theorem?

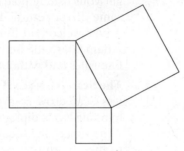

Solidifying Ideas

6. **Operating on the triangle.** Using a straightedge, draw a random triangle. Now carefully cut it out. Next amputate the angles by snipping through adjacent sides. Now move the angles together so the vertices all touch and the edges meet. What do you conclude about the sum of the angles of a triangle? Try this procedure with triangles having different dimensions.

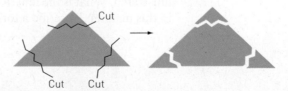

7. **Excite your friends about right triangles.** Describe the proof of the Pythagorean Theorem to someone who has never seen it before. Try to get him or her inspired and intrigued by math! Record the event and the various reactions.

8. **Easy as 1, 2, 3?** Can there be a right triangle with sides of length 1, 2, and 3? Why or why not? Can you find a right triangle whose side lengths are consecutive natural numbers?

9. **Sky high (S).** On a sunny, warm day, a student decides to fly a kite on the college green just to relax. His kite takes off and soars. He lets all 150 feet of the string out and attracts a crowd of onlookers. There is a slight breeze, and a spectator 90 feet away from the student notices that the kite is directly above her. Unlike a real kite, this math-question kite has the string going in a straight line from the student to the kite. How high is the kite from the ground?

10. **Sand masting (H).** The sailboat named Sand Bug has a tall mast. The backstay (the heavy steel cable that attaches the top of the mast to the back, or stern, of the sailboat) is made of 130 feet of cable. The base of the mast is located 50 feet from the stern of the boat. How tall is the mast?

11. **Getting a pole on a bus.** For his 13th birthday, Adam was allowed to travel down to Sarah's Sporting Goods store to purchase a brand new fishing pole. With great excitement and anticipation, Adam boarded the bus on his own and arrived at Sarah's store. Although the collection of fishing poles was tremendous, there was only one pole for Adam and he bought it: a five-foot, one-piece fiberglass "Trout Troller 570" fishing pole. When Adam's bus arrived, the driver reported that Adam could not board the bus with the fishing pole. Objects longer than four feet were not allowed on the bus. In tears, Adam remained at the bus stop holding his beautiful five-foot Trout Troller. Sarah, seeing the whole ordeal, rushed out and said, "Don't cry, Adam! We'll get your fishing pole on the bus!" Sure enough, when the same bus and the same driver returned, Adam boarded the bus with his fishing pole and the driver welcomed him aboard with a smile. How was Sarah able to have Adam board the bus with his five-foot fishing pole without breaking the bus line rules and without cutting or bending the pole?

12. **The Scarecrow (ExH).** In the 1939 movie *The Wizard of Oz*, when the brainless Scarecrow is given the confidence to think by the Wizard (by merely handing him a diploma, by the way), the first words the Scarecrow utters are, "The sum of the square roots of any two sides of an isosceles triangle is equal to the square root of the remaining side." An isosceles right triangle is just a right triangle having both legs the same length. Suppose that an isosceles right triangle has legs each of length 3. What is the length of the hypotenuse? Is the Scarecrow's assertion valid? This question illustrates the true value of a diploma without studying.

13. **Rooting through a spiral.** Start with a right triangle with both legs having length 1. What is the length of the hypotenuse? Suppose we draw a line of length 1 perpendicular to the hypotenuse and then make a new triangle as illustrated. What is the length of this new hypotenuse? Suppose we continue in this manner. Describe a formula for the lengths of all the hypotenuses.

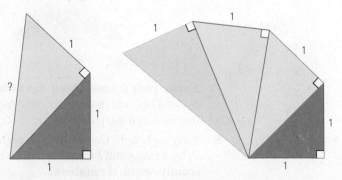

14. **Is it right? (H)** Suppose someone tells you that she has a triangle with sides having lengths 2.6, 8.1, and 8.6. Is this a right triangle? Why or why not? Is there an angle in the triangle larger than 90°? Justify your answer.

15. **Train trouble (H).** Train tracks are made of metal. Consequently, they expand when it's warm and shrink when it's cold. When riding in a train, you hear the clickety-clack of the wheels going over small gaps left in the tracks to allow for this expansion. Suppose you were a beginner at laying railroad tracks and forgot to put in the gaps. Instead, you made a track 1 mile long that was firmly fixed at each end. On a hot day, suppose the track expanded by 2 feet and therefore buckled up in the middle, creating a triangle.

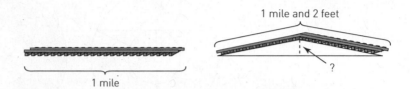

1 mile and 2 feet

?

1 mile

Roughly how high would the midpoint be? Now you may appreciate the click-clack of the railroad track.

Creating New Ideas

16. **Does everyone have what it takes to be a triangle?** Suppose a friend comes up to you and says, "Hey, I just made a triangle with sides of length 2431; 5642; and 3210." How would you respond to him? What basic fact about triangles do you conclude from this dialogue with your friend?

17. **Getting squared away.** In our proof of the Pythagorean Theorem, we stated that the second figure is actually two perfect squares touching along an edge. Prove that they are indeed both perfect squares. It will be useful for you to use the puzzle pieces from your kit to build the first big square again and carefully notice how the pieces all fit together.

18. **The practical side of Pythagoras.** Suppose you are building a patio and you want to make certain that the sides of your patio meet at right angles. Using the converse of the Pythagorean Theorem that appears at the end of Mindscape 22, give a practical and easy method to check that the angle between two adjacent sides is 90°.

19. **Pythagorean pizzas (H).** You have a choice at the local pizza place: For the same price you can get either one large pizza or both a small and a medium. How can you determine which way you get more by using just the Pythagorean Theorem and knowing the diameters of the different sizes of pizza? Describe an easy procedure to figure out which deal to choose.

20. **Natural right (S).** Suppose r and s are any two natural numbers where r is bigger than s. Using the converse of the Pythagorean Theorem that appears at the end of Mindscape 22, show that the triangle having side lengths equal to $2rs$, $r^2 - s^2$, and $r^2 + s^2$ is actually a right triangle. Are there infinitely many different right triangles having all sides of integer lengths?

Further Challenges

21.   Well-rounded shapes. Suppose we have two circles having the same center. The small one has radius r, and the large one has radius R. Let's now consider two shapes. The first is the doughnut-like region between the two circles. The second is a disk whose diameter is the length of the line segment whose endpoints are on the large circle and whose center point touches the small circle at its north pole. What are the areas of these two shaded regions? How do their sizes compare with each other? (We note that the formula for the area of a circle of radius r equals πr^2.)

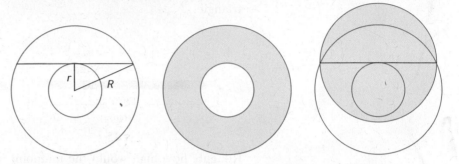

22.   A Pythagorean Theorem for triangles other than right triangles. Suppose we have a triangle that is not a right triangle. For example, consider:

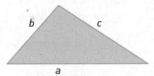

Now, if we drop a perpendicular line from the top vertex down to the side of length a and we cut that side into two pieces of lengths, a' and a'', we would have:

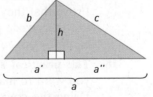

It turns out that $a^2 + b^2 = c^2 + 2aa'$. Use the Pythagorean Theorem with the two right triangles in the preceding picture to produce two equations. Subtract one equation from the other and notice that the h^2 terms drop out. Now, deduce the preceding formula (remember that $a = a' + a''$, so you can solve for a''). Once you have proved this formula, show what happens if the angle between sides a and b is 90°. Notice that you have actually proved the "converse" of the Pythagorean Theorem: If a triangle has sides with lengths a, b, and c satisfying $a^2 + b^2 = c^2$, then it is a right triangle.

In Your Own Words

23.   With a group of folks. In a small group, discuss and actively work through the proof of the Pythagorean Theorem. After your discussion, write a brief narrative describing the proof in your own words.

For the Algebra Lover

Here we celebrate the power of algebra as a powerful way of finding unknown quantities by naming them, of expressing infinitely many relationships and connections clearly and succinctly, and of uncovering pattern and structure.

24. **Double trouble.** Suppose you know a right triangle has legs of length a and b, and hypotenuse of length c. Verify that a triangle with sides of length $2a$, $2b$, and $2c$ still satisfies the Pythagorean Theorem.

25. *K***-ple trouble.** Suppose you have a right triangle that has legs of length a and b, and hypotenuse of length c. Let k be a fixed natural number. Verify that a triangle having side lengths ka, kb, and kc also satisfies the Pythagorean Theorem.

26. **Padding around.** You have a rectangular patio with a diagonal measurement of 20 feet. Suppose the length of the patio is y and the width is x. Find an equation that relates x and y. What's the area of the patio if the $x = 12$ feet?

27. **Pythagoras goes the distance.** Plot the points $(5, 0)$ and $(0, 12)$ in the xy-plane. Use the Pythagorean Theorem and your sketch to find the distance between the points $(5, 0)$ and $(0, 12)$ in the xy-plane. Add the points $(0, 0)$ and $(5, 12)$ to your sketch and find the distance between $(0, 0)$ and $(5, 12)$.

28. **Ahoy there! (H)** Your exotic sailboat, which you named the *Pythagoras*, has two sails. Each one is the shape of a right triangle. The smaller sail has legs of length 1 and 3 meters and hypotenuse of length x meters. The larger sail has one leg whose length equals the length the hypotenuse of the shorter sail. The larger sail has a hypotenuse of length $\sqrt{26}$ meters. Find the lengths of all the sides of both triangular sails.

A VIEW OF AN ART GALLERY

Using Computational Geometry to Place Security Cameras in Museums

Angelo Hornak/© Corbis

The Guggenheim Museum in New York City

> *The interior is infinite, all the way to the mystery of the inmost, the charged point, a kind of sum total of the infinite.*
>
> PAUL KLEE

We now move from the Pythagorean Theorem, a geometric gem that has been around for over 2000 years, to a geometric question that has been around for just 20 years. This new geometric question asks how many eyes are necessary to view an entire art gallery—all at once.

A Look at a Gallery

When we walk around an art gallery, we study and enjoy the works hanging on the walls. Normally we don't pay much attention to the floor plan. Floor plans, however, are exactly what we wish to study now. If we consider a gallery floor plan, and assume the gallery is one big open space with no interior walls, then we might see something like the diagram on the next page. We will assume that each wall is flat (so the Guggenheim is out).

Since we are assuming that our museum consists of one large room with no interior walls or partitions, the surrounding walls form a jagged "loop" that does not cross through itself. In other words, the building is completely sealed, and there is only one gallery on the inside. Let's call any figure that is made up of straight line pieces that are connected end to end to form a loop a *polygonal closed curve*. The corners where walls meet are called *vertices*. So, the floor plan of a gallery made up of straight walls and having just one big open space on the inside is an example of a polygonal closed curve—notice that the gallery may have many nooks and crannies.

Top view of an interestingly shaped gallery

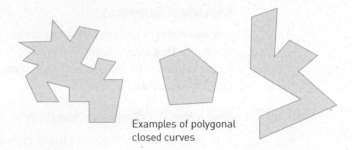

Examples of polygonal
closed curves

Finding Points of View

A sad truth is that today museums need security systems to guard their valuable
works of art. How can one keep a watchful eye on things? Suppose our goal is
to strategically place video cameras such that every point of the gallery will be
viewed by at least one camera. In an attempt to be unintrusive, suppose we wish
to mount cameras only in the corners of the gallery. Each camera is equipped with
a wide-angle lens that enables the camera to view everything inside the V formed
by the two adjacent walls the camera is housed between.

It is possible that the corner formed by the two walls actually sticks into the
room. In this case, the camera is equipped with a special fish-eye type lens, giving
it nearly panoramic vision and a sweeping view of everything visible between one
wall of the V and the other.

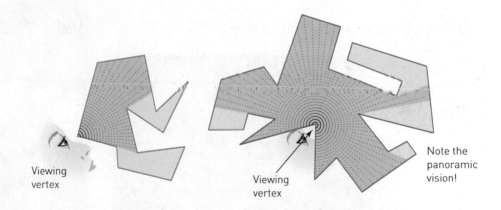

Shading shows the area that
can be viewed from the camera
located at the vertex with the eye.

Viewing
vertex

Viewing
vertex

Note the
panoramic
vision!

Our Main Question

Many lines of sight with cameras
at *every* corner

How many cameras are needed to ensure that every point in the gallery
is seen by some camera?

Of course, the answer depends on how complicated the floor plan is.
However, no matter how many walls and bends it has, if we place one camera at every corner, then we are able to keep a watchful eye on every point
in the gallery. But cameras are expensive, and perhaps a camera in every
corner is not necessary. For example, in the gallery illustrated what is the
largest number of cameras we can remove so that every point inside is still
visible from some camera? This issue leads to the following question.

Modified Question

For an arbitrary gallery having v corners (vertices), how many cameras are needed to ensure that every point in the gallery is seen by some camera?

Pared down to its geometrical essence, the question becomes the "Klee Art Gallery Question."

The Klee Art Gallery Question

For an arbitrary polygonal closed curve in the plane with v vertices, how many vertices are required such that every point in the interior of the curve is visible (directly in the line of sight) from at least one of these chosen vertices?

This question represents an issue in an area of study known as *computational geometry*—an area of mathematics dealing with efficient methods of computing things involving geometric objects. These methods often lead to improved algorithms, and thus there is much cross-pollination between computational geometry and computer science. Before answering the Klee question, let's first mention the famous art name associated with this question. Paul Klee was an imaginative abstract artist whose work is world-renowned. However, this question is named after Victor Klee from the University of Washington, who in 1973 first posed it casually in conversation. Victor Klee's work in combinatorics and computational geometry is also world-renowned. In fact, Victor Klee's gallery question has led to a tremendous body of interesting work in combinatorial geometry.

Paul Klee, Hauser-Enge, 1939, Swiss. Watercolor and charcoal.

Paul Klee, *Castle in a Garden,* 1919.

Let's now consider various art gallery shapes to build some insights and make some guesses. For each example, mark the fewest vertices required so that every point inside is directly in the line of sight of at least one of your marked vertices.

Where would you place
your cameras in these
gallery floor plans?

An Untangling Example—The Comb Gallery

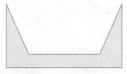

2-tine comb gallery

Let's look at a floor plan that resembles a comb. Each tine is a thin triangle that opens onto one long hall. Notice that each long narrow triangular hallway needs its own camera somewhere. But how many cameras do we need? Well, we could place two cameras in a number of ways to view the entire interior. One camera alone cannot do the job since it is impossible for one camera to see down both triangular hallways. So, the smallest number of cameras needed for this 2-tine comb is two. Notice that the comb has six vertices. What is the fewest number of cameras needed for the 3-, 4-, and 5-tine combs? How many vertices does each comb have? What is the relationship between vertices and cameras? What can we conclude about the minimum number of cameras needed for comb-shaped galleries?

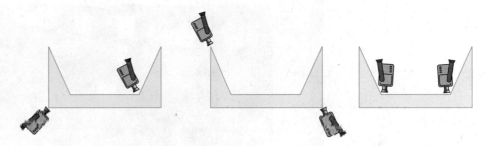

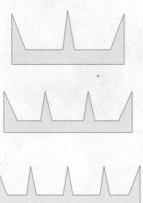

If you answered the previous questions, you've discovered that we need exactly three cameras for the 3-tine comb, which has 9 vertices; four cameras for the 4-tine comb, which has 12 vertices; and five cameras for the 5-tine comb, which has 15 vertices. Of course, we could make comb-shaped galleries with any number of tines. Such a gallery with t number of tines has $3t$ vertices and requires t cameras. So, from these examples, we conclude that a gallery with v vertices may need as many as $v/3$ cameras. We will now discover that that number of cameras will *always* be enough, no matter what the shape of the gallery—thus answering the Klee Art Gallery Question with the following theorem.

The Art Gallery Theorem.

Suppose we have a polygonal closed curve in the plane with v vertices. Then there are $v/3$ vertices from which it is possible to view every point on the interior of the curve. If $v/3$ is not an integer, then the largest number of vertices we need is the biggest integer less than $v/3$.

This result was first formulated and proved by Vasek Chvátal, from Rutgers University, a couple of days after Victor Klee raised his off-the-cuff but interesting question. A few years later Steve Fisk, from Bowdoin College, gave a different proof of the Art Gallery Theorem. It is his argument that we will now think about together. On the *Heart of Mathematics* Web site, you can find a program that allows you to construct your own art galleries and experiment with camera placements.

Creativity Happens

Creative ideas can arise anytime and anywhere as long as our minds are open. The story of how Steve Fisk came up with the idea for his proof is a wonderful and funny real-life illustration of this maxim. Here Professor Fisk recounts the episode:

> *Here's the quick story. My wife and I were traveling across Asia during 1975–1976, and during our one-month stay in Iran I visited the math library at the University of Tehran. I read Chvátal's article — or rather I looked at it, for the proof seemed so complex that I didn't want to read it. I did like the result, though. A few weeks later we were traveling on a bus between Herat and Kandahar in Afghanistan. It was hot, dusty, and noisy with squawking chickens on the roof. I was thinking about it — well I was probably dozing off actually — and I thought of the proof.*

Courtesy of Steve Fisk

Professor Steve Fisk and his future wife, Judy Lloyd, in Asia.

Proving the Art Gallery Theorem

Let's investigate why $v/3$ vertices are always enough to see every point inside a polygonal curve. Where do we begin? The difficulty is that we have no idea what the polygonal curve looks like. That is, we don't know the floor plan of the gallery. It may have hundreds or thousands of complex hallways and corners. What should we do? When faced with a challenging issue in life, we should always try to begin with an easy case and warm up to the more difficult task.

What is the simplest polygonal curve (gallery floor plan)? The answer is a *triangle*. Notice that the triangle has three vertices and that from any one vertex we are able to view every point in the triangle. So only one vertex is needed to view the entire interior, and, happily, $v/3$ in this case is 3/3, which equals one vertex. Thus, the theorem is true for any triangular polygonal curve. We also notice that, in this simple case, we can use any of the three vertices as a location for our camera. That was easy. Now that we've tackled the simple, let's move to the more complex.

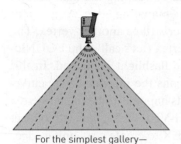

For the simplest gallery— the triangle—any vertex can be used to view the entire museum.

Divide and Conquer

Our arbitrary polygonal curve gallery will not always be a triangle, of course. What do we do with a more complex shape? The answer is to divide it into a whole bunch of triangles—that is, convert one complicated situation into a bunch of easy ones. Let's now see how to divide the inside of the polygonal closed curve into triangles.

We build triangles in the interior of the polygonal curve by adding edges from one vertex to another with the only restriction being that two edges cannot cross. The vertices of these triangles must all be vertices of the original polygonal curve. Notice that there are many different ways to create such triangles.

> *Life Lesson*
>
> Often in life, hard questions are made up of many easy pieces.

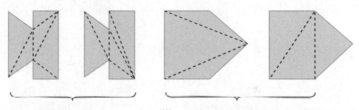

Two examples of two different triangulations of the same polygonal gallery.

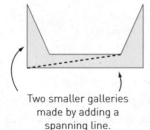

Two smaller galleries made by adding a spanning line.

How do we know that we can always divide the polygonal gallery into triangles in this way? In practice, whenever we draw a polygonal curve, it's easy to break its interior into triangles, but we need a method that will always work. Our strategy is first to do something a little easier than dividing our gallery into triangles. Instead, we just divide the inside of the polygonal curve into two pieces by drawing in a line that spans the inside of the polygonal curve between two vertices. We would then have two easier problems to solve—in this case, two smaller polygonal closed curves that we want to divide into triangles. In each of those two smaller curves, we will draw spanning lines again. Pretty soon our pieces will be so small that they must be triangles. Let's see how to find a spanning line. Remember, if we can break a hard problem into two easier problems, we have a better chance of solving it.

Illuminating the Dividing Line

Start at any vertex and notice that two edges emanate from it and that they form an angle. Actually, they form two angles, one on the interior of the polygonal curve and

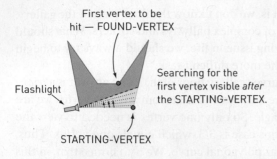

First vertex to be
hit — FOUND-VERTEX

Flashlight

Searching for the
first vertex visible *after*
the STARTING-VERTEX.

STARTING-VERTEX

one on the exterior. From the vertex we point a flashlight along one of the interior edges. The flashlight shines against the edge and illuminates the vertex at the other end—let's call that vertex the STARTING-VERTEX. We now start sweeping the light around the interior angle toward the other emanating edge. At all times we look to see whether the light ever strikes another vertex. The first time it does, we connect that vertex (let's call it the FOUND-VERTEX) to the vertex where the flashlight is located. In this case, we have located a line that spans the interior of the curve.

If, as the flashlight swings, the first vertex the light hits happens to be the far vertex of the other emanating edge (let's call it the ENDING-VERTEX), then we connect the STARTING-VERTEX to the ENDING-VERTEX to create a spanning edge.

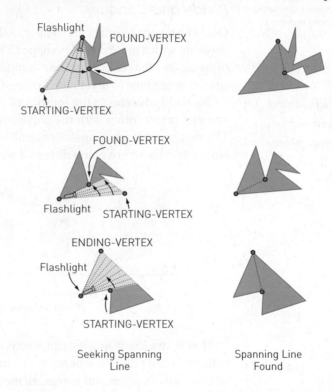

Flashlight

FOUND-VERTEX

STARTING-VERTEX

FOUND-VERTEX

Flashlight   STARTING-VERTEX

ENDING-VERTEX

Flashlight

STARTING-VERTEX

Seeking Spanning
Line

Spanning Line
Found

Repeat Until Triangles Abound

We have now broken the interior of the curve into two pieces, creating two polygonal closed curves. If either or both of these curves are not triangles, then we repeat the process just described until the entire polygon has been divided into triangles. Recall two important features of this triangulating process:

A triangulation

• Each vertex of each triangle is a vertex of the original polygonal curve.

• For any triangle, the entire triangle is visible from any one of its three vertices.

A Dash of Color

Next, we are going to assign a color to each vertex of the polygonal curve—red, yellow, or blue. We may color the vertices any way we wish, as long as we abide by one rule: The three vertices of any one individual triangle must have different colors. That is, every single triangle in the triangulation must have one red vertex, one yellow, and one blue. How can we color the vertices following this rule? Try some examples and figure out how to do it.

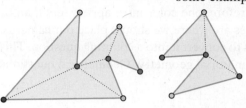

The answer is just to pick one of the triangles and color each of its vertices a different color. At least one of the edges of that triangle is shared with another triangle. This new triangle has two of its vertices colored (the two from the common edge that the triangles share). This leaves one uncolored vertex. We color it the third color. We repeat this procedure, one triangle at a time. Since all the vertices are on the polygonal curve, this process colors all the vertices as required.

Red Eye

Suppose we start with *any* gallery and triangulate the polygonal gallery and color its vertices as described above. Suppose we put cameras at every red vertex. Would we be able to see every point inside the polygonal gallery? Think about this question, and try to provide both an answer and an explanation.

The answer is, "Yes, we'd be able to see everything." Why? Select a random point inside the polygonal curve. It must be in one of our triangles. We know that that point is visible from each of the three vertices of the triangle. By our coloring rule, there must be a vertex of this triangle colored red (remember that every triangle must have its vertices colored with different colors). From that red vertex we can see the point! Therefore, every point inside the polygonal curve is visible from the red vertex. Would we be able to see every point inside the polygonal curve from just the yellow vertices? Why? Would we be able to see every point inside the polygonal curve from just the blue vertices? Why?

We have now discovered that every point inside the polygonal curve is visible from any set of vertices all having the same color. We are now going to count! How many red vertices are there? How many yellow vertices? How many blue vertices? The answers can be given succinctly: "We have no idea." Okay, well, that isn't too informative. What can we say about the number of vertices of various colors? Let's call the number of red vertices R, the number of yellow vertices Y, and the number of blue vertices B. Even though we do not know what R, Y, or B equals, we do know something about all of them together. What does $R + Y + B$ equal?

Since every vertex of the polygonal curve was colored one of the three colors and we are told that there are a total of v vertices, we know that $R + Y + B = v$. This is a bit of progress. We have three numbers (R, Y, and B) that add up to v, so they can't all three be bigger than $v/3$. That is, at least one of them must be less than or equal to $v/3$. So, we have just discovered that there must be a color that is used less than or equal to $v/3$ times. We have already seen that, if we look just from the vertices colored with that color, we can see the entire interior of the polygonal curve. So, we are always able to see every point inside from less than or equal to $v/3$ vertices—which is exactly what we wanted to prove.

The Strategy in Review

Let's review the strategy of our proof. We first reduced the difficult issue into an enormous number of less difficult issues (we divided our gallery into triangles).

Then we colored the vertices of each triangle with three different colors. We discovered that every interior point in the polygon would be visible if viewed from all the vertices of the same color. Finally, we counted the number of vertices of each color and realized that we cannot have each of the three colors occurring more than one-third of the time. Therefore, one color must appear less than or perhaps equal to $v/3$, and those vertices are where we should place our cameras. The key to this geometry question was to convert it into a counting question. This technique is the main theme of combinatorial geometry: Geometrical questions are answered by counting.

Mirror, Mirror on the Wall…

We have considered both ancient and modern questions and their answers. Often it's interesting to look into the future and contemplate those questions that remain unanswered. Here is one such geometry question that relates to our art gallery discussion. No one knows the answer.

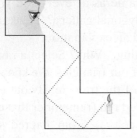

Seeing around corners in a mirrored room.

Suppose we have a gallery in the shape of a polygonal closed curve, but this time it is a gallery of mirrors. That is, each wall is completely covered by a mirror. Reflect on this image if you will. Since the walls are all mirrored, we can see around some corners by looking through reflections off various walls. Guarding a mirrored gallery might need far fewer video cameras. In fact, from a single point inside the gallery (not necessarily at a vertex), we might be able to see every other point inside the gallery by using the mirrors. Some people believe there is always at least one point in the gallery from which every other point inside is visible, but no one knows for certain.

Visibility Within a House of Mirrors Question.

For an arbitrary polygonal closed curve having mirrored sides, must there exist a point from which every point inside the polygonal curve is visible?

We can find such vantage points in some examples. But how can we prove that a point of visibility *always* exists for *all* polygonal curves—no matter how complicated the curve? Will the solution involve some of the analysis used by the Greeks; or will it follow from an insightful observation by some eighth graders using computer graphics software; or will the solution involve new techniques; or will someone find a mirrored polygonal closed curve that contains no point of complete visibility? As with all unanswered questions, we will not know for sure until someone's creative ideas lead us to a complete answer.

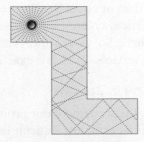
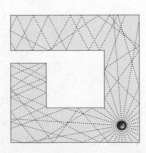

A Look BACK

IF WE HAVE a polygonal closed curve in the plane with v vertices, then there are $v/3$ vertices from which it is possible to view every point on the interior of the curve. The Art Gallery Theorem can be proved by dividing the interior into triangles, coloring their vertices, and then considering the view at vertices of the same color.

We began our exploration of this topic by drawing and examining examples to develop insights. Then, faced with potentially complicated closed curves, we broke this complicated situation into many manageable pieces—triangles. We looked at examples and then attacked a difficult issue by dividing it into many simpler pieces.

These life strategies are among our most helpful guides when our curiosity and wonder lead us to an encounter with the unknown.

Life Lessons

Ground your understanding in examples.

•

Look for patterns.

•

Divide and conquer.

MINDSCAPES Invitations to Further Thought

*In this section, Mindscapes marked (**H**) have hints for solutions at the back of the book. Mindscapes marked (**ExH**) have expanded hints at the back of the book. Mindscapes marked (**S**) have solutions.*

Developing Ideas

1. **Standing guard.** Draw the floor plan of a gallery with three vertices. What shape do you get? What's the smallest number of guards you need?

2. **Art appreciation.** State the Art Gallery Theorem.

3. **Upping the ante.** How many guards do you need for a gallery with 12 vertices? With 13 vertices? With 11?

4. **Create your own gallery.** Use the program on the *Heart of Mathematics* Web site to create at least three interesting galleries and place cameras at as few vertices as possible. Sketch or print out your floor plans.

5. **Keep it safe.** At what vertices would you place cameras so that you use as few cameras as possible and so that each point inside the curve is visible from a camera?

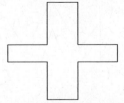

Solidifying Ideas

6. **Klee and friends.** Explain to a friend the statement of the Art Gallery Theorem and its proof. What are his or her reactions and questions?

7. **Putting guards in their place.** For each floor plan, place guards at appropriate vertices so that every point in the museum is within view.

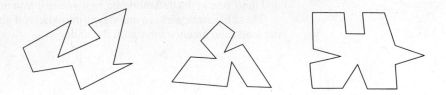

8. **Guarding the Guggenheim.** The Art Gallery Theorem does not tell us how many guards are needed to guard museums that have curved walls. For each floor plan, place the minimum number of guards at vertices so that every point in the museum is within view.

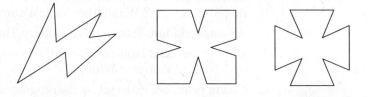

9. **Triangulating the Louvre (H).** Triangulate the floor plans by adding straight segments that do not cross each other yet span the insides and extend from one vertex to another.

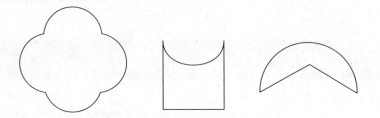

10. **Triangulating the Clark.** Triangulate the floor plans by adding straight segments that do not cross each other yet span the insides and extend from one vertex to another.

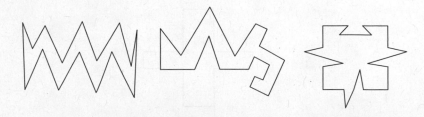

11. **Tricolor me (ExH).** For each triangulation, color the vertices red, blue, or green so that every triangle has all three colors.

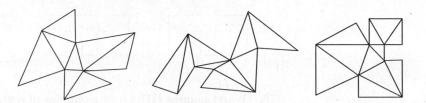

12. **Tricolor hue.** For each triangulation, color the vertices red, blue, or green so that every triangle has all three colors.

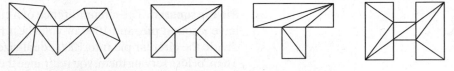

13. **One-third.** Write the number 6 as a sum of three natural numbers in several different ways, and, in each sum, circle a number that is less than or equal to 2.

14. **Easy watch.** Draw a floor plan of a museum with six sides that needs only one guard to view the entire gallery.

15. **Two watches (S).** Draw the floor plan of a museum with 10 sides that needs exactly two guards to view the entire gallery.

Creating New Ideas

16. **Mirror, mirror on the wall.** Consider the floor plans of the rooms below.

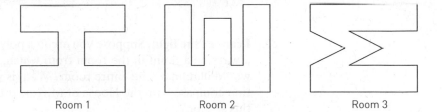

Room 1   Room 2   Room 3

Suppose that all the walls are covered with mirrors. For each room, find a point inside from which all other points are visible. Use illustrations showing the reflections to justify your answers.

17. **Nine needs three (H).** Draw a floor plan for a museum having nine sides that needs exactly three guards to watch the entire gallery. Show the placement of the guards in your drawing.

18. **One-third again (ExH).** If a natural number is written as the sum of three natural numbers, show that one of the numbers in the sum must be less than or equal to one-third of the original natural number.

19. **Square museum (S).** If a museum has only right-angled corners, how many guards are necessary? It turns out that you can guard such museums with

only one-fourth as many guards as corners. In the examples below, where would you place the guards to guard the museum?

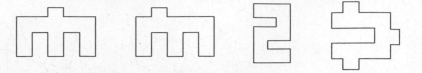

20. **Worst squares (H).** Draw examples of museums with only right-angled corners having 12 sides, 16 sides, and 20 sides that require three, four, and five guards, respectively.

Further Challenges

21. **Pie are squared.** The circumference of a circle of radius r is $2\pi r$. Suppose you have a round pie, and 1000 of your best friends come over for dessert. You divide the circular pie into 1000 equal-size pieces by cutting from the center. Then, before serving them, you rearrange the pieces by putting the first piece with the curved part up, the next piece right next to it with curved edge down, the next one with curved side up, alternating until all 1000 pieces are arranged. The shape you have constructed is almost an exact rectangle except that its top and bottom edges are each made of 500 just slightly curved tiny segments that came from the edge of the pie. How long is the rectangle? How wide is it? What is its area? Why is this story a convincing demonstration that the formula for the area of a circle is correct?

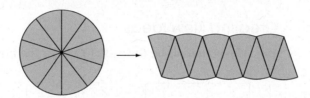

22. **I can see the light.** Suppose you are in a polygonal mirrored room. Will there always be a point in the room from which, if a light were placed, that light would illuminate the entire room? What is your guess? How does this question compare with the House of Mirrors question mentioned at the close of this section?

In Your Own Words

23. **Personal perspectives.** Write a short essay describing the most interesting or surprising discovery you made in exploring the material in this section. If any material seemed puzzling or even unbelievable, address that as well. Explain why you chose the topics you did. Finally, comment on the aesthetics of the mathematics and ideas in this section.

24. **With a group of folks.** In a small group, discuss and actively work through the statement and proof of the Art Gallery Theorem. After your discussion, write a brief narrative describing the theorem and proof in your own words.

25. **Creative writing.** Write an imaginative story (it can be humorous, dramatic, whatever you like) that involves or evokes the ideas of this section.

26. **Power beyond the mathematics.** Provide several real-life issues—ideally, from your own experience—that some of the strategies of thought presented in this section would effectively approach and resolve.

For the Algebra Lover

Here we celebrate the power of algebra as a powerful way of finding unknown quantities by naming them, of expressing infinitely many relationships and connections clearly and succinctly, and of uncovering pattern and structure.

27. **Less than.** You've triangulated your polygon and assigned a color (red, blue, or yellow) to each vertex so that each triangle has all three colors appearing. Let r equal the number of red vertices, b equal the number of blue vertices, and y equal the number of yellow vertices. If $r < 10$, $b < 8$, and $y < 12$, what's the largest number of vertices the polygon can have?

28. **Greater than.** You've triangulated your polygon and assigned a color (red, blue, or yellow) to each vertex so that each triangle has all three colors appearing. Let r equal the number of red vertices, b equal the number of blue vertices, and y equal the number of yellow vertices. If $r > 6$, $b > 7$, and $y > 10$, what's the smallest number of vertices the polygon can have?

29. **Counting the colors.** Your polygon has 40 vertices. Thirty percent have been colored red, 20% yellow, and the remainder blue. Determine the number of vertices of each of the three colors.

30. **Only red.** Twelve of your polygon's vertices have been colored red. If this equals 15% of the vertices, how many vertices does your polygon have?

31. **Totaling triangles.** If a polygon has n sides, it can be shown that any triangulation has $n - 2$ triangles. Suppose you have a triangulated polygon and notice that the number of sides in the polygon plus the number of triangles equals 100. How many sides does your polygon have?

4.3 THE SEXIEST RECTANGLE

Finding Aesthetics in Life, Art, and Math Through the Golden Rectangle

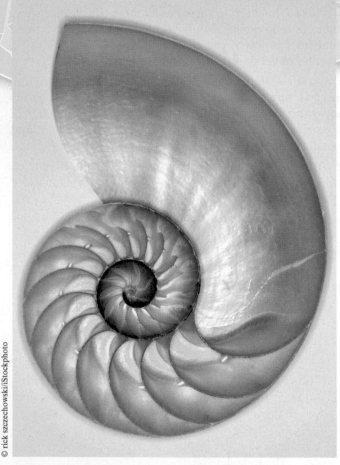

© rick szczechowski/iStockphoto

Are you into it?

> *Geometry has two great treasures: one is the theorem of Pythagoras; the other, the division of a line into extreme and mean ratio. The first we may compare to a measure of gold; the second we may name a precious jewel.*
>
> JOHANNES KEPLER

On our journeys through various mathematical landscapes we have become conscious of the issue of aesthetics—in particular, the intrinsic beauty of mathematical truths. We're discovering that mathematics is not just a collection of formulas tied together by algebra but is instead a wealth of creative ideas that allows us to investigate, explore, and discover new realms. Now, however, we wonder if mathematics can be used to discover structure behind the aesthetics of art and nature.

Rectangular Appeal

In our discussion of Fibonacci numbers we asked the following geometrical question that begs to be asked again: What are the dimensions of the most attractive rectangle—the rectangle we might imagine when we close our eyes on a dark starry night and dream of the ideal rectangle? When someone says *rectangle*, we think of a shape. What shape is it? From the rectangles given here, choose the one you find most appealing:

Given these choices, a high percentage of people think that the second rectangle from the left is the most aesthetically pleasing—the one that captures the true spirit of rectangleness. That rectangle is referred to as the *Golden Rectangle*. It is the length of the base relative to the length of the height that makes it a Golden Rectangle.

What precisely is the ratio of base to height that produces the Golden Rectangle? Recall that, in our conversations about numbers, we found a ratio that was an especially attractive number. The ratio arose in our discussions of the Fibonacci numbers, and we denoted it by the Greek letter phi, φ. It was called the *Golden Ratio* because it satisfied the symmetrical equation of ratios:

$$\frac{\varphi}{1} = \frac{1}{\varphi - 1}.$$

Specifically, we found that the Golden Ratio, φ, is the number $(1 + \sqrt{5})/2 = 1.618\ldots$

You may want to glance back at the Fibonacci discussion in Section 2.2 and revisit the relationship $\varphi/1 = 1/(\varphi - 1)$. (The Greek letter φ used to denote the Golden Ratio was introduced in the past century to honor the famous ancient Greek sculptor Phidias, much of whose work appears to involve the Golden Ratio.)

The Golden Ratio gives us the satisfying relationship of height to width for those rectangles that many deem extremely pleasing to the eye. The precise mathematical definition of a Golden Rectangle is any rectangle having base b and height h such that

$$\frac{b}{h} = \varphi = \frac{1 + \sqrt{5}}{2}.$$

We have already discovered how the Fibonacci numbers and the Golden Ratio appear in nature's spirals. Do the proportions of the Golden Ratio make the Golden Rectangle especially attractive and, if so, why? These questions have given rise to heated debate and much controversy. In 1876, Gustav Fechner, a German psychologist and physicist, conducted a study of people's taste in rectangles—a taste test—and found that 35% of the people surveyed selected the Golden Rectangle. So, although the Golden Rectangle seems likely to win an election, we would not expect the outcome to be a landslide.

The Golden Rectangle in Greece

The Greeks appear to have been captivated by the proportions of the Golden Rectangle as evidenced by its frequent occurrence in their architecture and art. As a classic illustration, consider the magnificent Parthenon in Athens, built in the 5th century BCE.

The Parthenon today is pretty run-down—in fact, it's in ruins. However, perhaps you're a step ahead of us, guessing that the big rectangle contained in the Parthenon is a Golden Rectangle. Actually, if we measure the sides and do the division, we will see that the rectangle is not a Golden Rectangle! So what's the point? Well, when the Parthenon was built, it was much fancier—in particular, it had a roof. Imagine now that the roof is in place. If we form the rectangle from the tip of the rooftop to the steps, we will see a nearly perfect Golden Rectangle.

Another example of the Golden Rectangle in Greek sculpture is the Grecian eye cup. The one pictured is inscribed inside a perfect Golden Rectangle.

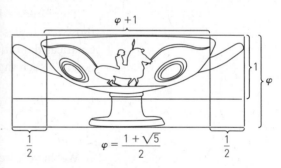

$$\varphi = \frac{1 + \sqrt{5}}{2}$$

It remains an unanswered question whether Greek artists and designers intentionally used the Golden Rectangle in their work or chose those dimensions solely based on aesthetic tastes. In fact, we are not even certain that such artists were consciously aware of the Golden Rectangle. Although we will likely never know the truth, it is romantic to hypothesize that the Greeks were not conscious of the Golden Rectangle, because this then shows how aesthetically appealing its dimensions are and that we are naturally attracted to such shapes. Some people, however, believe that the occurrence of Golden Rectangle proportions is simply coincidental and random. While some believe that ancient Greek works definitely contain Golden Rectangles, others believe that it is nearly impossible to measure such works or ruins accurately; thus, there is plenty of room for error. In the preceding pictures, all the superimposed rectangles are perfect Golden Rectangles. Was their presence random or deliberate? Are Golden Rectangles really there? What do you think?

The Golden Rectangle in the Renaissance

It appears that mathematicians in the Middle Ages and the Renaissance were fascinated by the Golden Rectangle, but there is much question as to whether this enthusiasm was shared by artists of the time. Leonardo da Vinci was a math enthusiast, but did he know about the Golden Rectangle? Did he deliberately use it in his work? While historians debate such issues, let's take a look at Leonardo's unfinished portrait of St. Jerome from 1483. In the reproduction on below, we have superimposed a perfect Golden Rectangle around the great scholar's body.

Intentionally or otherwise, Leonardo selected proportions that were aesthetically appealing, and such dimensions resemble those of the Golden Rectangle. Although we are not certain whether Leonardo intentionally used the Golden Rectangle, we do know that 26 years later he was aware of its existence. In 1509, Leonardo was the illustrator for Luca Pacioli's text on the Golden Ratio titled *De Divina Proportione*. It was famous mainly for the reproductions of 60 geometrical drawings illustrating the Golden Ratio.

Leonardo da Vinci *St. Jerome,* 1480.

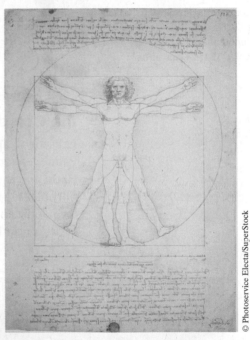

Leonardo da Vinci's illustration for Luca Pacioli's *De Divina Proportione.*

The *Divine Proportion* is a synonym for the Golden Ratio. In fact, many people, including Johannes Kepler, referred to the Golden Ratio as the Divine Proportion, or as the *Mean and Extreme Ratio*. Sometimes imaginations ran a bit too wild. Pacioli claimed that one's belly button divides one's body into the Divine Proportion. If you're not ticklish, you can easily check that this is not necessarily true.

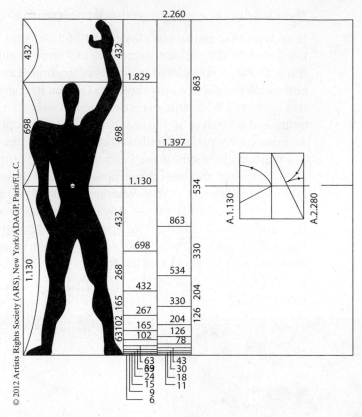

Note the Fibonacci-like pattern in Le Corbusier's 1946 *Modular Man.* Proportional system: $6 + 9 = 15$, $9 + 15 = 24$, and so on.

The Golden Rectangle and Impressionism

Let's now leap ahead about 300 years to the creative age of French Impressionism. Painter Georges Seurat was captivated by the aesthetic appeal of the Golden Ratio and the Golden Rectangle. In his painting *La Parade* from 1888, he carefully planted numerous occurrences of the Golden Ratio through the positions of the people and the delineation of the colors. The use of the Golden Ratio in works of art is now known as the *technique of dynamic symmetry*.

George Seurat's *La Parade*, 1888.

ABCD, FGHJ, EBIK are all golden rectangles; we also note that $\dfrac{GE}{EA} = \dfrac{EA}{FE} = \varphi$.

The Golden Rectangle in the 20th Century

In the 20th century, artists were still fascinated with the beautiful proportions of the Golden Rectangle. French architect Le Corbusier believed that people are comforted by mathematics. In this spirit, he deliberately designed this villa to conform with the Golden Rectangle.

© 2012 Artists Rights Society (ARS), New York/ADAGP, Paris/F.L.C.

Le Corbusier was one of the architects involved in the design of the United Nations Headquarters in New York City. Here we again see the influence of the Golden Rectangle in this monolithic structure.

© dibrova/iStockphoto

United Nations

Finally, we note that the Golden Rectangle appears often in other art forms, including musical works. As an illustration, consider the work of French composer Claude Debussy. In his 1894 work "Prelude to the Afternoon of a Faun," he deliberately placed numerous ratios of musical pulses (called *quaver units*) that approximate the Golden Ratio.

$$f \quad\quad f \quad\quad\quad\quad f \quad\quad\quad\quad f$$
Bar 19 Bar 28   Bar 47   Bar 70

$$19 + 28 = 47$$

←———— 81 Seconds ————→

←———— 129 Seconds ————→

$$\frac{129}{81} = 1.592$$

Quaver units for "Prelude to the Afternoon of a Faun." Note:

$$\frac{817}{515} = 1.5864 \ldots \approx \varphi$$

Why the Appeal?

Why do we see proportions conforming to the Golden Ratio in so many works of art? To answer this question, let's return to Le Corbusier's villa and notice that the living area creates a large square, whereas the open patio on the left has a rectangular shape. Look what happens when we compare the proportions of the whole villa to the small rectangular patio:

Le Corbusier, Villa.

Patio turned on its side and enlarged.

Both are Golden Rectangles!

This rectangular similarity is actually a fundamental and beautiful mathematical property of the Golden Rectangle. This property might explain why the Golden Rectangle is so aesthetically pleasing.

To examine this property in general, let's picture a Golden Rectangle with base equal to $(1+\sqrt{5})/2$ and height equal to 1, so that $b/h = \varphi = (1+\sqrt{5})/2$

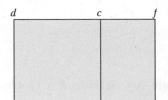

We now divide this Golden Rectangle *aefd* into a square (*abcd*) and a smaller rectangle *befc*. The smaller rectangle is formed by removing that largest square from the Golden Rectangle. We will soon prove that it was the Golden Ratio proportions of the Golden Rectangle that automatically made the smaller rectangle, *befc*, golden!

An Unexpected Rectangle

The fact that a Golden Rectangle comprises a square and a smaller Golden Rectangle may well explain its aesthetic appeal. This "self-proliferation" feature represents an attractive regenerating property: If we look at the smaller Golden Rectangle and now remove the largest possible square inside it, we are left with an even smaller Golden Rectangle. Can you visualize continuing this process of removing the square and getting another even smaller Golden Rectangle forever? There is, in some sense, a self-similarity property at work here: At any stage in this process, no matter how small the Golden Rectangle is, when we chop off the biggest square possible, we have created an even smaller Golden Rectangle. We will observe a similar situation when we consider fractals.

Why is this surprising mathematical fact true? It comes from the pleasing algebraic relationship that the Golden Ratio satisfies:

$$\frac{\varphi}{1} = \frac{1}{\varphi - 1}.$$

The Golden Rectangle Within a Golden Rectangle.

If a Golden Rectangle is divided into a square and a smaller rectangle, then the small rectangle is another Golden Rectangle.

Proof

Let's begin with our picture of a Golden Rectangle.

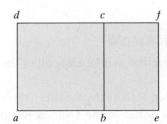

As before, we might as well declare that the length *ad* is 1 unit, and *ae* has length φ. To show that rectangle *befc* is a Golden Rectangle, we must show that the ratio of its longer side to its shorter side, that is, *ef/be*, is φ. So we will need the lengths of the sides of the smaller rectangle. Well, *ef* is easy to figure out: It equals *ad*. So *ef* = 1. What is *be*?

We note that *be* is just *ae* minus *ab*. So,

$$be = ae - ab.$$

But $ae = \varphi$, and $ab = 1$. So,

$$be = \varphi - 1.$$

So, the ratio

$$\frac{ef}{be} = \frac{1}{\varphi - 1}.$$

But recall our pleasing identity:

$$\frac{\varphi}{1} = \frac{1}{\varphi - 1}.$$

Therefore, *ef/be* equals φ, and the small rectangle *befc* is indeed a Golden Rectangle. This observation completes our proof.

Constructing Your Own Golden Rectangle

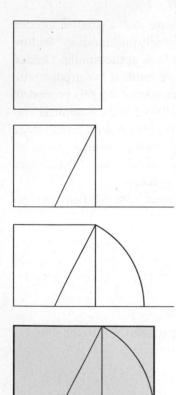

Perhaps you are now convinced that the Golden Rectangle is aesthetically intriguing and downright cool. You want one for yourself. Sure, you can call 1-800-COOL-REC and order one (operators are standing by), but why waste your money? We can make a perfect Golden Rectangle ourselves for free. It may appear that such a perfectly proportioned rectangle would be complicated to create. Not so. In fact, it's easy to construct a perfect Golden Rectangle. Here's how: First we build a square.

Next, we connect the midpoint of the base of the square to the northeast corner of the square with a straight line segment. We then extend the base of the square with a straight line segment off to the east, like a landing strip. We now have a picture that looks like the figure to the left.

Now we draw part of a circle whose center is the midpoint of the base and whose radius extends to the northeastern corner of the square. We note where the circle portion hits the landing strip. The line segment drawn inside the square from the midpoint to the northeastern corner is actually a radius of the circle arc drawn. We now have the picture to the right.

Next, we construct a line perpendicular to the landing strip and passing through the point where the circle hit the landing strip. We then extend the top edge of the square to the right with a straight line until it hits the perpendicular line just drawn. Finally, we erase the excess landing strip to the right of the arc, giving us the diagram shown here.

That was pretty easy. Now take a look at that big rectangle we just constructed (we made ours a bit darker). Do you find yourself drawn to that tall, dark, and handsome rectangle? If so, it's all right, because that rectangle is a perfectly precise Golden Rectangle.

Why This Procedure Produces a Golden Rectangle

We begin by recalling the final picture of our construction and labeling all of the vertices.

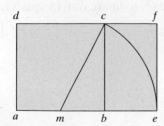

To prove that the rectangle *aefd* is really a Golden Rectangle, we must show that the length of *ae* divided by the length of *ad* is equal to the Golden Ratio $(1 + \sqrt{5})/2$. So, we want to prove that

$$\frac{ae}{ad} = \frac{1 + \sqrt{5}}{2}.$$

The size of the rectangle is not important. What matters is the ratio of the two sides. We can call the length of *ad* 1 unit and note that this now completely determines the length of everything else in the rectangle. Given this agreement, our goal is to figure out what the length of *ae* is. Notice that *ae* is just *am* plus *me*. If we can find *am* and then *me*, then we will have *ae*, since *ae* = *am* + *me*. Remember that we started with a square, and *m* bisected the bottom side. So *am* = *mb* = 1/2. Great—all we need to do is find *me*.

The truth is that the length of me is mysterious. Let's see if we can find another line segment having the exact same length as *me*. Examine the preceding picture and find another line that has the same length as *me*. Try this before reading on.

Did you guess *mc*? If so, great. Note that both *mc* and *me* are radii for the same circle, so the segments must have the same length. Instead of finding the length of *me*, let's find the length of *mc*. Why is this quest easier? The answer is that *mc* is part of a right triangle. In fact, it is the hypotenuse of the triangle *mbc*. Notice that we already saw that *bc* is equal to 1 and *mb* is equal to 1/2. Thus, using the Pythagorean Theorem, we can figure out the length of *mc*. Why not try to figure it out on your own before reading on?

Here we go:

$$(1)^2 + \left(\frac{1}{2}\right)^2 = (mc)^2.$$

That is,

$$1 + \frac{1}{4} = (mc)^2 \quad \text{or} \quad \frac{5}{4} = (mc)^2.$$

Notice the 5 making its debut in this discussion. This development is great news since we want a $\sqrt{5}$ at some point. In fact, note that to solve for *mc* we need to take 1/ the square root of both sides, but, because the length *mc* is positive, we have

$$mc = \frac{\sqrt{5}}{2} \text{ (because } \sqrt{4} = 2\text{)}.$$

Remember that *mc* has the same length as *me*, so,

$$me = \frac{\sqrt{5}}{2}.$$

Therefore,

$$ae = \frac{1}{2} + \frac{\sqrt{5}}{2} = \frac{1 + \sqrt{5}}{2}.$$

Now for the big finish:

$$\frac{ae}{ad} = \frac{\left(\dfrac{1 + \sqrt{5}}{2}\right)}{1} = \frac{1 + \sqrt{5}}{2} = \varphi.$$

Life Lesson

Often in life when faced with a difficulty, it is valuable to look for something else that is comparable, but easier to resolve.

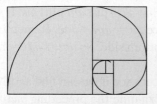

So, we have a Golden Ratio, which proves we've constructed a perfect Golden Rectangle.

Golden Spirals

We close with one last aesthetically pleasing construction. Let's take a Golden Rectangle and start drawing successive squares. Within each square, we will draw a quarter of a circle having a radius equal to the side of the square. If we do this, we get a spiral.

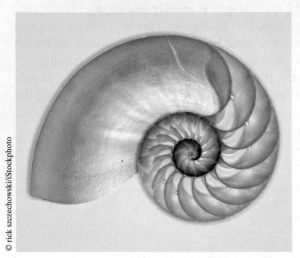

This spiral closely approximates the *logarithmic spiral*, and it occurs in nature in various forms, such as the nautilus sea shell. The natural and aesthetic beauty of this spiral may be described mathematically. We first consider the center of the spiral. By the center we mean that point at which the spiral spins around infinitely often—the point that the spiral is heading toward. How can we locate the very center of the spiral?

Locating the center is surprisingly simple. We need only draw a diagonal in the largest Golden Rectangle from the northwest corner down to the southeast corner and then draw the diagonal in the next largest Golden Rectangle from its northeast corner to its southwest corner.

These two diagonals intersect at the precise center of the spiral. You may also have observed another unexpected fact: All analogous diagonals on all subsequent pairs of Golden Rectangles lie on the first two diagonals. This follows from the fact that each rectangle has exactly the same proportions. Thus, we see structure and beauty in the construction of the Golden Rectangle and the associated spiral.

What makes the curve of the spiral so appealing? Here is a mathematical observation that may account for its appeal. Select any point on the spiral and connect that point with the center of the spiral. Now draw the line that passes through that chosen point on the spiral but just grazes the curve of the spiral (such a line is called a *tangent line*). Notice the angle made by these two lines (the tangent at the point and the line connecting the point to the center). These angles are nearly the same, no matter which point on the spiral you selected.

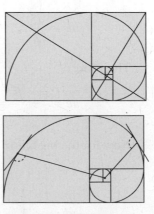

Angles are nearly equal.

On the *Heart of Mathematics* Web site, you can find a program to generate these spirals and thus create your own works of art.

We'll now close our discussion of the Golden Rectangle, but not forever. Several other examples of Golden Rectangles occur in surprising places; but for them we will have to wait until we talk about the Platonic solids.

A Look BACK

A RECTANGLE is a Golden Rectangle if the ratio of its base to its height equals the Golden Ratio. If we remove the largest square from a Golden Rectangle, the small remaining rectangle is itself another Golden Rectangle. Thus, we can create a sequence of smaller and smaller Golden Rectangles. This sequence of Golden Rectangles leads to spirals that occur in nature.

We can build a Golden Rectangle by starting with a square and elongating it by using a simple geometric procedure. We can verify that the ratio of base to height is the Golden Ratio by applying the Pythagorean Theorem.

Art, aesthetics, geometry, and numbers all meet in the Golden Rectangle. Its appealing proportions have appeared in art throughout history and we can also find them in nature. Do the mathematical properties of the Golden Ratio somehow create the beauty of the Golden Rectangle? Some ideas span the artificial boundaries of subjects—in this case from the algebra of numbers (the Golden Ratio) to the geometry of rectangles (the Golden Rectangle). Seeking connections across disciplines often leads to new insights and creative ways of understanding.

Life Lessons

Take ideas from one domain and explore them in another.

MINDSCAPES   Invitations to Further Thought

*In this section, Mindscapes marked (**H**) have hints for solutions at the back of the book. Mindscapes marked (**ExH**) have expanded hints at the back of the book. Mindscapes marked (**S**) have solutions.*

Developing Ideas

1. **Defining gold.** Explain what makes a rectangle a *Golden Rectangle*.

2. **Approximating gold.** Which of these numbers is closest to the Golden Ratio? 1.16; 1.29; 1.62; 1.98.

3. **Approximating again.** Which of the following objects most closely resembles a Golden Rectangle? A 3 × 5–inch index card; an 8.5 × 11–inch paper; an 11 × 14–inch paper; an 11 × 17–inch paper.

4. **Same solution.** Why does the equation $\varphi - 1 = \dfrac{1}{\varphi}$ have the same solution as the equation $\dfrac{\varphi}{1} = \dfrac{1}{\varphi - 1}$?

5. **X marks the unknown (ExH).** Solve each equation for x:

  a. $\dfrac{2x}{1} = \dfrac{1}{x-1}$ b. $\dfrac{x}{3} = \dfrac{2}{x-4}$ c. $\dfrac{3x}{2} = \dfrac{1}{x+1}$

Solidifying Ideas

6. **In search of gold.** Find at least three examples of Golden Rectangles in your surroundings. If possible, include photographs or sketches and estimates of the ratio of base to height for each example.

7. **Golden art.** In the masterpiece *Paris Street, Rainy Day* by Gustave Caillebotte (1877) shown below, find as many Golden Rectangles as you can.

Gustave Caillebotte, *A Paris Street, Rainy Day*, 1877.

© Fine Art Images/SuperStock

8. **A cold tall one?** Can a Golden Rectangle have a shorter base than height? Explain your answer.

9. **Fold the gold (H).** Suppose you have a Golden Rectangle cut out of a piece of paper. Now suppose you fold it in half along its base and then in half along its width. You have just created a new, smaller rectangle. Is that rectangle a Golden Rectangle? Justify your answer.

10. **Sheets of gold.** Suppose you have two sheets of paper, an unmarked straightedge, and a pair of scissors. Explain how you can use one of the sheets of paper and the straightedge to construct a perfect Golden Rectangle on the other sheet. (*Hint:* You may cut the first piece of paper.)

11. **Circular logic? (H).** Take a Golden Rectangle and draw the largest circle inside it that touches three sides. The circle will touch two opposite sides of the rectangle. If we connect those two points with a line and then cut the rectangle into two pieces along that line, will either of the two smaller rectangles be a Golden Rectangle? Explain your reasoning.

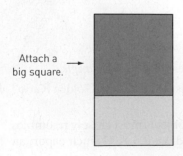

Attach a big square. →

12. **Growing gold (H).** Take a Golden Rectangle and attach a square to the longer side so that you create a new larger rectangle. Is this new rectangle a Golden Rectangle? What if we repeat this process with the new, large rectangle?

13. **Counterfeit gold?** Draw a rectangle with its longer edge as the base (it could be a square, it could be a long and skinny rectangle, whatever you like, but we suggest that you do not draw a Golden Rectangle). Now, using the top edge of the rectangle, draw the square just above the rectangle so that the square's base is the top edge of the rectangle. You have now produced a large new rectangle (the original rectangle together with this square sitting above it). Now attach a square to the right of this rectangle so that the square's

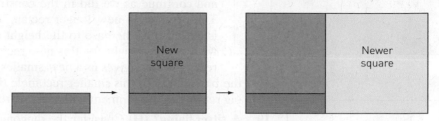

left side is the right edge of the large rectangle. You've constructed an even larger rectangle.

Now repeat this procedure—that is, append to the top of this huge rectangle the largest square you can and follow that move by attaching the largest square you can to the right of the resulting rectangle. Start with a small rectangle near the bottom left corner of a page and continue this process until you have filled the page. Now measure the dimensions of the largest rectangle you've built and divide the longer side by the shorter one. How does that ratio compare to the Golden Ratio? Experiment with various starting rectangles. What do you notice about the ratios?

14. **In the grid (S).** Consider the 10×10 grid at left. Find the four points that, when joined to make a horizontal rectangle, make a rectangle that is the closest approximation to a Golden Rectangle. (*Challenge:* Suppose the rectangle can be tilted.)

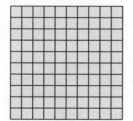

15. **A nest of gold.** Consider the figure of infinitely nested Golden Rectangles on page 269. Suppose we remove the largest square, and, with the rectangle that remains, we enlarge the entire picture so that its size is identical to the original rectangle. How will that enlarged picture compare to the original figure? Explain your answer.

Creating New Ideas

16. **Comparing areas (ExH).** Let G be a Golden Rectangle having base b and height h, and let G′ be the smaller Golden Rectangle made by removing the largest square possible from G. Compute the ratio of the area of G to the area of G′. That is, compute Area(G)/Area(G′). Does your answer really depend on b and h (the original size of G)? Are you surprised by your answer?

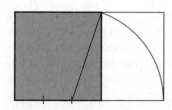

17. **Do we get gold?** Let's make a rectangle somewhat like the Golden Rectangle. As before, start with a square; however, instead of cutting the base in half, cut it into thirds and draw the line from the upper right vertex of the square to the point on the base that is one-third of the way from the right bottom vertex. Now use this new line segment as the radius of the circle, and continue as we did in the construction of the Golden Rectangle. This produces a new, longer rectangle, as shown in the diagram. What is the ratio of the base to the height of this rectangle (that is, what is base/height for this new rectangle)? Now remove the largest square possible from this new rectangle and notice that we are left with another rectangle. Are the proportions of the base/height of this smaller rectangle the same as the proportions of the big rectangle?

18. **Do we get gold this time? (S)** We now describe another construction of a different type of rectangle. It is exactly the same as the Golden Rectangle except that, instead of starting with a square, we begin with a rectangle whose base

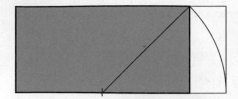

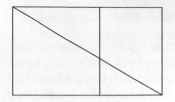

is twice as long as its height. Now connect the midpoint of the base to the upper right vertex with a line, and use this line as the radius of the circle and continue as we did in the construction of the Golden Rectangle. This produces a new, longer rectangle, as shown in the diagram. What is the ratio of the base to the height of this new big rectangle (that is, what is base/height for this new rectangle)? Now remove the original rectangle. This gives us a new, smaller rectangle. Are the proportions of the base/height of this smaller rectangle the same as the proportions of the big rectangle? Experiment with starting rectangles of differing proportions.

19. **A silver lining? (H)** Consider the diagonal in the Golden Rectangle shown here and draw in the largest square possible. Notice that one edge of the square cuts the diagonal into two pieces. What is the ratio of the length of the entire diagonal to the length of the part of the diagonal that is inside the square? That is, compute the length of the entire diagonal divided by the length of the part of the diagonal that is inside the square. Surprised?

20. **Cutting up triangles.** Draw any right triangle. Find a way of cutting up that triangle into four identical triangles such that each one is identical in shape and proportion to the original large triangle except that it is scaled down to one-fourth the area.

Further Challenges

21. **Going platinum.** Determine the dimensions of a rectangle such that, if you remove the largest square, then what remains has a ratio of base to height that is twice the ratio of base to height of the original rectangle.

22. **Golden triangles.** Draw a right triangle with one leg twice as long as the other leg. This triangle is referred to as a *Golden Triangle*. Suppose that one leg has length 1 and the other has length 2. What is the length of the hypotenuse? Next draw a line from the right angle of the triangle to the hypotenuse such that the line is perpendicular to the hypotenuse. Now cut up the larger of the two new right triangles into four triangles (see Mindscape 20, "Cutting up triangles"). Show that all five triangles are the same size and are Golden Triangles. We will use this neat cutting up of the Golden Triangle in Section 4.4, "Soothing Symmetry and Spinning Pinwheels."

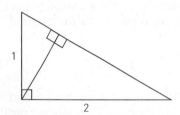

In Your Own Words

23. **Personal perspectives.** Write a short essay describing the most interesting or surprising discovery you made in exploring the material in this section. If any material seemed puzzling or even unbelievable, address that as well. Explain why you chose the topics you did. Finally, comment on the aesthetics of the mathematics and ideas in this section.

24. **With a group of folks.** In a small group, discuss and actively work through the geometric construction of the Golden Rectangle. After your discussion, write a brief narrative describing the construction in your own words.

25. **Creative writing.** Write an imaginative story (it can be humorous, dramatic, whatever you like) that involves or evokes the ideas of this section.

26. **Power beyond the mathematics.** Provide several real-life issues—ideally, from your own experience—that some of the strategies of thought presented in this section would effectively approach and resolve.

For the Algebra Lover

Here we celebrate the power of algebra as a powerful way of finding unknown quantities by naming them, of expressing infinitely many relationships and connections clearly and succinctly, and of uncovering pattern and structure.

27. **Special K.** As a student at the University of Ketchikan in Ketchikan, Alaska, you decide to create a new logo for the school's dogsled team, the Special Ks. Your design involves a Golden Rectangle with its height of length φ and its base of length 1. You block off a square as shown in the figure, and then draw a K using the diagonals of the square at the bottom and the smaller rectangle at the top. Find the area of the red square and the area of the black rectangle. Also find the length of the lower diagonal of the K. (You'll need the Pythagorean Theorem from Section 4.1.)

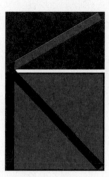

28. **Special x.** Find all values of x satisfying the equation $5 - x = \dfrac{-1}{x} + 4$. Are any of the values of x special? Explain. (*Hint*: rewrite the equation and compare it to others in the section that use a different symbol for the variable.)

29. **In search of x.** Solve each equation for x:

$$\frac{x}{2} = \frac{1}{x+2} \qquad \frac{x^2}{2} = \frac{4}{x} \qquad 3x^2 = \frac{81}{x}$$

30. **Adding a square.** Your school's Healthy Eating garden plot is a rectangle measuring 2 meters by x meters. You decide to expand the area for more organic vegetables by adding a large square plot against the side measuring x meters. Write an expression in terms of x that gives the total area of the expanded garden. If the total area is 24 square meters, find x.

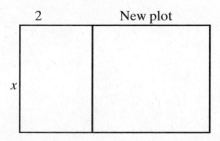

31. Golden Pythagoras (H). If you have a Golden Rectangle with base length $\varphi = \dfrac{1+\sqrt{5}}{2}$ and height length 1, the Pythagorean Theorem from Section 4.1 tells you that the length of the diagonal of your rectangle is $\sqrt{1^2 + \left(\dfrac{1+\sqrt{5}}{2}\right)^2}$. Simplify this expression as much as possible.

SOOTHING SYMMETRY AND SPINNING PINWHEELS

Can a Floor Be Tiled Without Any Repeating Pattern?

© serts/iStockphoto

© furnau┌pa/iStockphoto

© Fernando Alonso Herrero/iStockphoto

> *The mathematical sciences particularly exhibit order, symmetry, and limitation; and these are the greatest forms of the beautiful.*
>
> **ARISTOTLE**

Symmetry! (*Top*: Turkish mosaic.
Bottom right: Alhambra relief.
Bottom left: Moroccan mosaic.)

Understanding the world often comes down to discovering pattern and order. When we perceive that the orbits of the planets are ellipses or that crystals are made of orderly arrangements of molecules, we feel that we have detected something significant about the structure of nature. Our sense of beauty often centers on balance and harmony, but sometimes we're moved when something departs from an expected order. People have lived with regular patterns for centuries, but only recently have we discovered patterns that straddle order and chaos. These patterns without symmetry have a haunting, unsettling beauty of their own. And again we find that an exploration driven by abstract intrigue ends in ideas that have an uncanny ability to describe our real physical world.

Before considering chaos, let's consider order. What makes the artistic works on the previous page so alluring? The answer, in short, is symmetry.

Symmetry

We begin by describing what symmetry is in general, and then we define its meaning more precisely. By pinning down the meaning of symmetry, we allow ourselves

249

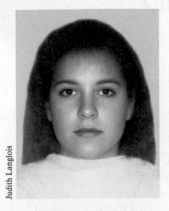

Judith Langlois

An average computer-generated face made from 32 actual faces

© Mihail Orlov/iStockphoto

to develop variations of it, including a whole new world of patterns that paradoxically combine order with chaos.

We are drawn to the symmetrical. Recently, psychologists performed experiments to show that facial symmetry is an important component of our notion of beauty and influences our choice of a mate. How does one study human symmetry and beauty? Judith Langlois of The University of Texas at Austin wanted to know which faces were considered beautiful, so she showed people different photographs of faces and recorded their reactions. We might think that the most beautiful faces would be extreme in some way—extremely thin, or extremely chiseled, or extremely something else. But the faces most often chosen as beautiful were extremely . . . well, average and symmetrical. Average—not extreme. Dr. Langlois created the most symmetrical and most average faces she could by creating composites of many photographs of people. The averaged face was completely symmetrical and was the one most preferred. Thus, the love of symmetry may literally be in our genes.

When we turn from faces to patterns in floor tiles, wall coverings, and paintings, again we are drawn toward the symmetrical. In most classical patterns, the symmetry is soothing. However, modern patterns, such as the Penrose Pattern and the Pinwheel Pattern, shown on pages 252 and 253, combine symmetry with chaos to create a disturbing yet hypnotic dissonance.

What Is Symmetry?

We sense symmetry when small portions of a pattern are repeated elsewhere in the overall pattern. Let's imagine that M.C. Escher's symmetry drawing has been extended indefinitely to cover the entire plane. Each area has a symmetry of its own, but the symmetry of the whole pattern lies in the regularity with which they appear over the whole plane. Imagine that we cover the entire plane with a sheet of rigid plexiglass. Suppose we trace every area onto the plexiglass. Now imagine sliding the plexiglass over one/unit to the right. What would we see? We'd notice that each area on the plexiglass matches up perfectly with an area on the plane and each plexiglass area matches a plane area. The plexiglass experiment confirms our intuition that the pattern has symmetry.

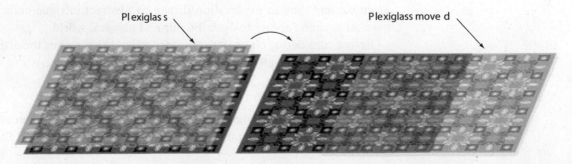

Plexiglas s   Plexiglass move d

© Mihail Orlov/iStockphoto

To refine this idea, we'll say that *a rigid symmetry* of a pattern in the plane is a motion of the plane that preserves the pattern and does not shrink, stretch, or otherwise distort the plane. In other words, it is a motion of a plexiglass copy of the pattern that ends up in a position that again exactly matches the pattern. A rigid symmetry could be a shift, a rotation, a flip, or any combination of these.

By tracing the different patterns onto transparencies, you can try to discover the various rigid symmetries (or lack thereof) in the illustrations at the opening of this section.

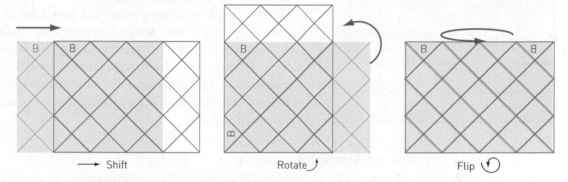

Shift   Rotate   Flip

A Symmetry of Scale

One day in Wonderland, Alice found a cake helpfully labeled, "Eat Me," which she dutifully did. She quickly grew to five times her height. Although it is not explicitly mentioned in *Alice in Wonderland*, surely she was standing on a square-tiled floor. When she had finished her growth spurt, she looked down at the distant floor expecting to feel giddy because of her new altitude. However, much to her surprise, she had no such sensation. Instead, by squinting, she saw that the original square tiles of the floor could be amalgamated in groups of five by five, so that these bigger super-tiles formed a pattern in the plane that looked, from her new height, exactly like the original pattern.

Taking another bite of the cake, she grew to five times her new height. But again, the super-tiles could be grouped to form super-super-tiles whose pattern of covering the plane looked identical to the original pattern. No matter how much cake Alice ate—and she packed it away—the pattern, when properly grouped, produced an identical pattern but on a larger scale.

Being mathematically inclined, Alice noted that this feature of the pattern revealed a symmetry of scale. Being in Wonderland, she wondered if symmetry of scale was related to the cool pictures of self-similar objects that she saw while thumbing through Chapter 7 of this book. She had, however, already learned that, after discovering an idea, one should pin it down with a definition.

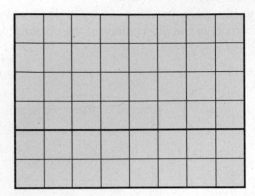

We'll say that a pattern in the plane has *a symmetry of scale* or *is scalable* if the tiles that make up the pattern can be grouped into super-tiles that still cover the plane and, if scaled down, can be rigidly moved to coincide with the original pattern.

Several simple patterns possess this new kind of symmetry. The square pattern and the design composed of equilateral triangles (see page 260) clearly illustrate the concept of duplicating the same pattern at a larger scale by grouping the original tiles together to create super-tiles. This duplication of pattern through grouping seems to indicate a high level of organization, regularity, and symmetry. Or at least so it appears.

A Strange Question

Suppose we have a pattern that is so regular that it has a symmetry of scale. Must that pattern also have rigid symmetries?

Anyone's first attempts at making a pattern with a symmetry of scale involve putting the tiles down in a highly regular pattern. But one of the most remarkable recent developments in geometry shows us that the world is more interesting and varied than we might first guess. Scalable patterns can be so chaotic that they have no rigid symmetries at all! These scalable patterns with no rigid symmetries are bizarre and intriguing modern creations.

Chaotic Patterns

In the mid-1960s, Robert Berger, in his Ph.D. thesis in applied mathematics at Harvard University, constructed some examples of patterns that have no rigid symmetries yet have a scalable property. These examples combined thousands of

different shapes of tiles, and they opened the door to an exploration of what types of exotic patterns in the plane might be possible. A decade later, Roger Penrose constructed famous patterns, called *Penrose Patterns*, with no rigid symmetries using only two tile shapes, referred to as *kites* and *darts*.

Here is a picture of a pattern created using Penrose tiles. No covering of the plane by Penrose kites and darts has any rigid symmetries; nevertheless, these Penrose Patterns do not look completely chaotic. Part of the reason for their somewhat stable appearance is that, in any Penrose Pattern, every tile occurs in one of 10 possible orientations in the plane. But even that amount of regularity can be absent in some truly exotic patterns.

Penrose tiling made a brief appearance in British bathrooms when Kleenex Quilted toilet paper used the pattern in its design. Sir Roger Penrose was not amused and sued Kimberly-Clark for copyright infringement. His spokesman, David Bradley, stated the case forcefully:

> So often we read of very large companies riding rough-shod over small businesses or individuals. But when it comes to the population of Great Britain being invited by a multinational to wipe their bottoms on what appears to be the work of a Knight of the Realm without his permission, then a last stand must be made.

Kimberly-Clark, flushed with embarrassment but not to be wiped from the scene, took the plunge and re-released Kleenex Quilted with a less mathematically intriguing feel.

In 1994, John Conway of Princeton University and Charles Radin of The University of Texas at Austin described another exotic tiling of the plane called the *Pinwheel Pattern* using one single triangular tile. This Pinwheel Pattern not only has a symmetry of scale and has no rigid symmetry, but, in addition, the tiles occur in infinitely many orientations, thereby generating the sense of a pinwheel spiraling out of control from any vantage point. The Pinwheel Pattern gives the impression of a bewildering and disturbing jumble. Only by looking carefully can we find the way to group the triangles by fives to produce the super-tiles that demonstrate its symmetry of scale.

Construction of the Pinwheel Pattern

How can we construct such a pattern? Where would we start to build a pattern that has a symmetry of scale but no rigid symmetry? The answers are that we think ahead and build in both scalability and lack of rigid symmetry during our construction.

For a pattern to have a symmetry of scale, we must be able to group the tiles to create super-tiles of the same shape as the original tiles but larger. In the case of the Pinwheel Pattern, the basic tile is pretty simple. It is a right triangle with one leg twice as long as the other, which we'll call a *Pinwheel Triangle*. By the Pythagorean Theorem we see that a Pinwheel Triangle with legs of lengths 1 and 2 has a hypotenuse equal to $\sqrt{5}$. Those specific proportions are chosen because five identical Pinwheel Triangles can be assembled to form a larger Pinwheel Triangle—that is, another right triangle with one leg twice as long as the other. We'll call that large Pinwheel Triangle made up of five smaller Pinwheel Triangles a *5-unit Pinwheel Triangle*.

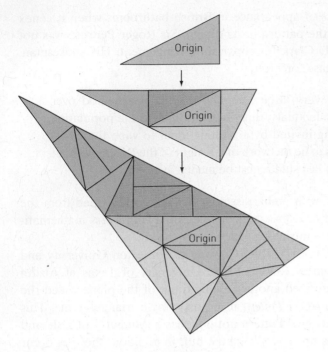

The phrase "5-unit" reminds us how many of the original-size tiles were used to make a super-tile. Each of the five triangles in the 5-unit Pinwheel Triangle is in a particular position, and we label those positions 1, 2, 3, 4, and Interior, as illustrated. We will always arrange the five triangles making up a 5-unit Pinwheel Triangle in exactly this way, which we call the *T-arrangement*, recognizing the T whose stem is the edge between triangles 4 and Interior, and whose top is an edge of triangle 3.

Building Super-Tiles

We are now ready to describe how to tile the entire plane with the Pinwheel Pattern. We begin with a single Pinwheel Triangle. See Then we surround it with four other identical Pinwheel Triangles to create a 5-unit Pinwheel Triangle in the T-arrangement such that our original triangle is the Interior triangle of the 5-unit Pinwheel Triangle. This strategy specifies for us how to lay the first five triangles. After those five tiles have been laid, how would we decide where to lay the next 20 tiles?

The answer requires us first to note that, after we lay the first five tiles, we have created one 5-unit Pinwheel Triangle. We now surround this 5-unit Pinwheel Triangle with four other 5-unit Pinwheel Triangles in the T-arrangement. We have just created a 25-unit Pinwheel Triangle such that our original 5-unit triangle is in the Interior position of the 25-unit Pinwheel Triangle. Each of the four 5-unit Pinwheel Triangles that we just put down are made up of five Pinwheel Triangles in the T-arrangement. Thus, we have just arranged 20 tiles.

We can determine the location of the next 100 tiles by surrounding the 25-unit Pinwheel Triangle that we just created by four other 25-unit Pinwheel Triangles in the T-arrangement, thus creating one 125-unit Pinwheel Triangle. Notice in the illustrations of this process that the successively larger triangles cover increasingly larger regions around the original tile so that the process can be continued forever to cover the entire plane. Also, observe how the increasingly larger Pinwheel super-Triangles spin around in different orientations as they grow, giving the pattern its name. After continuing this process forever, the resulting pattern in the plane is the Pinwheel Pattern.

A Symmetry of Scale

Let's group the triangles of the Pinwheel Pattern into groups of five using the T-arrangement so that the 5-unit Pinwheel Triangles form

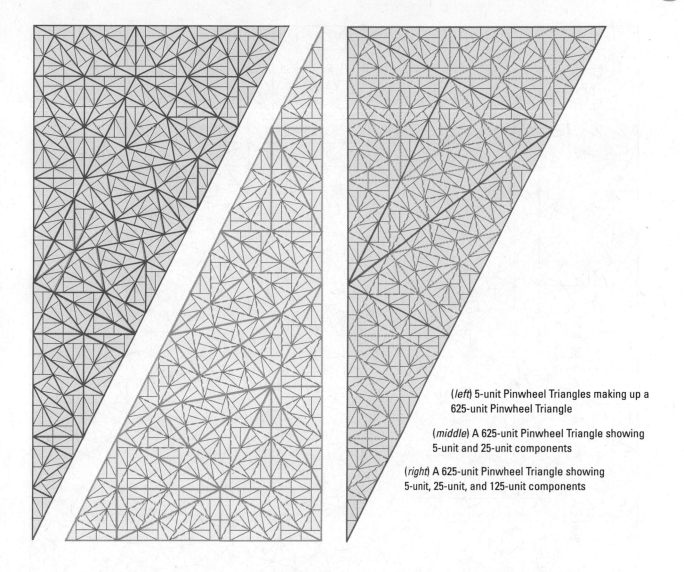

(*left*) 5-unit Pinwheel Triangles making up a 625-unit Pinwheel Triangle

(*middle*) A 625-unit Pinwheel Triangle showing 5-unit and 25-unit components

(*right*) A 625-unit Pinwheel Triangle showing 5-unit, 25-unit, and 125-unit components

an identical Pinwheel Pattern in the plane but at a larger scale. To begin, we hunt down the original triangle. Then we find the four surrounding triangles that make up the first 5-unit Pinwheel Triangle. That 5-unit Pinwheel Triangle is the first of the 5-unit super-tiles that will make up a Pinwheel Pattern in the plane but at a larger scale. The next stage of the construction of the Pinwheel Pattern consists of placing four more 5-unit Pinwheel Triangles in the T-arrangement to construct a 25-unit Pinwheel Triangle. Those four 5-unit Pinwheel Triangles are the next four of the 5-unit super-tiles.

The construction proceeds by surrounding the 25-unit Pinwheel Triangle by four more 25-unit Pinwheel Triangles, each of which is composed of five 5-unit Pinwheel Triangles.

Those five 5-unit Pinwheel Triangles in each of the four new 25-unit Pinwheel Triangles are the next 20 5-unit super-tiles. This process can be continued to

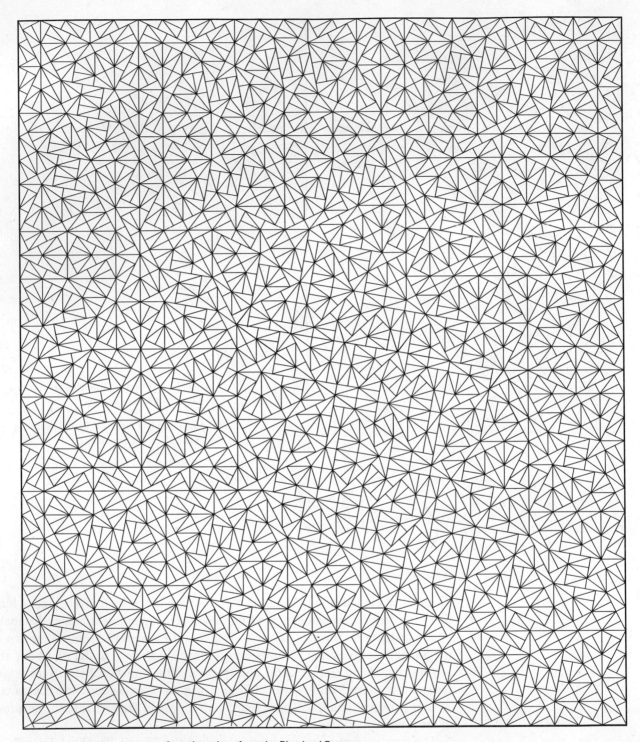

A random piece from the Pinwheel Pattern

amalgamate all the triangles in the Pinwheel Pattern into groups of 5-unit super-tiles. These create a Pinwheel Pattern of the plane that is identical, up to scaling, to the original Pinwheel Pattern.

The Pinwheel Pattern on page 256 looks jumbled and disorderly. In fact, it's not easy to pick out the super-tiles from the jumble. But they are there, and we ask you now to take a blue marker and trace the super-tiles, each comprising five of the original Pinwheel Triangles. After you have traced these super-tiles, you will have drawn a new, blue pattern of the plane with tiles five times as large as the original ones. Now take a green marker and group the blue tiles in fives to create super-super-tiles, each containing 25 of the original tiles. These green super-super-tiles form yet another Pinwheel Pattern of the plane. Of course, this process of creating larger and larger super-tiles can continue indefinitely. The only way to get a real sense of this amalgamation process and its subtlety is by actually drawing the super-tiles in. Take the time to explore and enjoy the gentle structure of this tundra of triangles.

Different Groupings?

In many patterns that possess a symmetry of scale, it is possible to group the tiles in several different ways to demonstrate the symmetry of scale. For example, consider the square pattern. We could group the tiles into two-by-two squares of four red tiles to create 4-unit super-tiles that again tile the plane in the square pattern. However, we could also shift the 4-unit groupings diagonally down to the right one unit, for example, to create a different way of grouping the tiles into 4-unit super-tiles. A lower-right tile in the first grouping would become an upper-left tile in the alternative grouping. Thus, in the case of this pattern of squares, many different ways exist to group the original tiles into 4-unit super-tiles to create a new square pattern that demonstrates its symmetry of scale.

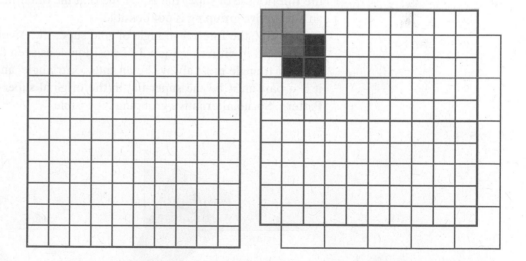

However, as you identified the 5-unit super-tiles in the Pinwheel Pattern, you may have noticed that you had no choice as to how to group the five triangles together to create the super-tile. In fact, there is only one way to group them together.

Uniqueness of Scaling.

There is only one way to group the Pinwheel Triangle into super-tiles to create a Pinwheel super-pattern in the plane.

The symmetry of scale of the Pinwheel Pattern shows us that there is a way to group the Pinwheel Triangles into fives to create a Pinwheel super-pattern of the plane. We want to show now that it is impossible to group the triangles into 5-unit, T-arrangement super-tiles in any other grouping.

Here, again, is our 5-unit super-tile labeled 1, 2, 3, 4, and Interior. Would it be possible for an Interior triangle to be part of a 5-unit group in some position other than the Interior position? *Hint:* The answer is no. To see why, we simply consider all the possibilities.

Let's look at the Pinwheel Pattern, find any 5-unit super-tile, and locate the Interior triangle of that super-tile. Now let's explore the possibility that, by grouping the tiles differently, the Interior triangle might be a part of a different 5-unit super-tile. That different super-tile would, of course, overlap the first super-tile, and the Interior triangle would coincide with a triangle in position 1, 2, 3, or 4 of the other super-tile. Here we see what would happen if triangle 1 of a supposed purple super-tile B were to coincide with the Interior triangle of a gold super-tile A. As we see, the other tiles do not fit. So we know that the Interior triangle of gold super-tile A cannot be part of another grouping that puts it in position 1 of purple super-tile B.

Could the Interior triangle of gold super-tile A be in position 2 of a supposed purple super-tile B? Again, by laying the position-2 tile of purple super-tile B over the Interior tile of super-tile A, we see that the other tiles do not line up, so such an alternative grouping is not possible.

Likewise, the Interior triangle of gold super-tile A could not be part of purple super-tile B in either position 3 or position 4, as seen here. Consequently, each Interior triangle must always be an Interior triangle, and the super-tile of which it is a part must be the super-tile in the original super-pattern of the Pinwheel Pattern. So, no alternative grouping is possible.

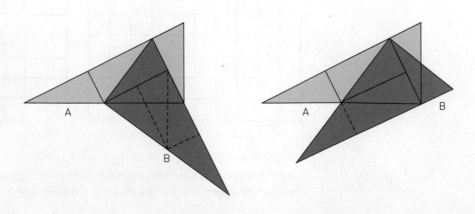

We now know that it is impossible to group the Pinwheel Triangles into 5-unit, T-arrangement super-tiles in any other grouping. We can use that fact to see that we also cannot group the Pinwheel tiles into 25-unit super-super-tiles in any other arrangement. Likewise, there is no alternative way to group the tiles of the Pinwheel Pattern into 125-unit super-super-super-tiles, and so on. Thus the groupings of the tiles in the Pinwheel Pattern into super-tiles are unique. This observation, in turn, demonstrates that its symmetry of scale is unique and completes our proof.

Rigid Symmetries?

We now turn our attention to the rigid symmetries of the Pinwheel Pattern. We will soon discover that the proof of the uniqueness of grouping provides a key insight into resolving the rigid symmetry issue. We have seen that the Pinwheel Pattern enjoys a symmetry of scale. Let's now explore other symmetries that the Pinwheel Pattern might exhibit. Photocopy an image of the Pinwheel Pattern onto a transparency, and move the transparency in search of translational, rotational, or reflectional symmetries of the Pinwheel Pattern. Mark a tile as your base tile. Move the transparency so that the base tile lies above some other tile in the pattern. Now look at the other tiles on your transparency. Do they all lie directly over an identical tile in the picture? If not, try again. You can, in fact, get large groups of tiles around your moved base tile to line up. But do all the triangles on your transparency coincide with a triangle on the page?

Perhaps that's enough trying, because we could try until the proverbial cows come home and we would never get the transparency pattern to cover the entire original pattern. It cannot be done! The Pinwheel Pattern has no rigid symmetries at all!

No Rigid Symmetry.

The Pinwheel Pattern has no rigid symmetries.

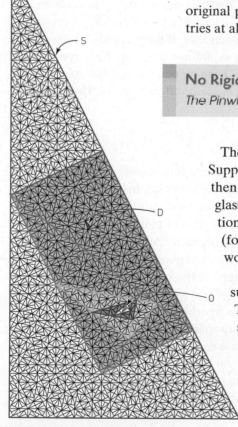

A 3125-unit Pinwheel Triangle

The key to this observation is the uniqueness of the grouping process. Suppose the Pinwheel Pattern had a rigid symmetry. If this were the case, then we would be able to trace the Pinwheel Pattern onto a piece of plexiglass, label the original tile of the Pinwheel Pattern O (for Original position), shift the plexiglass so that the triangle O covers a different triangle D (for Destination position), and every Pinwheel Triangle on the plexiglass would coincide with a Pinwheel Triangle on the plane.

Now let's consider the grouping process again. Let's first construct 5-unit super-tiles to form a new Pinwheel Pattern in the plane by 5-unit Pinwheel Triangles. Next let's group those 5-unit triangles by fives to create 25-unit super-super-tiles that form a new Pinwheel Pattern of the plane by 25-unit super-super-tiles. We can continue this process as long as we wish, making the super-tile with triangle O in it become so large that it contains as big an area around it as we wish.

Thus, if we continue this process long enough, we will engulf both triangles O and D in one really big super-duper-tile. Let's do sufficiently many groupings so that both triangles O and D are in the same super-duper Pinwheel Triangle S (for Super-duper).

Notice that our supposed rigid symmetry rotated and slid all the original triangles of the Pinwheel Pattern so they exactly coincided with other original Pinwheel Triangles. So, if we outlined with a marker the super-duper-triangles created by grouping, then the supposed rigid symmetry would take each such super-duper-triangle onto another super-duper-triangle. Therefore, super-duper-triangle S would be shifted to some other super-duper-triangle T. But remember that there is only one way to group the triangles in a Pinwheel Pattern to construct Pinwheel Patterns of a larger scale. If we could move S to T, then we would have constructed two different groupings of the original tiles into super-duper-tiles—the one grouping that contains S and the other grouping that contains T. Since S and T overlap, they could not be two different super-duper-triangles in the same grouping—they would represent two different ways to group the original tiles into super-duper-tiles. This statement contradicts the *Uniqueness of Scaling* observation, and thus our supposed rigid symmetry cannot possibly exist. Therefore, we conclude that the Pinwheel Pattern can have no rigid symmetry.

The previous argument may be challenging to understand. But once we master the basic idea of the overlapping large triangles and remember the uniqueness of scaling, the concepts fall into place.

Life Lesson

It requires time and effort to make ideas and strategies our own.

A Deduction from the Proof

The proof shows us a feature of patterns that possess symmetry of scale and *do* have rigid symmetries: Namely, there must be more than one way to group the tiles to create the super-tiles. We just showed that, if there is only one grouping of tiles into super-tiles, then that pattern would not have any rigid symmetries. Take the square pattern or the equilateral triangle pattern, for example, and see how to group the tiles into super-tiles in several different ways. This variety is shown in the illustration.

Exotic Patterns in Nature

Patterns with no rigid symmetries are fascinating objects in their own right, and, as often happens with intriguing mathematical insights, they may also have implications in nature. In this case, these exotic patterns may reflect the geometric structure of a certain class of materials called *quasicrystals*. The connection between patterns without rigid symmetries and the structure of quasicrystals was first noticed in 1984 when scientists observed a diffraction pattern in the quasicrystal alloy $Al_{5.1}Li_3Cu$ that showed an unusual fivefold symmetry. The similar look of this quasicrystal and this aperiodic Penrose Pattern suggested the connection. As with many other developments in mathematics, perhaps these exotic patterns will be among those that at first seem rare and bizarre and later are found everywhere. These exotic patterns will become better understood as more examples are developed and more subtle properties of them are explored.

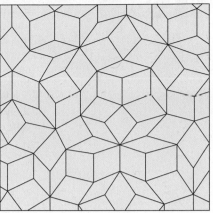

Patterns in Three Dimensions

As an epilogue to the issues of patterns in the plane, we take a brief, pictorial look at patterns in tiling 3-dimensional space. Cubes, of course, will do the job in an orderly manner full of symmetries of all kinds. Here is a surprising pattern constructed by Colin Adams of Williams College that fills space with identically shaped, fattened solid knots. The drawing of the knotted tile below left is the work of Pier Gustafson and illustrates Adams's finding.

Finally, below right is a picture of a pattern in space that has no rigid symmetries. This pattern is a 3-dimensional analogue of the Pinwheel Pattern in a plane. The picture has a wonderfully jumbled look that may better reflect reality than the orderly, perhaps artificial, neatness of the cubical filling of space.

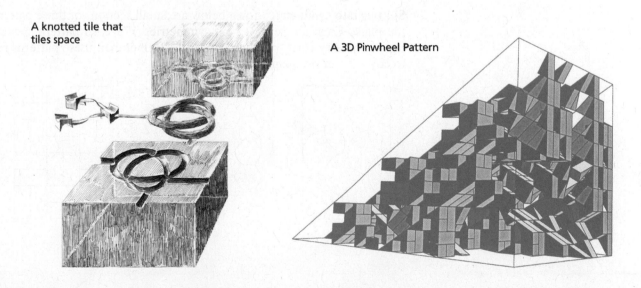

A knotted tile that tiles space

A 3D Pinwheel Pattern

A Look BACK

PATTERNS IN THE PLANE can display at least two types of symmetries—rigid symmetries and symmetries of scale. Surprisingly, some patterns have symmetry of scale but do not have any rigid symmetries at all. The Pinwheel Pattern is constructed by building increasingly larger super-tiles of the same triangular shape. The tiles can be grouped into super-tiles in only one way. That uniqueness of grouping is the key to showing that the Pinwheel Pattern has no rigid symmetries.

   We developed these exotic patterns by starting with a vague idea of symmetry and then pinning it down. Once we defined our concept more precisely, we could identify different types of symmetries and develop new patterns that we had not known before.

Life Lessons

Specifying the meaning of a familiar term or notion can open our eyes to new possibilities and ideas.

MINDSCAPES   Invitations to Further Thought

*In this section, Mindscapes marked (**H**) have hints for the solutions at the back of the book. Mindscapes marked (**ExH**) have expanded hints at the back of the book. Mindscapes marked (**S**) have solutions.*

Developing Ideas

1. **To tile or not to tile.** Which of the following shapes can be used to tile the entire plane?

2. **Shifting into symmetry.** Shown below are small sections of three patterns in the plane. Each has several rigid symmetries. For each pattern, describe a rigid symmetry corresponding to a shift. (Remember, these patterns repeat to cover the entire plane.)

 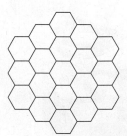 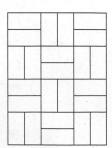

3. **Flipping over symmetry.** For each pattern in Mindscape 2, describe a rigid symmetry corresponding to a flip. Which patterns have more than one flip symmetry?

4. **Come on baby, do the twist.** Which patterns in Mindscape 2 have a rigid symmetry corresponding to a rotation? Do any of the patterns have more than one rotational symmetry?

5. **Symmetric scaling (ExH).** Each of the two patterns below has a symmetry of scale. For each pattern, determine how many small tiles are needed to create a super-tile. How many are required to build a super-super-tile?

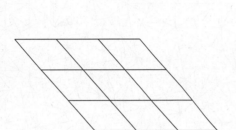 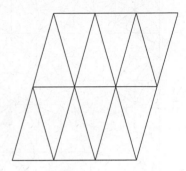

Solidifying Ideas

6. **Build a super.** Draw a 1, 2, $\sqrt{5}$ right triangle in the center of a page. Draw four identical right triangles around it to create a similar right triangle that is a 5-unit super-tile in the T-arrangement.

7. **Another angle.** Look at the 5-unit super-tile you created in Mindscape 6. Measure the angle between the line determined by the base of the original triangle and the base of the 5-unit super-tile.

8. **Super-super.** Surround your 5-unit super-tile with 20 more right triangles, each identical to the original triangle, to construct a 25-unit super-super-tile.

9. **Expand forever (H).** If you continue the process of Mindscapes 7 and 8, why can you cover the whole plane?

10. **Triangular expansion.** Take an equilateral triangle. Surround it by three other identical equilateral triangles to create another equilateral triangle. Which way is it facing?

11. **Expand again.** Take your 4-unit equilateral triangle and surround it with 12 equilateral triangles to create a 16-unit super-triangle. Which way is it oriented?

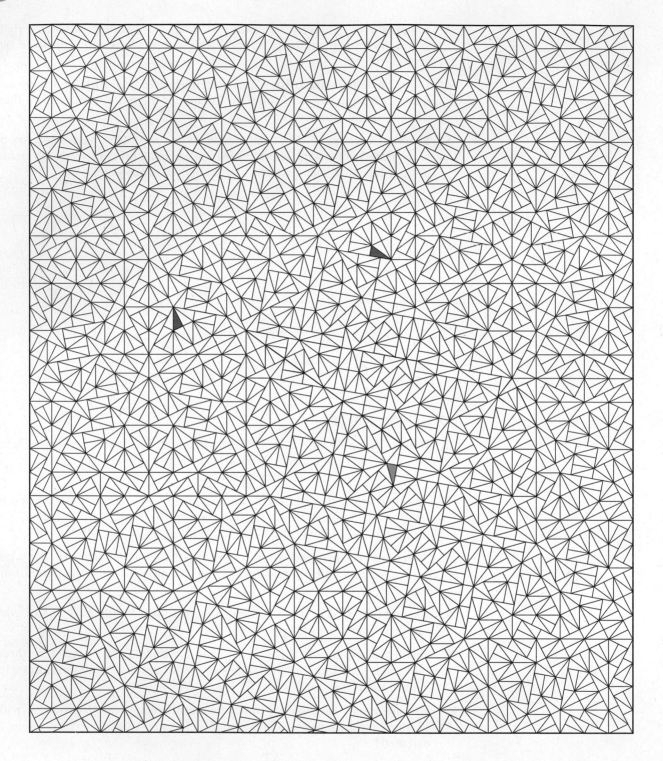

12. One-answer supers. Here is a Pinwheel Pattern. For each filled-in tile, outline the surrounding tiles that create the 5-unit super-tile and the 25-unit super-tile of which it is a part. There is only one correct answer.

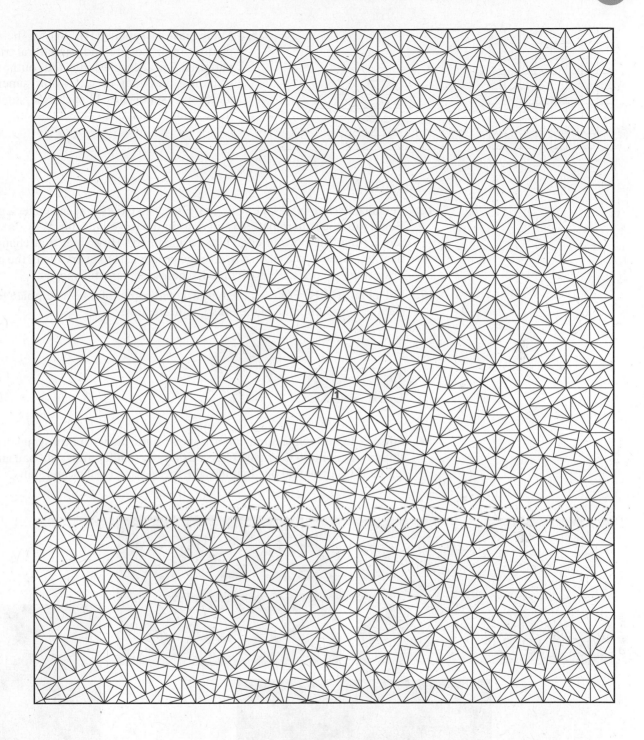

13. **Close to periodicity (S).** If you cover this Pinwheel Pattern with a transparency and shift it to make the ✳ tile cover tile 1, how many tiles near the moved ✳ tile line up before they don't?

14. **Golden periodicity (H).** Can you construct a periodic pattern that covers the plane and is made from the 1, 2, $\sqrt{5}$ Golden Triangles?

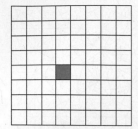

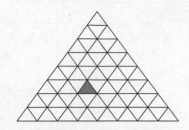

15. **Many answer supers (H).** Shown here are pictures of the square and equilateral triangle patterns. For each tile indicated, outline surrounding tiles that create 4-unit super-tiles and 16-unit super-tiles. In each case, show that there is more than one correct answer.

Creating New Ideas

16. **Personal Escher (S).** Make your own Escher-like tiles that can cover a plane. Begin with the square pattern and deform the top edge a bit. Why does that require that the bottom edge also be modified? How does that change all the tiles? Note how a change to an edge propagates to every tile in the plane. Distort the square tiles to become Escheresque, fanciful shapes.

17. **Fill 'er up? (ExH)** For each tile below, could copies of it be assembled to create a pattern filling the plane? If so, indicate how.

18. **Friezing.** In the architectural patterns shown here, can you find a symmetry that shifts every point to the right? Can you find one that flips the right half and the left half? Can you find one that flips the frieze over? Or flips and shifts?

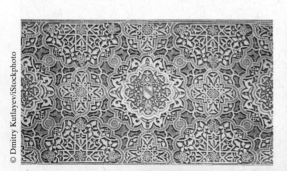

19. **Wallpapering.** In the patterns shown below, can you find a symmetry that shifts every point to the right? Can you find one that flips the right half and the left half? Can you find one that flips the pattern over? Or flips and shifts? Can you rotate the pattern?

© Eduard Andras/ iStockphoto

© Jens Carsten Rosemann/iStockphoto

© Jens Carsten Rosemann/iStockphoto

© Bill Noll/iStockphoto

© tulcarion/iStockphoto

20. **Commuters? (S)** Take the square pattern covering the plane. Consider two symmetries. One flips the pattern about a diagonal line through the central square. The other rotates the pattern 90° clockwise. If you first perform the flip and then the rotation, do you get the same symmetry as you would if you were to first do the rotation and then do the flip?

Further Challenges

21. **Penrose tiles.** Roger Penrose constructed two tiles that can be used to cover the plane only in aperiodic ways. One tile is called a *kite*, one is called a *dart*. Show how to construct a tiling using these tiles. Notice that each tile contains both a blue and red arc. The only rule for assembling the tiles is that the ends of blue arcs must touch only the ends of other blue arcs and ends of red arcs must touch only the ends of other red arcs. Given this rule and using many copies of the kites and darts, show how to construct a tiling using these two tiles.

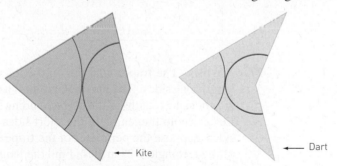

Kite

Dart

22. **Expand forever.** Why does any shape that can be assembled to form a larger version of itself that surrounds it on all sides give a means for covering the whole plane with tiles of that shape?

In Your Own Words

23. **With a group of folks.** In a small group, discuss and actively work through the construction of the Pinwheel Pattern and prove that it has no rigid symmetries. After your discussion, write a brief narrative describing the construction and proof in your own words.

24. **Creative writing.** Write an imaginative story (it can be humorous, dramatic, whatever you like) that involves or evokes the ideas of this section.

For the Algebra Lover

Here we celebrate the power of algebra as a powerful way of finding unknown quantities by naming them, of expressing infinitely many relationships and connections clearly and succinctly, and of uncovering pattern and structure.

25. Super total. Recall that the Pinwheel Triangle has sides of length 1, 2, and $\sqrt{5}$. The figure below shows a super triangle made of five Pinwheel Triangles. Find the lengths of the sides of the super triangle.

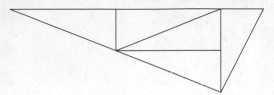

26. Sororitile. As a community service project, your sorority offers to replace the tile in the main room of your town's recreation center. The floor plan is given below. (The room is a 30 by 30 square with a 10 by 10 square cut out of one corner.) If you plan to use 1 by 1 foot square tiles in the most basic pattern (see the figure associated with Mindscape 15), how many tiles will you need?

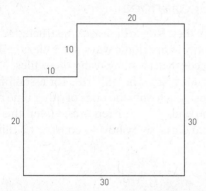

27. XY-tiles. The trapezoidal tile on the left has one short side and three equal-length long sides. The tile on the right is a rectangle whose long and short sides are the same lengths as the corresponding long and short sides of the trapezoid. Let x denote the length of the short sides and y denote the length of the long sides. Suppose the perimeter of the trapezoid is 32.5 inches and the perimeter of the rectangle is 25 inches. Find the lengths of the sides of both tiles.

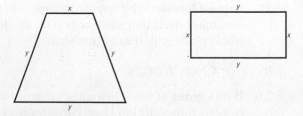

28. School spirit. Your dorm bathroom is tiled using white hexagonal tiles arranged in a honeycomb pattern (see the middle figure associated with Mindscape 2). One weekend, you and your roommates decide to replace all the white tiles with purple and gold tiles in honor of your school colors, with some green tiles thrown in as accents. (One of your roommates is from New Orleans.) You need a total of 7200 tiles. You want three times as many gold tiles as green tiles, and 1200 more purple tiles than twice the number of green tiles. How many tiles of each color do you need?

29. **T-total (H).** Suppose you start with one small tile. You create a super tile by surrounding the small tile with t additional small tiles identical to the original. Then you create a super-super tile by surrounding your super tile with t additional super tiles. Write an expression for the total number of small tiles you have in your super-super tile. Suppose this total is 81. Find the value of t.

4.5 THE PLATONIC SOLIDS TURN AMOROUS

Discovering the Symmetry and Interconnections Among the Platonic Solids

The Calligrapher and Mathematician: Johann I. Neudorfer and His Son (1561) by Nicolas Neufchatel.

> *God eternally geometrizes.*
> PLATO

Symmetry and regularity lie at the heart of classical beauty. We have an instinctive affinity for symmetrical objects—things that can be turned or reflected to return to their original shapes. The sphere is the ultimate in symmetry. From any vantage point, the sphere looks the same. If we forgo the graceful constant curvature of the sphere and consider objects with flat sides, then how symmetrical and graceful can they be? Here we examine such flat-sided objects and explore their robust symmetries that have intrigued people for thousands of years.

We explore symmetric solids—referred to as regular, or Platonic, solids—by imagining ourselves holding them, building them, being inside them, cutting them. These "solids" are solid. They are real things for us to hold and enjoy. We study them by first moving from the qualitative to the quantitative—how many edges, faces, and vertices does each solid have? Then we record our observations and find coincidences. In life, coincidences are flashing lights signaling us to look for reasons and relationships that show a deeper structure. We find surprising connections among the regular solids that turn these separate objects into a coherent collection.

Symmetry in the Plane

As always, we consider a simple case with which to ground ideas. Before tackling objects in space, let's quickly consider symmetric shapes in the plane. In the plane, the most symmetrical object is the circle. But if we restrict ourselves to objects with straight sides, the most symmetrical objects are the *regular polygons*—polygons having all sides of equal length and all angles of equal measure. For any natural number *n*, there is a regular polygon having *n* sides. These regular polygons have lots of symmetry.

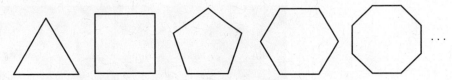

∞ - gon?

We see that there are infinitely many different regular polygons, and the more sides a regular polygon has, the more it looks like a circle—the most symmetric of all shapes in a plane. Perhaps we could view the circle as an infinitely many-sided polygon.

What 3D Objects Are Analogous?

We now turn from figures in a plane to solids in space. We will discover a surprising difference between the regular polygons in two dimensions and the analogous objects in three dimensions. For now, we will just think about solids in space.

The sphere is the most symmetrical of solids in space because it remains unchanged when revolved about any line through its center or reflected through any plane through its center. Building a sphere isn't easy. If we wanted to construct a solid with lots of symmetry, we might decide to settle for something that is less symmetrical than the sphere but that can be more easily built. Specifically, suppose our solid has flat sides and straight edges. What properties would such solids have in order to be as symmetrical as possible?

Symmetric Solids with Flat Faces: Platonic Solids

Certainly, all the faces of a solid should be the same and should be symmetrical themselves; that is, the faces should be identical regular polygons. But how will the faces fit together? We would like every corner (vertex) of the solid to look exactly like any other vertex. Thus, we require that the number of edges emanating from any vertex of the solid always be the same. So, for a solid made up of flat sides to be as symmetrical as possible, its faces should be identical regular polygons, and the number of edges coming out of any vertex of the solid should be the same for all vertices. Such symmetric solids are called *regular* or *Platonic solids*.

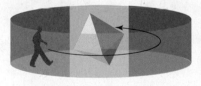

A regular solid looks the same from different points of view.

Different points of view.

The cube is the most familiar of the regular solids; its classic structure is well known. Every side of the cube is a square and from any vertex of the cube, three edges come together. Can you think of other regular solids? How many other regular solids do you think there are? The Pythagoreans pondered these same questions thousands of years ago. Remember that there are infinitely many regular polygons in the plane. It seems reasonable to guess that there will be infinitely many regular solids in space. On the following pages are 3D pictures of five regular solids: tetrahedron, cube, octahedron, dodecahedron, and icosahedron.

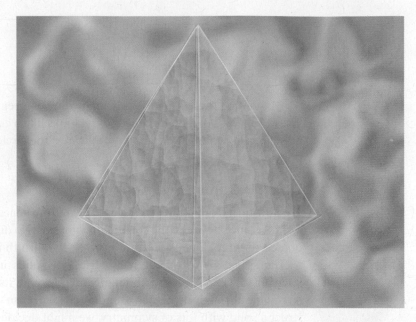

Tetrahedron in 3D (use your *Heart of Mathematics* 3D glasses to view these images)

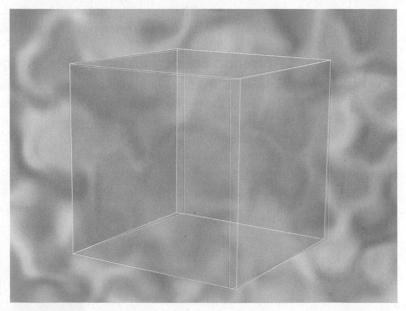

Cube in 3D

Can you think of another one? No, you can't. Why? Because even though there are infinitely many regular polygons in the plane, there are only *five* regular solids in space. What a dramatic difference between the geometry of the plane and the geometry of space! Why are there only five? If you cannot wait to learn why, you can take a peek at the "Feeling Edgy" section in Chapter 6. The argument ironically relies on features of solids that are not solid! For now, let's just accept the fact that there are only five regular solids. The regular solids are regular since all the faces of each are regular polygons identical in size to each other

Octahedron in 3D

Dodecahedron in 3D

Icosahedron in 3D

face of the regular solid, and the number of edges emanating from each vertex is the same. Each Platonic solid looks identical from the vantage point of each face.

The Mystical Allure of Platonic Solids

For centuries, the Platonic solids were associated with mystical powers. Even though the great Greek mathematician Euclid wrote about these regular solids,

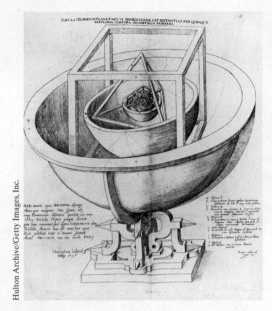

Kepler's unsuccessful planetary
effort

Life Lesson

**Don't be afraid to
make mistakes while
creating new ideas.**

they are named after Plato because he apparently attempted to relate them to the fundamental components that he believed made up the world. The Pythagoreans, who knew there were exactly five regular solids and no others, held them in awe.

Johannes Kepler shared this sense of awe for the Platonic solids. Although Kepler is remembered best for his laws of planetary motion, he was proudest of his book *Mysterium Cosmographicum*. In it he proposed a theory to explain the structure of the solar system. At the time Kepler lived, there were six known planets. Kepler showed that it is possible to take the five regular solids, put one inside the other, and have the sizes of inscribed and circumscribed spheres about these solids reveal the sizes of the orbits of the planets.

Kepler believed that a mystical power of the five regular solids dictated the planets' orbits. Sadly, his theory was refuted in 1781 when another planet, Uranus, was discovered. Unfortunately, there still are only five regular solids. We know that the regular solids have no relation to the orbits of the planets, but we also know that the only way to understand and make discoveries is to try new things and not be afraid to be wrong. These qualities cannot be separated from Kepler's genius.

Hold Them in Your Hands: First, The Tetrahedron

Before we examine the Platonic solids closely, please explore these shapes a bit on your own. To do that, you must have a set of them. So, before proceeding, you will need to make a complete set of the five regular solids. You will find the necessary materials and instructions in your kit.

The first and simplest regular solid is the *tetrahedron*. The tetrahedron is made up of four identical equilateral triangles. We assemble them to make a triangular-based pyramid, the tetrahedron. A tetrahedron that we hold and see is imperfect. A real tetrahedron exists only in the mind and is the idea that this physical model suggests. The real tetrahedron is perfect—without blemishes or flaws. It has four identical faces, four vertices, and six edges. We can hold a model with our hands, but we can conceptualize the real tetrahedron in our minds, turn it about, and see it from the inside out or in other ways.

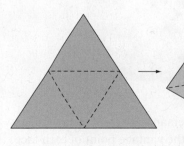

Think about the tetrahedron and understand its geometry so completely that you could answer any question about it. For example, suppose we were inside a tetrahedron sitting on one face with a vertex directly above our heads. What would we see as we looked up? Try to visualize the tetrahedron in several ways. Hold it in your mind. Are the edges sharp or dull? Are the vertices dangerously pointy? Feel it. Cut it up. Explore it. What questions arise in getting to know it? Learn to rely on the model in your mind.

Now rotate the tetrahedron and understand all its possible symmetries. Do not be satisfied with general impressions. Move from having a feeling about symmetry to being able to define exactly what symmetry is. Think about how you would describe all symmetries of the tetrahedron.

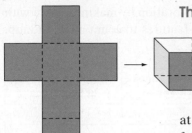

The Other Four

Next we move to the *cube* — the most familiar of the regular solids. One incarnation of cubes is dice. Their regularity allows us to expect with equal likelihood that any one of the six faces will land up after a die is thrown. Cubes are roughly the shape of some rooms. This fact is helpful in trying to visualize the cube from the inside. If you are in a room right now, look at the corners. You see three sides coming together at the corners. You see the faces meeting at right angles.

The next regular solid is the *octahedron*, which is constructed from eight identical equilateral triangles. It has the appearance of two bottomless pyramids stuck together.

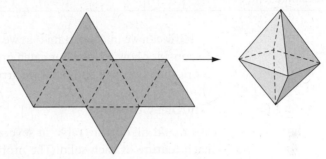

The *dodecahedron* is the only solid with pentagonal faces. It is made up of 12 identical pentagons.

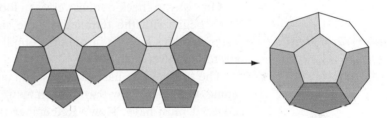

The Platonic solid with the most faces is the *icosahedron*. It is constructed from 20 identical equilateral triangles. Notice that, at each vertex of the icosahedron, five triangles meet.

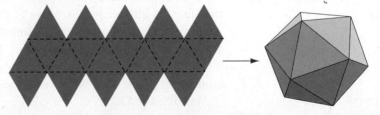

From the Qualitative to the Quantitative

After you have made your own set of Platonic solids, you can begin to explore their nuances. Let's leap beyond qualitative impressions toward quantitative properties.

In the case of the regular solids, what natural features can we count? How many faces, vertices, and edges are there? How many sides does each face have? How many edges come together at each vertex? How many faces come together

Life Lesson

In life, quantitative exploration can often bring into focus hidden texture and richness.

at each vertex? Let's be systematic about this exploration by making a table with the regular solids noted down one side and the features to count along the top. Now fill in the chart of features on the next page.

	Number of Vertices	Number of Edges	Number of Faces	Number of Faces at Each Vertex	Number of Sides of Each Face
Tetrahedron					
Cube					
Octahedron					
Dodecahedron					
Icosahedron					

How can we make certain that we are not miscounting? It is always worthwhile to consider how to avoid or catch errors. One strategy is to do the same thing in more than one way and see if we arrive at the same answer.

Double Counting

We could fill out the table in several different ways. First, we could just count each feature of each solid. The problem is, how can we be certain that we have not failed to count, say, an edge, or that we have not counted a vertex more than once? Although tackling the tetrahedron may be simple, that icosahedron has a heck of a lot of edges, vertices, and faces.

One way to check our answers is to look for relationships among the numbers. Let's begin with the tetrahedron, the simplest figure. Examining simple cases often gives us insights into more complicated situations and may lead to unexpected discoveries.

The tetrahedron has four faces, and each face is a triangle. Once we know that, we can figure out how many edges it must have. How? Remember that each of the four faces has three edges. So, our first guess is that there must be three edges for each of the four faces for a total of 4 × 3, or 12, edges. Of course, this method of counting is wrong, because each edge of the tetrahedron is counted twice. Each edge touches two different faces, so we've counted each edge twice.

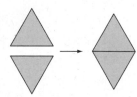

Because each edge has been counted exactly twice, we know that 4 × 3 gives us a number that is exactly twice too big. Hence, we know that the number of edges is (4 × 3)/2 = 6. Of course, with the tetrahedron, it is relatively easy to count the number of edges. Still, we have learned a lesson that can be applied to the more complicated figures.

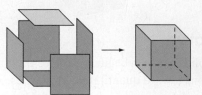

Let's try this line of thought with the other figures. The cube has six faces. Each face is a square. So there must be (6 × 4)/2, or 12, edges on a cube. We arrive at this answer because there are 4 edges on each of the 6 squares, so there are 6 × 4 edges before we glue the squares together to make the cube. Once we make the cube, we glue one edge of one square to exactly one edge of another square. This gluing then makes the number of edges on the cube exactly half of the number of the 6 × 4 original edges of the squares. Thus, there are 12 edges on a cube.

The octahedron has 8 faces, each one a triangle. Therefore, an octahedron has $(8 \times 3)/2 = 12$ edges. The dodecahedron has 12 faces, each one a pentagon. Therefore, a dodecahedron has $(12 \times 5)/2 = 30$ edges. The icosahedron has 20 faces, each one a triangle. Therefore, an icosahedron has $(20 \times 3)/2 = 30$ edges.

Now let's look at the table from the previous page and check the number of vertices in a similar way. Can you devise a way of calculating the number of vertices of a regular solid once you know how many faces it has, how many sides each face has, and how many faces come together at each vertex? After you have devised the method, use it to check your entries for the number of vertices for each figure.

Coincidence? We Don't Think So

Look at the previous table again. Do you notice any coincidences? Coincidences occasionally are just coincidences, but if we see several numbers appearing more than once, we might wonder whether those entries are somehow related. What coincidences appear in our table of vertices, edges, and faces?

Observations

The number of faces of the cube equals the number of vertices of the octahedron; the number of vertices of the cube equals the number of faces of the octahedron; and the number of edges of the cube and octahedron is the same. Why?

Let's think about these observations while examining our models. The number of faces of the cube equals the number of vertices of the octahedron. Could we put those two apparently coincidental numbers together geometrically? Can we explain geometrically why those coincidental numbers are not at all coincidental?

Visualizing the Big Picture

Let's think big. Stand in the middle of a room — preferably a small square room, since such a room is most nearly a cube. As you stand in this room look up, down, ahead, behind, left, and right. Those six directions are the six faces of the cube in which you stand. We want to understand the coincidence of the octahedron

Put on your 3D glasses and enjoy.

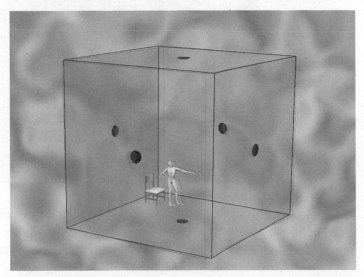

having the same number of vertices as the cube has faces. So, look at each face of the cube — that is, each wall plus the floor and ceiling. If you placed a dot (vertex) on each face of the cube, where would be the most natural place on each face for it? Look up. Where would you put a vertex? There is really only one natural place—the middle, right in the middle of the ceiling. Imagine a dot in the center of the ceiling. Now look at the center of each of the four walls, and don't forget the floor. Place a dot on each. How many vertices have you placed? Six. One for each face of the cube in which you stand.

Now that you have placed some vertices, try drawing edges between them in an attempt to make the skeleton of a solid. Which vertices would you connect to create an edge of another solid? Look at the center of the ceiling. Remember that, at the center of the ceiling, you have placed a vertex in your imagination. The ceiling is surrounded by four walls, and each wall also has a vertex in its center. It seems natural to connect the ceiling vertex with each wall vertex. Don't you feel compelled to draw those four connecting lines—those edges? The edges connect the ceiling with each wall because each wall abuts the ceiling. It seems reasonable that faces that touch should have their centers joined by an edge. Do you see in your mind's eye the four edges from wall centers to the ceiling center? You are standing so that you see those four edges rising like a tent or a pyramid over your head.

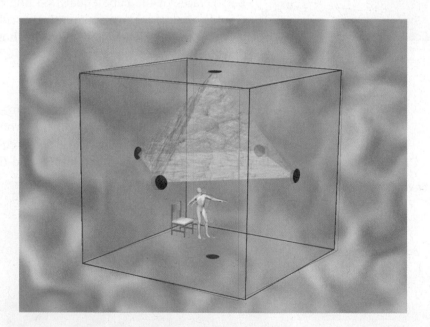

Now join consecutive wall-centers together to create a square "encircling" your midsection and forming the base of the pyramid. Do you see those four sides going from wall-center to wall-center enclosing you in a square embrace? Now complete the picture with edges connecting the wall-centers with the center of the floor where your feet are planted. Do you see those four edges creating an upside-down pyramid on the same central diamond base?

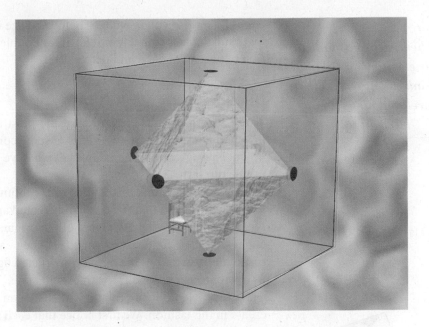

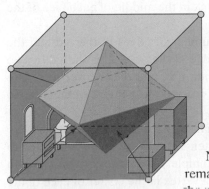

©Corbis

You now have the skeleton—the edges—of the octahedron. Do you see why the edges of the cube are in one-to-one correspondence with the edges of the octahedron? Each edge of the cube is an edge where two faces abut. But each of those two faces has a vertex in its center, and those two face-centers are connected with an edge of the octahedron we created. Those two edges naturally correspond. Each cube edge has a natural octahedron edge and vice versa. So there are the same number of cube edges as octahedron edges.

Now we have the edges of an octahedron. It remains only to fill in the triangular faces to complete the octahedron. It is easy to fill those triangles in with our mind's eye, but let us think as we do so. Do you see that each corner of the cube is associated with one triangular face of the octahedron? Do you see that the fact that there are three faces coming together at each vertex of the cube leads to the fact that the figure we are creating has triangles for faces?

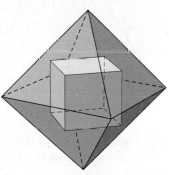

We have just seen that the very essence of the cube somehow includes in its being the essence of the octahedron. These figures are closely linked. What happens if we start with an octahedron and imagine the same process? Can we float in the center of the octahedron? Of course we can in our mind's eye. We can move two pyramids of Cheops, one upside down, to create in our mind a wonderful octahedron. We can float inside that octahedron and begin to place a vertex in the center of each of the eight triangular faces. The vertices in the centers of the triangular faces can be joined to create

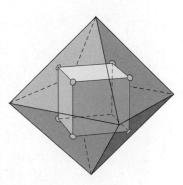

edges to enclose you in what? Try to visualize it! Do you see yourself surrounded by that cube suspended amid the great double pyramid? Do you see the connection between the cube and octahedron?

Dualing Solids

These Platonic solids are paired at the most basic level of their essential defining characters. They are two views of some common essence. The one entails the other. They are our first examples of what we will soon call duality, but they will not be our last.

Let's test the limits of the idea and continue to gain insight. In this case, we started with one geometric solid and created another. We put a vertex in the middle of each face. When two faces of the original solid shared a common edge, we connected their centers with an edge. Then each vertex of the original solid corresponded to a face of the newly constructed figure. This process will always allow us to create a new solid from an existing one.

Now let's look at the relationship between the old solid and the new one we create in this fashion. We first notice that each face of the old gives a vertex of the new. That is easy to see, because we put a vertex in the middle of each face of the old solid.

Next we notice that there are the same number of edges in the old and the new solids. This equality is true since every edge of the original is shared by two faces. Each of those two faces had a vertex put in it, and the two vertices were connected by an edge. Each edge of the old is crossed by one edge of the new. So they are in one-to-one correspondence.

Finally, we look at each vertex of the original solid. Each face that meets there has a vertex of the new figure placed in its center. These vertices are connected to create a polygonal face of the new solid. For example, look at the corner of a room and imagine it as a vertex of a cube. If we put a vertex in the center of each of the walls that come together at that corner and connect them, we get a triangular face of the new solid we are creating. So, in fact, we know how many sides the new-polygonal face has by seeing how many faces come together at the original vertex. That is, each face of the solid we are creating from the cube is a triangle. Of course, after we noticed that we had created an octahedron, we knew that each face was a triangle. But now we see a relationship between the triangles of an octahedron and the number of faces that come together at each vertex of a cube.

If we now take our new solid and perform the same process of constructing a new solid from it, it is no coincidence that we get a solid of the same type that we started with.

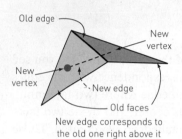

New edge corresponds to the old one right above it

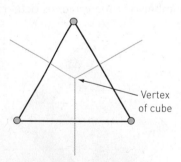

The vertex of the cube abuts the corresponding octahedron face edge.

Duality

This process of creating one solid from another is so appealing that it deserves a name. Since we see great connections between the two solids—the faces of one correspond to the vertices of the other, and the edges correspond—the name chosen to refer to this phenomenon is *duality*. This name suggests that the two solids form a pair. It also suggests that the "opposite" features of the two solids correspond: The faces of one correspond with the vertices of the other. A polygonal

3D art: natural pairings between the Platonic solids

3D images showing the duality between the octahedron and the cube

face is the highest dimensional external feature of the solid. It corresponds to the lowest dimensional feature of the dual object, the vertex. The edge is in the middle of the vertex-edge-face hierarchy, and it corresponds to the dual solid's edge because it is also in the middle.

What does duality generate from each solid? The cube gives an octahedron. An octahedron gives the cube back. A dodecahedron gives the icosahedron; and the icosahedron gives the dodecahedron back. What does the tetrahedron give back as its dual? Itself! The tetrahedron is *self-dual*.

3D images depicting the duality between the dodecahedron and the icosahedron

Amorous Solids

For over two millennia, the regular solids have been Platonic. Surely with all this duality going on, the time has come to rename certain pairs the *amorous solids*. The cube and octahedron have a physical intimacy that is not purely Platonic. The dodecahedron and the icosahedron are closely entwined. And the self-dual tetrahedron has a relationship with itself that surely goes beyond the bounds of Platonism. We hereby begin a campaign to rename the Platonic solids the more lively and appropriate amorous solids to reflect their physical beauty and their habit of congregating in dual pairs.

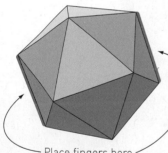

Place fingers here

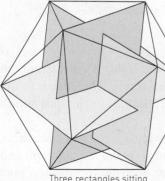

Three rectangles sitting in the icosahedron

Our Golden Promise

At the end of our journey through the aesthetic world of the Golden Rectangle we said we would see Golden Rectangles occurring in surprising places. It turns out that they occur in Platonic solids. To illustrate, we examine the icosahedron. Consider the drawing and note the rectangle inside the icosahedron. Then examine your physical model of the icosahedron and put your fingers on the opposite sides of the rectangle. Feel the rectangle between your fingers.

In fact, if that were a perfect icosahedron, that rectangle would be a perfect Golden Rectangle! Can you find other Golden Rectangles inside the icosahedron? Consider the picture at left. All three of those rectangles are Golden Rectangles. Do you see how they all appear to be perpendicular to each other? They actually are! In fact, this is the way that Luca Pacioli constructed the icosahedron in his book *De Divina Proportione*. He observed that, if we begin with the three Golden Rectangles shown in the diagram at the left, then attaching vertices together to make triangles will result in 20 equilateral triangles, thus creating an icosahedron. We ask you to think more about this construction in the Mindscapes.

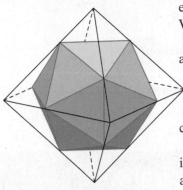

The octahedron surrounding and protecting the icosahedron

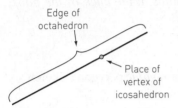

Edge of octahedron

Place of vertex of icosahedron

Notice in the preceding illustration that the rectangles are all linked in an exotic manner. Can you figure out what is unusual about this group of rectangles? We'll describe this remarkable property in the next chapter.

One last golden fact. We know that the icosahedron and the dodecahedron are dual, that the cube and the octahedron are dual, and that the tetrahedron is dual with itself. Thus, it appears that there is no connection between the octahedron and the icosahedron. We know for sure they are not dual. Well, these two solids turn out to have an interesting connection. We describe this now, but caution you that this special phenomenon is not duality!

Observe that there are 12 edges on the octahedron and 12 vertices on the icosahedron. If we are careful, it is possible to place a small icosahedron inside a larger octahedron in such a way that each vertex of the icosahedron touches exactly one edge of the octahedron. It is a delicate placement. Look at your octahedron and see where you can place the vertices of the icosahedron on the edges of the octahedron. Visualize the icosahedron within the octahedron before reading on.

Notice that the vertices of the icosahedron do not touch the edges of the octahedron exactly in the middle of the edges. The vertices touch the edge somewhere off center. If we take any edge of the octahedron in the preceding figure and cut it where the vertex of the icosahedron touches that edge, we would have two line segments, one longer than the other.

If we now let these lengths determine a rectangle, then we get a perfect Golden Rectangle. We not only are able to fit an icosahedron inside an octahedron in a special way, but, in doing so, we cut up all the edges of the octahedron into pieces that exhibit the Golden Ratio! Life is full of amazing connections and mathematics can help us see them!

A Look BACK

THE FIVE REGULAR, or Platonic, solids are the tetrahedron, the cube, the octahedron, the dodecahedron, and the icosahedron. No other solids can be created with identical, regular polygonal faces meeting together so that the number of edges emanating from any vertex of the solid is the same. The regular solids come in dual pairs where the vertices of one correspond to the faces of the other. The tetrahedron is self-dual. We saw surprising relationships among the Platonic solids, the Golden Rectangle, and the Golden Ratio. Different mathematical ideas are interconnected. The deep relationships between seemingly distinct mathematical objects are at the heart of mathematics and are the core of its beauty and elegance.

Our strategy for discovering interesting properties of the Platonic solids was to take a quantitative look at their features and then look for patterns. We counted the vertices, edges, and faces of each regular solid and saw some surprising coincidences. Exploring those seeming coincidences led us to discover, among other things, the idea of duality.

To understand ideas or objects, a good first step is to experience them as directly as possible. Going from a qualitative to a quantitative view allows us to see far more detail and opens a whole new world of understanding. Quantitative measurements allow us to see patterns that might not otherwise be apparent. Patterns and coincidences are signposts signaling relationships and interconnections. These techniques can help us analyze the unknown and think of new ideas in every arena of life.

MINDSCAPES ▸ Invitations to Further Thought

*In this section, Mindscapes marked (**H**) have hints for solutions at the back of the book. Mindscapes marked (**ExH**) have expanded hints at the back of the book. Mindscapes marked (**S**) have solutions.*

Developing Ideas

1. **It's nice to be regular.** What makes a polygon a *regular* polygon? Sketch six different regular polygons and three nonregular ones.

2. **Keeping it Platonic.** What makes a solid a *regular* (*Platonic*) solid?

3. **Count 'em up.** How many faces, edges, and vertices are there in a cube? How many are in a tetrahedron?

4. **Defending duality.** Explain why the cube and the octahedron are duals of each other.

5. **The eye of the beholder.** Suppose you have models of the Platonic solids that are not transparent. Below are sketches of three such solids drawn from different viewpoints. Identify each one and explain the perspective from which each solid was viewed.

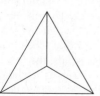 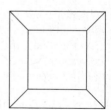 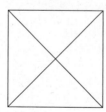

Solidifying Ideas

6. **Build them.** Using toothpicks and pieces of plastic hose, strings and straws, posterboard, mahogany, or your kit, make a complete set of the Platonic solids for yourself.

7. **Unfold them.** For each of the Platonic solids, draw a picture showing how to unfold it in various ways so it lies flat on the plane, and how to refold it to create the solid again.

8. **Edgy drawing (H).** Draw pictures in the plane that show the edges of each of the regular solids in the sense that each edge in your drawing corresponds to an edge on the solid and each one joins another one whenever the corresponding edges on the solid touch. For example, you could inflate each regular solid until it becomes a sphere with curved edges on it and then draw the stereographic projection.

9. **Drawing solids.** Draw each solid by completing the beginnings of the drawings shown here.

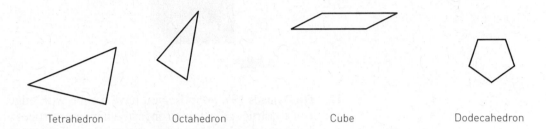

Tetrahedron   Octahedron   Cube   Dodecahedron

10. **Life drawing.** Draw the regular solids using your physical models. Include shading and perspective. You may title your work *Regular Still Life*.

11. **Count.** For each of the regular solids, take the number of vertices, subtract the number of edges, add the number of faces. For each regular solid, what do you get?

12. **Soccer counts (ExH).** Look at a soccer ball. Take the number of vertices, subtract the number of edges, add the number of faces. What do you get? This counting can be tricky, so think of a systematic method of accomplishing it.

13. **Golden Rectangles.** Take your toothpick or straw model of an icosahedron and place it on the table. Now prop it up so that only one edge is resting on the table. Locate the Golden Rectangle spanning the edge on the table and the top edge. Locate the pair of vertical edges that form two sides of a Golden Rectangle. Locate a pair of horizontal edges halfway up from the table to the top edge that form two sides of a Golden Rectangle. If you are dexterous, carefully weave pieces of string to construct those three Golden Rectangles. Measure the base and height of one of those rectangles and then divide the height into the base and see how it compares to the Golden Ratio.

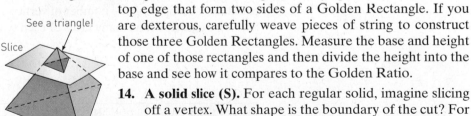

See a triangle!

Slice

14. **A solid slice (S).** For each regular solid, imagine slicing off a vertex. What shape is the boundary of the cut? For example, slicing a vertex off a tetrahedron gives a triangular cut.

15. **Siding on the cube.** Suppose we start with the edges of a cube (that is, its skeleton). Now, on each square face, we glue four triangles together as shown at right. Sketch a drawing of this figure. Count the number of its vertices, edges, and faces.

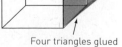

Four triangles glued to the edges of the cube

Creating New Ideas

16. **Cube slices (H).** Consider slicing the cube with a plane. What are all the different-shaped slices we can get? One slice, for example, could be rectangular. What other shaped slices can we get? Sketch both the shape of the slice and show how it is a slice of the cube.

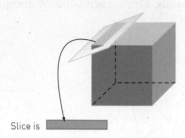

Slice is

17. **Dual quads (S).** Suppose you have a cube with edges of length 1. Suppose you construct an octahedron inside it whose vertices are in the centers of the faces of the cube. How long are the edges of that octahedron?

18. **Super dual.** Suppose you take a cube with edges of length 1 and construct an octahedron around it by making the center of each triangle of the octahedron hit at a vertex of the cube. How long are the edges of the octahedron?

19. **Self-duals.** Suppose you have a tetrahedron having each edge of length 1. Construct a tetrahedron inside whose vertices are in the centers of the faces of the original tetrahedron. How long are the edges of that inscribed tetrahedron?

20. **Not quite regular (ExH).** Suppose you allow different numbers of triangles to come together at different vertices of a solid. Show how to produce solids with arbitrarily large numbers of triangular faces.

Further Challenges

21. **Truncated solids.** Slice off all the vertices of each of the regular solids to produce new solids that have two different types of sides. Fill in the chart by counting or computing the number of vertices, edges, and faces each solid now has. Also describe how many faces of each type the truncated solid has.

Solid (pretruncated)	Number of Vertices	Number of Edges	Number of Faces
Tetrahedron			
Cube			
Octahedron			
Dodecahedron			
Icosahedron			

22. **Stellated solids.** Take each regular solid and replace each face by a collection of equilateral triangles (one attached to each face) to produce a new solid that looks like a star:

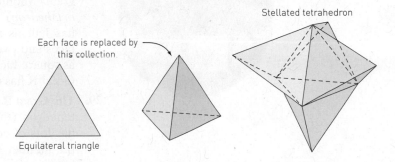

Stellated tetrahedron

Each face is replaced by this collection

Equilateral triangle

Fill in the chart by counting or computing the number of vertices, edges, and faces of each.

Solid (prestellated)	Number of Vertices	Number of Edges	Number of Faces
Tetrahedron			
Cube			
Octahedron			
Dodecahedron			
Icosahedron			

In Your Own Words

23. **Personal perspectives.** Write a short essay describing the most interesting or surprising discovery you made in exploring the material in this section. If any material seemed puzzling or even unbelievable, address that as well. Explain why you chose the topics you did. Finally, comment on the aesthetics of the mathematics and ideas in this section.

24. **With a group of folks.** In a small group, discuss and actively work through the idea of duality of the Platonic solids. After your discussion, write a brief narrative describing duality in your own words.

25. **Creative writing.** Write an imaginative story (it can be humorous, dramatic, whatever you like) that involves or evokes the ideas of this section.

26. **Power beyond the mathematics.** Provide several real-life issues—ideally, from your own experience—that some of the strategies of thought presented in this section would effectively approach and resolve.

For the Algebra Lover

Here we celebrate the power of algebra as a powerful way of finding unknown quantities by naming them, of expressing infinitely many relationships and connections clearly and succinctly, and of uncovering pattern and structure.

27. **Soccer solid.** A soccer ball has a total of 32 sides consisting of regular pentagons and hexagons, with eight fewer pentagons than hexagons. How many sides does it have of each type?

28. **How do you pronounce that?** There is a semi-regular solid with 26 faces composed of squares, hexagons, and octagons called the great rhombicuboctahedron (*rhombi-cube-octahedron*). Let's just call it the Great R. It has half as many octagonal faces as square faces and two-thirds as many hexagonal faces as square faces. Find the number of faces the Great R has of each type.

29. **The Great R returns.** The great rhombicuboctahedron (the Great R) from Mindscape 28 has lots of edges and vertices. Let E denote the number of edges and V denote the number of vertices. The values of E and V satisfy the equations $E + V = 120$ and $5E - 2V = 264$. Find the values of E and V.

30. **Rhombi-what?** There is a semi-regular solid with 62 faces called (ironically) the small rhombicosidodecahedron (*rhombi-cosi-dodecahedron*). We'll call it the Small R. It has half again as many square faces as triangular faces, and three fifths as many pentagonal faces as triangular faces. Find the number of faces the Small R has of each type.

31. **Small again.** Let E denote the number of edges and V denote the number of vertices for the solid Small R from Mindscape 30. The values of E and V satisfy the equations $E + 2V = 240$ and $3E - V = 300$. Find the values of E and V.

THE SHAPE OF REALITY?

How Straight Lines Can Bend in Non-Euclidean Geometries

Does space curve?

> *If there is anything that can bind the heavenly mind of man to this dreary exile of our earthly home and can reconcile us with our fate so that one can enjoy living—then it is verily the enjoyment of the mathematical sciences and astronomy.*
>
> **JOHANNES KEPLER**

Mathematics can help us understand the cosmic, the unapproachable, and the mysterious. Nothing is more cosmic and mysterious than the entire universe. For thousands of years, people have pondered the fundamental question: What is the shape of our universe?

In any attempt to understand the world around us, it is only natural to wonder about the geometry of our physical existence. Of course, our universe is incredibly vast, and our experience is limited by time and space. Thus our question is by no means easy. Does space bend or curve? What does it even mean for space to bend or curve? Since we do not see space bending or curving around us, our initial sense is that space is flat. However, because we exist on such a microscopic scale compared to that of the entire universe, perhaps we don't sense the reality of the "big picture."

Let's apply some of our techniques of analysis and try to discover the shape of space.

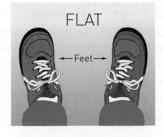

A First Sketch of the "Big Picture"

How do we start to understand the geometry of something so large that it seems beyond our capacity to comprehend it? Start by looking at something smaller.

First, we look at the ground under and just around our feet. What do we see? Flat.

Thus, it seems reasonable to guess that our world around us is flat like a plane. This guess is not completely crazy, especially to those who live in Kansas. The world around us does tend to look pretty flat. This observation led people throughout history to study the flat plane and its rich geometry. However, it turns out that Earth is shaped like a ball. This nontrivial fact illustrates two important points. First, there is no pressing need for the Flat Earth Society; and second, what we observe locally may not accurately depict what is occurring on a larger scale.

Before taking on the whole universe, perhaps we should consider the geometry of the next simplest realm: the sphere.

A Next Sketch: The Geometry of a Sphere

What is the shortest distance between two points? In a flat, unobstructed world, that shortest distance is always a straight line. But in New York City the shortest distance from 5th Ave. and 42nd St. to 8th Ave. and 38th St. is not a straight line. What path does the crow take? If the crow drove a taxi, he would have to follow the grid of streets to find the shortest path. So the shortest paths between two points—the "straight lines"—depend on the shape of the space where we live. We live on Earth. So what are the "straight lines" on Earth? Let's travel around and see.

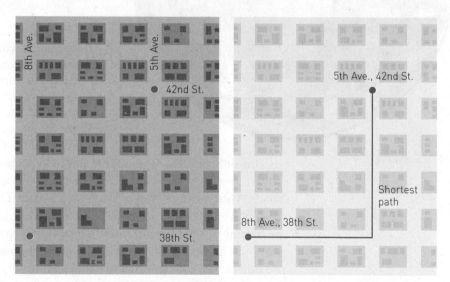

New York City

Since Earth is round, we do not live on a plane; yet many travelers live in a plane a good deal of the time. Pilots have a great attachment to fuel and hate to run out of it at 35,000 feet in the air. Thus, airplanes go from place to place along the shortest routes possible. Pilots, like crows, know that Earth is round and choose their routes accordingly. Let's see what those routes are. The best method for bringing this point home would be for you to now take a nonstop plane trip from Chicago, Illinois, to Rome, Italy. We'll wait patiently, but you had better send us a postcard.

Path shown = 5300 miles

Rubber band's
path = 4800 miles

The arc path is part
of a "great circle"

Traversing the Globe

Chicago and Rome are both at the latitude of nearly 40° north. You might think that the shortest route from Chicago to Rome would be to stick to the 40° latitude line the whole way. Let's measure how far that route would be. We will do this by measuring distances on a globe and using the scale to tell the mileage. If we take out a tape measure and place it along the 40° latitude line, we see that the distance is 5300 miles. Is there a shorter route? If we're flying the plane, we had better find out.

A good, though messy, way to find the shortest route involves using a greased globe and a rubber band. We take the globe and grease it until it is so slippery that nothing, including the rubber band, will stick to it. We next put two pins in the globe, one at Chicago and one at Rome and then stretch a rubber band over the two pins. We first hold the rubber band down so it sits on the latitude line. Then we let go. Does it stay on the 40° latitude line? We don't think so. In fact, it slides up to a shorter route. Instead of staying on the latitude line, the rubber band finds a genuinely shorter route. Notice that the route heads north and goes over Labrador and Dublin, Ireland, before heading back south on its way to Rome.

Is this route really shorter? Let's measure. We place our measuring tape along our new route and measure about 4800 miles. This new route saves about 500 miles!

Let's take a closer look at this rubber-band route. If we extend the route, we get a *great circle* that is as big as possible going around the whole globe—that is, a circle whose center is at the center of the globe and whose circumference is as long as the equator.

Indeed, the shortest path between any two points on the globe is always on a great circle that contains them. The segments of great circles are the shortest distance between two points on the globe. Why?

Why Great Circles Are the Way to Go

Chicago

Rome

Let's think about why the great circle segments are the shortest paths. If Earth were hollow, the shortest path from Chicago to Rome really would be a straight line inside Earth. So for our purposes, let's imagine a straight-line tunnel connecting Chicago and Rome burrowing right through the planet. The shortest route on the surface would deviate as little as possible from that straight, underground Chicago–Rome tunnel.

Let's notice something about circles and lines. If we take two points and make a circle that contains them, then bigger circles are flatter and therefore stay closer to the straight line between the two points. So, among the paths that stay on circles, taking the biggest circle on the globe containing Chicago and Rome—that is, the *great circle*—will stay closest to the straight line. The latitude circle, being smaller, bends out more from the straight line and is therefore longer. "Straight lines," that is, the shortest paths on Earth, are great circle segments.

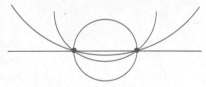

The larger the circle,
the "flatter" the segment.

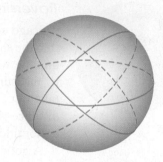

Latitude is longer

Distances in a Different World

We live on Earth, which is essentially a ball, but how about a bug on the wall? If our bug doesn't fly, its world is in the shape of the walls. So, when it sees its dining destination on some other wall, it has some serious calculations to perform. What is the shortest distance from here to dinner? Take a guess. A good guess would be the path shown in the figure below.

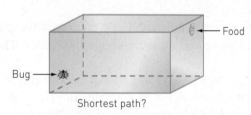

Shortest path?

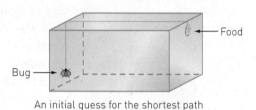

An initial guess for the shortest path

Life Lesson

Think about some simple cases.

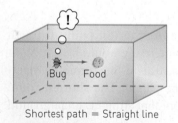

Shortest path = Straight line

Where to cross the edge?

Is this guess the shortest path? Suppose dinner is on the same wall. This situation is easy. The bug simply walks in a straight line. The bug is off to a great start.

How about if dinner is on an adjacent wall? It is pretty clear that the bug needs to go straight to the boundary edge and then straight on the next wall to its dinner. The question is, Where on that edge should it cross? Describe a method for locating the best crossing place.

Notice that, if the walls were at some angle other than 90°, the distances from the bug to the crossing point and the crossing point to dinner do not change. So let's consider a different question. Suppose the bug is on an open door, and its dinner is on the wall.

"Hinged door"

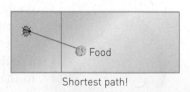

Shortest path!

Imagine that, as the bug is considering its shortest route, someone comes along and closes the door. Suddenly the question becomes much easier. Now the bug and its dinner are on the same wall, and the bug simply proceeds in a straight line. Now the bug is on a roll.

Let's now return to the scenario where the bug's dinner is on the opposite wall. How will it figure out the shortest route? Having experienced the closing of the

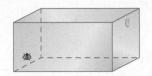

Can you think of other ways of unfolding the walls, keeping the food and the bug in the same relative locations?

door, surely our bug cannot resist the idea of unfolding the walls. The problem is that there are many ways to unfold them. Which one should the bug choose? The straight lines from bug to dinner vary in length, depending on the route.

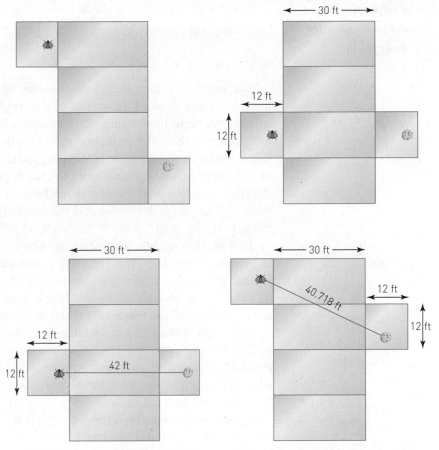

Some straight paths are shorter than others.

Shortest path

What to do?

One method of deciding would be to unfold the room in all possible ways and measure the straight-line distances. We've seen several different room-unfolding possibilities. Notice that different unfolding scenarios result in different placements of the food and the bug. Visualize the reassembly of the flattened rooms and verify that the relative positions of the bug and the food are always the same. Once we find the shortest flattened path, we can draw the straight line on the unfolded model and then refold it. In this case, the shortest path takes the bug over five walls—a dramatic departure from our first guess.

Now we have a better sense of shortest paths and straight lines in various worlds, including our own earthly sphere. Putting these straight lines together allows us to explore some basic geometry that captures the essence of the graceful curvature of the sphere. Let's put three straight lines together to make a triangle.

Curvy Geometry
Triangles on a Sphere

Draw any triangle in the plane. Add up the three angles. The result is 180°.

Cut off angles.   Sum is 180.

But we've seen that, in different realms, we have different ideas of straightness. Let's now explore the angles of a triangle made out of straight lines on a sphere. We begin by drawing a large triangle on a sphere. For example, let's put one vertex on the north pole, one vertex on the equator at 0° longitude, and the third vertex on the equator at 90° longitude. The edges of this triangle consist of two longitudinal segments from the north pole to the equator and a segment that goes one-fourth of the way around the equator. What are the angles at each vertex? Each one is 90°. So what is the sum of the three angles of that triangle on the sphere? 90° + 90° + 90° = 270°. Yikes!

This result is slightly disconcerting. The sum of the angles of this triangle on the sphere is not 180°, but 270° (90° too much). Is it possible that all triangles on the sphere have angles that sum to 270°? Let's see.

Take the big triangle above and break it into two by drawing in the longitudinal segment from the north pole down to the equator at 45°.

Now each of the half-sized triangles has angles of 90°, 90°, and 45°. That sum is 225°, 45° more than the 180° we would have in a triangle on the flat plane. This result is stranger still, since not only do the angles fail to add up to the comfortable 180° we know and love, but now we see that on a sphere, different triangles have different sums of angles.

As always, we must look for patterns. Is there any regularity among our measurements? The big triangle had 270°, 90° too much. When we divided it in half, each half had 225°, 45° more than 180°. Did you notice that the total surplus of angle for the two smaller triangles stayed at 90°? In other words, when we took the big triangle and measured the surplus angle bigger than 180°, we got 90°. Then, when we divided the big triangle into two smaller triangles, each of the halves had a surplus of 45°, or 90° altogether.

Suppose we start with any triangle on a sphere and cut it in half by bisecting one of the angles. What is the relationship between the angles of the original triangle and the angles of each of the two subtriangles? Well, the new angles •° and ••° add up to 180° since they are on a straight line. So, the total excess of the two triangles must be equal to the excess for the original big triangle.

Notice what happens if we take a small triangle on the sphere. What is the surplus of its angles? Not very much. A small part of a sphere is basically flat, so the angles of a triangle there will have almost exactly the same angles as the angles of a triangle on a flat plane. It seems that larger triangles have greater excess in the sum of their angles than small triangles do. Furthermore, if a large triangle is divided into smaller triangles by adding edges, since all the added angles created are along straight lines or divide existing angles, the total excess of all the

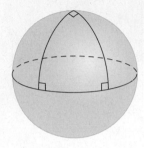

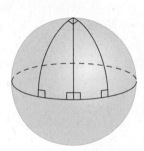

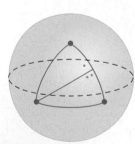

•° + ••° = 180°

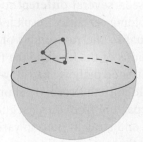

The sum is just a smidge greater than 180°.

subtriangles making up a bigger triangle must be the same as the excess of the big triangle. What corresponds to the excess? It turns out that the excess increases as the area of the triangle increases. Thus, we see that the sum of the angles of a triangle on a sphere will always exceed 180° but that small triangles will just barely exceed 180° and large triangles will exceed 180° by a more substantial amount.

Extra Degrees Through Curvature

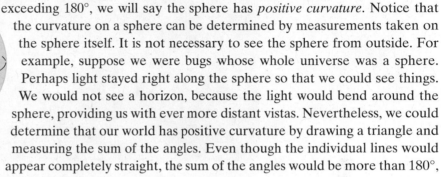

Curvature of the sphere causes "straight lines" to bow out a bit.

The sum of the angles of any triangle on a sphere exceeds 180° because of the curvature of the sphere. And, since every triangle on a sphere has a sum of angles exceeding 180°, we will say the sphere has *positive curvature*. Notice that the curvature on a sphere can be determined by measurements taken on the sphere itself. It is not necessary to see the sphere from outside. For example, suppose we were bugs whose whole universe was a sphere. Perhaps light stayed right along the sphere so that we could see things. We would not see a horizon, because the light would bend around the sphere, providing us with ever more distant vistas. Nevertheless, we could determine that our world has positive curvature by drawing a triangle and measuring the sum of the angles. Even though the individual lines would appear completely straight, the sum of the angles would be more than 180°, clinching the positive-curvature claim.

We now have one space, the plane, where all triangles have angles that add up to 180°. That constant sum is our benchmark, so we will say the plane has *zero curvature*—it is flat. We saw another space, the sphere, with positive curvature where all triangles have angles that add up to more than 180°. Surely we cannot resist asking the question, "Is there a space with *negative curvature*—that is, where the sum of the angles of a triangle is less than 180°?"

Geometry on a Saddle

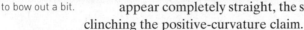

Horseback riding provides us not only with a sore bottom but also with an interesting geometrical opportunity. The surface of a saddle has an appealing shape and provides a surface ripe for experiments using rubber bands and butter. Suppose we place three pins as shown in the diagram on the next page, one near the front of the saddle and two near the stirrups. (This is a poor time to actually sit on the saddle.) We now grease the saddle with butter and put a rubber band around every pair of pins. The rubber bands will slide to the shortest distances between pins. So we will have a rubber band triangle on the surface of the saddle. We now wish to measure the angles. If you don't happen to have a saddle handy, estimate using the angles in the diagram, and compare the sum of the angles to 180°. The sum of the angles of this triangle is less than 180°. Of course, the rubber-band method will not always work on a saddle because the line between some pairs of points, like the center front to the center back of the saddle, would leave the surface and float in the air. However, rubber bands are good tools for finding the shortest distances between some pairs of points on the saddle. Whether you use rubber bands or another method for finding the shortest distances between points, triangles on the saddle will have

3D picture of a saddle (Use your
Heart of Mathematics 3D glasses
to view.)

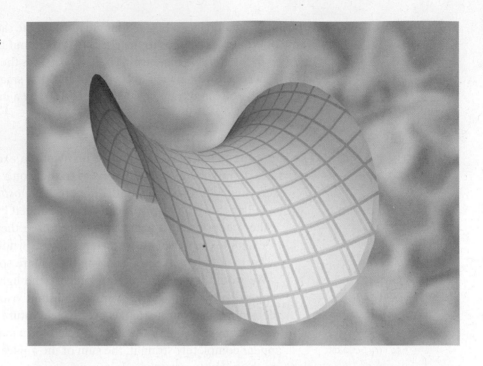

Triangle on the surface of a saddle
(Angles sum to less than 180°.)

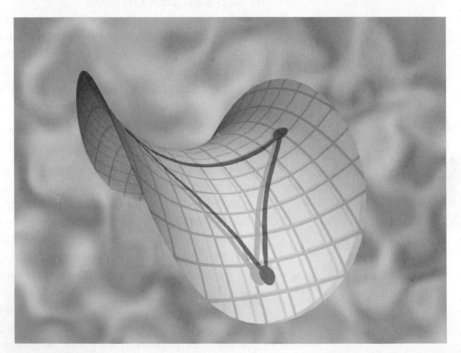

Sum of the angles
is less than 180°.

sums of angles less than 180°, because the sides of the triangles curve inward and
thus cause those angles to shrink. So this space has negative curvature and is an
example of the exotic world known as *hyperbolic geometry*.

We have seen three different types of geometry: plane geometry, which we say
has zero curvature (all triangles have angle sums of exactly 180°); spherical geom-

etry, which we say has positive curvature (angle sums vary depending on the size of the triangle, but always exceed 180°); and hyperbolic geometry, which we say has negative curvature (angle sums vary depending on the size of the triangle, but always are less than 180°). It certainly appears as though hyperbolic geometry is exotic and foreign to our real-world existence, which brings us back to our original question: What is the shape of our universe?

The Shape of Our Universe

We have just caught glimpses of three types of geometry: planar, spherical, and hyperbolic. Which type models our universe? Think of an experiment that we could perform to answer this. (*Hint*: What property distinguishes the three?)

Let's measure the angles of triangles. Suppose we make a big triangle and measure its angles. If the sum of those angles equals 180°, then we would conjecture that our universe has zero curvature. If the sum of those angles exceeds 180°, then we'd guess that our universe has positive curvature. If the sum of those angles is less than 180°, then we'd guess that our universe is curved negatively. Would anyone actually attempt this experiment? Yes!

The great mathematician Carl Friedrich Gauss tried this experiment in the early 1800s. He formed a triangle using three mountain peaks near Göttingen, Germany: Brocken, Hohenhagen, *und* Inselsberg. He had fires lit on each mountain top (Smokey the Bear would not have been amused) and used mirrors to reflect the beams of light to form a triangle having side lengths roughly 43, 53, and 123 miles. He carefully measured the angles of the triangle and added them up. His sum was within 1/180 of a degree of 180°. That small difference could easily have been caused by errors in measurement.

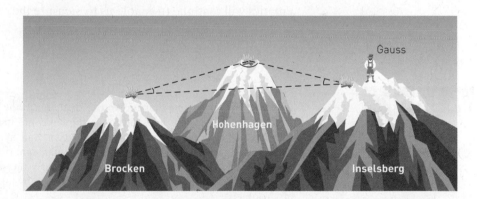

This evidence certainly leads us to think that our universe is neither positively nor negatively curved and that the universe is flat. What is the problem with this conclusion? Think about this question in view of our spherical geometry observations. Recall that, in spherical geometry, if we have a small triangle, then the triangle is nearly flat, and thus the sum of its angles is nearly 180°. Thus, although Gauss's triangle was big, compared to the entire universe it wasn't even a speck. Thus, on such a microscopic scale, it is not surprising to see that the evidence points to a geometry having zero curvature. We would need an enormous triangle to detect the existence of any actual curvature. Is this experiment even practical?

And if it were, would anyone actually attempt it? The answer to the first question is *possibly* and to the second is *yes*.

Today scientists believe that the universe exhibits two important properties. The first is that it is *homogeneous*, which basically means that any two large sections of space will look the same—of course, here "large" needs to be LARGE. The second is that the universe is *isotropic*, which means that, as we look around, things look about the same in every direction. It turns out that we can find geometrical objects that are homogeneous and isotropic that are either planar, spherical, or hyperbolic. This fact leads to a question of great interest to scientists today: Does the universe have zero, positive, or negative curvature?

A large group of scientists now believes that the universe is negatively curved—that is, that the geometry of the universe is actually the exotic hyperbolic geometry suggested by the saddle. In fact, a conference was held in October 1997 at Case Western Reserve University that brought together 20 cosmologists and 5 mathematicians to discuss the possible shape of the universe and ways to measure its curvature. NASA very recently used *MAP*—the Microwave Anisotropy Probe—to measure microwave background radiation, which is a residue of the "big bang." They have determined that within the limits of instrument error the universe is, in fact, flat. European scientists have sent up the Planck Probe. This probe should be able to make even more careful measurements of the variations in microwave radiation. These modern experiments capture the spirit of Gauss's attempts to measure the curvature of the universe.

So, what is the shape of our universe? In the first edition of this text, we wrote, "Although many experts believe it may be hyperbolic and negatively curved, no one knows for certain." However, the new evidence appears to favor the theory of a flat universe. Twenty-first century science and technology together with mathematics will continue to enable us to measure the curvature of our vast space and understand its subtle and beautiful geometry.

A Look BACK

SPACE CAN HAVE various shapes. We can distinguish how space bends by examining the shortest paths—straight lines, although they may not necessarily be straight. Three different kinds of geometry are planar, spherical, and hyperbolic. The flat plane, round sphere, and saddle are good models for planar, spherical, and hyperbolic geometries, respectively. On a very small scale all look nearly the same, and thus we have not yet been able to determine the shape of our universe by taking measurements of our local environment.

The distinguishing feature of the three different geometries is their curvature. If a space has zero curvature (the sum of the angles of any triangle is exactly 180°), then the space is flat. If a space has positive curvature (the sum of the angles of any triangle exceeds 180°), then the space is spherical. Finally, if the space has negative curvature (the sum of the angles of any triangle is less than 180°), then the space is hyperbolic.

When we wish to consider big issues it is often valuable to start with simple and familiar models or examples and build from there. By identifying both similarities and differences in our various examples, we can often discover the underlying structure that determines the general case.

Life Lessons

Start with the simple and build from there.

•

When you don't know what to do, consider everything.

•

Look for patterns.

MINDSCAPES   Invitations to Further Thought

*In this section, Mindscapes marked (**H**) have hints for solutions at the back of the book. Mindscapes marked (**ExH**) have expanded hints at the back of the book. Mindscapes marked (**S**) have solutions.*

Developing Ideas

1. **Walking the walk.** Here are three walks from corner X to corner Y in a city. The first walk is seven blocks long. How long are the other two? If you only travel east or north, how long is any other walk from corner X to corner Y?

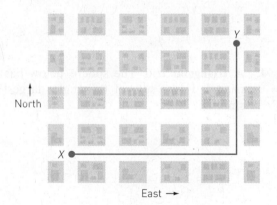

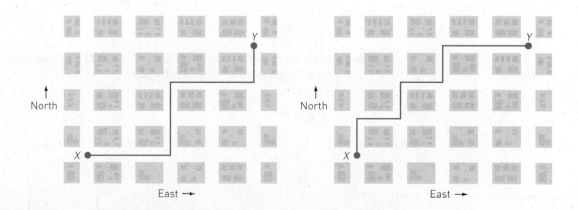

2. **Missing angle in action.** The triangles below are drawn in the plane and the numbers represent the degrees of the angles. In each figure, compute the unknown angle(s).

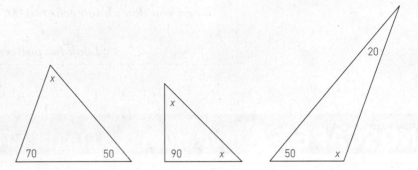

3. **Slippery X.** A triangle is drawn on a sphere. Can you determine the size of the angle x? Why or why not?

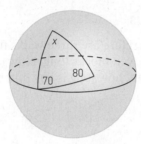

4. **A triangular trio.** The sphere below has three triangles on it. For which triangle is the sum of the angles largest? For which triangle is the sum smallest?

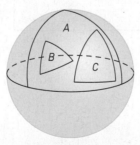

5. **Saddle sores.** The triangle at right is drawn on a saddle surface. Can the angle x be as large as 90°? Why or why not?

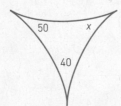

Solidifying Ideas

Travel agent. In each of the following three Mindscapes, get a globe and trace the shortest paths between the pairs of cities. For each pair on the left, find the location on the right that is on the shortest path between them.

6. Austin, Texas–Tehran, Iran   Reykjavik, Iceland

7. Williamstown, Massachusetts–Beijing, China   Denali, Alaska

8. Austin, Texas–Beijing, China   Near the north pole

Latitude losers (H). In each of the following three Mindscapes, you are given a pair of cities that are on the same latitude. Fill in the table by measuring the distance from city to city, first staying along the latitude and then measuring the distance taking the great-circle route.

	City Pair	Latitude Distance	Great-Circle Distance
9.	Beijing, China–Chicago, Illinois		
10.	Mexico City, Mexico–Bombay, India		
11.	Sydney, Australia–Santiago, Chile		

Triangles on spheres. For each of the following five Mindscapes, on a globe, draw triangles whose vertices are the following sets of cities. For each such triangle, measure the sum of the three angles of the triangle.

12. Minneapolis, Minnesota; Austin, Texas; Williamstown, Massachusetts.

13. **(S).** Panama City, Panama; Nome, Alaska; Dublin, Ireland.

14. Quito, Ecuador; Monrovia, Liberia; Thule, Greenland.

15. Quito, Ecuador; Bangkok, Thailand; the south pole.

16. Wellington, New Zealand; Moscow, Russia; Rio de Janeiro, Brazil.

17–21. **Spider and bug.** For each pair of points on the boxes, describe the shortest path from one point to the other.

17.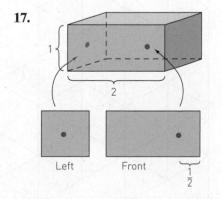

18.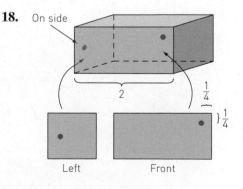

19.

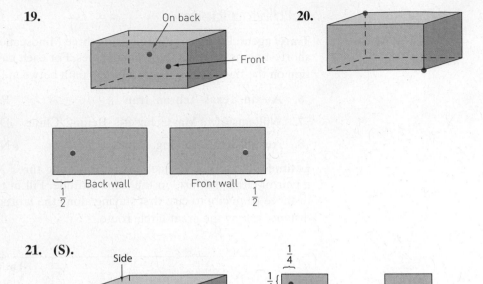

20.

21. (S).

22. Becoming hyper. Professor William Thurston found a neat way to build a model of hyperbolic geometry. Photocopy many equilateral triangles. (Enlarge the sheet given here on a copier.) Cut them out and tape them together so that seven triangles meet at each vertex. You will have to bend the triangles to fit

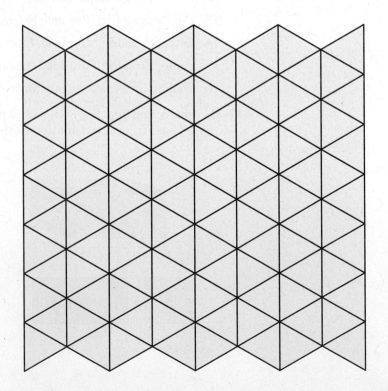

them together. Continue attaching the triangles so that seven come together at each vertex. You will notice that your model will become floppy. The larger you make it, the floppier and more accurate your model will be.

23. **Deficit angles (H).** Draw a big triangle on your floppy sheet constructed in Mindscape22. Span several of the pieces by flattening a section on the ground and drawing a straight line, then flatten another section and draw another straight line, and then complete the triangle in the same way. There is a lot of squashing involved. Now measure the three angles and add them up. What do you get?

24. **Same old.** Go to a vertex on your floppy plane. Look at the pattern of all the triangles that you can reach from there passing through at most two triangles. Now go to another vertex and do the same. Are the patterns the same or different?

25. **Gauss II.** Try Gauss's experiment. Select three tall objects that are reasonably far away from one another (for example, three buildings or trees) and measure the angles between them. What are your measurements? Sum the angles.

Creating New Ideas

26. **Big angles (H).** What is the largest value we can get for the sum of the angles of a triangle drawn on a sphere? Experiment with larger and larger triangles and compute the largest sum value.

27. **Many angles (S).** Draw three different great circles on a sphere. How many triangles have you made on the sphere? Compute the sum of the angles of all the triangles22. Draw another group of three different great circles and answer the questions again. What do you notice? Make some conjectures.

28. **Quads in a plane.** Measure the sum of the angles of the quadrilaterals below. Why is the sum of the angles of any quadrilateral in a plane the same?

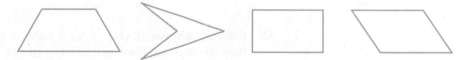

29. **Quads on the sphere.** Below are quadrilaterals on spheres. Measure the sum of the angles of each quadrilateral. Make a conjecture about the relationship between the sum of the angles of a quadrilateral and its area.

30. **Parallel lines (ExH).** On a plane, if you draw a line and then choose a point off the line, there is one and only one line that goes through that point and misses the line. Is this true for a sphere? Take a line on a sphere (which, remember, is a great circle) and take another point. How many lines—that is, great circles—can go through the point and miss the first great circle altogether? Explain your findings.

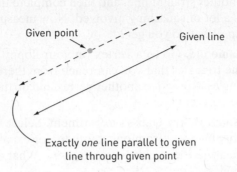

Given point   Given line

Exactly *one* line parallel to given
line through given point

31. **Floppy parallels.** On the floppy plane you constructed in Mindscape 22, draw a line and then choose a point some distance off the line. How many lines can go through the point and miss the first line altogether even if they are extended indefinitely?

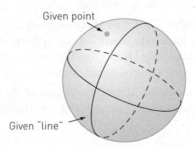

Given point

Given "line"

32. **Cubical spheres (ExH).** Take a cube. Put a point in the middle of each face. Now draw the straight lines to the middles of each of the sides of that face, thus producing a plus (+) sign on each face. The kinked line that goes from the center of one face to the center of an adjacent face forms a bent edge on this cubical world. Thus we have created eight "bent" triangles whose vertices are the centers of the faces of the cube. Now what is the sum of the angles for each of those triangles? What is the sum of all the angles of all the triangles? Compare your answer to the answer to Mindscape 27.

33. **Tetrahedral spheres.** Let's do a similar calculation for the tetrahedron. Put a vertex at the center of each face of a tetrahedron and connect adjacent faces over the center of each edge. Answer all the questions in Mindscape 32.

34. **Dodecahedral spheres.** This Mindscape is the same as the previous two, except start with a dodecahedron.

35. **Total excess.** Using the observations from the previous Mindscapes and Mindscape 27, make a conjecture about the total excess of sums of angles of triangles that cover up polyhedra as described.

Further Challenges

Geometry on a cone (H). In Mindscapes 36 through 39 we will consider the following construction: Take a piece of paper and cut out a pie-shaped piece of angle *z*. Now put the two ends of the cut-out piece together to construct a cone. Let's see what happens if we look at triangles that surround that cone point. To draw a triangle, we have to determine what a straight line is on the cone. That is fairly easy, since the cone is made from a piece of paper that was originally flat. Take two points on the cone, flatten the cone in such a way that both points are on the flat part, and connect them with a straight line. A triangle is just made of three straight lines. First measure the angles in a triangle that does not go around the cone point.

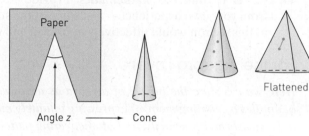

36. **What is the sum of the three angles? Why?** Consider the more interesting case of a triangle that goes around the cone point. Draw the three sides separately by flattening in different ways.

37. **What is the sum of the angles of your triangle?** Is the sum the same for all triangles that go around the cone point? Let's try some more experiments, this time by removing thinner and fatter pie slices before making our cone.

38. **Removing a slice of the pie.** Complete the following table by making the cones, drawing triangles around the cone points, measuring the angles of the triangles, and adding them up.

Angle of Pie Removed	Sum of Angles of Triangles	Difference from 180°
30°		
60°		
90°		
180°		

39. **Conjuring up a conjecture.** Make a conjecture about the relationship between the angle of the slice removed to make the cone and the excess above 180° of the sum of the angles of a triangle that goes around the cone point.

40. **Tetrahedral angles.** What is the sum of the angles around each vertex of the tetrahedron? For each vertex, compute 360° minus the sum of the angles at that vertex. Multiply that number by four since there are four vertices of the tetrahedron. What do you get? Are you surprised?

In Your Own Words

41. **Personal perspectives.** Write a short essay describing the most interesting or surprising discovery you made in exploring the material in this section. If any material seemed puzzling or even unbelievable, address that as well. Explain why you chose the topics you did. Finally, comment on the aesthetics of the mathematics and ideas in this section.

42. **With a group of folks.** In a small group, discuss and actively work through the relationship between the sum of the angles of a triangle on a sphere and a plane. After your discussion, write a brief narrative explaining the relationship in your own words.

43. **Creative writing.** Write an imaginative story (it can be humorous, dramatic, whatever you like) that involves or evokes the ideas of this section.

44. **Power beyond the mathematics.** Provide several real-life issues—ideally, from your own experience—that some of the strategies of thought presented in this section would effectively approach and resolve.

For the Algebra Lover

Here we celebrate the power of algebra as a powerful way of finding unknown quantities by naming them, of expressing infinitely many relationships and connections clearly and succinctly, and of uncovering pattern and structure.

45. **Taxi total.** You're in Manhattan for a job interview. A taxi picks you up from your hotel and takes you x blocks north, then $x/2$ blocks east, then $x + 3$ blocks north, then $x/6$ blocks east to your destination. Write an expression in terms of x that gives the total number of blocks traveled. Simplify the expression as much as possible. If the total number of blocks you traveled is 19, determine how many blocks you traveled north between the time you entered the cab and the time the cab first turned east.

46. **Angle x.** A triangle in the Euclidean plane has angles measuring x, $2x$, and $3x$ degrees. Find x.

47. **Measuring moment (H).** While lounging around your room one day, avoiding your history reading, you notice a tissue box and decide to do some measuring. The box has dimensions l, w, and h, measured in inches. Your measurements reveal that the perimeters of the faces have the following lengths: $2l + 2w = 26$, $2l + 2h = 24$, and $2w + 2h = 14$. While it might not help you with your history assignment, solve these equations simultaneously to find l, w, and h.

48. **Is x big enough?** A triangle drawn on the sphere has angles measuring 80, 90 and x degrees. If x satisfies the equation $x^2 - 30x + 200 = 0$, find x. (There's only one value for x that works. What is it and why?)

49. **Negative x?** A triangle is drawn on a surface having negative curvature. Two of the angles measure 80 degrees. If x is the measure of the third angle, and x satisfies the equation $x^2 - 25x + 100 = 0$, find x. (There's only one value for x that works. What is it and why?)

4.7 THE FOURTH DIMENSION

Can You See It?

Max Weber, Interior of the Fourth Dimension, 1913.
National Gallery of Art

Interior in the Fourth Dimension
(1913) by Max Weber.

> *listen, there's a hell of a
> universe next door: let's go!*
>
> e.e. cummings

The very phrase *fourth dimension* conjures up notions of science fiction or even the supernatural. The fourth dimension sounds eerie, romantic, mysterious, and exciting; and it is. Physicists, artists, musicians, and mystics all conceptualize the fourth dimension differently and for different purposes. The mystique of the fourth dimension appeals to all who contemplate it.

The fourth dimension lies beyond our daily experience. So, visualizing, exploring, and understanding it seem at first impossible. Such understanding would require us to develop an intuition about a world that we will never see. Nevertheless, that understanding is within our reach. We will succeed in building insights without experience. We will become at home in an environment that we cannot touch, see, or otherwise sense. Successful explorations of an unfamiliar realm often begin by delving into the depths of the familiar. The fourth dimension provides a dramatic testament to the power of developing a concept through analogy.

Building Up from the Familiar

There is no physical direction for us to look in to see the fourth dimension; our everyday world is too confining. But we have to start somewhere, so we'll start with what we know and understand. We will get to the fourth dimension by starting with zero and moving up.

Making a Point

What is 0-dimensional space? What would a 0-dimensional world be like? In a 0-dimensional space there are zero "degrees of freedom." That means there would be no room for anything. No information is required for us to identify a particular location in this 0-dimensional space. What must this space look like? The answer is *a point*. In 0-dimensional space, the entire universe is just a point. A point has no length, height, or width—so, there are zero degrees of freedom.

Life Lesson

*By looking closely
at what we already
know, we may discover
something new.*

•
└ 0-dimensional space.

If you lived there,
you'd be home by now.

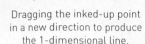

Dragging the inked-up point in a new direction to produce the 1-dimensional line.

We now take the extremely confined 0-dimensional world and build a more spacious one. Suppose we have a remarkable ink pad: If we ink up an object and drag it, the ink leaves a trail behind the object. What kind of geometric object would we create if we were to take 0-dimensional space (the point), ink it up on this ink pad, and then drag it right and left? We would sweep out a line.

Getting in Line

One-dimensional space is a line. For convenience, let's suppose that our line is marked with numbers. If someone lives at a certain point on the line, what is the minimum amount of information we require to locate that person? The answer is just one number, namely, the number on the line where that person lives. If we think about this number as an address that identifies where this person lives, we need only the house number, since there is only one street (namely, the line). So we need only one piece of information to identify precisely any point on the line. Therefore, we say that the line is 1-dimensional space.

My house

$\frac{3}{2}$

My address is $\frac{3}{2}$: one piece of information and you know where I live implies 1-dimensional space!

We have just moved from 0-dimensional space to 1-dimensional space. We now wish to move another dimension higher. Suppose we take the line and ink it up all over and drag it in a new direction. Notice that our moving the line east or west just places more ink on the line we've already created. To create something new, we need to drag it in a different direction. Suppose we now drag the line in the north–south direction (perpendicular to the original line). We sweep out a plane. The plane is 2-dimensional space.

Inked-up line dragged vertically.

Produces the 2-dimensional plane.

Home on the Plane

The plane is 2-dimensional; if we draw two numbered axes for reference, as we often do, then to find out precisely where someone lives in this space we need only two pieces of information: a vertical number (representing the street) and a horizontal number (the address on that street). These two numbers pinpoint the person's house. If our friend told us only her vertical position, we would know only what street she lived on but would not know for sure which house was hers. We need another piece of information, in this case, her horizontal address. Given these two numbers, we can locate our friend's point (house) in the plane. Thus the plane is 2-dimensional space.

The plane

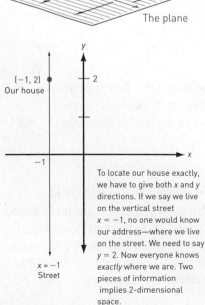

(−1, 2)
Our house

y

2

−1

x

$x = -1$
Street

To locate our house exactly, we have to give both x and y directions. If we say we live on the vertical street $x = -1$, no one would know our address—where we live on the street. We need to say $y = 2$. Now everyone knows *exactly* where we are. Two pieces of information implies 2-dimensional space.

Again, we observe a pattern of progression: By dragging an inked-up 1-dimensional space in a new direction, we build 2-dimensional space. We are developing a paradigm for moving up dimensions. Our next step is to describe the process for creating 3-dimensional space from 2-dimensional space.

Space—The Final Frontier?

Following the ink-and-drag analogy, we ink up the entire 2-dimensional plane and drag it in a new direction different from the east–west and the north–south directions. We choose a direction perpendicular to the plane. Suppose we drag the inked-up plane in the up–down direction. Then we sweep out space as we know it. To locate a point in this space, we need three pieces of information, and we call this 3-dimensional space. This space is the world in which we live. We have three degrees of freedom: north–south, east–west, and up–down. Once again, we note how a completely inked-up 2-dimensional plane dragged in a new direction results in 3-dimensional space.

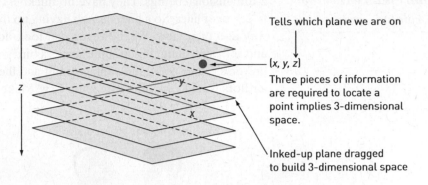

Tells which plane we are on

(x, y, z)

Three pieces of information are required to locate a point implies 3-dimensional space.

Inked-up plane dragged to build 3-dimensional space

The Fourth Dimension

We have now come to the moment we've been eagerly awaiting. Do we now have enough experience to make a first attempt at describing 4-dimensional space? Let's try to follow our established pattern. First we ink up all of 3-dimensional space. This inking would be both challenging and messy to do in practice, so we'll just perform it in our minds. We ink up every single point in 3-dimensional space including every molecule of air. It may be useful to think of 3-dimensional space as a sponge that can hold ink everywhere. Now we drag that inked-up space in a new direction—let's say in a direction perpendicular to all three of the original directions of 3-dimensional space. At first glance, this last step may present something of a challenge.

What does it mean to select a direction perpendicular to all the directions of the space in which we live!? A fourth direction that is different from the three we already know? Where is it? We are unable to point to it, but if we ignore that fact for the moment and drag the inked-up copy of our 3-dimensional space in that new direction, what we sweep out is 4-dimensional space. Each point in that 4-dimensional space can be located by four pieces of information. To make this foreign idea meaningful, we must develop some insight and clarity into the notion of a "new direction," one that is different from the three we all know and love.

"New" direction

Dragging all of three dimensions into a "new" direction. This figure cannot be drawn here!

Understanding Through Analogy

When the going gets tough, the smart stop going! In other words, instead of searching hither and yon for this mysterious fourth direction, we will attempt to understand it by looking at the dimensions with which we are comfortable. This fourth direction exists entirely in the mind. It is a virtual direction. Why not become familiar with it through a process of analogy? We seek to understand how objects fit into this 4-dimensional space just as we understand them in 2-dimensional and 3-dimensional space. Reassessing our basic notion of "inside" will lead to unexpected realizations.

We begin our process of analogy by stepping back and considering the simpler world of two dimensions—the plane of this piece of paper for example. Imagine that in this world live two authors—Ed and Mike. Thus, they themselves are 2-dimensional beings. They have no thickness at all. They cannot lift their heads to see over lines. So a line, though having no thickness at all, is able to block their way and their view. We, however, can look down onto the page—their world—and see everything. It is in some sense, a bird's-eye, panoramic view of their entire universe. Think about this vantage point and then draw what the two 2-dimensional authors would look like from your view. Make a quick sketch before reading on.

How authors appear to us.

A line blocks the authors' view of each other.

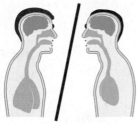

Ed   Mike

We immediately notice that, looking at this 2-dimensional world from our vantage point above it, we can see inside each 2-dimensional person and examine his inner organs. To a 3-dimensional being, the insides of a 2-dimensional person are exposed and appear open. Notice, however, that, when Ed and Mike look at themselves, they are unable to see each other's insides; their bodies are completely sealed by their 1-dimensional skin. Their 1-dimensional skin is able to separate their insides from their outside world.

All I see is skin; he's completely sealed up! I can't see his brain.

Why are 3-dimensional people able to see the 2-dimensional people's inner organs? The answer is that we are able to look down at their entire 2-dimensional universe from a third direction and thus are able to see everything: insides and outsides simultaneously. They cannot hide anything from us. In fact, if they wanted us to perform internal surgery, and give them their much-needed lobotomies, we could perform it without even cutting their skin open. We could simply reach in from the third dimension and touch their insides.

Imagine the great frustration we would experience in failed attempts to describe our extra degree of freedom to these 2-dimensional authors. Three-dimensional people certainly have an overall aerial view of the entire 2-dimensional world that 2-dimensional people could not easily comprehend.

An Analogy Taking Us to the Fourth Dimension

Suppose now that a 4-dimensional creature is looking "down" at us. Since it would be looking at our 3-dimensional world from a new and different direction, it would have a view of our world analogous to what we have of a 2-dimensional world. It could see everything—inside and outside—simultaneously. Our inner organs would be open and in full view of the 4-dimensional creature.

Let's call the 4-dimensional creature Fredddd (notice Fredddd's 4D nature!… don't groan, it's cute). Is it really possible for Fredddd to see our insides since they are completely covered all around by our skin? Remember the analogy of the 2-dimensional beings. They appeared to themselves as completely sealed all around *in their world*. However, their 1-dimensional skin does not block the view of 3-dimensional eyes. Similarly, although our 2-dimensional skin separates our insides from our 3-dimensional outside world, it would not block Fredddd's view from his 4-dimensional vantage point. Think again about the 2-dimensional creatures with their 1-dimensional skin that did not block our view.

We can think about Fredddd's incredible "panoramic" view and attempt to develop a sense of 4-dimensional space from our 3-dimensional world via analogy to our attempts to describe our world to the 2-dimensional-plane folk.

Are Safes Safe Across Dimensions?

Suppose now that the 2-dimensional authors, Ed and Mike, wrote a most successful and incredibly well-received 2-dimensional math book, *The Spleen of Mathematics*. Instead of cash royalties, in their contract negotiations they agreed to receive 2-dimensional bars of gold (actually Golden Rectangles, but that is another story). Now they wish to put their gold bars in a sealed vault so they will safely sit until Ed's and Mike's retirement. What would a sealed vault look like in 2-dimensional space?

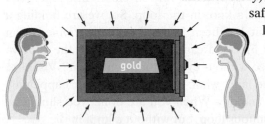

Here is their 2-dimensional sealed vault. Notice that, when the 2-dimensional-plane authors look all around it, they see it is completely sealed up. All the sides are solid, and it is impossible for anyone to steal their gold. To make sure, they even stand guard and watch the vault and surrounding area intensely.

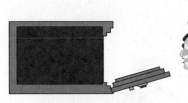

Our 2-dimensional vault—sealed ALL around!

A 3-dimensional person could trivially remove the gold without breaking the sealed vault and without the authors seeing it happen. We can simulate this theft with a Post-It® note playing the role of their gold. The 3-dimensional person just reaches down to the page right over the Post-It note and peels it off the 2-dimensional page. Suppose the authors decide to forget about their retirement and take out the gold now. They break the sealed vault, and what do they see? Their nest egg has vanished into thin 2-dimensional air! Think about how they must feel. They saw with their own eyes that the vault was completely sealed *all around the entire time*; there was no way to get in. The disappearance appears simply impossible.

Our gold is gone! But the vault was *never* broken! Impossible!

Let's return to our 3-dimensional world and to our 3-dimensional book. Imagine that we placed this book (a 3-dimensional object that is worth its weight in gold) into a sealed box. We examine the box along all its sides and convince ourselves that it is completely sealed all around. Now, working by analogy, we invite you to describe how Fredddd can remove our hidden treasure without opening the sealed box and without our seeing him take it. By doing so, we are building some insights and intuition into the abstract realm of the fourth dimension.

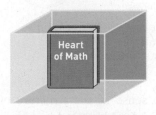

Our book inside a COMPLETELY SEALED box. Can't get to it.

Headache

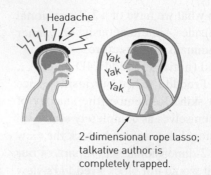

2-dimensional rope lasso; talkative author is completely trapped.

Life Lesson

Don't close your mind to ideas that first appear counterintuitive.

Lift a piece of the rope out of the plane and into the third dimension.

"Part of the rope has vanished or was cut!?"

I'm out!

Use your 3D glasses to really see the lifting rope.

Lassoing in Low Dimensions

Returning to the 2-dimensional world of the authors, suppose that one of the authors (the loud one) annoys the other author to his breaking point. Almost crazed, the aggravated author decides to rid himself of his chattering co-author once and for all. As he does not believe in murder, he decides to take some rope in two dimensions and encircle his co-author with a lasso. Although the aggravated author can still hear the rantings of the loud author, it is impossible for the lassoed author to escape (he has no sharp objects to cut the rope). Think now of ways to help the talkative author to escape without touching him. Remember that the lasso is made of pliable rope.

To rescue the author, we reach down and lift a little bit of the rope up into the third dimension (out of the plane). What would the quiet author, who is watching all this, see from his perspective? What would it look like to him? It would appear as though a piece of the rope was cut off and removed. Is the rope really cut? No, it's just that we moved a piece of it out of sight into the third dimension, and thus it appears that a section of the rope disappeared. If we now allow the rope to fall back into the plane, it would seem as if the missing piece of rope had magically reappeared.

Now, back to our 3-dimensional world. Suppose we have a piece of real rope in our hands. The rope's ends are joined together to form a loop. However, on closer inspection, we notice that there is a knot in the loop. So, we are holding a piece of rope that has been knotted and then joined. Can you explain how Fredddd can help you unknot the rope in 4-dimensional space without ever actually cutting it? Visualize what you would see and what might actually be happening. Remember to argue by analogy. When you are done, you should be holding the rope, still a continuous loop, but without a knot in it.

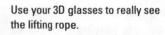

Knotted rope loop in three dimensions

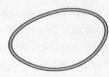

Unknotted rope! No cutting.

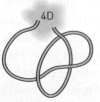

Freddddd lifts a piece of the rope up into the fourth dimension. Is the rope cut? No! But from our 3-dimensional vantage point, it's open. So, . . .

Move loop

Untwist

Now Freddddd lowers back the piece of rope he was holding in the fourth dimension ——→ we see the rope "fuse" together magically.

These mind experiments enable us to develop a sense of the structure of 4-dimensional space. This type of analogy appears in Edwin Abbott's classic book *Flatland*, written in 1884.

Visualizing Cubes

To illustrate further the notion of dimensionality, we now consider the geometry of space by constructing and visualizing a 4-dimensional cube. We proceed, as always, to warm up by building cubes in all the lower dimensions first. This method will allow us successively to build a cube in the next higher dimension using the cube constructed in the previous dimension. Notice that we are again examining dimensionality sequentially.

The 0-dimensional cube is pretty simple—*everything* is just that one point. Thus, the 0-dimensional cube is a point. We now ink up the 0-dimensional cube and drag it one unit in a new direction. This dragging produces a line segment that is actually a 1-dimensional cube. If we ink up the line segment and drag it one unit in a direction perpendicular to the previous one, we get a 2-dimensional cube (also known as a square). If we ink up the entire 2-dimensional cube and drag it one unit in a direction perpendicular to the previous ones, we construct a 3-dimensional cube (also known as a cube).

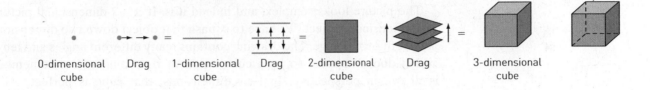

| 0-dimensional cube | Drag | 1-dimensional cube | Drag | 2-dimensional cube | Drag | 3-dimensional cube |

The last cube is actually just a drawing—an artist's rendition—of a 3-dimensional cube. It is impossible to draw a perfect and complete 3-dimensional cube on 2-dimensional paper. But we can draw a representation of it, showing perspective, which our 3-dimensional eyes can see and interpret correctly. For example, consider the following drawing, but view it solely as a shape on the page rather than as a diagram of a 3-dimensional object. What do we literally see? Notice that some lines of this drawing are longer than others, and some of the angles are huge while others are very small. These properties are certainly not those of a 3-dimensional cube. On a 3-dimensional cube, all the edges are the same length, and all the angles are perfect right angles. So, why does the drawing look so funny? Because, it's just an artist's rendition of the cube, not the cube itself. We are creatures who see and understand 3-dimensional objects instinctively, so we make the adjustments and interpret the picture in the appropriate manner. We "see" all the angles as right angles and all the edges as equal even though they are not drawn that way. Our mind compensates automatically to what our eyes actually see.

Can you now describe a process to construct a 4-dimensional cube and draw a picture of it? Again, as always, we proceed by analogy. Let's take the 3-dimensional cube (note that it is a solid object) and ink it up. We don't just ink up its faces, but we ink up every point of the cube, even the points inside, Perhaps it is useful to think of the solid 3-dimensional cube as a cubical sponge. This way we may be able to visualize the ink being inside all the points of the

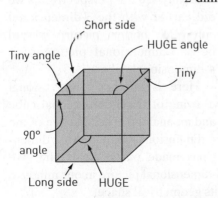

A figure in a plane—certainly not symmetrical!

Life Lesson

Look at things in new ways.

solid cube. We now drag it one unit in a direction perpendicular to the previous ones, and we have a 4-dimensional cube. Since we are moving the 3-dimensional cube in a direction different from the previous ones, the 3-dimensional cube never hits itself as it is dragged. Of course, this dragging is performed entirely in our imagination. Here is one view of the 4-dimensional cube, also known as a *hypercube*.

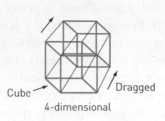

Cube → 4-dimensional ← Dragged

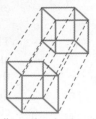

4-dimensional cube with new edges dotted to illustrate the "dragging" in the "new" direction.

The picture looks complex, and indeed it is. It is a 2-dimensional picture of a 4-dimensional object. We have to squash that object down two dimensions to have it fit on the page. The drawing contains many different angles and lengths, and it doesn't appear to be a cube made of right angles. But remember, it is drawn in perspective. In four dimensions, that cube is perfect: All the edges are the same length, and all the angles are right

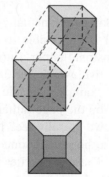

"Top" view of a 3-dimensional cube

angles. Our difficulty is that we do not have eyes that perceive four dimensions, so we cannot automatically see the perspective as we did earlier with the 3-dimensional cube. A better picture would be a 3-dimensional picture of a 4-dimensional cube.

Here is another 2-dimensional drawing of the 3-dimensional cube and an analogous perspective of the 4-dimensional cube in three dimensions made out of soap film. We must try to create a 3-dimensional picture in our minds to get a better sense of its geometrical shape.

Beyond the Fourth

We can now even build a 5-dimensional cube. How? Just remember: When life gets you down, just ink it up and drag it! Ink up every point of the 4-dimensional cube and drag it one unit in a direction perpendicular to the previous ones, and, voilà, we have a 5-dimensional cube. We can use this successive reasoning to consider the geometrical structure of dimensions greater than four. Through careful analysis we are able, within our mind, to gain geometrical insights into worlds that we are unable to see with our eyes.

A soapy 4-dimensional cube in three dimensions

5D cube: Just a 4D cube dragged in a new direction. Dotted edges are the new ones added. Put on your *Heart of Mathematics* 3D glasses to get a hint of 5D!

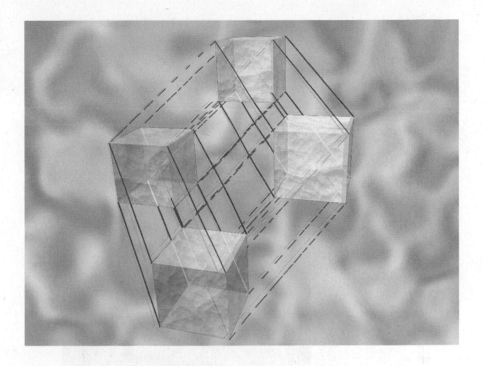

Cubes—Unplugged and Unfolded

Before leaving these cubes, we return to the issue of the perspective drawing of the 3-dimensional cube on the paper and the inability to draw a perfect 3-dimensional cube. The truth is that we can depict the 3-dimensional cube on paper in such a way that all its edges are exactly the same size and all its angles are perfect right angles. This seems impossible since we need all three dimensions to build the 3-dimensional cube. The secret is to draw the cube *unfolded*. Thus, we could draw it as shown at right.

We must understand how to assemble it: We fold it up along the drawn edges and join the outer edges together in the correct manner. Here is the unfolded 3-dimensional cube; we have marked the pairs of edges that need to be joined.

This construction suggests that perhaps we could unfold a 4-dimensional cube in 3-dimensional space in such a way that all its edges are of the same length and all its angles are right angles. All we would have to do is join pairs of faces together. Indeed, on page 316 is a drawing of an unfolded 4-dimensional cube in 3-dimensional space. To assemble it, the external faces would have to be paired up and joined. This unfolded 4-dimensional cube was the inspiration for Salvador Dali's 1954 painting *The Crucifixion, Corpus Hypercubicus*, in which Dali attributes religious significance to the fourth dimension.

Does the Fourth Dimension *Really* Exist?

From a practical point of view, higher dimensional space definitely does exist and is used frequently, for example, in big business and high finance. We illustrate this fact by answering the question, "How do I use the fourth dimension to make money?" Suppose we are in the business of manufacturing dental floss, and we

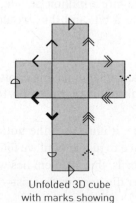

Unfolded 3D cube. Now all the sides are perfect squares.

Unfolded 3D cube with marks showing edges to be joined.

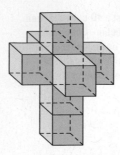

Unfolded 4D cube
in three dimensions

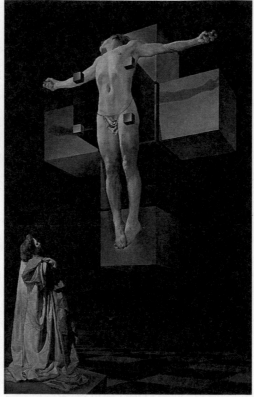

Salvador Dali, Fundacio Gala-Salvador Dali, Artists Rights Society (ARS), New York 2012

wish to make lots of profit through wholesome dental hygiene. What decisions would we have to make? Here are some basic ones:

- *Where do we have our factory?*

  If it's in the middle of nowhere, the rent is inexpensive, but then we incur high shipping costs. If we're in the big city, we have high rents but almost no shipping costs.

- *How big should our factory be?*

  The bigger the factory, the faster we can produce dental floss, but the greater the rent.

- *How many workers do we hire?*

  If we hire a few, we achieve less productivity; if we hire a million people, they would just squeeze into the factory, couldn't move a bit, and thus would not produce anything.

- *How much capital should we borrow?*

  There are tax advantages in realizing some debt and also in having capital on hand, but if we incur dramatic debt, we may go bankrupt in our attempts to pay off our loans.

Even though this scenario oversimplifies the issue, it illustrates the notion of four degrees of freedom, four variables, which we have to juggle until we find levels for each that maximize overall profit. In other words, this problem lies within the fourth dimension, and it demonstrates that higher dimensions do exist—even in the world of big business.

Life Lesson

Open your mind to the practical utility of abstract notions.

Marcel Duchamp captures the fourth dimension as time in *Nude Descending a Staircase #2* (1912). Philadelphia Museum of Art.

Is It Time?

Many believe that time is perhaps the fourth dimension. Although time does play a dimensionlike role, we do not enjoy the freedom of motion through time that we have in the other dimensions; for example, it is difficult for us to move backward in time.

In fact, any attribute may be used to represent dimensions in a *model* of space. For example, we may use sound, color, or temperature as dimensions. That is, to locate a unique point in 6-dimensional space we can identify a point in 3-space (three coordinates) and then also identify a pitch (say C-sharp), a color (purple), and a temperature (35°F). These attributes precisely describe a point in a 6-dimensional world.

Is It Physical?

Okay, but does the fourth dimension exist physically? Certainly the fourth dimension may exist physically, but, just like the 2-dimensional beings who are unable to see the third dimension, perhaps we are unable to see that extra direction of freedom. Perhaps we are living on just a slice of a 4-dimensional universe. In fact, modern scientific theories describe the universe as having many physical dimensions beyond three. At this point, however, we are unable to perceive them directly. But we have already seen that the fourth dimension does exist in our minds. We can describe it, draw it, navigate around in it, and enjoy it. The fourth dimension is a real world that we can explore and view with aesthetic appreciation all beyond the sensory world but within the world of the mind.

Before moving on, we pose one last question that leads to the next stop on our mathematical tour: What happens if we are allowed to *bend* space?

FOUR-DIMENSIONAL SPACE is a world we construct in our minds by following the pattern created by the 0-, 1-, 2-, and 3-dimensional spaces that we already know. Four-dimensional creatures could pluck this text from a 3-dimensional locked safe as easily as we can lift a paper clip off a table top. To enclose valuables in four dimensions, we need a hypercube—the 4-dimensional analog of a cube. We can visualize 4-dimensional space by looking at 2- or 3-dimensional "pictures" or by unfolding objects like hypercubes.

We developed our ideas of the fourth dimension by looking at the familiar and finding patterns and analogies. The fourth dimension as developed here is not a vague wonderland, ill-defined and idiosyncratic for each individual. It has a specific shape and definite form. We followed the pattern of sweeping a lower dimensional space to create the next higher dimension. This pattern among the known spaces was sufficiently clear that we could proceed to follow the pattern even when physical models were no longer available to confirm our understanding.

One method for opening our minds to ideas imperceptible to our immediate senses is the use of analogy. Starting with familiar objects or ideas, we can look for patterns and then extend those patterns to create new ideas. We can ask ourselves, "If these patterns were extended, what properties would the next object on the list have?" Once we have described that next object, we have created a new idea, whether the next object can actually be constructed physically or not. Reasoning by analogy is a powerful method for creating new ideas.

Life Lessons

**When a question seems too difficult to answer,
try to answer an easier, related question.**

•

Find patterns among familiar ideas.

•

Create new ideas by analogy.

MINDSCAPES   Invitations to Further Thought

*In this section, Mindscapes marked (**H**) have hints for solutions at the back of the
book. Mindscapes marked (**ExH**) have expanded hints at the back of the book.
Mindscapes marked (**S**) have solutions.*

Developing Ideas

1. **At one with the universe.** Below is a sketch of a 1-dimensional space. Identify
   the points $x = 3$, $y = -5/2$, and $z = 2.25$.

   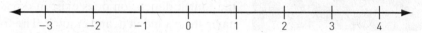

   How does the fact that you need only *one* number to identify a particular
   point relate to the dimension of the space?

2. **Are we there yet?** Why does the information "$x = 4$" *not* specify a unique
   point in the plane? What does this say about the dimension of the plane?

3. **Plain places.** Plot the following points in the plane: $(2, 1)$, $(-1/2, 0)$, $(\pi, -3)$.
   Is there a point in the plane for which you need exactly three numbers to
   specify its location?

4. **Big stack.** If you take a huge number of sheets of paper and stack them up,
   what do you get? What is the dimension of the structure you built?

5. **A bigger stack.** If you take a huge number of different 17-dimensional spaces
   and stack them up, what do you get? Guess its dimension.

Solidifying Ideas

6. **On the level in two dimensions.** Pictured in the chart below are level slices of
   objects in two dimensions. That is, we took a 2-dimensional object and made
   several parallel slices with a line at different levels. What are the objects?

	Level 1	Level 2	Level 3	Level 4	Level 5
Object 1	•	• •	• •	• •	•
Object 2	——	• •	• •	• •	•
Object 3	——	• •	• •	• •	——
Object 4	•	•	•	•	•

7. **On the level in three dimensions (S).** Pictured in the chart below are level slices of objects in three dimensions. That is, we made several parallel slices with a plane at different levels. What are the objects?

	Level 1	Level 2	Level 3	Level 4	Level 5
Object 1	•	○	○	○	•
Object 2	•	△	△	△	▲
Object 3	●	○	○	○	●
Object 4	•	⬭	○ ○	⬭	•

8. **On the level in four dimensions.** Pictured in the chart below are level slices of objects in four dimensions. That is, we made several parallel slices with 3-dimensional space at different levels. What are the objects?

	Level 1	Level 2	Level 3	Level 4	Level 5
Object 1	•	⊖	⊖	⊖	○
Object 2	⬚	⬚	⬚	⬚	⬚
Object 3	•	•	•	•	•
Object 4	•	△	△	△	▲

9. **Tearible 2's.** In the pictures below, describe how you would remove the gold bar from the various barriers just by bending and moving the walls in the 2-dimensional plane. Indicate which pictures require that you tear the barrier to remove the bar.

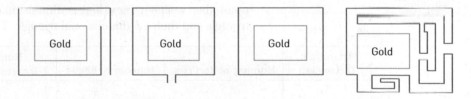

10. **Dare not to tear?** For the figures in the Tearible 2's that required tearing to remove the gold, describe how you could remove the gold without tearing by using the third dimension.

11. **Unlinking (H).** Using the fourth dimension, describe how you would unlink the pictured pairs of objects.

12. **Unknotting.** Describe how you would unknot the following knots using the fourth dimension.

13. **Latitude.** Some 4-dimensional person has an object. She shows you 3-dimensional cross-sectional slices of it. The slices look like a circle in one level of 4-space with increasingly smaller circles at each level above and below until they end at a point at the top and a point at the bottom. What is the object?

14. **Edgy hypercubes (H).** Produce drawings of the regular cube, the 4-dimensional cube, the 5-dimensional cube, and the 6-dimensional cube.

15. **Hypercube computers (ExH).** Parallel processing computers use 4, 5, and higher dimensions by locating a processor at each vertex of the cube. One processor sends information to processors that are attached to it by an edge. The distance between two vertices is defined to be the minimum number of edges required to create a path from one of the vertices to the other. What is the longest distance between vertices on a 3-dimensional cube? 4-dimensional cube? 5-dimensional cube? In general, an n-dimensional cube?

Creating New Ideas

16. **N-dimensional triangles (ExH).** We saw how to build cubes in all dimensions; how about triangles? A 0-dimensional triangle is just a point. A 1-dimensional triangle is a line segment; you know what a 2-dimensional triangle looks like; a 3-dimensional triangle is a tetrahedron. What is the pattern? We take the triangle we just created and then add a new point in the next dimension "above" the triangle. If we draw new edges from the vertices of the triangles to our new point, then we have a next higher dimensional triangle. Sketch a 4-dimensional triangle and then a 5-dimensional triangle. Fill in the following table.

Triangle's Dimension	Number of Vertices	Number of Edges	Number of 2-Dimensional Faces	Number of 3-Dimensional Faces
1				
2				
3				
4				
5				
n (in general)				

17. **Doughnuts in dimensions.** Suppose we have a mysterious object in four dimensions. If we slice the object in half with a 3-dimensional slice, we'd see the surface of a hollow doughnut.

If we take a 3-dimensional slice just above or below that first slice, we'd see the surface of another doughnut—this one thinner than the first.

As we slice higher (and lower) we see doughnuts whose waistlines get thinner and thinner, until finally we just see a circle.

What is the 4-dimensional object if we know the level 3-dimensional slices? Why is the answer "a circle of spheres"?

18. **Assembly required (S).** As promised in the preceding section, here is your chance to glue together a 4-cube. Draw a picture of the unfolded 4-dimensional cube as eight 3-dimensional cubes. Indicate in your drawing which faces get glued together to reassemble the 4-cube in four dimensions.

19. **Slicing the cube.** Take a 3-dimensional cube balancing on a vertex and imagine slicing it with many parallel planes starting with this one. Sketch the various types of level curves, that is, cross-sectional slices, we'd see. For example, the first few would look like triangles that are increasing in size. Continue sketching the slices and make sure to include all the shapes we'd see.

Move plane down and slice cube.

20. **4D swinger.** The plane is just a half line swung around a point. Three-dimensional space is a half plane swung around a line.

Swing a half plane around
a line and make a 3D space.

Suppose we make a circle with a dot inside it on the half plane. When we swing the half plane around to make 3-dimensional space, the circle and the point produce objects in 3-dimensional space.

Describe the objects. What pair of objects do we produce if we swing only the point around but keep the circle fixed?

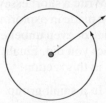

Swing around and
sweep out a plane.

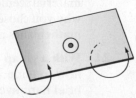

Now what is 4-dimensional space? It is half 3-space swung around a plane. Hard to see? Yes—but try. What do we get if we take a pair of linking circles in the half 3-space and swing it around to make 4-dimensional space? What if we take a sphere in the half 3-space with a point inside? Swing the point around, but leave the sphere still. Do we get a sphere linked with a circle? Explain your answer as best as you can.

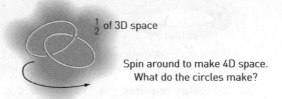

$\frac{1}{2}$ of 3D space

Spin around to make 4D space.
What do the circles make?

Further Challenges

21. **Spheres without tears.** A sphere is the set of points at a fixed distance from a given point. A sphere in 1-dimensional space is just two points. To make a sphere in 2-dimensional space, we take two copies of the sphere in 1-dimensional space, fill each one in (color in all the points between the points), and then glue the outer edge of one of the filled-in spheres in 1-dimensional space to the outer edge of the other (this requires us to bend each sphere out a bit). This process produces a circle (a sphere in 2-dimensional space). Generalize this procedure to produce a sphere in 3-dimensional space (include pictures), and then use this method to describe how to construct a sphere in 4-dimensional space.

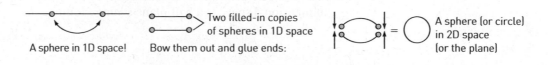

A sphere in 1D space!   Two filled-in copies of spheres in 1D space   Bow them out and glue ends: $=$ A sphere (or circle) in 2D space (or the plane)

22. **Linking.** Start with two linked circles in 3-dimensional space. Make one of the circles the equator of a sphere in 4-dimensional space by constructing increasingly smaller circles as we rise and descend in levels of 4-dimensional space. Can we pull the sphere and circle apart in 4-dimensional space, or are they linked? Explain your answer.

In Your Own Words

23. **Personal perspectives.** Write a short essay describing the most interesting or surprising discovery you made in exploring the material in this section. If any material seemed puzzling or even unbelievable, address that as well. Explain why you chose the topics you did. Finally, comment on the aesthetics of the mathematics and ideas in this section.

24. **With a group of folks.** In a small group, discuss and actively work through how to untie a knot in 4-dimensional space. After your discussion, write a brief narrative describing the process in your own words.

25. **Creative writing.** Write an imaginative story (it can be humorous, dramatic, whatever you like) that involves or evokes the ideas of this section.

26. **Power beyond the mathematics.** Provide several real-life issues—ideally, from your own experience—that some of the strategies of thought presented in this section would effectively approach and resolve.

For the Algebra Lover

Here we celebrate the power of algebra as a powerful way of finding unknown quantities by naming them, of expressing infinitely many relationships and connections clearly and succinctly, and of uncovering pattern and structure.

27. **Vertex values.** The Math Club wants to host a party on the theme "Hypercube!" As part of the decorations, they want to build a model of a 6-dimensional cube using Styrofoam balls as vertices. If a cube in n dimensions has 2^n vertices, how many balls does the Math Club need?

28. **Edge expenditures.** The Math Club's "Hypercube!" party planning requires they obtain sufficient dowels to serve as edges in their 6-dimensional hypercube decoration described in the previous Mindscape. If a cube in n dimensions has $n2^{n-1}$ edges, how many dowels does the Math Club need?

29. **Cubical calculation.** An ordinary 3-dimensional cube has six square faces, twelve edges, and eight vertices. In a 4-dimensional cube, the number of square faces is twice the number of square faces in the 3-dimensional cube plus the number of edges in the 3-dimensional cube. The number of edges in the 4-dimensional cube is twice the number of edges in the 3-dimensional cube plus the number of vertices in the 3-dimensional cube. How many square faces are there in a 4-dimensional cube? How many edges are there?

30. **Triangular totaling.** A tetrahedron is a 3-dimensional "triangle." It has four faces, each a 2-dimensional triangle, and six edges. It turns out that a "triangle" in $n - 1$ dimensions has $n(n - 1)/2$ 2-dimensional faces and $n(n - 1)(n - 2)/6$ edges. (See Mindscape 16.) Compute the number of edges and the number of 2-dimensional faces there are in a 5-dimensional triangle.

31. **Summing square.** After the Math Club built their 6-dimensional hypercube model, club president Edwinna decided to calculate how many square faces (2-dimensional faces) this monster model had. If a cube in n dimensions has $(n)(n - 1)2^{n-3}$ 2-dimensional faces, how many such faces does the Math Club's model have?

CHAPTER FIVE

Contortions of Space

5.1 Rubber Sheet
Geometry

5.2 The Band That Wouldn't
Stop Playing

> *The whole of mathematics is nothing more than a refinement of everyday thinking.*
>
> ALBERT EINSTEIN

Most objects in our everyday world are more or less rigid. Cars, buildings, and tables are of a fixed size and shape. Consequently, the common building blocks we use to describe the shapes in our world include the firm templates of the straight line, triangle, circle, and sphere. However, some objects in our world do bend, stretch, or distort—rubber bands, waterbeds, and our waistlines. In our mind's eye, we can imagine the rigid objects of our world as stretchable and amorphic. That image gives us a whole new potential reality to explore. In such an unreasonably contortable universe, we discover things that surprise and amuse us. These discoveries lead us to new insights into our real world, including unraveling some of the mysterious behavior of DNA, the stuff of life itself.

We begin our exploration of contortable space through some classic examples of flexible possibilities, including removing a stretchable vest without first taking off one's coat. Such examples lead us to a notion of equivalence by distortion. This idea of equivalence captures some geometrical aspects of objects but is far more flexible than the rigid equivalence of "congruence" that is discussed in high school geometry. Some features of objects do not change with bending or stretching, and it is these we describe and explore.

Counting basic components of surfaces leads us to intriguing, twisted objects that initially defy our intuition. For example, a relationship among vertices, edges, and faces of random doodles on rubber surfaces helps us prove the rigid fact that there are only five regular solids. Another application of this flexible geometry is meteorological: some pair of points opposite each other on Earth must at any given moment have the exact same temperature and pressure.

As we have already mentioned, a great way to stumble on new insights is to first find different ways of describing a situation. Also, seeking the essential and ignoring the irrelevant, we can represent the very core of an issue free from distracting and unnecessary baggage. Even in an abstract world created in our imagination, we can look for patterns and describe similarities and differences. The tools of thought that are illustrated in this contortable world are most powerfully applicable in our everyday lives.

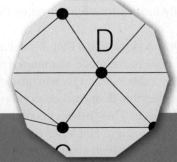

5.3 Knots and Links

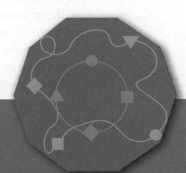

5.4 Fixed Points, Hot Loops, and Rainy Days

Discovering the Topological Idea of Equivalence by Distortion

© Fine Art Images/SuperStock

Vincent Van Gogh, *Wheat Field with Cypresses,* 1889. Can you imagine a world that can be stretched and twisted?

> *The moving power of mathematical invention is not reasoning but imagination.*
>
> **AUGUSTUS DE MORGAN**

Have you ever noticed that a rubber band can be stretched to resemble a circle, a square, a triangle, and, in fact, any polygon or any distorted version of a polygon? From a "rubber geometry" point of view, all these shapes are equivalent since we can deform any one into any other.

Compare this new view of geometric equivalence with that of classical, Euclidean geometry, in which two

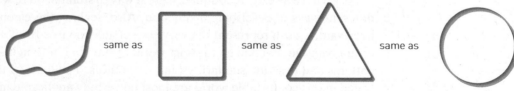

objects are basically the same if one can be rigidly moved to the location of the other, with the two matching perfectly.

As much as we try, we cannot stretch a rubber glove to look like a Ping-Pong ball or a hollow inner tube. Now let's imagine that all objects are made of material as malleable (and even more) than a rubber glove. Then, rubber geometry will provide us with an idea of geometric equivalence that is more flexible than Euclidean geometry yet captures some of the geometric character of objects. This "rubber geometry" has evolved into a branch of mathematics known as *topology*. The word *topology* was coined in 1847 and derives from the Greek word *topos*, which means "surface." Topology is a mind-stretching subject that frees us from conventional geometry and allows us to appreciate geometric characteristics of objects that remain unchanged even when the objects are stretched, shrunk, distorted, or contorted.

Life Lesson
Seek the essential.

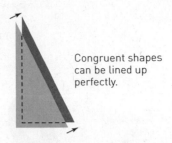

Congruent shapes can be lined up perfectly.

Fun with Rubber

Fooling around with rubber can be fun; but besides its entertainment value, it can stretch the imagination. To introduce the idea of elastic space, we present three short tales. (Well, since they are rubber stories, they really can be stretched into three tall tales.) Solve each challenge before reading the solutions.

First Challenge—The Divestment

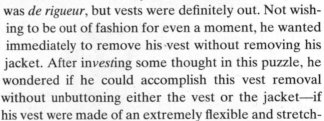

One of the authors frequently wore a natty suit complete with vest. One day he read in *Vogue* that vests were considered old-fashioned, even stodgy in some circles. He studied the article carefully and noted that the suit jacket was *de rigueur*, but vests were definitely out. Not wishing to be out of fashion for even a moment, he wanted immediately to remove his vest without removing his jacket. After in*vest*ing some thought in this puzzle, he wondered if he could accomplish this vest removal without unbuttoning either the vest or the jacket—if his vest were made of an extremely flexible and stretchable rubberlike material. Experiment in your mind's eye or in your clothes closet.

Second Challenge—Grandma's Flat Tire

While the first author was taking off his vest, the other author happened on an elderly woman whose car had been incapacitated by a flat tire. He first carefully advised her about how to use the tire iron. Soon the 86-year-old grandmother had pulled off the tire and removed the tube—being guided throughout by the author's numerous and helpful suggestions. She showed him the rather large puncture in the tube. In fact, the hole was about 5 inches across.

The silver-haired matron was patting the perspiration from her brow and catching her breath from the exertions of jacking up the car while our math hero said, "Do you think it is possible to turn the inner tube completely inside out? Just reach in through that hole and pull on the insides and see what happens."

The grandmother indulged the author and did as instructed. At last she stood, covered with grime from head to toe.

"You have just discovered a topological fact about the torus!" cried our impassioned author with glee.

"Is my tire almost fixed?" asked the grandmother.

"Whoops, got to get to class," said our fearless author as he nimbly dodged the tire iron that the grandmother apparently knew how to use after all. Challenge: Was she able to turn the inner tube inside out?

Third Challenge—the Ring

We are given a thin sheet of extremely flexible and stretchable rubber with two holes. An expensive gold ring passes through the two holes. Is it possible to remove the ring from one of the holes? Your challenge is twofold: Try to answer this question, and then write a funny story using the question as the punch line.

The Divestment Solution

To illustrate the notion of topology, we first turn to "The Divestment" challenge and examine how, if the vest is made of very distortable rubber, we can remove the vest without even unbuttoning the jacket or the vest.

Because he is not allowed to unbutton the vest, the vest has the same characteristics as a pullover sweater vest made of rubber. As the vest is extraordinarily

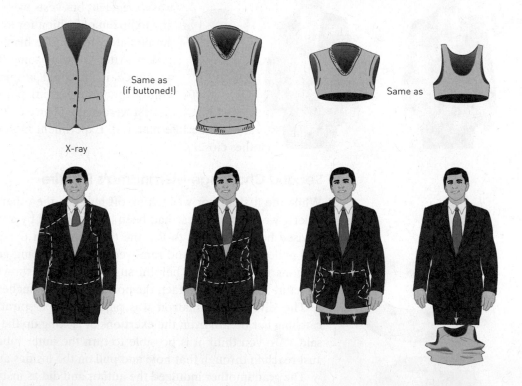

flexible, he can shrink the bottom part of the vest upward without changing its topological properties. (Remember, stretching or shrinking will not change an object in this new type of geometry.) In fact, he can shrink the bottom of the vest

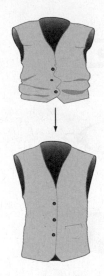

so that its bottom seam comes within an inch of the head hole and the arm holes. The vest now appears to be more like a stretchable tank top. Notice that all this shrinking can take place without the vest ever leaving the cozy confines of the suit jacket.

He can now stretch one shoulder strap down through the jacket sleeve to his hand, pass the strap around his hand, and shrink the strap back up the sleeve to its original size. If he then moves the other shoulder strap down and around the other arm, we see that the straps are no longer over his shoulders. Thus by wiggling, he can slide the entire rubber garment down to his feet right through the bottom of the jacket. He can then pick up the tank top and stretch it all out again to make the unfashionable original vest.

Despite the fact that he bent, stretched, and shrunk the vest, at every stage of the "divestment" the vest remained *equivalent by distortion* to the original one.

Grandma's Flat Tire Solution

To refine this notion of equivalence by distortion, we return to Grandma's flat tire and show how we can turn an extremely distortable inner tube, with a hole, inside out. In this sequence of figures, notice how we begin with an inner tube that is red on the inside and blue on the outside. We then carefully stretch and deform that hole and, once the dust settles, we are left with the red surface on the outside and the blue surface on the inside.

By the way, this feat can physically be accomplished with a regular inner tube having a 5-inch-diameter puncture. If you can get your hands on an inner tube, give it a try.

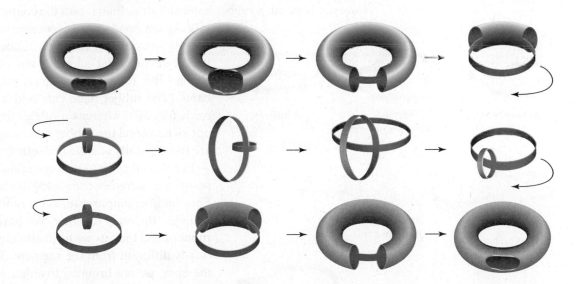

The Ring Solution

The counterintuitive fact is that the ring *can* be removed from one of the holes. Here is a sequence of stretches that accomplishes this surprising feat.

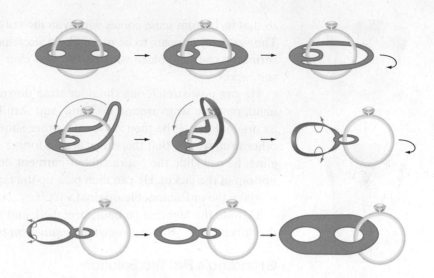

Equivalence by Distortion

Let's further develop our intuition regarding rubber sheet geometry by comparing various objects and determining what kinds of deformations are allowable and what are not. Since we wish to preserve, protect, and defend the fundamental structure of our object during a distortion, we will not allow breaking or tearing it. We want to maintain the integrity of the object in the sense that points that were close to one another in one position remain close throughout the distortion. As an analogy, think of the object as being made of molecules that grip adjacent molecules. If the object is stretched, each molecule still grips the same molecules. However, if we cut a rubber band and straighten it out, the continuity of the cycle is broken; a bond between connecting molecules has been destroyed. So, cutting a rubber band to get a line segment is not allowed, since doing so would change the fundamental structure of the rubber band. (The rubber band can hold a bunch of index cards together, whereas a rubber line segment cannot.) Therefore, the rubber band and the rubber line are topologically *inequivalent* objects.

In a similar spirit, we do not allow gluing together points that were not connected to each other before. For example, suppose we take a rubber line segment and glue the ends together. We have just created a rubber band that, as we have already noted, is structurally different from the segment. When we attach the ends, we are bringing together points that were not touching in the original segment. In the analogy of molecules, we are not allowed to create connections between molecules that did not exist before.

So, summing up, we'll say that two objects are *equivalent by distortion* if we can stretch, shrink, bend, or twist one, without cutting or gluing, and

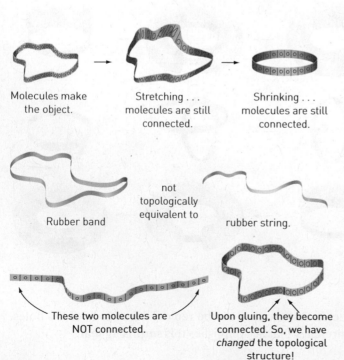

Molecules make the object.

Stretching . . . molecules are still connected.

Shrinking . . . molecules are still connected.

Rubber band

not topologically equivalent to

rubber string.

These two molecules are NOT connected.

Upon gluing, they become connected. So, we have *changed* the topological structure!

deform it into the other. The technical topological term for this notion is *isotopic*— a term you will find useful at topological cocktail parties. Cheers!

Undisturbed by Distortion

What are some features of an object that remain intact no matter how much it is distorted? A circle has the property that, if we remove any point from it, what remains of the circle is still one piece. Any distortion of the circle retains this property. This feature of a circle allows us to deduce, for example, that we could not stretch it to make it look like a circle with a feeler attached, because there is a point on the circle with a feeler that, if removed, actually breaks the object into two pieces.

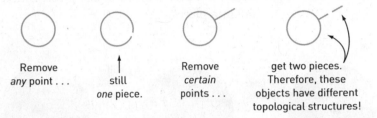

Remove *any* point . . .   still *one* piece.   Remove *certain* points . . .   get two pieces. Therefore, these objects have different topological structures!

However, if we remove any two points from a circle, the circle falls into two pieces. This feature allows us to deduce that a circle cannot be distorted to look like a theta curve—that is, an object in the shape of the capital Greek letter *theta*—because there are pairs of points that, when removed, leave the theta curve in one piece.

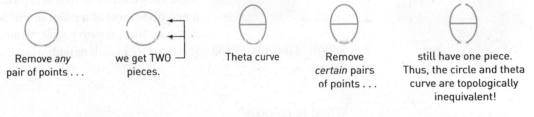

Remove *any* pair of points . . .   we get TWO pieces.   Theta curve   Remove *certain* pairs of points . . .   still have one piece. Thus, the circle and theta curve are topologically inequivalent!

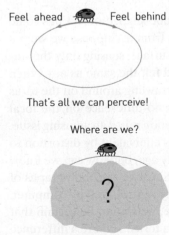

Feel ahead   Feel behind

That's all we can perceive!

Where are we?

?

What are we on?

Let's imagine ourselves as a small bug crawling along a circle; our entire sensory apparatus allows us to detect only the immediate neighborhood of the circle on which we crawl—not the space off it. So, wherever we stop, we can feel ahead and behind, and that's all. Our neighborhood feels the same as a slightly distorted line segment.

Suppose now that we are crawling along a curve, but we are not told whether we are crawling around on a circle or on a theta curve. Could we tell on which curve we are marching, purely by sensing our immediate neighborhood?

At most places, we could not tell the difference between the neighborhoods of a point on the circle versus a point on the theta curve; however, if we were standing at one of the two points where the horizontal bar attaches to the circle on the theta curve, we would feel a different surrounding. We would notice that there are three directions in which we could travel from that point. The number of segments emanating from a point would be preserved under any distortion.

Three different directions to move. There's no such intersection on a circle. So, this can't be a circle!

Therefore, the circle and the theta curve are not equivalent by distortion, because the theta curve contains a point (in fact, two points) from which three different segments emerge, whereas every point in the circle is like every other point on the circle in its having only two directions. So, this local difference in features distinguishes these two curves.

Feeling Our Way on Surfaces

To gain a deeper understanding of the idea of equivalence by distortion, we now examine various surfaces. Let's consider a sphere—that is, the boundary of a ball—and attempt to find features of it that would be preserved under distortions. If we remove one or two points from a sphere (or even more, of course), the object left over remains in one piece. But, if we cut out any loop, the sphere will fall into pieces.

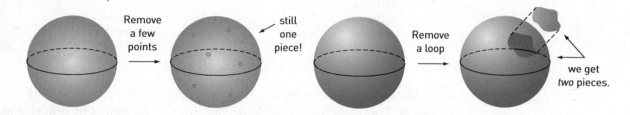

Observe that, if we were a bug on a sphere and could sense only the neighborhood immediately around us, every neighborhood would seem like every other one. In fact, every point on a sphere has a neighborhood surrounding it that is equivalent by distortion to a neighborhood of a point in the flat plane. Any object that has the property that around every point there is a small neighborhood equivalent to a small neighborhood on the plane is called a *surface*.

Neighborhood looks like:

What Is a Torus?

A *torus* is the boundary of a doughnut. This inner-tube-shaped object has some similarities to the sphere, as well as some differences from it. Suppose we were a bug and we lived on a torus. As we crawled over the surface, sensing only the surface itself, we would notice that every neighborhood felt the same as any other. In fact, we could not tell the difference between crawling around on the torus and crawling around on the sphere. There is no difference on the local level—they are both surfaces. In fact, this question raises an amusing issue. How do we know that the surface of Earth is equivalent by distortion to a sphere? Everyone learns this fact at an early age, but how can we know this and prove this by personal experience? Sure, we see photographs of Earth taken from outer space, but perhaps they were altered by computer. The point is that, as we travel along the surface, we cannot be certain that we are walking around on a sphere rather than on a torus! Is there a difference between a sphere and a torus? If so, how can we tell the difference?

Torus

Inside is hollow.

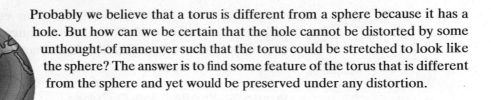

Probably we believe that a torus is different from a sphere because it has a hole. But how can we be certain that the hole cannot be distorted by some unthought-of maneuver such that the torus could be stretched to look like the sphere? The answer is to find some feature of the torus that is different from the sphere and yet would be preserved under any distortion.

Perhaps this is actually how Earth is.

Is a Torus a Sphere?

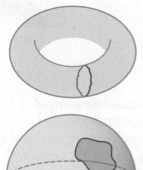

Look at the red circle on the torus pictured on the left. Suppose we cut the torus around that circle. Notice that the cut torus is still one piece, though now it's a tube. We have not separated the torus into two pieces. Suppose that we could somehow distort the torus to make it look like the sphere. Then, after the distortion, that red circle would become a red loop on the sphere. But cutting any loop on the sphere divides the sphere into two pieces. So, we see that, on the torus, the red loop does not cut the object into two pieces. After just stretching and distorting the surface, the red loop cuts the object (now the sphere) into two pieces, which is impossible. Therefore, we cannot distort a torus to look like a sphere; thus, these surfaces are topologically different objects.

We close this section with two intriguing questions. We hope that you first spend some time drawing pictures, visualizing objects in your mind's eye, and talking about these questions with others before reading the answers.

Puzzle 1—Holy Doughnuts

One piece!!

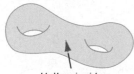

Hollow inside

The torus we just considered had one hole. Let's now consider tori with more than one hole. A two-holed torus is just the surface of a two-holed doughnut. Notice that the inside is hollow. It is easy to see that the figures below are all equivalent by distortion to the original two-holed torus. Our question here is, What happens if the two holes are linked?

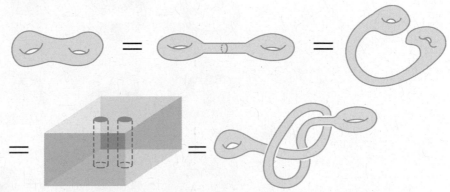

Specifically, is the linked two-holed torus equivalent by distortion to an unlinked two-holed torus? If so, draw a sequence of pictures showing how to stretch and distort one to get the other. If not, provide a reason why the linked holes cannot be unlinked.

Puzzle 2—Knotted Jell-O Cubes

Jell-O is beautifully translucent and wiggly, the perfect substance to inspire creative topological thoughts. Suppose we take two cubes of Jell-O, and, in each cube, we drill two holes so each cube would be some *solid two-holed torus*—that is, a two-holed torus whose inside is filled with Jell-O. In the first cube we just drill

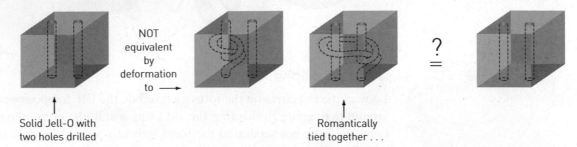

NOT equivalent by deformation to →

Solid Jell-O with two holes drilled

Romantically tied together . . .

two holes straight down to get an ordinary solid two-holed torus made of Jell-O. However, in the second cube, one of the holes is knotted. Although we will not prove it, these two Jell-O cubes are not equivalent by distortion; there is no way to unknot that knotted hole. We wish to consider a third Jell-O cube with two holes, one of which is knotted and circles around the unknotted hole. Question: Is this knotted Jell-O solid two-holed torus equivalent by distortion to the ordinary (unknotted) Jell-O solid two-holed torus? If so, draw sketches that show how to deform one into the other. If not, provide a reason why they are not equivalent. (You may use the previous fact about the first knotted Jell-O mold.)

Solutions—A Picture Is Worth a Thousand Words

The surprising and counterintuitive answer to both questions is, yes, they are equivalent! We now illustrate a sequence of moves that depicts the distortions needed to get from one object to the other.

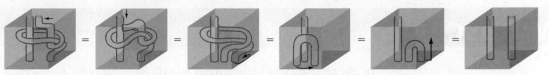

Down the hole!

A Twisted Final Thought

Path walked on outside.

Cannot get to the inside . . .

without crossing an edge!!

Each of the surfaces we've examined thus far has two sides, an inside and an outside—as we expect. If you are standing on the outside surface of a tube, for example, and you walk along the surface and return to your starting point without crossing an edge, then you will still be on the same side of the surface. Is it possible to make a surface so strange that we could start at one point, walk along a path never crossing an edge, and return to the same point, but find ourselves on the other side of the surface? This might seem impossible. But, as usual, questioning our intuition and our expectations reveals a whole world of fascinating possibilities. We will grapple with these disorienting ideas in the next section.

A Look BACK

RUBBER SHEET GEOMETRY, the nickname for topology, centers on a flexible notion of equivalence. Two objects are equivalent by distortion if we can stretch, shrink, bend, or twist one, without cutting or gluing, and deform it into the other. These distortions alter some geometric properties of the objects but preserve others. Looking for these invariant or unchanged features is an effective way to determine when two objects cannot be made equivalent by distortion. If we cut along any loop on a sphere, we divide the sphere into two pieces; however, there are some loops on a torus that when cut do not separate the torus into two pieces. Thus, the torus is not equivalent by distortion to the sphere. However, surprising pairs of objects are equivalent by distortion.

Our primary strategy for investigating this idea of rubber sheet geometry is to try things physically when possible or to draw lots of pictures. We must be open-minded to avoid narrow thinking, and we must seek specific reasons when we believe some distortion is impossible. Rubber sheet geometry forces us to identify essential features of objects and to ignore others.

In real life, we group people or objects together on the basis of characteristics they share. However, a great way to find new ideas and connections is to choose criteria for distinguishing among things. In this way, we can see old objects and ideas in a new light. We can explore this new perspective by performing physical or mental experiments, comparing and contrasting using our new criteria.

Exploring the consequences of an alternative reality can lead to creative insights.

Life Lessons

Seek the essential.

•

Ignore the irrelevant.

•

Seek novel ways to compare and contrast objects and issues.

MINDSCAPES   Invitations to Further Thought

*In the section, Mindscapes marked (**H**) have hints for solutions at the back of the book. Mindscapes marked (**ExH**) have expanded hints at the back of the book. Mindscapes marked (**S**) have solutions.*

Developing Ideas

1. **Describing distortion.** What does it mean to say that two things are *equivalent by distortion*?

2. **Your last sheet.** You're in your bathroom reading the liner notes for a newly purchased CD. Then you discover that you've just run out of toilet paper. Is a toilet paper tube equivalent by distortion to a CD?

3. **Rubber polygons.** Find a large rubber band and stretch it with your fingers to make a triangle, then a square, and then a pentagon. Are these shapes equivalent by distortion? What other equivalent shapes can you make with the rubber band? Can you stretch it to make a rubber disk?

4. **Out, out red spot.** Remove the red spot from the letters at right. For each letter, how many pieces result? Are the original letters equivalent by distortion?

5. **De-vesting.** Try "The Divestment" challenge. Find a stretchy vest or tank top, put it on, then put on a roomy jacket or sweatshirt. Now see if you can remove the vest without removing the jacket.

Solidifying Ideas

6. **That theta (S).** Does there exist a pair of points on the theta curve whose removal breaks the curve into three pieces? If so, the existence of those two points would provide another proof that the circle is not equivalent by distortion to the theta curve. Why?

7. **Your ABCs (H).** Consider the following letters made of 1-dimensional line segments:

ABCDEFGHIJKLMNOPQRSTUVWXYZ

Which letters are equivalent to one another by distortion? Group equivalent letters together.

8. **Puzzled?** Find the topology puzzle in your kit. Read the kit instructions for the puzzle and then solve the puzzle. Describe your solution in words and pictures.

9. **Half dollar and a straw.** Suppose we drill a hole in the center of a silver dollar. Would that coin with a hole be equivalent by distortion to a straw? Explain why or why not.

10. **Drop them.** Is it possible to take off your underwear without taking off your pants? You may assume you are wearing rubber undies. Explain why or why not. Include pictures.

11. **Coffee and doughnuts (H).** Is a standard coffee mug equivalent by distortion to a solid doughnut? Explain why or why not.

12. **Lasting ties.** Tie a thin rope around a friend's wrists, and then tie another one around your own, with the ropes linked as pictured. Can you unlink yourselves without cutting the ropes? Why or why not? (This challenge is one you should definitely make physical!)

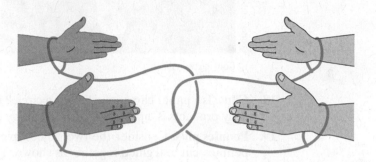

13. **Will you spill? (S).** Suppose you rest a glass of water in the palm of your hand. Is it possible to rotate the palm of your hand 360° without spilling the water or dislocating your arm?

14. **Grabbing the brass ring.** Suppose a string attached to the ceiling passes through a metal, nondistortable frame and then is tied to a brass ring, as illustrated in the picture. Is it possible to remove the brass ring without cutting the string?

15. **Hair care.** Is a regular comb equivalent by distortion to a regular hair pin? Explain why or why not.

16. **Three two-folds.** Take three pieces of paper and fold each in half and then in half again, as shown in the figure below. Suppose you cut out a semicircle from each paper as shown. Are those cut sheets of paper equivalent? Explain why or why not.

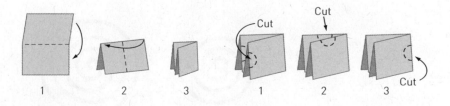

17. **Equivalent objects.** Group the objects in this photograph into collections that are equivalent by distortion.

Courtesy of the authors

Cooking utensils.

18. **Clips.** Is a paper clip equivalent to a circle? If not, to what other small stationery products is a paper clip equivalent by distortion?

19. **Pennies plus.** Consider the two objects pictured here. One is made of two pennies, cut and glued together as shown. The other is a thickened plus sign. Are these two objects equivalent by distortion?

20. **Starry-eyed.** Consider the two stars below. Are they equivalent by distortion?

21. **Learning the ropes.** Pictured below are two ropes, coiled differently. Are the ropes equivalent by distortion?

22. **Holy spheres.** Consider the two spheres shown. Each has four holes on its surface. Ropes are looped through the holes in different ways. Is the first

sphere with its two rope loops equivalent by distortion to the second sphere with its rope loops?

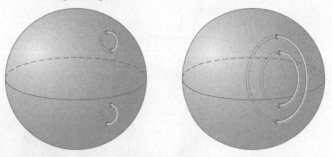

23. **From sphere to torus.** The following sequence of drawings takes a sphere and deforms it into a torus. Does this sequence describe an equivalence by distortion? Why or why not?

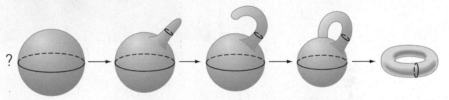

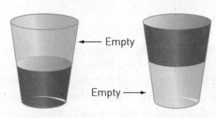

24. **Half full, half empty.** One glass is half filled with cranberry juice as illustrated. Another glass is also half filled with juice, although it is the top half (ignore such pesky issues as gravity). Are the two half-filled glasses with their contents equivalent by distortion? Explain why or why not.

25. **Male versus female.** Consider the male and female symbols at left. Assuming they are made out of 1-dimensional lines and curves, are the symbols equivalent by distortion? Explain why or why not.

Creating New Ideas

26. **Holey tori.** Are these two objects equivalent by distortion? If so, demonstrate the distortion with a sequence of pictures; if not, explain why not.

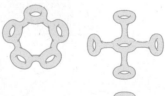

27. **More holey tori (H).** Are these two objects equivalent by distortion? If so, demonstrate the distortion with a sequence of pictures; if not, explain why not.

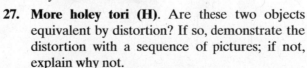

28. **Last holey tori.** Are these two objects equivalent by distortion? If so, demonstrate the distortion with a sequence of pictures; if not, explain why not.

29. **Beyond the holey inner tube.** Suppose you are given a two-holed torus with a large puncture in its side. Is it possible to turn it inside out? Explain why or why not.

Puncture

30. **Heavy metal.** Carefully examine this picture of a metal puzzle. Can you unlink the two pieces using distortion? You may assume that the metal is rubber.

©SuperStock/SuperStock

Metal Ring Puzzle.

31. **Rings around the ring.** Return to "The Ring" challenge given on page 328. Draw a red circle around one of the holes on the rubber sheet and a blue circle around the other. Now redraw the sequence of moves that unlinks the ring from one of the holes, but on each figure now include how the two colored circles get deformed. What is the result? Are both colored circles still looped around the ring?

32. **The disk and the inner tube (ExH).** Suppose you have a rubber disk with two holes and you glue a tube from one hole to another. Is that object equivalent by distortion to an inner tube with a puncture? Explain why or why not.

Holes

33. **Building a torus (S).** Suppose you are given a rectangular sheet of rubber. How could you glue the various edges together to build a torus? Indicate which edges get glued to which and how the edges match up.

34. **Lasso that hole.** Consider the first two tori on the left on page 341. Both have two punctures on their sides. On the first torus, a rope is looped through the two holes but does not go around the hole of the torus. On the second,

the rope is looped around the hole of the torus. Is it possible to distort the first torus to look like the second? How about if the rope looped around the hole twice, as shown in the rightmost torus below?

35. **Knots in doughnuts.** We are given two solid doughnuts: one with a worm hole drilled as shown and the other with a knotted worm hole drilled as shown. Are these two-holed doughnuts equivalent?

Further Challenges

36. **From knots to glasses (ExH).** Take the thickened knot and then add a solid tube, as illustrated below. Is it possible to distort this new object into a pair of eyeglass frames?

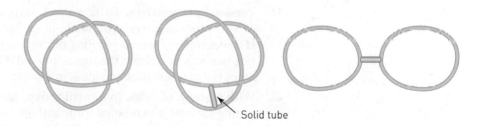

Solid tube

37. **More Jell-O.** Suppose we take a cube of Jell-O, drill two holes in it, and glue a Jell-O tube between them, as shown. Show that this Jell-O object is equivalent to the Jell-O cube on the right with two holes.

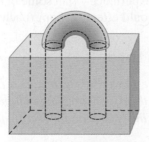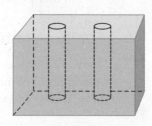

38. **Fixed spheres (H).** We are given two spheres made of glass. They are not distortable, they are rigidly fixed in space—one inside the other as indicated—and they cannot be moved. The middle sphere floats miraculously in midair without any means of support. A rubber rope hangs from the inside ceiling of the big sphere to the roof of the smaller sphere, as illustrated. The rope has a knot in it. Is it possible to unknot the rope without moving or distorting the two spheres?

39. **Holes.** Is a torus equivalent to a two-holed torus? If not, carefully justify your answer. (Consider the strategy used in "Is a Torus a Sphere?" on page 333 but with more than one cut.)

40. **More holes.** Is a two-holed torus equivalent to a three-holed torus? If not, carefully justify your answer. Can you generalize your observations and arguments?

In Your Own Words

41. **Personal perspectives.** Write a short essay describing the most interesting or surprising discovery you made in exploring the material in this section. If any material seemed puzzling or even unbelievable, address that as well. Explain why you chose the topics you did. Finally, comment on the aesthetics of the mathematics and ideas in this section.

42. **With a group of folks.** In a small group, discuss and work through the reasoning for how a two-holed torus with its holes linked can be distorted into an unlinked two-holed torus. After your discussion, write a brief narrative describing the method in your own words.

43. **Creative writing.** Write an imaginative story (it can be humorous, dramatic, whatever you like) that involves or evokes the ideas of this section.

44. **Power beyond the mathematics.** Provide several real-life issues—ideally, from your own experience—that some of the strategies of thought presented in this section would effectively approach and resolve.

For the Algebra Lover

Here we celebrate the power of algebra as a powerful way of finding unknown quantities by naming them, of expressing infinitely many relationships and connections clearly and succinctly, and of uncovering pattern and structure.

45. **Tea time.** Your grandmother has a very large collection of tea cups, sugar bowls, and mugs. Some have a one-holed handle, some have two holes, some have none. She wants you to organize them onto three shelves so that each shelf contains only vessels that are equivalent by distortion. She has n^2 cups with a one-holed handle, $40 - n$ with two holes, and $30 - n$ with no holes. If she has 150 items altogether, determine how many she has of each type.

46. **Ratio-rama.** These two shapes are equivalent by distortion. (Do you see why?) The left shape has squares for its boundaries and the right shape has isosceles right triangles for its boundaries. Let x denote the side length of the inner square and y denote the base (and height) of the inner triangle. If the side length of the outer square is $4x$ and the base (and height) of the outer triangle is $3y$, find the ratio of the area of the outer square to the area of the inner square and the ratio of the area of the outer triangle to the area of the inner triangle.

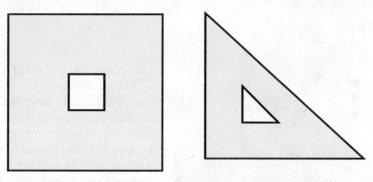

47. **How many holes? (H).** It's funny bagel day in the dining hall. Three shapes of bagels are being served, each shape with a different number of holes. To discover the number of holes in each of the three shapes, find the three values of n that satisfy the equation $n^3 - 5n^2 + 6n - 0$.

48. **Curvilinear.** In one xy-plane, sketch the graph of the equation $y = x + 2$; in another xy-plane, graph the equation $y = x^2$. Ignoring the axes, are the two graphs equivalent by distortion?

49. **Holey snowflakes.** Your little brother likes to make paper snowflakes by folding white paper and cutting lots of holes. He was very busy one January afternoon and made 60 snowflakes. Twenty percent have 16 holes, 25% have 18 holes, 30% have 15 holes, and the rest have 12 holes. Determine how many snowflakes he has of each type.

THE BAND THAT WOULDN'T STOP PLAYING

Experimenting with the Möbius Band and Klein Bottle

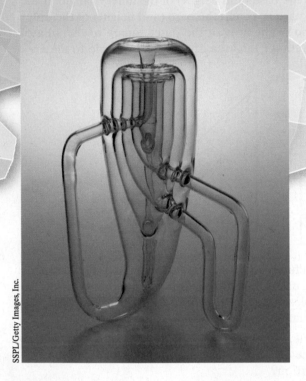

SSPL/Getty Images, Inc.

A sphere, the surface of a doughnut, and the surface of a two-holed doughnut are all different surfaces; it is impossible to distort any one to look like the others. However, each of these surfaces does have two sides—an inside and an outside—in the sense that, if we were bugs crawling on the outside surface and we walked along the surface and returned to our starting point without crossing an edge, we would still be on the outside of the surface. Does every surface have two sides in that sense? Maybe any surface automatically has two sides. But maybe not.

To help us think about the two sides of sidedness, we will build a physical model and then explore it through various experiments. In particular, we'll see what happens when we cut it up in various ways. We may encounter some surprising results. That feeling of surprise, however, means that there is more to understand. So we'll keep experimenting and looking at the results in different ways until they change from mysterious to simply neat. Even after we understand it completely, the Möbius band remains intriguing.

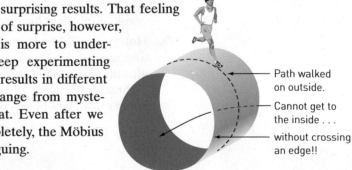

Path walked on outside.

Cannot get to the inside . . .

without crossing an edge!!

The Möbius Band

The Möbius band has been a crowd-pleaser for decades. After we have held a Möbius band and explored its endless edge, we experience an eerie sense of oneness: one edge, one side, one, one, one.

Looking at a picture is not good enough. You need a Möbius band, and you need it now. To make a Möbius band, get a strip of paper about 11 inches long and about 1 inch wide. First, bring the two ends of the strip together to make a loop. Now give

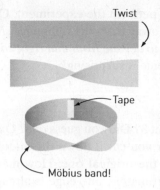

Twist

Tape

Möbius band!

one end of the strip a half twist and again bring the ends together. Tape the ends together. You now have a band with a half twist in it. You have a *Möbius band*.

What makes the band you hold so Möbius? It's that twist, of course. The half twist creates unexpected wonders. Let's explore. Hold the Möbius band; touch its edge; notice its simple elegance. As you slide it through your hands you may feel the urge to explore it a little further. What has the twist done besides create such haunting beauty? Some basic physical experiments will lead us to some startling discoveries.

Experiment 1—Life on the Edge

Start marking here

Let's start with an easy experiment. Take a red marker and begin coloring around the very edge of the Möbius band. We are envisioning your marker as soaking a little of the band as you slide along. Keep going until you get back to where you started. By the way, how many edges are there? Let's count them. There is the one you just colored, and . . . Surprise! What explains such a strange phenomenon?

Experiment 2—A Walk on the Wild Side

Start marking here

Let's now use a blue pen to count the number of sides. Put the pen in the middle of a side of the Möbius band and start drawing a line right down the middle all the way around the band. Keep drawing without lifting the pen. Keep going until you return to where you started. You have now drawn a circle on the Möbius band, and, of course, that circle, being one piece, all lies on the same side of the band. But wait. Peek on the opposite side of the band. What do you see? How could that circle be on the back side, too? The back side is the same side as the front side. How is that possible?

Experiment 3—Making the Cut

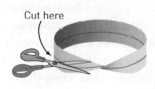

Cut here

So, the Möbius band seems to be deficient in edges and sides—only one of each, but abundant in mystery and intrigue. What is going on here? Before uncovering any explanations, let's try another experiment. Take a pair of scissors and cut the band lengthwise down the center core. (In fact, your blue pen line provides a guide for cutting.) Before physically executing this experiment, try it first in your mind's eye and make a guess as to what you will see. *Any guess is fine, but please guess something!* You may want to record your guess because guesses and even mistakes often lead to insights and an appreciation for hidden subtlety. Now cut the Möbius band along the center blue core line. What do you see?

Life Lesson
Make guesses.

Experiment 4—Hugging the Edge

Build another 11-inch × 1-inch Möbius band. We're going to cut it again lengthwise. But this time, instead of cutting right down the center, cut so that your scissors hug (stay close to) the right edge as you slide the Möbius band along while you cut. That is, cut so that the scissors are always 1/3 inch away from the right edge. Before proceeding, think about the result of this experiment in your mind's eye and make a guess about what you'll end up with when you've finished cutting.

Cut, staying $\frac{1}{3}$ inch away from the right edge of the strip with the scissors.

Remember: First think, then guess, then cut. Now perform the experiment. Cut the band lengthwise, but keep the scissors 1/3 inch away from the right edge. *No matter what happens, don't give in to the temptation to deviate from that right edge where your scissors currently are cutting!* The outcome is surprising!

Some Experimental Results

How many pieces did you end up with in experiment 3? Did you guess *one*? One, one, one. One edge, one side, and one piece after you cut it down the middle. Notice how the new narrow band is longer than the original one. How many edges does your long twisted strip now have? Fortunately, one edge is already marked in red. The other edge is basically marked in blue if you stayed right in the middle of the blue marked line while you cut. So, how many edges does this new band have? Two edges. In fact, check for yourself that the twisted strip you now hold has two half twists, two edges, and, you guessed it, two sides. Just one cut and we doubled everything! Why? Given all these observations, were you surprised by the outcome of experiment 4?

How can we understand why these peculiarities occur? We can see that the Möbius band behaves as it does, but to understand it we need to look at the band in different ways.

When trying to understand an object or an idea, it is often helpful to consider different descriptions of it. Frequently it is helpful to create a model—not necessarily a physical model but a representation that helps us to see the object from a certain point of view. In this case, let's think about how the Möbius band is constructed and use that knowledge to help us construct a flat representation of the twisted band.

Recall that we construct the Möbius band using a strip of paper. We attach the ends together in a special way, namely, with a half twist. To represent this process, let's draw a strip and indicate how the ends of the strip are to be glued to create the Möbius band. We'll mark the short sides with arrows to indicate that, to make the Möbius band, we must bring those short sides together and glue them to each other in such a way that the arrows are pointing in the same direction. Notice that the twisting is now captured in the directions of the arrows. In fact, as long as we understand that those arrows have to line up, we can think about the Möbius band without actually gluing the ends. We just understand that our picture means that those ends are, in fact, to be glued in the appropriate manner to create the Möbius band. The picture can be viewed as a kit with instructions to build the Möbius band (some assembly required).

Using this representation of the Möbius band, let's take another look at some of the results we observed.

Life Lesson

Look at alternative representations.

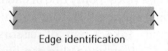

Edge identification

Glue so arrows line up.

A Möbius band!
(Assembly required . . .)

Making the Cut—Experiment 3 Revisited

Cutting down the center line would look like the diagram here. As the scissors reach the right edge, the cut returns where it started. How can we now see why we get only one piece after cutting? Let's place a small bug on the surface and have it walk around. As the bug crawls toward the right and crosses the upper part of the right side, it doesn't fall off—there is no cliff. In fact, it doesn't even

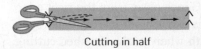

Cutting in half

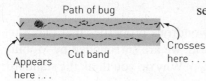

Path of bug

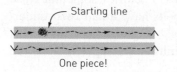

Crosses here . . .

Appears here . . .

Cut band

Starting line

One piece!

see an edge; it just sees a seam. Remember the gluing. It just sees the band continuing. As it crosses that upper-right edge, where does it reappear? Visualize this activity and remember the arrows.

The bug steps out on the lower-left edge (on the back) and continues to walk across until it reaches the lower-right edge and then continues on the upper-left edge (now in front) until it returns to where it started. How many pieces are there? One, since there is no piece of the band that was not touched by the bug! This demonstrates why cutting a Möbius band down the middle results in only one piece. In fact, our analysis provides a way to determine the length of the new band and the number of twists it has (if you don't believe us, just take a look at Mindscape 11 at the end of this section). Who said a math proof shouldn't have bugs in it? Okay, so much for our cutting sense of humor; let's see if this method can be used to analyze the cutting experiment in which we hugged the right shoreline.

Hugging the Edge—Experiment 4 Revisited

In this cutting experiment, we keep the scissors 1/3 inch from the right (bottom) edge. As we cut, we come to the right edge of the strip. Where do the scissors move beyond this point? The scissors continue at the upper-left side and cut to the right. However, now, as the band is twisted, the rule of staying 1/3 inch from the bottom side requires us, as in our diagram, to stay 1/3 inch from the top side. The key lies in the twisting. We continue to cut, and when we reach the right edge of our strip we return back to the lower left where we started our cut. How many pieces do we now have?

If we place a ladybug on the upper strip and have it walk to the right, it will walk until it gets to the right edge, and then it emerges . . . where? Once it moves beyond the lower-left edge, it continues to travel, passing the lower-right edge, which brings it back to the upper-left edge and then back to its starting place. So, the top and the bottom strips are connected to make one large strip. Did the ladybug traverse everything? No! That middle piece was untouched by the ladybug and thus is a second piece. This model demonstrates that, when we cut the Möbius band in this manner, we get two pieces. This analysis also shows much more, as you will again discover in the Mindscapes.

??

Where do we come out?

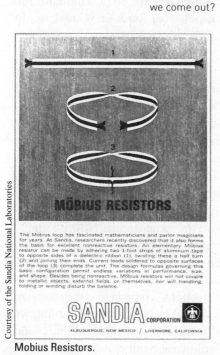

Courtesy of the Sandia National Laboratories

MÖBIUS RESISTORS

The Möbius loop has fascinated mathematicians and parlor magicians for years. At Sandia, researchers recently discovered that it also forms the basis for excellent nonreactive resistors. An elementary Möbius resistor can be made by adhering two 1-foot strips of aluminum tape to opposite sides of a dielectric ribbon (1), twisting these a half turn (2) and joining their ends. Current leads soldered to opposite surfaces of the loop (3) complete the unit. The design formulas governing this basic configuration permit endless variations in performance, size, and shape. Besides being nonreactive, Möbius resistors will not couple to metallic objects, external fields, or themselves, nor will handling, folding or winding disturb the balance.

SANDIA CORPORATION

ALBUQUERQUE, NEW MEXICO / LIVERMORE, CALIFORNIA

Mobius Resistors.

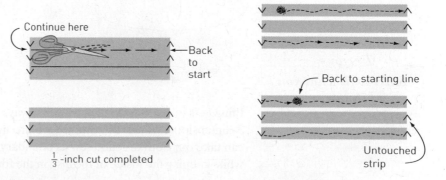

Continue here

Back to start

$\frac{1}{3}$-inch cut completed

Back to starting line

Untouched strip

Möbius Bands Abound

Although Möbius bands have a great many wondrous properties that may surprise us, it turns out that, in our everyday world, we are surrounded by Möbius bands. Where? Just look at any recycling symbol. Why do you think the standard logo is a Möbius band?

Möbius bands appear in some surprising places. Factories sometimes use Möbius bands as conveyor belts. Some conveyor belts have a half twist underneath them so that the belt wears evenly on both "sides" at the same time. This twist extends the life of the belt. In fact, such a Möbius belt was patented by the B.F. Goodrich Company.

One-Sided Surfaces

The mysterious features of the Möbius band arise from its single-sidedness, but, of course, the Möbius band has an edge to it. Is it possible to construct a surface that has only one side and no edge at all? A sphere has no edge, but if we walk on one side of the surface, we can never get to the other side. Now we are asking if there is a surface without any edges such that we can start at one point, walk along the surface, and arrive at the opposite side of our starting point—just as we did on the Möbius band. Think about this disorienting question before moving on.

She can never get to the inside from the outside; he can never get to the outside from the inside. The sphere has no edges but two sides!

The answer is yes. One strategy for constructing such a surface would be to start with a Möbius band and somehow get rid of its edge. Since it is not clear how to accomplish this task, we examine similar, related situations in the hopes of discovering a useful idea.

Suppose we have a sphere with its northern polar cap removed. The sphere with its northern polar cap removed has a boundary, which we'll call the Arctic Circle. Also, the polar cap itself has a circle boundary. We could eliminate the arctic circle boundary by putting the northern polar cap back on. Could we do something similar to eliminate the boundary of the Möbius band, thus creating a surface with no boundary?

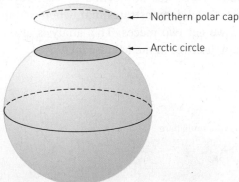

Cloth disk

Sewing

Eventually we get stuck.

Northern polar cap

Arctic circle

Well, let's try to add a cap as we did on the Arctic Circle. Recall that the Möbius band has only one edge. Let's take a distortable disk (a truly floppy disk) and attach the disk to the Möbius band along the edge of the Möbius band. To make it physical, let's suppose we have a cloth Möbius band and a cloth disk whose boundaries have the same lengths. Now we simply start sewing the edge of the disk along the edge of the Möbius band. We do fine for a while, but eventually something gets in our way. In fact, we will always get stuck if we remain in our pedestrian 3-dimensional world. If, however, we are more liberal in our choice of habitat, we can take our cloth disk and sew its boundary on to the boundary of the Möbius band while keeping the inside of the disk in the fourth dimension out of the 3-dimensional

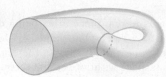

Klein bottle

space in which the Möbius band lives. This construction does eliminate the boundary of the Möbius band. However, this surface, known as the *projective plane*, is a bit unsatisfying in that it is too abstract and hard to visualize in our mind's eye. Is there a one-sided surface that is easier to visualize?

The Klein Bottle

Although we do not justify it here, the fact is that no one-sided surface without an edge can be constructed entirely in 3-dimensional space. Nevertheless, we can effectively describe an elegant one-sided surface known as the *Klein bottle*. The rules for constructing a Klein bottle are simple, and the resulting surface can be attractively modeled. To construct the Klein bottle, take a rectangle and glue one pair of opposite sides together without a twist, creating a tube. Next glue the other pair of opposite sides to each other, putting in a half twist as in the Möbius band.

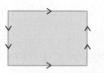 Gluing . . .

Arrows don't
match up!

Move inside so
that arrows line up.

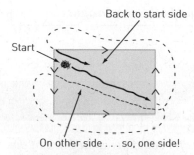

Start

Back to start side

On other side . . . so, one side!

When we follow the directions for constructing a Klein bottle, we soon discover that the Klein bottle passes through itself. We can construct a non–self-intersecting Klein bottle in the fourth dimension, or we can construct the Klein bottle in three dimensions having one circle-shaped self-intersection. The neat property of a Klein bottle is that its inside is the same as its outside! In the rectangular-edge matching model, we can see the one-sidedness by just following a path over the twisted edge and seeing that we switch sides. Alternatively, we can look at the 3-dimensional model and trace our finger lengthwise along the tube and see that, when we continue around, we find ourselves on the inside. What would happen if you tried to fill a Klein bottle with your favorite beverage? Is it inside? Is it outside? Cheers!

Besides the amazing hidden secrets we uncovered in the mysterious Möbius band and the convoluted Klein bottle, we also saw the value of the diagrams with the arrows. They enabled us to create, understand, and explain phenomena that first appeared to be mystifying. In fact, similar diagrams can be used to describe and analyze any surface.

The Klein bottle has many interesting mathematical properties, and it is also an especially intriguing work of art. Artists have created graceful renditions of the Klein bottle out of stone, glass, and metal. Perhaps this section will inspire you to make a Klein bottle of your own.

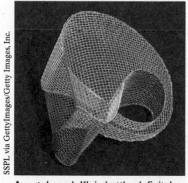

A metal-mesh Klein bottle: definitely
can't hold liquid!

A crystal Klein bottle

A Look BACK

THE MÖBIUS BAND has one edge, one side, and remains in one piece after being cut lengthwise down the middle. It demonstrates that a surface can be one-sided. A Klein bottle is a surface without an edge that also has only one side; however, the Klein bottle cannot be constructed entirely in 3-dimensional space without passing through itself.

To better understand the Möbius band and the Klein bottle, it is useful to represent those surfaces as a construction model with construction rules. Both the band and the bottle can be described as a rectangle or a square with edges glued together according to the assembly instructions. Examining deconstructed versions of the Möbius band and the Klein bottle lets us visualize relationships that may be difficult to see in the assembled versions.

Often what first appears surprising and mysterious can be explained clearly by thinking about the situation in a completely different manner. By doing so, we see another side to the issue, and we begin to build greater understanding and deeper insights.

Life Lessons

Build physical models and perform experiments as a way to make new discoveries.

•

Look for alternative means of describing a situation to help explain surprising outcomes and build new insights.

MINDSCAPES Invitations to Further Thought

In this section, Mindscapes marked (**H**) *have hints for solutions at the back of the book. Mindscapes marked* (**ExH**) *have expanded hints at the back of the book. Mindscapes marked* (**S**) *have solutions.*

Developing Ideas

1. **One side to every story.** What is a Möbius band?

2. **Maybe Möbius.** How can you look at a loop of paper and determine if it's a Möbius band?

3. **Singin' the blues.** Take an ordinary strip of white paper. It has two sides. Color one side blue and leave the other side white. Now use the strip to make a Möbius band. What happens to the blue side and the white side?

4. **Who's blue now?** Take an ordinary strip of white paper. It has two sides. Color one side blue and leave the other side white. Now give one end of the strip two half twists (also know as a full twist). Tape the ends together. Do you get a Möbius band?

5. **Twisted sister.** Your sister holds a strip of paper. She gives one end a half twist, then she gives the other end a half twist in the same direction, then she tapes the ends together. Does she get a Möbius band?

Solidifying Ideas

6. **Record reactions.** Explain to a friend how to make a Möbius band and how to cut it up in the various ways described in this section. Make your friend guess the outcomes before he or she does the cutting. Explain why they work as they do. Record your friend's reactions and thoughts.

7. **The unending proof.** Take a strip of paper and write on one side: "Möbius bands have only one side; in fact while." Next turn it over on its long edge and write "reading *The Heart of Mathematics*, I learned that." Now tape the strip to make a Möbius band. Read the band. This provides another proof that the Möbius band has only one side.

8. **Two twists.** Take a strip of paper, put two half twists in it, and glue the ends together. Cut it lengthwise along the center core line. What do you get? Can you explain why?

9. **Two twists again.** Take a strip of paper, put two half twists in it, and glue the ends together. Now cut it lengthwise while hugging the right edge. What do you get?

10. **Three twists (H).** Take a strip of paper, put three half twists in it, and glue the ends together. Cut it lengthwise along the center core line. What do you get? Find an interesting object hidden in all that tangle.

11. **Möbius length (S).** Use the method on page 346 and 347 for identifying edges of a Möbius band to find out how long a band we get when we cut the Möbius band lengthwise along the center line. Give the length in terms of the length of the original Möbius band.

12. **Möbius lengths.** Use the edge identification diagram of a Möbius band to find the lengths of the two bands we get when we cut the Möbius band by hugging the right edge. Give the lengths in terms of the length of the original Möbius band.

13. **Squash and cut.** Take a Möbius band and squash it flat on the table. Cut it like a buzz saw. What do you get?

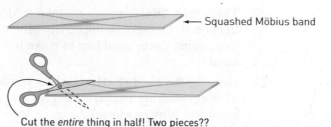

← Squashed Möbius band

Cut the *entire* thing in half! Two pieces??

14. **Two at once.** Take two strips of paper and put them on top of each other. Twist them together as though you were making a Möbius band, tape the tops together, then tape the bottoms. What do you have? What do you get if you cut it lengthwise down the center core if you keep the two bands together?

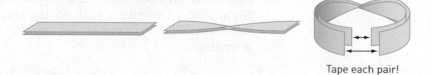

Tape each pair!

15. Parallel Möbius. Is it possible to have two Möbius bands of the same length situated parallel in space so that one hovers over the other exactly 1/8 inch away? Explain why or why not.

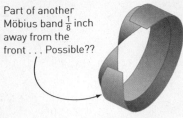

Part of another Möbius band $\frac{1}{8}$ inch away from the front . . . Possible??

16. Puzzling. Suppose you have a collection of jigsaw pieces as shown below. They can be put together to form a strip. Can they be assembled into a Möbius band? Can you explain why or why not?

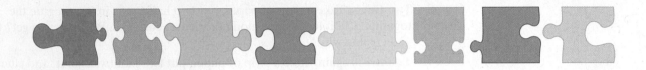

17. Möbius triangle. Make a 1-inch-wide Möbius band, lay it like a circle on a table, and carefully flatten it (thus making three folds). What shape do you have? What is the shortest flattened Möbius band that you can make? Please build one.

18. Thickened Möbius. Imagine a Möbius band thickened so the edge is as thick as the side. We'll call this a *thickened Möbius band*. How many edges does it have?

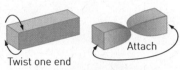

Twist one end   Attach

19. Thickened faces. How many faces (sides) does a thickened Möbius band have (see Mindscape 18)?

20. Thick then thin. Suppose we take a Möbius band, thicken it to make a thickened Möbius band (see Mindscape 18), and then shrink the original face to make it an edge. Is this new object a Möbius band?

21. Drawing the band (ExH). Imagine you have a Möbius band made of cloth with a drawstring around the edge. Can you draw the drawstring completely together? Why or why not?

22. Tubing (H). Suppose we take two Möbius bands and make a small hole in each. We then glue to each hole a tube connecting the two bands. How many edges does this new object have? How many sides?

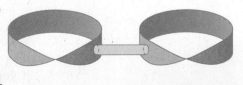

23. Bug out (ExH). Suppose you are a ladybug on the "outer" surface of the Klein bottle. Describe a path on the surface of the bottle that you can travel that would get you to the "inside" where a special gentleman bug is waiting.

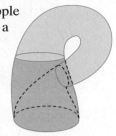

24. **Open cider.** Consider the Klein bottle half filled with apple cider, as illustrated. Describe how you could pour out a glass of cider without opening the bottle.

25. **Rubber Klein (S).** Suppose you have a rectangular sheet of rubber. Carefully illustrate how you would associate and then glue the edges of the sheet together to build a Klein bottle. Draw a sequence of pictures illustrating the construction.

Creating New Ideas

26. **One edge.** Using the method on page 347 for identifying edges of a Möbius band, prove that the Möbius band has only one edge.

27. **Twist of fate (S).** Using the edge-identification diagram of the Möbius band, prove that, when you cut a Möbius band lengthwise down the center, you have two half twists.

28. **Linked together.** Using the edge-identification diagram of the Möbius band, prove that, when you cut the Möbius band by hugging the right edge, the two pieces you get are interlocked.

29. **Count twists.** Using the edge-identification diagram of the Möbius band, determine the number of half twists each band has after you cut the Möbius band while hugging the right edge.

30. **Don't cross.** Can you draw a curve that does not intersect itself and that goes around a Möbius band three times without crossing over the edge? Experiment and explain your answer.

31. **Twisted up (H).** Suppose you are given a band of paper with a lot of twists in it. How can you tell without counting whether you have an even number of half twists or an odd number? Can you deduce a general fact about what you would have if there are an odd number of half twists and what you would have if there are an even number of half twists? Try. (*Hint.* Experiment by drawing on various physical models.)

32. **Klein cut.** Consider a rubber rectangular sheet with the edges associated as to create a Klein bottle. Suppose we make two cuts as indicated. How many pieces would we have? What would those pieces look like?

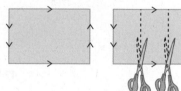

33. **Find a band.** Find a Möbius band on the surface of a Klein bottle. (The answer to the Rubber Klein story [Mindscape 25] may help.)

34. **Holy Klein.** Show that the figure on the left is equivalent to a Klein bottle with a hole.

35. **Möbius Möbius.** Show that the Klein bottle is two Möbius bands glued together on their edges.

Further Challenges

36. **Attaching tubes.** Consider a Möbius band with two small holes. Suppose we connect these holes with a tube. We could glue the tube to the holes in two

different ways, as illustrated. Are these pictures equivalent? (*Hint:* Slide one end of the tube around the band.)

37. **Möbius map (H).** Using felt-tip color pens that soak through both sides of the paper, draw a map on the Möbius band that has five countries, each of which touches the other four; that is, each and every country shares a border with the four other countries. Similarly, draw a map on the Möbius band that has six countries, each of which shares a border with the five other countries.

38. **Thick slices.** Thicken a Möbius band and then carefully cut (slice) the band along two adjacent sides straddling a common edge. How many pieces are you left with? If there is more than one piece, what is their relationship to one another?

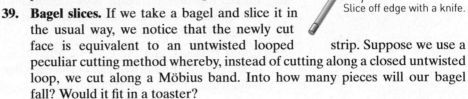

39. **Bagel slices.** If we take a bagel and slice it in the usual way, we notice that the newly cut face is equivalent to an untwisted looped strip. Suppose we use a peculiar cutting method whereby, instead of cutting along a closed untwisted loop, we cut along a Möbius band. Into how many pieces will our bagel fall? Would it fit in a toaster?

Slice off edge with a knife.

40. **Gluing and cutting.** Consider a rectangular sheet of rubber with its edges identified, as shown here. Suppose we make two cuts as indicated. How many objects would we be left with? Describe the objects.

In Your Own Words

41. **Personal perspectives.** Write a short essay describing the most interesting or surprising discovery you made in exploring the material in this section. If any material seemed puzzling or even unbelievable, address that as well. Explain why you chose the topics you did. Finally, comment on the aesthetics of the mathematics and ideas in this section.

42. **With a group of folks.** In a small group, discuss and work through the reasoning for how we get two interlocked strips when we cut the Möbius band 1/3 inch off the right edge. After your discussion, write a brief narrative describing the method in your own words.

43. **Creative writing.** Write an imaginative story (it can be humorous, dramatic, whatever you like) that involves or evokes the ideas of this section.

44. **Power beyond the mathematics.** Provide several real-life issues—ideally, from your own experience—that some of the strategies of thought presented in this section would effectively approach and resolve.

For the Algebra Lover

Here we celebrate the power of algebra as a powerful way of finding unknown quantities by naming them, of expressing infinitely many relationships and connections clearly and succinctly, and of uncovering pattern and structure.

45. **On the edge.** Suppose you make a Möbius band with a strip of paper x inches long. (Assume there is no overlap of the taped edges.) How long is the edge of your Möbius band? Suppose you cut your band in half down the middle as described in Experiment 3 in this section. How long is the edge now?

46. **Möbius triple.** You have three Möbius bands. One has an edge with length in inches equal to x. The other two have edges of length $2x + 3$ inches and $5x - 4$ inches. If the three lengths total to 60 inches, find the length of each edge.

47. **Möbius decorating.** Your dorm decides to decorate for Homecoming by creating oodles and oodles of colorful Möbius bands to hang from the light fixtures in the lounge. You create a total of 300 Möbius bands. Twenty percent are purple, 35% are gold, 25% are green, and the rest are evenly divided between red and blue. Find the number of Möbius bands of each color.

48. **Möbius prep (H).** You have a large rectangular sheet of paper, x inches by y inches, with y measuring the longer side. You want to cut this sheet into rectangular strips, ½ in by 6 inches, for making Möbius bands. You know that y is a multiple of 6 and x is a natural number. What values of x and y will allow you to create exactly 72 bands? (*Hint:* Feel free to use guess-and-check to determine your answer.)

49. **Painting Möbius.** A fraternity has a large canvas banner shaped like a Möbius band the entire surface of which is painted with the school colors. The band was created using canvas that measured 3 feet by 20 feet, with no overlap when the ends were stitched together. One year, the Board of Trustees decides to change the school colors to all new colors, so the frat boys have to repaint the entire band. How many square feet of canvas must be covered with paint?

5.3 KNOTS AND LINKS

Untangling Ropes and Rings

Alexander Cuts the Gordian Knot by Jean-Simon Berthelemy (1743–1811).

© World History Archive/Alamy Limited

> *God is like a skilful Geometrician.*
>
> **SIR THOMAS BROWNE**

When we think of flexible, bendable, and contortable objects, at some point we think of string. Loops of string can be knotted or looped together with other loops of string. This mundane, everyday observation actually leads to interesting discoveries with important consequences. Here we investigate the twisted world of knots and links and find that knots, at the most fundamental level, play a role in the creation of life itself.

Physical experimentation is an excellent way to become acquainted with the variety and intrigue of knots. As we hold and try to disentangle a loop of knotted string, we sometimes wonder whether we are making matters better or worse. This question leads us to the important idea of measuring the complexity of a position of the string. By deciding on some way to gauge the complexity of a knot, we can determine whether we are making progress toward disentangling it. The strategy of specifically measuring complexity is applicable broadly in and outside of mathematics. It allows us to tackle complicated issues a step at a time.

A Twisted DNA Tale

The fiber of life itself is encoded in strands of DNA entwined intricately in its famous double helix. These double strands herald life's message in the watery world of our cells. They jostle among molecular building blocks floating in their surroundings and attract pieces with their geometrical and magnetic charms.

At the supreme moment in the life of a DNA molecule, it begins its magical splitting. The ladder rungs split apart, and one strand peels away from the other, beginning

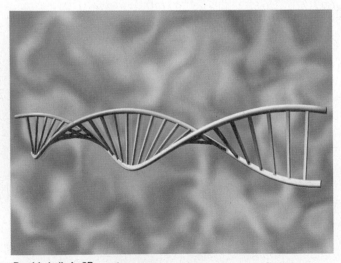

Double helix in 3D—put on your Heart of Mathematics 3D glasses to view.

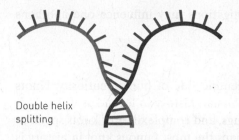

Double helix
splitting

DNA twisted
and knotted

Pulling the strands apart would
cause physically impossible spinning.

the wondrous process of self-replication. This image of the two connected sides of the ladder peeling apart to go independently off on their own to gather materials and ultimately create duplicates is amazing.

Unfortunately, it can't be true; the picture we just painted is impossible. Strands entwined in a double-helical fashion would be linked, tangled, and crumpled. For example, take two pieces of string and twist one around the other to form a double helix. Remember that the helix is twisted thousands of times. If we start at one end and attempt to pull the two strands apart, the other end would have to spin around thousands of times, and the complicated tangle would disrupt the process in midstream. The DNA strands would have to be strong and agile enough to untangle and untwist themselves, and chemists assure us that such acrobatics are beyond the physical abilities of DNA molecules. Some DNA molecules have their ends attached to one another, thereby forming a double-helical loop. In this case, the two strands are linked, and they would be impossible to pull apart without breaking some bonds.

What can we conclude? We are forced to conclude that, during the replication phase of DNA, the two long strands do not stay individually connected. They must frequently break apart, cross over the other side of the ladder, and reassemble themselves. The geometrical linking of the two sides of DNA requires this process of cutting, moving, and reconnecting during the process of splitting and separating. Bolstered by the mathematical certainty that a process of breaking and reconnecting must be happening, molecular biologists are seeking and have isolated enzymes whose special function is to allow unlinking. DNA and its reproductive habits provide a powerful, current application of the theory of knots and links.

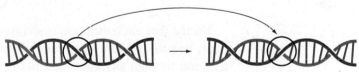

During replication, strands break and rejoin.

The theoretical study of knots and links has been useful to biologists in deducing more subtle conclusions about DNA as well. An enzyme acts on DNA locally, meaning that its influence is restricted to the part of the DNA strand that it touches. One application of the abstract study of knots emerged when Dewitt Sumners, a knot theorist, was able to prove in the abstract that only certain types of knots in DNA could result from the localized action of an enzyme. Substantial amounts of money were spent to do molecular biological experiments to show the same thing that Sumners's theoretical work implied without any laboratory experimentation. Such revelations have led scientists and knot theorists to

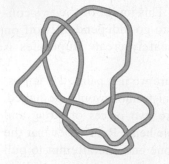

collaborate in the hopes that such investigations may influence one another's fields of inquiry.

The Gordian Knot

Knots have not been confined to the scientific side of human curiosity. Knots and links have long played their parts in human history in legend as well as science. Perhaps the intricate tangles, crossings, and complexities of knots speak to a gnarled core of the human psyche. Perhaps the most famous knot in history is the Gordian knot.

The Gordian knot held a mythical sway over the Phrygian people in the 4th century BCE. "The one who can untie the Gordian knot will be lord of all Asia," the legend maintained. When Alexander the Great arrived in Phrygia and was confronted with the fiendish knot, he took his great sword and cut the knot in two, thereby fulfilling the prophecy.

Had he studied knot theory instead of conquering the known world, he might well have saved his edge and been able to untie the knot without breaking the spirit of the unknotting legend—a breach of etiquette that, no doubt, weighed heavily on his mind. Although we may never be asked to untie a knot to become king of the known world, we remain fascinated by knots, which play a significant role in the structure of life.

Gordian knot

Knots We May Know

We begin this discussion with pictures of some knots we may know.

Unknot   Trefoil knot   Figure-eight knot   Square knot   Granny knot

Notice that each example is a closed curve rather than a strand having free ends, because a rope with free ends can always be untied simply by pushing a free end through any knottiness in the string. So, although we talk about taking a length of string and putting a knot in it, in the context of knot theory, unless we glue the free ends together, we have not actually created a real knot.

Let's return to the pictured knots. The first of these is the simplest knot of all. It is a round circle, and its only problem is that it is not really knotted. In fact, it is unknotted. Mathematicians no doubt thought long and hard about what to call that circle in the context of knot theory. After deep consideration, it was decided that any unknotted curve would henceforth be referred to as the *unknot*. Although a round circle is an unknot, more complicated curves that can be untangled to become a round circle are also examples of the unknot. For example, we can untangle the curve at the lower right to see that it is in fact the unknot in disguise.

Once we have gotten beyond the unknot, we can start looking at some knotted knots, the simplest of which has three crossings and three lobes when pictured.

Push through   Not a real knot

A free end means "not knotted."

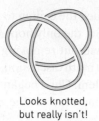

Looks knotted,
but really isn't!

The 3's involved give it its name, the *trefoil knot*. No matter how hard we try to move the trefoil knot around, we cannot just bend and stretch it, without cutting it, to make it become the unknot. It is genuinely knotted.

The next knots in the sequence of pictures are the *figure-eight knot*, the *square knot*, and the *granny knot*. The figure-eight knot is called the figure-eight knot because it can be drawn to look somewhat like the numeral 8. The square knot is called a square knot because it can be drawn to look somewhat square. The granny knot is called a granny knot because it apparently looks somewhat like someone's granny. These knots are relatively simple in the sense that each can be drawn with just a few crossings. However, more complicated knots can require any number of crossings.

An Unsolved Knotty Challenge

An unsolved challenge is to distinguish one knot from another. Suppose we see a photograph of a complicated knot. Could we describe a process for rearranging the knot to put it in a position in which a new photograph of it has the smallest possible number of crossings? In particular, if it is the unknot, could we devise a strategy for disentangling it to make it look like the circle? No one has yet found an effective way to answer these questions in general. Perhaps we could think of some way to count crossings or count the pattern of overs and unders that could tell us definitely whether a pictured knot is actually the unknot. No one knows how to do that yet. Many questions about knots have yet to be answered, but other facts about knots are known and some are quite surprising. We consider one such insight in the following story.

Not Math to Knot Math

The road to mathematical understanding is not without its major impasses and potholes. This story is about a student who hit one pothole too many.

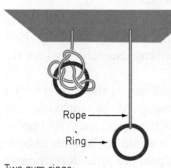

Rope →

Ring →

Two gym rings:
one tangled, one not.

Chris was a conscientious student in a course that used this book as a text. Sure, Chris wasn't a real math fan (yet), but Chris would read *The Heart of Mathematics: An invitation to effective thinking* and would try to work through the assigned homework. The text was not as bad as Chris had expected. Yes, some of the corny jokes and comments were slightly lame, but it made for some unusual reading. Chris's mind was spinning with notions of prime numbers, different sizes of infinities, the fourth dimension, and the Klein bottle. Chris was experiencing mathematical thinking. Then there was Chris's roommate, Pat.

Pat was in the same math course as Chris but gave up and used the text to prop up the CD player. Pat had lots of interesting tattoos and was pierced in a variety of unusual places (including Detroit). Pat's mind was spinning, but sadly the spinning was not caused by new intellectual ideas. Pat could not see the practical value of discussing, arguing, and debating various counterintuitive mathematical notions with friends. Pat, who was unable to participate in these lively gatherings (remember, Pat's text remained under the CD player, unread), decided instead to go to the gym to work out on the rings.

There Pat was faced with a practical problem: Although one ring hung straight and free, the other rope had become hopelessly knotted and entangled through the ring as it hung from the high ceiling. Pat could not reach the ceiling to remove the snared rope, unknot, untangle, and then reattach it. So, the only hope of practicing the iron cross was to untangle the knotted rope without removing it from the ceiling. At this moment, Chris, having finished a lively math discussion, walked in and was inspired to help Pat work out this knotty problem. Chris began encouraging Pat by yelling out some of the life lessons Chris had read in *The Heart of Mathematics*:

Life Lessons

Experiment to discover new insights.

•

Measure the complexity of a problem.

•

Never give up.

Pat thought a bit and saw that one of the most obvious measures of the complexity of this situation was the number of crossings appearing in the rope on its tangled way to the ceiling. If one could always reduce the number of crossings by one, then, Pat realized, it would be possible to untangle the rope; because, if Pat could reduce the number of crossings first by one, then Pat could further reduce the new number by one more and so on until there were no crossings left. After untangling the rope without removing the rope from the roof or the ring from the rope, the rings would hang straight. Pat was so happy to discover such a creative idea that Pat felt uplifted, returned to the math course, removed the text from under the CD player, read it, aced the course, got a great job, and got a new tattoo that read "Math Rocks."

Untangling a Knotted Rope—Reducing Complexity

Life Lesson

Reduce the complexity of a problem.

Upon first inspection, it may appear that one couldn't unknot and unlink the tangled and knotted rope without removing it from the ceiling. However, Pat's approach is effective, and it works often in many settings: Namely, measure the complexity of a situation and work to reduce it.

Suppose we have a complicated tangle such as the one in the story. Further suppose that a way exists to remove one of the crossings of the tangle and that this method does not depend on the idiosyncrasies of the tangle itself. By removing this crossing, we would have another tangle, but it would not be quite as complicated, because it has one less crossing. If we repeat this procedure, then we reduce the number of crossings again by one. Repeating this process until there are no crossings left leaves us with an untangled rope with a loop.

How do we know we can always remove a crossing from our tangled and knotted rope? Start where the rope is attached to the ceiling and move down toward the ring. Find the first place where that hanging rope crosses over or under some other part of the rope or the ring. The crossing is made of two parts: the part on the rope from the ceiling and the other part that crosses it. Notice that we can remove that crossing by pulling that crossing piece up toward the ceiling and then all the way around the bottom of the tangled mess. Now the rope has one less crossing. We repeat. We find the first crossing we come to as we proceed down the rope from the ceiling and remove that crossing in the same manner as before. We repeat again and again; this leads us to an untangled and unknotted rope. In the Mindscapes you'll have the chance to physically try this with some string. You will discover that it is actually easier to do by just playing with the string rather than by strictly following this procedure.

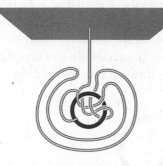

Links and Chains

We all know what a "chain" is: It is an object that is constructed from some number of closed loops that may be knotted either individually or about one another. What people may not know is that mathematicians call such a collection of loops a *link*. A trivial example of a link is a collection of round circles that do not interact at all. This link is boring; it's called the *unlink*.

The simplest example of a link where some real linking goes on is the two-component link that looks like two pieces of a chain. Notice that each piece is an

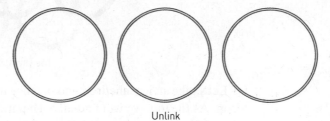

Unlink

unknotted curve, but we cannot disassemble the links to become a collection of two circles that do not interact. This example is a genuine link, but links among curves are not always so straightforward and present us with many challenges.

The Linking Challenge

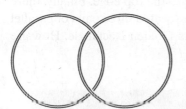

Is it possible to link three rings together in such a manner that they are indeed linked, yet if we remove any one of the rings, the other two remaining rings become unlinked? It seems that the answer must be no. Consider, for example, the link shown here. The three circles are linked. If we remove the middle circle, the other two become unlinked; however, if we remove the right or left circle, the other two circles remain linked. What we want to know is whether we can link the three rings so that the removal of *any* (not just a particular) ring leads to unlinking of the remaining two. The answer seems to be *no*. Do you agree?

The Borromean Rings—They Either Hang Together or They Don't Hang at All

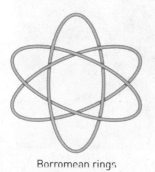

Borromean rings

The answer to the linking challenge is "Yes, there is a way to link three rings such that the removal of any one will automatically unlink the other two." An example of such a linking is known as the *Borromean rings*. The Borromean rings slightly resemble the Olympic symbol. Notice that, if we pretend that the red ring is gone, then the blue ring is completely behind the green ring. Therefore,

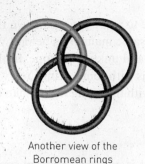

Another view of the
Borromean rings

the blue ring and the green ring are unlinked if the red ring is removed. Similarly, notice that we get unlinking if the blue ring is removed and also if the green ring is removed.

The delicate linking of the Borromean rings turns out to be a natural wonder. The Borromean rings are not just some abstract idea of the mind; here nature provides us with a tactile example of this intriguing pattern of linking in the regular solids—in particular, the icosahedron, the 20-sided solid. Recall that, if we take two edges of the icosahedron that are parallel and directly across from each other, they form two sides of a Golden Rectangle. The Golden Rectangle is constructed simply by using the parallel sides as the short sides of the rectangle and completing the rectangle with line segments that go directly through the icosahedron.

The Olympic rings

Let's take an icosahedron and balance it on a table on one edge of the icosahedron. At the top, we will find an edge parallel to the bottom. Also, there is a pair of horizontal edges exactly halfway from the table to the top edge. Finally, there is a third pair of edges that are vertical. So, we have located three pairs of parallel edges. Now let's complete each pair of edges into a Golden Rectangle. How are these three related? Please look at them.

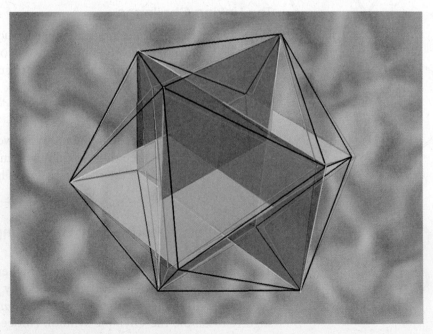

3D version of the icosahedron with golden Borromean rings

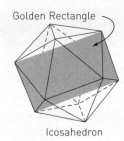

Golden Rectangle

Icosahedron

The edges of the three rectangles form a set of Borromean rings. The icosahedron has in it a natural set of three mutually perpendicular Golden Rectangles whose edges form the Borromean rings. Three ideas—the Golden Rectangle, the Platonic solids, and the Borromean rings—all come together in the icosahedron; they exemplify the interconnectedness and beauty of our geometric universe.

Other Knotty and Linking Challenges

The Borromean rings are three closed curves such that the removal of any one curve allows the remaining two to be separated from each other. Is it possible to construct a set of four, five, or more closed curves such that as a group they are all linked, and yet, if we remove any one curve from the group, the remaining curves all become unlinked? Try it in the Mindscapes.

Knot theory is a vast subject, full of challenging, unsolved problems. We end this section by repeating the most basic question in knot theory that remains unanswered. We hope you will enjoy turning your mind toward this simple-sounding question that has stumped mathematicians for many years.

Unanswered Question

Given two pictures of knots, is there a practical procedure, for example, a computer program, that would tell us whether one knot could be deformed into the other?

A Look BACK

THERE ARE MANY EXAMPLES OF FLEXIBLE, elastic knots and links in the natural world. DNA molecules with their twisted, linked structure demonstrate that studying knots and links is important for understanding ourselves. Knots and links hold many surprises. A tangled gymnast's ring can be untangled, but the Borromean rings cannot. The Borromean rings form a codependent group. Together they cannot be disentangled; yet after any one of them has been removed they become independent—that is, they completely fall apart.

Our basic strategy for dealing with knots and links is first to experience them directly with physical models and then to think about how to measure their complexity. To untangle a tangled gymnast's ring we measure its complexity by counting crossings and then demonstrate a procedure for removing a crossing. By repeatedly simplifying the tangled rope a little at a time, we can eventually untangle it altogether.

The technique of incremental solution allows us to deal with complex situations. Instead of being frozen into inaction by the sheer size and difficulty of a problem, we can look for a way to measure progress step by step. Small, but definite, forward motion, repeatedly applied, will get us anywhere.

Life Lessons

Get direct experience.

•

Measure complexity.

•

Reduce complexity.

MINDSCAPES · Invitations to Further Thought

*In this section, Mindscapes marked **(H)** have hints for solutions at the back of the book. Mindscapes marked **(ExH)** have expanded hints in the back of the book. Mindscapes marked **(S)** have solutions.*

Developing Ideas

1. **Knotty start.** Which of the following knots are mathematical knots?

2. **The not knot.** What is the unknot?

3. **Crossing count.** Count the crossings in each knot below.

4. **Tangled up.** Is the figure below a knot or a link?

5. **Ringing endorsement.** What are the Borromean rings?

Solidifying Ideas

6. **Human knots.** Find four friends. Have the five of you join hands as shown. Without unhanding anyone, have the group move around and attempt to unknot. Once the group has tried in earnest, have everyone regroup and take their original positions. Now, switch two crossings, as shown, and try to unknot. Write about the event and your findings.

   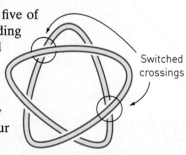

   Switched crossings

   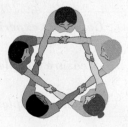

7. **Human trefoil.** What is the minimum number of people you need to make a human trefoil knot? Get some close friends and try it. Draw a diagram showing the human knot.

8. **Human figure eight.** What is the minimum number of people you need to make a human figure-eight knot? Get some close friends and try it. Draw a diagram showing the human knot.

9. **Stick number (ExH).** What is the smallest number of sticks you need to make a trefoil knot? (Bending sticks here is not allowed.)

10. **More Möbius.** Make a Möbius band with three half twists. Cut it lengthwise along the center core. What kind of knot have you made?

11. **Slinky (H).** Take a Slinky, lengthen one of its ends, and push it through and attach it to the other end. Is that a knot?

12. **More slink.** Take a Slinky, and this time weave an end up and down through the twists and then attach it to the other end. Is this knotted?

13. **Make it.** Use a piece of string or an extension cord to determine if this illustration is a knot or the unknot.

14. **Knotted (S).** Take an unknotted loop. Tie a knot in the middle as illustrated. Is this now a knotted loop? Explain why or why not.

15. **Slip.** Take an unknotted loop and put a *slip knot* on it: Why does the knot go away without cutting? Should it be called a *slip unknot*?

16. **Dollar link.** Take two paper clips and a dollar and fasten them as illustrated. Now pull the ends of the dollar so as to straighten it out. Report what happens to the paper clips.

17. **Knotted loop.** Take some string and put a loop at the end of it. Now knot the string through and around the loop as much as you want, and tie the unlooped end to a chair or table. Attempt to untangle the string without untying it from the chair. (*Hint:* It is easier to just play around with the knot than to follow the mathematical algorithm presented in the story.)

18. **Borromean knot.** Is it possible to switch some number of crossings in the picture of the Borromean rings on page 366 to make it a trefoil knot? If so, what is the minimum number of crossing switches that you need? A crossing switch just changes an undercrossing to an overcrossing or an overcrossing to an undercrossing.

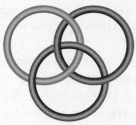

Another view of the
Borromean rings

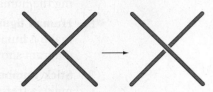

A crossing switch

19. Unknotting knots (H). In each of the two knots at right, unknot the knot by switching exactly one crossing. Show the crossing you switched and how the new object becomes the unknot.

20. Alternating. A picture of a knot is an *alternating picture* if as you follow along the string you see an undercrossing, then an overcrossing, then an undercrossing, and so forth until you return to where you started. Of the five simple knots below, which are alternating?

Unknot   Trefoil knot   Figure-eight knot   Square knot   Granny knot

21. Making it alternating. Consider the knot on the left. Notice that the crossings do not indicate if they are overcrossings or undercrossings. Select the crossings so that the picture of the knot is alternating.

22. Alternating unknot. Draw an alternating picture of the unknot that has at least two crossings.

23. One cross (H). Prove that any loop with exactly one crossing must be the unknot.

24. Two loops (S). Is there a picture of two linked loops that has exactly three crossings? Could all three crossings be created by one loop crossing the other?

25. Hold the phone. Disconnect the wire from the phone to the wall. Tangle it up and plug it back in. Can you untangle it without unplugging the wire?

Creating New Ideas

26. **More unknotting knots.** In these two knots, find the fewest number of crossing changes needed to create a new unknotted object. Illustrate the crossing changes and how the new loop is unknotted.

27. **Unknotting pictures (S).** Suppose you are given a picture of a complicated knot. Show that you can always change that knot to the unknot if you are allowed to switch as many crossings as you wish. How would you proceed systematically?

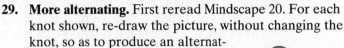

28. **Twisted.** Suppose we are given a figure consisting of two twisted strings that are connected by noninteracting curved arcs, as shown. What property of the twist would determine if it is a knot or a link? Experiment!

29. **More alternating.** First reread Mindscape 20. For each knot shown, re-draw the picture, without changing the knot, so as to produce an alternating drawing of the knot.

30. **Crossing numbers.** Suppose you are given pictures of two knots. If they have a different number of crossings, then must the knots be different knots? If so, explain; if not, provide different pictures of the same knot with different numbers of crossings.

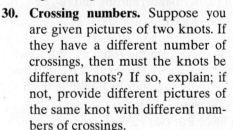

31. **Lots of crossings.** Suppose you are given a picture of a knot. Can you move the knot, without cutting, so that a picture of the new position has more than any specified number of crossings? If so, how; if not, why?

32. **Torus knots (H).** Can you draw a trefoil knot on a torus without having the loop cross itself? If so, draw a picture and then show that the loop you drew is a trefoil knot.

33. **Two crosses.** Prove that any loop with exactly two crossings must be the unknot.

34. **Hoop it up.** Show that every knot can be positioned such that all of it lies in the plane as disjoint curves except for some unknotted, semicircular hoops all the same size.

35. The switcheroo. Pictured below is a way of combining two knots by putting one after the other. Show that you can deform this combined knot so that the rightmost knot will now be on the left side.

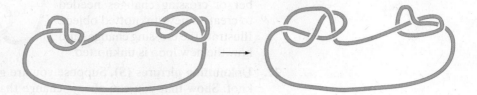

Further Challenges

36. 4D washout. Why is the study of knots and links vacuous if you allow yourself to move in 4-dimensional space?

37. Brunnian links (H). Link four loops together in such a way that they are all linked and yet the removal of any one loop will unlink all the others. This generalization of the Borromean rings is called the *Brunnian links*. Can you link five loops together in this fashion? Can you link any number together in this way? (*Hint:* The loops need not be round circles. Make the Borromean rings out of rubber bands, cutting and re-gluing one. Now move them around.)

38. Fire drill (ExH). A fire starts in your fifth-floor room. All you have available is a collection of slippery three-foot loops. They are strong, but short. You have a post in the room around which you can link one. How can you arrange the loops so that you can climb down? (The loops are too small and slippery to tie together by doubling them up and using them as if they were small pieces of rope.)

39. Bing links. Put two rubber bands together as pictured below (this is known as a *Bing link*). Now make another identical pair and put them around a pole with four long extensions, as pictured at right. Without removing them from the pole, maneuver the rubber bands so that each rubber band touches only one extension.

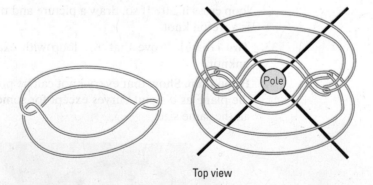

Top view

40. Fixed spheres again. We are given two spheres that are made of glass and thus are not distortable. They are rigidly fixed in space, one inside the other as illustrated, and they cannot be moved! The middle sphere floats miraculously in midair without any means of support. There is a rubber rope that hangs from the inside ceiling of the big sphere to the roof of the smaller sphere, as shown. The rope has a knot in it. Show that it is possible to unknot the rope without moving or distorting the two spheres. (This question appeared in Section 5.1, Mindscape 38.) Suppose now we add another straight rope as shown. Can you unknot the top rope without introducing a knot in the bottom rope?

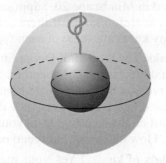 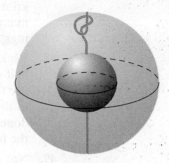

In Your Own Words

41. **Personal perspectives.** Write a short essay describing the most interesting or surprising discovery you made in exploring the material in this section. If any material seemed puzzling or even unbelievable, address that as well. Explain why you chose the topics you did. Finally, comment on the aesthetics of the mathematics and ideas in this section.

42. **With a group of folks.** In a small group, discuss and work through the reasoning for the solution of the tangled ring story. After your discussion, write a brief narrative describing the ideas in your own words.

43. **Creative writing.** Write an imaginative story (it can be humorous, dramatic, whatever you like) that involves or evokes the ideas of this section.

44. **Power beyond the mathematics.** Provide several real-life issues—ideally, from your own experience—that some of the strategies of thought presented in this section would effectively approach and resolve.

For the Algebra Lover

Here we celebrate the power of algebra as a powerful way of finding unknown quantities by naming them, of expressing infinitely many relationships and connections clearly and succinctly, and of uncovering pattern and structure.

45. **Tree house tangle.** Your little sister needs help tying knots in a rope at 1-foot intervals for climbing up to and down from her tree house. She has a thick rope 12 feet long that, when knotted properly, will reach exactly to the ground. She needs a foot of rope to tie to the floor of the tree house and eight inches of rope for each of the six climbing knots. What's the distance from the ground to the floor of the tree house?

46. **Box o' knots (H).** There is a box of knotted bunches of string in your math instructor's classroom. The number of trefoil knots in the box is x and the number of figure-eight knots is y. You take all the knots out of the box and display them so each looks like the trefoil or figure-eight knot illustrated in Mindscape 20. If there are a total of 21 knots and 74 crossings, how many knots are there of each type?

47. **New knots in the box.** Your math instructor has a new box of knots. The number of unknots in the box is x, the number of trefoil knots in the box is y, and the number of figure-eight knots is z. You take all the knots out of the box and display them so each looks like the unknot, trefoil or figure-eight knot illustrated in Mindscape 20. Suppose you also know there are twice as many unknots as trefoil knots. If there are a total of 36 knots and 72 crossings, how many knots are there of each type?

48. **Rinky dink links.** The local Half-Dollar Store sells cheap key rings (two for a half-dollar) that are easy to link together. You purchase a bunch of rings and use them – three at a time – to create a bunch of Borromean rings as shown in the tri-colored figure in the text (not the all-green figure). When you're done, you count all the crossings in all your sets of Borromean rings and find the total is 24. How much did you spend on rings at the Half-Dollar Store?

49. **Not another box of knots?!** Yes, your math instructor has yet another box of knots. This time, the number of unknots is x, the number of trefoil knots is y, the number of figure-eight knots is z, the number of square knots is w, and the number of granny knots is v. After carefully studying the figures of these knots in the text, write an expression that represents the total number of crossing that occur in all of the knots in the box when they are displayed like the figures in the text.

FIXED POINTS, HOT LOOPS, AND RAINY DAYS

How the Certainty of Fixed Points Implies Certain Weather Phenomena

© Fine Art Images/SuperStock

Starry Night (1889) by Vincent van Gogh.

> *What science can there be more noble, more excellent, more useful . . . than mathematics.*
>
> BENJAMIN FRANKLIN

Times are changing. Things are moving. The world is in flux. With all this movement, it is comforting to find settings where stability is absolutely required—not by statute or custom but by the higher law of mathematics. Here we'll examine various instances where things must remain fixed and consider some surprising applications and consequences regarding the climate on Earth.

Often the road to insight begins with intentional failure. For example, suppose we seriously attempt a truly impossible task such as constructing an example that contradicts a true theorem. We will fail. But looking carefully at why we fail can lead us to understand why the theorem is true. This strategy of intentional failure is a potent generator of ideas.

Sometimes the ideas are just small pieces of a larger puzzle. We can analyze what is happening there, or there, or there, but we do not have a global picture that incorporates all the information. We now need to step back and seek some organizing structure that puts the local information together to form a global whole.

The Rubber Break-Up Challenge

Consider two disks: One is red and one is blue, but they are identical in size. The only difference between the two disks, besides their color, is that the red one is made of extremely stretchable, shrinkable, and distortable rubber, whereas the blue one is made of rigid material, so it is unbending and cannot be distorted. We are handed these disks, the red on top of the blue. Since they are the exact same size, each point of the red disk touches one and only one point of the blue disk, and, similarly, each point of the blue disk touches one and only one point of the red disk. So, we have paired the points of the red disk with the points of the blue disk—each one has a partner. In other words, we have found a one-to-one correspondence between the points of the two disks.

Red point touches
blue point beneath it.

Of course, the world is in flux. Suppose the red points start to dance around. That is, suppose we now lift up the red disk, stretch, shrink, fold, bend, twist and otherwise distort it (but no cutting), and then place it back on top of the blue disk. The only rule is that the red disk cannot hang off the blue one. That is, the red disk must sit completely on the blue, but in any way, shape, or form that you wish. Plainly, we have mixed up and destroyed many original pairs of partners of points. In fact, there may be some blue points that have no red points above them (they are points without partners), and there may be some blue points that have several red points above them (those blue points have several partners). There may be other blue points that have exactly one red partner, but that partner may not be its original corresponding one.

Here we see a blue
point without a red partner.

A single pairing of a red and a blue point—but not an *original* pairing.

Here we see three red points on top of one blue . . . competition!

Here is a challenge: Can you find a distortion of the red disk so that, once it is placed on top of the blue disk, no red point is paired with its original blue partner, that is, each original pair of partners had a break-up? Try to distort the red disk in an attempt to have every point on the red disk move from its original position.

Some of Our Attempts

We thought we'd share a couple of our simple attempts at finding a solution to the preceding challenge. Our first try was to twist the red disk on the blue one like a record on a record player (check a history book for a description of a "record

Life Lesson
Discovering a global structure can be a considerable challenge but often is the key to true understanding.

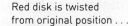

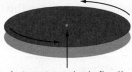

Red disk is twisted
from original position . . .

but center point is fixed!

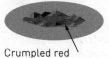

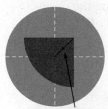

Rotate red disk
around here.

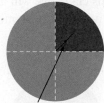

One of these four red
points at the midpoint
remained fixed!

player"). The problem is that the center point of the red disk would remain fixed, so the center pair of corresponding points would remain. Foiled!

Next we tried folding the red disk in half and then in half again to make a four-fold quarter piece. We then placed the piece smack-dab in the middle of the blue disk. Notice that the original center of the red disk (now the corner of the quarter piece) is not touching the center of the blue disk. Also, lots of points have been shuffled around. It may appear that we had met the challenge. But suppose we connect the center of the blue disk to the corner of the quarter red piece with a line segment and mark the midpoint of that segment. Because we folded the red disk twice, there are four red points below that midpoint. If we now rotate the quarter red piece around that midpoint, it matches up perfectly with the upper-right quarter of the blue disk and, therefore, one of the red points under that midpoint was never moved; it remained fixed. Thus, that pairing of points is an original pairing. Foiled again!

Crumpled red
mess . . . fixed points??

Finally, in frustration, we just crumpled up the red disk randomly and dropped it on the blue one. Certainly that must do it!?

It's Hard to Accomplish the Impossible

Give up? So did we. The annoying part is that the challenge sounded so darned doable. Amazingly enough, however, it is not. That is, it is impossible to stretch, rotate, distort, fold, and squash the red disk and place it back on top of the blue disk in such a manner that each point on the red disk moves. This surprising fact is a mathematical result known as the *Brouwer Fixed Point Theorem*.

The Brouwer Fixed Point Theorem.

Suppose two disks of the same size, one red and one blue, are initially placed so that the red disk is exactly on top of the blue disk. If the red disk is stretched, shrunken, rotated, folded, or distorted in any way without cutting and then placed back on top of the blue disk in such a manner that it does not hang off the blue disk, then there must be at least one point on the red disk that is fixed. That is, there must be at least one point on the red disk that is in the exact same position as it was when the red disk was originally on top of the blue one.

This result has many applications within the abstract mathematical realm of topology and other areas. In fact, recently scientists have been realizing that such a result has applications to human physiology. The Brouwer Fixed Point Theorem may help us understand the behavior of heart muscles and heart attacks.

Why is this theorem true? Since we have no idea what distorting we will do, we need to demonstrate that this property holds without knowing what the distortion looks like. This task sounds nearly impossible, but a clever idea saves the day. (*Caution*: The proof ahead is challenging.)

The Proof

In the beginning, we have one disk on top of another, and the points of the blue disk are paired up in a one-to-one manner with the points on the red disk. Each blue point gazes up and memorizes its special red partner point. As the red disk is stretched, bent, folded, rotated, compressed, and otherwise distorted, each blue point carefully watches the movements of its special red point. When the distorting process finally terminates and the mangled red disk is laid to rest on the blue, each blue point sees exactly where its special red partner is now located. We now wish to show that there *is* some red point that finds itself sitting exactly on top of its corresponding blue point. To verify this claim, we will assume the opposite—that is, we will now assume that each red point has been moved off its blue mate. We will now see this assumption is impossible.

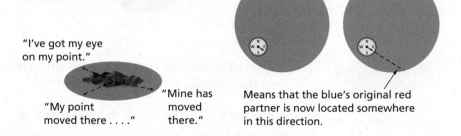

"I've got my eye on my point."

"My point moved there . . . ."

"Mine has moved there."

Means that the blue's original red partner is now located somewhere in this direction.

So strong is the attraction of each point on the blue disk toward its original red mate that each blue point has an invisible pull toward its red point's current location. This magnetic-like pull can be detected with a special compass. When we place this little compass at any particular point on the blue disk, it points in the direction of where its original red partner is now located. In fact, we may imagine placing a tiny compass on every single blue point; each compass would point in the direction of its corresponding red point. Notice that since we didn't cut the red disk, nearby blue points will have their compasses pointing in similar directions.

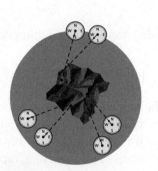

Nearby points have compasses pointing in roughly the same direction.

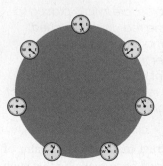

Arrows along the boundary circle always point into the blue disk.

Moving a Compass Along Circular Paths—The Winding Number

Let's consider what happens to the compass arrow as we move around the circular boundary of the blue disk. At each point around the boundary circle, the compass arrow points toward the destination of its associated red point. Of course, we don't know where that red point is; however, we do know that the destination is somewhere in the blue disk. Consequently, if we start at the top of the boundary circle and move once around the circle, the compass arrow will wind around exactly once. Following the same procedure, do this experiment for yourself. A good way to see it is first to just keep the compass arrow constantly pointing toward the middle of the circle. This experiment shows that the compass arrow circles once around the compass as we move around the boundary of the circle.

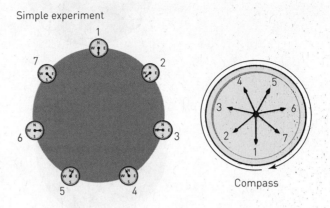

Simple experiment

Compass

Now what happens when we do the same process, this time on a circle that is not quite the boundary circle of the blue disk but is just barely inside it (see page 376)? Well, notice that at each position the compass arrow is pointing in almost exactly the same direction as it was pointing from the nearby point actually on the boundary. So, as we go around, we will still see the compass arrow circling exactly once. We call the number of times the compass arrow winds around as we travel along a circle the *winding number*. So, we see that, along the circular boundary of the blue disk and on circles just within the boundary, the winding numbers are all 1; the arrow circles once around the compass.

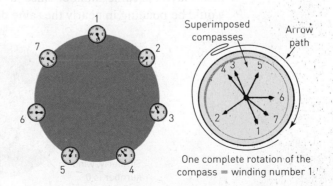

Superimposed compasses

Arrow path

One complete rotation of the compass = winding number 1.

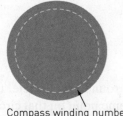

Compass winding number traveling around this slightly smaller circle??

Let's make another observation. If we take a tiny circle around the center of our blue disk, then, as we move around that tiny circle, all the destination points stay in almost the same location. Consequently, the compass arrow will be pointing in nearly the same direction for all the points in that tiny circle. So, in going around the tiny circle while pointing to the destination points, the compass arrow does not wind around the compass at all. It has winding number 0.

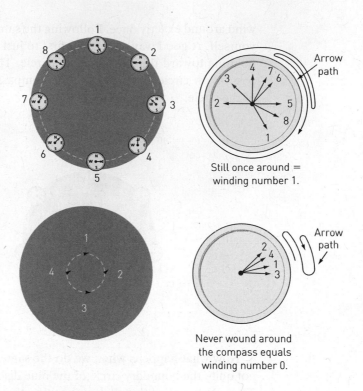

Still once around =
winding number 1.

Never wound around
the compass equals
winding number 0.

Various Winding Numbers

The question now is, what happens as we consider the winding numbers for the circles of different radii, all centered at the center of the blue disk, as we move from the boundary circle toward the center of the blue disk? Near the boundary, the winding number is 1. Near the center, the winding number is 0. So, somewhere along the way, the winding number had to change. But how is that possible? Whenever a particular circle has a certain winding number, all nearby circles must have the same number since, at each location nearby, the compass arrow would be pointing in nearly the same direction.

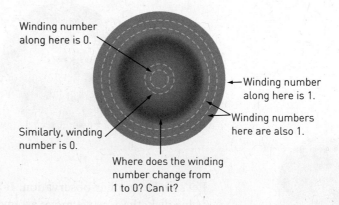

Winding number along here is 0.

Winding number along here is 1.

Winding numbers here are also 1.

Similarly, winding number is 0.

Where does the winding number change from 1 to 0? Can it?

The only way the winding number can change is if around some circle, the winding number *does not exist at all*. In other words, at some point on that circle, the compass does not know in which direction to point. That situation would occur only if the red point is exactly above its original blue point—if that original red/blue pair of points remained together. This romantic and happy ending demonstrates that there must be a fixed point, and our theorem is proven to be true.

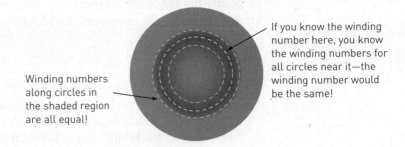

Winding numbers along circles in the shaded region are all equal!

If you know the winding number here, you know the winding numbers for all circles near it—the winding number would be the same!

Meteorological Fixed Points

We now illustrate the power of this circle of ideas with an application to the weather. The north and south poles are always cold; however, they are not always equally cold. Air pressure varies from time to time and place to place. These sentences just about exhaust our personal knowledge of meteorology. But we do know something about the weather that practicing meteorologists might not know.

The Meteorology Theorem.

At every instant, there are two diametrically opposite places on Earth with identical temperatures and identical barometric pressures.

Let's clarify this statement. At any instant in time, every location on Earth has both a temperature and a pressure. Every point on Earth has exactly one point that is diametrically opposite, or antipodal. The *Meteorology Theorem* says that, if we have a globe before us with the temperatures and the pressures written at each point, we can always find a pair of antipodal points on the globe where the temperature and pressure at one point are identical to the temperature and the pressure at the other point.

Although we will not give a proof of the *Meteorology Theorem*, the fundamental idea is to adopt the winding number argument used in the Brouwer Fixed Point Theorem. In trying to understand the *Meteorology Theorem*, however, the first question we may ask is, Forget the pressure—I have enough pressure as it is. Why is there a pair of antipodal points where the temperatures are the same? The answer to this more basic question is contained in the proof of the *Hot Loop Theorem.*

The Hot Loop Theorem.

If we have a circle of variably heated wire, then there is a pair of opposite points at which the temperatures are exactly the same.

Warm-Up Exercise

Before we read the proof of the Hot Loop Theorem, let's try to defeat the theorem by building a counterexample. Draw a circle and label the points around the circle with temperatures. Try to label it in such a way that no opposite pair of points has the same temperature. Do this before reading on, because your attempt provides the essence of the idea behind the result.

Result

When you undertake this challenge, you will learn several things.

1. You cannot label each point since there are infinitely many (in fact, uncountably many) points on a circle. So, you label just some of the points and assume that the temperatures between labeled points vary smoothly.

2. Temperatures vary continuously. That is, if it is 50° at one point, the temperature around that point must be *close* to 50°.

3. You will learn that you cannot defeat the Hot Loop Theorem.

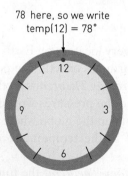

$x = 1.5$

$x' = 7.5$
opposite of x

78 here, so we write
temp(12) = 78°

The Proof of the Hot Loop Theorem

Let's label the points on the circle as a clock is labeled—that is, 12 at the top, then 1, 2, 3, and so on. Assume that points in between these numbers are labeled with decimals. For each point x on the circle, we'll call the opposite point x'. So, for example, if $x = 1.5$, then $x' = 7.5$.

For each point x on the circle, let us call the temperature at that point temp(x). In other words, temp(x) is the temperature at point x. For example, if the temperature at point 12 is 78°, then we would write temp(12) = 78°. So, we seek to locate a point x on the circle where temp(x) = temp(x') (the temperature at point x is equal to the temperature at the point opposite x).

We will start at 12 and consider the difference between the temperature there and the temperature at the opposite point—that is, temp(12) − temp(6). As we move around the clock, we consider temp(1) − temp(7), temp(2) − temp(8), and so on.

Notice what happens to this difference as we move around the circle. Suppose that temp(12) − temp(6) is a positive number. That means that temp(12) is bigger than temp(6). Points near 12 will still be hotter than points near 6. Perhaps, when we get to 1, temp(1) − temp(7) will still be positive; that is, the temperature at 1 is larger than the temperature at 7. Keep going. At 2, perhaps temp(2) − temp(8) is still positive; that is, temp(2) is greater than temp(8). Likewise, temp(3) might be larger than temp(9). Keep going. It might even be that temp(4) is larger than temp(10). Maybe temp(5) is still larger than temp(11).

But look out! Could temp(6) be larger than temp(12)? No! Remember that we started the whole discussion assuming that temp(12) is larger than temp(6). But, as we move halfway around the circle, we see that we are switching the order of subtraction. So, if temp(12) − temp(6) is positive, then temp(6) − temp(12) is negative.

Therefore, somewhere between 12 and 6 the difference must change from positive to negative, and, when it does, it is 0. At that point x, temp(x)

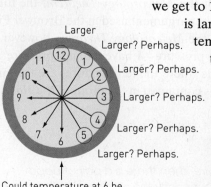

Larger
Larger? Perhaps.
Larger? Perhaps.
Larger? Perhaps.
Larger? Perhaps.
Larger? Perhaps.

Could temperature at 6 be
higher than temperature
at 12? Perhaps??

$= \text{temp}(x')$. That means that the temperature at x equals the temperature at the opposite point x', and we have established the validity of the Hot Loop Theorem!

From Loops to Spheres—The Meteorology Theorem

The Hot Loop Theorem tells us that there are lots of opposite (or antipodal) points on Earth where the temperatures are the same. Take any great circle at all—for example, the equator. There must be a pair of antipodal points on the equator that have the same temperature. Likewise, for every longitudinal circle there is a pair of opposite points that have the same temperature.

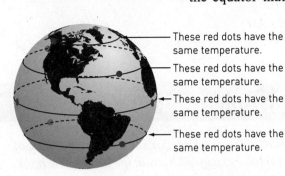

These red dots have the same temperature.

These red dots have the same temperature.

These red dots have the same temperature.

These red dots have the same temperature.

By the same reasoning, on the equator there is at least one pair of opposite points where the pressures are the same. So have we proved the Meteorology Theorem? No. A point on the equator whose temperature equals the temperature of the opposite point may not be a point whose pressure is the same as the opposite point's pressure. The temperatures and the pressures form independent sets of numbers around the equator. We require a slightly more elaborate line of reasoning to deduce the Meteorology Theorem, but we won't present that proof here.

We close this section with the secure feeling that even in an ever-changing world we are certain, using abstract mathematical ideas, that some things will remain the same.

A Look BACK

THE BROUWER FIXED POINT THEOREM reveals the surprising fact that no matter how we bend, stretch, or deform a disk and put the result within the bounds of where it started, there will always be at least one point that has not moved. The Hot Loop Theorem implies that there are always two opposite points on Earth that have identical temperatures or pressures.

The key to understanding these issues lies in examining the situation point by point but then reasoning about the whole. In the case of the Brouwer Fixed Point Theorem, we consider the winding numbers of a compass arrow as it travels from point to point around various circles. In the Hot Loop Theorem, we consider the differences in temperatures of opposite points. We can try to turn and fold a disk to contradict the Brouwer Fixed Point Theorem. We can try to put numbers on a circle in an attempt to contradict the Hot Loop Theorem. Our failures give us insights into why the theorems are true. These theorems are difficult to prove because they require us to put together many bits of information into a structured whole. In one case, we must look at how circles of points behave, and, in the other, we must look at opposite points on the loop.

Failures can be excellent teachers. If we believe something is true, a good way to prove it is to try hard to show that it is false and see where the arguments break down. Looking at the cause of a failure often points the way to success. Sometimes understanding a complicated situation requires that we gather and organize many small bits of information to see a global pattern. When we have mounds of microscopic information, we can step back, squint, and see whether the details coalesce into a coherent whole.

Life Lessons

> *Attempt impossible tasks and carefully observe the failure.*
>
> •
>
> *Learn from failed attempts.*
>
> •
>
> *Collect information at the micro level and analyze it at the global level—*
> *act locally, think globally.*

MINDSCAPES Invitations to Further Thought

In this section, Mindscapes marked (H) have hints for solutions at the back of the book. Mindscapes marked (ExH) have expanded hints at the back of the book. Mindscapes marked (S) have solutions.

Developing Ideas

1. **Fixed things first.** What does the Brouwer Fixed Point Theorem assert?

2. **Say cheese.** You're making an open-faced cheese sandwich. You are very pleased to discover that your perfectly square slice of processed American cheese product fits exactly over your perfectly square slice of bread. In a moment of inspiration, you decide your sandwich will be even more perfect if you pick up the slice of cheese, rotate it one-quarter of a turn, and then place it back down on the bread. Once you make this final adjustment, will there be a point on the cheese slice that is in exactly the same place it was before you rotated the cheese? If so, describe that point; if not, explain why not.

3. **Fixed flapjacks.** You're making pancakes and thinking, as always, of the Brouwer Fixed Point Theorem. You flip over one of your perfectly circular pancakes so that it falls back exactly onto its original place on the griddle. What point or points of the pancake are in the exact same position they were in before the flip?

4. **Pointing out opposites.** Look at the circle at right. For each colored point, use a pen of the same color to mark the corresponding antipodal point on the circle.

5. **Loop around.** What does the Hot Loop Theorem assert?

Solidifying Ideas

6. **Fixed on a square.** Does the Brouwer Fixed Point Theorem hold true if the disks are replaced by two square sheets? Explain why or why not.

7. **Fixed on a circle.** Does the Brouwer Fixed Point Theorem hold true if the disks are replaced by two circles (that is, just the boundaries of the disks without including the inside)? Explain why or why not.

8. **Winding curves (H).** In each drawing below is a disk with a marked green circle. The disk was then stretched, twisted, shrunk, and generally distorted and then restored to its original form. Of course, the green circle was stretched, twisted, shrunk, and generally distorted as well. We depict the green circle after the distortion by the analogously marked yellow curve. For each picture, compute the winding number of the yellow image as you travel once around the original green circle.

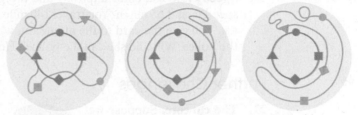

9. **Winding arrows.** In each drawing below we have a circle (equator) drawn with arrows as indicated. Compute the winding number of the arrows as you travel once around each of the equators.

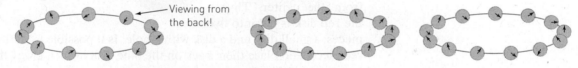

Viewing from the back!

10. **Under pressure (S).** Must there be two antipodal points on the equator with the same barometric pressure?

11. **Not the equator.** Must there be two opposite points with the same temperature on any small circle drawn on Earth?

12. **Home heating (H).** Prove that there are two points somewhere in your room that are exactly 5 feet apart and have *precisely* the same temperature.

13. **Polar populations.** Suppose at each point on Earth you count the number of people within 1000 miles of that point. Are there two antipodal points that have the same number of people within 1000 miles?

14. **Lighten up.** Must there exist two antipodal points on Earth where the amounts of sunlight (lightness) are identical?

15. **Shot disk (H).** Suppose we have two disks, one red and one blue, and we remove the center point from the red and place that punctured disk on top of the blue. If we now distort the red disk and place it back on the blue, must there be a point on the punctured red disk that remains fixed?

Creating New Ideas

16. **Lining up (H).** Suppose we have two line segments having identical length, one red and one blue. We begin with the two segments positioned one on top of the other. We then take the red segment, stretch, bend, shrink, and distort it with-

out cutting, and place it on the blue line so that it does not hang off at either end. Must there be a point on the red line that is in its original position?

17. **A nice temp.** Must there be two antipodal points on Earth where the temperature is 62°? Explain.

18. **Off center (ExH).** Suppose we have two disks, one red and one blue, and we remove one point from the red (not necessarily the center point) and place the punctured red disk on top of the blue. If we then distort the red disk and place it back on the blue, must there be a point on the punctured red disk that remains fixed?

19. **Diet drill.** Suppose someone weighs 160 lbs. and decides to go on a diet. After three months, the person weighs 149 lbs. Did this person weigh 154.5 lbs. after one and a half months? Did the person weigh 154.5 lbs. at some time within that three-month period? Explain your answer.

20. **Speedy (S).** You enter a tollway and are given a toll card at 12:00 noon. The speed limit is 65 m.p.h. You travel 140 miles on the tollway and then exit at 2:00 P.M. You give the card to the attendant, who looks at it and immediately calls the police. Why did she call the police? How can she prove you broke a law?

Further Challenges

21. **The cut core.** Suppose we have the red and blue disks as before. Now, however, on the red disk we cut out a small circle from the center. Thus, the red disk is cut into two 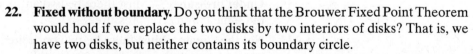 pieces: a small disk and a disk with a hole. Is it possible to deform these two red pieces and place them back on the blue disk in such a way that every red point is moved?

22. **Fixed without boundary.** Do you think that the Brouwer Fixed Point Theorem would hold if we replace the two disks by two interiors of disks? That is, we have two disks, but neither contains its boundary circle.

23. **Take a hike (ExH).** A hiker decides to climb up Mount Sanitas. There is only one trail to the top. He starts at the base of the mountain at 8:00 A.M. Saturday. He climbs, stops, rests, backtracks a bit, but finally gets to the summit by 5:00 P.M. that evening. The next morning at 8:00 a.m. he begins to hike down. Again he stops, rests, backtracks (in fact returns back to the top because he left his tent up there), but finally gets back to the bottom by 5:00 p.m. that evening. Must there exist a precise point (altitude) on the trail with the property that, at the very moment he crossed that point, his watch showed the exact same time going up as it did coming down? Carefully explain your answer.

In Your Own Words

24. **Personal perspectives.** Write a short essay describing the most interesting or surprising discovery you made in exploring the material in this section. If any material seemed puzzling or even unbelievable, address that as well. Explain why you chose the topics you did. Finally, comment on the aesthetics of the mathematics and ideas in this section.

25. **With a group of folks.** In a small group, discuss and work through the arguments involved in the proofs of the Brouwer Fixed Point Theorem and the Hot Loop Theorem. After your discussion, write a brief narrative describing the ideas in your own words.

For the Algebra Lover

Here we celebrate the power of algebra as a powerful way of finding unknown quantities by naming them, of expressing infinitely many relationships and connections clearly and succinctly, and of uncovering pattern and structure.

26. **Two temps.** At midnight last night, points A and B on the surface of Earth had exactly the same temperature and barometric pressure. If we treat the values of T and P as numbers without units, the values satisfy the equations $T + P = 91.80$ and $T - P = 32.20$. Find the values of T and P.

27. **Beat the drum circle.** Your university marching band has a tradition at the end of every football halftime show. They form a large circle in the field with the 50-yard line passing through the center of the circle, and then march counterclockwise while playing the school fight song. If the bass drummer starts on the 50-yard line, and the band marches 180 degrees around the circle, where does the bass drummer end up?

28. **Look before you leap.** There is a fire in your high-rise dorm one night. You have to jump from a fourth-story window onto a circular fire safety net. While contemplating your chances of surviving the fall, you notice that the firefighters holding the net are rotating—going clockwise and then counterclockwise. You become mesmerized watching the motion of one firefighter's red hat. It moves 1/4 of the way around the circle clockwise, then 1/3 of the way counterclockwise, then 1/2 of the way clockwise, then 1/8 of the way counterclockwise. How far has the firefighter in the red hat moved from his original position? Express your answer as a fraction of the distance around the circle clockwise or counterclockwise. (Historical aside: One of the most popular of the fire safety nets was called the *Browder life net* after Thomas Browder, who invented and patented the device in 1887. Wonder if he knew Felix Brouwer?)

29. **Finding y.** Each graph below corresponds to a function whose y values lies between 0 and 2. Which functions have an x value that gives a corresponding y value of 1? Which have an x value that gives a corresponding y value of 0.5?

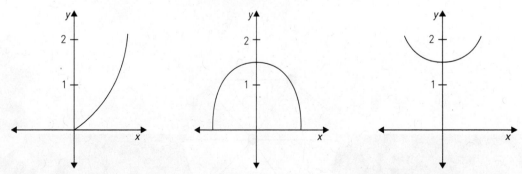

30. **For all y?** Look at the graphs for the previous Mindscape. Which functions have the property that you could pick any value, call it b, between 0 and 2 on the y-axis and find a value on the x-axis, call it a, so that the b is the value corresponding to a?

CHAPTER SIX

Modeling Our World through Graphs

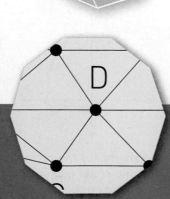

6.1 Circuit Training

6.2 Feeling Edgy?

> *Acting allows me to tell a lot of stories. You know, start at the beginning, finish at the end, and tell everything in between. Modeling is just an image.*
>
> **CAMERON DIAZ**

Connections are everything—it's often all about who you know and who knows you. Beyond personal connections, the Internet connects server to server around the globe; emails connect senders and receivers; airline route maps depict connections between cities; electronic circuits connect nodes and components; even to chart the spread of a contagious disease we might need to show when an infected person has contact with others. When we strip away all the extraneous issues, each of the previous scenarios involve objects and connections between pairs of those objects, whether the objects are computers, cities, electronic components, or even germ-infested individuals. Many familiar situations can be better understood and resolved if we focus on the fundamentals. In many cases, the most basic features consist of a collection of objects together with a list of which pairs are connected and which pairs are not. When we draw diagrams highlighting these connections, we find structure and patterns that are at once simple and subtle. We also find we can answer some intriguing and practical questions by modeling complicated situations through focusing on objects and connections. With a solid model in place to explore, the insights are inevitable.

The first scenario we will explore involves a city with several bridges joining four landmasses. By mentally shrinking the massive landmasses to tiny points, and by letting curves between those points represent the corresponding bridges, we produce a line drawing that clearly reveals fundamental connections—such a drawing is known as a *graph*. This drawing allows us to answer questions about the ways in which we could travel along the various bridges in the city, leading to an important result with applications from mail delivery to snow plowing and everything in between.

Analyzing these simple drawings of points and curves in other ways reveals surprising and otherwise hidden relationships. Just *counting* the numbers of points, curves, and regions in a drawing has implications ranging from showing that these can't be more than five regular solids to showing which computer circuits can be printed on a single-layer chip. This chapter not only celebrates some important modern ideas of mathematics and the power of mathematical modeling, but also one of the greatest lessons of mathematical thinking: focus on the essential.

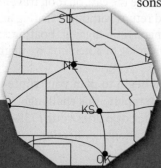

6.3 Plane Old Graphs

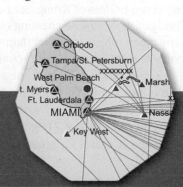

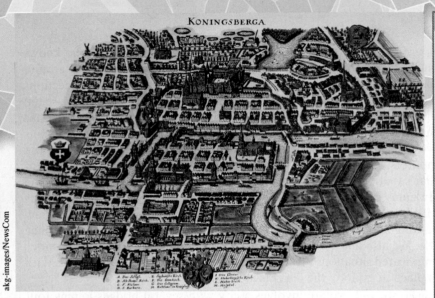

The seven bridges of Königsberg.

> *"A problem was posed to me about an island in the city of Königsberg, surrounded by a river spanned by seven bridges, and I was asked whether someone could traverse the separate bridges in a connected walk in such a way that each bridge is crossed only once. I was informed that hitherto no one had demonstrated the possibility of doing this, or shown that it is impossible."*

LEONHARD EULER IN A LETTER TO GIOVANNI MARINONI

You were there—the year: 1735; the place: Königsberg, Prussia. Königsberg was *the* place to be seen in 18th century Europe and thus a popular Spring Break destination and the filming locale for the video series *Prussians Gone Wild*. Actually, the island in the middle of the city was called Kneiphof, which translates to "pub yard"—yes, every hour was a "happy" one. As our story opens, you find yourself sitting in a Sternbucks coffee house, nursing a Venti Mocha Espresso Cappucino with a mint cinnamon twist and pondering the challenge of the day, which is well-known since it is written on a sandwich board outside the Sternbucks. It offers a prize of two wienerschnitzels and a dachshund for anyone who can leave the café on foot, traverse all seven bridges of Königsberg exactly once each, and return without revisiting any bridges or otherwise cheating (that includes swimming, flying, digging, boating, or any means of motion other than walking). Given the meaning of Kneiphof, we might view this challenge as a *pub-crawl*.

You spread a map of Königsberg on the table and begin to plan your route in the hopes of winning the valuable prizes. As you contemplate this challenge, you soon find yourself discovering ideas about connections. Little does your 1735-self imagine that by the 21st century those identical ideas about connections would have led to riches far greater than wiener dogs.

The Königsberg Bridge Puzzle
Connections

An excellent method to generate important new ideas is to start with a specific question, discover an answer, and then isolate new concepts from your solution. Those new concepts often can be applied to many situations far beyond the original question. One of the most famous mathematicians of all time, Leonhard Euler (pronounced "*oiler*," reminding us of his many *slick* mathematical discoveries), used that exact strategy on the Königsberg Bridge Puzzle. In solving that one puzzle, he not only found a method to solve any similar question that involved any configuration of bridges, but his solution also opened the door to an exploration of connections which became the mathematical subject now known as *graph theory*.

Again imagine yourself with your map of Königsberg as you turn your mind toward finding a route over all the bridges. Notice the essential features of the city—the rivers, the landmasses, and the bridges. In planning your route, one of the most important steps is to realize that most of the details on the map are irrelevant to the particular puzzle at hand. So let's take the important step of ignoring irrelevant features so that we focus on the essential features of our situation.

The original map shows many roads and buildings in the city, but for the purpose of determining whether we can find a route that traverses each bridge exactly once, the roads and buildings on any particular landmass are irrelevant. So let's simply ignore those features of the map and generate a simpler map.

Life Lesson

Build from the concrete.

Life Lesson

Ignore the irrelevant; focus on the essential.

Life Lesson

Simplify whenever possible.

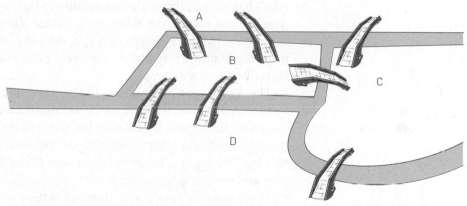

The shapes of the landmasses don't matter either, so we can simply replace each landmass by a dot. In fact, the only relevant features are the connections between the landmasses. That is, we focus on which landmass is connected to which other one by a bridge. Each bridge creates a connection between two landmasses.

Now we are well on our way to isolating the essential features of the Königsberg Bridge Puzzle. We have a collection of places (landmasses now denoted as dots) and some pairs of them are connected (bridges now denoted as arcs between pairs of dots)—and these places and connections are all that matter for the challenge at hand. We could name the places A, B, C, and D. And we could describe the connecting bridges by writing which pair of places each bridge connects. So the seven bridges of Königsberg can now be denoted as AB, AB, AC, BC, BD, BD, and CD.

After we've stripped away all the unnecessary diversions, we could restate the Königsberg Bridge Puzzle as follows: Suppose we have four dots named A, B, C, and D, and we have seven connections among them, namely, AB, AB, AC, BC, BD, BD, and CD. Can we start at some dot and choose connections to take us from dot to dot in such a way that we use every connection exactly once *and* we return to the dot at which we started? Notice that the starting place does not matter; that is, if we could solve the puzzle starting at some dot we could also solve it starting from any other dot. Why is this true?

It might, at first, appear that we have not really made any progress toward solving the puzzle, but we have isolated the essential ingredients, and that is an enormous step forward. In fact, those essential ingredients—dots and connections—are the essential ingredients that comprise a modern mathematical area called *graph theory*. A *graph* is simply a set of *vertices* (that is, the dots) together with a collection of connections of pairs of the vertices (that is, the lines or arcs connecting pairs of dots). The connections are called *edges*. So the graph associated with the Königsberg Bridge Puzzle has vertices A, B, C, and D and edges AB, AB, AC, BC, BD, BD, and CD. Our simple picture that just shows the vertices and edges has all the information we require to tackle the Königsberg Bridge Puzzle. Now we can restate the Königsberg Bridge Puzzle as follows: Can we start at some vertex in the Königsberg graph, then choose edges to take us from vertex to vertex in such a way that we use every edge exactly once *and* we return to the vertex at which we started?

The Königsberg graph shows us that there are four landmasses (represented by the vertices) and it shows which landmasses are connected by bridges (represented by the edges). By the way, the order of the vertices that describe an edge does not matter; in other words, CD and DC both mean the same thing, namely, a "bridge" (or edge) connecting landmasses C and D. Since two bridges connect A to B and two bridges connect B to D, we simply list AB twice and BD twice, but we don't make any attempt to distinguish the two AB bridges from one another. For example, we don't care which AB is the bridge nearer D. We make no distinction between the two AB edges, because an edge is just a connection between its vertices. No other feature of an edge matters, such as how or where we draw it.

Stepping Out

If we take a walk over the edges of the Königsberg graph, we can represent that walk by the ordered sequence (list) of edges that we traversed. For example, if we just go from A to B to D to C to B and back to A, we could represent that walk by listing the edges (AB)(BD)(DC)(CB)(BA) in the given order. Notice that it doesn't matter which of the two BD edges we choose, as each one accomplishes the same task of getting us from B to D.

Suppose we take a walk around the Königsberg graph and return back to where we started. If we write down the ordered sequence of edges involved, notice that each edge that we traverse must start at the point at which the previous edge ended. So in our example (AB)(BD)(DC)(CB)(BA), after we traveled along one of the edges from A to B, the next edge of course started at B and ended at some other point, in this case D. Then we took an edge from D to C, then journeyed on the edge from C to B, and then went from B to A using the edge we didn't use the first time. Since we returned to the place where we started, the first letter and the last letter will coincide.

Let's forget the parentheses and just look at the letters that are written down in our list of the edge-sequence, namely, ABBDDCCBBA. Notice that every letter appears an *even* number of times, because every internal letter is the end of one edge and therefore, in turn, the beginning of the next edge, so every internal letter appears in pairs, while the first and last letters are the same (because we return to where we started), so that letter appears in pairs as well. The list of edges of such a circuit is the Noah's Ark of graph theory: Every letter appearing comes in pairs, in this example: BB, DD, CC, BB, and of course our starting and ending location AA.

This observation about letters appearing an even number of times lets us solve the Königsberg Graph Puzzle. Why? Well, let's look at the list of all seven edges of Königsberg. They are AB, AB, AC, BC, BD, BD, CD. If we walked over all the Königsberg edges *each exactly once* in any order at all, those seven pairs of letters would be the ones describing our walk. It might start (AB)(BC)(CD)... and so on. But if each edge in the Königsberg graph (that is, each given pair of letters such as AC) appeared exactly once in our walk, then the total number of A's would be 3 (an odd number), because we can see that there are exactly three A's in our seven edges—namely one A each in the edges AB, AB, and AC, and no other A's in any of the other edges. Similarly, the total number of B's would be 5, the number of C's would be 3, and the number of D's would be 3.

But we saw that if we *were* able to take a walk over the edges—traversing each edge exactly once—and return to where we started, then when we recorded the edges in the order that we walked over them, each letter would appear an even number of times. But this even number of appearances of each letter is impossible for the Königsberg graph because we just noticed that each letter appears an odd number of times on our list of bridges. Thus we conclude that it is *impossible* to start at one location, traverse each and every edge exactly once, and return to our starting point.

This observation settles the Königsberg Graph Puzzle and thus settles the Königsberg Bridge Puzzle, definitely proving that it is impossible to walk over each bridge just one time. Please think through the reasoning and "bridge" the ideas of the argument together for yourself until every step makes sense. Dominoes may help.

The Domino Theory

One way to look at our reasoning is to consider dominoes. Remember that each edge is represented by two letters that denote the two vertices flanking the edge. Suppose you represent each edge by a domino and put the two letters of the vertices involved at either end of the domino. You need to see that if you take *any*

Life Lesson
Look for patterns.

Life Lesson
An assumption that leads to a contradiction must be false.

walk over a sequence of edges and come back to where you start, then in the written sequence of edges, each vertex letter must appear an even number of times. Taking a walk over the edges corresponds to putting the dominoes in a long row in which touching dominoes must have the same letters on their touching ends. So every letter would appear in touching pairs, except for the first and last letters, which would be the same and therefore create another pair.

Next we must realize that if we used each of the seven edges of Königsberg exactly once, then there would be vertex letters appearing an odd number of times. In other words, if we take seven dominoes, one for each bridge, then on those seven dominoes, various letters appear an odd number of times. In fact, *every* vertex letter would appear an odd number of times—A three times, B five times, C three times, and D three times; however, if just one of those letters appeared an odd number of times that would be sufficient to know that walking over every edge exactly once and returning to our starting point would be impossible. Think of the dominoes. If we put them down in one long string that started and ended in the same letter, then every letter would have to appear an even number of times.

| A | B | B | D | D | C | C | B | B | A |

Following Our Ideas Beyond Königsberg

When we solved the Königsberg Graph Puzzle, we actually came upon an idea that can be employed to answer many similar questions as well. One of the most effective strategies of thinking is to take existing ideas and see how those ideas can be applied in different contexts. It's a little like recycling, except instead of returning used cans to create new cans, we turn over used ideas in our minds to create new ideas.

A key idea in solving the Königsberg Graph puzzle involved noting how many edges were incident with (attached to) each vertex. We'll call the number of edges incident to a vertex V the degree of V, or deg(V) for short. So in the graph at the left, the degree of C is 4 and the degree of D is 6. Also, deg(A) = deg(B) = 2 and deg(E) = 4.

In the Königsberg bridge graph, the degree of A is 3, as are the degree of C and the degree of D. The degree of B is 5. Having any vertex with odd degree made it impossible to complete the Königsberg walk successfully.

Let's see whether it would be possible to take a walk over some other collections of bridges that connect various landmasses, keeping in mind this idea of degree. Let's just suppose we have a collection of islands named A, B, C, D, E, and F and we have some bridges that connect various pairs of those islands. Suppose the bridges are labeled: AB, AD, BC, BD, BE, BF, BF, CD, CE, DE, and DF. As with the Königsberg Bridge Puzzle, we model this scenario using a graph. Let's see whether it would be possible or impossible in this graph to walk over each edge once and return to where we started.

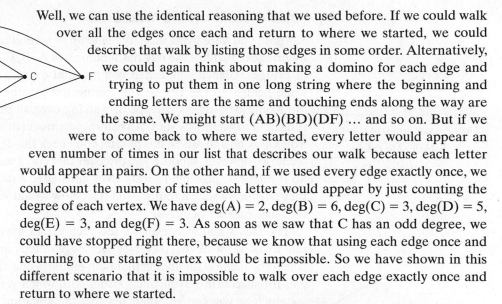

Well, we can use the identical reasoning that we used before. If we could walk over all the edges once each and return to where we started, we could describe that walk by listing those edges in some order. Alternatively, we could again think about making a domino for each edge and trying to put them in one long string where the beginning and ending letters are the same and touching ends along the way are the same. We might start (AB)(BD)(DF) … and so on. But if we were to come back to where we started, every letter would appear an even number of times in our list that describes our walk because each letter would appear in pairs. On the other hand, if we used every edge exactly once, we could count the number of times each letter would appear by just counting the degree of each vertex. We have deg(A) = 2, deg(B) = 6, deg(C) = 3, deg(D) = 5, deg(E) = 3, and deg(F) = 3. As soon as we saw that C has an odd degree, we could have stopped right there, because we know that using each edge once and returning to our starting vertex would be impossible. So we have shown in this different scenario that it is impossible to walk over each edge exactly once and return to where we started.

In fact, we now have an easy way to determine if we *cannot* walk over a given collection of bridges and come back to where we started. If there is any landmass with *an* odd number of bridges (corresponding to a vertex with odd degree in the graph model), our special kind of walk is impossible. By the way, notice that walking over bridges without walking over the same bridge twice is equivalent to tracing the graph in one pen-stroke without tracing over the same edge twice and without lifting your pen. So now you can quickly look at a picture of a graph and tell when you can't trace it in one stroke and return to where you start. For example, it is impossible to trace these graphs in one pen-stroke without going over the same edge twice. Can you see why?

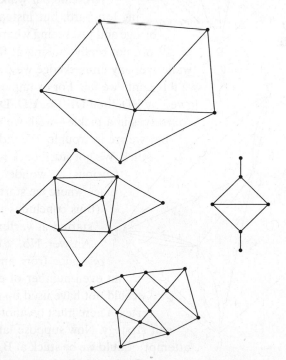

Take a Walk on the Wild Side

Suppose you were living in a city made up of several landmasses, some connected by bridges. Perhaps the city has landmasses A, B, C, D, E, and F and a long list of bridges. Glancing at a map you see that every landmass can be reached by some bridge, so the city is truly connected together. Now someone gives you the Königsberg-type challenge of walking over all the bridges exactly once and returning to where you start. As before, you model this configuration with a graph. You know how to attack this puzzle: you check the degree of each vertex. In this scenario, however, suppose you discover that *every* vertex *does* have *even* degree. So it *might* be possible to walk over each edge exactly once and return, but is there *definitely* such a route? And if so, how would you discover that walk given the large number of edges?

The fact is that whenever we see each vertex having even degree, there *definitely* will be a walk that traverses each edge exactly once and returns to where we start. The method for finding that path employs one of the most important strategies for success in solving many issues with which you are faced: namely, first do as much as you can and then reassess the situation at that point. Don't worry about completing the whole task when you begin. You may not know how to complete the entire task at the start, so instead of waiting until you can imagine a complete solution, first just do what you can.

Let's apply that reasoning to a specific example. Suppose there are six islands A, B, C, D, E, and F and there are many bridges AB, AB, AC, AF, BC, BC, BD, BD, CD, CE, CF, DE, DE, DE, DF, DF, EF, and EF. We create our graph model and first observe that every vertex has even degree: deg(A) = 4, deg(B) = 6, deg(C) = 6, deg(D) = 8, deg(E) = 6, deg(F) = 6.

To create our walk, let's apply the technique of simply not thinking too hard, but instead simply walking over any available unused bridge and just seeing where we end up. Let's start at vertex A and simply set off, not really looking at the picture or even thinking. Instead let's just walk over any unused edge we come to and continue until we get stuck; that is, we'll try until we fail. For example, we might start by walking over edge AB, followed by BD, then DF, FE, ED, DC, CB, BA, AC, CF, FA. At this point we find ourselves in a pickle—well, we're not actually in a vinegar-soaked vegetable, but we are in trouble. We did not journey over *all* the edges before getting stuck. We got stuck back where we started with no exit edges unused.

Curious, we wonder: Why did we end up getting stuck back at the place where we started? This question now requires some thought (thus concludes the "no thinking" part of our analysis). If we started at vertex A, could we have ended up stranded at B? Answer: No. When we first arrived at B, we would have used *one* edge from among all the edges at B. But recall, there are an even number of edges at B (vertex B has even degree), so we could not have used up the *last* edge at B when we first arrived at that vertex. There must be another edge incident with B that we can use as an exit strategy. Now suppose later we returned to B (as we did in our failed attempt). Could we be stuck at B at that time (that is, is it possible that there are no more B-edges)? No. Because at that point we have used only 3 of B's edges

Life Lesson
Don't wait for the whole solution. Do what you can.

and that is not an even number. So there must be another unused edge at B that we can now use to exit that landmass. In other words, since there are an even number of edges incident with B, every time we come *into* B there must be some unused edge available to use as an exit.

So suppose we start at A and just continue going over unused edges for as long as we can; that is, until we are stuck because there are no unused edges available. Then given that B has even degree, we see that we can never get stuck at B—there's always an unused exit strategy. Of course, applying this same even-reasoning to any letter other than our starting point A shows that we can never get stuck at C or D or E and so on for the same reason we couldn't get stuck at B—each of those vertices has even degree. The only place we can get stuck is back at A, because by starting at A, we used up one of the edges at A to make our initial exit off that vertex, thus leaving us with an odd number of unused edges at A. Since there are only a limited number of edges altogether, eventually we must be unable to go on, and at that moment we must find ourselves back at vertex A.

So let's assess our progress: The good news is that we have taken a pleasant walk, never duplicating edges, and we returned to where we started. The bad news is that we did not traverse all the edges. But we should celebrate this partial victory; that is, going over some of the edges and getting back to where we started without doubling back on a previously traveled edge. Our challenge now is: How can we alter our path so that we go over even more edges? Answer: Let's take our current walk and splice in another walk over unused edges.

Extending Our Journey by Adding a *Loop-d-Loop*

We first notice that at each vertex, the number of unused edges is even. We know this will always be the case because every vertex started out with even degree; then we used an even number of edges at each vertex in our first walk. So there must be an even number of unused edges remaining at each vertex. This even number might be zero, as is the case at vertex A in our example.

By examining our first attempt at creating a walk on our graph, we discover that there are some edges at vertex B that we never used. Suppose we simply start at B and travel over unused edges for as long as we can—again without thinking, simply taking any edges that were not previously used during our initial walk that took us from A back to A. For example, we could take edge BC, then CE, then ED, and then DB. At that point we are stuck again. There are no more unused edges that start at B. Notice that in taking unused edges starting from B, we ended up back at B when we finally got stuck. Why? The same reason as before: Every vertex has even degree (remember to look only at the edges not used on the first walk). So the only way to get stuck is back at B where we started. We've just created a little loop from B back to B.

Now we can simply take the walk we had before, which was (AB)|(BD)(DF)(FE)(ED)(DC)(CB)(BA)(AC)(CF)(FA), but when we get to B (at the location denoted by |), we'll just splice in the loop-d-loop detour walk (BC)(CE)(ED)(DB) before proceeding. So our new, longer walk that starts and ends at A is: (AB)(BC)(CE)(ED)(DB)(BD)(DF)(FE)(ED)(DC)(CB)(BA)(AC)(CF)(FA).

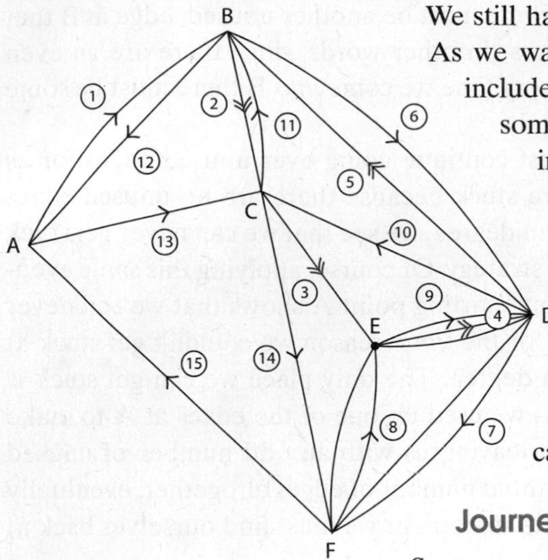

We still haven't traveled along all the edges, but now we know what to do. As we walk along our current walk that starts and ends at A (that now includes our B loop-d-loop), we notice that when we get to E we find some unused edges. We can just start at E and without any thinking simply take any unused edges until we get stuck. In this case, we could walk along (EF)(FD)(DE). Now we can splice in that short circuit (AB)(BC)(CE)(EF)(FD)(DE)(ED)(DB) (BD)(DF)(FE)(ED)(DC)(CB)(BA)(AC)(CF)(FA) and realize that we've succeeded in our mission. We found a way to create a walk going over each edge once. Although we were working with a specific example, in fact, our method works in general as well. Let's try to state the most general circumstances in which we can find such a walk over the edges in a graph.

Journeying Along Graphs

Suppose we have a graph: a collection of vertices together with a collection of edges that connect pairs of vertices (each edge is written as the pair of vertices it connects). A graph is said to be *connected* if we can walk from any vertex to any other vertex by a sequence of edges in which the end of any one edge in the sequence is the beginning of the next edge in the sequence. All the graphs we've used as models and examples in this section are connected graphs.

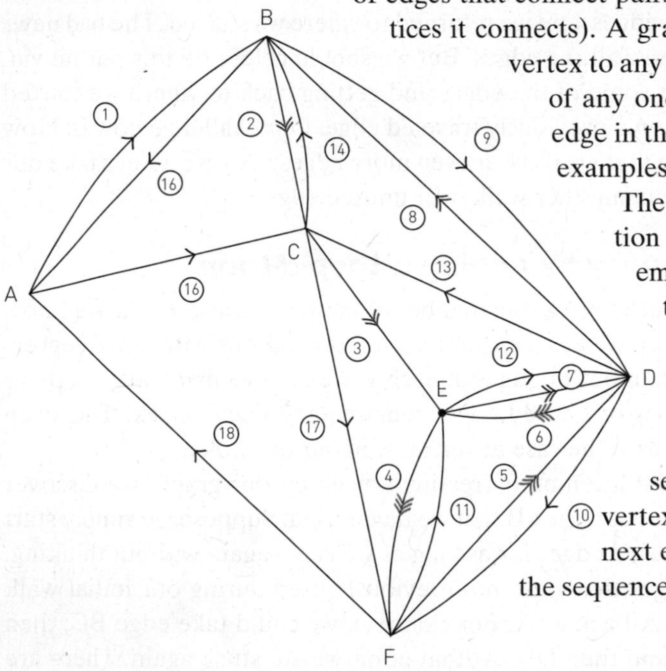

The Königsberg Bridge Puzzle led us to explore the question about sequences of edges on a graph. Since the mathematician Euler explored this question, it is traditional to use his name in talking about the kinds of paths we have been discussing. So this exploration gives rise to a definition that captures the essence of the walks we have been discussing.

We define an *Euler circuit* of a graph to be an ordered sequence of the edges of the graph such that the second vertex of each edge in the sequence is the first vertex of the next edge, every edge in the graph appears exactly once in the sequence, and the first and last vertices are the same.

Euler's Circuit Theorem.

A connected graph has an Euler circuit if and only if every vertex has even degree.

So now we can quickly solve any puzzle that asks whether we can trace a given graph without lifting our pencil or going over the same edge twice while ending up where we started. We know a quick method for determining whether it is possible—just check to see whether every vertex has an even number of edges emanating from it. And we even know a method for finding such a tracing when

Life Lesson
Definitions should follow from insights.

it exists—we first just do the best we can and then splice in a loop-d-loop circuit to augment our first attempt at tracing the given graph, and if we are still not done, we splice in another loop-d-loop circuit and so on until we have gone over every edge of the graph.

A Look BACK

We analyzed the question of when we can walk over a collection of bridges that connect various landmasses without going over the same bridge twice. This question was prompted by the famous Königsberg Bridge Puzzle from the year 1735. Landmasses connected by bridges were abstracted into the mathematical idea of a graph consisting of vertices and edges that connect vertices in pairs.

We found that a graph has an Euler circuit (that is, it can be traced starting anywhere and ending where we start without lifting our pencil or going over the same edge twice) if and only if it is connected and every vertex appears an even number of times in the list of edges. The study of graph theory goes far beyond Königsberg and leads to important insights into computer science, sociology, neuroscience, epidemiology, social networks, and many other applications.

Our strategy for developing this topic was to build from the concrete. We analyzed a specific question. Our method was to ignore the irrelevant and isolate essential ingredients from the situation. Our method of solution included looking for patterns and building insights from failed attempts. Then we took insights that we came to and extended them to apply to related situations. The strategies of investigation that we used here can generate insights and resolve issues far beyond mathematical ones.

Life Lessons

Analyze specific questions. Then generalize.

•

Find the essential ingredients.

•

Take a solution to one question and see where else it can be applied.

MINDSCAPES ⟩ Invitations to Further Thought

*In this section, Mindscapes marked (**H**) have hints for solutions at the back of the book. Mindscapes marked (**ExH**) have expanded hints at the back of the book. Mindscapes marked (**S**) have solutions.*

Developing Ideas

1. **Map maker, map maker make me a graph.** Represent this map using a graph, with a vertex (dot) for each landmass and an edge (line or arc) for each bridge.

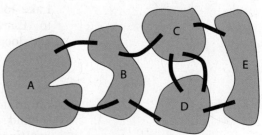

2. **Unabridged list.** Represent each landmass from Mindscape 1 as a vertex and give each vertex a letter. Represent each bridge from Mindscape 1 as an edge of your graph using a pair of letters as in the text. Make a list of the edges. What are the degrees of each vertex of your graph?

3. **Will the walk work?** Does your graph from Mindscapes 1 and 2 have an Euler circuit? That is, can you take a walk around the town shown in the map in Mindscape 1, cross each bridge exactly once, and return to where you started? If yes, describe such a walk by listing the edges. If not, explain why not.

4. **Walk around the house.** Is it possible to traverse this graph with a path that uses each edge exactly once and returns to the vertex at which you started? In other words, does the graph have an Euler circuit? If so, find one. If not, explain why not.

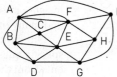

5. **Where's Waldoskova?** Do some research to find the location of Königsberg on a modern map. Does it still have the same name?

Solidifying Ideas

6. **Walk the line.** Does this graph above have an Euler circuit? If so, find one. If not, explain why not.

7. **Walkabout.** Does this graph have an Euler circuit? If so, find one. If not, explain why not.

8. **Linking the loops.** In this map, the following walks can be taken from various starting points: CAADDFFC, FCCBB CCEEF, DCCBBEEBBAAD. Can these walks be spliced together to create one walk that starts on landmass C, crosses each bridge exactly once, and then returns to C? If so, find such a walk. If not, explain why not.

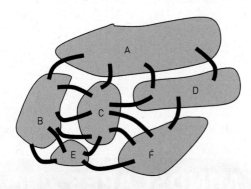

9. **Scenic drive. (S)** Here is a map of Rockystone National Park. One scenic drive is Entrance to Moose Mountain to Rockystone Lake to Lookout Below to Entrance. Can you add loop-d-loops to this drive to obtain a trip that traverses each road in the park exactly once and returns to the entrance? If so, find such a trip. If not, explain why not.

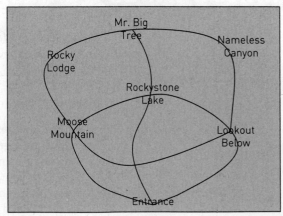

10. **Under-edged. (H)** Does this graph have an Euler circuit? If so, find one. If not, add the fewest number of new edges until such a circuit is possible. (Remember that edges can be curved.)

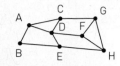

11. **No man is an island.** The country of Pelago consists of six islands. Create a graph to model the islands and bridges of Pelago. What is the degree of each vertex of your graph? Does your graph have an Euler circuit? Why or why not?

12. **Path-o-rama.** For each graph below, determine if the graph has an Euler circuit. If such a path is possible, present it. If not, explain why.

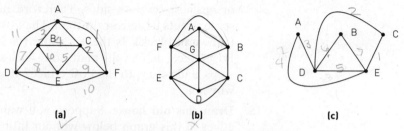

(a)   (b)   (c)

13. **Walk around the block.** Create a graph of the buildings on your campus that you use frequently, with the buildings (dorm, classroom, library, gym, etc.) as vertices, and edges joining vertices when they have a direct walkway or road connecting them. Does this graph have an Euler circuit? If so, find one. If not, explain why not.

14. **Walking the dogs.** Your dogs, Abbey and Bear, love to walk in the town park, modeled with the graph below. Based on the drawing, what do the vertices in this graph represent? What do the edges represent? Abbey and Bear recently learned about Euler circuits and try to traverse each edge exactly once during their walk (as they visit many trees more than once). Will they be successful?

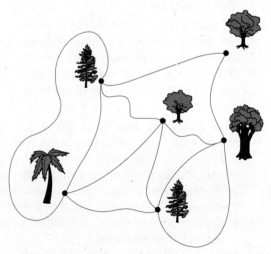

15. **Delivery query.** The next time you see a postal worker delivering mail, ask her how she plans her route. In particular, ask if she is fortunate enough not to have to retrace her steps down streets where she has already delivered the mail.

Creating New Ideas

16. **Snow job. (ExH)** Shown here is a map of the tiny town of Eulerville. The streets are white; the blocks are orange. After a winter storm, the village snowplow can clear a street with just one pass. Is it possible for the plow to start and end at the Town Hall (shown in black), clearing all the streets without traversing any street more than once? If so, find such a route. If not, explain why not.

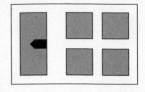

17. **Special delivery. (ExH)** Julia is the letter carrier for the town of Eulerville shown in the previous Mindscape. The streets are white; the blocks are orange. Assume that any street with a block on each side will have homes or business on both sides. Therefore Julia must walk along both sides of some streets to deliver the mail. The Post Office is located in the Town Hall (shown in black). Is there a route for Julia to use so she can start at the Post Office, deliver the mail along each street without having to retrace her steps, and return to the Post Office? If so, find such a route. If not, explain why not.

18. **Draw this old house.** Suppose you wanted to trace out the edges in this graph below without lifting your pen from the paper so that you draw each edge exactly once but with a new twist: You don't care if you return to the point where you started. Is it possible? If so, do it. If not, explain why you can't.

19. **Path of no return.** Consider this map showing a town with six landmasses and ten bridges. If you started on land mass C and walked as far as possible without going over the same bridge twice, where could you possibly get stuck? Why? What if you started on landmass D? What is special about C and D?

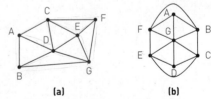

20. **Without a trace.** Is it possible to trace out either of these graphs without crossing an edge more than once? (Don't worry about ending up where you start.) If yes, describe the path. If no, explain.

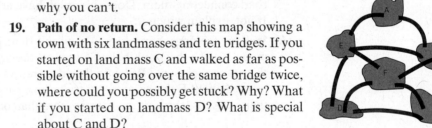

(a)   (b)

21. **New Euler.** In the three previous Mindscapes, you were presented with graphs that had no Euler circuit because they had vertices with odd degree (an odd number of incident edges). But in three of the four graphs, you could find a path that traversed each edge exactly once. Such a path is called an *Euler path*. Each of your Euler paths started and ended at a vertex of odd degree. Did this have to happen for these graphs? If you had more than two vertices of odd degree, could an Euler path exist?

22. **New edge—new circuit.** Look at the graph for Mindscape 18. Draw your own copy of this graph and add a second edge from E to F. Does the new graph have an Euler circuit? If you wanted to, could you start and end that circuit at vertex E, with the new edge being the last edge you used?

23. **New edge—new path.** Review your work for Mindscape 22. If you start your Euler circuit starting at E and end without using the new edge, what have you done? How does this relate to the question in Mindscape 18?

24. **Path to proof.** Suppose you have a connected graph in which every vertex has even degree except for two vertices with odd degree. If you add an edge between the two odd-degree vertices, what can you say about the resulting graph? Apply the reasoning from Mindscapes 22 and 23 to deduce that the original graph must have an Euler path that starts at one odd-degree vertex and ends at the other. Test your reasoning on graph (b) for Mindscape 20.

25. **No Euler no how.** Look at graph (a) for Mindscape 20. If you add one edge between two of the odd-degree vertices, does the resulting graph have an Euler circuit? Does this lead you to a conjecture about how many odd-degree vertices a graph can have and still have an Euler path?

Further Challenges

26. **Degree day. (S)** For each graph below, determine the degree of each vertex. (An edge that begins and ends at the same vertex is called a *loop*. Such an edge is counted twice when determining the degree of a vertex.) Then for each graph, compute the sum of all the degrees of the vertices in that graph. Count the number of edges in each graph. What do you notice?

 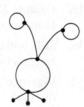

27. **100 degrees of proof.** Review your work for Mindscape 26. Can you make a conjecture that relates the number of edges in any graph to the degrees of all the vertices in that graph? (You might discover a result known as the Handshake Theorem. Can you explain why it's called that?)

28. **Degrees in sequence.** Can you draw a graph that has six vertices with degrees 5, 4, 3, 2, 2, 2? (This list is called the *degree sequence* of the graph.) Is there more than one way to draw such a graph? What does the sum of the numbers in the degree sequence equal? Can you draw a graph with degree sequence 3, 3, 2, 1, 1? What about 4, 3, 2, 1, 1?

29. **Even Steven.** Review your work in Mindscape 28 to make a conjecture about the number of vertices of odd degree that a graph can have. Prove your claim.

30. **Little League lesson. (H)** You are in charge of scheduling the baseball games for your town's Little League. There are 11 teams in your league. Usually you play ten games in a season, but some of the coaches want to extend the season to 13 games. So every team would play 13 games (thus playing more than one game against some of the teams). How will you explain to the coaches why this plan can never work?

In Your Own Words

31. Personal perspectives. Write a short essay describing the most interesting or surprising discovery you made in exploring the material in this section. If any material seemed puzzling or even unbelievable, address that as well. Explain why you chose the topics you did. Finally, comment on the aesthetics of the mathematics and ideas in this section.

32. With a group of folks. In a small group, discuss and actively work through the reasoning involved in proving that a connected graph with all vertices of even degree has an Euler circuit. After your discussion, write a brief narrative describing the methods in your own words.

33. Creative writing. Write an imaginative story (it can be humorous, dramatic, whatever you like) that involves or evokes the ideas of this section.

34. Power beyond the mathematics. Provide several real-life issues—ideally, from your own experience—for which some of the strategies of thought presented in this section would provide effective methods for approaching and resolving them.

For the Algebra Lover

Here we celebrate the power of algebra as a powerful way of finding unknown quantities by naming them, of expressing infinitely many relationships and connections clearly and succinctly, and of uncovering pattern and structure.

35. Finding V. Suppose a graph has n vertices and $(1/2)n(n - 1) = 15$ edges. How many vertices does the graph actually have?

36. Still looking. (H) Suppose a graph has n vertices and $(1/2)n(n - 1) = 45$ edges. How many vertices does the graph actually have?

37. Pigeon count. Mary runs every morning. She follows an Euler circuit through the park near her house. Along the first edge she sees one pigeon, along the second edge she sees three, along the third edge she sees five, and so on. If there are ten edges total in Mary's run, how many pigeons does she see? Find an expression that gives the total number of pigeons she would see if there were n edges in her run.

38. Edges vs. vertices. Someone chalked a graph outside your dorm. You observe that the number of edges is twice the number of vertices. You also see that the number of edges times the number of vertices is equal to 30 plus four times the number of vertices. Let V be the number of vertices in the graph and E the number of edges. Write an equation showing the relationship between V and E and another equation involving EV. Use these two equations to find the value of V.

39. Family values. Your teacher claims to have a family of graphs where the graph with n vertices has $(1/2)n(n - 1)$ edges, and n can be any natural number. How many edges does the graph with 100 vertices have? What about the graph with two vertices? With one vertex? Draw examples of these last two graphs.

FEELING EDGY?

Exploring Relationships Among Vertices, Edges, and Faces

> *Inspiration is needed in geometry, just as much as in poetry.*
>
> ALEKSANDR PUSHKIN

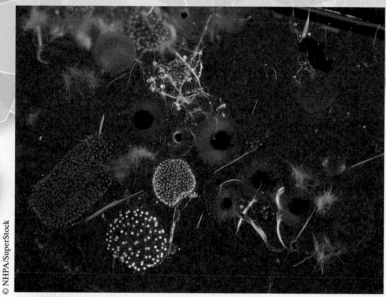

Radiolaria—microorganisms whose skeletons sometimes form regular solids.

In our discussion of the regular (Platonic) solids, we boldly asserted that there are only five of them. This statement seems just plain wrong given that there are infinitely many regular polygons in the plane. We said we would later prove that there are exactly five regular solids, and later is now. The regular solids are hard, rigid bodies. Their essence includes their flat faces, each a regular polygon. Surprisingly, we can prove that there are only five regular solids without even thinking about them as solids. The realm of the distortable contains all the ideas that limit the number of regular solids. Thinking about stretching and distorting allows us to uncover hidden structure within the rigid world of the regular solids.

We discover properties of rigid solids by first doodling in the plane and looking at what we draw. Looking carefully and quantitatively at something as simple as a random doodle gives a key insight about the regular solids. Once again, the key to unlocking a difficult idea is to observe a common phenomenon clearly and quantitatively.

Discovering by Doodling

"Do I have to draw you a picture?!" This exclamation is often heard in conversations involving an exasperated and impatient pontificator trying to get across some idea to a poor soul who would rather be somewhere else. Nevertheless, the answer is, "Yes."

Few methods for communicating fundamental ideas are more productive and successful than drawing a picture. Pictures not only reveal structure and pattern but often show more than what the artist intentionally put into them. The discovery of such new information is an important feature of the power of pictures. To illustrate this idea and to move us toward a better understanding of the Platonic solids, let's try a simple example. Please take out a blank piece of paper, and remember the artistic adage: "A picture is worth a thousand words."

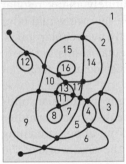

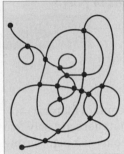

18 dots
17 regions } Combining = ⑤
33 edges + 2 = ⑤

Put the paper in front of you and place a pen on it. Close your eyes. Now pick up the pen and draw sweepingly, randomly on the page without lifting the pen or running off the edge. Don't draw too much, because later we're going to ask you to count various parts of your work of art. Once you've had enough, stop drawing, lift the pen from the paper, and open your eyes. You have created your own masterpiece—you're another Picasso. Shown at left is a sample of the genre.

We, the authors, will now attempt to demonstrate our ability to predict your actions across the void of time and space. Specifically, we will divine some features of the random drawing you just created. Please give us a moment to concentrate. Something is coming into focus . . . yes . . . yes . . . we see a squiggly object. Right? *Amazing.*

Please now accentuate each place where the curves cross each other in your drawing by placing a big dot at each crossing point. Also, draw dots at the points where you started and stopped. We now confidently assert that your drawing divides the paper into various regions.

Count on It—Detecting a Pattern

Okay, you may not yet be impressed with our clairvoyance, but hold on. We are detecting a pattern. We notice that each of your big dots has exactly four curves coming out from it—except for the starting and ending dots. Are we right? If we aren't, then we must wonder if you really closed your eyes and drew randomly. If not, please start over again.

Hold on—we see something else. In our minds, we see that you have not just drawn a random squiggle after all. You have drawn exactly two fewer edges than dots and regions combined. Now why in the world did you do that?

Please check our clairvoyance: Count the number of dots, regions, and edges (an edge is a segment, curved or not, connecting two dots). Don't forget to count the one region on the outside—that is, the one containing the edge of the paper. (How did we know there was exactly one of these exterior regions?) Notice that the number of dots plus the number of regions is exactly two more than the number of edges!

We have duly demonstrated our powers to read your mind and see through your eyes. However, these magical illusions are really mathematical feats. When you drew your squiggly curve, you unknowingly created some relationships among parts of your picture. These relationships are consequences of the topology of the paper and its idealized counterpart, the 2-dimensional plane. These relationships are topological in nature in that they are not associated with the particular size or shapes in your drawing. Let's discover why the number of dots plus the number of regions is always two more than the number of edges. As always, we begin with an easy case.

Building up Pictures

On a sheet of paper, let's draw a dot (also known as a *vertex*) and then draw a line segment from there, making a dot at the end of the segment. Now count the number of segments we have drawn. That is not too difficult—it is one. How many dots have we made? Two. How many regions are there on the page? One—everything around the segment. In this simple example, the number of dots plus the number of regions is three, which is indeed two more than the number of edges.

So far, so good. Now draw another line segment that starts at one of the dots and goes in any direction. Draw a dot at its end. Count again.

Vertices	=	3
Edges	=	2
Regions	=	still 1

Next, join up the two end dots to make a triangle. Count again.

Vertices	=	still 3
Edges	=	3
Regions	=	2 (inside the triangle and outside the triangle)

Continue adding segments that start at an existing dot and either go to an existing dot or just stick off some way and end with a new dot. New edges may not cross or touch any point of the existing diagram except for the vertices. Notice that, if we add a segment that connects two existing vertices, we create another region as we add an edge. If the new edge just sticks out somewhere, we do not create another region, but we do add another vertex. So, whenever we add an edge, we also add either one more region or one more vertex. We will continue to have two fewer edges than vertices and regions combined no matter how many edges we add to our picture. The objects we are drawing, consisting of edges connected together, are called *connected graphs*.

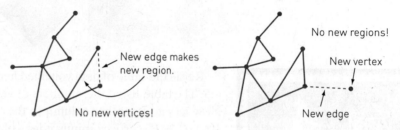

The Pattern Revealed

By adding one edge at a time, we can draw connected graphs as complicated as we wish. We could in theory draw one with 1,234,867 edges having 746,346 vertices. The thing we now know for sure is that any connected graph with 1,234,867 edges and 746,346 vertices will have exactly 488,523 regions. We are certain of the number of regions because, as we draw the figure one edge at a time, the number of vertices minus the number of edges plus the number of regions always equals 2, even if we draw millions of edges.

We were not the first to notice this fact about doodling. About 350 years ago, René Descartes observed this same phenomenon; however, he did not prove it. It took another 200 years until our hero Leonhard Euler came along and provided a complete and rigorous justification why this conjecture is always true. Euler figured out how to take the observation and prove it as fact—it is a theorem. Let the letter V stand for the number of vertices in the graph, E the number of edges, and F the number of regions into which the graph divides the plane. (Why do we use the letter F? We'll *face* that question soon.) The outside region does count as a region.

The Euler Characteristic Theorem.

For any connected graph in the plane, $V - E + F = 2$, where V is the number of vertices, E is the number of edges, and F is the number of regions.

All this formula says is that when we add an edge to a connected graph, we also add either one vertex or one face. So, we have already shown why this theorem is true. Writing our insight in this form allows us to give it a fancy name that will memorialize Euler through the ages. The quantity $V - E + F$ is called the *Euler Characteristic* and is a simple insight from which we can draw important conclusions.

Only Five Regular Solids

In our discussion of the Platonic solids, we stated that the universe contains only five regular solids (the tetrahedron, cube, octahedron, dodecahedron, and icosahedron). We can now give a complete justification of this statement. As we will discover, this result can be deduced as a surprising consequence of the Euler Characteristic.

Leonhard Euler stamp.

Recall the table of data you filled in regarding the regular solids in Section 4.5. The table listed the numbers of vertices, edges, and faces in each solid. Now let's add another column to the table. The new column will compute $V - E + F$, vertices minus edges plus faces for each of the regular solids. (This connection with the solids clears up why we used F to represent regions.) For each of the regular solids, let's compute $V - E + F$, giving us the following table:

	Number of Vertices	Number of Edges	Number of Faces	$V - E + F$
Tetrahedron	4	6	4	2
Cube	8	12	6	2
Octahedron	6	12	8	2
Dodecahedron	20	30	12	2
Icosahedron	12	30	20	2

Our answer to $V - E + F$ is 2 for every solid. This answer is like a mystical mantra for the universe: 2. Remember that we already showed that the Euler Characteristic for any connected graph in the plane is 2. Is this repeated answer of 2 a coincidence? Unexpected similarities in seemingly different settings suggest hidden relationships.

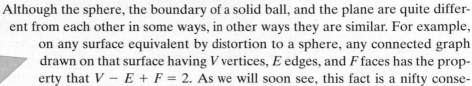

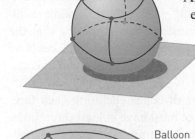

A Plane–Sphere Correspondence

Although the sphere, the boundary of a solid ball, and the plane are quite different from each other in some ways, in other ways they are similar. For example, on any surface equivalent by distortion to a sphere, any connected graph drawn on that surface having V vertices, E edges, and F faces has the property that $V - E + F = 2$. As we will soon see, this fact is a nifty consequence of the Euler Characteristic Theorem.

Suppose we draw a connected graph on a rubber balloon. Let's further suppose that the air hole, where we blow up the balloon, is located in the middle of a face (so the air hole does not pass through an edge or vertex).

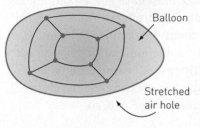

Balloon

Stretched air hole

Let's place our rubber balloon surface on a flat plane so that it is resting with the air hole (tied in a knot) sticking up. Now let's unknot the air hole and carefully stretch out the balloon (remember it's malleable rubber) so that it's flat on the plane. This activity takes a great deal of visualization in our mind's eye. Another way to think about this procedure is to project the balloon onto the plane using a stereographic projection. (Recall our discussion of stereographic projection from our geometric look at infinity in Section 3.5.) We now have a connected graph on the plane, and the number of vertices, edges, and faces of the connected graph are equal to the number of vertices, edges, and faces of the original graph on the balloon. Notice that the face that had the air hole is now the region that entirely surrounds the graph.

We know from the Euler Characteristic Theorem that the number of vertices minus the number of edges plus the number of faces of the connected graph in the plane is equal to 2. But since these numbers are exactly the same as the corresponding numbers on the balloon surface, we must have that $V - E + F = 2$ for any surface that is equivalent by distortion to a sphere. This stereographic projection process explains why all those numbers from our table combined to give 2: All the regular solids are equivalent to a sphere. Notice how we used the notion of a one-to-one correspondence here to see that the number of vertices, edges, and faces on the balloon equals the number of vertices, edges, and regions of the corresponding graph on the plane.

Only Five and No More—the Proof

We now know something about the relationships among vertices, edges, and faces of connected graphs drawn on a sphere or anything distortable to a sphere, such as the regular solids. Our challenge now is to use this Euler Characteristic relationship to show that there are only five regular solids.

Five Platonic Solids.

There are only five regular solids.

The question is, Could there be a regular solid that we have not thought of? If there were, then that *MYSTERAHEDRON* would satisfy the formula $V - E + F = 2$ just as all the regular solids do. That fact will lead to its demise and show us that the mysterious *MYSTERAHEDRON* is actually a *NONEXISTAHEDRON*. In other words, it cannot exist!

Part 1—Finding Relationships

Let's suppose that the *MYSTERAHEDRON* has V vertices, E edges, and F faces. Besides the Euler Characteristic formula, $V - E + F = 2$, some additional relationships rise among V, E, and F, because we are supposing that the *MYSTERAHEDRON* is a *regular* solid. Recall that in a regular solid all faces have the same number of sides, and all vertices have the same number of edges emanating out.

Let's use the letter s to stand for the number of edges enclosing each face. Notice that s has to be at least three, because a face must have at least three sides; otherwise, it would not be a real face. (Remember that the face with the smallest number of sides is a triangle.) Next we can count the number of edges, E, in a clever way. Multiply the number of faces, F, by s, the number of sides on each face, and notice that each edge is counted exactly twice because each edge is on exactly two faces. So sF divided by 2 gives the number of edges; that is, $E = sF/2$, or $F = 2E/s$. As an illustration, for the cube we see $s = 4$, $F = 6$, so

$$\frac{4 \times 6}{2} = 12 = \text{number of edges.}$$

In a regular solid, the number of edges that come together at any vertex is the same as the number that come together at any other vertex. Let's call that number of edges emanating out of a vertex c. Notice that c must be at least three. (Why must this be? Think about it and draw some pictures to convince yourself.)

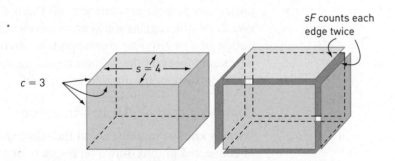

$c = 3$ $s = 4$ *sF* counts each edge twice

Since each edge has two ends, we could count the number of edges by multiplying the number of vertices, V, by c and dividing the result by 2 since cV is the number of edge ends, rather than the number of edges. So, $E = cV/2$, or $V = 2E/c$. For the cube, we see that $c = 3$ and $V = 8$, so

$$\frac{3 \times 8}{2} = 12 = \text{number of edges.}$$

These two formulas, $F = 2E/s$ and $V = 2E/c$, result from the idea of carefully counting things.

Part 2—Using the Relationships

Now we are in business, because we know that $V - E + F = 2$, and we know what V and F are equal to. Substituting, we get

$$\frac{2E}{c} - E + \frac{2E}{s} = 2$$

or, equivalently,

$$E\left(\frac{2}{c} - 1 + \frac{2}{s}\right) = 2$$

or, equivalently,

$$E\left(\frac{2}{c} + \frac{2}{s} - 1\right) = 2.$$

The positive number E multiplied by $(2/c + 2/s - 1)$ equals the positive number 2. So, $(2/c + 2/s - 1)$ must also be a positive number. Therefore, $2/c + 2/s$ must be larger than 1.

Since both c and s are 3 or greater, there are not many possible values for c and s that ensure that $2/c + 2/s$ is larger than 1. Notice that if either c or s were as large as 6, the number $2/c + 2/s$ would be 1 or less. So, neither c nor s can be as large as 6. Therefore, the only possibilities are as follows:

$c = 3, s = 3$ (which gives the tetrahedron)

$c = 3, s = 4$ (which gives the cube)

$c = 3, s = 5$ (which gives the dodecahedron)

$c = 4, s = 3$ (which gives the octahedron)

$c = 4, s = 4$ (which makes $2/c + 2/s = 1$, which is not larger than 1)

$c = 4, s = 5$ (which makes $2/c + 2/s = 9/10$, which is not larger than 1)

$c = 5, s = 3$ (which gives the icosahedron)

$c = 5, s = 4$ (which makes $2/c + 2/s = 9/10$, which is not larger than 1)

$c = 5, s = 5$ (which makes $2/c + 2/s = 4/5$, which is not larger than 1)

This proof demonstrates that there are only five regular solids in the entire universe and no others.

Concluding Thoughts

There are many proofs of this theorem. This proof has the advantage that it does not use geometry in the usual sense. For example, it does not refer to the interior angles of regular polygonal faces. So, in fact, we have proven that we cannot build even a distorted solid with faces all having the same number of sides and the same number of edges emerging from each vertex.

Perhaps we can empathize with Johannes Kepler for believing that this limit on the number of these aesthetically pleasing regular solids has a mystical significance for the structure of the universe. In fact, these solids do have physical implications that influence such things as the structure of certain molecules and some living organisms as well.

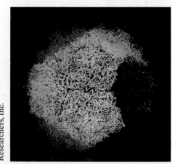

James M. Hogle/Photo Researchers, Inc.

Dodecahedral shell of a polio virus.

A Look BACK

THE EULER CHARACTERISTIC records a relationship among the number of vertices, edges, and regions created by a connected graph in the plane, namely, $V - E + F = 2$. Connected graphs on a balloon, a sphere, or a solid share the same property since we can create a one-to-one correspondence between figures on a balloon, sphere, or solid and those in the plane. Other relationships among the vertices, edges, and faces of a Platonic solid arise from their regularity. These relationships combined with the Euler Characteristic prove that there are only five regular solids.

We can discover the Euler Characteristic relationship by counting features of random doodles in the plane. Then we confirm our observations by building connected graphs one edge at a time and seeing that the $V - E + F = 2$ relationship continues to hold with each new added edge. Thinking about stretching and distorting allows us to see why drawings in the plane correspond naturally in a one-to-one manner with figures on a balloon or sphere, or the regular solids.

Looking at simple objects in a quantitative way lets us see patterns that might not otherwise be apparent. When we have found a pattern in one domain, we can increase its scope and power by seeking the same or related patterns in different settings. Great ideas are rare and should be applied in as large an arena as possible.

Life Lessons

Never underestimate the power of simple counting.

•

Expand an idea to different settings.

•

Cultivate ideas and reap great harvests.

MINDSCAPES ⟩ Invitations to Further Thought

In this section, Mindscapes marked (H) have hints for solutions at the back of the book. Mindscapes marked (ExH) have expanded hints at the back of the book. Mindscapes marked (S) have solutions.

Developing Ideas

1. **What a character!** What expression gives the Euler Characteristic?

2. **Count, then verify.** What are the values of V, E, and F for the graph shown? Compute $V - E + F$.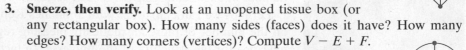

3. **Sneeze, then verify.** Look at an unopened tissue box (or any rectangular box). How many sides (faces) does it have? How many edges? How many corners (vertices)? Compute $V - E + F$.

4. **Blow, then verify.** Inflate a balloon and use a marker to draw the graph of the peace symbol shown on its surface. Find the values of V, E, and F. Compute $V - E + F$.

5. **Add one.** Find the values V, E, and F for the graph shown. Now add a vertex in the middle of one of the edges. How do the values of V, E, and F change? Do the new values still satisfy the Euler Characteristic formula?

Solidifying Ideas

6. **Bowling.** What is the Euler Characteristic of the surface of a bowling ball? Explain why.

7. **Making change.** We begin with the graph pictured at right. Each of the graphs that follow is a modification of it. For each graph compute the change in the number of vertices, edges, and regions.

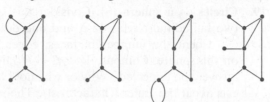

8. **Making a point.** Take a connected graph and add a vertex in the middle of an edge, making two edges out of the one. What happens to $V - E + F$? Explain why.

9. **On the edge (H).** Is it possible to add an edge to a graph and reduce the number of regions? Why or why not? Is it possible to add an edge and keep the same number of regions? Why or why not?

10. **Soap films.** Consider the following sequence of graphs generated by soap bubble films. As the vertical film shrinks, it passes through the unstable position in the third picture and then jumps to the last. Compute $V - E + F$ at each stage. What are the changes in the V, E, and F counts as we move from the second to the third picture?

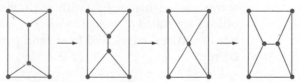

11. **Dualing.** What is the relationship between the Euler Characteristic for a regular solid and its dual? (See Chapter 4, Section 5.)

12. **No separation (S).** Suppose we have a connected graph that does not separate the plane into different regions. Prove that the number of edges is one less than the number of vertices.

13. **Lots of separation.** Suppose we are told that a connected graph cuts the plane into 231 regions. How many more edges than vertices are there?

14. **Regions (H).** Suppose we are given a connected graph made up of 151 edges. Is it possible that the graph cuts the plane into 153 regions? If not, what is the largest number of regions possible? (You may use loops.)

15. **Psychic readings.** Someone is thinking of a connected graph in the plane. It is made of 36 edges and cuts the plane into 18 regions. How many vertices are there?

16. **An odd graph.** Is it possible to draw a connected graph in the plane with an odd number of regions, an even number of vertices, and an even number of edges? If so, draw one; if not, explain why not.

17. **Another odd graph.** Is it possible to draw a connected graph in the plane with an odd number of regions, an odd number of vertices, and an even number of edges? If so, draw one; if not, explain why not. What makes your response to this scenario different from the one in the previous Mindscape?

18. **Circular reasoning.** Create a connected graph as follows. Draw a large circle and then place dots (vertices) at various points on the circle. Check that Euler's formula works for your graph. What do you notice about the number of vertices and edges in your graph? Will this result happen no matter how many vertices you put on your circle? Explain.

19. **Circles on a sphere (S).** Consider this sphere. It has two latitudinal circles on it and one dot on each circle. Count the number of faces, edges, and vertices on this sphere. Compute $V - E + F$. Did you get the answer you expected? Why or why not? What hypothesis about the Euler Characteristic Theorem does this question pertain to?

20. **More circles.** Consider the sphere described in Mindscape 19 but add an edge that connects the two dots, as shown. Now count the number of faces, edges, and vertices. Compute $V - E + F$ again. Why is this answer different from the previous calculation? What hypothesis about the Euler Characteristic Theorem do Mindscapes 19 and 20 illustrate?

21. **In the rough (S).** Count the number of facets, edges, and vertices of this diamond. Compute the Euler Characteristic for this gem.

22. **Cutting corners (H).** The following collection of pictures shows the regular solids with their vertices cut off. Such objects are called *truncated solids*. For each truncated solid, count the number of vertices, edges, and faces, and verify that the Euler Characteristic is correct.

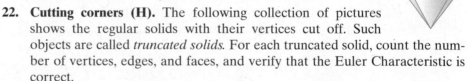

23. **Stellar.** The following collection of pictures shows the regular solids where each face was replaced by a vertex that sticks out and that attaches to the solid with edges to each of the vertices of the original face. Such objects are called *stellated solids*. For each stellated solid, count the number of vertices, edges, and faces, and verify that the Euler Characteristic is correct.

24. **A torus graph (ExH).** The Euler Characteristic $V - E + F$ can be applied to other surfaces besides spheres. We can compute it for tori, two-holed tori, and yet more complicated figures; however, for those surfaces the 2 in $V - E + F = 2$ is replaced by other numbers. Draw a connected graph on a torus that both circles the hole and goes through the hole of the torus. Compute $V - E + F$ for the graph. Try several other different graphs that go around the torus both ways and again compute $V - E + F$.

Graph must go around both ways, as marked.

25. **Regular unfolding.** Each graph below represents some regular solid that has been "unfolded," or projected down onto the plane. Determine which regular solid each graph represents.

Creating New Ideas

26. **A tale of two graphs.** Suppose we draw a graph that has exactly two pieces; that is, it is not connected but instead consists of two connected parts (such as the one shown). What is $V - E + F$ for that two-component graph?

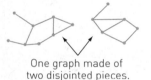

One graph made of two disjointed pieces.

27. **Two graph conjectures (S).** Can you conjecture a new formula for $V - E + F$ for graphs with exactly two pieces? Try various examples first.

28. **Lots of graphs conjecture.** Can you conjecture a new formula for $V - E + F$ for graphs with exactly three pieces? Try various examples first. Can you conjecture a new formula for $V - E + F$ for graphs with exactly n pieces? Search for a pattern.

29. **Torus count.** Three hollowed, triangular prisms were used to make the torus shown here. Carefully count the number of vertices, faces, and edges for this prismatic torus. Compute the Euler Characteristic for this torus.

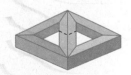

30. **Torus two count (H).** Carefully count the number of vertices, faces, and edges for the two-holed torus shown here. One way to view this two-holed torus is as two copies of the previous picture, with one side removed from each and then the open edges glued together. This operation is called the *connected sum*. Compute the Euler Characteristic for this two-holed torus.

Remove both faces and glue edges together.

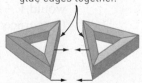

31. **Torus many count.** Using the preceding calculations and the notion of a connected sum, compute the Euler Characteristic for a three-holed torus. (First think about how to put together a three-holed torus using a two-holed and a one-holed torus.) Make a conjecture as to what the Euler Characteristic is for a torus with four holes, five holes, and, in general, h number of holes.

32. **Torus tours.** Consider the rubber rectangular sheet shown here with the edges identified to make a torus. Assuming that the edges are glued together as indicated, how many different edges would there be (*different!*)? How many different vertices would there be? How many faces? Compute $V - E + F$. (Make sure you are not counting the same things more than once!)

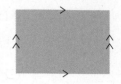

33. **Tell the truth.** Someone said that she made a two-holed torus by gluing 60 triangles together along their edges. Once it was all glued, this person claimed that there were 30 vertices on the object. Is she telling the truth? Explain why or why not.

34. **No sphere.** Suppose we have a sphere built out of 60 triangles. Why can't all the vertices have the same number of triangles coming into them? (*Hint:* Count edges by multiplying by something and then dividing by 2, and count vertices by a similar multiplying and dividing technique.)

35. **Soccer ball.** A soccer ball is made of pentagons and hexagons. Each pentagon is surrounded by five hexagons. Use the Euler Characteristic to figure out how many pentagons and how many hexagons are necessary to construct a soccer ball.

Further Challenges

36. **Klein bottle.** Using the diagram here for building a Klein bottle (as described in Section 5.2), count the number of vertices, edges, and faces. Use this data to conjecture the Euler Characteristic for the Klein bottle.

37. **Not many neighbors.** Show that every map has at least one country (oceans count as countries) with five or fewer neighbors—that is, countries that share a border with it. (*Hint:* Put a vertex in the center of each country and join two vertices if the two countries share a border. This procedure produces a graph with one vertex for each country. The number of edges emanating from a vertex equals the number of neighboring countries. Suppose each country has six or more neighbors. What's the smallest number of edges the graph would have? How many faces would there have to be to make the Euler Characteristic hold?)

38. **Infinite edges.** Suppose we consider a connected graph in the plane, but now we allow edges to go from a vertex out to infinity. What would $V - E + F$ be for such a graph? Why? (Experiment and discover the pattern.)

In Your Own Words

39. **Personal perspectives.** Write a short essay describing the most interesting or surprising discovery you made in exploring the material in this section. If any material seemed puzzling or even unbelievable, address that as well. Explain why you chose the topics you did. Finally, comment on the aesthetics of the mathematics and ideas in this section.

40. **With a group of folks.** In a small group, discuss and work through the reasoning for why any connected graph in the plane satisfies the relation that the number of vertices minus the number of edges plus the number of regions is equal to two. After your discussion, write a brief narrative describing the method in your own words.

41. **Creative writing.** Write an imaginative story (it can be humorous, dramatic, whatever you like) that involves or evokes the ideas of this section.

42. **Power beyond the mathematics.** Provide several real-life issues—ideally, from your own experience—that some of the strategies of thought presented in this section would effectively approach and resolve.

For the Algebra Lover

Here we celebrate the power of algebra as a powerful way of finding unknown quantities by naming them, of expressing infinitely many relationships and connections clearly and succinctly, and of uncovering pattern and structure.

43. **Rewriting Euler.** Rewrite the Euler Characteristic formula to give V in terms of E and F. Then write E in terms of V and F, and write F in terms of V and E.

44. **Euler assistance.** Your planar graph drawing has n^2 vertices, $n(n + 4)$ edges, and $6n - 10$ regions. Use Euler's formula to help you find the number of vertices, edges, and regions in your drawing.

45. **Decorated dodecahedron.** Your precocious little niece takes a Sharpie and writes a number on each face of your dodecahedron. She writes a 1 on one face, a 3 on another face, a 9 on another face, a 27 on another face, an 81 on yet another face, and so on. What's the largest number she writes on a face? Give an expression for the sum of all the numbers she writes. Find the sum.

46. **Stick figures. (H)** You draw n stick figures in the plane. Each one looks like the figure shown here. Assuming your figures don't touch or share vertices, into how many regions have you divided the plane? Write an expression for the total number of vertices minus the total number of edges plus the number of regions. Does it simplify to equal 2? If you add edges to your drawing without adding vertices, you can connect your figures. How many edges do you have to add to make Euler's formula apply?

47. **The other football.** You have a solid with $n(n^2 - 1)$ vertices, $n^3 + 6n + 2$ edges, and $8n$ faces. Use the Euler Characteristic to solve for n. Find the number of vertices, edges, and faces of your solid. (Can you think of a solid with these values?)

6.3 PLANE OLD GRAPHS

Drawing in the Plane and Coloring Maps

Lots of planes flying in essentially the same plane

> *"I found I could say things with color and shapes that I couldn't say any other way—things I had no words for."*
>
> GEORGIA O'KEEFFE

An electronic circuit fundamentally consists of nodes that are connected in various patterns. If we wanted to create a huge model of such a circuit, we could take a bunch of nodes and connect various pairs of them with extension cords. That complicated set of wires would physically demonstrate which nodes are connected to which others. An important question for manufacturing computer chips is whether that network of wires can be translated into a line drawing on a piece of paper. In other words, could the nodes and the connections between them be drawn on a page such that the lines connecting nodes never cross each other? If the circuit can be placed in the plane, then the circuit can be manufactured more cheaply. It is not always possible to draw a given circuit in the plane. Sometimes we can and sometimes we can't. Our challenge is to determine what makes the difference. That is, which graphs are planar and which are not?

Another collection of important images we study and use (both on the page and on our GPS screens) are maps. But a map is not truly attractive until its regions (its connected countries, states, or counties) are colored. And a clever coloring scheme makes it easy to distinguish where one region ends and another begins by making sure that different regions sharing a common border are colored differently. The question is: How many different colors must we have to be *certain* that we can color every possible map in this smart manner? That question stumped mapmakers, mathematicians, and many people in the general public for a hundred years before the answer was definitively proved. We now know the answer: no matter the map, we need only four colors! This surprising recent fact is a celebrated result of living mathematicians.

To appreciate and analyze these puzzles about drawings and our world, we use a potent combination of effective methods of thinking. One method is to try some specific examples, and when we have understood those particular cases, we realize that we can generalize our insights so they are applicable to many other areas as well.

414

The question of which line drawings can be drawn in the plane has important physical applications to computer chip manufacturing. Map coloring questions have direct applications to cartography. And the ability to analyze these questions develops mental strengths that are applicable to every challenge of life.

The Gas-Water-Electricity Puzzle

Laying the Groundwork with Pipes and Cables

The *Gas-Water-Electricity Puzzle* is a classic question about three houses and three utility companies. The three houses each need water, natural gas, and electricity. Is it possible to create a line drawing on a paper connecting each utility company to each house so that none of the lines cross on the page?

A related challenge is the *Five-Station Model Train Puzzle:* Is it possible for a model train designer to build a layout on a tabletop with five stations and with direct connections between every pair of stations without using any bridges, tunnels, or crossing tracks?

Of course, in the real world, utility companies run underground pipes and overhead wires to avoid the tragedy of star-crossed cables. And model train sets come with bridges and tunnels. But when we focus on the essential features of the two scenarios, we realize that the fundamental questions they pose can be modeled using graphs. By representing houses and utility companies by vertices, and the utility lines by edges, the Gas-Water-Electricity Puzzle asks if we can redraw the graph below on the left so that no edges cross each other (to remove all the *edge crossings*). For the Five-Station Model Train Puzzle, if the stations are the vertices and the tracks are the edges, the puzzle asks if we can redraw the graph below on the right to remove all the edge crossings. (Note: In each graph, there are numerous points on the page where two edges cross. These points are *not* vertices. The vertices in the graphs are represented by the enlarged dots.)

When faced with any question, one of the first steps toward finding a solution is to strip away extraneous information so that you can focus on the underlying important issues. The step of taking the Gas-Water-Electricity Puzzle and the Five-Station Model Train Puzzle and turning them into questions about graphs is an example of *mathematical modeling*. We have abstracted the essence of those puzzles into questions about graphs. That step allows us to ignore the distractions of the particular puzzles such as wondering where the natural gas comes from or making sure the tiny plastic trees provide enough shade for the neighborhoods of the electric train display. The strategy of creating a mathematical model of a real-world situation is one of the most powerful ways for resolving difficult challenges.

House 1   House 2   House 3

Gas lines   Water lines   Electric lines

Life Lesson

Focus on essentials.

$K_{3,3}$

K_5

The two graphs that model our two puzzles are famous and have descriptive names. The graph on the right is called the *complete graph on five vertices*. It is 'complete' because each of the five vertices is joined by an edge to each of the other vertices. It's denoted K_5. The "K" does not stand for some ancient misspelling of the word "complete," but rather honors Kazimierz Kuratowski, a Polish mathematician who made important contributions in this field (although we are not 100% certain if the K stands for his first or last name). The graph on the left is called the *complete bipartite graph $K_{3,3}$*, because, although not every pair of vertices is connected by an edge, there are two collections of vertices—one collection of three (the three utility companies) and another collection of three (the three houses)—such that every vertex from one collection (every utility company) is connected by an edge to every vertex in the other collection (every house). But there are no edges between vertices that are in the same collection; that is, no two houses are connected by an edge, nor are different utility companies connected by an edge.

Remember that our challenge is to keep the appropriate pairs of vertices joined by edges as originally described, and only change *how* those graphs are drawn in an attempt to draw them without any edges crossing. Here are attempts to remove at least some crossings in each graph. Notice that in each try, we did not entirely succeed since each new drawing still has a rogue edge crossing:

unwanted edge crossings

These drawings of the graphs are great attempts. In each case, we now have only one unfortunate crossing of edges. The remaining question is: Can we move the edges around in a more clever way such that each drawing has *no* crossings? Notice that moving edges does not change the graph *structure*; all the same pairs of vertices are still joined by edges even though those edges now might be longer or more curvy than before. We've just drawn the edges differently. So the redrawn graphs still model the same *connections* given in the two puzzles.

The questions posed in our two puzzles ask us whether it is possible for us to draw the two graphs that model the puzzles such that no edges cross. These two particular challenges lead us to a more general question about which graphs *can* be drawn in the plane and which graphs *cannot* be drawn in the plane without any edge crossings.

This question about "drawability" leads us to a new definition. We say a graph is *planar* if it can be drawn on a sheet of paper (that is, in the plane) such that no edges meet or cross except at vertices. The graph on the left may not look planar at first, because two of its edges are crossing in the picture; however, it is planar because we can redraw one edge to remove the crossing, getting a new drawing of the same graph, shown in the middle, where no edges cross. We can slide vertices and edges around a bit as well to obtain a more symmetric drawing of this graph, as shown.

Note that each of the previous drawings represents the same graph in the sense that each of these three graphs consists of four vertices, with each pair of vertices joined by an edge. This graph is the "complete graph on four vertices": K_4. A graph is planar if it's *possible* to draw it with no edge crossings. Being planar does not imply that *every* drawing of the graph avoids edge crossings—you could always draw a graph in such a way that edges cross each other. (Try it! Making your edges wiggly helps create lots of crossings.) To earn the title of *planar* graph it must be **possible** to create a planar drawing of the same graph (that is, a graph having the same number of vertices, with the same pairs of vertices connected).

Here are two more examples of planar graphs, but in each case we have drawn them with one or more edge crossings. Take a moment right now to redraw our not-so-planar drawings to create a planar drawing of each graph.

Can the Gas-Water-Electricity and Model Train Puzzles Be Solved? That Is, Are $K_{3,3}$ and K_5 Planar?

We still haven't solved the original two puzzles, but we can now restate them in the equivalent but more fancy graph theory way: Are $K_{3,3}$ and K_5 planar graphs? Play with the drawings for a while, moving edges around to see if you can draw all the edges without any crossings. Do you think such a planar drawing is possible for either of the given graphs?

You're right! The answer is "No." No matter how hard you try, there is no way to lay out the train tracks so that five stations can each be connected to the other four without having at least one bridge or crossing. Here's one way to demonstrate that K_5 is not planar.

The graph K_5 consists of five vertices, with an edge connecting each pair. So let's start simply by first considering just three vertices. They must be joined in pairs, so any planar drawing would consist of some sort of triangle-like shape such as:

We now need to add two more vertices, each with edges to all the other vertices. Let's work with the drawing on the right. Our argument will be similar for the drawing on the left. The fourth vertex must be either inside the "triangle" or outside. Here are the results (the new vertex is white).

Notice that if we color in the vertex, slide the four vertices around, and straighten out the edges, both of these drawings can be made to look like the following. (Again we see a planar drawing of K_4.)

But now we need to add a fifth vertex and draw an edge from it to each of the existing four vertices. If we put the fifth vertex on the outside, there is no way to draw the edge from it to the vertex in the middle without crossing one of the edges making up that outer triangle. (See figure on the next page on the left.) Likewise, if we put the fifth vertex in one of the inner triangles, there is no way to draw an edge from it to the vertex outside that triangle (See figure on the next page on the right.)

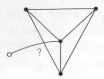

Thus, because we considered all possible placements of that fifth vertex, we conclude there is no way to draw K_5 without edge crossings, and so K_5 is not planar.

In our thinking above, notice that as we added new vertices, we had to add more and more edges. This leads to the question: Is there a connection between the number of vertices and the number of edges on any planar graph? The answer is, "no, because of *silly* planar graphs." By a silly planar graph, we mean one that has loops (edges connecting a vertex to itself) or multiple edges (several edges that connect the same pair of vertices). Notice that we can take any planar graph with a fixed number of vertices, and create a new and different planar graph by just adding a bunch of loops and multiple edges.

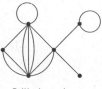

Planar   Still planar!

Therefore, with a fixed number of vertices, the number of edges a planar graph can have is unbounded. But what if we don't consider these silly planar graphs; that is, what if we just consider planar graphs that do not contain any loops or multiple edges? We'll now discover that if a friend has a planar graph with no loops and no multiple edges and does not show it to us, then if she just tells us the number of vertices her graph has, we can deduce the maximum number of edges in her drawing.

Not-so-plane Euler Enters the Picture

Remember Euler and his remarkable Characteristic from Section 6.2 (*Feeling Edgy?*). Euler discovered that for any graph drawn in the plane with no edge crossings, if the graph has V vertices, E edges, and F regions, then $V - E + F = 2$.

Notice that K_4 has 4 vertices, 6 edges and, in the plane drawing here 4 regions. (Remember to count the outer unbounded region.) Thus $4 - 6 + 4 = 2$, confirming the Euler Characteristic for this graph.

As we'll now see for ourselves, we can use the Euler Characteristic to obtain the maximum number of edges a planar graph with no loops and no multiple edges can have, just knowing the number of vertices it has. That is, suppose we are told that a particular graph has a certain number of vertices *and* is planar. Then we can tell (without even seeing the graph) the maximum number of edges that graph has.

The process has two steps. We will start with a plane drawing of the graph with no loops and no multiple edges and uncover a relation between the number of edges and the number of regions. Then we will bring Euler's Characteristic formula in to convert the relation involving edges and regions into a relation involving edges and vertices.

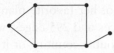

We derive our general results by thinking through a particular example. This graph drawing has $V = 6$, $E = 7$, and $F = 3$. We also notice that every region has at least three edges as part of its boundary. This observation will hold true for any plane graph drawing with no loops or multiple edges. The only way to create regions with only one or two edges involves loops or multiple edges.

So each region created by a graph with no loops or multiple edges has at least three edges on its boundary. Now, suppose you take a walk around the boundary of each region, as if inspecting the fences represented by the edges. As you walk along, imagine you have your hand running along the edges on the boundary, and that you count each edge as you pass it in each region. So in our example graph above, from inside the triangular region you count three edges; inside the square you count four, and for the outer region you count seven. Wait . . . seven?!? Yes, because you count each edge as your hand runs along it, so the right-edge sticking out into the outer region gets counted twice—once as you walked on one side of it to get to the end and pivot around that lonely vertex and then again as you walked back to the rest of the graph along the other side of that edge. (You could call this last edge a *peninsula edge*.)

Now, add up all the edge-counts you found over all regions. In our example, we get $3 + 4 + 7 = 14$. This number is exactly *twice* the number of edges in the graph, $14 = 2 \times 7 = 2E$. More generally, for any plane drawing of a graph, the total sum of boundary edges over all regions will count every edge exactly twice, because we really count each side of each edge. We know that each edge is counted exactly twice because each edge is either a peninsula and so is counted twice in our system, or the edge is on the boundary of two different regions and so is counted twice. We also know that for any plane graph with no loops and no multiple edges, every region has at least three edges on its boundary, so our sum total of edge sides must be at least $3F$. Thus we obtain the inequality $2E \geq 3F$. This impressive inequality arose simply from careful counting and thinking.

Enter Euler! Because our graph is planar, the drawing satisfies the equation $V - E + F = 2$. We solve this equation for F to obtain $F = 2 - V + E$, then substitute this equation for F into the inequality $2E \geq 3F$ to obtain

$2E \geq 3(2 - V + E)$. Distributing the 3 yields

$2E \geq 6 - 3V + 3E$. Subtracting $2E$ from both sides yields

$0 \geq 6 - 3V + E$. Subtracting $6 - 3V$ from both sides yields

$3V - 6 \geq E$. Equivalently,

$E \leq 3V - 6$.

And we have a beautiful upper bound on the number of edges we can have in a planar graph with no loops or multiple edges. One caveat: this inequality doesn't work for tiny planar graphs with only one or two vertices, so we also have to insist that V be at least 3.

Life Lesson
Use what you know.

A Ceiling for Edges.

For any planar graph with no loops or multiple edges having a vertex count of V (with V being equal to at least 3) and an edge count of E, we have that

$E \leq 3V - 6$.

So, for example, if our friend says she's thinking of her favorite planar graph with no loops or multiple edges and it has 100 vertices and 295 edges, we know, without even seeing her graph, that either she has trouble counting or it must be April Fools'. Because *if* her graph had 100 vertices (so $V = 100$), then the number of edges E must satisfy: $E \le 3V - 6 = 3(100) - 6 = 294$. So an edge-count of 295 is impossible for any such planar graph—and we know that something is wrong with her counts of vertices and edges of her graph without even seeing the graph itself!

Are $K_{3,3}$ and K_5 Planar Redux?

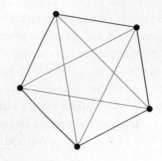

We can use the previous theorem that gave us an upper bound on the edge count to quickly show that K_5 is not planar. K_5 has five vertices and ten edges (five in the pentagon (blue) and five in the star (green) (see figure at left). If it *were* planar, $V = 5$ and $E = 10$ would have to satisfy the edge-vertex inequality $E \le 3V - 6$. But $3(5) - 6 = 9$, and 10 is not less than or equal to 9. Therefore we conclude that K_5 is not planar.

But we already showed this geometrically by considering all the possible placements of the fifth vertex. Does our edge-vertex inequality allow us to discover anything we didn't already know for certain? For instance, does it demonstrate that $K_{3,3}$ is not planar as it did for K_5? (You can try this now. How many vertices does $K_{3,3}$ have? How many edges? Do these values violate the inequality $E \le 3V - 6$ the way K_5 did?) We'll recycle the ideas that led to that inequality to discover another edge-vertex inequality that will enable us to show that $K_{3,3}$ is not planar. Remember that $K_{3,3}$ is the complete bipartite graph with six vertices, three on top and three on the bottom:

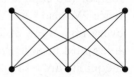

In any plane drawing of a bipartite graph without multiple edges (loops are never permitted in a bipartite graph), each region will have at least *four* edges on its boundary. We see this fact because the boundary of a region will be formed by edges that must go from top to bottom to top to bottom to top. Thus, each region must be surrounded by an even number of edges; so three-sided, triangular-shaped regions are impossible. If we now repeat our walk-around-the-regions-and-total-up-the-edges process on our planar bipartite graph, we obtain the inequality $2E \ge 4F$, because each region has at least four edges making up its boundary (rather than three in the general case we considered before). Substituting $F = 2 - V + E$ from Euler's Characteristic formula again, we obtain

$2E \ge 4(2 - V + E)$.

$2E \ge 8 - 4V + 4E$.

$0 \ge 8 - 4V + 2E$.

$4V - 8 \ge 2E$.

$2V - 4 \ge E$. Equivalently,

$E \le 2V - 4$ if every region has at least four edges and there are no loops or multiple edges.

This relationship between the number of edges E and the number of vertices V is another beautiful upper bound on the number of edges in terms of the number of vertices for these graphs. Once again, we have to insist that we have at least three vertices or this bound doesn't apply. But now we know that for any bipartite graph with at least three vertices and no multiple edges, we must have $E \leq 2V - 4$.

We now return to the graph $K_{3,3}$. It has six vertices and nine edges. If it *were* planar, then $V = 6$ and $E = 9$ would have to satisfy our new inequality. But $2(6) - 4 = 8$, and 9 is not less than or equal to 8. So $K_{3,3}$ cannot be planar—and that's a mathematical proof of our earlier conjecture.

Life Lesson
Seek the essential.

Kuratowski's Deep Insight

Why have we focused so much on K_5 and $K_{3,3}$? What's so special about these graphs? In fact, they are, in essence, the building blocks of *all* non-planar graphs. The fact is that in any non-planar graph, we can find a "copy" of K_5 or $K_{3,3}$ (or both). The tricky part is—what do we mean by "copy"? In some cases, a non-planar graph will contain an exact copy of K_5 or $K_{3,3}$, as the following graphs do. We know these graphs cannot be drawn without edge crossings, because doing so would implicitly create a planar drawing of K_5 or $K_{3,3}$—something we know is impossible.

 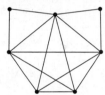

But look at the graph below. It's almost the same as $K_{3,3}$, except an extra vertex (shown in red) has been added to an edge, dividing that edge into two new edges. In effect, we have replaced one edge with a new vertex of degree two and two new edges.

This new graph is not the same as $K_{3,3}$ (since it has a different number of edges and different number of vertices than $K_{3,3}$ has), and it does not contain an exact copy of $K_{3,3}$, but like $K_{3,3}$, it is not planar. Why? Again, suppose this new graph could be drawn in the plane without edge crossings. Then if we just remove the red vertex and restore the divided edge into a single edge, we would have a planar drawing of $K_{3,3}$, which we know is impossible. Thus, this new graph is also not planar.

What about these graphs?

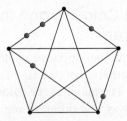 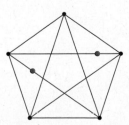

These graphs resemble K_5, but neither is, nor contains an exact copy of, K_5. Yet we know they are not planar, for if we could draw either without edge crossings, we could obtain a planar drawing of K_5 simply by removing the red vertices and restoring the original edges—which, again, is impossible.

So it's clear that if we add these red "freckle" vertices to $K_{3,3}$ or K_5, we will obtain a non-planar graph. And any graph that contains an exact copy of $K_{3,3}$ or K_5, or a copy with one or more added freckle vertices, must also be non-planar.

What's truly amazing is that $K_{3,3}$ and K_5 are the *only* obstacles in the way of planarity. That is, if a graph does not contain either an exact copy of $K_{3,3}$ or K_5 or a copy of one of those with freckles, then you can draw that graph in the plane. In other words, every non-planar graph, that is, every graph that is impossible to draw in the plane, must contain either an exact copy of $K_{3,3}$ or K_5 or a copy of one of those with freckles. This theorem is very difficult to prove. It was discovered by Kuratowski himself and was published in 1930.

Freckled graphs are called *subdivisions* of the original graph, because a freckle vertex "subdivides" an old edge into two new edges. Note also that freckle vertices are ordinary degree-two vertices in the new graph; it's just fun to call them freckles.

A graph G   One subdivision of G   Another subdivision of G
             (added vertices are red)   (added vertices are red)

Kuratowski's Theorem (1930).

A graph is planar if and only if it does not contain a copy of $K_{3,3}$ or K_5 or a subdivision of $K_{3,3}$ or K_5.

This surprising theorem is motivated by the two puzzles posed at the beginning of the section: The Gas-Water-Electricity Puzzle and the Five-Station Model Train Puzzle. One of the most effective strategies of mathematics and of thinking well in general is to identify the truly essential features of a situation from among a jumble of diverting clutter. Kuratowski's Theorem tells us that the two essential questions when trying to figure out whether a graph can actually be drawn in the plane are: Is $K_{3,3}$ in it? Is K_5 in it? That's it. The graph may have billions of vertices and millions of edges, but the existence or non-existence of a subdivision of $K_{3,3}$ or K_5 will tell the entire story of whether it is planar.

Map and Graph Coloring and the Infamous Four Color Theorem

Perhaps surprisingly, another application of planar graphs arises when we consider the cartographer's conundrum of picking colors for the countries or states on a map, globe, or GPS image. The mapmaker's coloring challenge is to color any potential world map in such a way that different regions that share a common border, that is, a part of their boundary, have different colors. The big question is: What is the minimum number of colors that always suffices to color any potential world map?

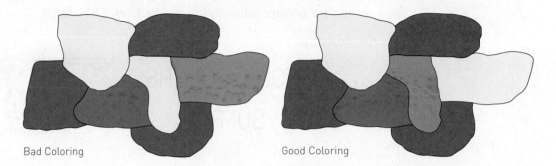

Bad Coloring   Good Coloring

Often when we think of a question, we soon realize that the question is not completely clear. For example, when we talk about "any potential world map," are there any hidden assumptions about what a world map could look like? In this case, one important hidden assumption is that each country is a connected piece. If the countries were allowed to come in many different pieces, like island nations, and yet all those pieces are to be of one color since they are part of the same country, then no fixed number of colors would enable us to color any possible map.

Here's a challenge to help illuminate that last claim. Suppose we allow countries to have many pieces, some of which might be sort of islands entirely inside other countries. Can you draw a map in which six countries each share a border with each of the other five countries? Can you draw a map in which ten countries each share a border with each of the other nine countries? The implication of these challenges involving disconnected countries is that we must confine our coloring question to maps in which each country is one connected piece.

So our revised map-coloring question is: What is the minimum number of colors that always suffices to color any potential world map in which each country is connected and different countries sharing a common border have different colors?

The question about the minimum number of colors required to color any map turned out to be one of the most famous mathematical challenges for a hundred years. In 1852, Francis Guthrie made a conjecture that became known as the Four Color Conjecture: namely, he conjectured that four colors would suffice for any map with connected countries. However, he was unable to prove that his assertion was true in general.

The Four Color Conjecture was famous partly because of its interesting history of errors and failures to solve it. In 1879, Alfred Kempe published what he thought was a proof of the Four Color Conjecture. However, in 1890 Percy Heawood discovered and pointed out a flaw in Kempe's reasoning. Peter Guthrie Tate published a different erroneous "proof" in 1880 and its error was exposed in 1891.

Because the Four Color Conjecture is easy to state and also because some erroneous proofs are relatively easy to discover and easy to be fooled by, the general public embraced the Four Color Conjecture as perhaps the most famous unsolved question in mathematics for a century. Many amateur mathematicians worked hard on this conjecture, often finding one of several subtly flawed proofs.

In 1976 mathematicians Kenneth Appel and Wolfgang Haken from the University of Illinois finally proved the *Four Color Theorem*. The result was celebrated with the slogan 'Four Colors Suffice' appearing many, places, including on postage cancellation marks from Urbana, IL. The proof involved the

FOUR COLORS

SUFFICE

extensive use of computer checking of thousands of combinatorial cases. This role of computer verification was a new and controversial aspect of mathematical proof and, to this day, many people remain uncomfortable with the idea that a mathematical theorem can be considered proved even though it requires far more steps than could be conceivably undertaken by any human being.

Six Colors Suffice

The Four Color Theorem is far too difficult to prove in one section of this book, but we can prove the Six Color Theorem, which is a great start. For one thing, it will show that no matter how complicated the map, a small, fixed number of colors (just six) is all that's needed to color the map so that no two countries that share a border have the same color. You might have thought that some complicated maps would require 100 different colors. To start work on such a question, our first step is to find a way to translate the question about maps into a related question about graphs. This translation from a map-coloring question into a question about graphs is yet another example of mathematical modeling and how modeling can allow us to solve challenges that would otherwise be too difficult.

For the houses and utilities question, we replaced each house and each utility with a vertex and each connecting line with an edge to create a graph. Similarly, in tackling the coloring conundrum we can isolate the essential issues by ignoring many of the complicated features of an actual map. Let's just put a single vertex *in* each state or country and, if two states or countries share a border, then put an edge *between* their two vertices by drawing a line that crosses the border. In that way, irrelevancies such as the shapes of the states or countries can be ignored. Each vertex of the graph represents a region (country or state) and each edge denotes which regions share a border.

Let's look at a map of the lower forty-eight states of the United States. We can put a vertex in each state and draw an edge between any two states that share a border. Now our original challenge of coloring each state can be transformed into the challenge of coloring each vertex in this associated graph in such a way that no two vertices that are connected by an edge have the same color. This strategy of transforming one question into a related question is, as we have seen before and will see again now, often powerful.

Life Lesson
Use simple models to understand complex situations.

Life Lesson
Transform one question into a related question.

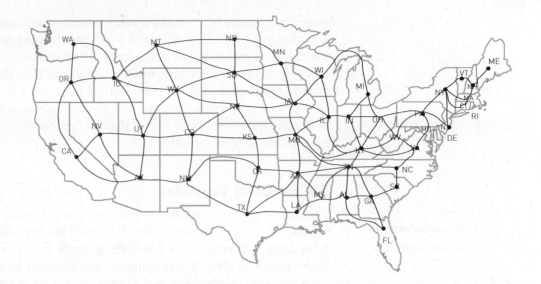

Coloring the regions of a map is the same as coloring the vertices of the related graph. Since that related graph is drawn on the map, it must be a planar graph. So we have translated our map-coloring question into a question about coloring the vertices of a planar graph.

The Four Color Theorem or the Six Color Theorem can now be translated into a statement about coloring the vertices of a planar graph. Translated, it states that given a planar graph, using only four (or six) colors, it is possible for you to color the vertices such that no two vertices joined by an edge receive the same color. To prove the Six Color Theorem, we'll need a useful fact about maps and planar graphs.

Every Map Has a Country with Five or Fewer Neighbors

The first surprising fact about maps is that no matter what map you draw, there will always be at least one country that has five or fewer neighbors. Try it! Just try to draw a map on a sheet of paper in which *each* country has six or more neighbors. Try as you might, alas, you are doomed to failure. Here's why.

Counting a Degree with Just One Hand.
Every planar graph with no loops or multiple edges must contain at least one vertex of degree less than or equal to five.

Life Lesson
Build on what you know.

We establish this claim as follows. You may recall from a few paragraphs ago, that one consequence of the Euler Characteristic (that is, $V - E + F = 2$ for any planar graph) was that any planar graph with no loops or multiple edges satisfies the condition $E \leq 3V - 6$. Suppose that you could draw a graph in which every vertex had degree 6. Then since every edge has two ends, the number of edges E in that graph would be $E = 6V/2 = 3V$. But then the number of edges E is not less than or equal to $3V - 6$ as the inequality guarantees, so not every vertex in a planar graph with no loops or multiple edges can have degree 6. If vertices have even higher degrees than 6, then matters get even worse. So we conclude that every planar graph with no loops or multiple edges must have at least one vertex of degree five or less, which means that every possible map has at least one country with five or fewer neighbors.

The Six Color Theorem

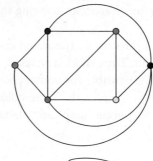

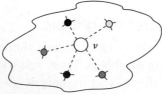

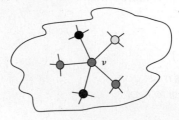

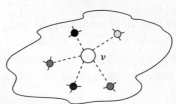

Since every possible map has at least one country with five or fewer neighbors, we can use this fact to verify the Six Color Theorem. We will work our way up from maps with only a few countries to maps with more and more countries. Every map with six or fewer countries can certainly be colored with six colors, since we could simply color each country with a different color. So that's a good and easy start.

As usual, let's translate our insight into a statement about planar graphs with no loops or multiple edges. In fact, from now on let's just assume that our planar graphs have no loops or multiple edges so we don't have to repeat that condition each time. If we have a planar graph with six or fewer vertices, then we can six-color the graph, meaning that we can color each vertex in such a way that no edge has both its vertices colored with the same color. Again, that's easy; we just color each vertex with a different color. The figure here shows one example. We could get away with fewer than six colors on this sample graph, but with six crayons in our Crayola box, why not use them all?

Now let's consider a planar graph that has 7 vertices. We know from our previous result that at least one of those vertices, let's just call it v, has degree 5 or less. Let's momentarily remove that vertex v and the edges that contain v. (The figure to the left shows only the removed vertex v and its remaining neighbors. All other vertices and edges are left out for clarity.)

Then we have a graph with only 6 vertices. Give that graph a 6-coloring, which we know we can do since we just established that. Then look at the vertices that were neighbors of v. Since there are 5 or fewer of them, at least one of our six colors was not used on v's neighbors. Now we can add v and its edges back into the graph and color v with that color.

So now we know that we can find a 6-coloring for any planar graph with 7 vertices. Now let's consider a planar graph that has 8 vertices. We know that at least one of those vertices, call it v, has degree 5 or less. Let's again momentarily remove v and the edges that contain v. (The figure to the left shows only the removed vertex v and its remaining neighbors. All other vertices and edges are left out for clarity.)

Then we have a graph with 7 vertices. Give that graph a 6-coloring, which we know we can do since we just proved it. Now look at the vertices that were neighbors of v. Since there are 5 or fewer of them, at least one of our six colors was not used on v's neighbors. Now we can add v and its edges back into the graph and color v with that color.

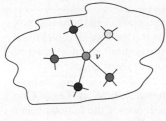

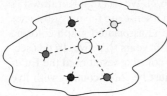

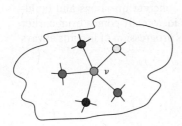

So now we know that we can find a 6-coloring for any planar graph with 8 vertices. Now let's consider a planar graph that has 9 vertices. We know that at least one of those vertices, call it v, has degree 5 or less. Remove v and the edges that contain v. (The figure to the left shows only the removed vertex v and its remaining neighbors. All other vertices and edges are left out for clarity.)

Then we have a graph with 8 vertices. Give that graph a 6-coloring, which we know we can do since we proved that before. Then look at the vertices that were neighbors of v. Since there are 5 or fewer of them, at least one of our six colors was not used on v's neighbors. Now we can add v and its edges back into the graph and color v with that color.

So now we know that we can find a 6-coloring for any planar graph with 9 vertices.

You may notice that the last three paragraphs were basically identical. Also, notice that we could copy the previous paragraph and tweak the appropriate numbers to show that any graph with 10 vertices can be 6-colored and then we could copy the paragraph again to prove that any planar graph with 11 vertices can be 6-colored, and so on forever. Since there will be no stopping this process, we conclude that every planar graph with any number of vertices can be 6-colored.

Therefore, we have proved the Six Color Theorem.

The Six Color Theorem.

Given a map consisting of connected regions printed on a sheet of paper the regions can be colored using only six different colors such that any two different regions sharing a common border will have different colors.

Life Lesson

Look for patterns.

As a mathematical aside, the thinking technique of working our way up step by step to graphs with an ever-increasing number of vertices is an example of a technique of proof known as *mathematical induction*. It just means that we repeat the same argument over and over again to conclude that what we are trying to prove works for graphs with larger and larger numbers of vertices. Proving the Five Color Theorem is not much more difficult than what we just did, but the proof is a bit technically complicated, so we will not include that analysis here. However, the proof that only four colors suffice is considerably more complex and involves many advanced techniques from combinatorics and graph theory as well as the theory of computation.

The Four Color Theorem took a hundred years to prove. Many mistakes were made along the way. New ideas had to be discovered and new techniques invented in order to finally settle that long-standing conundrum. The proof of the Four Color Theorem is a testament to the perseverance and imagination of human beings. Sometimes questions that defy our best efforts for a long time can ultimately be solved through clear thinking and focused effort.

A Look BACK

SOME GRAPHS CAN BE DRAWN IN THE plane without extra edge crossings and some cannot. We started with two puzzles, the Gas-Water-Electricity Puzzle and the Five-Station Model Train Puzzle, and found the essential questions involved were questions about whether or not two particular graphs could be drawn in the plane without crossing edges.

The graphs involved in the puzzles were the complete bipartite graph $K_{3,3}$ and the complete graph on five vertices, K_5. We showed that neither $K_{3,3}$ nor K_5 is planar. One method of pinning down a proof of their non-planarity made use of the famous Euler Characteristic formula from the previous section (Section 6.2). We saw that Kuratowski's Theorem stated that those two special graphs $K_{3,3}$ and K_5 in a sense are the only obstacles to planarity of a graph. Then we introduced maps and the associated graph coloring questions. The famous Four Color Theorem took about 100 years to prove. In this section, we gave a proof of the Six Color Theorem. This map coloring result used the Euler Characteristic to prove that in any possible map, there must be some country with five or fewer neighbors.

This section illustrated the strategy of starting from concrete questions and building upon ideas that we had previously developed. Every idea really comes from earlier ideas and incrementally develops those ideas further. This progression of insights always leaves us to think: What new ideas are still to come?

Life Lessons

Build from what you have and move forward.

•

Model complex situations with simpler constructs.

MINDSCAPES Invitations to Further Thought

*In this section, Mindscapes marked (**H**) have hints for solutions at the back of the book. Mindscapes marked (**ExH**) have expanded hints at the back of the book. Mindscapes marked (**S**) have solutions.*

Developing Ideas

1. **Don't be cross.** Here is a drawing of a graph with four vertices and an edge joining every pair of those vertices plus two additional edges and vertices. Redraw the graph so that none of the edges cross.

2. **"De Plane! De Plane!" (S)** Is the graph given in Mindscape 1 planar? Explain.

3. **Countdown (H).** For the graph drawing shown, count the number of vertices, edges, and regions. Verify that $V - E + F = 2$.

4. **Connect Connecticut.** Here is a map showing the counties of Connecticut. Create a graph by placing one vertex (dot) in each county and then drawing an edge (line or arc) between two counties if they share a common boundary.

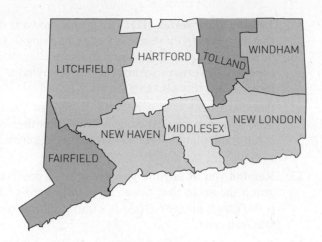

5. **Criss-Cross.** Is it possible to redraw the graph shown so that none of the edges cross? In other words, is the graph planar? Draw it without edges crossing or explain why you can't.

Solidifying Ideas

6. **Don't cross in the edge.** Each of the graphs drawn below is actually planar. Redraw each one so that none of the edges cross.

  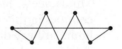

7. **Hot crossed buns.** Each of the graphs drawn below is actually planar. Redraw each one so that none of the edges cross.

8. **Starring the pentagon.** Is it possible to redraw the graph shown so that none of the edges cross? In other words, is the graph planar? Draw it without edges crossing or explain why you can't.

9. **Spider on a mirror.** Is it possible to redraw the graph shown so that none of the edges cross? Draw it without edges crossing or explain why you can't.

10. **One more vertex.** The graph here is drawn to show that it's planar (there are no edge crossings). Can you add a new vertex v to the graph and draw an edge from v to each of the original vertices without creating any edge crossings? Do it or explain why you can't.

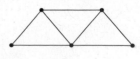

11. **Yet one more vertex (H).** The graph shown is drawn to show that it's planar (there are no edge crossings). Can you add a new vertex v to the graph and draw an edge from v to each of the original vertices without creating any edge crossings? Do it or explain why you can't.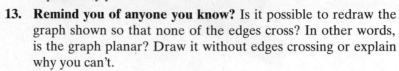

12. **Familiar freckles.** Is it possible to redraw the graph here so that none of the edges cross? In other words, is the graph planar? Draw it without edges crossing or explain why you can't.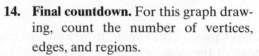

13. **Remind you of anyone you know?** Is it possible to redraw the graph shown so that none of the edges cross? In other words, is the graph planar? Draw it without edges crossing or explain why you can't.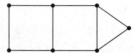

14. **Final countdown.** For this graph drawing, count the number of vertices, edges, and regions.

15. **Euler check-up.** Use your answer to the previous Mindscape to check Euler's formula: $V - E + F = 2$. Does it work? If not, recheck your counting from the previous Mindscape!

16. **Euler second opinion.** For the graph drawing shown here, count the number of vertices, edges, and regions. Then check Euler's formula: $V - E + F = 2$.

17. **Coloring Connecticut.** Return to the map in Mindscape 4. The Four Color Theorem asserts that it is possible to color the counties in this map using at most four colors so that counties that share a boundary get different colors. Is it possible to accomplish this with fewer than four colors? If so, describe such a coloring. If not, explain why not.

18. **Westward Ho!** Here's a map of some states in the western U.S. Describe how you would create a graph model of this map in which each vertex represents a state, then draw it. The Four Color Theorem asserts that it is possible to color the states in this map using at most four colors such that states sharing a boundary get different colors. Is it possible to accomplish this coloring with fewer than four colors? If so, describe such a coloring. If not, explain why not.

19. **A colorful museum.** This figure shows the floor plan of a museum (doors and windows have been omitted). Describe how you would create a graph model of this floor plan in which each vertex represents a room, then draw it. Is it possible to assign a paint color to each room using only three colors so that rooms with shared walls get different colors? If so, describe such a coloring. If not, explain.

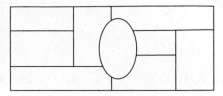

20. **Limit of 5.** Start drawing a planar graph. Keep adding vertices, and as many edges as you can but avoiding edge crossings. When you get up to about ten vertices, stop and see if your graph has a vertex of degree five or less (a vertex with at most five edges joined to it).

Creating New Ideas

21. **Starring the hexagon.** Is it possible to redraw this graph so that none of the edges cross? In other words, is the graph planar?

22. **Kowabunga Kuratowski (ExH).** Show that this graph is not planar by finding a subdivision of K_5 in it.

23. **Kool Kuratowski.** Show that this graph is not planar by finding a subdivision of $K_{3,3}$ in it.

24. **Getting greedy. (H)** Suppose you are asked to color the vertices of this graph in the following way: You must color the vertices one-by-one in numerical order and no two vertices joined by an edge can be the same color. You are given the list of colors red, yellow, blue, green, purple, and orange. The coloring rule is to look at this ordered list of colors, and select the first color on the list that satisfies the criteria above. So, for example, vertex #1 would be colored red. Following the rules, how many colors did you use? (A method that uses the first available item on a list to complete a series of tasks is an example of a *greedy algorithm*.)

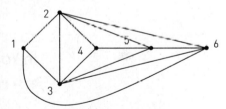

25. **Stingy rather than greedy.** By coloring the vertices in the graph from the previous Mindscapes in any order, is it possible to use fewer colors than you used in the previous Mindscape and still have no two vertices of the same color joined by an edge?

Further Challenges

26. **Getting more colorful.** Graphs don't have to be planar to have their vertices colored. How many colors would you need to color K_5 such that any two vertices joined by an edge get a different color? What's the smallest number of colors you need to color $K_{3,3}$ such that any two vertices joined by an edge get a different color? Are you surprised at your answer?

27. **Chromatic number.** The chromatic number of a graph is the smallest number of colors you need to color each vertex such that any two vertices joined by an edge get a different color. Looking at the previous Mindscape, what's the chromatic number of K_5? What's the chromatic number of $K_{3,3}$?

28. **Chromatic questions (ExH).** Find the chromatic number for each graph below.

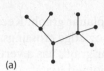

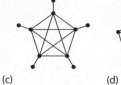

(a)   (b)   (c)   (d)

29. **Chromatically applied.** There are eight radio stations broadcasting in your region of the state. The graph shows each station as a vertex, with an edge drawn between pairs of stations that are too close to be assigned the same radio frequency. How many different frequencies do you need in order to insure that there will be no interference between stations?

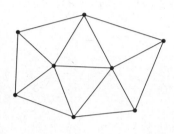

30. **Do four colors really suffice?** Suppose this map shows the provinces of a country. A peculiarity of this country is that it has one province comprised of two separate pieces, the striped regions. Thus when coloring this map, a cartographer must use the same color on these two separate pieces. Is it possible to complete the coloring using only four colors? Does this suggest a need to clarify the Four Color Theorem?

In Your Own Words

31. **Personal perspectives.** Write a short essay describing the most interesting or surprising discovery you made in exploring the material in this section. If any material seemed puzzling or even unbelievable, address that as well. Explain why you chose the topics you did. Finally, comment on the aesthetics of the mathematics and ideas in this section.

32. **With a group of folks.** In a small group, discuss and actively work through the reasoning involved in proving the Six Color Theorem. After your discussion, write a brief narrative describing the methods in your own words.

33. **Creative writing.** Write an imaginative story (it can be humorous, dramatic, whatever you like) that involves or evokes the ideas of this section.

34. **Power beyond the mathematics.** Provide several real-life issues—ideally, from your own experience—for which some of the strategies of thought presented in this section would provide effective methods for approaching and resolving them.

For the Algebra Lover

Here we celebrate the power of algebra as a powerful way of finding unknown quantities by naming them, of expressing infinitely many relationships and connections clearly and succinctly, and of uncovering pattern and structure.

35. **Planar puzzle.** Someone chalked a planar graph on the sidewalk in front of the student union. It has twice as many edges as vertices and twelve fewer regions than three times the number of vertices. How many vertices does the graph have? Will the chalker be called to the Dean's Office?

36. **More puzzling.** Three of your friends tell you they've seen a graph chalked in front of the math building. Leonhard says the graph is planar, with no loops or multiple edges. Kazmir says the graph has 50 vertices. Kenneth says the graph has 150 edges. If you know that Leonhard and Kazmir are correct, what's your reaction to Kenneth? If you know Leonhard and Kenneth are correct, what's your reaction to Kazmir?

37. **X marks the spot.** A drawing of a planar graph has V vertices, E edges, and F regions. Suppose you know a number x such that $V = x^2 - 14x$, $E = x^2 - 12x - 8$, and $F = x^2 - 11x - 36$. Find the number of vertices, edges, and regions in your graph drawing. Draw a graph with these values.

38. **Maximizing E. (H)** Suppose a planar graph has twelve vertices and no loops or multiple edges. What's the largest number of edges it can have?

39. **Vertex z.** A drawing of a planar graph has V vertices, E edges, and F regions. Suppose you know a number z such that $V = z^3 - 6z^2 + 6$, $E = z^3 - 5z^2 + 20$, and $F = 2z^2 - 5z - 8$. Find the number of vertices, edges, and regions in your graph drawing.

6.4 NETWORKING

Using Graphical Models to Find the Shortest, Closest, and Cheapest

© Robert Churchill/iStockphoto

How edgy is your own social network?

Computer networks, social networks, transportation networks—no matter which webs ensnare us, we're faced with networks. The central common feature shared by each of these objects that we call networks is the connections between things. Computer networks connect computers, social networks connect people, transportation networks connect cities or airports. The fundamental information that a network records is which pairs are connected. Joe and Jane are friends on Facebook. American Airlines offers direct flights from Los Angeles to Dallas and from Dallas to Boston. A road map tells the distances between cities. Basic questions about these real-world networks include: For marketing, how many friendship connections between Jim and Josie? For easy travel, how many stops from Los Angeles to Boston? For a delivery service, how short a distance must we travel to deliver a set of packages? To answer these questions about networks, we can use a strategy that we have seen many times before: We focus on the essentials.

In all networks, the essentials are the objects and their relationships with one another (the connections). When we boil the network down to those two essential ingredients, we realize that any network can be modeled by capturing its core features in a *graph*. A mathematical model is a structure that captures one or more key aspects of a real-world scenario. In future chapters we will see mathematical models used to explore population growth (in Chapter 7), predict the future (in Chapters 8 and 9), and measure risk and rewards (in Chapter 10). Here we'll discover that graphs provide a very useful type of model for representing structures that involve connections. In earlier sections we saw how graphs can model several real-world puzzles like the Königsberg Bridge Puzzle and the Gas-Water-Electric Puzzle. In this section, we will see how graphs can model many real-world scenarios, from Facebook to traveling salespeople, and can help us to answer optimization questions about a wide variety of different networks since all these networks are essentially graphs.

The strategy of modeling real-world scenarios by graphs allows us to make better sense of many complex situations by focusing our attention on the important connections involved. Modeling real-world questions using graphs is so important that one of the yet-unsolved questions in this area has a $1 million prize offered for its solution. So reading this section might make you… *a millionaire!* Stay tuned.

Facebook

Few Friends Today; A Googolplex Tomorrow

Think back to the first few minutes after you created your original Facebook page. You probably started out with just a few friends, and some of those friends were probably friends with each other. The following diagram might represent those early moments after you initially opened up your account:

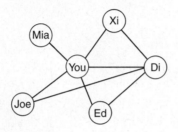

You and your friends are represented by large dots and your friendship connections are represented by line segments or arcs. We can simplify the diagram somewhat to obtain:

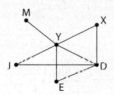

This figure is a drawing of the graph model for your initial Facebook friendship relations. The dots, your Facebook friends, are the vertices of this graph and each edge indicates that the two people at the ends of that edge are Facebook friends. Notice that we could have drawn the graph differently.

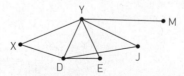

Because this new drawing shows the same vertices (representing you and your friends) with the same friendship connections, it is a drawing of the same graph, even though the drawing looks different. The essential features of the graph model are you and your friends (the dots or vertices) and the specified pairwise connections between them (the edges). A drawing of this model is just a way to visualize it and to make the connections clear, and there are many different ways to render a drawing that captures those same interconnections.

Of course, it wasn't long after you launched your Facebook page that you had a hundred friends, many of whom were friends with each other. If you tried to draw a graph to model this scenario, you would have 101 vertices (you and your hundred closest friends) and probably thousands of edges showing a web of friendship connections. There would be 100 edges joined to your vertex alone—certainly a complicated picture. But a computer can store all these vertices and edges without even one microchip breaking a sweat. Many graphical models of real-world situations are complicated to draw but are easy for a computer to store and exploit.

This partial American Airlines route map is a graph that models the flights that they offer. The vertices are the cities, and the edges represent the existence of a direct flight. So this graph captures information about which cities are connected by American Airlines flights to the Caribbean.

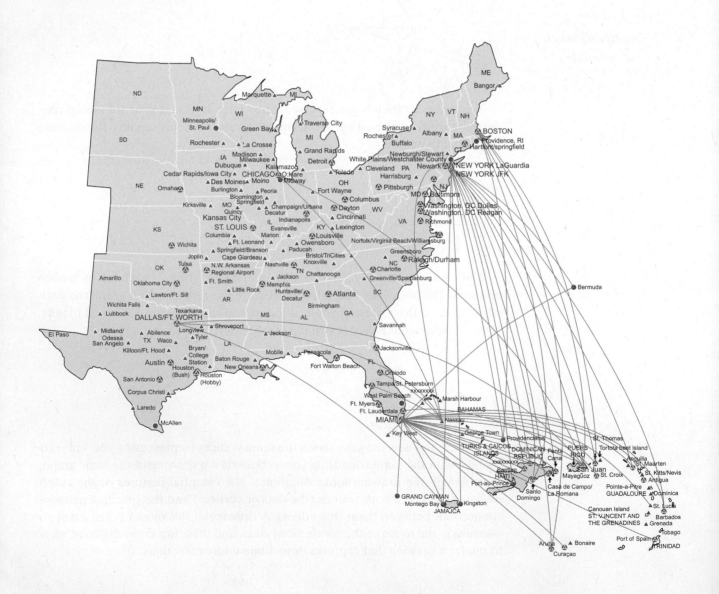

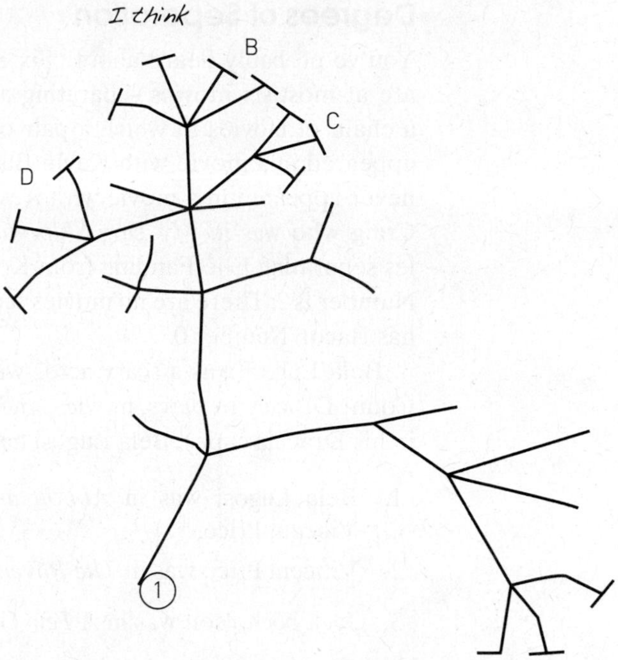

I think

B

C

D

1

Life Lesson

Different situations may have common structure.

Biologists use graphs to represent the relationships among species. Here is naturalist Charles Darwin's first sketch of an evolutionary tree, which appears in his First Notebook on Transmutation of Species (c. 1837).

A concept map shows ideas and connections between ideas. One good way to study a topic is to draw a concept map that makes clear to you what ideas are connected to what other ideas. Here is a concept map for finding a summer job.

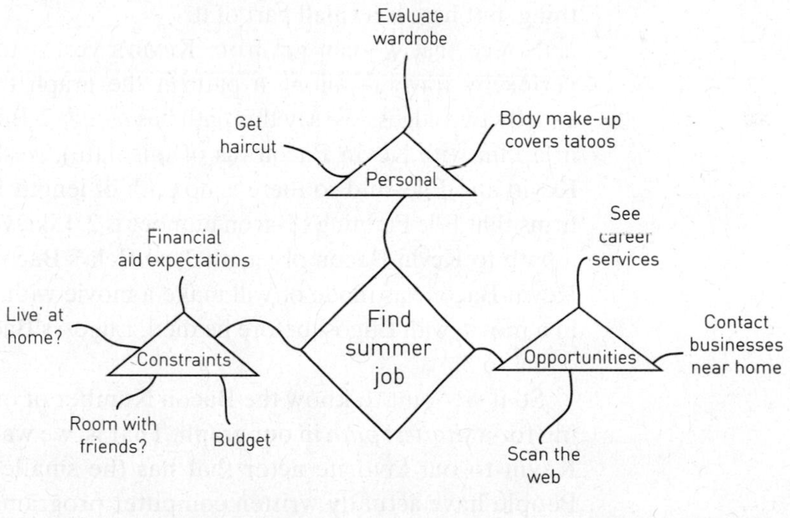

Evaluate wardrobe

Get haircut

Body make-up covers tatoos

Personal

Financial aid expectations

See career services

Live at home?

Find summer job

Constraints

Opportunities

Contact businesses near home

Room with friends?

Budget

Scan the web

By looking at examples, we see that many scenarios in the real world can be described as networks of objects or ideas that are connected. Such networks are well modeled by graphs. They help clarify basic structure and can also help us answer questions including the number of stops necessary to fly from Nome, Alaska to Naples, Florida or the connections of infected people who could spread a deadly disease or how far Hollywood actors are separated from Kevin Bacon.

Degrees of Separation

You've probably heard about "Six Degrees of Kevin Bacon": that is, that there are at most six movies separating any Hollywood actor from Kevin Bacon via a chain of movies in which a pair of actors both appeared. Any actor who has appeared in a movie with Kevin Bacon has Bacon Number 1. Elle Fanning has never appeared in a movie with Kevin Bacon, but she was in *Déjà vu* with Polly Craig who was in *My Dog Skip* with Kevin Bacon, so this chain of two movies separating Elle Fanning from Kevin Bacon means that Elle Fanning's Bacon Number is 2. There are no movies separating Kevin from himself, of course, so he has Bacon Number 0.

Bela Lugosi was a scary actor who died in 1956 and was famous for playing Count Dracula in plays, movies, and beyond (the great beyond … he was buried in his Dracula cape). Bela Lugosi has a Bacon Number of 3 because:

1. Bela Lugosi was in *Abbott and Costello Meet Frankenstein* (1948) with Vincent Price;

2. Vincent Price was in *The Raven* (1963) with Jack Nicholson; and

3. Jack Nicholson was in *A Few Good Men* (1992) with Kevin Bacon.

So even semi-ancient, scary, dead actors have modest Bacon Numbers.

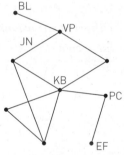

We can model Six Degrees of Separation from Kevin Bacon with a graph. Let each actor be a vertex and draw an edge between any pair of actors who have appeared in a film together. This graph has thousands of vertices and millions of edges, so we won't even try to draw the entire thing, but here is a small part of it:

Notice that we can get from Kevin's vertex to Elle's vertex by traveling along a path in the graph that has exactly two edges. We say this path has *length 2*. Because Elle has never appeared in a film with Kevin Bacon (as of this date), we know there is no edge joining Kevin and Elle, and so there is no path of length 1. Thus, this shortest path confirms that Elle Fanning's Bacon number is 2. Likewise, notice that Bela Lugosi has a path to Kevin Bacon of length 3, so Bela's Bacon Number is at most 3. Unless Kevin Bacon has made or will make a movie with some really old actor who was in a movie with Lugosi before he died, Lugosi's Bacon Number is and will remain equal to 3.

So if we want to know the Bacon Number of our favorite actor, we are looking for a *shortest path* in our graph. That is, we want to find a path of edges from Kevin to our favorite actor that has the smallest possible number of edges. People have actually written computer programs available on the Internet to find the Kevin Bacon number and path of "film links" for any actor you want. This information is extremely useful and interesting—especially if you happen to be Kevin Bacon. Finding the shortest path in a graph modeling airline routes will find you the flights with the fewest layovers. Finding the shortest paths in a graph modeling the spread of an infectious disease might help pinpoint the source of an outbreak. There are good algorithms for finding shortest paths in

Life Lesson

Focusing on underlying structure allows you to answer many questions at once.

Paul Erdös (1913—1996)

a graph, which you may enjoy if you study graph theory in a future course.

If you are intrigued by the idea of Bacon Numbers, your inner mathematician will be happy to know that many years before Kevin even starred in his first film, math fans had a similar idea and coined the term Erdös Number. Paul Erdös was a prolific and highly influential Hungarian mathematician. He traveled widely, collaborated with hundreds of colleagues, and was the author or co-author of over 1500 papers. Mathematician Caspar Goffman is thought to be the first to use the term "Erdös number" in a paper published in 1969. If you have published a paper with Erdös, your Erdös Number is 1. If you have not published with Erdös, but you have published a paper with someone who has, then your Erdös Number is 2. Of course, if you are Erdös, your Erdös Number is 0. Occasionally, the criteria for calculating Erdös numbers are eased a bit, as in the case of homerun king Hank Aaron. He is considered to have Erdös number 1 because he and Erdös autographed the same baseball.

The authors of this textbook have small Erdös numbers. Burger wrote a paper, "On the decomposition of vectors over number fields," with Jeffrey D. Vaaler, who wrote a paper, "Multiplicative functions and small divisors" with Erdös (as well as another mathematician named Alladi). So Burger's Erdös Number is 2. Starbird wrote a paper, "Products with a metric factor," with Mary Ellen Rudin, who wrote a paper, "A non-normal box product," with Erdös. So Starbird's Erdös Number is also 2. Thrilling! The authors are still waiting for an agent to call asking us to appear in a blockbuster movie so that we can get a Bacon Number. (We hope Kevin will be in it.)

The Bacon Number and Erdös Number are examples of a phenomenon called the Six Degrees of Separation, which basically states that almost any two people on Earth can be connected by a chain of six or fewer people, each one of whom has personally met the next person in the chain. Think about how long the chain would be to get from you to more or less anyone else reading this book. Here's how:

1. You probably had a mathematics teacher, who probably went to some presentation about mathematics;

2. That presenter probably went to some other presentation about mathematics given by someone;

3. That someone probably went to a presentation by one of the authors of this book.

4. The author probably spoke at some conference attended by some other teacher;

5. That teacher probably attended some other conference attended by another teacher;

6. And that teacher was a mathematics teacher of the other reader.

It is very possible that there might be an even shorter chain. Thus all *Heart of Mathematics* readers are connected and form a tight-knit family.

You are closely linked with essentially every person on the planet by a short chain of these direct connections. It is comforting to know that the world is so

Life Lesson

Quantifying impressions can alter perspectives.

small; however, it is also a bit frightening. This insight means that a person infected with a deadly disease has been in a room with a person who was in a room with a person who was in a room with a person who was in a room with a person who was in a room with a person who was in a room with you. We can only hope that some of these folks along the line covered their mouths as they sneezed.

From finding a short chain between pairs of people, we now turn to a different challenge whose goal is efficiency. When we seek to assemble disco balls at a lively party, our minds naturally spin toward graph theory to minimize extension cords.

Powering Disco Balls

Your family is planning a big outdoor party. There will be eight tables under a big tent, arranged as follows:

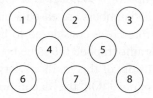

To set the perfect mood, they want to have a spinning disco ball hanging above each table. This means, among many other things, that they need to run extension cords so that there is electricity over each table. Being the family expert in mathematical thinking, your help is enlisted to find an efficient arrangement of cords. You know there is an outdoor electrical outlet near Table #1 and that between any two tables you need at most one extension cord. Assuming you can connect a cord to as many other cords as you wish, what's the smallest number of extension cords you need to have electricity at each table?

Did you get 7? Create a graph to model your proposed set-up, with the tables as vertices and the extension cords as edges. Maybe you drew something like this:

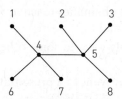

The graph you created to model the arrangement of extension cords to get electricity to each disco ball must have had two properties. First, the graph must be *connected*, that is, each vertex in the graph can be reached from any other vertex in the graph by following a path of edges in the graph. If the graph were not connected, not all the disco balls would be connected to the power source and those balls would not be able to spin. The second property your graph must have is that it contains no *cycles*; that is, you can't start at a vertex, traverse a sequence of edges without going over the same edge twice, and get back to where you started. If your graph had a cycle, you would be able to remove at least one extension cord and still leave all the disco balls powered and spinning.

Graphs with these two properties occur so often that they have a special name. A *tree* is a graph that is connected and contains no cycles. Here are some more examples to illustrate trees.

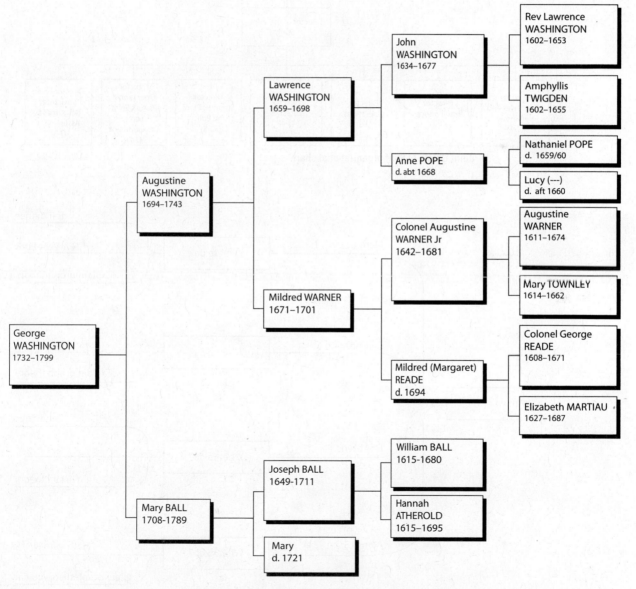

These graphs are trees.

These are not trees
G is not connected; H has cycles.

Trees occur frequently as graph models. Here are some more examples:

George Washington's ancestral tree.

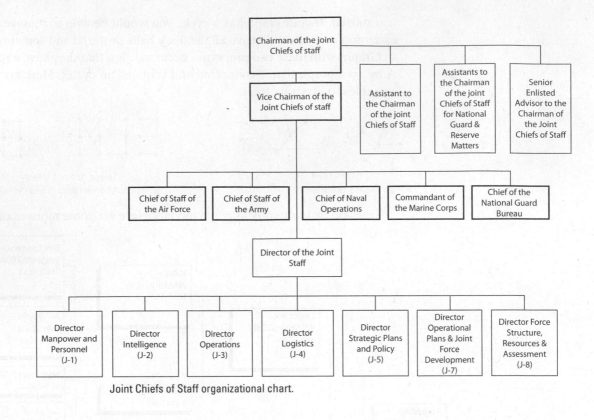

Joint Chiefs of Staff organizational chart.

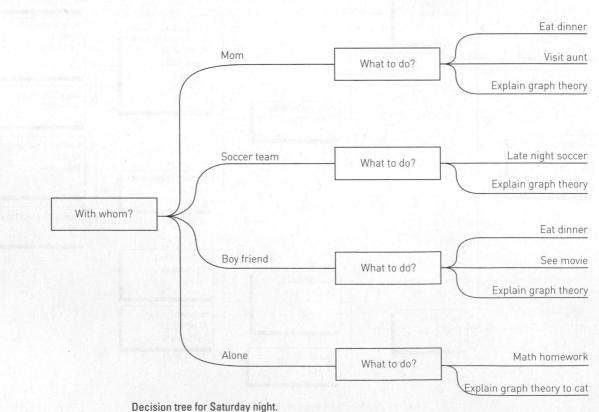

Decision tree for Saturday night.

Spanning Trees

The strategy of creating graphs as models of situations in the real world often allows us to find the best way to accomplish a task. Frequently when we are trying to minimize cost, we find trees springing up all around us. Putting in cable lines is a great example. Suppose you are building an apartment complex and want to lay cable lines in an

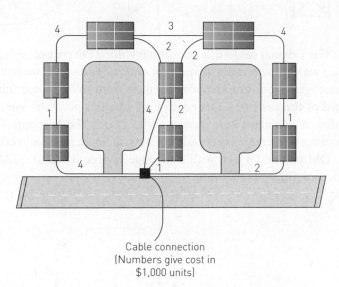

efficient way. The cable company offers the following schematic showing estimated costs for installing various segments of the cable lines. Your company always watches the bottom line, so you want to choose the segments that will allow cable access in each apartment building, but with no unnecessary connections. Which segments would you choose? Why?

To simplify the situation, we can begin by first investigating the issue while ignoring the costs; that is, we can explore how to hook every apartment

Cable connection
(Numbers give cost in
$1,000 units)

up with cable lines without having unnecessary connections. We can model this scenario with a graph: let the buildings be our vertices and the edges be the possible cable lines. Include a vertex for the cable connection at the street as well. Now our simplified challenge has been replaced with a question about our graphical model. Our goal now is to select edges from the graph to produce a connected graph that contains every vertex.

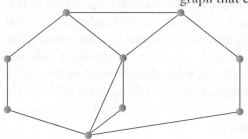

The result will be a graph that includes all the vertices in our original graph but with just enough edges from that original graph to be connected. If we had selected so many edges that those edges created a circuit, then we would be able to discard at least one of those edges while still reaching every vertex. So the graph we select will be a tree that contains all the vertices of the original graph. Such a tree is called a *spanning tree* of the original graph. Here are a couple of spanning trees in the cable-line graph. The edges of the spanning trees are shown in red.

Life Lesson

When faced with a difficult task, do a related, easier task first.

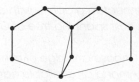 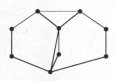

Most graphs have many different spanning trees. Here are some more examples of graphs, in each of which we have highlighted in red just one of several possible spanning trees. In each graph locate for yourself one or two different spanning trees. Notice that for each graph, you can travel from any vertex to any other vertex by following a path of edges in the given spanning tree, but there are no cycles in the spanning trees.

  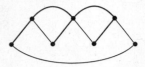

The original cable-line question included the cost of laying various cable lines. Can we choose a cable layout that *minimizes* cost? If we include costs on each of the edges in our graph model, then we want to find a spanning tree such that the sum of the costs of all its edges is as small as possible. We're looking for what's called a *minimum cost spanning tree*. Here's the graph model for the cable-line question with edge costs added. Can you find a minimum cost spanning tree?

Did you find a tree with total edge cost equal to $17,000? There's more than one possibility. Here's one:

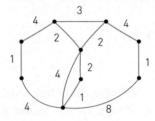

How did you create your minimum cost spanning tree? Did you have a method or did you just guess? One reasonable method would be to start by picking the cheapest edge in the graph. Then you could pick the cheapest remaining edge as long as that edge does not create a cycle. By continuing to always take the cheapest edge that does not create a cycle, you will eventually produce a spanning tree. Surprisingly, this method of merely always choosing the cheapest edge without creating a cycle will always produce a spanning tree whose total cost is less than or equal to the cost of any possible spanning tree. Joseph Kruskal, Jr., discovered this method in 1956. This method is often used to minimize costs in many situations that can be modeled by a graph. Here is Kruskal's strategy more formally.

Life Lesson

Think about how you think about a specific question to generate general methods.

Kruskal's Algorithm.

Given a connected graph G with a cost on each edge, to find a minimum cost spanning tree of G do the following:

1. *Choose an edge of minimum cost to start.*
2. *Among all available edges that you haven't already chosen and that do not create a cycle, pick an edge of minimum cost and add it to your selected edges.*
3. *Repeat Step 2 until you create a spanning tree.*

Let's apply Kruskal's algorithm on the graph below:

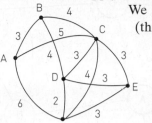

We choose the edge DF first because it is the cheapest (the smallest number). Then we can choose several of the edges with cost 3, as long as we don't create a cycle. One possibility is to choose edge DC, then edge DE, and then edge AB. At this point we can't choose any of the other cost-3 edges without creating a cycle, so we look at cost-4 edges. We could add DB or BC without creating a cycle. Let's take BC. We now have a spanning tree with total edge cost of 15, shown with bold edges below.

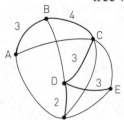

Why does Kruskal's algorithm always produce a minimal cost spanning tree? To verify this fact, let's start with any spanning tree T at all, and we will show that its cost is greater than or equal to the cost of the Kruskal tree. Our strategy is to modify T step by step without ever increasing its cost until it becomes the Kruskal tree.

Let's think about the edges being put on the Kruskal tree in the order prescribed in the Kruskal algorithm. Find the first edge e in the Kruskal tree that is not in the spanning tree T. Since that Kruskal edge e is not in T, if e were added to the spanning tree T, a circuit would be created. That circuit would have to contain a non-Kruskal edge f (because the Kruskal tree, being a tree, has no circuits). The cost of the edge f must be greater or equal to the cost of the Kruskal edge e, because e was the cheapest edge available when it was added. So if we substitute e for f in T, the total cost of the new T would stay the same or go down. If we continue through the edges of the Kruskal tree, repeating this process of modifying the spanning tree T, then when we are finished, we will have turned T into the Kruskal tree without increasing its total cost. Therefore, no tree T could have started with a lower total cost than the Kruskal tree has.

Kruskal's algorithm is very convenient because at each step we make the best available choice (the cheapest edge that doesn't create a cycle) without regard to what lies ahead, and yet that process leads to the best possible result. The real-world challenge of choosing the cheapest cable lines to connect buildings led us to model that situation with a graph and then to realize that finding minimal cost spanning trees would give the answer. Then we saw that clear thinking led to Kruskal's algorithm, which neatly solved our dilemma. Minimal cost spanning trees help us to efficiently connect a whole set of objects. We now turn our attention to another question about efficiency—a question best driven home by a road trip.

The Hamiltonian Road Trip

There's a smile on Hallie Hamilton's face. She's planning her first road trip as a dining Hall Culinary Consultant for StudyGrub Food Service. Hallie will be visiting eight colleges and universities whose dining halls serve delicious meals designed and supplied by her company. She'll meet with the dining hall staff and students at each school, and she looks forward to hearing everyone's upbeat and good-natured ideas about how to improve the dining experience for all. To avoid the potential

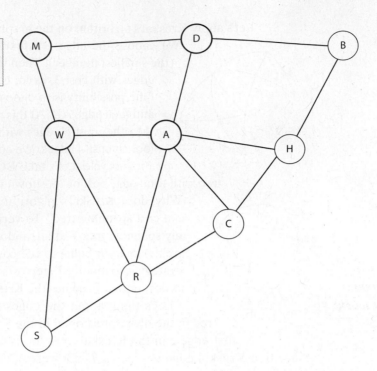

boredom of seeing the same sight twice during her trip, Hallie hopes to find a route that goes through each college town exactly once before returning home. Looking at a map, she is overwhelmed by all the information in front of her.

Fortunately, Hallie knows how to use a graph to capture the key features relevant to her trip: the colleges and the primary roads between them. She draws a graph showing each college as a vertex, with an edge between each pair for which a reasonable driving route exists. She simplifies the driving routes to look like straight lines, even though the roads are curvy and picturesque.

The graph has nine vertices (eight colleges: A, B, C, D, M, R, S, W and Hallie's home town H) and fourteen edges. Because Hallie wants to visit each college only once, she is looking to traverse selected edges in the graph in such a way that she visits each vertex only once. Unlike the case of the Königsberg Bridge Puzzle in Section 6.1, she does not care about traversing each edge in the graph. She needs to visit every vertex, but would like to avoid traveling over any edge that takes her back to a vertex she has already visited. Look at the graph for a few minutes and see if you can find a route that starts at vertex H (home), visits every other vertex only once, and returns to vertex H.

In fact, there are several answers. Here's a list of edges that works:

HB, BD, DM, MW, WS, SR, RC, CA, AH

We see that every vertex is visited exactly once, except the starting and ending vertex. In contrast, recall that in an Euler circuit every edge is traversed exactly once (see Section 6.1). In Hallie's trip, she will visit each vertex once, but there are many edges in the graph that do not appear on the list, so this circuit is not an

Euler circuit. Instead, we call a circuit that goes through every vertex exactly once and returns to its starting vertex a *Hamiltonian cycle* (or *Hamiltonian circuit*) after the 19th century Irish mathematician, physicist, and astronomer William Hamilton.

In 1857, Hamilton invented the *Icosian game*, the goal of which was to find a path on the edges of the 12-sided dodecahedron that visits each vertex exactly once and returns to the starting vertex (see Section 4.5 for a detailed description of the dodecahedron).

This solid has graphical representations that show all the edges and vertices, though such drawings usually distort the faces.

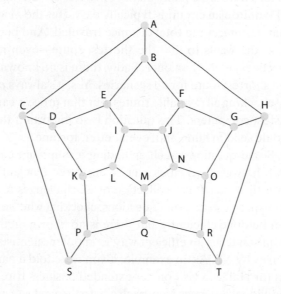

See if you can find a Hamiltonian cycle in this graph. (Does it matter what vertex you start with. Why or why not?)

Is There an Easy, Euler-like Answer?

In Section 6.1 we saw that an Euler Circuit in a graph is a circuit that traverses each edge exactly once and returns to the starting vertex. We figured out the precise conditions under which a graph will have an Euler circuit: namely, when it is connected and every vertex has even degree.

It seems reasonable that there must be an equally simple characterization of those graphs that have a Hamiltonian circuit. The completely counterintuitive reality is that we still don't know whether such a simple characterization exists or not! Mathematicians are still looking for a simple group of conditions that describe exactly those graphs having a Hamiltonian circuit. There are some conditions that describe some families of graphs with a Hamiltonian circuit, but they mostly boil down to requiring that the graph have lots of edges. There are also some conditions that make such a circuit impossible. You can think of some of these conditions for yourself, and we will explore a few of these in

Life Lesson

Seek simplicity.

Life Lesson

Not all questions have simple answers.

the Mindscapes. We can all look forward to future insights that may someday completely describe which graphs do have a Hamiltonian circuit and which ones do not.

The Traveling Salesperson Challenge and the Questionable "Equation" P = NP?

Hallie's business trip and Hamiltonian circuits are closely related to a famous question called the Traveling Salesperson Problem. Like Hallie, the traveling saleswoman wants to visit each city in her sales territory as efficiently as possible. There are usually plenty of routes for driving between the cities, so finding a Hamiltonian circuit is typically easy. But the *new* challenge is to find the route that *minimizes* the total distance traveled. And because this is a real-world question, she wants to find the shortest route—even if it means passing some cities (vertices) more than once. Sadly, there is no known efficient algorithm for finding the shortest route for all scenarios. Yes, it's always *possible* to find a shortest route by checking all available routes, but that process can take lots of time, even on the fastest computers. This question is so important that finding an efficient method would lead to fame and, even better, fortune.

Some questions, such as finding a minimum cost spanning tree or a shortest path between two cities, can be answered efficiently. Computer scientists denote with the letter **P** the collection of all challenges for which an efficient solution is known. For some other questions, checking whether a proposed solution is correct can be done efficiently, but there is no known method for finding such a solution from scratch in an efficient way. Computer scientists denote this collection of challenges by **NP**. As an example, if you were told a particular order of a hundred cities for Hallie to visit on her extended business trip, it would be easy to add up the individual distances from each city to the next city and answer the question, "Is the total distance for that trip less than 4000 miles?" However, it might be very difficult to determine whether some order of visiting the cities would be less than 4000 miles. There are an enormous number of ways to order 100 cities, so if you had to check the total distance for each possible ordering of the cities, it would take you longer than the history of the universe even on the fastest computers in existence. The number of orderings is itself a number with 158 digits!

One of the biggest unsolved questions in computer science and mathematics is whether these two collections of challenges are really the same: Does **P** (easy to solve challenges) = **NP** (easy to check but *apparently* hard to solve challenges)? In other words, maybe there exist efficient ways of answering the questions in **NP** but we just haven't found them yet. Determining whether **P** = **NP** is one of the seven Millennium Problems designated by the Clay Mathematics Institute in Cambridge, Massachusetts. Settling the **P** = **NP** Challenge would win you a prize of $1 million. So finding an efficient way to solve the general Traveling Salesperson Problem could lead you to that million-dollar prize.

Life Lesson

Thinking clearly can make you rich.

A Look BACK

WE'VE SEEN THAT GRAPHS ARE USEFUL TOOLS FOR MODELING real-world situations that involve people, cities, organizations, or any circumstances in which objects and pairwise connections are involved. By focusing on the basic features of a scenario, we can often answer questions about how to optimize our efficiency and effectiveness. This strategy of isolating essential features is what mathematical modeling is all about: a model can capture the basic features or relationships on which we want to focus, allowing us to answer questions or make wise predictions. Our world is full of models—maps, diagrams, equations, graphs—all of which help us to better understand and navigate in our complex world.

Our strategy of investigating the topic of modeling the real world with graphs was to ignore extraneous diversions, thereby revealing the central issues of objects and connections.

Life Lessons

Clear the clutter.

•

Focus on the essential.

•

Use simple cases to build up to more complicated ones.

MINDSCAPES Invitations to Further Thought

*In this section, Mindscapes marked (**H**) have hints for solutions at the back of the book. Mindscapes marked (**ExH**) have expanded hints at the back of the book. Mindscapes marked (**S**) have solutions.*

Developing Ideas

1. **Up close and personal.** Create a graph to model your friendship connections with a few (up to six or eight) of your friends. You and your friends are the vertices, with an edge between any two who are friends.

2. **Network lookout.** Find an example of a network online or in print.

3. **Tree house.** Find a spanning tree in the graph below.

4. **Hamiltonian holiday (S).** You are interning for a nonprofit organization that works with similar agencies in Asia. Your boss is planning to visit the cities in the graph below. He knows you've studied graph theory and so asks you to find a Hamiltonian circuit that starts in Beijing, travels to each of the other cities only once, and returns to Beijing. Is there more than one such route?

5. **Home style.** Create a graph to model the rooms in your house as a network. Let each room be a vertex, and draw an edge between two vertices if there is a doorway or archway between the rooms. Treat hallways as separate rooms. Feel free to ignore closets. (If your house has more than one story, use just the rooms on one floor to avoid the confusion of stairways.) How many vertices does your graph have? How many edges?

Solidifying Ideas

6. **Six degrees or less.** Suppose this graph is a model for your mom and her Facebook friends, along with some of their friends, and their friends' friends. The people are the vertices, with an edge between two people if they are friends on Facebook. Let the *Mom number* of a person be the smallest number of edges you need to travel in the graph to get from that person to your mom. So your mom has Mom number 0 and because you're her friend (right!?), you have Mom number 1. How many vertices in the graph represent people who

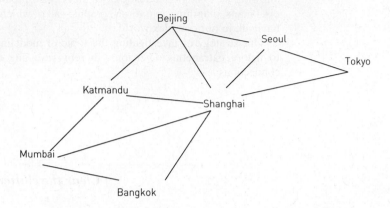

have Mom number 2? Is there anyone with Mom number 3? Can you find four people who are all Facebook friends with each other?

7. **Degrees of you.** Find ten willing friends or dorm-mates. Draw a graph in which each of you is a vertex, with an edge between any two people who are in a class together. Note the vertices that are distance one away from you (people with whom you share a class). Note the vertices that are distance two away from you (people you don't have a class with, but who you have a classmate in common with). Are there any vertices left? Are there any vertices you can't reach using a path of edges in the graph?

8. **Campus shortcut.** Find a map of your campus and create a graph model that shows the main buildings or other locations you visit regularly. Each location is a vertex, with an edge between two vertices for which there is an easy campus route between the locations that doesn't pass through a third location. If one of the vertices is your dorm, find a shortest path from your dorm to each of the other vertices. (Measure shortness in terms of the number of edges in your graph, not the distance each edge represents.)

9. **Arborist lesson.** Which of the graphs below are trees? Explain.

10. **Seeking spans.** Find a spanning tree in each of the graphs below.

  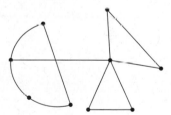

11. **Span or not? (H).** Each graph below has some edges marked in red. For each graph, determine whether the red edges (and incident vertices) form a spanning tree. Explain.

 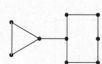

12. **Minimum cost.** Find a minimum cost spanning tree in the graph below.

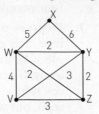

13. **Kruskal cost (H).** Use Kruskal's algorithm to find a minimum cost spanning tree in this graph. List the edges in the order in which you choose them. Is there more than one such tree? Explain.

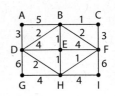

14. **Span or no span.** Does the graph below have a spanning tree? Explain.

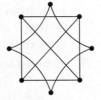

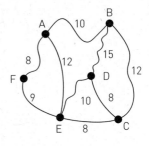

15. **Paving savings.** The graph shows the roads in a rural county in your state. All the roads are gravel, but the county plans to pave certain roads to make travel easier. They want to pave only enough roads so that people can travel from each town to any other town on paved roads, even if they can't always take the shortest route. The value on each edge shows the cost of paving that road in $10,000 units. Which roads would you choose to pave in order to keep the cost as low as possible? Explain your answer.

16. **Hamilton Study.** Look at the graph you drew to model part of your campus in section 6.4, Mindscape 8. Is there a Hamiltonian circuit in your graph? Why or why not?

17. **Business trip redux.** Look back in the section and find the graph Hallie created to help plan her business trip. Find a Hamiltonian circuit that differs from the route given in the text.

18. **Handling Hamiltons.** For each graph below, find a Hamiltonian circuit or explain why one doesn't exist.

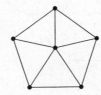
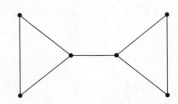

19. **Road trip.** You are checking out graduate programs in basket weaving. The best schools are in the eight cities represented by vertices in the graph here, one of which is your home vertex H. Is there a way you can leave home, visit each city exactly once, and return home? Why or why not?

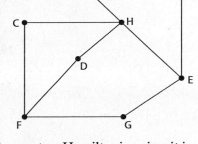

20. **Back to Hallie's trip.** Look back in this section and find the graph Hallie created to help plan her business trip. If you want to create a Hamiltonian circuit in this graph, are there some edges you must use? Why? Is it possible to create a Hamiltonian circuit that uses the edge AW? What about AD? Explain.

21. **Solve the Icosian Game.** Find a Hamiltonian circuit on the dodecahedron. Use the graph model given in the text on p. 447.

Creating New Ideas

22. **Hunt for Hamilton (S).** A large island country has all its cities on the coast. All the airline routes between cities fly over the interior of the island. The graph shows the cities as vertices and the airline routes as edges. Can you find a Hamiltonian circuit in this graph? Why or why not?

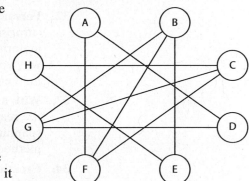

23. **Has no Hamilton.** Give some characteristics that make it impossible for a graph to have a Hamiltonian circuit. Explain your answer. (See Mindscapes 6.4.18 and 6.4.19 for ideas.)

24. **Cubing Hamilton (ExH).** Can you find a Hamiltonian circuit on the edges of a cube? That is, can you start at one corner of a cube, traverse edges so that you visit each of the other seven corners exactly once, and return to your starting corner?

25. **Hamiltonian path.** A Hamiltonian path is a path in a graph that visits each vertex exactly once, without necessarily returning to the starting vertex. Which of the graphs in Mindscape 18 have Hamiltonian paths?

26. **Sorry, no path.** Give some characteristics that make it impossible for a graph to have a Hamiltonian path. Explain your answer.

Further Challenges

27. **Unique minimum.** Suppose you have a connected graph with costs on the edges with no two edges having the same cost. Could this graph have more than one minimum spanning tree? Explain.

28. **Many minima.** How many different minimum cost spanning trees are there in the graph? (We consider two spanning trees to be different if there is at least one edge in one tree that's not in the other.)

29. **Guaranteed span.** What condition(s) must be put on a graph in order to guarantee that it has a spanning tree?

30. **Leafing out (ExH).** A vertex in a tree that has degree one is called a leaf. Draw a tree with exactly two vertices. Draw a tree with three vertices. (These aren't very interesting trees!) Draw a tree with five vertices and another with seven. Count the number of leaves in each of your trees. Make a conjecture about the minimum number of leaves a tree with at least two vertices must have. Can you prove your conjecture is true?

31. **Edge count.** Look at all the trees you drew in the previous Mindscape. Count the edges in each tree. How does the number of edges relate to the number of vertices in the tree? Draw some more trees and see if your relation holds. Make a conjecture.

In Your Own Words

32. **Personal perspectives.** Write a short essay describing the most interesting or surprising discovery you made in exploring the material in this section. If any material seemed puzzling or even unbelievable, address that as well. Explain why you chose the topics you did. Finally, comment on the aesthetics of the mathematics and ideas in this section.

33. **With a group of folks.** In a small group, discuss and actively work through the reasoning involved in using Kruskal's algorithm to find a minimum cost spanning tree. After your discussion, write a brief narrative describing the methods in your own words.

34. **Creative writing.** Write an imaginative story (it can be humorous, dramatic, whatever you like) that involves or evokes the ideas of this section.

35. **Power beyond the mathematics.** Provide several real-life issues—ideally, from your own experience—for which some of the strategies of thought presented in this section would provide effective methods for approaching and resolving them.

For the Algebra Lover

Here we celebrate the power of algebra as a powerful way of finding unknown quantities by naming them, of expressing infinitely many relationships and connections clearly and succinctly, and of uncovering pattern and structure.

36. **Dollars and cents.** Your spanning tree has three edges with costs x^3, $x^2 - x$, and $2x$, all measured in dollars. The total cost of your tree is $7x$. Find the cost of each edge in your tree.

37. **Adding up.** Your spanning tree has four edges with costs z, $z/2$, $z/3$, and $z/4$, all measured in Euros. The total cost of your tree is 50 Euros. Find the cost of each edge.

38. **Small edges.** The two cheapest edges in your spanning tree have costs $54/(x+2)$ and $24/(x+5)$ and a total cost of 8. What is the cost of the cheapest edge? (You may need the quadratic formula.)

39. **Vertex search (H).** Your graph has a Hamiltonian circuit with $(x+67)/(x+4)$ edges. There are $(x+32)/(x+2)$ edges that are not in the Hamiltonian circuit, and a total of 17 edges in the graph. How many vertices does the graph have?

40. **Binary gossip tree.** You told a secret to two of your friends. Then they each told two of their friends. Then those friends each told two of their friends. And then those friends each told two of their friends. Model this gossip trail with a graph, using vertices for the people and an edge between each two people who shared the secret. (Assume there are no repeated people amongst all the friends.) How many people hear the secret in the last stage of this process? In all, how many now know the secret?

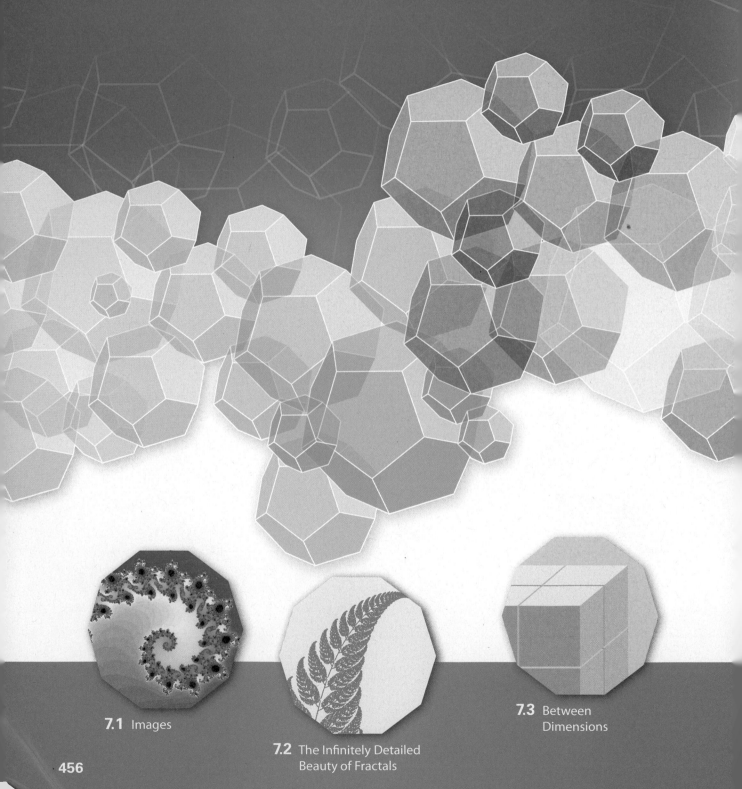

CHAPTER SEVEN

Fractals and Chaos

7.1 Images

7.2 The Infinitely Detailed Beauty of Fractals

7.3 Between Dimensions

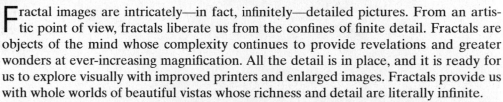

God has put a secret art into the forces of Nature so as to enable it to fashion itself out of chaos into a perfect world system.

Fractal images are intricately—in fact, infinitely—detailed pictures. From an artistic point of view, fractals liberate us from the confines of finite detail. Fractals are objects of the mind whose complexity continues to provide revelations and greater wonders at ever-increasing magnification. All the detail is in place, and it is ready for us to explore visually with improved printers and enlarged images. Fractals provide us with whole worlds of beautiful vistas whose richness and detail are literally infinite.

The images that follow form the starting point for our investigations into fractals and mathematical chaos. We will examine the images and search for patterns. In the process we will discover that the whole picture resembles small parts of the picture but on a larger scale. Once we see this self-similarity at different scales, we investigate processes that can generate such features; we find that repeated applications of simple transformations result in these beautiful images.

Fractals have significance beyond the images we can produce. They show us that simple processes, repeated many times, lead to objects of enormous complexity. Many mathematical models of natural phenomena use simple processes that are repeated many times. These models attempt to predict the weather, the stock market, and population size. These models reflect the chaotic complexity of our world and suggest that our ability to predict the future is severely limited.

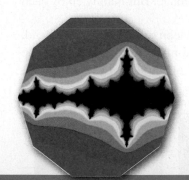

7.4 The Mysterious Art of Imaginary Fractals

7.5 The Dynamics of Change

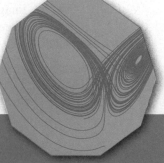

7.6 Predetermined Chaos

Viewing a Gallery of Fractals

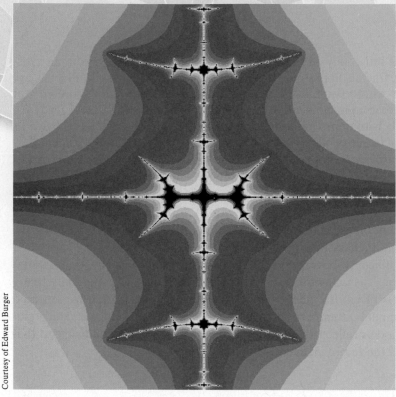

Courtesy of Edward Burger

A fractal

Not chaoslike, together crushed and bruised, But, as the world harmoniously confused; Where order in variety we see, And where, though all things differ, all agree.

ALEXANDER POPE

A natural first step in the study of fractals is simply to look at and enjoy some. Look for patterns and similarities within each image. But mainly enjoy each picture for itself.

Real or fake? These spectacular vistas have never seen the light of day. Instead, these images were created using a simple, repeated mathematical process. The images compel us to wonder whether nature relies on similar processes to produce itself.

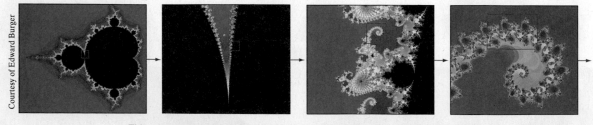

Courtesy of Edward Burger

This spectacular Mandelbrot Set provides an infinitely detailed tribute to Benoit Mandelbrot's contribution to the development of a whole new field of study. If graphic lines were infinitely fine and microscopes were infinitely powerful, you could increase the power indefinitely, focus on increasingly

© Michey/Age Fotostock America, Inc.

© Karin Sebelin/Age Fotostock America, Inc.

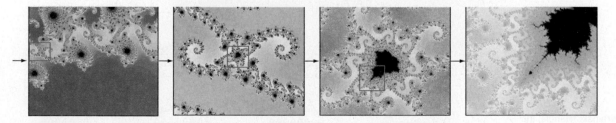

small sections, and see at each magnification new intricate details that possess a structure similar to the larger sections. As we will see in Section 7.4, the Mandelbrot Set is an intellectual triumph that spans mathematics and art.

Real, fake, healthful, or tasty? Answer: Real and healthful, certainly. Tasty, to some. Self-similar, definitely. The whole bunch of broccoli looks much like a sub-bunch, which looks much like a sub-sub-bunch.

Courtesy of Edward Burger

A bunch looks like . . .

a sub-bunch, which looks like . . .

a sub-sub-bunch.

Here they are together—note the relative size differences, although they look similar in the above photographs.

In the Sierpinski Triangle, the whole is identical to a part, which is identical to a sub-part, and so on—forever.

How many Quackers do you see? This Quacked Wheat box exemplifies a process of creating a picture within a picture within a picture, which is the basic construction of many incredible fractal images.

Sure we don't think of ferns as the breakfast of champions, but could this image arise from a Quacked Wheat box? Yes. The whole fern is not a fern at all but merely a Quacker-esque picture within a picture.

Gaston Julia never saw graphic renderings of the infinitely complex sets he conceived more than 70 years ago.

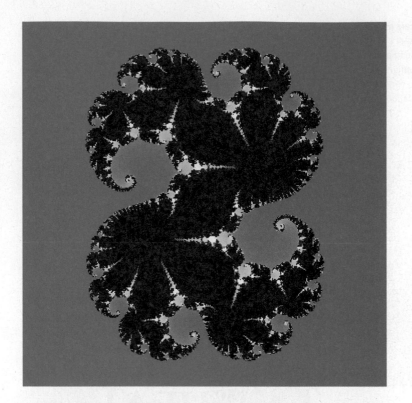

Courtesy of the authors

More of Julia's sets

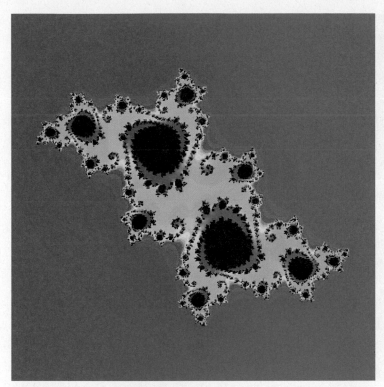

Yet another of Julia's sets

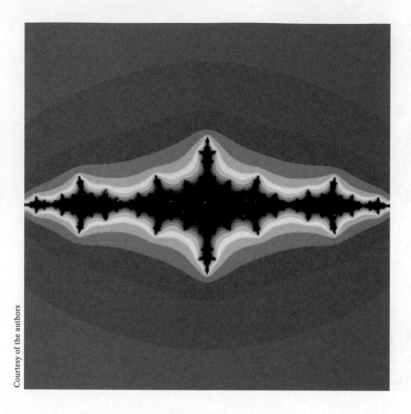

Courtesy of the authors

A Look BACK

These images are all infinitely intricate but arise from simple processes. They are the icons of fractals and chaos. We now move to discover and to understand their origins, structures, and nuances.

Life Lessons

Look for patterns.

MINDSCAPES Invitations to Further Thought

In this section, Mindscapes marked (H) have hints for solutions at the back of the book. Mindscapes marked (S) have solutions.

Seeing Things

Self-seeking. Each image in the preceding fractal gallery exhibits self-similarity to some extent. Here we ask you to uncover it.

1. **The incredible shrinking duck.** On the Quacked Wheat box, outline the sub-picture that is an identical, but reduced, copy of the whole picture. Roughly what fraction of the height is the reduced picture compared to the whole?

2. **Multiplicity (S).** In the Sierpinski Triangle, outline three sub-figures that are identical but reduced copies of the whole figure. For each sub-figure you outlined, compare its width, as a fraction, to that of the whole Sierpinski Triangle.

3. **Different sizes.** In the fern, find three reduced copies of the whole picture—one about 85% as large as the whole figure, the other two much smaller—that make up the whole picture except for the stem.

4. **Blooming broccoli.** In the bunch of broccoli, find three sub-bunches that look nearly identical to the whole bunch. Which would you rather eat, the whole bunch or the sub-bunch?

5. **Not quite cloned.** In the Mandelbrot Set shown, the whole is not identical to any sub-part. Find some sub-parts that nevertheless look similar to some yet smaller sub-parts.

6. **Julia's descendants.** Some of the Julia Sets pictured on pages 462–464 hang together in one piece, and some are made up of many pieces. For each Julia Set that is not connected, find two reduced copies of the whole Julia Set that make up the whole picture.

7. **Maybe moon.** What features of the fractal forgeries of the cratered vista make it look realistic?

8. **Exposing forgeries.** What features of the fractal forgery of the cratered vista expose it as a fake?

9. **Nature's way.** Find some examples of self-similarity in nature.

10. **Do it yourself (H).** Draw a figure that contains several reduced copies of itself.

In Your Own Words

11. **With a group of folks.** In a small group, discuss the images in this section. After your discussion, write a brief narrative describing some common themes that the pictures from this section share and how these images might be produced.

12. **Creative writing.** Write an imaginative story (it can be humorous, dramatic, whatever you like) that involves or evokes the images and ideas of this section.

7.2 THE INFINITELY DETAILED BEAUTY OF FRACTALS

How to Create Works of Infinite Intricacy Through Repeated Processes

Sierpinski Garden, 1997, Khaldoun Khashanah.

Sierpinski Garden (1997) by Khaldoun Khashanah. Studying and teaching fractals inspired Professor Khashanah to paint his interpretation of the landscape along the Delaware River.

> *I coined* fractal *from the Latin adjective* fractus. *The corresponding Latin verb* frangere *means to break: to create irregular fragments . . . how appropriate for our needs!*
>
> BENOIT MANDELBROT

Some artists are renowned for their attention to detail, producing pictures that are incredibly intricate. Yet only within the realm of mathematics can images be created that are literally infinitely intricate. But what does *infinitely detailed* mean? Could such an image be drawn with the finest penpoint? Of course not. But it is possible to describe such images so precisely that, for any point on our canvas, we can determine whether that point is in the design or not. These images can be drawn to any degree of detail that our finest printers can muster. But the totality of their intricacy is fully present only in our mind's eye. Every power of magnification reveals yet further detail. In some cases, we will understand the image so well that we can literally draw an enlarged version of the tiniest parts magnified a million, a billion, or even a trillion times. Other images reveal unthought-of variations and surprises at ever-increasing levels of magnification. These mysterious images give us new vistas to discover as we explore them in an ever-increasing depth, yet continue to hold deeper secrets for others to uncover later.

These images of the mind are created by following a pattern that is repeated over and over, forever. Barbers and hairstylists make us all familiar with this kind of process when they use two mirrors to enable us to see the backs of our heads, and we see hairdo after neatly cut hairdo receding into the distance. Feedback loops of video images generated by pointing a video camera at the screen of a monitor displaying the video feed are another example. Feedback processes create great pictures of infinite detail and intrigue. Let's begin exploring infinitely intricate images by looking at some specific examples.

466

Ad Infinitum

Example 1—Parallel Mirrors

Take a hand mirror and hold it facing a regular mirror. What do you see?

What you are seeing is the result of a repeating process. When you look in the big mirror, you see yourself holding a small mirror. But what can be seen within that small mirror? You see a reflection of the small mirror in the big mirror. But what is the reflection of the small mirror reflecting?

Light travels at finite speed; there are limits to the reflecting capabilities of mirrors; and eyes have limited acuity. So, the image we see within these real mirrors is not infinitely complex. However, we know what image we would see in an idealized pair of mirrors. That image would have infinite detail, because each image of the hand mirror would contain in it an image of a yet smaller hand mirror, and so on forever. Infinitely reflecting mirrors provide a good, concrete way to think about images with infinite detail.

Parallel mirrors are a real-world example of a repeating process that has a self-similarity property.

Example 2—Video Feedback

Here's another physical example. Take a video camera and set it up so that a monitor displays the image that the video camera sees. Video stores often set up cameras in this fashion. Now point the camera at the monitor. What do you see? The camera is taking a picture of the monitor. But the monitor is displaying a picture of what is being seen by the video camera. But the video camera is seeing the monitor, so the monitor displays a picture of the monitor, whose picture is the picture of the monitor, whose picture is a picture of the monitor, and so on . . .

Again, the physical constraints of pixels prevent video feedback from actually having infinite complexity, but in the abstract we can think of an infinitely detailed feedback image displayed on the monitor.

Example 3—Eat Your Wheat

Sometimes your day starts with infinite complexity—especially if you eat Quacked Wheat. On the cereal box, we see a Quacker holding a box of Quacked Wheat, in which we see a smaller picture of the Quacker holding a smaller box of Quacked Wheat, in which we again see the Quacker holding the Quacked Wheat, and so forth . . .

The parallel mirror, video feedback, and the Quacked Wheat box all represent infinitely detailed themes. In each of these illustrations, parts of the picture look identical to larger parts of the picture, but at a different scale. These pictures display self-similarity. We have now seen several examples of a similar theme. Our next step is to isolate the core ideas from these illustrations.

Life Lesson
Isolate key ideas from examples.

Richard Megna/Fundamental Photographs

This 19th-century Vaishravana Mandala from Tibet, used for meditation, captured some essence of self-similarity long before Quacked Wheat ever existed.

Self-Similarity

Consider broccoli. Broccoli is a food possessing a legion of virtues for health and well-being. We know, however, that some people do not like broccoli, and we are going to help those people by pointing out a feature of broccoli that will demonstrate that a little broccoli goes a long way. In fact, a little piece of broccoli is similar in structure to the entire head. The picture below shows a bunch of broccoli, and underneath it is an enlarged picture of a small piece of it.

Underneath the enlarged picture is an even more enlarged picture of an even smaller piece. Notice how similar those three pictures look. So, if you want to convince your mother or doctor (or both) that you are eating your broccoli, just eat the tiny, untasteable piece in the last picture, send them the blown-up photo, and they will be impressed with your diligent adherence to a healthful diet. *Bon appetit!*

Coastlines are wiggly. When we look at the coastline of the United Kingdom on a world globe, we will see a wiggly outline. Now look at a large map of the United Kingdom alone. We discover that each of the wiggles of the globe-level view has been magnified to reveal a finer construct of

Photographs of various pieces of broccoli at various magnifications

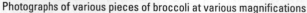

Same three broccoli pieces, now accompanied by a tastier treat to show the scaling

At each magnification we see wiggles on wiggles—at all scales we see self-similarity.

wiggles at a smaller scale. Now let's look at a map of Wales. Again, we see that the smaller wiggles comprise finer wiggles still. At each level of magnification, the wiggles have some resemblance to the image at previous magnifications—a real-world illustration of self-similarity across scales.

Coastlines and broccoli both capture the flavor of *self-similarity*—that is, the characteristic of looking the same as or similar to itself under increasing magnification. An object that has this distinctive self-similarity feature is referred to as a *fractal*. Coastlines and broccoli are examples from nature. We will soon see some purely geometrical fractals that so eerily resemble natural objects we will have to wonder whether the physical processes that produce self-similarities in nature might be similar to iterative processes that produce self-similarities in geometric examples.

A Process of Repeated Replacement

Pictures with infinite intricacy can be produced by a process of repeated replacement. We start with a picture made of line segments, replace each line by a particular jagged line made up of several shorter segments, then replace each of those shorter segments by the same, but smaller, jagged line, and so on. This procedure, if continued forever, produces objects of infinite detail. In 1904, Helge von Koch first described this type of spiky replacement. Let's see this process in action.

Koch's Kinky Curve

Let's start with a line segment and replace the segment with four equal-length segments each one-third as long as the original. Thus we have created a line segment with one kink in the middle. But, in this chapter, we never stop. Let's do it again. Replace each segment with four segments each one-third the length of the previous segments, as before. We now have kinks on the kinks. Let's continue and continue and continue, forever. The object we end up with cannot be drawn completely, because it has infinitely many wiggles. Let's look at this *Koch Curve*

Creation of the Koch Curve

under increasing magnification. Take the part of the curve that came from one of the first four parts, and enlarge it to be three times its size. The enlarged version is identical to the entire Koch Curve. The Koch Curve exhibits exact self-similarity. Looking at a part of the Koch Curve, you cannot tell if you are seeing it actual sized or magnified a billion times. Either way, it looks completely, exactly the same.

The Koch Curve is one example of an infinite replacement process. Let's ground our understanding of this process by looking at other specific examples that use infinite replacement—namely, the Sierpinski Triangle, the Menger Sponge, and, more generally, a special collage-making procedure.

Sierpinski Triangle

Waclaw Sierpinski first described his fractal triangle in 1916. We start with a filled-in triangle and replace it with a triangle with the upside-down middle triangle removed, as illustrated. This process leaves three filled-in triangles; each is a smaller version of the original triangle. We have therefore established a process of replacing a filled-in triangle with three smaller triangles contained in it. So, we can take the result of that first replacement process and replace each of the three smaller triangles with three yet smaller triangles. To each of the nine filled-in triangles now present, we can apply the same process. We continue forever.

Stage 0   Stage 1   Stage 2   Stage 3   Stage 4

Notice that the resulting object, known as the *Sierpinski Triangle*, looks exactly like any one of the three sub-triangles that comprise it, except for scaling. It also looks exactly like any one of the nine filled-in sub-sub-triangles at the next stage, and so on. The construction guarantees this self-similarity because the process of replacement is identical for each sub-triangle, as it was for the whole triangle.

Menger Sponge—A 3-Dimensional Example

Self-similar objects need not be confined to the 2-dimensional world of the plane. In 1926, Karl Menger described a neat example of a 3-dimensional fractal that has come to be known as the *Menger Sponge*. Here a solid cube is replaced by the 20 solid sub-cubes as shown. It's as though we got frustrated with a Rubik's cube and punched out its core on all faces. Continuing by replacing each of those 20 sub-cubes with 20 sub-sub-cubes yields 400 tiny cubes. This process can be continued forever, and the resulting object is known as the Menger Sponge. Notice that the whole Menger Sponge is precisely the same, up to scaling, as each of the 20 original sub-cubes, and it is also precisely the same as each of the 400 sub-sub-cubes at the next stage, and so on, forever.

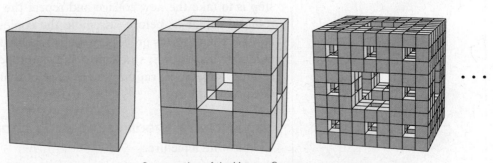

Construction of the Menger Sponge

© DeAgostini/SuperStock

Getting a Collage Education

Aren't collages attractive? A collage is made up of separate images artistically grouped together. Although the collage gathers together many individual pictures, these pictures give one overall impression. Here we are going to do an experiment in collage-making that illustrates the ideas of self-similarity and repeating processes.

In making collages, some people prefer to use just one picture, make several reduced copies of it, and assemble them to create a collage based on one initial theme. In other words, they take their favorite picture (probably of themselves), make several smaller copies of it, and then assemble them artistically on the page.

Some people (such as the authors) go to extremes. The collage they just created is so attractive that they decide that it should be the initial picture to be used to craft yet another collage. So, they take a photograph of the entire collage and basically start the process all over. They make the same number of copies as before, each copy reduced in size as before. They then assemble the copies into a collage by configuring the copies in the way they did previously, and lo and behold, they have a new collage.

Courtesy of the authors

Courtesy of the authors

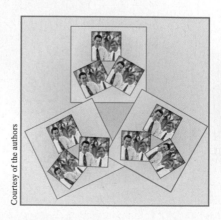

Courtesy of the authors

Once started on this road to collage making, who could stop? The next step is to take the new collage and repeat the whole process. Then make reduced copies as before, assemble the copies as before, and create yet another collage. This process could be repeated 100 times, 1000 times, or a million times. But why stop there? Why not repeat the process forever? The resulting object will capture the essence of that collage. Let's pin this idea down and explore it.

Instructions for Creating Repeated-Image Collages

1. Start with a picture.
2. Make some number of copies, each reduced by a specified amount.

3. Position each reduced picture on a page in a specified location, creating a new picture on the page (which is really a collage of pictures).

4. Start with the resulting picture and return to step 1, repeating this cycle forever.

We'll elaborate and illustrate each of these collage-making instructions so that anyone following the steps will create the same work of art.

1. Start with a picture.

2. Make a specific number of copies: 5, 10, or any number. Each copy is reduced by a specific amount. For example, copy 1 may be 1/3 the original size, copy 2 may be 1/2 the original size, copy 3 could be 3/4 the original size, copy 4 could be 1/1000 the original size, and so on. Each of the copies is reduced by a specific amount and will be reduced by that same amount each time.

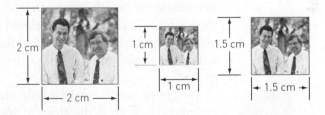

3. Position each reduced copy of the picture on a page in a particular place at a particular angle to create a new picture. It could even be flipped over. For example, the center of copy 1, positioned vertically, might be at the point 3 inches to the right and 4 inches up from the bottom of the page. Copy 2, rotated 20° counterclockwise, might be centered at the point 5 inches to the right and 6 inches up. Copy 3 might be centered . . . and so on. Notice that some of the pictures may even overlap one another.

4. This new picture is then used to start the whole process again. That is, exactly the same number of copies of this picture are made, each reduced to the same sizes as before.

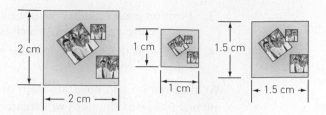

They then are positioned in exactly the same places as the first copies were positioned. The result is another picture that is a collage of a collage. As you see, it really consists of reduced copies of reduced copies of the original picture. Repeat this whole process again and again, each time using the result of the previous creation as the initial picture for the next collage.

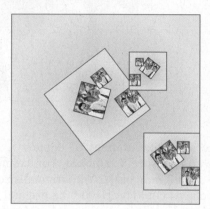

After you repeatedly apply these instructions, you will find that the resulting collages start to settle down in the sense that you see slighter and slighter differences between the resulting collage and the previous one. In other words, the picture we see after the 100th step will look almost identical to the picture we see after the 101st step. However, the full intricacy of the whole infinite process is not actually present except in our minds as we imagine that the process has been repeated infinitely many times.

Intriguing images can be constructed using this collage-generating mechanism. In fact, you can go to the *Heart of Mathematics* Web site to create your own fractals from your own initial images using this collage method. See for yourself how easy it is to become a fractal artist—just keep clicking the "repeat" button!

In addition, this method of generating images offers some surprises, including the reality that even artistically challenged people can create spectacular images. Before we explore such tantalizing possibilities, let's see some of the images we can construct using this infinitely iterated collage method. As a familiar illustration, we return to the Sierpinski Triangle.

Sierpinski Triangle Encore

We start with a filled-in equilateral triangle and then make three smaller copies, each half as tall and half as wide as the original. We position them to create

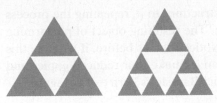

a three-triangle collage as shown at left. These three reduced copies and their positions specify this collage-generating instruction set. We take the resulting picture, make three reduced copies, and place them in the same arrangement again. We continue forever. The resulting image, the Sierpinski Triangle, has infinitely many holes. Notice if we take the Sierpinski Triangle and perform the operation on it (make three reduced copies and arrange them as specified), we will reconstruct exactly the Sierpinski Triangle.

Barnsley's Fern

Perhaps the images and objects constructed so far appear somewhat formal and sterile. So, here we see how Michael Barnsley developed a fractal that looks uncannily lifelike. This beautiful and lifelike fern is actually created by using a collage-making process, and, in fact, it is rather simple in that only four reduced copies of an original image are needed to produce this intricate and naturalistic result. It differs from our previous collage-making method in that some of the reduced copies shrink the length and the width by different amounts; in fact, one of the four reductions squashes the width down to zero. Let's see this process in action.

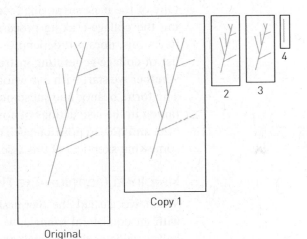

Original

Copy 1

2   3   4

We'll begin with a simple caricature of a fern. First, we draw the four reduced copies separately so we can see how they are shrunk. The first copy is 0.85 as large as the original in height and in width. The second copy shrinks the width to 0.3 its original width and the height to 0.34 its original height. The third copy shrinks the width to 0.3 its original width and the height to 0.37 its original height. The fourth copy shrinks the width to zero and shrinks the height to 0.16 the original height. These copies are positioned on the page as shown at left. Notice that the third copy is flipped over as well as rotated. Also notice that the copies overlap, which is allowed.

As always, let's take the original picture and repeat the collage-making process (make four reduced copies and position them) to make a collage. Next, let's take that image and apply the collage-making instructions to it, repeating the process once, twice, three times, and infinitely often. The resulting object of performing this operation infinitely is *Barnsley's Fern*. Notice that, as before, if we take this intricate image and perform the operation on it (make four reduced copies and arrange them as specified), we will reconstruct Barnsley's Fern exactly again.

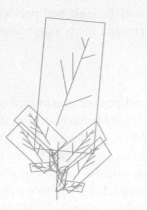 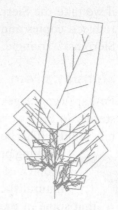 

Barnsley's Fern

Artistic Ability Unnecessary

One of the most surprising features of this method of creating images by following the collage-making process is that the final outcome, after infinitely many iterations, does *not* depend on the original, initial image. If we have a specific list of collage-generating instructions, the final collage picture will be the same whether we start with our whole page blackened, a single dot, or the *Mona Lisa*. The form, beauty, and substance of the final, delicate picture are completely contained in the instructions of how many copies to make, how much to shrink each one, and how to position each one rather than in the initial image. You might be somewhat skeptical of this assertion, so let's see why it's true.

Sierpinski Triangle—The Final Bow

When we created the Sierpinski Triangle using the collage method, we started with an equilateral triangle, we made three smaller copies, each half as tall and half as wide as the original, and we placed them in three positions on the page. Let's follow these exact same instructions, but this time let's start with a square instead of a triangle. Notice that the first figure does not look too much like the Sierpinski Triangle; however, let's take that image and apply the collage-making instructions to it. Now we are getting closer. Let's take the resulting picture, make three reduced copies, and place them in the specified positions again. We now have a blocklike Sierpinski Triangle. After several more repetitions of this process, notice how the images are becoming much more like the Sierpinski Triangle. If we look very closely at the last image, we will still see tiny squares making up the image; however, the impression now is of triangles, not squares. Continue forever. The resulting image is *exactly* the Sierpinski Triangle.

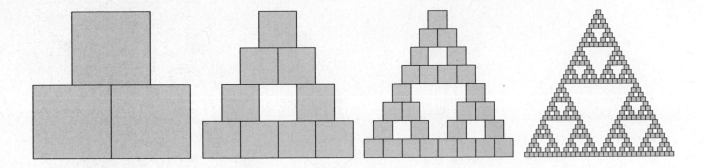

Just for fun, we show the first five applications of the Sierpinski Triangle collage-making instructions, one starting with a dot, one starting with a squiggle, and one even starting with the Sierpinski Triangle. Notice that, in each case, the resulting object is becoming the Sierpinski Triangle. These examples suggest that the collage-making instructions, not the starting image, determine the final picture. The starting image is irrelevant.

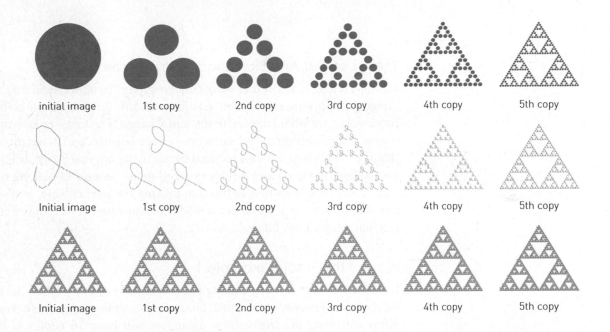

Barnsley's Fern Regrown

Let's reconstruct Barnsley's Fern from a different initial drawing. Let's try it starting with something that doesn't resemble a fern at all—for example, a limousine. We make the four reductions exactly following the instructions as before, arrange those reduced images as specified, and repeat the process. Notice how that mechanistic symbol, a car, transforms itself into the ecologically sensitive fern. If we were philosophically inclined, we would say that the study of fractals is helping humanity create a bridge between industrial coldness and natural beauty—lucky for you, we are not so inclined.

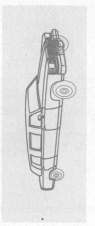  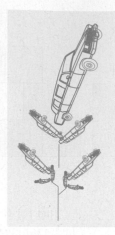 

It's Not Where We Begin, It's How We Get There.

If one repeatedly applies a specific set of collage-making instructions, starting with any initial image, the infinite collage process will produce the same result.

The Proof that Any Start Leads to the Same Result

It is heartening to know that we cannot go wrong. To understand why any starting image produces the same result, let's just carefully go through the collage-making process and see what happens to the initial image. The first step is to make several copies of the starting image, each one reduced in size, and place those copies on the page. Each one of those reduced copies of the original image, being part of the first collage, is copied and further reduced in the process of making the next collage. Each copy is further reduced in making the next collage, and so on. So, at any stage, the collage we create is made of some number of reduced copies of the original image. Let's be more specific.

Idea of the Proof—Example 1

If the instructions call for making four copies, for example, then the first collage will have four copies of the initial image, although they may overlap one another. After following the instructions again, we will have 16 copies of the original image, since the first collage was made of four copies of the original image and that whole collage was copied four times. After the next stage, there will be 64 copies, and in general there will be 4^n copies of the original image making up the nth-stage collage. But since at each stage the copies are reduced in size, all copies of the original image become tinier and tinier with subsequent repetitions of the collage-making process. Soon they become so tiny that what we see after many steps is not determined by the original images, each copy of which is so tiny that we can't see it anyway; the picture is determined by the locations of 4^n greatly reduced copies of the original image. The position of each tiny, multiply reduced copy of the original image is determined by the instructions that tell us where to put each copy.

Idea of the Proof—Example 2

As a matter of fact, if we start with a single point, even the shrinking factors are not necessary. In the case where the collage-making instructions call for four copies, the first "collage" would simply consist of 4 points positioned according to the location instructions. Those 4 points would be duplicated four times to create 16 points positioned per instructions, then duplicated four times to create 64 points, and so on. Notice that after 100 steps, there will be 4^{100} points, each placed where a reduced image would have been placed had we started with an image. But, after reducing any image 100 times, it would be indistinguishable from a point anyway. So, we've just shown that any starting picture produces the same final collage.

Modeling the World

If we want to model a picture with an infinitely detailed fractal approximation, there is a simple way to do it. Just sketch the goal picture and cover it with a set of images that are reductions of an initial picture to define the collage-making instruction set. This process of putting the pieces together generates some amazing artificial images that suggest reality. Examine the two images on the next page, and guess which is an authentic photograph taken by a photographer, and which is a fractal imposter created by a mathematician. The fractal fake is strongly convincing, and repetition

© Markus Genn/Age Fotostock America, Inc.

© Andrey Stenkin/iStockphoto

of simple rules lies at the heart of the fractal image. In fact, if we throw in a bit of randomness every once in a while, then the resemblance to reality and nature is eerie. Such fractal images are so striking that we can't help but wonder whether the processes by which natural wonders are created might similarly involve repetitious applications of simple rules together with a bit of randomness. This theme of

Image originally created by IBM Corporation.

© olikli-3d-rf/Alamy Limited

Image originally created by IBM Corporation.

© Alamy Creativity/Alamy Limited

Image originally created by IBM Corporation.

© Karin Sebelin/Age Fotostock America, Inc.

randomness suggests an alternative method of constructing fractals. (By the way, the real photograph by a photographer is the image on the left.)

The Chaos Game

Repeating the collage-making instructions creates intricate images, but another way to generate such delicate images is by random luck. We now illustrate this assertion by describing what is often referred to as the *Chaos Game*.

Let's number the three vertices of an equilateral triangle 1, 2, and 3. We start at any vertex and randomly choose a number: 1, 2, or 3. (For example, we could roll a regular die and let 1 or 6 mean 1; 2 or 5 mean 2; and 3 or 4 mean 3). Whatever number we generate, we move halfway from where we are toward that numbered vertex and make a dot and remain at that point. We then roll again to determine the next number and move halfway from where we are toward that numbered vertex and make a dot. We repeat, repeat, repeat, creating a sequence of dots. Computers can easily be programmed to repeat this process quickly. As we make the dots we will notice that the marks do not resemble a jumbled random collection of points filling up the triangle, but instead they begin to describe a particular pattern. If we continue for many steps, we will see the Sierpinski Triangle materialize right before our eyes. As we continue forever, the dots we draw will fill in the Sierpinski Triangle in increasingly finer detail. In fact, if you go to the *Heart of Mathematics* Web site, you can simulate this Chaos Game and watch the fractal materialize. Let's think about why this process will generate the Sierpinski Triangle.

Remember that each of the three sub-triangles in the Sierpinski Triangle is identical to the whole Sierpinski Triangle, except smaller. So, taking the whole Sierpinski Triangle and moving each point to the point halfway toward the top vertex, for example, just reduces the entire Sierpinski Triangle to the upper sub-triangle. Similarly, if we start with the whole Sierpinski Triangle and move halfway toward either of the other two vertices, the whole Sierpinski Triangle gets shrunk to become exactly one of the three sub-triangles. This observation tells us that, if we start with a point on the Sierpinski Triangle and move halfway toward any vertex, we will land on another point of the Sierpinski Triangle. So, although our process of generating the dots is random, we are certain that the dots we make are part of the Sierpinski Triangle.

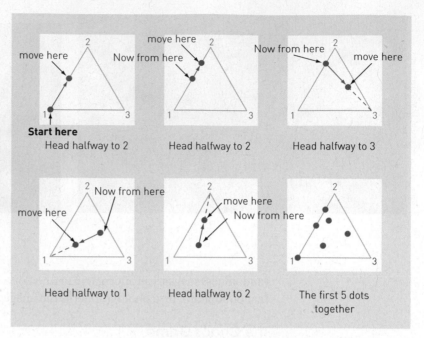

First five stages if the random numbers were 2, 2, 3, 1, 2, . . .

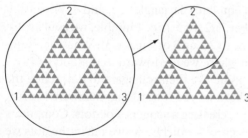

We now claim that the dots will tend to fill the *entire* Sierpinski Triangle in the sense that, as we follow the sequence of dots, we get arbitrarily close to any point in the whole Sierpinski Triangle. By getting close to each point of Sierpinski's Triangle, the pointillistic drawing we are creating will evolve into the whole Sierpinski Triangle in ever-increasing detail. So, we need to ask ourselves why this random sequence of points will tend to fill in the whole Sierpinski Triangle.

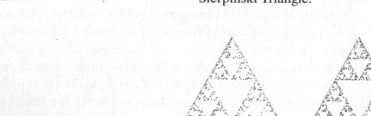

500 dots   1000 dots   1500 dots

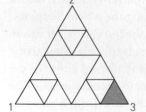

How will we visit points in the shaded lower-right, lower-right triangle?

Let's begin with an easy question. Suppose we start with the lower-left vertex. Is it possible that we will *never* visit a point in the upper triangle? Well, that is possible if we *never* randomly generate a 2. But, if we are randomly generating infinitely many numbers all from the collection 1, 2, and 3, and each number is just as likely to arise as any other, as a practical matter, we will eventually generate a 2. As soon as we do, no matter where we were before, we will move into the upper triangle. Similar arguments convince us that we will eventually land somewhere in each of the three sub-triangles. But notice that each sub-triangle is itself made of three sub-sub-triangles. Can we be certain of visiting each of those nine sub-sub-triangles? For example, will we visit the lower-right lower-right sub-sub-triangle? What kind of "dice throws" would get us into that small sub-sub-triangle? Well, all we have to do is randomly generate two 3's in a row. If we are randomly picking 1, 2, or 3 infinitely often, sometime along the way we will pick two 3's in a row, and, whenever we do that, we will find ourselves in the lower-right lower-right sub-sub-triangle.

How will we visit points in the shaded lower-right triangle?

If we consider one or two more levels, then we will begin to see the rationale for why our dots fill the whole Sierpinski Triangle. For example, why are we basically certain to occasionally land in the lower-right sub-sub-sub-triangle in the lower lower-left sub-sub-triangle in the lower-left sub-triangle? Or, another way to ask this question is, What sequence of random numbers will guarantee that we will land in that sub-sub-sub-triangle? We can check that any sequence of the form 3, 1, 1 will land us in that sub-sub-sub-triangle. But, if we pick numbers infinitely many times, that specific sequence will eventually come up—in fact we will see that sequence infinitely many times in our endless list of random digits. If you are intrigued by this notion of randomness, then you'll be happy to know that we will consider this concept in greater detail in the next chapter . . . stay tuned.

At this point we can see three things:

- For any sub-sub-sub- . . . -sub-triangle in the whole Sierpinski Triangle, there exists some specific sequence of numbers that will take us there.

- If we generate numbers randomly infinitely many times, then that specific sequence of numbers will eventually appear.

- Every such sequence generated takes us to a point in the Sierpinski Triangle.

So, every tiny nook and cranny of the Sierpinski Triangle will eventually be visited (but no other points will be hit), and the Sierpinski Triangle will be drawn to that precision.

Even Random Roads Lead to Fractals

The Chaos Game illustrates how a random process can lead to a predictable picture. Similar random processes give an alternative way to construct infinitely detailed models of the world and other fractal images arising from the collage-making process. In fact, there are theories that the stock market and even one's heart rate exhibit this type of behavior. Thus we see that chance, together with some simple rules, leads us to the infinitely intricate world of fractals, a world that quite possibly overlaps with our own physical world. However, some of the most beautiful examples of fractals—as we'll see in Section 6.4—arise out of abstract mathematical ideas. Before turning to those abstract works of fractal art, in the next section we will take a closer look at the intricate fractals we've created here and wonder if they have length, area, or volume.

A Look BACK

INFINITELY DETAILED images can be created by repeating simple processes infinitely many times. These fractal images can be created by feedback loops, by iterated collage constructions, or by random processes. Following a set of collage-making instructions will lead to the same image no matter what picture we start with. Starting at a point and randomly moving halfway toward one of three points sketches the Sierpinski Triangle pointillistically. All these fractal images are infinitely detailed in the sense that they display more intricacy at every increased magnification. These models lead to compelling pictures that seem to capture some of nature's vistas.

Examples of fractals in everyday life suggest the idea of pictures that can have infinite detail. Parallel mirrors, video feedback loops, and cereal boxes all create images that would be infinitely detailed if they were carried to their logical extremes. All these are generated by repeated applications of a process. We can repeat other processes, for example, repeatedly replacing part of a picture (such as a line segment) with a modified version (such as a bent segment) or repeatedly replacing a collage by a collage of itself. By using variations of the idea of repeating a process, we create intriguing images.

Pushing ideas to an extreme often leads to new and unexpected insights and often clarifies essential features. Looking at extremes is a means of applying ordinary objects or ideas in uncommon ways. After we have an idea, we can pin it down and explore its consequences by looking for extensions, variations, and alternative methods.

Life Lessons

Simple repeated processes can lead to complex and interesting outcomes.

•

Take ideas and look for extensions, variations, and alternatives.

•

Explore ideas by carrying them to the extreme.

MINDSCAPES   Invitations to Further Thought

In this section, Mindscapes marked (**H**) *have hints for solutions at the back of the book. Mindscapes marked* (**ExH**) *have expanded hints at the back of the book. Mindscapes marked* (**S**) *have solutions.*

Developing Ideas

1. **A search for self.** What does self-similarity mean?
2. **Desperately seeking similarity.** Which of these objects display self-similarity at different scales: broccoli, cupcake, coastline, fern, your left foot?
3. **Too many triangles?** At stage 0, the Sierpinski triangle consists of a single, filled-in triangle. (See page 471.) At stage 1, there are three smaller, filled-in triangles. How many filled-in triangles are there at stage 2? How many at stage 3? What's the pattern? How many triangles are there at stage 4? How many will there be at stage n?

4. **Counting Koch.** Look at the early stages of the Koch curve on page 470. The top figure is stage 1; it has four line segments. The next figure is stage 2. How many line segments does it have? How many line segments do you think there are at stage 3? (Count them to check your answer!) What's the general pattern?

5. **Argyle art.** Take four line segments and place them to form the X configuration as shown in the figure. Make four copies half the size of the original and arrange them in the X formation shown on the right. (Note that each half-size X aligns with one of the branches of the larger X). Follow these instructions to create the third stage of the collage.

Solidifying Ideas

6. **Nature's way.** Find several examples of objects in nature that, like broccoli, display self-similarity at different scales.

7. **Who's the fairest?** Can you position three mirrors in such a way that in theory you could see infinitely many copies of all three mirrors? Either describe the positions of the mirrors or describe why this feat is not possible.

8. **Billiards and mirrors.** On an idealized, square billiard table, you want to hit a shot that traces a perfect diamond. To line up your shot, you take four mirrors and position them in the centers of each side wall. Before you remove the mirrors, you make your shot directly along the line from the center of one mirror to the center of the next. As you look into the facing mirror, how many billiard balls do you see?

9. **MTV.** You've become a rock star and consequently feel that there can never be too many of you. So you make a video of infinitely many appearances of yourself. Do it! Arrange a TV, a video camera, and yourself as described in

"Example 2—Video Feedback" (page 467) to produce a picture of infinitely many copies of yourself and your guitar admiring infinitely many copies of yourself and your guitar. If you have a decent voice, you can sing on the video; otherwise, call Milli Vanilli.

10. **Photo op.** Suppose you arrange two mirrors facing each other at a slight angle, as shown. Place a camera parallel to one of the mirrors. Snap the picture. The picture will contain many increasingly smaller pictures of the camera. Will they be arranged going off to the right, the left, or up?

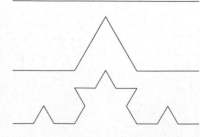

11. **How many me's?** Arrange mirrors and camera as in Mindscape 10, but put yourself in the picture right above the camera. Suppose the mirrors are about 5 feet apart, and the shutter is opened 1/1000th of a second. Light travels at about 186,000 miles per second, and a mile is 5280 feet. If the photographic paper or the pixels were sufficiently detailed and sensitive, roughly how many of you will you see in the photo? Is that enough?

12. **Quacker, Quacker, Quacker.** Suppose you had the job of shooting the photograph for the next Quacked Wheat box, and the picture was to be one of you holding a Quacked Wheat box. Of course, the picture on the Quacked Wheat box you are holding should be of you holding a Quacked Wheat box, and so on. Suppose you have a camera but no mirror; however, you do have a photocopier that can reduce images to any size. How can you create the picture for the box?

13. **Sierpinski hexed (S).** Take an equilateral triangle and remove a hexagon from the center as shown at right, leaving three equilateral triangles with sides 1/3 the length of the sides of the original triangle. Repeat the process of removing a hexagon from each sub-triangle. Repeat, repeat, repeat, forever. Make a sketch of what remains after one step, two steps, and three steps. Describe the locations of infinitely many points that are left after infinitely many steps. Let's call the remaining points *Sierpinski Dust.*

14. **The Kinks (ExH).** Koch's kinky curve is created by starting with a straight segment and replacing it with four segments, each 1/3 as long as the original segment. So, at the second stage the curve has three bends. At the next stage, each segment is replaced by four segments, and so on. How many bends does this curve have at the third stage? The fourth stage? The nth stage?

15. **Four times (H).** Draw a picture on a square piece of paper. Now make a collage by reducing your picture to 1/4 its length and width and making four copies.

Divide another square sheet of paper into four squares, and center each copy of the picture in each square to make the second stage of the collage. Making the four reduced copies and positioning them in the four squares constitute the collage-making instructions. What does your collage look like at the third stage? The fourth stage? Sketch your guess of what the collage looks like at the final infinite stage.

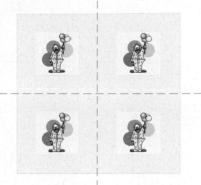

16. **Burger heaven (S).** Sketch a picture of a hamburger, and make three reduced copies, two reduced to 1/4 the original length and width, one reduced to 1/2 its original size. Position them as shown to create the second stage of the collage. Follow these same instructions and sketch the third stage and the fourth stage, and sketch a guess of what the infinite stage looks like.

Second stage of burger collage

17. **Ice cream cones.** Draw a picture of an ice cream cone, and follow the same instructions as in Mindscape 16. Sketch the third stage and the fourth stage, and sketch a guess of the infinite stage. How similar are your results compared to the burger heaven results?

18. **Sierpinski boundary.** Take the boundary of a triangle. Make three reduced copies, each reduced to 1/2 the length and width. Arrange them as you did when making the Sierpinski Triangle. Repeat the process and draw the result.

19. **Catching Z's (H).** Take a Z. Put in nine smaller Z's, as shown, to create the second stage. If the smaller Z's are 1/6 as long as the large one, roughly how long is the line through the Z's at the third stage if the line through the original, big Z is 24 centimeters long?

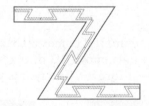

20. **Replacement pinwheel.** Take a 1, 2, $\sqrt{2}$ right triangle. Divide it into five identical similar right triangles, as shown at right. Repeat the process in each triangle. Draw the picture that you get after three steps.

21. **Koch Stool (ExH).** Start with a line segment, mark it into three equal pieces, and replace the middle piece with a small square missing its bottom. Now, for each of the five smaller line segments, repeat this procedure. Draw the next two iterations of this process. If we repeat this process forever, we can call the final result a *Koch Stool*.

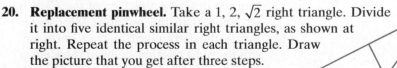

22. **Koch collage stool.** Given the Koch Stool described in Mindscape 21, explain how you can create the figure using the collage method.

23. **Sierpinski shooting.** Suppose that you were playing the Chaos Game to create the Sierpinski Triangle and you rolled 1, 1, 3, 2. Shade in the sub-sub-sub-sub-triangle in which you would land. How about if you rolled 1, 3, 3, 1? How about 3, 2, 2, 2?

24. **Sierpinski target practice (H).** What sequence of numbers will land you in the sub-sub-sub-triangle labeled *a* if you are playing the Chaos Game to create the Sierpinski Triangle? What sequence would land you in *b*? How about *c*?

25. **Cantor Set.** Start with the interval [0, 1]. Build a 1-dimensional collage by making two copies of the interval, each shrunk to 1/3 the original length, and putting them at the ends of the interval. Using these collage-making instructions, draw the next two stages of the collage process. The result of doing this process infinitely many times is called the *Cantor Set*.

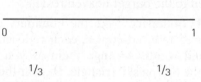

0 ────────────────────────────── 1

¹/₃ ──────── ¹/₃ ────────

Creating New Ideas

26. **Cantor luck (H).** Start with the point 0. Flip a coin. If it comes up heads, move 2/3 of the way toward 1, if it comes up tails, move 2/3 of the way to 0 from wherever you are at the time. Repeat forever. The points you find are drawing a picture of the Cantor set defined in the previous Mindscape. Verify that any point in the Cantor Set will move to another point in the Cantor Set under the coin-flipping-and-moving process. Notice that the points will approximate the whole Cantor Set as you continue the process.

27. **Cantor Square (S).** Take a square. Make four copies, each reduced to 1/3 the length and width of the original. Position these four copies in the corners of the

original square. Repeat these collage-making instructions. Draw the next two stages of the construction. We will call the result of doing this process infinitely many times the *Cantor Square*.

28. Cantor Square shrunk. Take the Cantor Square. Describe why applying the collage-making instructions to it (making four reduced copies and putting them in the corners of the original square) results in the exact same Cantor Square that you started with.

29. Cantor Squared. Draw the four corners of a square, label them 1, 2, 3, 4, and start at one corner. Randomly choose a number from 1 to 4. Move 2/3 of the way to the corner whose number comes up from wherever you are at the time. Repeat forever. Verify that, with this process, any point in the Cantor Square will move to another point in the Cantor Square. Notice that, as you continue the process, the points will approximate the whole Cantor Square.

30. Hexed again. Suppose you start with the three vertices of an equilateral triangle labeled 1, 2, 3. Start at any vertex and randomly choose numbers 1, 2, or 3. When you choose a number, move two-thirds of the way to that numbered vertex from where you are. Show that the points you generate will result in the Sierpinski Dust described in the Sierpinski hexed Mindscape 13.

31. Pinwheel spun. Take a 1, 2, $\sqrt{5}$ right triangle. Divide it into five identical right triangles. Repeat the process in each triangle. Draw the picture that you get after three steps. Describe the relationship between this fractal construction and the construction of the Pinwheel Pattern that was a nonperiodic covering of the plane described in Section 4.4.

32. Antoine's necklace. *Antoine's necklace* is a delicate necklace indeed. The first stage is a solid doughnut, and the second stage results from replacing it by eight linked sub-doughnuts. Each of these eight sub-doughnuts is in turn replaced by eight sub-sub-doughnuts, and so on, forever. Draw the next stage of Antoine's necklace.

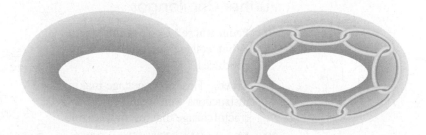

33. Menger jacks. For the game of jacks, let's imagine constructing a six-pointed metal jack whose ends are square. Begin with a solid cube. Make seven reduced copies, each reduced to 1/3 the size in each dimension, and position the seven little cubes in the original cube to create a jack. Draw the result of performing the collage-making instructions again.

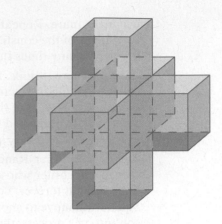

34. **A tighter weave.** Look back at the Tight Weave story in Chapter 1 (page 9). There you constructed a pattern by dividing a purple square into a 3 × 3 grid of nine squares, making the central square gold, and repeating the process in each of the remaining eight squares. The purple remaining after repeating that process infinitely often is sometimes called a *Sierpinski Carpet*. Here let's modify the process. Instead of dividing the original square into a 3 × 3 grid and making the central one gold, suppose we divide the square into a 5 × 5 square and make the central one gold. Then take each of the 24 remaining squares, divide it into a 5 × 5 grid and make the middle one gold, and continue this process forever. Starting with a square as stage 1, draw the first three stages of this replacement process.

35. **A looser weave.** Let's modify the carpet-designing process yet again. Instead of dividing the original square into a 3 × 3 or 5 × 5 grid and making the inner one gold, suppose we divide the square into a 4 × 4 square and make the central 4 squares gold. Then take each of the 12 remaining squares, divide each of them into a 4 × 4 grid and make the central four gold, and continue this process forever. Starting with the square as stage 1, draw the first three stages of this replacement process.

Further Challenges

36. **From where?** Look at this fractal. What collage instructions would produce it?

37. **Treed.** Describe collage-making instructions that would result in a fractal collage that resembles a tree.

38. **Flaky (H).** Describe collage-making instructions that would result in a fractal collage that resembles a snowflake.

39. **How big a hole?** In the Tighter weave (34) and Looser weave (35) Mindscapes, how much of the total pattern in each is gold after infinitely many steps?

Cantor Set

0   1

40. **4D fractal.** Describe a fractal in 4-dimensional space that is analogous to the Menger Sponge. How many sub-hypercubes would you divide a hypercube into? How many would you remove to create the next stage?

In Your Own Words

41. **Personal perspectives.** Write a short essay describing the most interesting or surprising discovery you made in exploring the material in this section. If any material seemed puzzling or even unbelievable, address that as well. Explain why you chose the topics you did. Finally, comment on the aesthetics of the mathematics and ideas in this section.

42. **With a group of folks.** In a small group, discuss the infinite collage-making process. After your discussion, write a brief narrative describing the process and why any starting picture yields the same result if you continually repeat the same collage-making instructions.

43. **Creative writing.** Write an imaginative story (it can be humorous, dramatic, whatever you like) that involves or evokes the ideas of this section.

44. **Power beyond the mathematics.** Provide several real-life issues—ideally, from your own experience—that some of the strategies of thought presented in this section would effectively approach and resolve.

For the Algebra Lover

Here we celebrate the power of algebra as a powerful way of finding unknown quantities by naming them, of expressing infinitely many relationships and connections clearly and succinctly, and of uncovering pattern and structure.

45. **Graphing Koch.** Plot the following points in the xy-plane:

    $A = (0, 0)$, $B = (6, 0)$, $C = (9, 4)$, $D = (12, 0)$, $E = (18, 0)$

    Draw line segments from A to B, B to C, C to D, and **so on.** What fractal does your figure resemble the first stage of? (It's not an exact copy—that is impossible if we want all whole number coordinates for the points).

46. **Koch adds life.** Look at the points given in the previous Mindscape. Find an equation for the line that passes through points A and B and an equation for the line that passes through the points B and C.

47. **White out.** Each stage of the Sierpinski triangle is a figure composed of white and green triangles (as depicted on page 471). Mindscape 3 focused on the number of green (filled-in) triangles at each stage, but in this Mindscape we're going to focus on the white triangles. Counting these bad boys is a little trickier, but you can verify for yourself that at stage n there are $1 + 3 + 3^2 + \ldots + 3^{n-1}$ white triangles. It turns out that this sum can be simplified to equal $\dfrac{3^n - 1}{2}$. Confirm that these two quantities are equal for Stage 4: that is, let $n = 4$ and compute the sum and the quotient, verifying that the two answers are equal.

48. **Measuring the length of your Koch.** The Koch stool is presented in Mindscape 21. If the starting line segment (at Stage 0) has length 1, then the total length of all the stool segments at Stage 2 is 5/3 because each of the five segments is 1/3 the length of the original segment. At Stage n, the total length of all

the stool segments is $(5/3)^n$. Compute the total length of the Koch stool segments for $n = 2$, 3, and 4.

49. **Counting Koch segments (H).** Suppose you start with a line segment of length 1 and begin to construct a Koch stool (See Mindscape 21). At each stage, the line segments are $1/3$ the length of the segments in the previous stage. Give an expression for the length of a line segment at Stage n of the stool. The previous Mindscape tells us that at Stage n, your Koch stool has total length $(5/3)^n$. Use this formula to help you determine how many segments you have at Stage n.

BETWEEN DIMENSIONS

Can the Dimensions of Fractals Fall Through the Cracks?

Are these Swiss cheese-like
Platonic solid fractals
3-dimensional or . . .?

©1991 John C. Hart, Electronic Visualization Laboratory,
University of Illinois at Chicago/Courtesy University of Illinois

> *. . . without dimension; where
> length, breadth, and height,
> And time and place are lost;
> where eldest Night And chaos,
> ancestors of nature,
> hold Eternal anarchy.*
>
> JOHN MILTON

What is dimension? In our journey through geometry in Chapter 4, we developed an intuitive idea of the dimension of space as measuring how many degrees of freedom we have. But this intuitive meaning of dimension seems a bit limiting and difficult to apply if we consider objects more intricate than the line, the plane, and so on. Can we sensibly associate a dimension to any object, or is dimension an idea that applies only to rather regularly shaped things?

To explore this question we begin, as usual, by examining carefully the simple and familiar with an eye toward understanding the more complex. In regular objects we see patterns associated with dimension that suggest how to extend the idea of dimension to irregular objects.

Easy Does It

We begin by looking at the familiar. Here are some warm-up questions to get you in the mood for what is ahead. Try to answer all of them before reading on.

- How many dimensions does a line have?
- Can you give an example of an object that is 2-dimensional?
- The cube is ???-dimensional. What does ??? equal?

You will not be shocked by the answers we are about to give to these questions. The line is 1-dimensional. One of the simplest examples of a 2-dimensional object is a filled-in square. The cube is 3-dimensional. Okay, not too tricky. It seems as though we have some reasonable notion of dimensionality: If the object just has length but no width or height, then it is 1-dimensional; if it has length and width but no height, then it's 2-dimensional; if it has length, width, and height, then it's 3-dimensional. That's that.

Life Lesson
Start simply.

493

Well, that *is* that as long as we're considering only simple shapes. But what about the coastline of Norway, which isn't so simple? Suppose we consider (as we tend to do in this chapter) objects of infinite intricacy. For example, what is the dimension of the Koch Curve? On the one hand, since it's basically made out of 1-dimensional line segments, it might seem to just have length, so it's reasonable to think that it's 1-dimensional. On the other hand, it has infinitely many bends and corners. If we look at it, it's so incredibly jagged that we might call it fuzzy—like mold on stale bread—and it seems to have some slight thickness. Thus, perhaps it is 2-dimensional. But wait: It cannot be 2-dimensional since it's just lines, yet it cannot be just 1-dimensional since its infinite fuzziness seems to take up a bit of area. What is the dimension of the Koch Curve?

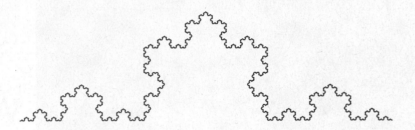

Our conflicting thoughts are leading us to a strange possibility: Perhaps the dimension of the Koch Curve is some number that is bigger than 1, but less than 2. Huh? Does it make sense to have a dimension between 1 and 2? The answer is that first we need to pin down our idea of dimension. Let's examine simple things carefully and search for patterns.

What Is Dimension?

A straight line is 1-dimensional. A filled-in square is 2-dimensional. A solid cube is 3-dimensional. Somehow we intuitively know these facts. The problem is that our intuition fails us when we look at the infinitely intricate. Let's try to devise, without appealing to our intuition, a systematic means of determining the dimensions of the line, the square, and the cube—dimensions we already know. If we could accomplish this task, then perhaps we could apply the method to compute the dimensions of more exotic objects.

Let's begin with the filled-in square. Suppose the length of each side is 1. How many copies of this square can be assembled to produce a larger square? Well, four will do, or nine would work as well. If we use four squares to build a big square, we notice that the new square is twice as long as the original square. So, to make a square two times as large as the original square (that is, if we set the scaling on a copier machine to print edges twice as large), we need four of the original-size squares to make the larger square.

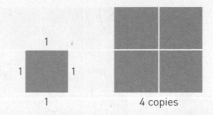

4 copies

What if we consider a cube having each side equal to 1? How many copies of this cube could be used to build a cube each of whose sides are twice as long? The answer is that we require eight cubes to build a cube with edges two times as large as the first cube. So, if we had a 3-dimensional photocopier, we would scale things by two to go from the first cube whose sides are 1 to the larger cube whose sides are 2.

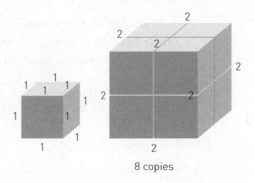

8 copies

Let's now consider a straight-line segment having length 1. Of course, two copies of this line produce a line that is twice the size of the original. So, in this case, if we set the copying machine to make an image twice the size, we could construct that enlarged image with just two copies of the original segment.

1   2 copies — twice as long as first segment

In Search of a Pattern

Let's organize our findings in a chart such as the one below.

Original Object	Dimension of the Object	Scaling Factor to Make a Larger Copy	Number of Copies Needed to Build the Larger Copy
Line	1	2	$2 = 2^1$
Square	2	2	$4 = 2^2$
Cube	3	2	$8 = 2^3$

3 copies

Life Lesson
Look for patterns.

Do you see a pattern? It turns out that, if we write the number of copies needed to build the next larger copy of the original object as the scaling factor raised to some power, then that exponent is equal to the dimension of the object.

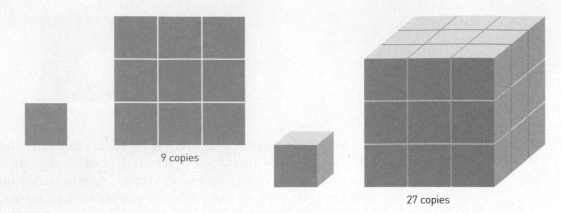

9 copies

27 copies

Now that we see a pattern, let's confirm our theory by scaling by a factor of 3 instead of 2 to see whether the pattern continues to hold. Here is the chart we construct for a scaling factor of 3:

Original Object	Dimension of the Object	Scaling Factor to Make a Larger Copy	Number of Copies Needed to Build the Larger Copy
Line	1	3	$3 = 3^1$
Square	2	3	$9 = 3^2$
Cube	3	3	$27 = 3^3$

So, our hypothesis about the relationship between the scaling factor and the number of copies needed to construct a scaled-up version is confirmed. Raising the scaling factor to the power of the dimension gives us the number of copies needed to construct that larger-scale version of the object.

Testing Our Conjecture

Let's test our hypothesis with some other objects and see if we can use this procedure to compute their dimensions.

Let's consider a filled-in equilateral triangle with all sides of length 5. How many copies of this triangle are needed to make an equilateral triangle with sides of length 10?

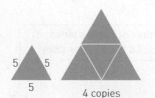

5   5
5   4 copies

We need four copies to build an equilateral triangle with sides twice as long. If we take the number of copies we need and write that number as the scaling factor raised to some power, then we see $4 = 2^2$. We need nine copies to build an equilateral triangle with sides three times as long. Once again, if we take the number of copies we need and write that number as the scaling factor raised to some power, then we see $9 = 3^2$. So, we see that the exponent we need is 2. Since we believe that a filled-in triangle is 2-dimensional, we see once again that the exponent is telling us the dimension.

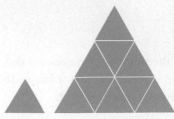

9 copies

Using this method, figure out the dimension of a filled-in rectangle whose base is 2 and height is 1. First, we must figure out how many copies of this rectangle are needed to build a similar rectangle that is larger. Again, we need four copies to construct a rectangle with sides twice as long. Thus, if we write the

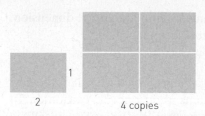

1

2

4 copies

number of copies we need as the scaling factor raised to some power, then we again see $4 = 2^2$.

So, the dimension of the rectangle is again the power 2. Since we have observed a relationship between dimension and this exponent, let's crystallize these observations into an official definition of dimension.

Life Lesson

Let examples lead to definitions.

Dimension Finally Defined

Suppose we have an object for which N copies can be assembled to construct a larger version of it that has been scaled by a factor of S. We now define the *dimension*, which we call d, of the object to be the power to which we have to raise S to have it equal N. That is, d is the number such that

$$S^d = N.$$

We already know that this definition of dimension conforms to our intuitive sense of dimension in the cases of the line segment, the square, the cube, the filled-in triangle, and the filled-in rectangle, so it is a plausible distillation of our intuition. We now want to apply our clear understanding of the known to analyze the unknown.

From the Ordinary to the Exotic

We have worked hard to discover a precise definition of dimension. Now let's test our wings and compute the dimensions of more interesting and intricate objects. We begin with the Koch Curve.

Example 1—Koch Curve Dimension

Recall that the Koch Curve is the fractal we get by iterating a certain process infinitely often. We begin with a line segment and then cut it into three equal pieces. We then replace the middle piece with an inverted **V** made of two segments, each having the same length as the middle piece we removed. We now repeat this procedure with each of the four smaller line segments. We continue this process, creating an object of infinite intricacy and self-similarity.

What is its dimension? This was the question that started our entire discussion. Our first, strange guess was that it would be some number between 1 and 2. Think with us now as we compute the dimension of a Koch Curve.

How do we compute dimension? We must first ask how many copies of the Koch Curve are required to build a larger version of the Koch Curve. How many would you guess? One answer is four. Imagine gluing four copies together to produce a big version of the Koch Curve.

Now we come to a more challenging question: How much do we have to scale the original Koch Curve to enlarge it to the bigger one we just built? It is a worthwhile exercise to ponder this question a bit.

The answer is that we scale by 3. Notice that, if the original Koch Curve had a horizontal length of 1, then our enlarged version would have a horizontal length of 3. Thus we see that the scaling factor is indeed 3. So, what is the dimension d? According to our definition of dimension, the dimension d of the Koch Curve is the power d such that

$$3^d = 4.$$

Finding the numerical value of d involves logarithms, which we will not review here. As a practical matter, to compute logarithms we just need a calculator. To solve for d, use a calculator that has an LN ("natural log") key. Specifically, d equals ln 4/ln 3. If we enter this number into a calculator we get this result:

$$d \approx \frac{1.38629}{1.09861} = 1.26185.\ldots$$

On your calculator check that 3 raised to the 1.26185 ... power equals 4. So, the dimension of the Koch Curve is 1.26185. It is indeed bigger than 1-dimensional and yet smaller than 2-dimensional. Notice how the infinite intricacy of the Koch Curve creates an interesting fuzziness that increases the dimension ever so slightly.

Example 2—Sierpinski Dimension

Let's now return to the Sierpinski Triangle. What is its dimension? Notice that three copies of the Sierpinski Triangle can be assembled to create a larger version and that the larger version is twice the size of the original one (the scaling factor in this case is 2). Thus, the dimension d of the Sierpinski Triangle is the number such that

$$2^d = 3.$$

We can compute this dimension using a calculator by computing

$$\frac{\ln 3}{\ln 2}$$

and we see that

$$d \approx \frac{0.47714}{0.30103} = 1.58496.\ldots$$

So, again we see that the dimension of this fractal is greater than 1 but smaller than 2.

These two examples give us the tools by which we can compute the dimensions of many fractals that were constructed using the collage-making procedure—as long as the shrinking factors can be determined. We can use a calculator to find that

$$d = \frac{\ln N}{\ln S}$$

where, again, N is the number of copies required to make a larger version of the original object and S is the scaling factor required to enlarge the original object to the size of the larger version.

A Look BACK

WE CAN TAKE THE IDEA OF DIMENSION and extend it to encompass and describe many of the interesting fractal images we constructed. Surprisingly, the dimensions of fractal objects are not confined to the natural numbers 1, 2, or 3 but instead fall somewhere between these dimensions. Perhaps this fact is not so surprising, because a fractal is cloudlike: It is difficult to see where it begins and where it ends. Fractal dimension captures and in some sense measures the beautiful cloudlike essence of fractals.

Our strategy for developing the idea of fractal dimension is to start with observations of familiar objects. For familiar, regular objects like cubes and triangles, we find a relationship connecting a scaling factor with the number of copies needed to produce similar objects at a larger scale. That insight allows us to give a more encompassing definition of dimension. Our extended idea of dimension applies to more complicated objects like fractals.

Starting with simple and familiar cases and understanding them deeply allows us to reach beyond what we know now. Patterns among familiar objects and ideas can then be fit on unfamiliar terrain to show features not previously evident. By defining carefully an idea that we cull from common experience, we have a beacon for exploring the unknown.

Life Lessons

Start with the simple and familiar.

•

Look for patterns.

•

Apply patterns to new settings.

MINDSCAPES · Invitations to Further Thought

*In this section, Mindscapes marked **(H)** have hints for solutions at the back of the book. Mindscapes marked **(ExH)** have expanded hints at the back of the book. Mindscapes marked **(S)** have solutions.*

Developing Ideas

1. **Parallel grams.** At right we see a parallelogram. How many copies do you need to build a parallelogram that is twice as large? How many copies do you need to build one three times as large?

2. **Moving on up.** Here's a triangle along with some larger versions. What is the scaling factor for each of the larger copies?

3. **Bigger rug.** Here's a picture of the Sierpinski Carpet (see p. 9). How many copies do you need to build a version three times as large?

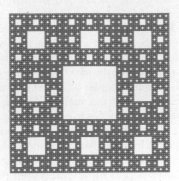

4. **Dimension connection.** What equation relates the dimension d of an object to the number of copies N needed to construct a larger version scaled by a factor S?

5. **Divining dimension.** If a fractal object requires $N = 3$ copies to construct a version scaled up by a factor of $S = 2$, what is the dimension of the object?

Solidifying Ideas

6. **Stay inbounds.** Give two consecutive integers that bound the fractal dimension of (a) the Mandelbrot Set, (b) the Menger Sponge, and (c) the Cantor Set.

7. **Regular things (H).** Find the fractal dimension of these two objects using the definition of fractal dimension from this section.

8. **More regular things.** Find the fractal dimension of a parallelogram.

9. **Any right triangle.** Take any right triangle. It can be broken into four similar triangles, each one having edges that are 1/2 the length of their corresponding edges in the big triangle. Use this fact to show that, using the dimension definition, any filled-in right triangle has dimension 2.

10. **Sierpinski carpet (S).** Compute the fractal dimension of the purple part of the Sierpinski carpet—that is, the fractal constructed in Chapter 1: "A Tight Weave" (page 9).

11. **Koch Stool.** Compute the fractal dimension of the Koch Stool described in Section 7.2, Mindscape 21.

12. **Cantor Set (H).** The Cantor Set was constructed by taking a unit segment, making two copies each of length 1/3, and putting those copies at the two ends of the unit segment. Repeating this collage-making process infinitely often results in the Cantor Set. What is the dimension of the Cantor Set?

13. **Cantor reduced.** Suppose you take a unit interval, make two copies each shrunk to 1/4 its length, and position them at the ends of the unit interval. Repeating this process infinitely results in a fractal similar to the standard Cantor Set. What is its dimension?

14. **Long Koch (ExH).** The first stage in the construction of the Koch Curve is a line segment of length 1. In the second stage that segment is replaced by four segments, each of length 1/3. So, the length at the second stage is 4/3. What is the length of the third stage? What is the length of the fourth stage? What is the length of the nth stage in the Koch construction? What would you say is the length of the final Koch Curve?

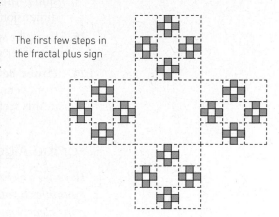

The first few steps in the fractal plus sign

15. **Plus (H).** This fractal plus sign is self-similar in that, if it is reduced by a factor of 1/3, four reduced copies can be put together to create it again. What is its dimension?

Creating New Ideas

16. **Tinier triangles (S).** Suppose you make something similar to the Sierpinski Triangle, but this time you make three copies, each reduced so that the sides are only 1/3 of the original's sides' lengths, and then you position the reduced triangles at the three corners. Notice that this process has the same result as removing a hexagon from the triangle, as we did before. Continue this collage-making process infinitely. What is the dimension of the resulting collage?

17. **Menger Sponge (ExH).** Compute the fractal dimension of the Menger Sponge.

18. **Thinning.** Take a square. Make two copies, each reduced to half size in length and width, and position them in diagonal corners. Repeat the process infinitely often to create a fractal. What is the fractal?

19. **Not much.** What is the fractal dimension of the fractal in the previous Mindscape?

20. **Koched (H).** Create a Koch-like curve with fractal dimension ln 5/ln 4.

Further Challenges

21. **Find a fractal.** Describe a fractal having dimension ln 8/ln 3.
22. **Find a 1.5 fractal.** Describe a fractal having dimension 1.5.

In Your Own Words

23. Personal perspectives. Write a short essay describing the most interesting or surprising discovery you made in exploring the material in this section. If any material seemed puzzling or even unbelievable, address that as well. Explain why you chose the topics you did. Finally, comment on the aesthetics of the mathematics and ideas in this section.

24. With a group of folks. In a small group, discuss the meaning of fractal dimension. After your discussion, write a brief narrative describing the fractal dimension of the Koch Curve in your own words.

25. Creative writing. Write an imaginative story (it can be humorous, dramatic, whatever you like) that involves or evokes the ideas of this section.

26. Power beyond the mathematics. Provide several real-life issues—ideally, from your own experience—that some of the strategies of thought presented in this section would effectively approach and resolve.

For the Algebra Lover

Here we celebrate the power of algebra as a powerful way of finding unknown quantities by naming them, of expressing infinitely many relationships and connections clearly and succinctly, and of uncovering pattern and structure.

27. Varying dimensions. Using any method, including guess-and-check, find all the real numbers x that satisfy each equation:

$$x^5 = 243 \qquad 2^x = 1024 \qquad x^3 = -1 \qquad (x^5)/(x^3) = 36$$

28. Dimensional thinking. The text defines the dimension of an object as follows. Suppose you have an object for which N copies can be assembled to construct a larger version of the object that has been scaled by a factor of S. The dimension of the object, called d, is the power we have to raise S to in order to obtain N. Thus, S, d, and N satisfy the equation $S^d = N$.

Suppose it takes 125 copies of an object of dimension 3 to make a larger version. What's the scale factor being used? Suppose it takes 64 copies of a different object to create a copy twice as large. What's the dimension of this object?

29. Power play (H). Using any method, including guess-and-check, find all the real numbers x that satisfy each equation:

$$x^3 = -8 \qquad 4^x = 1024 \qquad 5^x = 625 \qquad x^7 = -1 \qquad (x^3)/(x^5) = ¼$$

30. Marching madness. The marching band wants to create a half-time show with a fractal theme. For their first drill number on the football field, they want to present their version of the Sierpinski Carpet (see Chapter 1, Story 6). The director has designated the 81-member drum line for this drill. The drummers will form a square block with 9 rows of 9 players each. To indicate the gold portions of the carpet, certain drummers will put gold shower caps on their heads. Create a 9-by-9 grid to represent the drummers. Indicate which drummers should put on shower caps to create the best approximation you can to a Sierpinski Carpet.

31. **More marching madness.** For the second drill in the fractal-themed half-time show, the marching band will present Stages 0 through 3 of the Koch curve. At each stage, every segment of the curve has the same number of students on it. There are 256 students in the band. How many students will there be on a segment of the Stage 3 curve?

Infinitely detailed beauty—a hint of things to come. . .

THE MYSTERIOUS ART OF IMAGINARY FRACTALS

Creating Julia and Mandelbrot Sets by Stepping Out in the Complex Plane

God as the Architect of the Universe (c. 1230). Is the universe a Benoit Mandelbrot-like fractal?

akg-images/NewsCom

> *. . . what the imagination seizes as beauty must be truth—whether it existed before or not.*
>
> JOHN KEATS

The infinite iterative processes we will examine here do not involve the collage-generating theme described and exploited in the previous two sections but instead an imaginative motion of points in the plane. The abstract mathematical objects resulting from these moving points were first studied more than 80 years ago—long before computer technology allowed us to generate images of them. Only in the past 30 years, with the development of powerful computers and graphics software, have we become able to see images of these abstract conceptions. The pictures not only surprised and excited people around the world but also led to an entirely new branch of mathematical inquiry.

Visualization is a powerful means of discovering structure. These fractal images are like a Rorschach test of mathematics. We see patterns within patterns, nuances, similarities, and differences with new intricacies emerging at every level of magnification. Questions about what we see or don't see are still among the unsolved problems of mathematics.

Julia and Mandelbrot Sets

The first collection of objects we examine are called *Julia Sets*, named after the French mathematician Gaston Julia. In the 1920s Julia discovered these sets (that is, collections) as natural objects in his investigation of imaginary numbers. He studied some of the many unusual properties these Julia Sets possess even though he could not create or imagine the incredibly rich and intricate pictures of them that tantalize us today. The visual renderings of these sets were created by sophisticated computers more than 50 years after Julia's initial work. These images startle and delight all who see them. The beauty and complexity of the Julia Sets are great surprises that add to the interest in Julia's original mathematical work.

A drawing of Gaston Julia, who tragically had his nose blown off in World War I.

Below, and on the next page, are some images of various Julia Sets. When we magnify one part of the set, we see a smaller set with a structure similar to that of the original one. This self-similarity produces images of endless intrigue, because every magnification exposes new wonders. While you enjoy these spectacular pictures, we invite you to think about how one could ever describe or create such infinitely detailed objects.

In the late 1970s Benoit Mandelbrot discovered a revolutionary means of studying an aspect of all such Julia Sets. He described what is now known as the *Mandelbrot Set*, and in 1980 the image of his set was first created using special computer graphics software. The Mandelbrot Set still conceals many mysteries and is the object of continuing investigation. You may well join the crowd of individuals who consider the Mandelbrot Set a supremely intricate and beautiful icon of mathematical art.

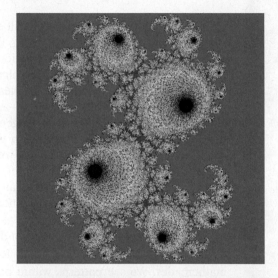

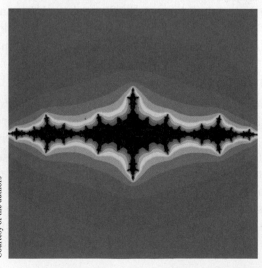

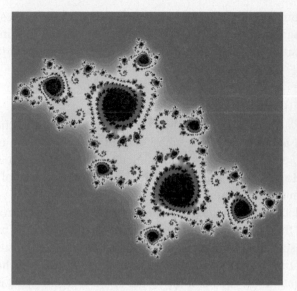

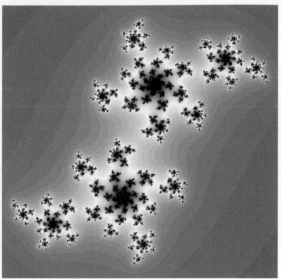

Fathering Fractals

Benoit Mandelbrot pioneered a whole new area of study concentrated on fractals, images of infinite detail, such as these Julia Sets and his Mandelbrot Set. He truly is the father of fractals, since, as we have mentioned, he invented the word *fractal*, which he coined from the Latin *fractus*, meaning broken or irregular. He applied the term to all sets having infinite detail, such as the collage-generated sets discussed in the previous two sections and the Julia Sets we will look at here. *Fractal* does not have a rigorous and precise mathematical definition, but it conjures up the ideas of similarity at different scales and of infinitely detailed intricacy.

Benoit Mandelbrot—the father of fractals

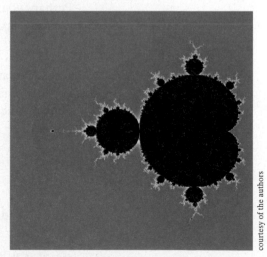

The Mandelbrot Set

We now turn our attention to the basic question: How were these incredibly detailed and imaginative sets created? To understand their origin, we follow the fertile imagination of some great thinkers and enter the world of imaginary numbers.

A Process in the Plane

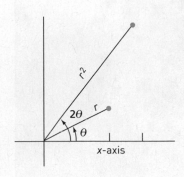

We first describe a process by which we start with one point in the plane and use it to move to another. This procedure is used to generate the elegant Julia and Mandelbrot Sets. The process is not difficult, but it may seem artificial and strange at first. However, we will see that it is a natural example of performing arithmetic with imaginary numbers.

Suppose we select a point in the plane and draw the line segment from that point to the origin. We then measure the length of that segment and the angle moving counterclockwise starting from the positive x-axis to the segment. Using this point, we will now describe a second point. The new point we get will have distance from the origin equal to the square of the length of the segment—that is, the square of the distance of the original point to the origin—and will be in the direction of twice the angle of the original segment. So, in short, we double the angle and square the length. Sounds like an unnatural procedure to us . . . at least for the moment. We'll describe this procedure more fully in a few paragraphs.

Points in the plane can be interpreted as imaginary numbers, and the process described in the last paragraph actually corresponds to squaring an imaginary number. We next present a brief look at imaginary numbers and how they are multiplied and added together. For the purpose of constructing Julia and Mandelbrot Sets, however, the geometric process just mentioned is really all we need to know. Interpreting the process as imaginary arithmetic provides the inspiration and motivation for the significance of this particular process of taking one point in the plane and generating another one.

Life Lesson

If something is an anomaly in your worldview, expand your view to make it fit.

Imaginary Numbers

Perhaps you have already been introduced to the imaginary number i. The unreal number i is defined to be the number satisfying the following amazing property:

$$i^2 = -1.$$

If we take any real number from the number line and square it, we never get a negative number. Thus we immediately see that this new number i is *not* a real number, and so we call it an *imaginary number*. Since $i^2 = -1$, many write i as

$$i = \sqrt{-1}$$

More generally, we define *complex numbers* to be numbers of the form

$$a + bi,$$

where a and b are real numbers. So, for example, the following are all examples of complex numbers:

$$2 + 3i, \ 5.7i, \ 1 + i, \ 2\pi - 12i, \ -5 + \sqrt{3}i.$$

All we'll need to do with the complex numbers is add them and multiply them. First we'll just figure out how to perform these operations; then we'll see a way to visualize both addition and multiplication. The visualization of these two operations is all that is needed to understand how the beautiful Julia and Mandelbrot Sets are created.

An Addition Complex

The natural guess as to how to add two complex numbers is correct. For example, $2 + 5i$ plus $11 + 4i$ equals $13 + 9i$.

Here is the rule for adding complex numbers:

$$
\begin{array}{r}
(a + bi) \\
+(c + di) \\
\hline
(a + c) + (b + d)i
\end{array}
$$

so,

$$\boxed{(a + bi) + (c + di) = (a + c) + (b + d)i}$$

In other words, the approach is not exotic. We add the first two numbers together and then add together the two numbers in front of the i's.

A Multiplication Complex

Multiplying complex numbers takes a bit more time to figure out. Here's how it works: The key is to make certain that every term in one complex number gets multiplied by every term in the other. It is similar to the "FOIL" ("first, outer, inner, last") method introduced in some algebra classes. We'll multiply $2 + 5i$ and $11 + 4i$ (remembering the crucial fact that $i^2 = -1$ and noting how it is used in the third line):

$$
\begin{aligned}
(2 + 5i)(11 + 4i) &= (2 + 5i)11 + (2 + 5i)4i \\
&= (22 + 55i) + (8i + 20i^2) \\
&= (22 + 55i) + (8i - 20) \\
&= (22 + 55i) + (-20 + 8i) \\
&= (22 - 20) + (55 + 8)i \\
&= 2 + 63i.
\end{aligned}
$$

Let's adapt the previous example to find a formula for multiplying complex numbers in general. Notice how all the steps below parallel the steps in the previous example.

$$
\begin{aligned}
(a + bi)(c + di) &= (a + bi)c + (a + bi)di \\
&= (ac + bci) + (adi + bdi^2) \\
&= (ac + bci) + (adi - bd) \\
&= (ac + bci) + (-bd + adi) \\
&= (ac - bd) + (bc + ad)i.
\end{aligned}
$$

So we see:

$$(a + bi)(c + di) = (ac - bd) + (bc + ad)i.$$

The final answer to the multiplication looks strange because $i^2 = -1$; but remember that we arrived at that answer simply by multiplying $(a + bi)$ by $(c + di)$ just as we were taught in algebra class, and then we used the fact that $i^2 = -1$ to simplify the answer. Therefore, it is easier to remember the *procedure* for multiplying complex numbers than to try to memorize the *formula*. For the purpose of making Julia and Mandelbrot Sets, all we care about is multiplying a complex number by itself. So, let's state this squaring as a special case:

$$(a + bi)^2 = (a + bi)(a + bi) = (a^2 - b^2) + (2ab)i.$$

Visualizing Complex Numbers

Recall that we visualized the real numbers by studying the real number line. The complex numbers can be visualized as a plane—the *complex plane*. The diagram represents the complex plane:

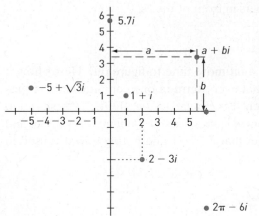

Notice that we have two axes: one horizontal (*x*-axis) and the other vertical (*yi*-axis). The number $2 + 3i$ can be found by moving over 2 in the horizontal direction and then moving up 3 in the vertical direction. The diagram depicts the numbers $2 + 3i$, $5.7i$, $1 + i$, $2\pi - 6i$, $-5 + \sqrt{3}i$ in the complex plane. So, in general, to find the point $a + bi$, we just need to go over to the number a on the x-axis and then move up or down to b on the yi-axis.

We can now think of a different way to understand addition and multiplication of complex numbers.

Here is an easy way to perform addition of complex numbers geometrically in the complex plane. Draw a line segment from each of the two numbers to the origin $(0 + 0i)$. Then make a parallelogram using these two lines as two adjacent sides. The vertex of the parallelogram that is across from the origin is the point that represents the sum of the two numbers. For example, returning to $2 + 5i$ plus $11 + 4i$, we draw the diagram shown here. Notice that the vertex across from the origin represents the complex number $13 + 9i$, which is the sum. So, the parallelogram gives a geometric way to add complex numbers—very cool.

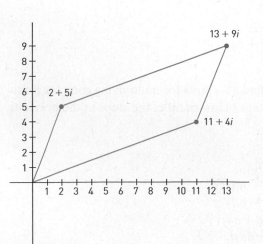

We can also multiply a complex number by itself geometrically in the complex plane, and, as you may have guessed, we will discover that the "process in the plane" described on page 508 actually corresponds to multiplying a complex number by itself. Let's look at this geometric view of multiplication. To multiply a number by itself, we first plot the number in the complex plane and connect it by a straight-line segment to the origin. We will take note of the pitch (or slope) of this line segment and its length.

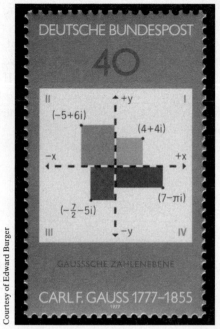

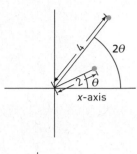

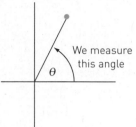

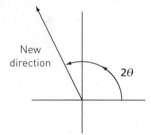

As we described in "A Process in the Plane," we can determine the line's pitch by considering the angle made by the line and the right half of the x-axis. Specifically, we consider the angle made by starting at the right half of the x-axis and moving counterclockwise until we reach the line segment.

We now consider a new direction: We start at the right half of the x-axis and move counterclockwise to an angle that is twice the angle we just found.

At that angle we move out from the origin to a length that is equal to the square of the length of the original line segment and place a point at that location. That is, we swing around counterclockwise by an angle that is double the angle of our line segment. This angle gives us a new pitch from the origin. We now draw a line segment from the origin in this new direction so that its length is equal to the length of our original line segment multiplied by itself. We place a point at the end of the segment. A little trigonometry would show that this new point is at the precise location of the square of the original complex number—that is, the result of multiplying the complex number by itself.

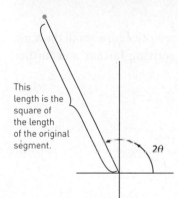

This length is the square of the length of the original segment.

Let's illustrate this elaborate procedure with an example. Using the arithmetic method just outlined, we can compute that $(2 + 3i)(2 + 3i)$ equals $-5 + 12i$. Now notice how we get to the same point from the geometric procedure just outlined. So, to find the square of a complex number geometrically, we look along a direction from the origin that is *double* the angle made by the original number and the right half of the x-axis and place a mark that is located at a distance from the origin that equals the *square* of the distance of the original number to the origin.

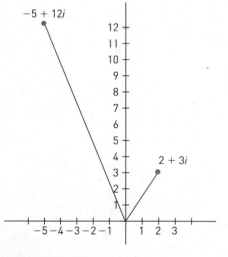

Squaring Practice and a Julia Set Warm-Up

We now have all the necessary mathematics to describe how the fabulously intricate Julia and Mandelbrot Sets are created. Those pictures result from an iterative process whereby we perform an operation and then repeat, repeat, repeat . . . forever. This process, however, is different from the collage method. To get us in the mood for this process and also to practice the arithmetic of complex

numbers we just described, let's consider some experiments and make some simple observations.

Let's take the number $2 + 3i$, square it, take that new number and square it, take that result, square it, and repeat, repeat. . . . What is the list of numbers we get? We already computed $(2 + 3i)(2 + 3i)$ and saw it equaled $-5 + 12i$. If we repeat this process with our new answer we get $(-5 + 12i)(-5 + 12i) = (25 - 144) + [2(-60)]i = 119 - 120i$. If we continue to repeat this process, here are the numbers we get (verify the third one for yourself):

Starting with $2 + 3i$, each number is the square of the number that preceded it.

$-5 + 12i$
$-119 - 120i$
$-239 + 28560i$
$-815616479 - 13651680i$
$665043872449535041 + 22269070348069440i$

$$\vdots$$

$(\pm$ a really huge number$) + (\pm$ another huge number$)i$

$$\vdots$$

Here is a rough plot of those same imaginary numbers. Notice how small the scale needs to be. Why? It's because those numbers are moving farther and farther from the origin.

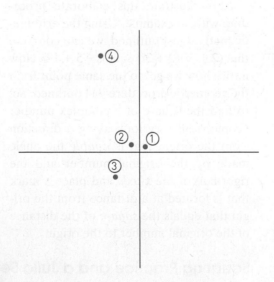

Let's try the same experiment starting with the number $0.5 - 0.8i$. Here is what we get (verify the first entry):

Starting with $0.5 + 0.8i$, each number is the square of the number that preceded it.

$$-0.39 + 0.8i$$
$$-0.48\ldots - 0.62\ldots i$$
$$-0.15\ldots + 0.59\ldots i$$
$$-0.33\ldots - 0.18\ldots i$$
$$0.07\ldots + 0.11\ldots i$$

$$\vdots$$

$$(\pm \text{ a really tiny number}) + (\pm \text{ another tiny number})i$$

$$\vdots$$

Here are those complex numbers plotted. Notice how large the scale is now; that's because we see that these complex numbers are zooming toward the origin, $0 + 0i$.

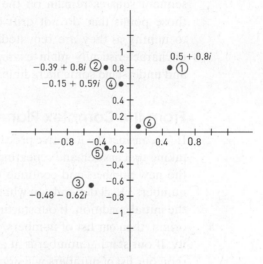

Finally, let's try this iterative squaring procedure starting with $3/5 + 4/5i$.

Starting with $3/5 + 4/5i$, each number is the square of the number that preceded it.

$$-7/25 + 24/25i = -0.28 + 0.96i$$
$$-0.84\ldots - 0.53\ldots i$$
$$0.42\ldots + 0.90\ldots i$$
$$-0.64\ldots + 0.76\ldots i$$
$$-0.17\ldots - 0.98\ldots i$$
$$-0.94\ldots + 0.33\ldots i$$

$$\vdots$$

$$(\pm \text{ a modest number}) + (\pm \text{ another modest number})i$$

$$\vdots$$

The graph of these points is quite interesting. Notice that they neither head off to the outer reaches of the complex plane nor spiral into the origin. In fact, they seem to stay along the circle centered at the origin having radius 1. Why? By using the Pythagorean Theorem we can compute that the distance between the original point $3/5 + 4/5i$ and the origin is equal to 1. Therefore, when we square it, the distance away from the origin of the new number will be 1^2, which is still 1. Similarly, all other squares will remain on the circle of radius 1.

Insights from Our Experiments

So, what do we discover from these experiments? We see that if a point outside the circle centered at the origin of radius 1 is repeatedly squared and squared again, the values drift off to infinity. If the point is inside the circle, then successive squaring makes the numbers head toward the origin. Finally, if we start with a point on the circle, then the subsequent squares remain on the circle. So, the circle is the boundary between those points that do not drift off to infinity and those points that do go off to infinity as they are repeatedly squared. This circle is actually a simple and uncharacteristically plain example of a Julia Set. We are now ready to face and understand some incredible, imaginative fractals.

From the Complex Plane to the Infinitely Beautiful

Let's consider the iterative process of starting with a complex number, squaring it, taking that answer, and repeating the process (so square that number, then square the new number, and continue to square successively). What happens to these numbers? It all depends on what the initial number was; that is, it all depends on the initial condition. If our starting number is at a distance greater than 1 from the origin, then our list of numbers, starting with that first one, will head off to infinity. If our starting number is at a distance less than or equal to 1 from the origin, then our list of numbers will not head off to infinity.

If we think of all the points in the complex plane as all the possible starting numbers, then we see that the circle of radius 1 centered at the origin is the interface between those starting numbers that result in their lists going off to infinity and those starting numbers that result in their lists not going off to infinity. This interface, or boundary, between these two classes of numbers is a Julia Set. Here is a picture of the "filled-in" Julia Set, where we color in all the starting numbers whose lists do not head off to infinity.

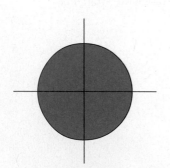

Adding Some Spice

To generate infinitely intricate Julia Sets we need to add one more twist to our process. Suppose we pick a particular complex number: Let's call it $a + bi$. So, $a + bi$ is a *fixed number*. Now suppose we begin with a starting complex number, square it, and then add our fixed number $a + bi$. We can now iterate this process: Take the answer we get, square it and then add $a + bi$, take that new answer, square it and add $a + bi$, and so on, forever. For example, suppose $a + bi = -1 + 0i$. If we start with the initial number $0 + 0i$, here is what we generate:

$$0 + 0i$$
$$(0 + 0i)^2 + (-1 + 0i) = -1 + 0i$$
$$(-1 + 0i)^2 + (-1 + 0i) = 0 + 0i$$
$$(0 + 0i)^2 + (-1 + 0i) = -1 + 0i$$

and so on . . .

If we start with the initial number $2 + 3i$, then we generate:

$$2 + 3i$$
$$(2 + 3i)^2 + (-1 + 0i) = (-5 + 12i) + (-1 + 0i) = -6 + 12i$$
$$(-6 + 12i)^2 + (-1 + 0i) = (-108 - 144i) + (-1 + 0i) = -109 - 144i$$
$$(-109 - 144i)^2 + (-1 + 0i) = -8856 + 31392i$$

and so on . . .

Julia Sets Exposed

Notice that this procedure is exactly the same as our original squaring procedure, except for the slight difference of adding on the number $a + bi$ at each stage after we square our number. It turns out that this slight difference leads to a world of difference. As before, given a certain starting number, our list of iteratively generated numbers will either drift off to infinity or not. In the preceding example, notice that the starting number $0 + 0i$ does not drift off to infinity, whereas the starting number $2 + 3i$ does. If we think of all points in the complex plane as all possible starting numbers, we see that some of them lead to lists that go off to infinity and some lead to lists that do not. The interface between these sets is called a *Julia Set*. Here is a picture of the filled-in Julia Set when $a + bi$ equals $-1 + 0i$. The boundary around this set is the Julia Set associated with the shifting number $-1 + 0i$.

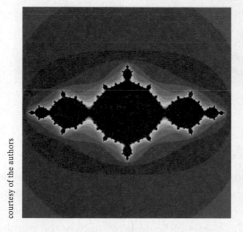

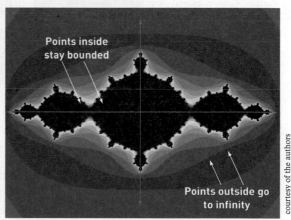

Points inside stay bounded

Points outside go to infinity

$(-1 + 0i)$ - Julia Set

courtesy of the authors

Notice that, when we did not add any number after squaring, our Julia Set was a boringly symmetric circle. However, when we added $-1 + 0i$ after each squaring operation, all of a sudden the Julia Set became an exotic fractal set. Remember what this set represents.

It turns out that, if we change the $a + bi$ (the fixed number we add after we square at each stage), then the associated $(a + bi)$-Julia Set (the boundary between the starting numbers that result in unboundedly larger and larger values and the starting numbers that result in values that remain bounded) changes dramatically. Here are some different Julia Sets associated with different values of $a + bi$. Make your own or see renderings of these using the neat program on the *Heart of Mathematics* Web site. It even allows you to zoom in and see these beautiful images at higher and higher magnification.

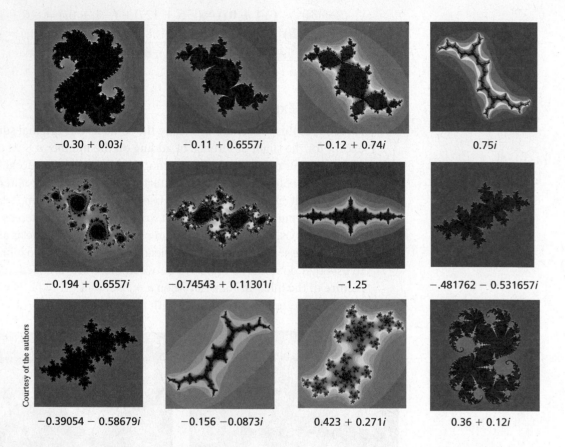

−0.30 + 0.03*i*	−0.11 + 0.6557*i*	−0.12 + 0.74*i*	0.75*i*
−0.194 + 0.6557*i*	−0.74543 + 0.11301*i*	−1.25	−.481762 − 0.531657*i*
−0.39054 − 0.58679*i*	−0.156 −0.0873*i*	0.423 + 0.271*i*	0.36 + 0.12*i*

Courtesy of the authors

Color-Coded Retreat

In the pictures above, the area outside the filled-in Julia Sets is depicted by several different colors. Remember that each starting point outside the Julia Set generates a sequence of complex numbers that eventually become larger and larger without bound. The colors indicate how quickly the numbers get larger. So, for example, starting with a complex number in the red or orange regions, its iterative list will go off to infinity faster than if the starting number were selected from the blue or yellow areas. So, the colors give a sense of how quickly the numbers get large in our iterative process.

That's how Julia Sets are made. There is one Julia Set for each complex number $a + bi$, and we call it the $(a + bi)$-Julia Set. We just fix $a + bi$ and then ask which complex numbers in the plane, when viewed as starting seeds for the squaring and then adding $a + bi$ iteration process, result in lists of numbers that get larger and larger without bound and which starting seeds result in lists of numbers that do not get arbitrarily large. The interface between these two collections is the ($a + bi$)-Julia Set. One fascinating feature of Julia Sets is that they reveal complicated structure at any magnification. They truly are infinitely detailed. Here is a Julia Set with a magnified area to show the fractal-like detail.

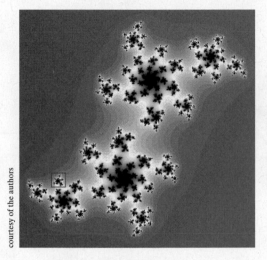

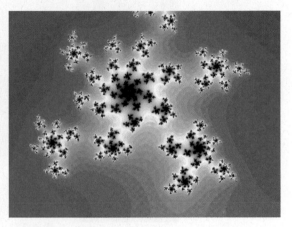

The ($-0.11031 + 0.67037i$)-Julia Set

An enlarged view of the area in the tiny red square

Mandelbrot Set

After being immersed in the (seemingly infinite) details of the construction of each Julia Set, it is time to step back and consider the universe of all ($a + bi$)-Julia Sets at once. We are now ready to construct the Mandelbrot Set. We begin by making a simple observation about each preceding filled-in Julia Set: Either the filled-in black region is just one piece, or it's made up of several smaller pieces. This basic distinction is the key to understanding the Mandelbrot Set. The Mandelbrot Set is an object that captures information about the collection of *all* ($a + bi$)-Julia Sets.

For each point $a + bi$ in the complex plane, we can go off and create the ($a + bi$)-Julia Set. In other words, we can go off and draw the Julia Set in which the number we add at each stage after squaring is $a + bi$. That filled-in ($a + bi$)-Julia Set will either be one piece or more than one piece. If the filled-in ($a + bi$)-Julia Set is one connected piece, then we declare that the point $a + bi$ is in the Mandelbrot Set. So the Mandelbrot Set is the collection of all complex numbers $a + bi$ with the property that the filled-in ($a + bi$)-Julia Set is just one piece.

So, for each point in and near the Mandelbrot Set, we can associate that point with a certain ($a + bi$)-Julia Set. If the point is *in* the Mandelbrot Set, then the

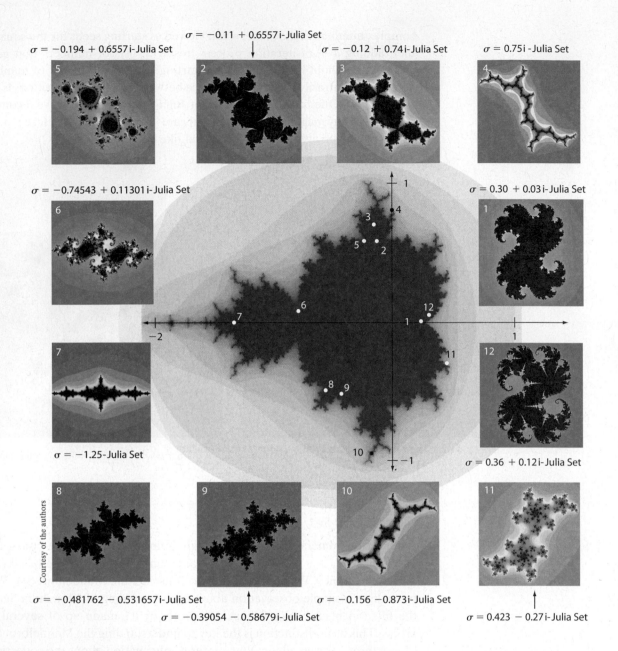

σ = −0.194 + 0.6557 i - Julia Set

σ = −0.11 + 0.6557 i - Julia Set

σ = −0.12 + 0.74 i - Julia Set

σ = 0.75 i - Julia Set

σ = −0.74543 + 0.11301 i - Julia Set

σ = 0.30 + 0.03 i - Julia Set

σ = −1.25 - Julia Set

σ = 0.36 + 0.12 i - Julia Set

Courtesy of the authors

σ = −0.481762 − 0.531657 i - Julia Set

σ = −0.39054 − 0.58679 i - Julia Set

σ = −0.156 −0.873 i - Julia Set

σ = 0.423 − 0.27 i - Julia Set

filled-in Julia Set associated with it will be *one* piece. If the point is *outside* the Mandelbrot Set, then the associated filled-in Julia Set will be made of *several* pieces. In the picture, we select various points in and around the Mandelbrot Set and show inlaid pictures of the associated filled-in Julia Sets.

Here are certain enlarged regions of the Mandelbrot Set to highlight its amazing fractal-like structure. Since there are infinitely many Julia Sets and both the Julia and Mandelbrot Sets are infinitely detailed at increasing magnifications, it is heartening to know that we can never run out of glorious images for calendars, posters, and screen savers.

courtesy of the authors

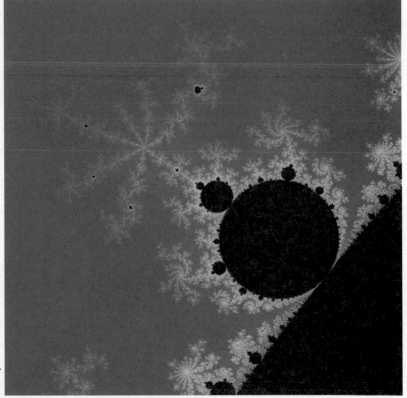

courtesy of the authors

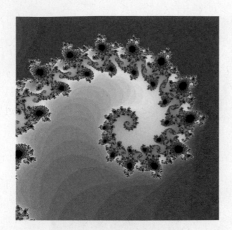

P. S. (Processes Besides Squaring)

We close this section by noting that there are other iterative processes we could have used besides squaring and then adding a fixed number. If we used other iterative processes, repeated forever, and then asked if the starting number leads to a list of numbers getting arbitrarily large or not, then we could construct other types of Julia Sets. We close this chapter with a Julia Set constructed in the same manner we have described except with more elaborate iterative processes (see page 504). Enjoy its wondrous beauty.

A Look BACK

WE ARE ABLE TO GENERATE truly imaginative images using the geometry and arithmetic of complex numbers. Every complex number leads to a different Julia Set. The $(a + bi)$-Julia Set captures those points in the complex plane that do not head off to infinity under the iterative process of repeated squaring and adding $a + bi$. The Mandelbrot Set includes all complex points whose Julia Sets are connected—that is, are in one piece. The Julia and Mandelbrot Sets are the icons of fractal art, and they arise from iterative processes of complex numbers.

Complex numbers arise as an abstract idea to create solutions to equations that have no real-number solutions. By conceiving of complex numbers geometrically as points in the plane, we gain new insight about their relationships. Visualization is a powerful tool for detecting patterns and suggesting relationships. The Julia Set began as an abstract idea about complex numbers, but, after seeing actual images of them, we discover new worlds of interest and new questions to ponder.

Finding a variety of ways to represent the same idea adds depth to our understanding. Visualizing ideas through images is a powerful way of noticing previously unseen patterns and relationships. By expressing the same idea in different ways, we not only see new sides of the idea but also can explore the connections among the different views. Those connections create new insights of their own.

courtesy of the authors

Life Lessons

Visualize the same idea in different ways to see relationships.

•

If something true appears anomalous, expand your worldview to encompass it.

•

Understanding is rock. Memory is sand. Build on rock.

MINDSCAPES   Invitations to Further Thought

*In this section, Mindscapes marked (**H**) have hints for solutions at the back of the book. Mindscapes marked (**ExH**) have expanded hints at the back of the book. Mindscapes marked (**S**) have solutions.*

Developing Ideas

1. **Not Raul.** Who was Gaston Julia?

2. **Use your imagination.** What does the number i equal? Simplify the following expressions: i^2, $(2i)^2$, $(2i)(3i)$.

3. **Complex plots.** The complex number $3 + 2i$ is plotted on the axes shown here. Redraw the axes on a sheet of paper and plot the following points: $2 - 3i$; $1.33i$; $0.5 + 0.25i$; $-\pi - i$.

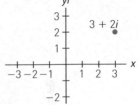

4. **Innie or outie?** If you pick a point $a + bi$ that is inside the Mandelbrot Set and use that point to build the $(a + bi)$-Julia Set, how many pieces will the Julia Set have?

5. **Outie or innie?** If you use the point $a + bi$ to build the $(a + bi)$-Julia Set, and the Julia Set has more than one piece, does the point you used lie inside or outside the Mandelbrot Set?

Solidifying Ideas

6. **Arithmetic.** Compute the following sums and products of complex numbers: $(2 + 3i) + (3 - 7i)$; $(-2 + 5i) + (4 + 6i)$; $(1 - 2i)(2 + 3i)$; $(-i)(4 - 2i)$; $(3 - 2i)^2$.

7. **More arithmetic (ExH).** Compute the following sums and products of complex numbers: $(1/2 + 3i) + (5/2 - i)$; $(-1 + i) + (3 + 2i)$; $(1 - \sqrt{2}i)(3 + \sqrt{2}i)$; $(3 - 2i)(3 + 2i)$; $(2 - \sqrt{5}i)^2$.

8. **Quick draw.** Two complex numbers are marked and labeled on each axis in the graphs shown. Geometrically compute their sums using the parallelogram rule. Geometrically compute their products by adding the angles to the two points measuring counterclockwise from the right half of the x-axis to find the direction and then multiplying their lengths to get the distance out from the origin. Check your answers by adding and multiplying the complex numbers algebraically.

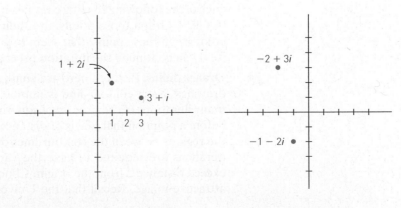

9.   **Quick draw II.** Two complex numbers are marked and labeled on each axis, as shown below. Geometrically compute their sums and products. Check your answers by adding and multiplying the complex numbers algebraically.

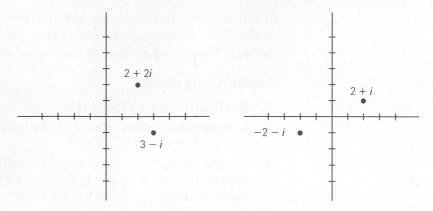

10. **Be square (H).** Let's get in the swing of squaring the imagination by squaring each of the following complex numbers: $1/2, 2 + 2i, -1 + i, 2 + i$. First compute the squares just using algebra. Then, for each one, draw it on the complex plane and measure its angle from the x-axis and its distance from the origin. Finally, measure the angle from the x-axis and the distance from the origin for the square. Did the angle double and distance square? Yes? Good. No? Oops!

11. **Squarer.** For each of the complex numbers $3 + 4i$ and $3/7 - (4/7)i$, find its square geometrically. Check your answers by computing the squares just using algebra.

12–15. **III iterate I (12:S).** Feeling queasy? Then compute the first three iterates of $z^2 + i$ for each of the following starting complex numbers z: $0 + 0i, i, 2 + 3i$, and $1 - i$. Graph the points and their iterates on the same graph, and color green those points that seem to you would be headed toward infinity if you continued the iteration process.

16–19. **III iterate II (18:S).** Compute the first three iterates of $z^2 + 0.27$ for each of the following starting complex numbers z: $0 + 0i, 0.5 + 0.5i, 2 + 3i, i$. Graph those points and their iterates on the same graph, and color green those points that seem to you would be headed toward infinity if you continued the iteration process.

20–23. **III iterate III.** Compute the first three iterates of $z^2 + (0.11 - 0.67i)$ for each of the following starting complex numbers z: $0 + 0i, 0.5 + 0.5i, 1 + 0i, 0 + 0.5i$. Graph those points and their iterates on the same graph, and color green those points that seem to you would be headed toward infinity if you continued the iteration process.

24. **Orange Julias.** Pictured here are some Julia Sets surrounded by contour drawings. Each contour line is numbered. The number represents how many iterations of the process of squaring and adding $a + bi$ are required before a point on that line is of distance 2 or more from the origin. Color the regions between the contour lines different colors to show how many iterations are required to have the various points outside the Julia Set exceed distance 2 from the origin. Choose your colors to create the most attractive image. Recall that the University of Texas at Austin's color is

burnt orange and the Williams College color is purple, so use burnt orange and purple prominently.

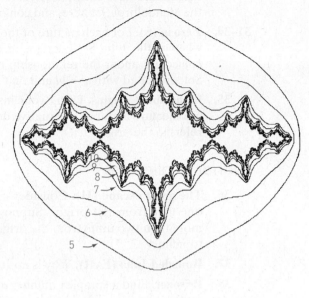

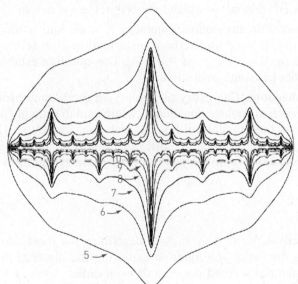

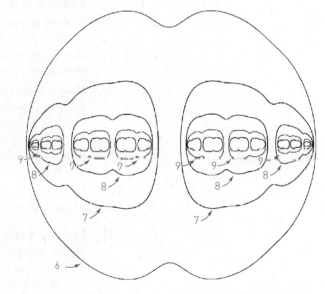

25. **Julia Webbed.** Visit the *Heart of Mathematics* Web site. Then, plug in some random complex numbers and create their associated Julia Sets. Print out your favorites and hang them in your room as psychedelic posters.

Creating New Ideas

26–28. **Great escape?** For each of the complex numbers $(-1 + i)$, $(0 + 0i)$, $(0.25 + 0.25i)$, guess whether the iterates of the number escape to infinity or not under the transformation $z^2 + (-0.2 + 0.65i)$.

29–30. Mandelbrot or not? (H). The images on the following page are little Julia Sets for each of several complex numbers. Which of these complex numbers should be in the Mandelbrot Set? Mark the points on the picture of the Mandelbrot Set here, and confirm your answers.

31–34. Zero in (S). For each picture of the Julia Sets in Mindscapes 29–30, note whether the Julia Set contains or does not contain the origin. Make a conjecture about the relationship between the origin being in the Julia Set associated with $a + bi$ and that Julia Set being connected.

35. Mandelbrot origins. Using your insights from Mindscapes 31–34, make a conjecture about an alternative description of the Mandelbrot Set that refers to the origin.

Further Challenges

36. The great escape (H). Consider $z^2 + a + bi$ where $a + bi$ is less than one unit from the origin. Suppose z is a complex number that lies more than two units from the origin. Show that iterates of z will go off to infinity.

37. Bounded Julia (ExH). Why is no Julia Set unbounded?

38. Prisoner. Find a complex number $a + bi$ that is fixed under the transformation $z^2 + (1 + i)$—that is, for which $(a + bi)^2 + 1 + i = a + bi$.

39. Always a prisoner. For any complex number $c + di$, find a complex number $a + bi$ that is fixed under the transformation $z^2 + (c + di)$—that is, for which $(a + bi)^2 + c + di = a + bi$. This question establishes that every Julia Set has some points in it.

40. Mandelbrot connections. Can every two points on the Mandelbrot Set that are close to each other be connected by a fairly small squiggly path? Look at pictures of the Mandelbrot Set and make a guess. (*Hint:* Don't be discouraged if you are unable to settle this question, since it is an unsolved problem. No one in the world knows whether the answer is yes or no.)

In Your Own Words

41. Personal perspectives. Write a short essay describing the most interesting or surprising discovery you made in exploring the material in this section. If any material seemed puzzling or even unbelievable, address that as well. Explain why you chose the topics you did. Finally, comment on the aesthetics of the mathematics and ideas in this section.

42. With a group of folks. In a small group, discuss what the Julia Set associated with the complex number $a + bi$ is. After your discussion, write a brief narrative describing which points are in the Mandelbrot Set in your own words.

43. Creative writing. Write an imaginative story (it can be humorous, dramatic, whatever you like) that involves or evokes the ideas of this section.

44. Power beyond the mathematics. Provide several real-life issues—ideally, from your own experience—that some of the strategies of thought presented in this section would effectively approach and resolve.

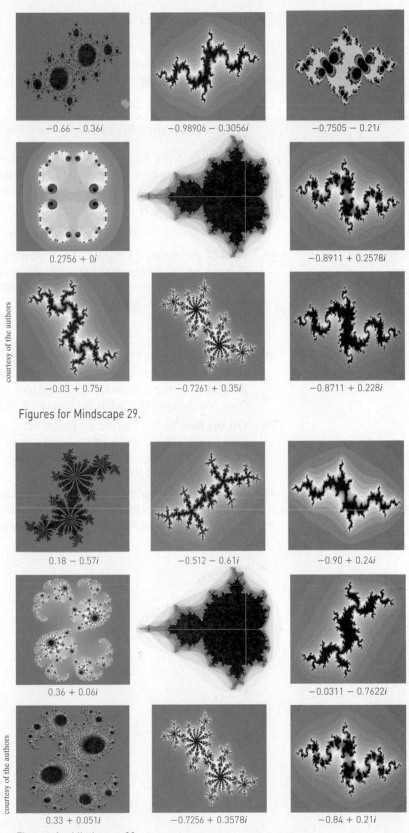

courtesy of the authors

$-0.66 - 0.36i$ $-0.98906 - 0.3056i$ $-0.7505 - 0.21i$

$0.2756 + 0i$ $-0.8911 + 0.2578i$

$-0.03 + 0.75i$ $-0.7261 + 0.35i$ $-0.8711 + 0.228i$

Figures for Mindscape 29.

$0.18 - 0.57i$ $-0.512 - 0.61i$ $-0.90 + 0.24i$

$0.36 + 0.06i$ $-0.0311 - 0.7622i$

$0.33 + 0.051i$ $-0.7256 + 0.3578i$ $-0.84 + 0.21i$

courtesy of the authors

Figures for Mindscape 30.

For the Algebra Lover

Here we celebrate the power of algebra as a powerful way of finding unknown quantities by naming them, of expressing infinitely many relationships and connections clearly and succinctly, and of uncovering pattern and structure.

45. **Quadratic complexity (H).** The quadratic formula gives us solutions to a general quadratic equation $Ax^2 + Bx + C = 0$ as follows: $x = \dfrac{-B + \sqrt{B^2 - 4AC}}{2A}$.

    (Notice that the formula gives two values for x, one with the plus sign and one with the minus sign.) Use this formula to find the solutions to the following equations. What kinds of numbers are these? (In the last equation, $B = 0$.)

    $$x^2 + x + 1 = 0 \qquad\qquad 2x^2 - 6x + 5 = 0 \qquad\qquad x^2 + 1 = 0$$

46. **Special products.** Compute the following products of complex numbers. What kind of numbers do you get for your answers? Can you give other examples where the product works out just as nicely? Can you see a pattern that explains what happens?

    $$(1 + i)(1 - i) \qquad\qquad (2 - i)(2 + i) \qquad\qquad (3 + 2i)(3 - 2i)$$

47. **Something's fishy.** The treasurer of the Marine Biology Student Association claims that the amount of money the association will have at the end of the year is one of the values of x (in \$) that satisfies the equation $20x^2 - 10x + 50 = 0$. (See Mindscape 45.) What do you think of his prediction? Explain.

48. **On the line.** In the xy-plane, sketch the graph of the line given by the equation $y = (3/2)x + 2$. Now imagine that your xy-plane is actually the complex plane. Verify that the point $2 + 5i$ lies on the line. Find three other complex numbers that lie on the line.

49. **A line with a complex.** Plot the points in the complex plane corresponding to the numbers $-1 + i$ and $4 - 2i$. Now imagine that your complex plane is really the good, old-fashioned xy-plane. Find an equation for the line through your two points.

THE DYNAMICS OF CHANGE

Can Change Be Modeled by Repeated Applications of Simple Processes?

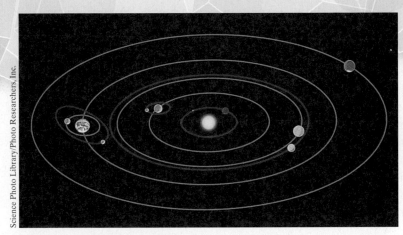

Science Photo Library/Photo Researchers, Inc.

Can we predict the location of the planets?

> *The mathematical phenomenon always develops out of simple arithmetic, so useful in everyday life, out of numbers, those weapons of the gods; the gods are there, behind the wall, at play with numbers.*
>
> **LE CORBUSIER**

In the previous sections of this chapter we have seen the detailed structure and rich nuance that arise from repeating a process again and again. Our focus up to this point has been visual in nature and that path led us to a beautiful world of fractals, in which the repeated application of a process generated infinitely intricate images. For the remainder of this chapter, we will apply this theme of repeating processes over and over to our everyday world in the hopes of uncovering rich nuance in the structure of our physical and natural universe.

Our world changes. Populations grow and shrink. The weather changes from day to day, season to season, and year to year. The positions of planets, moons, and stars evolve over time. Often we have some understanding of the influences that cause change. Applying this knowledge, we can fairly accurately predict the population, temperature, or positions of planets for the next day. Then, using those predicted conditions as the assumed starting points, we could predict the population, the weather, or the positions of planets for two days from our new starting point. Once more, taking the results of previous predictions as the starting point, we could take one more step of prediction. In this fashion, we can predict many days or years into the future. This process is a dynamic model of incremental prediction.

Dynamic models of incremental prediction are natural ways to describe change, and many important mathematical applications use them. As we will soon see, this technique of step-by-step change has intriguing and unexpected consequences. Among them are games, images, and a world of insight about our ability to accurately predict the future.

We will clarify our ideas about these repeating processes by looking at several different examples—bank accounts, positions of planets, populations, and games. From these disparate examples, we abstract and develop an idea of systems that change in a step-wise fashion.

Money

Suppose we deposit $1000 in a savings account that pays 5% interest compounded annually. After one year, our account will have a balance of 1.05($1000) = $1050. If we leave the money in another year, we will accrue 5% interest on all the money in the account. So, after two years we will have 1.05($1050) = $1102.50. If we leave our money in the account yet another year, we will earn 5% on what we now have; so, after three years we will have 1.05($1102.50) = $1157.63. The pattern is clear—not exciting, but clear. Take what we have at the beginning of the year, multiply that number by 1.05, and we have the amount of money we will end up with after one more year. As this process is repeated, we can just sit back and watch our money grow. Simple enough.

Planets

Suppose we know the locations of all the planets, asteroids, and other matter in the solar system. We know the speed and direction of each object; we know the masses of the Sun, planets, and asteroids; and we know about gravitational forces. Can we compute the locations and velocities of all those bodies after one revolution of Earth? Sure, pretty closely at least. How about after two revolutions? We could take our knowledge of the positions and velocities of the bodies after the first revolution and repeat the same calculations to compute the state of the solar system after two revolutions. And so on. Theoretically, we could continue year after year for thousands, millions, or even billions of years. This method can be used to make predictions, although how accurate our predictions will be is a matter for us to contemplate when we consider chaos in the final section of this chapter. The key is that we are able to take what we have, apply a rule, compute what we have next, and then repeat.

Population

How can we predict the population size after one year? We can take the current population size, estimate the number of births and deaths expected during the next year, and then compute the population size after one year. How can we predict the population after two years? We can just take our estimate for the population after one year, estimate the number of births and deaths, and again compute the new population size after one more year. How can we predict the population size many years from now? We'd just repeat, repeat, repeat.

Repeat, Repeat, Repeat

As we are seeing, models of the world often boil down to:

- Start somewhere.
- Apply a process and get a result.
- Apply the same process to that result to get a new result.
- Apply the process again to the new result to get a newer result.
- Repeat patiently and persistently, forever.

Computing compound interest, predicting the positions of the planets after time has passed, and estimating future population size all share the underlying theme of repetition. We have encountered this iteration theme before—not only in our journey through fractals, but in our discussion of the Fibonacci numbers. Recall that each Fibonacci number was obtained by adding together the previous two Fibonacci numbers. We took the results of what we had done before and used them to get the next number. We are now crystallizing the notion of an iterative procedure for predicting the future of changing, dynamic systems—a procedure that uses feedback from one step to produce the next.

Modeling

To build a mathematical model that reflects reality, we examine the real-world situation and cull from it some features we believe are important. We then attempt to capture their essence through mathematical relationships or processes that are analogous. For example, consider models for estimating population size.

How many people will there be on Earth in the year 2050? How would we begin to figure this figure? We might try to devise some reasonable predictions about how the population size will change from one year to the next given the number of people living at the beginning of a year, as well as their living conditions. Of course, developing realistic, predictive models is an extremely complicated issue, since so many factors influence population change. Therefore, as always, let's start simply.

One of the recurring themes of this book is the power of starting with a simple or an idealized case, building up insights, definitions, ideas, and methods, and then using that experience to handle the more challenging cases. In estimating population growth, let's start with a game that reflects some features of what makes populations grow or shrink.

Life Lesson
Start simply.

The Game of Life

Life was not invented in the 20th century, but the *Game of Life* was—by a mathematician, John Conway of Princeton University. Although the Game of Life models only one aspect of real life, it does reflect some of the dynamic development of populations over time. It may even be more interesting as a model of other phenomena, such as how fires propagate in a forest.

Conway's Game of Life is played on an infinitely extended grid. At each time interval, a square is either alive or not, depending on how many living squares surrounded it in the previous time interval. Thus, this model of population growth is focused on an organism's need for the right number of companions within its immediate vicinity. Here are the rules for determining which squares will be alive during the next generation:

- A living square will remain alive in the next generation if exactly two or three of the adjoining eight squares are alive in this generation; otherwise, it will die.

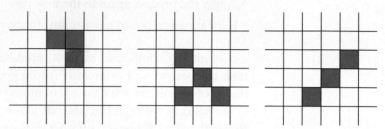

The red squares will be alive in the next generation.

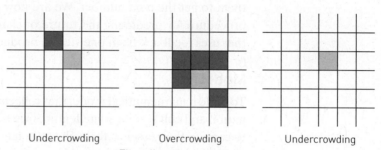

Undercrowding   Overcrowding   Undercrowding

The yellow squares will all die.

- A dead square will come to life if exactly three of its adjoining eight squares are alive; otherwise, it will remain dead.

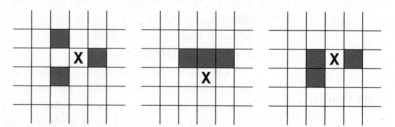

The currently dead **X** squares will come to life in the next generation.

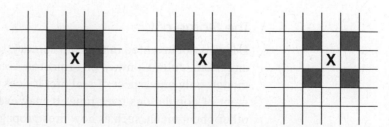

The currently dead **X** squares will remain dead in the next generation.

Let's look at the future generations for some initial sets of organisms.

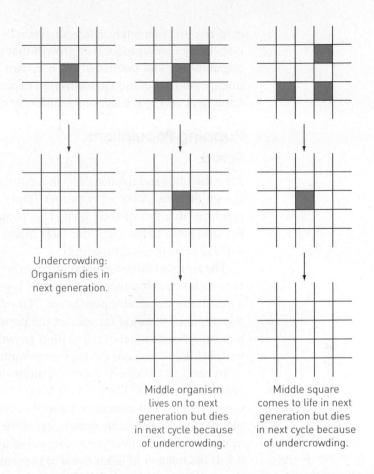

Undercrowding:
Organism dies in
next generation.

Middle organism
lives on to next
generation but dies
in next cycle because
of undercrowding.

Middle square
comes to life in next
generation but dies
in next cycle because
of undercrowding.

Slight differences in the initial sets of living cells sometimes evolve into dramatically different outcomes. Some initial sets give rise to a population explosion that grows without bound. Other populations die out quickly. Some become periodic—growing and shrinking in an infinitely repetitious, predictable way—or become stable. Others move across the grid like a migrant colony marching off forever onward. Here we see all these possibilities at play. For each figure, decide whether it gives rise to (1) a population explosion, (2) an extinction, (3) a stable pattern, (4) a periodic pattern, or (5) a migratory pattern. You can use the software on the *Heart of Mathematics* Web site to simulate the Game of Life and discover all kinds of surprising outcomes. Have fun!

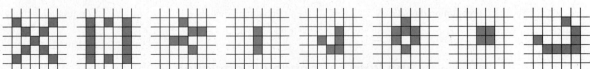

Choices: (1) population explosion; (2) extinction; (3) a stable pattern; (4) a periodic pattern; (5) a migratory pattern

The Game of Life provides a rough model of population development. Even with these simple rules, the potential patterns of development are intricate and fascinating. This game suggests that different initial population distributions give rise to completely different types of future behavior. Mathematical biologists study various models of growth that try to reflect different aspects of actual populations. The mathematical behavior of these models sometimes points to unexpected population patterns that occur in nature. Let's discover this phenomenon in action by considering a different population *pool*.

Popping Populations
Scrod

Like the birds and the bees, fish in a pond fall in love, copulate, and subsequently spawn little fish who, after proper schooling, themselves become amorous and produce little fish of their own. This romantic life cycle results in ponds full of fish and leads to the question: How many fish might we expect in the pond after several years of good breeding?

The simplest model of population growth might assert that the population doubles each year. However, such a model is grossly unrealistic, because the pond has a maximum sustainable population. If the fish population doubled each year, soon the population would far exceed the pond's capacity. Thus, we need to develop a more refined model of population growth that predicts that the fish population will shrink if it exceeds the maximum number that the pond can sustain.

In the 1840s Pierre François Verhulst developed a population model with this refinement in mind. The Verhulst Model represents each year's population not as a number of fish but instead as a density of fish in the pond—that is, as a *fraction* of the maximum sustainable population of the pond. So, for example, if there are half as many fish as the maximum sustainable number, we say the population density is 0.5. If the number of fish is equal to the maximum sustainable number, then we say the population density is 1.0. If in some year the population temporarily exceeds the maximum sustainable population, the population density might be 1.2, but then the population would drop the next year—there is an overpopulation of fish.

Life Lesson
> **There are many ways to say the same thing. Choose one that's useful.**

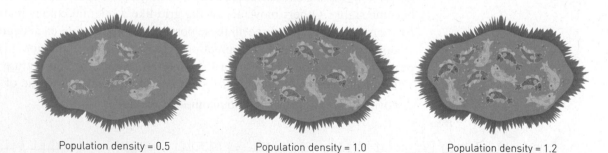

Population density = 0.5   Population density = 1.0   Population density = 1.2

The Rate of Change of Population

Verhulst reasoned that, if the population is small relative to the maximum sustainable population, then the fish will have lots of available room and resources and thus will happily reproduce at will. Consequently, the population will grow at a faster rate than it would if it were near the maximum sustainable population (in which case overcrowding and limited resources would inhibit amorous urges).

In the Verhulst Model, the rate at which the population increases (or declines) is proportional to how far away the population size is from the maximum sustainable number. Let's take a closer look at how the Verhulst Model works.

We first need to figure out the rate at which the population grows (or shrinks). To do this, let's write P_n for the population density at year n given as the fraction

$$P_n = \frac{(\text{the number of fish at year } n)}{(\text{the maximum sustainable population of fish})}.$$

Then $P_{n+1} - P_n$ is the change in population density from year n to the next year $n + 1$. That change divided by P_n gives the *rate at which the population is increasing or decreasing*. So, the rate at which the population is increasing or decreasing from year n to year $n + 1$ is equal to

$$\frac{P_{n+1} - P_n}{P_n}.$$

For example, suppose that, in a certain pond, the maximum sustainable population of fish is 5000. If in year 2 the fish population of the pond was 2500 and in year 3 the population was 3000, then we can compute the population densities as follows:

$$P_2 = \frac{2500}{5000} = 0.5 \quad \text{and} \quad P_3 = \frac{3000}{5000} = 0.6.$$

Thus, the change was

$$P_3 - P_2 = 0.6 - 0.5 = 0.1.$$

If we now take that change (0.1) and divide it by the population density we started with ($P_2 = 0.5$), we get 0.2 as the *rate at which the population increased*. Stated in plain English: The population increased by 20% from year 2 to year 3. Recall Verhulst's basic three-part assumption:

- If the population is far from the maximum sustainable population, we expect a large increase in population;
- If the population is not as far from the maximum sustainable population, we expect a lesser increase in population; and finally,
- If the population exceeds the maximum sustainable population, we expect a decrease in population.

Specifically, we notice that $1 - P_n$, the difference between the maximum sustainable population density, 1, and the population density at year n, Pn, measures how close the current population is to the maximum sustainable population. So, Verhulst's Model states that the rate of population change $[(P_{n+1} - P_n)/P_n]$ will be large when $1 - P_n$ is large, and the rate of population change will be small when $1 - Pn$ is small. That is, $[(P_{n+1} - P_n)/P_n]$ is proportional to $1 - P_n$.

We can phrase this observation in mathematical terms as follows: There is some fixed number such that $(P_{n+1} - P_n)/P_n$ will equal that fixed number multiplied by $1 - P_n$. For example, if we assume for the moment the fixed number is 0.7, we can write this idea as an equation:

$$\frac{P_{n+1} - P_n}{P_n} = 0.7(1 - P_n).$$

Multiplying through by the P_n term, we get

$$P_{n+1} - P_n = 0.7\, P_n(1 - P_n), \text{ or}$$
$$P_{n+1} = P_n + 0.7\, P_n(1 - P_n).$$

So, knowing the population density of one year allows us to find the population density for the next year. This formula, with all the n's and $n+1$'s running around, is confusing. Let's solidify the idea by working through a few specific examples.

Example 1: 1000 Fish

Let's assume that our pond has a maximum sustainable fish population of 5000. Suppose that the population this year is 1000 fish. Thus, we see that

$$P_1 = \frac{1000}{5000} = 0.2.$$

Since 0.2 is far from the maximum sustainable population density (which is 1), we would expect a good-sized increase in population in the next year. In fact, we can compute next year's population density (P_2) and see that it will be

$$P_2 = 0.2 + 0.7 \times 0.2(1 - 0.2) = 0.2 + 0.7 \times 0.2 \times 0.8 = 0.2 + 0.112 = 0.312.$$

So, the population for next year will be

$$0.312 \times 5000 = 1560 \text{ fish.}$$

The increase from the population level of 1000 to 1560 represents an increase in population of more than 50%—a pretty hefty increase.

Example 2: 4500 Fish

Suppose now that the initial population is 4500 fish. Thus, we have

$$P_1 = \frac{4500}{5000} = 0.9.$$

In this case, the population density is very close to the maximum sustainable number, so we would expect the population growth to slow down. Computing, we discover that next year's population density (P_2) will be

$$P_2 = 0.9 + 0.7 \times 0.9(1 - 0.9) = 0.9 + 0.7 \times 0.9 \times 0.1 = 0.9 + 0.063 = 0.963.$$

So, the population for next year will be

$$0.963 \times 5000 = 4815 \text{ fish.}$$

The population increase from 4500 to 4815 represents an increase of only 7%—a much smaller increase, just as we had expected.

Example 3: 7000 Fish

Finally, suppose the initial population is 7000 fish. So, we see that

$$P_1 = \frac{7000}{5000} = 1.4.$$

In this case, we have a population that exceeds the maximum sustainable population, so we would expect the population to decline. In fact, next year's population density (P_2) will be

$$
\begin{aligned}
P_2 &= 1.4 + 0.7 \times (1 - 1.4) \\
&= 1.4 + 0.7 \times 1.4 \times (-0.4) \\
&= 1.4 - 0.392 \\
&= 1.008.
\end{aligned}
$$

So, here we see that next year's fish population will be down to 5040. Notice that next year's population is smaller than the population of the year before, just as we had predicted. The population decrease from 7000 to 5040 represents a decrease of 28%.

Desperately Seeking the Constant

One key to accurately predicting future population size using the Verhulst Model is finding the constant number that best fits the situation. Finding a constant that leads to an accurate predictor is a difficult task. In our examples we used a constant of 0.7, but for a real pond the constant 2 or 0.01 or some other number might have more accurately predicted future populations.

Here are two instructive experiments we can perform with this little population model: The first is to fix a constant, like 0.7, experiment with different initial populations, and see what happens to the population in future years. The second experiment is to see what happens when we adjust the constant. In 1976, Robert May, a mathematical biologist, published a paper in the journal *Nature* titled "Simple mathematical models with very complicated dynamics." In it, May explored the surprising results of such experiments and helped to usher in a whole new perspective on predictability and chaos. We will explore these issues for ourselves in the next section.

A Look BACK

How CAN WE describe change? One method is to take it a step at a time. We can describe the current state and predict the next state. Then we can take that result and predict the next one. In this step-wise way, we can describe change over many steps—that is, over a long period of time. Both the Game of Life and the Verhulst Model illustrate this basic theme.

The real world is far too complex to understand and describe completely. Often our method of describing the world or predicting the next step is a simplification of reality. We just pick one or two important features and ignore the rest. Although we cannot expect such simplified models to be completely accurate, they may well reveal some aspects of the way in which the world will change. We start with an oversimplified model, expecting that we can modify it later.

To develop ideas, it is important to look at several examples of related phenomena in different settings and find common features. Often there are many ways to describe the same situation, so we can choose a method that is useful to us. By isolating essential features, we can develop descriptions that free us from distractions and allow us to focus on the core issues.

Life Lessons

Look at a variety of examples.

•

Start with simple cases.

•

Model a situation in various ways to emphasize different features.

MINDSCAPES　Invitations to Further Thought

In this section, Mindscapes marked **(H)** *have hints for solutions at the back of the book. Mindscapes marked* **(ExH)** *have expanded hints at the back of the book. Mindscapes marked* **(S)** *have solutions.*

Developing Ideas

1. **Any interest?** Suppose you put $1000 in a bank account that pays 3% interest compounded annually. How much money will you have in the account at the end of one year? At the end of two years? At the end of five years?

2. **Urban expansion.** In the year 2004 the village of Starburg has a population of 10,000. City planners predict that the population will grow at a rate of 7% per year. What is the population of Starburg expected to be in the year 2008?

3. **Pre-sushi.** Suppose a pond can sustain a maximum of 6000 fish. What is the population density when there are 3000 fish in the pond? What's the density when there are 4800 fish? What's the density when there are 7500 fish?

4. **A booby trap.** A small landmass off the Galapagos Islands can sustain a maximum of 50,000 blue-footed boobies. If the population density is 0.675, how many boobies are actually living on the island?

5. **Too many.** The island of Birdburg has an area of 17,000 square yards. If 50,000 cockatoos are distributed evenly on the island, then on average about how many birds are living on each square yard?

Solidifying Ideas

6. **Call your shots.** For each picture below, trace the path of the billiard ball in the sketch until it lands in a pocket. Assume that when the ball strikes a wall, it bounces off at the angles shown. What pocket do you call?

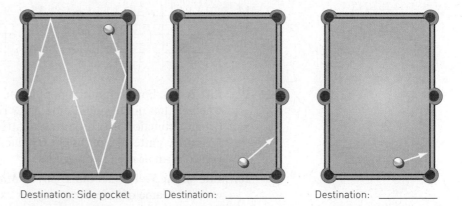

Destination: Side pocket   Destination: _____   Destination: _____

7. **Getting cornered.** Starting from the spot indicated on the billiard table at right, sketch two trajectories that result in the ball getting to the back left corner.

Want to sink the ball here

8. **Double your money (H).** Suppose you deposit $1000 in a savings account that pays 5% interest compounded annually. After one year, your account will have a balance of 1.05($1000) = $1050. By repeating this process, estimate the amount of time required to double your money.

9. **Too many (S).** Earth's surface is approximately 2,500,000,000,000,000 (2.5×10^{15}) square yards, including the oceans. The current population of Earth is about 7,000,000,000 (7×10^9) people. If the population were to grow at the rate of 1% per year, approximately how soon would there be a person for each square yard on Earth (including the oceans)? What can you deduce about the future rate of population growth?

10. **Rice bowl (H).** One day long ago, the Emperor of China wished to reward a clever servant. When asked what he wanted, the modest servant replied that he would be satisfied with one grain of rice on the first square of a checkerboard, two on the next, four on the next, eight on the next, and so on over the 64 squares of the checkerboard. Why did the Emperor behead the servant?

11. **Nature's way.** Find some examples, other than those mentioned in this section, of iterative processes at work in nature.

12–15. **The Game of Life (12:S).** For each initial population in the Game of Life, determine which cells will be alive after 1, 2, 3, and 4 generations. You can visit the *Heart of Mathematics* Web site to simulate the Game of Life if you wish.

12.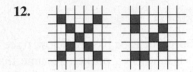

13.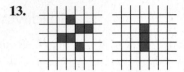

14.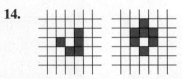

15.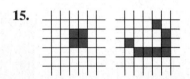

16–19. **Life cycles.** For each initial population in the previous Mindscape, guess whether that initial population gives rise to (1) a population explosion, (2) an extinction, (3) a periodic pattern, (4) a stable pattern, or (5) a migratory pattern. You might try one or two more generations to help you confirm or reject your guess.

20. **Life on the Web.** Visit our *Heart of Mathematics* Web site to experiment with the Game of Life. Record your experiments.

21. **Explosion.** Devise a new initial population in the Game of Life that leads to a population explosion. (*Hint:* Consider taking an initial population you know will explode and adding a small community far off that will not interfere with the explosion.)

22. **Extinction.** Construct a new initial population in the Game of Life that survives for several generations and then becomes extinct. (*Hint:* Consider taking an initial population you know and adding a small community far away that won't hurt.)

23. **Periodic population (H).** Construct a new initial population in the Game of Life that is *periodic,* that is, that returns to its original position after several generations. (*Hint:* Consider putting together a couple of periodic initial populations you know.)

24. **Programmed population.** On a computer or programmable calculator, enter this simplified version of the Verhulst equation, $y = 4x(1 - x)$, into its function memory. Start with $x = 0.245$ and compute $4x(1 - x)$ by pressing the function key. Take the result and again press the function key. Repeat 25 times and put the answers in the first column of a table. Now do the whole thing again, but this time start with the slightly different number $x = 0.246$. Enter the resulting answers in the second column of the table. The resulting answers start by being similar. Do they remain similar? We will explore your observation more in the next section.

25. **Programmed population: the next generation.** Using a computer or programmable calculator, input this simplified version of the Verhulst

equation, $y = 3.5x(1 - x)$. Start with $x = 0.245$ and compute $3.5x(1 - x)$ by pressing the function key. Take the result and again press the function key. Repeat 25 times and put the answers in the first column of a table. Now do the whole thing again, but this time start with the slightly different number $x = 0.246$. Enter the resulting answers in the second column of the table. Are you surprised at the difference between Mindscapes 24 and 25?

Creating New Ideas

26. **How many now?** Suppose that a population is modeled using the Verhulst Model, $P_{n+1} = P_n + cP_n(1 - P_n)$. Suppose the initial population measure P_1 is 0.25, and the next year's population measure P_2 equals 0.5. Find the value of the constant c. Using your answer, compute P_3.

27. **Fibonacci.** Fibonacci numbers are constructed using an iterative process. Explain how the results of one step in the process become the input for the next iteration of the process. Make a table with two columns. In the first column write the Fibonacci numbers where you start the process as usual with the first two numbers being 1 and 1. In the second column, do the same process of going from generation to generation; however, now start with 1 and 3. How are the columns similar and different?

28. **Fibonacci again.** For each of the two columns you created in Mindscape 27, make a new column. Next to the nth entry F_n in your existing column, enter F_{n+1}/F_n in the new column. What do you notice about the two new columns you created? Did the different starting seeds make a difference? What further experiments might you consider?

29. **Alien antenna (ExH).** Start with a **V**. Suppose each end sprouts a smaller **V**, and each of those ends sprouts an even smaller **V**. Draw the first four generations of this alien antenna. How many end points are there after four generations? How many would there be after n generations?

30. **Cobweb plots (ExH).** Here we visualize the simplified version of the Verhulst equation as we've seen in Mindscapes 24 and 25. That is, we consider the equation $y = cx(1 - x)$, in which c is a given constant value. The graph of this equation is a "sad-face" parabola that crosses the x-axis at $x = 0$ and $x = 1$. The peak of the parabola always occurs at $x = 0.5$. For example, let's consider the equation $y = 4x(1 - x)$ (so here, the constant c is equal to 4). If we start with $x = 0.2$, then we can use a calculator to find that $y = 0.64$. If we now repeat, then we plug in 0.64 for x and find that now $y = 0.9216$. If we repeat yet again, we plug 0.9216 in for x in the formula and find that the new $y = 0.2890. \ldots$

We can visualize this repeated process on the graph of our sad parabola by first drawing in the diagonal line, $y = x$, that goes right between the axes at an angle of $45°$. We now start at $x = 0.2$ on the horizontal axis and trace a path by going straight up until we reach the graph (this height yields our first y value, $y = 0.64$). From there we trace a path horizontally (left or right, in this case right) until we hit the diagonal line. Now we repeat the process again: We go up or down (in this case up) until we hit the parabola and that gives us the next y value, $y = 0.9216$. We then repeat again and again. We will generate a path with right-angle turns that go from the parabola to the diagonal line again and again. The first few steps are illustrated and we also include the process with two different starting values ($x = 0.3$ and also $x = 0.4$).

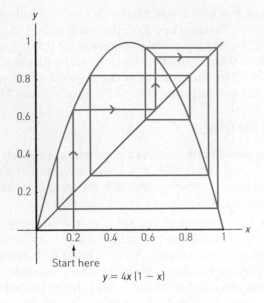

$y = 4x(1 - x)$

Start here

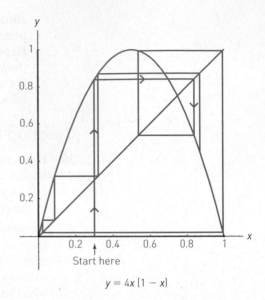

$y = 4x(1 - x)$

Start here

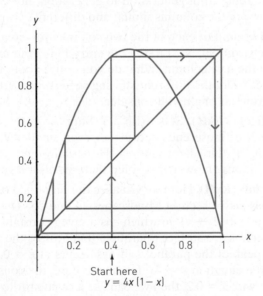

Start here
$y = 4x(1 - x)$

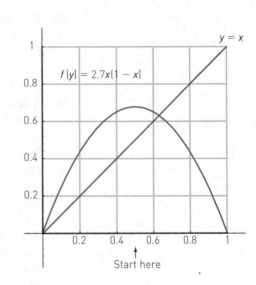

$f(y) = 2.7x(1 - x)$

$y = x$

Start here

Notice how different the paths look! The *Heart of Mathematics* Web site contains a program that allows you to see this repeated process for any starting value of *x*.

These paths are called *cobweb plots.* The cobweb plot records the results of a repeatedly applied transformation—namely, taking a value for *x*, applying the formula to find *y*, then taking that result as the next *x* and again applying the formula to find the next *y*, and so on.

Now consider the graph of $y = 2.7x(1 - x)$ given here. Start on the horizontal axis at 0.5 and carefully draw five iterations of the cobweb plot using a straightedge. Does the cobweb plot form a spiral? You can check your graph using the program on the *Heart of Mathematics* Web site. (*Hint:* To help make the drawing accurate, you should compute the *y* value when *x* equals 0.5 and then take that value and plug it back into the formula for *x*, repeat, etc.).

31. **More spiders (S).** Here is the graph of $y = 3.18x(1 - x)$. Start on the horizontal axis at 0.25 and draw five iterations of the cobweb plot as described in Mindscape 30. Does the path appear to be periodic or not (that is, does the path seem to repeat its steps from some point on or not)? You can check your graph using the program on the *Heart of Mathematics* Web site.

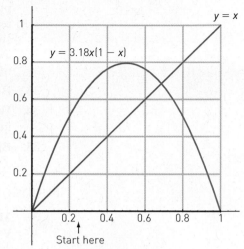

32. **Arachnids.** Here is the graph of $y = 4x(1 - x)$. Start on the horizontal axis at 0.25 and draw five iterations of the cobweb plot as described in Mindscape 30. Does the path keep going, or does it get stuck? You can check your graph using the program on the *Heart of Mathematics* Web site.

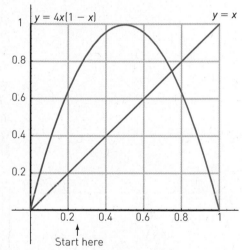

33–35. **Making dough.** Many delicious French desserts, such as Napoleons, are made of numerous thin layers of pastry. Once upon a time a French chef began, as usual, to make a masterpiece with a segment of dough one unit long. This time, however, he noticed that three grains of colored sugar—one red, one white, and one blue— were embedded in the pastry. He stretched the dough to twice its length and then folded it in half to produce a double layer of thinner pastry still one unit long. He again stretched the dough to double its length and folded it to produce four layers again one unit long. He repeated

the process: stretch to twice its length, fold to produce eight layers. He stretched and folded again and again. As our chef stretched and folded, he became fascinated with the movement of those three grains of sugar. In fact, he noticed that the red grain always returned to its original location. The white grain began somewhere, went to a different location after a stretch and fold, and then returned to its original location after one more stretch and fold. The blue grain rotated among three different places. The chef was so fascinated with the infinitely recurring itineraries of these grains of sugar that he stretched and folded the dough to such an extreme thinness that his dessert creation lived up to the high expectations for light, puffy French desserts and became the Napoleon of pastries.

33. **Red.** Show where the red grain of sugar may have been located initially.

34. **White (H).** Show where the white grain of sugar may have been located initially.

35. **Blue.** Show where the blue grain of sugar may have been located initially.

Further Challenges

36. **More cobwebs (H).** Consider the inverted **V** graph shown here and describe the points whose cobweb path leaves the unit square. (*Hint:* Any point in the interval (1/3, 2/3) leaves immediately. How about points in intervals (1/9, 2/9) and (7/9, 8/9)? Are there others that leave after more steps?)

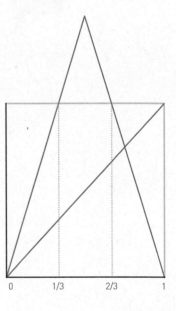

37. **Yet more cobwebs.** Given the inverted **V** graph below, find a starting point by trial and error with pictures such that the cobweb path intersects the diagonal in every 1/6th of the diagonal.

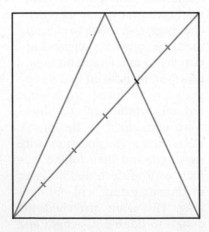

38. **Cantor's cuts.** Start with the unit interval [0, 1]. Remove the middle third of that interval and you will have [0, 1/3] and [2/3, 1]. Next, remove the middle third of each of those intervals. You will have [0, 1/9], [2/9, 1/3], [2/3, 7/9], and [8/9, 1] remaining. If you repeat the process of removing the middle third forever, the points that remain constitute the *Cantor Set*. Describe infinitely many points that remain in the Cantor Set.

39. **How much is gone?** In the construction of the Cantor Set, various intervals were removed. At the first step, the interval (1/3, 2/3) was removed. Then the intervals (1/9, 2/9) and (7/9, 8/9) were removed, and so on. Estimate the total length of the removed intervals. Use your answer to estimate the total "length" of the Cantor Set.

40. **How much remains?** Consider numbers that are not in decimal form (that is, base 10) but instead are represented in base 3. That is, every digit of the number is 0, 1, or 2. The first digit after the "decimal"point tells how many $1/3^1$'s you have; the next tells how many $1/3^2$'s you have; the nth place after the decimal point tells how many $1/3^n$'s you have. So, 0.212 (base 3), for example, represents $2(1/3) + 1(1/3)^2 + 2(1/3)^3 = 2/3 + 1/9 + 2/27$ or 23/27. Show that the points that remain in the Cantor Set are exactly those numbers whose base 3 decimal expansion can be written with only 0's and 2's. (Note that, just as $0.0999 \ldots = 0.1$ base 10, $0.0222 \ldots = 0.1$ base 3, so put all numbers in the ending 2's form rather than in the ending 0's form.) Since any sequence of 0's and 2's corresponds to a "decimal" number in the Cantor Set, show that there are more numbers in the Cantor Set than there are natural numbers. In fact, the cardinality of the Cantor Set is the same as the cardinality of the real numbers.

In Your Own Words

41. **Personal perspectives.** Write a short essay describing the most interesting or surprising discovery you made in exploring the material in this section. If any material seemed puzzling or even unbelievable, address that as well. Explain why you chose the topics you did. Finally, comment on the aesthetics of the mathematics and ideas in this section.

42. **With a group of folks.** In a small group, discuss the idea of iterative systems, particularly population models. After your discussion, write a brief narrative about systems that are described by repeated applications of a transforming process in your own words.

43. **Creative writing.** Write an imaginative story (it can be humorous, dramatic, whatever you like) that involves or evokes the ideas of this section.

44. **Power beyond the mathematics.** Provide several real-life issues—ideally, from your own experience—that some of the strategies of thought presented in this section would effectively approach and resolve.

For the Algebra Lover

Here we celebrate the power of algebra as a powerful way of finding unknown quantities by naming them, of expressing infinitely many relationships and connections clearly and succinctly, and of uncovering pattern and structure.

45. **Life happens.** A particular starting configuration in Conway's Game of Life has a number of living squares represented by m. A later generation has a number of living squares represented by n. The values m and n are both solutions to the equation $x^2 - 16x + 55 = 0$. Suppose you also know that $m < n$. Using this information, find the values of m and n.

46. **Perplexing population predictions.** Verhulst's model describes a relation between the population density one year (P_n) and the density the next year (P_{n+1}). Here's an example: $P_{n+1} = P_n + 0.3(1 - P_n)$. Thus, if $P_1 = 0.2$, then $P_2 = 0.2 + 0.3(1 - 0.2) = 0.2 + 0.3(0.8) = 0.2 + 0.24 = 0.44$. In other words, from Year 1 to Year 2, the model predicts that the population density of this community went from 0.2 to 0.44. Using the formula and the prediction found for Year 2, what does the model predict for Year 3? Using the answer you just found, what does the model predict for Year 4?

47. **More prediction peril.** Consider the previous Mindscape but now use the model: $P_{n+1} = P_n + 1.3(1 - P_n)$. This time, if $P_1 = 0.2$, then $P_2 = 0.2 + 1.3(1 - 0.2) = 0.2 + 1.3(0.8) = 0.2 + 1.04 = 1.24$. So with a much larger coefficient (1.3 instead of 0.3), the model predicts a much greater increase in population density from Year 1 to Year 2. hat does the model predict for Year 3? For Year 4?

48. **Cobweb parabola.** Mindscapes 30–32 describe cobweb plots that involve a parabola with equation $y = 4x(1 - x)$. As described there, this parabola opens downward ("sad-face") and intersects the x-axis at $x = 0$ and $x = 1$. Explain how these points of intersection can be found simply by setting $y = 0$ and solving for x.

49. **Two sad and one happy parabolas.** Building on your work for the previous Mindscape, determine where each of the following parabolas intersects the x-axis.

$$y = 6x(4 - x) \qquad y = x(2 - x) \qquad y = (5 - x)(1 - x)$$

Mindscape 30 points out that the parabola $y = 4x(1 - x)$ has its peak at $x = 0.5$, which is the point midway between the points where the curve intersects the x-axis. For each parabola above, find the value of x where the peak (or valley) occurs and find the corresponding value of y. Then sketch a graph for each parabola.

PREDETERMINED CHAOS

How Repeated Simple Processes Result in Utter Chaos

Smaller and Smaller (1956) by Escher, Maurits Cornelis.

© Fine Art Images/SuperStock

> *Chaos umpire sits, And by decision more embroils the fray By which he reigns: next him, high arbiter Chance governs all.*
>
> JOHN MILTON

We all know that the world is unpredictable and that we must expect the unexpected from time to time. But, somewhere deep in our minds, we might wonder whether the world is fundamentally orderly. Although we cannot find the key that organizes our surroundings and their meanings, perhaps a better theory, a better computer, or a better brain might do the trick. If we just had enough information, maybe we could predict the weather and the stock prices. Such predictions would keep us both dry and rich. We have made tremendous strides over the centuries in describing laws of nature. The accoutrements of modern life are a legacy of that insight. But how far can we expect these insights to take us? Are there limits to our ability to understand nature?

We now will discover another side of scientific progress—namely, its limitations. Many phenomena of the world are the result of repeatedly applied interactions at all scales—molecules jostling one another, animals reproducing, planets orbiting. In each case, inevitable minute inaccuracies in measuring the position, speed, or the laws of interaction are dramatically magnified over time, and our uncertainty about the future grows. It may not be surprising that there will be limits to our ability to accurately predict future outcomes due to our vagueness about the current conditions. What is shocking, however, is that uncertainty about the future grows even when there is *no* vagueness at all. Chaos may be a more fundamental feature of our world than we would prefer to believe.

Our journey through chaos will illustrate that direct experience is the best teacher. Ideas become real when we experience them firsthand, particularly if the experience is surprising. The feeling of surprise tells us that our understanding was

545

somehow not complete and invites us to seek a deeper understanding. Experience, then discomfort or surprise, then analysis comprise the path to understanding.

An Experiment—In Search of a Sign

We want you to experience something pretty surprising . . . but you'll need a calculator. So, before reading any more, put down this book (and no one will get hurt), find a scientific calculator that has a SIN (sine) key, and then pick up this book and continue reading.

With the help of the calculator, we're about to create a list of numbers. We're going to be relying heavily on the SIN (sine) key, so it will be the "key" player in generating our list of numbers. To begin, set the SIN key for angles in degrees (rather than radians). To check that your calculator is set properly, enter the number 180 and press the SIN key (or on some other calendars press SIN *then* 180). If you get an answer of 0, then you're all set (perfectly). If you see something like 0.801152 . . . , then your calculator is set on radians and you need to change it to degrees.

We're now ready to compute away. To record the results, make two columns with the numbers from 1 to 25 in the first column. Start the experiment by writing down a random decimal number between 0 and 1 in the second column next to 1. We are now ready to produce the remaining numbers in the second column by following these steps:

1. Type in your random number on the calculator.

2. Multiply it by 180.

3. Hit the SIN key and write down your answer in the second column next to 2. Keep all the decimal places.

4. Repeat steps 2 and 3 for each new number you generate. Continue to record your results until you have 25 numbers in the second column.

To illustrate this procedure (and to prove we actually did it ourselves), we selected the first number (the random one) to be 0.287. We multiplied it by 180 and then pressed the SIN key. The answer was 0.78434349, and this was our next number. We then multiplied it by 180 and hit the SIN key to get our third number. Our list follows, but we hope that you made up your own with your own first random number.

Pretty exciting, huh? Well, okay, so it's not so interesting . . . not yet, anyway. Now that you've made up your list, let's make up the exact list yet again, but this time something scary will happen!

Variation on a SIN Theme

Start by entering exactly the same first decimal number as before and repeat the process five times. The moment you record your fifth number, suppose your calculator makes a slight buzzing sound followed by the appearance of a cryptic symbol on its display and then goes blank. It turns out that your batteries have died. (To simulate this, shut off your calculator now.) Suppose you then rush

Life Lesson
Just do it. Make the effort to get experience.

1	0.287
2	0.78434349
3	0.62685095
4	0.92163865
5	0.24370037
6	0.69297498
7	0.82179119
8	0.53106712
9	0.99524087
10	0.01495068
11	0.04695167
12	0.14696871
13	0.44548493
14	0.98537011
15	0.04594498
16	0.14383972
17	0.43666283
18	0.98026879
19	0.06194772
20	0.19338834
21	0.57085546
22	0.97532697
23	0.07743501
24	0.24087690
25	0.68655271

out and replace the batteries. Well, now you've got to finish the list you started. But hey, why bother starting from the beginning? Instead, you decide to save some time and start with the fifth number (the last one you wrote down, before your calculator fainted). In fact, why bother typing in all those digits from that number? Instead, type in just the first six digits after the decimal point from the number and finish the list.

Try that scenario now. That is, record the first four numbers as they appear in your previous list and then for the fifth number, enter just the first six digits after the decimal point. Our fifth number is boldface in Table 1. Now, from that point, using that truncated number as the fifth number, complete the list and record your results. Now compare your new list to your original list. Go ahead and finish your list before looking at our results.

Table 1 shows what we got.

TABLE 1   SIN Experiment Repeated with Rounded Fifth Value

Original Time		This Time	
Step Number	Value	Step Number	Value
1	0.287	1	0.287
2	0.78434349	2	0.78434349
3	0.62685095	3	0.62685095
4	0.92163865	4	0.92163865
5	0.24370037	5	**0.243700**
6	0.69297498	6	0.69297412
7	0.82179119	7	0.82179273
8	0.53106712	8	0.53106303
9	0.99524087	9	0.99524212
10	0.01495068	10	0.01494674
11	0.04695167	11	0.04693932
12	0.14696871	12	0.14693033
13	0.44548493	13	0.44537699
14	0.98537011	14	0.98531226
15	0.04594498	15	0.04612653
16	0.14383972	16	0.14440412
17	0.43666283	17	0.43825728
18	0.98026879	18	0.98124664
19	0.06194772	19	0.05888135
20	0.19338834	20	0.18392805
21	0.57085546	21	0.54620504
22	0.97532697	22	0.98948314
23	0.07743501	23	0.03303366
24	0.24087690	24	0.10359211
25	0.68655271	25	0.31972972

Look at the original 25th number and compare it to the 25th number computed using the slightly truncated version of the fifth number. Why is there such an enormous difference between these two last numbers on our lists given that there was only an extremely tiny difference in our fifth numbers? One short answer is that sometimes very slight differences in initial conditions can have an enormous and dramatic effect on the final outcomes. A shorter answer is *chaos!*

Dueling Calculators

Here is a similar experiment that has even more surprising results. We borrowed a calculator from a student. The precision of the borrowed calculator was eight digits, whereas ours was nine. We entered the exact same decimal number between 0 and 1 on each calculator and began the previous process of multiplying by 180, hitting the SIN key, recording our answer, and repeating. Now, since we started with the same number, the answers should be the same, right? We repeated the process about 35 times. What do we observe? Why? Once again: *chaos!*

We started with the random number 0.7391. Table 2 shows what we got with successive trials. Huh? What could be more precise and accurate than the results we get from a calculator? Surely the answer we get is completely determined by what we punch in, and, indeed, it is. However, even with calculators we find that surprising developments can occur. As you can see, repeating a procedure like

TABLE 2 Experiment Using Dueling Calculators

Round	Our Calculator	Borrowed Calculator
1	0.7391	0.7391
2	0.73090122	0.7309012
3	0.74823571	0.7482357
4	0.71101516	0.7110152
5	0.78819629	0.7881963
6	0.61737461	0.6173746
7	0.93278107	0.9327811
8	0.20960844	0.2096084
9	0.61193460	0.6119345
10	0.93880464	0.9388047
11	0.19106878	0.1910686
12	0.56485729	0.5648568
13	0.97931362	0.9793139
14	0.06494242	0.0649414
15	0.20261016	0.2026071
16	0.59439941	0.5943916
17	0.95634610	0.9563533
18	0.13671325	0.1366909
19	0.41641387	0.4163499
20	0.96571997	0.9656678
21	0.10748581	0.1076489
22	0.33129584	0.3317792
23	0.86280720	0.8635739
24	0.41778307	0.4155936
25	0.96682764	0.9650479
26	0.10402551	0.1095849
27	0.32101951	0.3375106
28	0.84603979	0.8725123
29	0.46504041	0.3898922
30	0.99397488	0.9407660
31	0.01892734	0.1850170
32	0.05942697	0.5490674
33	0.18561266	0.9881424
34	0.55063050	0.0372430
35	0.98737654	0.1167356

multiplying by 180 and then pressing the SIN key gives grossly different answers after relatively few repetitions.

Suppose we are trying to predict the future of the population, the stock market, or the course of an epidemic or the weather. Our method for prediction may well involve an idea of how the system will appear after one step—a generation, a day, a year, or a second. Then the answer to that one-step prediction is fed back into our predictive engine, and the process is reapplied to generate a prediction for the state of the system after two steps. And so on. Based on our experience with the SIN key, our confidence in the state of a system after 35 steps should be extremely shaky. We are dealing with some strange observations, so let's try to understand them.

These examples illustrate how slight differences result in completely different outcomes after several repetitions of a process. This sensitivity is at the heart of mathematical chaos, a theme with profound implications for how well we can ever hope to know and understand the future.

Predicting Future Populations

Let's try another example, this one concerning population growth. Recall from Section 7.5 that, by making sensible assumptions about how fast populations will grow, we were led to a simple mathematical model for population growth. The Verhulst Model (page 532), sometimes referred to as the *logistic equation,* captured the reasonable idea that populations will increase in proportion to how much unused capacity exists in the environment. This assumption led Verhulst to the following relationship between populations from one year to the next. In this relationship P_n stands for the fraction of the sustainable capacity of the environment at year n—that is, the population density at year n. The Verhulst Model contains a constant that is selected to fit the specific environment being modeled. In the example, we use the constant 3:

$$P_{n+1} = P_n + 3P_n(1 - P_n).$$

Let's see what pattern of populations this model predicts from year to year. As an example, we started with the decimal number 0.058 for P_0 and computed future years' populations; that is, we plugged in 0.058 for P_n in the preceding equation and computed P_{n+1} (in this case, P_1). We did this calculation on two different calculators to see what might happen. Table 3 on the next page shows what we found.

One calculator is predicting that the population density at year 45 is 1.3159026, and the other calculator using the exact same formula is predicting that the population density at year 45 is radically different—specifically, 0.1293406901. So, if the maximum sustainable population of the pond is 10,000 fish, and we start with 580 fish, one calculator tells us we should expect 13,159 fish after 45 years (overpopulation), and the other calculator tells us we should expect only 1293 fish. So what's the answer? After 45 years, do we have a fish population that has exploded to 22 times our initial size or one that has merely doubled? We may well ask: Is one of these calculators broken? The answer is no. The enormous difference in the results arises from how many digits the two calculators carry before rounding off. The number of fish expected to be alive in a pond or the population of people on Earth might well be a crucial piece of knowledge for making policy decisions. Yet here we see that even a completely deterministic, simplistic model of population growth is susceptible to the problem of hypersensitivity. The sensitivity of this model to slight changes along the way makes it completely unreliable for predicting the future after 45 steps.

TABLE 3 Population Predictions on Two Different Calculators

n	P_n (calculator 1)	P_n (calculator 2)	n	P_n (calculator 1)	P_n (calculator 2)
1	0.22190800	0.2219080000	24	1.2938053	1.294691684
2	0.73990251	0.7399025186	25	0.15342458	0.1500870658
3	1.3172428	1.317242863	26	0.54308101	0.5327698814
4	0.063585171	0.0635851704	27	1.2875130	1.279548286
5	0.24221146	0.2422114600	28	0.17698247	0.2064616951
6	0.79284667	0.7928466660	29	0.61396151	0.6979674858
7	1.2855691	1.285569156	30	1.3249998	1.330394109
8	0.18421248	0.1842124572	31	0.033125663	0.0117309782
9	0.63504722	0.6350471407	32	0.12921072	0.0465110654
10	1.3303339	1.330333950	33	0.46675666	0.1795544243
11	0.011970483	0.0119705441	34	1.2134413	0.6214983234
12	0.047452057	0.0474522947	35	0.43644580	1.327212795
13	0.18305313	0.1830540180	36	1.1743283	0.0243697677
14	0.63168719	0.6316897517	37	0.56017204	0.0956974144
15	1.3296626	1.329663179	38	1.2993100	0.3553156723
16	0.014642343	0.0146402048	39	0.13262049	1.042515008
17	0.057926180	0.0579178125	40	0.47771737	0.9095474057
18	0.22163839	0.2216078311	41	1.2262278	1.156360173
19	0.73918284	0.7391012320	42	0.39400724	0.6139341425
20	1.3175575	1.317593034	43	1.1103038	1.324991176
21	0.062356531	0.0622179240	44	0.74289148	0.0331598544
22	0.23776111	0.2372584859	45	1.3159026	0.1293406901
23	0.78145341	0.7801591763			

These radical differences exist even when we are using the exact same mathematical model. The sensitivities of the mathematical model suggest that real populations may themselves have greater natural variations than we might intuitively have first thought. Once again, these examples indicate that, when we look at reality, we might find instances of unexpected sensitivity. Of course, a mathematical model is itself only an approximation of reality. Thus, we would expect the actual populations of fish to differ from the predicted numbers not only because of round-off problems but also because of inadequacies of the model. It is safe to say that the distant future is difficult to predict.

Predicting Planetary Positions

Using our knowledge of gravity and the masses of the planets, as well as their current locations and current velocities, we can predict where the planets will be in a year, and our answers will be almost correct. However, if we take that prediction, use the same method again for predicting locations of the planets after two years, we will be close to correct again, but slight differences in our estimates of the masses of the planets and their locations and velocities will make the two-year prediction a little less certain than the one-year prediction. After hundreds, thousands, or millions of years, the predictions will be completely different depending on whether we used one value for the mass of a planet or another value. The future locations of the planets in the solar system are also sensitive to initial conditions to such an extent that astronomers and mathematicians are currently debating whether the solar system is stable or whether after some time the planets might not attain increasingly eccentric orbits and literally fly apart. Although it's an ominous thought, don't count on that as a means of getting out of the final exam.

All these examples are leading us to appreciate that even determined and theoretically predictable mathematical models or physical systems can be so sensitive that we cannot effectively use them to predict the future. *Deterministic chaos* refers to the idea that wildly different futures can result from beginnings that are only minutely different, even in totally deterministic systems.

The Dawn of Chaos

Edward N. Lorenz, a meteorologist at the Massachusetts Institute of Technology, was the first to notice this idea of chaos—not in the weather but in systems of equations that describe the weather. In making his seminal discoveries about chaos, Lorenz used one of the most effective methods of investigation in history: discovery by accident.

Luck is a powerful means of discovery. It's something we all want to be attuned to. However, luck is not enough. Probably most of us see potentially significant accidents around us all the time, but, unlike Lorenz, we are not prepared to interpret the importance of them.

Life Lesson
Luck is a powerful means of discovery. Be attuned to it.

Chance favors the prepared mind.

LOUIS PASTEUR

In the 1960s, Lorenz had a primitive computer that he had programmed with equations designed to simulate global weather conditions. One day the program stopped, so he restarted it at an intermediate point. He wanted to save himself a little time, so, instead of copying in their entirety the decimal numbers that were describ-

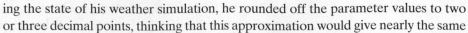

ing the state of his weather simulation, he rounded off the parameter values to two or three decimal points, thinking that this approximation would give nearly the same answers. (Does this story sound familiar?) He noticed that the answers he got were radically different from those he had gotten when he ran the simulation without the interruption. He investigated and discovered that the future weather predictions came out entirely differently when he rounded the parameters to two decimal places rather than using three or four. Just a tiny change in those parameters made a huge difference in his model's prediction of future weather. His serendipitous shortcut led to his crystallizing the idea of sensitivity to initial conditions in many mathematical models and in nature as well.

Pictured here is the Lorenz attractor associated with weather prediction models, whose attractive double spiral ushered in the era of chaos and fractals.

Lorenz attractor

"Chaos"—the Word

Some words have related yet significantly different meanings. The dictionary defines *chaos* as a "state of utter confusion and disorder; a total lack of organization or order." In 1975 James A. Yorke from the University of Maryland and his student Tien-Lien Li wrote a paper in the *American Mathematical Monthly* titled "Period Three Implies Chaos." They thereby applied this word to refer to a scientific concept whose meaning is related, but different. The chaos of mathematics, physics, and biology is not actually utterly disordered. In fact, scientific and mathematical chaos refers to systems, like the simple equations we have seen, that are ultimately completely deterministic and whose future states depend in a fixed and describable way on their current state. There are no uncertainties, no randomness, and no unknowns involved. However, as we have seen with our examples, these completely deterministic systems nevertheless display behavior that is surprisingly chaotic in appearance. They are chaotic in the sense that tiny variations in where we start or how many decimal digits we use in our calculations make an enormous difference in the final outcome.

We never know the current state of any physical system exactly, because there are always errors of measurement. But even when no measurement errors are involved, such as in calculating sines repeatedly, tiny errors due to rounding soon cause tremendous differences in the outcomes. The weather, populations of animals, and prices of goods are real-world phenomena for which we can develop mathematical models that try to predict the future. Unfortunately, these models are susceptible to the problem that slight errors in the beginning data and slight errors along the way propagate and expand. After a relatively short period of time, the models predict grossly erroneous outcomes.

Butterflies and Tornadoes

Because it was a meteorologist who first recognized the phenomenon of the dramatic sensitivity to initial conditions that inspired the mathematical notion of chaos, the parable of the *Butterfly from Brazil* has become an icon of chaos. Thus, we cannot resist giving our own rendition of the story of this famous, beautiful insect and its unfortunate effects on a girl and her dog in Kansas. Please read the boxed story below.

We might consider several responses to this classic butterfly tale—for example, "Can't that butterfly be found and stopped?" Although this question sounds silly, let's rephrase it in a more plausible way. Could we understand weather patterns so accurately that during a major drought we could perhaps set off a bomb over the ocean to cause a crop-saving rain on the drought area two weeks

The Butterfly from Brazil

On a sultry summer morning in the Amazon basin, a beautiful butterfly perched on the petal of an exotic purple flower. The still air and a lack of anything good on cable made the butterfly less energetic than was customary. The vision of her on that flower was indeed beautiful. Her eyelids, if butterflies actually had eyelids, drooped languidly over her eyes, and her mind, if butterflies actually had minds, slipped lazily from one inconsequential thought to another.

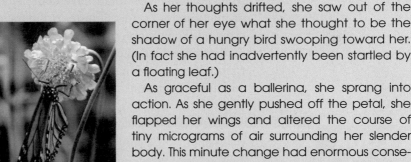

© Tammy Fullum/iStockphoto

As her thoughts drifted, she saw out of the corner of her eye what she thought to be the shadow of a hungry bird swooping toward her. (In fact she had inadvertently been startled by a floating leaf.)

As graceful as a ballerina, she sprang into action. As she gently pushed off the petal, she flapped her wings and altered the course of tiny micrograms of air surrounding her slender body. This minute change had enormous consequences. Air masses have some coherence, and they move somewhat in a pack. If air masses meet, a tiny difference can result in one mass sliding under another or over it—like the leading edge of a wedge. As the masses slide, they push neighboring air masses. The butterfly's flapping caused a small air mass to rise instead of fall, which in turn caused a bigger air mass to change its course, which made an even bigger air mass alter its course, and so on. The unpredictable event of a slight change in the air movement caused by the butterfly's flap led to other unpredictable events, which in their turn altered the patterns of increasingly larger bodies of air until two weeks later Dorothy was knocked unconscious when she was hit in the head by her flying dog Toto during a tornado in Kansas.

If that butterfly had had eyelids, perhaps she would not have been startled by the harmless leaf and would have remained still for several moments longer and would not have altered the course of those particular micrograms of air. Those molecules would have traveled a different path, other air masses would have been altered instead, and the good people of Kansas would have seen a beautiful, clear day instead of a devastating tornado raining cats and dogs (in particular, Toto). This concludes our version of the parable of the *Butterfly from Brazil*.

courtesy of the authors

later? During a war, could we set off a bomb over the ocean to cause a hurricane to wipe out the enemies' principal towns two weeks later? Edward Teller, sometimes referred to as the father of the atomic bomb, once said that the control of the weather would make atomic bombs look like children's toys. Is control of the weather more appropriately relegated to science fiction, or is it within the reach of future science fact? Today no one knows. We do know that little changes make dramatic differences, but we are unsure which changes lead to which differences.

These insights reveal that our potential to predict the future is severely limited. The Brazilian butterfly's wing flap is an example of a tiny change in current conditions that leads to massive changes as a consequence. If we had millions of temperature, pressure, and wind meters spread throughout the entire atmosphere, we could have a good sense of the weather now and also a pretty good way to predict the weather a few minutes or hours from now. But we would have some small uncertainty in the details of our prediction even a few minutes or hours later. As time passes, our uncertainties grow until quite soon our uncertainties outweigh our predictions. As a practical matter, since small uncertainties in our knowledge of the weather always exist, we cannot hope to ever have sufficient knowledge of the current state of the weather conditions at any one time to predict successfully the weather in several weeks.

A Look BACK

EVEN WHEN WE USE REASONABLY accurate mathematical models and reasonably accurate data, accurate long-term predictions become impossible due to the overabundance of accumulated error and uncertainty. Rounding off numbers in calculators as far out as the eighth or ninth decimal place leads to huge errors after only a handful of repeated applications of completely deterministic calculations. There is only one certainty here: Accumulated errors and uncertainty will always exist and, sadly, prevail. In a single word, *chaos!*

Performing experiments with calculators and computing future values of populations following the Verhulst Model give us direct experience with the idea of deterministic chaos. From these examples, we extrapolate the idea of systems that are sensitive to initial conditions and accumulated error.

Getting direct experience makes ideas much more real and immediate. As we observe the world, we sometimes notice unusual events. Often we simply ignore them, but by exploring the unusual, we can find whole new worlds. Today's unusual or extraordinary anomaly may well point the way toward tomorrow's main issue.

> *God has put a secret art into the forces of Nature so as to enable it to fashion itself out of chaos into a perfect world system.*
>
> IMMANUEL KANT

Life Lessons

Make the effort to gain experience—just do it.

•

Keep your eyes open for unusual occurrences.

•

Seek to understand anomalies.

MINDSCAPES   Invitations to Further Thought

In this section, Mindscapes marked **(H)** *have hints for solutions at the back of the book. Mindscapes marked* **(ExH)** *have expanded hints at the back of the book. Mindscapes marked* **(S)** *have solutions.*

Developing Ideas

1. **Does this thing come with a warranty?** If you use two different calculators to do the same calculations, are you guaranteed to always get exactly the same answer? Explain your answer.

2. **Root repeater.** Find a calculator with a square root key. Start with 0.999 and press the square root key over and over. What happens to the values? Do you eventually get 1? Do the experiment again. This time keep track of how many times you press the square root key to get an answer of 1. Now try starting with 0.9999. How many times did you need to press the square root key to get 1?

3. **Transforming experience.** The equation $y = 4x(1 - x)$ is an example of a *logistic transformation*. Find the value of y if $x = 0.5$. Find the value of y if $x = 0.437$.

4. **Up and over.** In this graph, we see a red diagonal and a blue inverted **V**. The point (0.4, 0.4) lies on the diagonal and is marked with a black dot. If you travel vertically up from this point, what point do you hit on the **V**? If you start at the point (0.4, 0.8) on the **V** and travel horizontally toward the diagonal, what point on the diagonal do you hit?

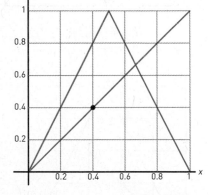

5. **Over and up.** Here is a simpler version of the diagonal-**V** diagram. One point is marked on the **V**. From this point, move horizontally and mark the point you hit on the diagonal. From this new point, move vertically and mark the point you hit on the **V**.

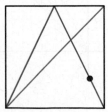

Solidifying Ideas

6–18.   The cobweb tent (ExH). Take a square in which the diagonal and an inverted **V** are drawn. Start at any point on the diagonal. From there, everything else is determined. Go vertically up or down as needed to head toward the inverted **V**. When you hit the inverted **V**, go horizontally right or left until you hit the diagonal. From there repeat the pattern going vertically until you hit the inverted **V** and then horizontally until you hit the diagonal. Repeat. Following this pattern creates the cobweb plot you have seen before (see Mindscape 30 on page 539–541 in the previous section for a full description of cobweb plots). Look at the following examples. These cobweb plots are the result of a repeated process, and they illustrate many of the ideas from this section. Mindscapes 6–18 all refer to this process. You can also produce these graphs using a program on the *Heart of Mathematics* Web site—the images are really cool!

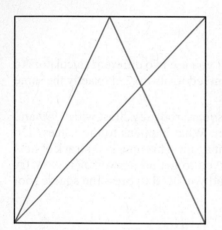

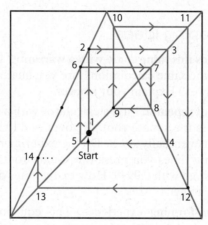

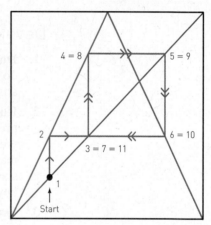

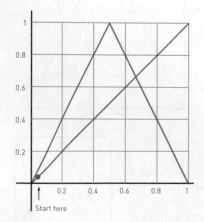

6.   Go. Start at the point marked on the diagonal at right, and draw the first four iterates of the cobweb plot. Use a straightedge and be as careful as possible to keep your lines vertical and horizontal.

7.   Staircase. Start at the point marked on the diagonal at the left, and draw the first four iterates of the cobweb plot. Use a straightedge and be as careful as possible to keep your lines vertical and horizontal.

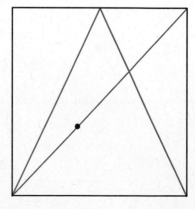

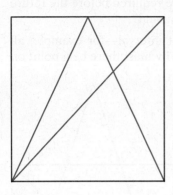

8. **How far?** Start at the point marked on the diagonal to the right, and draw the cobweb plot. How many steps are in the cobweb plot before the point quits moving?

9. **Points that quit (H).** In the diagram to the left, find a point whose iterations hit the diagonal in three different points and then remain fixed.

10. **Ups and downs.** Start at the point marked on the diagonal diagram at right below. Draw the cobweb plot until the path goes up to near the top of the diagonal and then down to near the bottom and repeats. This oscillation could go on forever.

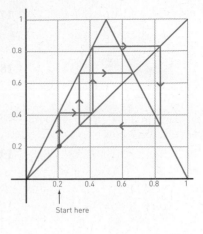

11. **Don't move (S).** Find the point on the diagonal other than 0 that does not move at all in its cobweb plot.

12. **Two step (S).** Find a point on the diagonal that goes up to the inverted **V**, over to the diagonal, down to the inverted **V**, and returns to the starting point. This point follows a repeating path that hits only two points of the diagonal and only two points of the inverted **V**.

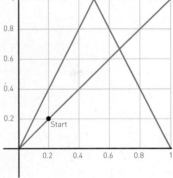

13. **Three step.** Find a starting point on the diagonal where the path hits the diagonal at three points. The path starts by going up to the inverted **V**, then right over to the diagonal, and down to the inverted **V**. It then goes left to the diagonal, up to the inverted **V**, and over to the right, at which point it has returned to its original position. So, this point repeats after three cycles; it has period 3.

14. **Four step.** Find a starting point on the diagonal where the path hits the diagonal at exactly four points. So, this point repeats after four cycles; it has period 4. (*Hint:* You might start near the lower left and create a sort of stair step. Try sliding the starting point and see what happens.)

15. **Grow up (H).** What part of the diagonal is covered if you start with all the points on the diagonal above the interval [0, 0.1] and apply the process once to each point?

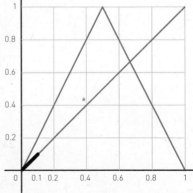

16. **Grow up again.** What part of the diagonal is covered if you start with all the points on the diagonal above the interval [0, 0.1] and apply the process twice?

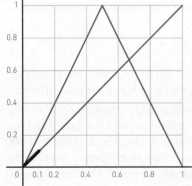

17. **A different patch (H).** Suppose you start with all the points on the diagonal above the interval $[0, 0.01]$. How many steps are required before the future positions of those points cover up the whole diagonal?

18. **Target practice.** Suppose we give you a target interval—for example, all the points on the diagonal above $[0.31, 0.32]$. Why must there be a point on the diagonal above the interval $[0.0, 0.01]$ that hits the target interval after some number of steps?

19. **Too high.** Consider a tent function that got too high and sticks out over the top of the unit square. Let's play the cobweb game and see which points stay within the unit square. Any point between 1/3 and 2/3 of the diagonal goes out immediately. Show that the point $(1/6, 1/6)$, which lies on the diagonal above the interval $[1/9, 2/9]$, leaves the square. How many steps are required? Likewise, show that the point $(5/6, 5/6)$, which lies on the diagonal above the interval $[7/9, 8/9]$, also leaves the square.

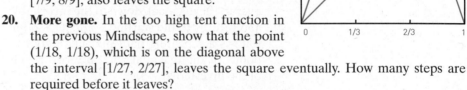

20. **More gone.** In the too high tent function in the previous Mindscape, show that the point $(1/18, 1/18)$, which is on the diagonal above the interval $[1/27, 2/27]$, leaves the square eventually. How many steps are required before it leaves?

21. **Too short.** Consider a short tent function. Will any point on the diagonal above the interval $[1/3, 2/3]$ ever hit the diagonal at a point above the interval $[0, 1/4]$? Why not?

22. **Where to?** Using the transformation $y = 3.5x(1 - x)$, calculate the first 30 values starting from 0.437 and from 0.438. Do the results stay fairly close to each other, or do they become quite different?

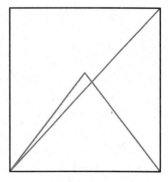

23. **Calculator slips.** Using the logistic transformation $y = 4x(1 - x)$, calculate the first 30 values starting from 0.437 and from 0.438. Do the results stay fairly close to each other, or do they become quite different?

24. **Just missed.** Find several examples in everyday life of tiny differences that led to huge differences in your own future—whether good, bad, or just different.

25. **Take stock.** Pick a stock. Some Web sites will display graphs about stocks. Look at the graph of the highs of that stock during each of the last 100 days. Next look at the graph of the highs of that stock during each of the last 100 weeks. Finally, look at the graph of the highs of that stock during each of the last 100 months. To what extent are these graphs similar? Do they seem to describe some chaotic features?

Creating New Ideas

26. Repulsive. If you take the number 1 and square it, then square the answer, then square the answer again, of course, you will get 1 again and again. So, 1 is a fixed point under that process. Show that every point near 1 gets increasingly distant from 1 under repeated squaring. We call 1 a *repeller*, since nearby points are repelled from it.

27. Attractive. Again consider the process of repeated squaring. If you take the number 0 and square it, you get 0 again. So, 0 is a fixed point under that process. Show that every point near 0 gets increasingly close to 0 under repeated squaring. We call 0 an *attractor*, since nearby points are attracted toward it.

28. Sierpinski attractor. Remember how the Sierpinski Triangle was created using a collage-making process. Consider the pictured image that is similar to it. Sketch the result of applying the Sierpinski collage-making instructions to this picture. Did the new image become more like the Sierpinski Triangle? If it did, then we would call the Sierpinski Triangle an attractor under the collage-making transformation.

29. Two step. Consider the transformation that takes any point x in the interval $[0, 1]$ to $(1 + \sqrt{5})x(1 - x)$. Compute the future values of 1/2 under repeated applications of this process.

30. Periodic attraction (S). Again consider the transformation that takes any point x in the interval $[0, 1]$ to $(1 + \sqrt{5})x(1 - x)$. As you saw in the previous Mindscape, the number 1/2 is a periodic point of period 2, meaning that it returns to itself after the process has been repeated twice. Compute 10 future values of the point 0.48 under repeated applications of this process. Do these points get closer to the repeated values of 1/2, or do they get farther away?

31. Periodic attraction. Again consider the transformation that takes any point x in the interval $[0, 1]$ to $(1 + \sqrt{5})x(1 - x)$. Compute 10 future values of the point 0.8 under repeated applications of this process. Do these points get closer to the repeated values of 1/2, or do they get farther away?

32. Four-peat (H). Consider the equation $y = 3.5x(1 - x)$. The point 0.5009 . . . is periodic. Compute future values until you find the period.

33. Nearly fourly. Consider the equation $y = 3.5x(1 - x)$. The point 0.36 is not periodic. Compute 12 future values under repeated applications of this equation. To what are they tending?

34. Tent attraction? Consider the point 0.4, which is of period 2 in the cobweb tent. Sketch the cobweb plot of the point 0.38. Do these values seem to be tending toward the periodic point's values or not?

35. Becoming periodic. The point 0.4 is of period 2 in the cobweb tent transformation. Find a point that takes two steps first and then becomes periodic with period 2.

Further Challenges

36. The Earth moved (ExH). Consider a transformation of the sphere that consists of moving each point by shifting its longitude by 30° to the west and by increasing its latitude by squeezing points upward away from the south pole and toward the north pole. This process keeps the north and south poles fixed and spirals points upward toward the north pole and away from the south pole. Describe the movement of the equator under the first two iterations of this process. What points are fixed under this transformation?

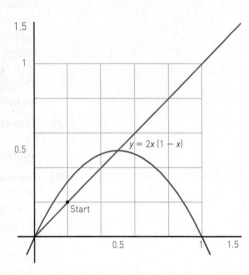

37. Poles apart. Consider the same transformation as in the previous Mindscape. For each fixed point, state whether points within 10 of latitude and longitude from each of the fixed points get increasingly close to that fixed point, making it an attractor, or increasingly far from that fixed point, making it a repeller.

38–40. Logistic cobwebs. Let's explore the relatively simple-looking transformation $y = rx(1 - x)$ in the following three Mindscapes. A *fixed point* is one where $x = rx(1 - x)$.

38. $r = 2$. For $y = 2x(1 - x)$, what points remain fixed? Draw the first 10 steps of the cobweb plot starting at the point marked. How does it relate to the fixed point?

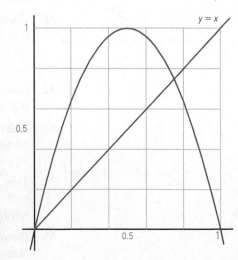

39. $r = 4$ (H). For $y = 4x(1 - x)$, sketch a cobweb pattern that indicates that there must be periodic points of period 3. (*Hint:* Start fairly near the lower-left part of the diagonal and create a stair-step pattern that ends near the top, thus pushing you back to where you started. Slide your starting point if necessary.)

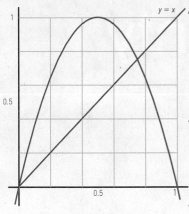

40. $r = 4$. For $y = 4x(1 - x)$, start at two random points near your periodic point of period 3 and compute the first 10 values under repeated application of the transformation. Do these values seem to be converging toward the plot of the period 3 point or not?

In Your Own Words

41. Personal perspectives. Write a short essay describing the most interesting or surprising discovery you made in exploring the material in this section. If any material seemed puzzling or even unbelievable, address that as well. Explain why you chose the topics you did. Finally, comment on the aesthetics of the mathematics and ideas in this section.

42. With a group of folks. In a small group, discuss the concept of mathematical chaos. After your discussion, write a brief narrative describing mathematical and natural systems that exhibit chaos.

43. Creative writing. Write an imaginative story (it can be humorous, dramatic, whatever you like) that involves or evokes the ideas of this section.

44. Power beyond the mathematics. Provide several real-life issues—ideally, from your own experience—that some of the strategies of thought presented in this section would effectively approach and resolve.

For the Algebra Lover

Here we celebrate the power of algebra as a powerful way of finding unknown quantities by naming them, of expressing infinitely many relationships and connections clearly and succinctly, and of uncovering pattern and structure.

45. V-quation. Look at the graph for Mindscape 4. The inverted "V" is composed of two line segments. Write down the coordinates (x, y) of the point at the top of the "V" and the coordinates of the two points where the "V" touches the x-axis.

46. Meeting up. Look at the graph for Mindscape 38. The red line has equation $y = x$. The blue curve has equation $y = 2x(1 - x)$. Set the right sides of these two equations equal to each other. Now solve for x. Use these values to find the points (x, y) where the line and the curve intersect.

47. We meet again. Look at the graph for Mindscape 39. The red line has equation $y = x$. The blue curve has equation $y = 4x(1 - x)$. Set the right sides of these two equations equal to each other. Now solve for x. Use these values to find the points (x, y) where the line and the curve intersect.

48. Calculating percentages (H). Return to Table 1 on page 547 showing the results of a calculator experiment. The table shows that rounding one value can lead to dramatic differences later on. What's the difference between the two values at Step 25? As a percentage, how much larger is the first value than the second value? In other words, let x denote the first (larger) value and y denote the second (smaller) value at Step 25. Write the ratio of x to y as a percentage.

49. Calculating percentages II. Repeat Mindscape 48 using the values from Step 45 in Table 3 on page 550 from this section.

Taming Uncertainty

Many, if not most, significant events in our lives arise from coincidence, randomness, and uncertainty. We meet friends and loved ones, we find intriguing opportunities, we fall into a profession or lifestyle. At a deeper level, the most basic interactions of molecules and subatomic particles are described in terms of probabilities and statistics. Nothing is more fundamental than chance. However, the uncertain and the unknown are not forbidding territories into which we dare not tread. There are ways to organize and understand them that can add meaning to our lives.

Probability enables us to better understand our uncertain world. It moves us from a vague sense of disordered randomness to a focused concept of measured proportion. Probability serves as the mathematical foundation of common sense, wisdom, and good judgment. Perhaps, too, it lets us view our world more truly as it is—a place where the totality follows rules of the aggregate while leaving individuals to their wild variation and unbridled possibilities.

We develop a measure of likelihood by looking at situations in which the future is uncertain but the possible outcomes are definite and easily described. Gambling games provide concrete and clear illustrations because dice and coins can teach us how to measure likelihood. We apply the principles we develop to measure the value of future possibilities, thus allowing us to weigh decisions involving the unknown future. Frequently, we must extrapolate the probable future from evidence from the past. This need presents us with the challenge of collecting and meaningfully interpreting data.

Surprises guide us in the study of the uncertain and the unknown. But we progress by considering concrete examples, by performing experiments to ground our theory in experience, and by looking at fallacies. Simple, clear cases let us develop principles that we can apply widely. So, even the uncertain and unknown are best understood by starting with the simple and the familiar.

8.4 Down for the Count

8.5 Drizzling, Defending, and Doctoring

CHANCE SURPRISES

Some Scenarios Involving Chance That Confound Our Intuition

Café Terrance at the place du Forum Arles, At Night, 1888, Vincent van Gogh. Analyzing what we see sometimes leads to surprising, counterintuitive results.

> *Chance, too, which seems to rush along with slack reins, is bridled and governed by law.*
>
> BOETHIUS

Many surprises lurk in the world of chance. We guess wrong because our intuition is untrained or mistrained. Each surprise is an enticement for us to find structure among the forces of the uncertain and unknown. In the following sections, we develop methods for analyzing chance scenarios. Let's begin with vintage TV.

Let's Make a Deal

Revisit the *Let's Make a Deal* scenario from Chapter 1 (Section 1.1, Story 7). The contestant selects one of three doors. The all-knowing host, Monty Hall, knowing the location of all prizes, opens another door to reveal one of the two mules rather than the lone Cadillac. The contestant now has the option to stick with the original guess or switch doors.

• *What we expect.* We probably expect that switching or sticking makes no difference. We might think that in either case the chance of getting the Cadillac is 1 out of 2, since there are two remaining closed doors.

• *Surprise.* Switching gives the contestant a 2 in 3 chance of winning the car. Why? The explanation is in Chapter 1 (Section 1.3, Story 7).

▶ Try It On the *Heart of Mathematics* Web site, you will find a program that allows you to play the game or simulate playing this game many times to confirm the *Let's Make a Deal* probability. If you want to make it physical, here is an experiment that verifies the probability live. From a deck of regular playing cards, remove three cards—a king and two aces. The king represents the Cadillac, and the aces represent the mules. Have a friend act as the dealer. The dealer shuffles these three cards and places them facedown on a table, side by side, without looking at them. Once the cards are on the table, the dealer peeks under each card so that the location of the king is known to the dealer but not to you. Point to a card. The dealer then turns over one of the other two cards to reveal one of the aces (the mules). You now have the chance to switch cards. Stick with your original guess and see what happens. Have the dealer shuffle the cards again and repeat the exact same scenario—don't switch—at least 30 times and see what fraction of the time you end up with the king (the car). Now, do the same experiment, but this time try switching and again record how often you find the king. After repeating this experiment several times, you will discover that about one-third of the time you find the king if you stick to your original guess, and about two-thirds of the time you will find the king if you switch.

Reunion Scene—Take One

Suppose we return to our 25th college reunion and we see an old classmate.

CLASSMATE: I have two clihildren.

WE: Wonderful! Is the *older* one a boy?

CLASSMATE: Yes! And . . .

(*At this point she chokes on an hors d'oeuvre and collapses.*)

Reunion Scene—Take Two

Suppose we return to our 25th college reunion and we see an old classmate.

CLASSMATE: I have two children.

WE: Wonderful! Is at least *one* of them a boy?

CLASSMATE: Yes! And . . .

(*At this point she chokes on an hors d'oeuvre and collapses.*)

 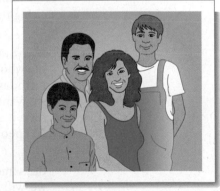

Naturally, we have some concern for the respiratory challenges of our former classmate; however, we probably would be more consumed with the following burning question: What is the chance that our classmate has two boys? Is the probability the same in both scenarios?

- *What we expect.* We probably expect that, in both cases, there is a 50–50 chance that she has two boys.

- *Surprise.* In Take One, the chance is exactly one-half (as expected); however, in Take Two the chance is only 1 in 3. Why? We'll see why in the next section.

▶ **Try It** You can simulate this situation with a deck of cards. Think of the boys as the black cards and the girls as the red cards. Shuffle the deck and remove cards from the top of the deck in pairs. First look at all the pairs whose first card is black. What fraction of those have both cards black? Now start over. Shuffle the cards and take them off in pairs; however, this time look at all pairs that contain at least one black card (disregard the pairs of two red cards). What fraction of those have both cards black? Does this experiment tend to confirm our original intuition or the surprising result?

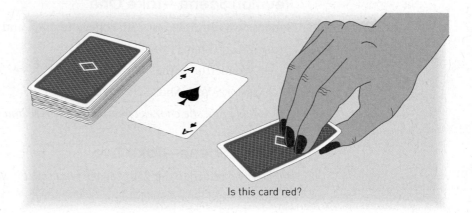

Is this card red?

The Birthday Question

How many people are needed in a room so that the probability of two people sharing the same birthday is roughly one-half?

- *What we expect.* We might expect that this experiment requires about 183 people. If 367 people were in the room, then we would be guaranteed at least two people who would share the same birthday, since 367 people can't all have different birthdays by the pigeonhole principle (see Chapter 2, Section 2.1). Therefore, if we want the chances of a matched pair of birthdays to be approximately 50–50, then it seems we would need about 183 people in the room (about half of the 367 people) for the chance of finding a shared birthday to be roughly 50%.

- *Surprise.* In a room containing only 23 people, the chance of two people sharing the same birthday is just over 50%. That is, in a random gathering of 23 people, we will more often than not find a pair of people with the same birthday. In a room with 183 people, the chance of finding a pair of people with the same birthday is over 99.999999%. Why? We'll see how to analyze this birthday surprise in the next section.

▶ **Try It** The next time you are in a room with 40 people or so, ask them for their birthdays and see whether you find a common birthday. Feel free to wager, if you are so inclined.

A Look BACK

MATTERS OF CHANCE can have satisfying explanations. However, we need to develop a sense of measuring uncertain events so that the experiences associated with the types of surprises discussed in this section come to appear natural and expected.

Life Lessons

Thinking about situations that jar our intuition can lead to new and important insights.

MINDSCAPES ⟩ Invitations to Further Thought

In this section, Mindscapes marked (H) have hints for solutions at the back of the book. Mindscapes marked (ExH) have expanded hints at the back of the book. Mindscapes marked (S) have solutions.

Developing Ideas

1. **Doors galore.** The 21st-century version of *Let's Make a Deal* has five doors instead of three. Two doors have cars behind them and the other three doors have mules. What percentage of doors have cars behind them?

2. **Birthday surprise.** How many people would you need to have in a room so that the chance that two (or more) of them share a birthday is over 50%?

3. **Opposite of heads.** Suppose you flip a coin 100 times, with 53 tosses landing heads up. What percentage of the tosses would be tails?

4. **Penny percent.** Suppose you flip a penny 50 times, with 28 tosses landing heads up. What percentage of the tosses would be heads? What percentage would be tails?

5. **Party time.** At a nephew's party, you decide to write down everyone's birthday. Here are your results:

Julia	Dec. 18	Isabel	Sept. 1
Max	Aug. 26	Colin	Dec. 31
Melinda	Oct. 1	Alexandra	Mar. 14
Zack	Jan. 17	Philip	Dec. 9
Drew	Apr. 18	Victoria	July 10
Margaret	June 16	Douglas	Oct. 31

What percentage of children have their birthdays in December? In February? What percentage have their birthdays in the same month as another child at the party?

Solidifying Ideas

6. **Flipping Lincoln.** Flip a penny 100 times and record how many pennies land heads up and tails up. What percentage of the pennies landed heads up?

7. **Flashing cards.** Shuffle a standard deck of 52 playing cards. Turn over the top 20 cards one by one and record how many are red. Of those cards you turned over, what percentage was black?

8. **King for a day.** Remove three cards from a deck of regular playing cards— a king and two aces. Shuffle these three cards and choose one at random. Record whether it's a king or ace. Repeat this experiment 20 times. What percentage of the time did you select the king?

9. **A card deal stick.** Remove three cards from a deck of regular playing cards— a king and two aces. Have a friend act as the dealer. The dealer shuffles these three cards and places them facedown on a table, side by side, without looking at them. Once the cards are on the table, the dealer peeks under each card so that the location of the king is known to the dealer but not to you. Point to a card. The dealer then turns over one of the other two cards to reveal one of the aces. Stick with your original guess, turn over that card, and record whether you chose the king. Have the dealer scramble the cards again and repeat the scenario—again, don't switch—and record the result. Repeat this experiment 50 times (you can get very quick at it). What percentage of the time did you choose the king?

10. **A card deal switch.** Remove three cards from a deck of regular playing cards—a king and two aces. Have a friend act as the dealer. The dealer shuffles these three cards and places them facedown on a table, side by side, without looking at them. Once the cards are on the table, the dealer peeks under each card so that the location of the king is known to the dealer but not to you. Point to a card. The dealer then turns over one of the other two cards to reveal one of the aces. Now switch your guess to the other facedown card. Turn over that card and record whether you chose the king. Have the dealer shuffle the cards again and repeat the scenario—that is, switch your guess each time after the dealer turns over an ace—and record the result. Repeat this experiment 50 times (you can get very quick at it). What percentage of the time did you choose the king?

11. **A card reunion—black first (S).** Using a shuffled deck of cards, remove cards from the top of the deck in pairs. For each pair where the first card is black, record whether the second card is red or black. After you have gone through 10 pairs, reshuffle the deck and repeat until you have recorded 50 cases where the first card of the pair is black. What fraction of those pairs had both cards black?

12. **A card reunion (H).** Using a shuffled deck of cards, remove cards from the top of the deck in pairs. For each pair where at least one of the cards is black, record whether both cards are black or one is black and one is red. After you have gone through 10 pairs, reshuffle the deck and repeat until you have recorded 50 cases where at least one card of the pair is black. What fraction of those pairs had both cards black?

13. **Birthday bash.** The next time you are in a room with 40 people or so, ask them for their birthdays to see whether you find a common birthday.

14. **Presidential birthdays (ExH).** Have two presidents of the United States shared a birthday?

15. **Vice-presidential birthdays.** Have two vice-presidents of the United States shared a birthday?

In Your Own Words

16. **Personal perspectives.** Write a short essay describing the most interesting or surprising discovery you made in exploring the material in this section. If any material seemed puzzling or even unbelievable, address that as well. Explain why you chose the topics you did. Finally, comment on the aesthetics of the mathematics and ideas in this section.

17. **With a group of folks.** In a small group, discuss the surprises involving chance found in this section. After your discussion, write a brief narrative describing the surprising features in your own words.

For the Algebra Lover

Here we celebrate the power of algebra as a powerful way of finding unknown quantities by naming them, of expressing infinitely many relationships and connections clearly and succinctly, and of uncovering pattern and structure.

18. **One Grecian urn.** A large urn in your kitchen is full of fruit. There are six apples, four oranges, and five pears. What percentage of the mixed fruit are apples? What percentage are not apples?

19. **Two Grecian urns.** Your math instructor keeps two urns on a table in front of the class. One is filled with red candies, the other with blue candies. There are 800 candies altogether in both urns. Fifty-six percent of the candies are in one urn. How many pieces are in the other urn?

20. **Pennies from heaven (H).** Residents of your dormitory decide to collect pennies as a fund-raiser for a charity. The donations are kept in a large container in the RA's room. One Saturday night, after an intense study session in the library, the RA decides to stand at the top of the stairs above the dorm common room and empty the pennies out so they rain down on the floor below. If 62% of the pennies land heads up, and 1900 pennies land tails up, how much money was collected for charity?

21. **Dating Penny.** The pennies described in the previous Mindscape attract the attention of a local coin collector who is curious about the years in which the pennies were minted. He discovers that 40% were minted in 1990 or later, 25% were minted in the 1980s, 25% in the 1970s, and 500 were minted before 1970. How many pennies were minted in 1990 or later?

22. **Changing your pants.** You have 20 coins in your pocket. Fifteen percent of the coins are pennies, 15% are nickels, 40% are dimes, and the rest are quarters. How much money do you have in coins?

8.2 PREDICTING THE FUTURE IN AN UNCERTAIN WORLD

How to Measure Uncertainty Using the Idea of Probability

Le Tricheur à l'As de Carreau (1635) by Georges de la Tour. (The Cheater with the Ace of Diamonds.) *Watch those hands!*

> *To be, or not to be . . .*
>
> SHAKESPEARE

What will be? How can we cope with the unknown—the uncertain future or unpredictable present? Some seek insight from tea leaves, the stars, or the entrails of sheep. Some gaze deeply into the translucent beauty of a crystal ball. Let's not. Instead, let's gaze deeply into the powerful world of transcendent ideas and take our vague view of the future and give it some structure. That is to say, let's construct a means to measure the possibilities for a future we cannot know. Quantifying the likelihoods of various uncertain possibilities is an impressively grand idea. How can we sensibly measure what we admit is unknowable?

We adopt a strategy used in previous investigations. We have already confronted numerous mysteries, including infinity and the fourth dimension. We uncovered their secrets by first understanding basic ideas deeply. Clarifying fundamental ideas enabled us to effectively develop precise notions and led us to new discoveries. Now we wish to delve into the uncertain and the unknown, so we seek examples where we have an intuitive sense of how to measure the likelihood of a future event. We look at those examples with the goal of finding patterns and techniques that can be applied more broadly. A careful examination of our intuition often leads to new insights and discoveries.

Likelihood in Everyday Life

The notion of likelihood is a major component of our everyday lives. How likely is it that a certain scenario will actually happen? What are the chances? As we will continue to see, often the answers to such everyday questions are surprising and counterintuitive.

Tomorrow it will either snow or not snow. Does this fact imply that there is a 50–50 chance of snow? If we are reading this book in Hawaii in June, then we

---

Life Lesson
Understand simple things deeply.

would not expect it to snow tomorrow. If, however, we are reading this book in Buffalo, New York, in June, then the answer is less clear. The point is, there is certainly a *chance* of snow, but is it *likely* to snow?

An amazing number of our actions and decisions are based on an intuitive sense of likelihood. In fact, "likelihood" often provides the foundation for what we think of as "common sense." Here is just a sample of some everyday issues and questions involving likelihoods, risks, and chances:

- Do you go to the dining hall for lunch the moment classes let out, or do you wait because you expect there to be long lines?

- While walking home at night, do you take the shortcut through the dark alley, or do you stick to the well-lit sidewalk? Why?

- When you are driving on a four-lane highway and are about to pass a car, you usually assume that the car will remain in its lane. Would you pass a car that was swerving in and out of its lane?

- Why do so many sexually active people practice safe sex?

- There are no nearby parking spaces. Do you park illegally in front of a store to run in for 5 minutes and take the chance of getting a ticket? What about 15 minutes? An hour?

Knowing how the future will unfold would be extremely valuable. We'd know which number will come up on the roulette wheel, which stock will skyrocket, and which numbered Ping-Pong balls will bubble out of the Lotto machine. We are constantly attempting to predict what will happen in our lives and act accordingly. To develop a measurement of likelihood, let's find some simple, concrete situations in which the future, though uncertain, presents clear, quantifiable alternatives.

On a Roll—the Measure of Likelihood

What are the chances?

Games of chance provide basic examples where the measurements of likelihood are reasonably clear. So let's measure uncertainties in the high-rolling domain of dice. Suppose we have an ordinary die with sides numbered from 1 to 6, and suppose it is a *fair die,* which means that no particular side is more likely to be rolled than any other. If we roll the die, what is the probability of rolling a 4? In other words, what number would you associate with the likelihood of a 4 coming up if you rolled a die once?

Probably you came up with 1/6. Why is 1/6 the probability of rolling a 4? Well, there are six possible outcomes of rolling a die. We could roll a 1, a 2, a 3, a 4, a 5, or a 6. All of these outcomes are equally likely with a fair die. Exactly one of these outcomes (rolling a 4) is the outcome whose likelihood we are assessing. Thus, there is exactly one way out of the six equally likely possible outcomes to roll a 4, and so there is a 1 in 6 chance of rolling a 4. The number 1/6 captures the idea that rolling a 4 is *one* of the *six* equally likely outcomes that are possible when the die is rolled.

Let's put our intuition to the test and experiment by rolling a die a bunch of times and recording the outcomes. We did some experimenting with 100 rolls of a die. Here are our results.

Number Appearing on Die	1	2	3	4	5	6
Times Rolled (out of 100)	18	16	20	17	15	14

We see that 17 out of our 100 rolls were a 4. Thus, 17/100 (or 0.17) of the time we saw a 4. This experiment seems to support our thinking that the probability of rolling a 4 is 1/6 or 0.1666. . . .

Let's now try to determine the probability of an even number appearing on the rolled die. As before, there are a total of six possible outcomes, each equally likely. However, now more than one outcome would lead to success (an even number): We could roll a 2, a 4, or a 6. Thus, there are a total of three different ways of rolling an even number. If we divide the total number of ways of rolling an even number by the total number of possible outcomes, we have 3/6, which equals 1/2, or 0.5. So the probability of rolling an even number is 1/2. This answer makes sense because half the numbers on the die are even; therefore, half the time we would expect to roll an even number.

A Measure of Likelihood—Probability

The concept of dividing the number of successful outcomes by the total number of possible outcomes provides us with a measure of likelihood. Notice that this fraction will always be a number between 0 and 1, where the closer this fraction is to 1, the greater our confidence that the successful outcome will actually occur, and the closer the fraction is to 0, the lower our confidence that the successful outcome will occur. Let's extend this concept of measuring likelihood into a precise definition.

Suppose we perform a certain activity in which there are only finitely many possible outcomes, any one of which is just as likely as any other to occur. Now we focus our attention on a specific collection of those outcomes (in the case of rolling a die, for example, we could think of the outcomes 2, 4, 6). A collection of outcomes is called an *event*. We then define the *probability of a particular event* to be the number of different outcomes in that particular event divided by the total number of possible outcomes. Let's say that again. Suppose that a certain activity (say, rolling a die) will result in a total of T possible outcomes, all of which are equally likely to occur (for rolling an ordinary, 6-sided fair die, T would be 6). We now consider a specific event, which we'll call E. (For instance, E might be the even numbers.) If we know that there are N different outcomes in the event E (in this case, N would equal 3, since there are three different outcomes that are even numbers), then we define the probability of the event E to be the number N/T. (In our example, this probability would be 3/6, or just 1/2.) So,

Probability.

The probability of the event E occurring =

$$\frac{N}{T} = \frac{(number\ of\ different\ outcomes\ in\ E)}{(total\ number\ of\ equally\ likely\ outcomes)}.$$

Notice that N is some number from 0 to T. Therefore, the smallest the probability could be is $0/T = 0$, and the largest the probability could be is $T/T = 1$. Observe that the larger the probability of an event, the more likely it is that an outcome in the event will occur.

Relative Frequency

As we repeat an experiment again and again, we can keep track of the number of times a particular outcome occurs. We can calculate the *relative frequency* of that particular outcome by dividing the number of times a particular outcome occurred by the number of times we repeated the experiment. In other words:

Relative Frequency.

Relative frequency of an outcome =
$$\frac{(\text{the number of times that outcome occurred})}{(\text{the total number of times the experiment was repeated})}.$$

For example, in our first die-rolling experiment, we saw that a 4 appeared in 17 out of 100 rolls. So, the relative frequency of rolling a 4 in this repeated die-rolling experiment is 17/100, which equals 0.17. The probability of rolling a 4 is equal to $1/6 = 0.1666\ldots$. Notice how close 0.17 is to $0.1666\ldots$ It seems reasonable that the more times we repeat an experiment and compute the relative frequency of an outcome, the closer that frequency should be to the actual probability of that outcome. This insight is known as the *Law of Large Numbers.*

Law of Large Numbers.

If an experiment is repeated a large number of times, then the relative frequency of a particular outcome will tend to be close to the probability of that particular outcome.

The Birth of Probability.

Probability started with dice. The French nobleman Antoine Gombauld, the Chevalier de Méré, was a famous 17th-century French gambler. He loved dice games. One of his favorites was betting that a 6 would appear at least once in 4 consecutive rolls of a die. After some time, Gombauld became bored with this game of chance and devised a new game by scaling up from one die to two dice. In the new game, he bet there would be at least one pair of 6's in 24 consecutive rolls of a pair of dice. Soon he noticed that he tended to lose with his new game. Bothered by this discovery, in 1654, Gombauld wrote a letter to the French mathematician Blaise Pascal, who, in turn, mentioned this problem to Pierre de Fermat. The two mathematicians solved the mystery.

They computed that the probability that a 6 would appear at least once in 4 consecutive rolls of a die is equal to 0.52. Since this probability is slightly greater than 0.5, over the long run, Gombauld would win slightly more often than he would lose. However, the probability of seeing at least one pair of double 6's in 24 consecutive rolls of a pair of dice is equal to 0.49. (We will verify both of these probabilities ourselves in Section 7.4.) Since this probability is slightly less than 0.5, Gombauld would lose more often on average, than he would win.

This observation by the gambler Gombauld and the answer given by Fermat and Pascal led to the birth of the study of probability. You may be amused by Pascal's view of humanity. In a letter to Fermat referring to Gombauld, Pascal wrote:

He is very intelligent but he is not a mathematician; this as you know is a great defect.

Analyzing Our Chance Surprises

Armed with the ideas of probability and relative frequency, let's take another look at the chance surprises from the previous section, as well as some additional dicey issues.

All Boys?

The two reunion scenarios in the last section asked what we can deduce from the slightly different dialogues: (1) her older child is a boy versus (2) at least one of her children is a boy. In each case, what is the probability that the speaker has two boys? Our analysis must begin with a careful listing of the possibilities.

A person with two children may have had first a boy then a girl, first a girl then a boy, first a boy then a boy, or first a girl then a girl. These are the four equally likely ways to have two children. Enumerating these possibilities helps us analyze the scenarios, whereas if we rely on vague intuition, we could easily be led astray.

In the first scenario, we know that her older child is a boy. So, either she had first a boy then a girl or first a boy then another boy. Thus, the probability of her having two sons is 1/2. However, in the second scenario, we know that the three possibilities are first boy then girl, first girl then boy, or first boy then boy. In only one of these three equally likely possibilities would she have two sons. Therefore, the probability of her having two boys is only one in three, or 1/3. The number of equally likely possibilities—3—divided into the number of those we are interested in—1—gives the probability.

More Dicey Issues

Given the genesis of probability, it seems only fitting that we roll some more dice. Suppose we now roll *two* fair dice. What is the probability of rolling "snake eyes" (rolling a sum of 2)?

Well, there's only one way of getting the numbers on the two dice to add up to 2: Each die must be showing a 1. So, there is only one outcome to yield "snake eyes." We now need to figure out the total number of possible outcomes. A reasonable guess is 11, because when we roll two dice, we see a total of 2, 3, 4, 5, 6, 7, 8, 9, 10, 11, or 12. The trouble with this guess is that not all these outcomes are equally likely. For example, we have already seen that there is only one way to roll a 2: Each die must be showing a 1. However, there are two different ways of rolling a 3: The first die could be a 1 and the second die could be a 2; *or* the first die could be a 2 and the second die could be a 1. These two possibilities are different outcomes. To get a clear picture of all the possible outcomes, it is better to color the dice different colors to distinguish one from the other. The table below illustrates all the possible outcomes of rolling two dice, a red one and a green one. Notice that there are a lot more than 11 different outcomes.

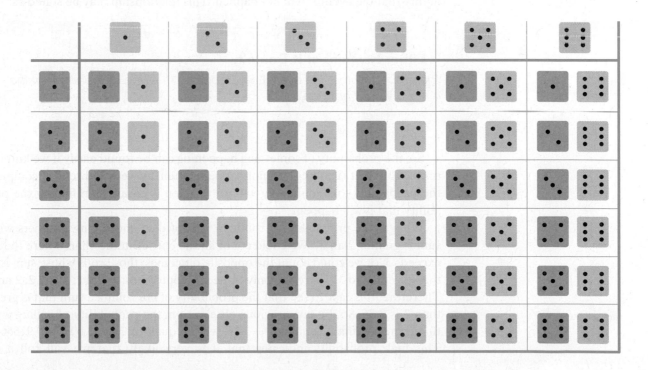

Notice that there are 36 possible, equally likely outcomes. Since only one of them produces "snake eyes," we conclude that the probability of rolling "snake eyes" is 1/36, which equals 0.0277.... This probability is pretty small, so we would not expect to see "snake eyes" too often. What is the probability of rolling a total of 4? What is the probability of rolling a total of 13? What is the probability of rolling a total of 7? Figure out these probabilities using the table.

The Probability of Success versus Failure

Let's think about the probability of an event *not* happening. What is the probability of rolling two dice and getting anything other than a total of 7? There are 30 outcomes that do not give us 7. Thus, the probability of not rolling 7 is 30/36, which equals 5/6 = 0.8333.... Therefore, it is likely that we will not roll a total of 7. How does this answer relate to the probability of rolling 7? Do you notice an interesting connection between these two probabilities? The probability of rolling a 7 is 6/36 = 1/6 = 0.1666.... When we add the probability of rolling a total of 7 to the probability of not rolling a total of 7, we get exactly 1. Take a few moments to extend this observation into a general principle. Once you have formulated a specific idea of the relationship between the probability that an event will happen and the probability that an event will not happen, continue reading.

When we roll a pair of dice, there are 36 equally likely outcomes. Six of these outcomes add up to 7, and the other 30 outcomes add up to something other than 7. Thus, the 36 total outcomes can be divided into the successes (6 outcomes) and the nonsuccesses (30 outcomes). So, the probability of getting 7 (6/36) plus the probability of not getting 7 (30/36) must add up to 1 (36/36). This insight lets us find the probability of an event if we know the probability that the event *won't* happen. The probability that an event E will happen is equal to 1 minus the probability that the event E will not happen. This relationship may be stated as:

It Either Happens or It Doesn't.
The probability that the event E *will happen* = 1 − *(the probability that the event* E *will not happen).*

So, the probability of something happening can be found easily if we know the probability that the thing will not happen. Often, as we will discover, computing the probability that something will not happen is easier than finding the probability that it will happen.

What is the probability of rolling a pair of dice and seeing numbers whose sum is greater than 3? We could count all the entries in the chart where the sum exceeds 3, or we could count the opposite outcomes: those rolls whose sum is less than or equal to 3. There are only three such outcomes: 1 and 1; 1 and 2; 2 and 1. Therefore, it is easy to see that the probability of *not* rolling a sum that *is* greater than 3 is 3/36. So, by our previous observation, the probability of rolling a sum that *is* greater than 3 equals 1 − 3/36, which equals 33/36 = 11/12 = 0.91666.... This high probability indicates that it is very likely that we will roll a sum greater than 3. This example also illustrates the power of looking at an issue in a different way. Looking at a situation from another perspective may lead to an easy and elegant solution.

Yahtzee

Maybe you have played the game Yahtzee, which is a dice-lover's dream since it involves rolling five dice at once. Players score points when several dice reveal the same number, so let's answer the following question:

The Yahtzee Pair Question.

What is the probability of rolling five dice and getting at least one matched pair; that is, having at least two of the dice showing the same number?

This question gives us a great excuse to employ the strategy of computing the opposite probability from what we desire. Instead of directly computing the probability of getting a pair the same, let's concentrate on computing the probability of *not* getting a matched pair, that is, the probability of rolling five dice and seeing five different numbers. We know that 1 − (the probability of not getting any matched pair) = (the probability of getting at least one matched pair). Since five dice are a lot to shake at once in our minds, let's instead start not as a high roller but as a low roller.

Starting with Two Dice

Let's first consider the warm-up challenge of finding the probability that rolling two dice will give a pair of the same number. The first die will come up with some number, so the question could be rephrased: What is the probability that the second die shows the same number as the first? Well, there are six possible numbers that can come up on the first die, and for the second die to match the first, it must land on that one and only number that agrees with the first die. So, the probability of a match is 1/6. To check our reasoning, let's again compute the probability, but this time let's use the chart that we used for the previous dice-rolling experiment.

Which entries on that chart correspond to the two dice being the same? The entries along the diagonal starting at the upper-left corner have both numbers the same. There are six entries on that diagonal, so the probability of the two dice being the same is 6/(6 × 6), which again, luckily, equals 1/6. We're definitely on a roll!

Not Different Numbers

It will be useful to consider a different point of view for computing the probability of rolling two dice and seeing the same number. This time, let's first find the probability of the opposite outcome—the probability that the two dice do *not* show the same number. It is this strategy that will enable us to solve the original Yahtzee Pair Question.

As before, there are 6 × 6 different possible, equally likely outcomes from rolling two dice. How many of these 36 possible outcomes produce different numbers? There are six possible numbers for the first die. However, once that first number is known, the second die must avoid that particular number like the plague, thus leaving only five possible numbers to ensure that the two dice are different. So, for every one of the six possible numbers for the first die, we have five possible numbers for the second die (all the numbers except the first die's number). This gives a total number of 6 × 5 = 30 outcomes in which the two numbers are different. Therefore, the probability of rolling two dice showing different numbers is

$$\frac{6 \times 5}{6 \times 6} = \frac{5}{6} = 0.83333 \ldots.$$

Using our previous relation between an event happening and it not happening, we conclude that the probability that two dice come up with the same number is equal to

$$1 - \frac{5}{6} = \frac{1}{6} = 0.16666 \ldots,$$

yet again confirming our previous computations. We're on our way to a winning streak.

Another Die

So, the probability that two rolled dice will each land on the same number is fairly low: 1/6. What if we roll three dice? What is the probability that two or more of the three come up with the same number? We examine this new situation with an eye toward finding a pattern that will allow us to answer the question for larger numbers of dice.

Let's again consider the opposite outcome; namely, that all three dice show different numbers. Let's first ask: How many possible triples of numbers are possible outcomes for rolling three dice? There are six possibilities for the first die, six possibilities for the second die, and six possibilities for the third. We can imagine creating a cubical graph to record each of those triple numbers. Instead of just a red die and a green die, we can add a yellow die. Therefore, there must be

$$6 \times 6 \times 6 = 216$$

possible triples of numbers that could arise when rolling three dice.

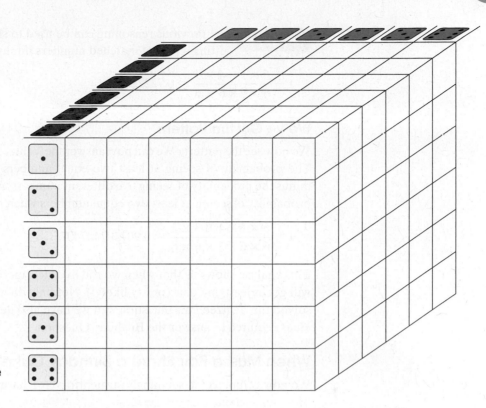

A 3-dimensional table of all possible outcomes of rolling three dice (not filled in!)

How many of these triples have the property that all three numbers are different? Theoretically, we could just look at our cubical chart and count; however, that would be difficult to see in practice. So instead let's think. The first die can show any number from 1 to 6, so there are six possibilities. To be different from the first, the second die can show any number *except* for the number of the first die, so the second die has five allowable numbers. The third die has to avoid the numbers of both the first and the second die, which leaves four allowable numbers. How many in all? Well, for each of the six possible numbers for the first die, we have any one of five possible numbers for the second die, and for each of those combinations, we have four possible numbers for the third die. Thus, there would be

$$6 \times 5 \times 4 = 120$$

possible triples of numbers in which all three numbers are different.

So, when rolling three dice, the probability of getting three different numbers is equal to

$$\frac{6 \times 5 \times 4}{6 \times 6 \times 6} = 0.55555. \ldots$$

Therefore, the probability of the opposite (having at least two dice out of the three share the same number) is

$$1 - \frac{6 \times 5 \times 4}{6 \times 6 \times 6} = 0.44444. \ldots$$

Can you now determine the probability of a match if four dice are rolled?

Try It   The previous reasoning can be used to show that the probability of getting a pair of matched numbers if four dice are rolled is equal to

$$1 - \frac{6 \times 5 \times 4 \times 3}{6 \times 6 \times 6 \times 6} = 0.72222\ldots.$$

We've Got the Pattern

We now see the pattern. We can now answer the Yahtzee Pair Question completely. The probability of seeing at least two equal numbers when rolling five dice is 1 minus the probability of seeing five different numbers when rolling five dice. So the probability of seeing at least two equal numbers when rolling five dice is:

$$1 - \frac{6 \times 5 \times 4 \times 3 \times 2}{6 \times 6 \times 6 \times 6 \times 6} = 0.907407407407\ldots.$$

This analysis shows us that when we roll five dice, more than 90% of the time, we will get at least one pair (pretty likely!). Not only do we now see the pattern that solved the Yahtzee Pair Question, but we have also developed all the probability ideas required to answer the Birthday Question.

When Must a Pair Share a Birthday Cake?

We now return to the seemingly straightforward, harmless question posed in the previous section:

The Birthday Question.

How many people are needed in a room so that the probability that there are at least two people whose birthdays are the same day is roughly one-half?

Let's make the assumption that it is equally likely to be born on one day as on any other—no day is more or less popular for celebrating birthdays. That is, the probability that someone is born on any given day, say December 9, is 1/365, since there are 365 days in the year (let's pretend we never leap) that are all equally likely candidates for one's birthday, and exactly one of them is December 9. Our strategy for answering this Birthday Question is to follow the path of ideas laid out in answering the Yahtzee Question.

Starting with Small Crowds

Let's first find the probability that two people share the same birthday. The first person has some birthday, so the question could be rephrased: What is the probability that the second person has the same birthday as the first person? Well, there are 365 possible days for a birthday, so the probability of that happening is 1/365. To check our reasoning, let's again compute the probability, but this time let's use a chart, as we did for the dice-rolling experiment.

How many different possible pairs of dates are there for the birthdays of two people? Well, there are 365 possibilities for the first person and 365 possibilities for the second person. We could make a huge chart similar to the two-dice chart, only this one would have 365 rows and 365 columns. Making the chart would be difficult, but figuring out how many outcomes are represented in the chart is easy: $365 \times 365 = 133,225$. Which entries in that chart correspond to the two people having the same birthday? The entries along the diagonal starting at the upper-left corner have both dates the same (starting with January 1 and ending with December 31). There are 365 entries on that diagonal, so the probability of the two people having the same birthday is $365/(365 \times 365)$, which again, much to our delight, equals 1/365.

Different Birthdays

Now let's first find the probability of the opposite outcome—the probability that they do *not* share the same birthday. It is this strategy that will enable us to solve the original Birthday Question.

As before, there are 365×365 different pairs of birthdays. How many of these 133,225 possible outcomes produce a pair of different birthdays? There are 365 possible and allowable birthdays for the first person. However, once that person's birthday is known, the second person must avoid that particular date as one would avoid stale birthday cake. So, for every one of the 365 possible dates for the first person, we have 364 possible dates for the second person (all the dates except the first person's date). This gives a total number of $365 \times 364 = 132,860$ pairs of dates in which the two dates differ. Therefore, the probability of two people having different birthdays is

$$\frac{365 \times 364}{365 \times 365} = \frac{364}{365} = 0.9972\ldots.$$

Using our previous relation between an event happening and it not happening, we conclude that the probability that two people have the same birthday is equal to

$$1 - \frac{364}{365} = \frac{1}{365} = 0.00273\ldots,$$

yet again confirming our previous computations.

A Few More People

In a room with only two people, the probability that they share a common birthday is extremely low. What if there were three people in the room? What is the probability that two or more of the three share the same birthday?

Let's again consider the opposite outcome, namely: All three people have different birthdays. Let's first ask: How many possible triples of dates are there for the birthdays of any three people? There are 365 possibilities for the first person,

365 possibilities for the second person, and 365 possibilities for the third person. Therefore, there must be

$$365 \times 365 \times 365 = 48{,}627{,}125$$

possible triples of dates.

How many of these triples have the property that all three dates are different? The first person's birthday can be any date, so there are 365 possibilities for that person. The second person's birthday can be any date, except for the date of the first person, so the second person has 364 possible dates. The third person has to avoid the dates of both the first and the second person, which leaves 363 possible dates. How many in all? As with the dice, we multiply these numbers together to discover that there are

$$365 \times 364 \times 363 = 48{,}228{,}180$$

possible triple dates in which the three dates are different.

So, when we have three people, the probability that they have three different birthdays is equal to

$$\frac{(365 \times 364 \times 363)}{(365 \times 365 \times 365)} = 0.9917\ldots.$$

Therefore, the probability of the opposite (having at least two people out of three share the same birthday) is

$$1 - \frac{(365 \times 364 \times 363)}{(365 \times 365 \times 365)} = 1 - 0.9917\ldots = 0.0082\ldots.$$

Although this probability is still extremely small and nowhere near the 0.5 probability that we seek, we do notice that having a birthday match with three people is about *three times* as likely as with two people. Can you now determine the probability of a match if four people are in the room?

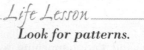 **Try It** The previous reasoning can be used to show that the probability of having a pair of matched birthdays among four people is equal to

$$1 - \frac{(365 \times 364 \times 363 \times 362)}{(365 \times 365 \times 365 \times 365)} = 0.01635\ldots,$$

which, although still nowhere near 1/2, is almost twice as large as the probability of finding a match among three people. We can continue to compute the probabilities in this manner. For example, a match among five people would have a probability of

$$1 - \frac{(365 \times 364 \times 363 \times 362 \times 361)}{(365 \times 365 \times 365 \times 365 \times 365)} = 0.0271\ldots.$$

We've Got the Pattern

We now see the pattern. If we continue for various numbers of people, we could produce the following chart.

Life Lesson
Look for patterns.

Number of People in the Room	Probability of at Least Two Sharing the Same Birthday
5	0.027 . . .
10	0.116 . . .
15	0.252 . . .
20	0.411 . . .
25	0.568 . . .
30	0.706 . . .
40	0.891 . . .
50	0.970 . . .
60	0.994 . . .
70	0.9991 . . .
80	0.99991 . . .
90	0.999993 . . .

It is truly surprising how quickly the probability heads toward 1. With only 50 people, it is almost a sure thing that there will be a match. With 90 people, we are essentially 100% confident of a match; yet 90 is a far cry from 366 people, which guarantees a match for sure. We also have an answer to our Birthday Question: The probability of a birthday match with 23 people is 0.5072. . . .

Retraining Our Intuition

If our intuition leads us astray, we need to look at the situation in different ways until not only our reason but also our intuition is convinced.

Why is the actual answer of 23 people so much lower than we first guessed? When dealing with many small probabilities simultaneously, our intuition often does not accurately correspond to reality. Before thinking about the Birthday Question, our intuition was probably influenced by some simple yet wrong reasoning. We might have reasoned that because 366 people are required to guarantee that two people will share the same birthday, then 183 people will be required for a 0.5 probability. Now we see that such reasoning is far from correct. Somehow, to make this birthday principle real to us, we must retrain our intuition.

A helpful technique in retraining the intuition is to try the birthday experiment in several actual gatherings to see that, in fact, pairs of people will share the same birthday. Another approach is to examine situations similar to the Birthday Question and discover the answer in the new setting. We could try analogous experiments in other settings, such as with cards, to experience the underlying principles at work. In the Mindscapes we invite you to try several. The surprising answer to the Birthday Question illustrates the power of analyzing simple cases carefully and then seeing how the principles apply to a harder case.

The Birthday Question is just one of several instances where probability and everyday intuition diverge. The "Cool dice" that you saw in Chapter 1 and that also appear in Mindscape 36 in this section present another cool counterintuitive fact that you can share with your friends (and use to get rich). The four funky dice required for this exercise are included in your kit that accompanies this text.

Our entire discussion of probability is implicitly based on a concept known as *randomness*. What is randomness? Is it synonymous with unpredictability? We will either visit the subtle notion of randomness in the next section . . . or not.

A Look BACK

PROBABILITY PROVIDES us with a quantitative method to analyze the uncertain and the unknown. It is a measure of the likelihood of an event, such as rolling a sum of 7 with two dice. For an activity (like rolling two dice) with only finitely many equally likely outcomes, the probability of a particular event is the number of different outcomes in that particular event divided by the total number of possible outcomes. So the probability of rolling a sum of 7 is 6/36 = 1/6. Using this basic definition and careful analysis, we can understand many probabilistic situations, some leading to surprising results. Perhaps the most famous and surprising example of unexpected probability is the Birthday Question, whose answer is that, in a group of 23 people, it is slightly more likely than not that two of them have the same birthday.

We develop ideas about probability by starting with familiar situations in which the probabilities are intuitively clear—for example, the simple cases of rolling dice. From those examples, we extrapolate the basic idea of probability. We then formulate a specific definition. Finally, we explore consequences of that definition and discover some surprising results.

Carefully analyzing simple and familiar events opens the door for us to understand more complex and puzzling situations. We can more easily see patterns and develop insights from simple and clear examples than we can from complex or muddy examples. So, focusing on the simple and familiar allows us to concentrate on uncovering the essential principles.

Life Lessons

Analyze simple things deeply.

•

Deduce general principles.

•

Apply them to more complex settings.

MINDSCAPES Invitations to Further Thought

In this section, Mindscapes marked (H) have hints for solutions at the back of the book. Mindscapes marked (ExH) have expanded hints at the back of the book. Mindscapes marked (S) have solutions.

Developing Ideas

1. **Black or white?** Your friend chooses his sartorial color scheme by putting all of his black and white T-shirts in a drawer, then closing his eyes and reaching into the drawer, and selecting a shirt. The probability that he wears a white T-shirt is 3/5. What is the probability that he wears a black T-shirt?

2. **Eleven cents.** You have a dime and a penny. Flip them both, noting whether each coin lands heads up or tails up. List all possible outcomes. Let E be the event that you get at least one head. List all the outcomes that give E. What is the probability that E occurs?

3. **Yummm.** You have a small bag of candy-coated chocolates that melt in your mouth; three are red, four are yellow, two are green, and five are blue. If you take a piece out of the bag at random, what is the probability it is green? What is the probability it is blue? What is the probability that you will eat it?

4. **Rubber duckies.** A game at a carnival has 75 rubber ducks floating in water. The ducks are numbered 1 to 75, with the numbers written on their undersides so they can't be seen. To play, you select a duck and see what number it has. If the number is less than 60, you win a consolation prize. If the number is at least 60 but less than or equal to 70, you win a stuffed duck. If the number is greater than 70 you win a giant, stuffed banana. What is the probability you win a stuffed duck? What is the probability you don't win a duck or a banana?

5. **Legally large.** What does the Law of Large Numbers assert?

Solidifying Ideas

6. **Lincoln takes a hit.** On your wall is a poster containing equal-sized pictures of each of the presidents of the United States. You take a dart, close your eyes, and throw it randomly at the poster. What is the probability that you will hit Lincoln?

7. **Giving orders.** Order the following events in terms of likelihood. Start with the least likely event and end with the most likely.

   • You randomly select an ace from a regular deck of 52 playing cards.

   • There is a full moon at night.

   • You roll a die and a 6 appears.

   • A politician fulfills all his or her campaign promises.

   • You randomly select the queen of hearts from a regular deck of 52 playing cards.

   • Someone flies safely from Chicago to New York City, but his or her luggage may or may not have been so lucky.

   • You randomly select a black card from a regular deck of 52 playing cards.

8. **Two heads are better.** Simultaneously flip a dime and a quarter. If you see two tails, ignore that flip. If you see at least one head, record whether you see one or two heads. Repeat this experiment 30 times. Calculate the number of double heads divided by 30. How close is this answer to the computed probability of having two boys in the second reunion scenario?

9. **Tacky probabilities.** Before doing the following experiment, think a bit, and then guess the probability and record it. Cup five identical, standard thumbtacks in your hands. Shake them, and then toss them slightly upward and let them fall onto a smooth, tiled floor. Count how many of the tacks land completely on their flat side and how many land resting against their points. Repeat this experiment 10 times, then use your data to estimate the probability of a tossed thumb tack landing point-side down.

Tack landing flat   Tack landing against the point

10. **BURGER AND STARBIRD.** Suppose you randomly select a letter from BURGER AND STARBIRD. Imagine writing these letters on Ping-Pong balls—one letter per ball—then putting them all in a barrel and removing one. What is the probability of  pulling out an R? What is the probability of pulling out a B? What is the probability of pulling out a letter that appears in the first half of the alphabet? What is the probability of pulling out a vowel?

11. **Monty Hall.** Read and rework the *Let's Make a Deal* scenario from Chapter 1, "Fun and Games." Work through the solution. Next, find a friend and simulate the *Let's Make a Deal* situation, keeping track of the outcomes under the two possible strategies—the switch strategy and the stick strategy. Perform the experiment approximately 40 times and record the results. Do the experimental data accord with the analysis of the probabilities? You can simulate *Let's Make a Deal* by visiting the *Heart of Mathematics* Web site.

12. **7 or 11 (S).** What is the probability of rolling a sum of 7 or 11 with two fair dice?

13. **D and D.** You simultaneously flip a dime and roll a die. Make a table of all the possible outcomes. What is the probability of seeing Roosevelt and a 4? Suppose now that someone else flipped and rolled, did not show you the result, but reported that the die shows a 2. What is the probability that the dime is showing tails? Justify your answer.

14. **The top 10 (ExH).** Suppose you have 10 marbles. They are marked with the numbers 1, 2, 3, 4, 5, 6, 7, 8, 9, or 10. They are placed in a jar, and you reach in and select one. What is the probability that the number you select has a factor of 3? What is the probability that the number you select is a prime number? What is the probability that the number you select is even? What is the probability that the number you select is evenly divisible by 13?

15. **One five and dime (H).** Someone simultaneously flips a penny, a nickel, and a dime. Make a list of all the possible outcomes. What is the probability of seeing three presidents? What is the probability of seeing exactly two presidents? Suppose now that you do not see the outcome, but you are told that a president is showing. Now, what is the probability of seeing three presidents? Suppose, instead, that you are told that Lincoln is showing. What now is the probability of seeing three presidents? Why do the answers differ?

16. **Five flip.** Someone flips five coins, but you don't see the outcome. The person reports that no tails are showing. What is the probability that the person flipped five heads?

17. **Flipped out.** We take a coin and flip it 10,000,000 times (okay, we have a lot of time on our hands). We notice that 6,010,375 times it landed on heads. What do you suspect about the coin?

18. **Spinning wheel.** A roulette wheel has 36 spaces marked from 1 to 36, half of which are marked red and half black. In addition, there are two green spaces

marked 0 and 00. What is the probability of the little ball landing on 13? What is the probability of it landing on a red spot?

19. **December 9.** Choose two people at random. What is the probability that they were both born on December 9?

20. **High roller (H).** Using two fair dice, what is the probability of rolling a sum that exceeds 4?

21. **Double dice.** You roll two fair dice. What is the probability you will roll a double (two 1's, two 2's, two 3's, and so on)?

22. **Silly puzzle.** After a professor explains the Birthday Question to her class of 20, she points out that the probability of having a birthday match in the class is around 0.4. A student raises her hand and states that she is certain that there will be a birthday match. She knows no one's birthday except her own. Explain why she was able to make this statement with such certainty.

23. **Just do it.** Find groups of roughly 35 people together (in a class, dorm, or dining hall) and have each person in turn shout out his or her birthday. Is there more than one pair of matches? Record your results.

24. **No matches (S).** Suppose 40 people are in a room. What is the probability that no two people share the same birthday?

25. **Spinner winner.** If you were to spin the wheel illustrated to the right, it is equally likely to stop at any point. You win if it stops on a space that is 6 or higher. Otherwise you lose. What is the probability of winning?

Creating New Ideas

26. **Flip side (S).** Someone flips three coins behind a screen and says, "I flipped at least two heads." What is the probability that the flipper flipped three heads?

27. **Other flip side.** Someone flips three coins behind a screen and says, "I didn't flip all tails." What is the probability that the flipper flipped all three heads?

28. **Blackjack.** From a regular deck of 52 playing cards, you turn over a 5 and then a 6. What is the probability that the next card you turn over will be a face card?

29. **Be rational (ExH).** Suppose someone has randomly selected two numbers from the set of the first one million natural numbers and used them to make a fraction. Reduce the fraction to its lowest terms. Is there a 0.5 probability that both the numerator and the denominator are odd numbers? Why or why not?

30. **Well red (H).** Someone shows you three cards. One is red on both sides, another is blue on both sides, and the last is red on one side and blue on the other. The cards are shuffled, and you are then shown one side of one card. You see red. What is the probability that the other side is blue? Is it 0.5? Explain.

31. Regular dice. Dungeons and Dragons players use dice in the shape of each of the regular solids (see Section 4.5). The faces are always numbered 1 through the number of total faces there are. You shake all five dice. What is the probability of your throwing a total of 6?

32. Take your seat. You decide to fly to California on EconoJet Airlines. You are randomly assigned a seat. Seats are numbered by row from 1 to 40 and in each row by A, B, C, or D, and amazingly, there is only one window seat in each row. The plane is boarded from the rear in groups of 10 rows at a time. What is the probability that you will be in the first group to board the plane? What is the probability that you get a window seat?

33. Eight flips. What is the probability of flipping a half dollar eight times and a head appearing at least once?

34. Lottery (S). The lottery in an extremely small state consists of picking two different numbers from 1 to 10. Ten Ping-Pong balls numbered 1 through 10 are dropped in a fish bowl, and two are selected. Suppose you bet on 2 and 9. What is the probability that you match at least one number? What is the probability you match both numbers?

35. Making the grade. What is wrong with the following statement? "The way I figure, the probability I get a 4.0 average this term is 0.2. The probability I get below a 4.0 average this term is 0.9." Explain. Given the statement, guess the person's actual GPA.

Further Challenges

36. Cool dice (ExH). Find the four dice from your kit (see the kit instructions). Show that, no matter which die someone picks, there is always another die such that the probability of rolling a higher number on the second die is greater than 0.5. That is, there is no best die. Order the dice A, B, C, D such that B beats A, C beats B, D beats C, and A beats D. This phenomenon is like a conference of sports teams in which B generally beats A, C generally beats B, and D generally beats C, but, because D has some weakness that one of A's strengths can take advantage of, A generally beats D. Play this dice game with several friends. What is the probability that the die with 6's and 2's will beat the die with 4's and blanks?

37. Don't squeeze. Five shoppers buy Charmin toilet paper. One Charmin out of 10 in this batch is defective—it's unsqueezable. You want to save everyone from this catastrophe, so you stop them at the door and ask to squeeze their Charmin. After squeezing 5 rolls, what is the probability that you have located 1 or more defective Charmins?

38. Birthday cards. Using a regular deck of 52 playing cards, select a card at random, record it, and then put it back in the deck. Shuffle the cards and select another card at random and record it, put it back, and so on. How many cards do you draw before you select the same card for the second time? Do this experiment several times. Calculate the probability of choosing 10 times and seeing 10 different cards.

39. Too many boys. Long, long ago and far, far away, an emperor believed that there were too, too many males and not enough females. To correct this wrong, the emperor decreed that, as soon as a woman gave birth to a male child, she would not be permitted to have any more children. If the woman gave birth to a female, she would be allowed to continue bearing children. What was the result of this decree? After the decree, what fraction of the babies will be male? Carefully explain your answer.

40. **Three paradox (H).** The correct probability of tossing three coins and having all three landing the same is 1/4. What's wrong with the following dubious reasoning?

> When we toss three coins, we know for a fact that two of the coins will land the same; therefore, we only have to get the third one to match. Thus, the probability is 1/2.

In Your Own Words

41. **Personal perspectives.** Write a short essay describing the most interesting or surprising discovery you made in exploring the material in this section. If any material seemed puzzling or even unbelievable, address that as well. Explain why you chose the topics you did. Finally, comment on the aesthetics of the mathematics and ideas in this section.

42. **With a group of folks.** In a small group, discuss and work through the details involved in the answer to the Birthday Question. After your discussion, write a brief narrative describing the reasoning in your own words.

43. **Creative writing.** Write an imaginative story (it can be humorous, dramatic, whatever you like) that involves or evokes the ideas of this section.

44. **Power beyond the mathematics.** Provide several real-life issues—ideally, from your own experience—that some of the strategies of thought presented in this section would effectively approach and resolve.

For the Algebra Lover

Here we celebrate the power of algebra as a powerful way of finding unknown quantities by naming them, of expressing infinitely many relationships and connections clearly and succinctly, and of uncovering pattern and structure.

45. **What's the big event?** The section defines the probability of the event E as follows. If T equals the total number of equally likely outcomes and N equals the number of these outcomes (all different) in the event E, then the probability of the event E equals N/T. For each given probability below, find the number of outcomes N.

$$\frac{N}{T} = \frac{3}{5} \text{ and } T = 100; \qquad \frac{N}{T} = \frac{5}{8} \text{ and } T = 48; \qquad \frac{N}{T} = 0.40 \text{ and } T = 75.$$

46. **Tallying totals.** For each given probability below, find the total number of outcomes T.

$$\frac{N}{T} = \frac{3}{8} \text{ and } N = 240; \qquad \frac{N}{T} = 0.35 \text{ and } N = 175; \qquad \frac{N}{T} = \frac{8}{15} \text{ and } N = 64.$$

47. **Algebra roulette.** Your math instructor has created a special roulette wheel. It has 32 spaces marked with x and four spaces marked with x^2. To play the game, you spin the wheel and earn M&Ms. If it stops on a space marked x, the number of M&Ms you win equals x. If it stops on a space marked x^2, the

number of M&Ms you win equals x^2. Your instructor tells you that x satisfies the equation $4x^2 + 32x = 80$. Determine how many candies you would win for each outcome of a spin of the wheel. What is the probability you will win x^2 M&Ms?

48. **Probability beans.** The jellybean jar on your math instructor's desk has 60 jellybeans in it – each jellybean is red, yellow, or green. There are twice as many red jellybeans as yellow ones, and four fewer green ones than yellow ones. If you pick a jellybean at random, what's the probability you get a yellow one?

49. **Picking up the tab (H).** Ed and Mike go out to lunch and decide to roll dice to determine who will pick up the tab. They each roll the same collection of dice. Ed's roll shows all 5's and 3's, with a total on all the dice of 21. Mike's roll show all 2's and 6's, with a total of 18. Mike loses, so he picks up the tab, but that's not the point here. Suppose the number of 5's in Ed's roll was the same as the number of 2's in Mike's roll, and the number of 3's in Ed's roll was the same as the number of 6's in Mike's roll. How many dice were Ed and Mike rolling? And who ordered the raw broccoli?

8.3 RANDOM THOUGHTS

Are Coincidences as Truly Amazing as They First Appear?

During the great Sammy Sosa–Mark McGwire home-run race of 1998, Mark McGwire tied a home-run record of 61 home runs on his own father's 61st birthday. *What an amazing coincidence!*

Coincidences are so striking because any particular one is extremely improbable. However, what is even more improbable is that no coincidence will occur. We saw in the Birthday Question that finding, in a room of 50, two people who share the same birthday is extremely likely, even though the probability of any particular two people having the same birthday is extremely low. If you were one of a pair of people in that room with the same birthday as someone else, you would feel that a surprising coincidence had occurred—as indeed it had. But almost certainly some pair of people in the room would experience that coincidence. Let's now delve into the mysterious world of coincidences.

Coincidences and random happenings easily befuddle our intuition. To expose them for what they are, we must describe them clearly and analyze them quantitatively. Looking at simplified situations will help us understand whence misleading impressions arise. As usual, we start with concrete examples.

A Deadly Coincidence

We began working on a first draft of this section for the first edition of this text during a two-week period in late June and early July of 1997. During that time, five famous people died—television celebrity Brian Keith (June 24), deep-sea diver Jacques Cousteau (June 25), actor Robert Mitchum (July 1), actor Jimmy Stewart (July 2), and news commentator Charles Kuralt (July 4). As we were writing about randomness, we began to ponder: What are the chances of five famous people dying during those two weeks? Isn't it strange that during the two weeks we were writing a section about coincidence such a public coincidence actually occurred?*

Having noticed this sad but interesting phenomenon, we decided to analyze and attempt to understand it. Is it strange or not that five famous people died during a two-week period? We know we should be sad, but should we be surprised? Contemplating this question brings up several of the main ideas associated with randomness, probability, and coincidence. The first is the meaning of *strange*. Presumably we mean that the event had a low probability of occurring. But to associate a probability with the event of five famous people dying, we are obliged to specify the total collection of possible occurrences with which to compare the death of the five. One possibility is to consider all deaths during that period and ask how likely it is that five of the people who died would be famous.

At least 52 million people died in 1997, which is an average of one million deaths per week worldwide. Thus, our question might be rephrased in a provocative way: Among the roughly two million people on Earth who die during any two-week period, what is the probability that five of them are famous? Already this phrasing of the question makes the fact of five famous deaths a little less surprising. But perhaps these five deaths would still be surprising if the total number of famous people is extremely low. How many famous people are there on Earth?

We just made some conservative guesses of the number of famous people in different categories: 300 singers; 600 actors; 600 sports figures; 1200 leaders; 150 scientists; 200 businesspeople; 200 artists and writers; and 750 miscellaneous people. So, let's say there are 4000 famous people in the world. Nearly all are famous for less than 40 years of their lives, so we estimate that at least 100 must die each year, which amounts to approximately two per week. Therefore, as a matter of fact, our five deaths are pretty much right on target for the average number of deaths of famous people for a two-week period. To check our figures, we looked in a world almanac that lists famous people who died between October 1995 and October 1996. The list consisted of 105 names, so our estimates appear to be reasonably accurate.

Actually, the coincidence of five famous deaths should probably be viewed in the context of all possible remarkable coincidences that might have happened during that two-week period. We didn't set out to look for a coincidence involving deaths of celebrities. The huge collection of all conceivable coincidences diminishes the significance of the five-death coincidence even further. When looked at in this way, many coincidences lose some of their luster.

Let's now see if we can come to understand the presence of coincidence in our own lives.

---

*As an even more eerie coincidence, while working on an early draft of this section for the 3rd edition back in June of 2009, within a matter of days, the following celebrities died: sidekick Ed McMahon, Queen of Pin-Up Farrah Fawcett, King of Pop Michael Jackson, Pitchman Billy Mays, Oscar Winner Karl Malden, 50s TV Star Gale Storm, Vegas Impressionist Fred Travalena. We're so concerned about the loss of (famous) lives as we move to future editions, that we refused to read the tabloids while writing this fourth edition.

Personal Coincidence

Think of the most amazing coincidences in your life. Perhaps you were walking in the airport in Chicago and ran into an old friend you hadn't seen in 10 years. Perhaps you thought about a car crash, and the very next day, a relative was in an automobile accident. Perhaps you and a college classmate independently decided to open a Starbucks, rather than go to grad school to study math. Perhaps your birthday is the same as a string of numbers in your social security number. Some remarkable coincidences have occurred in your life.

How unlikely are any of these events? The answer is that each one of them is extremely unlikely. However, let's look at the same situations from a different point of view. Let's consider the probability that you will *avoid* remarkable coincidences. Often, to better understand a possibility, it is valuable to consider the opposite possibility.

Each day, suppose you wake up in the morning and think of an event that has a one-in-a-thousand chance of happening that day. In other words, imagine that 1000 equally likely things might happen that day, of which one of them is the rare event that you are considering. Let's compute the probability of not one of those coincidences coming to pass during a year. The first day, you have a 999/1000 probability of not experiencing that coincidence (pretty likely it will not happen). Using the ideas developed in the previous section, we see that your chance of missing rare coincidences both the first and second days is $999^2/1000^2$ since you have 999 times 999 possible ways of not experiencing the coincidence during the two days and 1000 times 1000 total things that could happen during those two days. Your chance of missing out for any number of days is simply 999 raised to the power of the number of days divided by 1000 raised to the power of the number of days, which is the same as 999/1000 raised to the number of days. Using a calculator and taking 999/1000 to the 365th power, we see that missing out every day for a year has a probability of 0.69. So, your chance of experiencing one of your one-in-a-thousand coincidences during one year is 0.31—nearly 1/3 (one in three chances—not that unlikely). During a three-year period, your chance of missing every single day is 999/1000 raised to the 1095th power, a mere 0.33. In other words, the probability that during a three-year period your one-in-a-thousand event will occur at least once is a whopping 2/3 (two in three chances).

The probability that such a one-in-a-thousand event will happen at least once in 10 years is 0.97, and after 20 years the probability that at least one such unlikely event will happen to you is 0.9993. In other words, even if we select, in advance, each morning the one-in-a-thousand coincidence that we would count for that day, each of us is almost certain to experience that coincidence from time to time. Of course, in practice we would take note of a one-in-a-thousand coincidence even if it did not happen to be our particular coincidence *du jour*. Therefore, we see that it would be truly remarkable if we never experienced such a coincidence.

Life Lesson

To better understand a possibility, consider the opposite possibility.

Moral: Coincidence Happens.

How to Get Rich Quick as a Stock Whiz

Predicting the future is a feat full of folly. One method of beating the odds is to make many predictions and then declare success if a few of them materialize (and hope that people will just forget those that don't). Another is to cover all bets. Let's take a look at how we can predict the future in the stock market with impressive accuracy, from some people's point of view.

Let's take a list of 1024 investors and send them a letter on Monday. To 512 investors we write, "IBM stock will go up next week"; to the other 512 we write, "IBM stock will go down next week." The following week, we send 512 letters to the group for whom we were correct and write to 256 of them, "IBM stock will go up next week," and to the other 256 we write, "IBM stock will go down next week." At the end of that week 256 people will begin to pay attention to our ability to predict the future. We continue the pattern. After nine weeks, two people will have seen us predict the future nine times in a row. Now we ask them to send us a large check requesting our next week's prediction. We then send an "up" letter to one, a "down" letter to the other, and, to assuage our conscience, refund the payment from the person whom we misled.

Another up and down day on the market

In real life, we presume that such a scam is illegal, but it does happen inadvertently in another way. There are thousands of people who predict stock market activity. Some are correct sometimes. Suppose these stock analysts were literally flipping coins to make their predictions. Still, we would expect that someone would have a good track record if enough of them were flipping coins. The moral of this story is beware of investment counsel that says, "This expert correctly predicted the big crash of 1987." That person may well have done so, and that person no doubt believes that the prediction was not based on randomness but was, instead, based on special insight. How are we going to determine whether the truth rests in randomness among a lot of predictors who really don't have any special insight, or in the incredible instincts of a few special people? This question is essentially impossible to answer. Or is it? We leave these conundrums for you to ponder but do caution you not to take advice blindly.

Hey, Hey, We're the Monkeys

As we have said, although it is rare to see a particular coincidence within a random event, it is even more improbable to *never* see a coincidence. We further illustrate this idea with perhaps the most famous example of randomness.

Suppose we have a very large number of monkeys, each banging away randomly on his or her own word processor. Their typing is completely random. If we let them type indefinitely, would one of them at some point randomly type out Shakespeare's entire *Hamlet?*

The answer is yes. If we have enough monkeys typing long enough, we are bound to get *Hamlet.* Why? Because if we perform a random event enough times, we would expect to see any possible outcome, no matter how unlikely it may be. This result is known as the *Infinite Monkey Theorem* or, as we like to refer to it, *Hamlet Happens.*

The Infinite Monkey Theorem.

If we put an army of monkeys at word processors, eventually one will bash out the script for Hamlet.

This observation was first made by the astronomer Sir Arthur Eddington in 1929 while describing some features of the second law of thermodynamics. He wrote: "If I let my fingers wander idly over the keys of a typewriter it might happen that my screed made an intelligible sentence. If an army of monkeys were strumming on typewriters they might write all the books in the British Museum. The chance of their doing so is decidedly more favorable than the chance of the molecules returning to one half of the vessel."

Although *Hamlet* does happen, it does not happen often. Suppose 1,000,000 monkeys randomly typed at standard 48-key computer keyboards, and each typed one character per second. We would expect to wait more than

1,000,000,000,000,000,000,000,000,000,000,000,000,000,000,000,000,000

years before one of them typed "To be or not to be: that is the question."

There are many literary references to this random fact about monkeys and Shakespeare. One is from Douglas Adams's book *The Hitchhiker's Guide to the Galaxy*: "'Ford!' he said, 'there's an infinite number of monkeys outside who want to talk to us about this script for *Hamlet* they've worked out.'"

Our favorite story on this subject is a 1960s routine from the button-down mind of comedian Bob Newhart. In the routine, a lab technician monitoring infinitely many monkeys typing on typewriters is interviewed on a news program. His words were roughly: "Scientists have claimed that if you have enough monkeys typing randomly for enough time, one of them will eventually produce *Hamlet*. To test this theory, we have brought in a pack of monkeys and have been letting them type away for 72 days now. Let's see how they're doing. [He reaches over and pulls out the paper from one of the typewriters and reads.] 'To be, or not to be: that is the gezortenblatt.' Well, I guess we're not quite there yet—back to the studio."

This classic comedy routine illustrates another important point. If we let those monkeys type away, we will not only see *Hamlet,* but also any possible variation of it. Basically, if we have enough monkeys and enough time, we could generate every book ever written—although for this book only two monkeys were required.

This monkey business has more than just mere entertainment value. In 1993, George Marsaglia and Arif Zaman from Florida State University used this monkey fact to detect errors in computer programs that generate random numbers. Their basic strategy was to convert the numbers to letters and, roughly speaking, determine how likely it would be for the random-number generator to type a particular phrase, such as "To be, or not to be." Building a computer program to generate random numbers is a challenging task. However, as we've seen with the monkeys, within the purely random we occasionally will see familiar patterns. As one may expect, we might have to wait a long, long time to stumble across a

desired pattern. Thus, if you do attempt the monkey experiment, don't be discouraged if after typing away for 500 trillion quadrillion years, they only produce *Macbeth*.

Random Spots

Randomness often fools us. Below are six pictures, each containing 12 spots. Five of those pictures were drawn randomly, meaning that we used a random process to choose the location of each spot within its square. However, one of the collections of spots was drawn by a person who had not studied randomness whom we asked to draw 12 spots randomly in the square. Can you guess which one was drawn by the person who was trying to be random rather than by an actually random process?

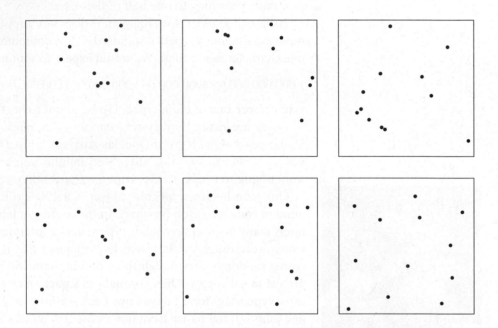

The answer is that the right bottom one was drawn by a person. Why would we guess that configuration was not randomly generated? Notice that in a sense the randomly located spots usually show clusters and areas with no spots at all. The bottom right image, on the other hand, has its spots spread evenly and carefully over the entire rectangle. Often people feel that random distributions should be evenly spread out. Of course, any distribution of spots could result from a random process; however, the randomly generated patterns usually exhibit more clustering (that is, "coincidences") and often have large empty spaces. A naive person simulating randomness tends to want to avoid having large open spaces.

Expect the Unexpected

If seeing random spots before your eyes is a bit disconcerting, then fear not, there is a terrific exercise that can help us retrain our intuition into randomness that only involves coin-flipping. This fantastic flipping experiment is one that you can try for yourself with a couple of friends or family members—who are unfamiliar with the subtleties of randomness. It's really great fun and in terms of randomness, really hits the spot (whoops, another spot before our eyes).

Give your two friends the following charge. You will leave the room and once you do, one of them will flip a coin 200 times and record the heads and tails in order as they occur. Independently (without consulting with the coin-flipper), the other person will pretend to flip a coin 200 times and just write down a random sequence of H's and T's that could have resulted. They are to decide which role each person will take (the real flipper or the fake flipper) but they are not to tell you who's who.

You now leave the room and let the flipping fly. When you are called back into the room, you are handed two sheets of paper each containing a run of 200 H's and T's. Your mission is to identify the genuine flipping sequence and expose the fraudulent flipping sequence. You can try this experiment yourself by looking at the following two sequences of H's and T's. Before reading on, which one do you think was generated by a person and which one records actual coin flips?

```
H T T H H H H H T H T T T H H H H T H H
H H H T T H T T T H T H H H H T H T H T
H T T H H H T T H T T H T T T T H T T T
T H T T T H T T H T H T H H H T T H H H
H H T H H H T H H T T H H H H H T H H H
H H T H T T H H T H T H T H T H T H T H
T H H T H T T T T T T H H T T H H H T T
T H H T H T H H T H H T H H H H T H H T
H T T T T H T H T T T H T H H T H T T H
T H H T H T H H T H T H H T H T T H T H
T H T T H H T H H T H H T H H T H H H T
H T H T H H T H T T T H T T T T T T T T
H T H T H T H H T H T H H H T T H H H T
H T H H H T T T H T H T H T H T H T T H
H H H H T H T H H T T T T H H T H T H H
T H H T H H T H H H H H H T T T H H T
H H T T H T H T H H H H T H T H H T H
T H H T T T T H I I H T T T T H T H H
T H T T T H H H H T T T T H H H T H T H
T T T H H T H T H H H H H T T H T T T H
```

A run of 200 H's and T's

```
H T T T H T T H H H I T T H T T H T T T
H H T H H H T T H T H H H T H T H T H T
T T H H T H T T T H H T T H T T H T H T
H T T T H H T H H H T H T T T T H H T H
H H H T T T H H T T T H T H T H T H H H
T H H H H H H T H H T H H T T H T H H H
T H H T H H H T T H T T H T H H T H H H
H T T T H T T H H H H T T T H T H T T H
H T T T H H H H H T T T H T T H H H H T
T H T T H H H T T T H T T T H T H T T T
T T H T H H T H H H H T T H T H T H T T
T H H T T H H H H H H T T T H T H T H T
H H T H H T H H H H T H T H T H H H H H
T H H T T H H H T T H T T H T H T T T H
T T H H H H T H T H H H T H T T T T H T
H T H T H H T T T H H H T H T H T T H T
T T H T H H H T T T T H T H H H T H H T
T T T H H H H T H H T T T H H H H T T H
H H H T H H H H T H T T H H H T H H H T
T T H T H H H T T H T H H H T H T H T H
```

Another run of 200 H's and T's

Can you tell which of the preceding is the random one? The answer is that the first one was generated with an actual coin and the second one was created by a person who deliberately attempted to write a random-looking list of H's and T's. How can you tell?

The answer is that we must learn to expect the unexpected. When probabilistically unsophisticated people attempt this challenge of acting randomly, they are generally reluctant to put too many H's in a row or T's in a row. It is natural to think that too many of one letter in a row is somehow unrandom; however, the coin has no such prejudice. Here are the same two sequences with runs of more than five H's in a row highlighted in red and runs of five or more T's highlighted in blue.

```
H T T H H H H H T H T T T H H H H T H H
H H H T T H T T T H T H H H H T H T H T
H T T H H H T T H T T T H T T T T H T T
T H T T T T H T T H T H T H H H T T H H
H H T H H H H T T H H H H H H T H H H T
H H T H T T H H T H T H T H T H T H T H
T H H T H T T T T T T H H T T H H H T T
T H H T H T H H T H H T H H H H T H H T
H T T T T H T T H T T T T H T H H T T H
T H H H T H T H H T H T H T H T H T T H
T H T T H T H H T H H T H H T H H T H T
H T H T H H H H T T T T T H T T H T T T
H T H T H T H H T H T H T H H T H H H T
H T H H H T H T T T T H T H T H T T T H
H H H H T H T H T T T T T H H T H T T H
T H H T H H T H H H H H H T T T H H H T
H H T T H T H T H H H H H T H T H H T H
T H H T T T T T T H T T H T T T T H T H
T H T T H H H H T T T T T H H H T H T H
T T T H H T H T H H H H T T T H T T T H
```

```
H T T T H T T H H H T T T H T T H T T T
H H T H H H T T H T H H H T H T H T H T
T T H H T H T T T H H T T H T T H T H T
H T T T H H T T H H T H T T T H H T H
H H H T T T H H T T T H T H T H H H H
T H H H T H T H H T H H H H T T H T H H
T H T H H H T T H T T H T H H T H H H
H T T T H H H T H H T T T H H T T T H
H T T T H H H T T H H H T T H T H H H T
T H T H H H T T T H T T H T T H T H H T
T T H T H H T H H H T H T T T H T H T
T H H T H H T H H H H T H T T H T H H T
H H T H H T H H H H H T H T H T H H H H
T H H T H H H H T H T T T H T T T H H H
T T H H H H T H T H H H T H T T T H T
H T H T H T H H T T T H H H T H T T H T
T T H T H H H T T T H T H H H T T H T
T T T H H H H T H T H T H H T T H H H H
H H H T H H H T H T H H H H T T T H H T
T T H T H H H T H H H T H H H T H T H H
```

Notice that the coin-generated sequence has many streaks of five or more of the same letter in a row. The human-generated sequence had only two streaks with five in a row, and no streak with six in a row. The reality is that the probability when flipping a coin 200 times of having at least one streak of six or longer is 96%. The probability of having at least one streak of five or longer is 99.9%. So the moral of the story is, yet again, to expect the unexpected—surprise!

History Does Not Matter

Confusion about randomness can cause serious trouble for gamblers. One of the most common mathematical crimes that gamblers commit is thinking that their luck must change. Suppose a gambler is playing a game involving flipping a fair coin in which he wins if the coin comes up heads and loses if the coin comes up tails. Unfortunately, 10 tails have turned up in a row, so he says to himself, "Surely my luck will change; the coin's got to come up heads *this time*. I'll bet my life savings." That kind of reasoning *does* have a transformative effect on the life of our gullible gambler. He arrived at the casino in a small $5,000 Buick, but will travel home in a huge $500,000 bus.

To demonstrate the fact that random events are totally blind to history, we performed the following computer simulation. We simulated flipping a coin 11 times in a row. Then we repeated that simulation 1,024,000 times. In other words, we simulated more than 11 million coin flips. The reason we performed this "11-coin-flips" simulation 1,024,000 times is that with that enormous number of trials, on average the laws of probability imply that we would expect about 1000 of those simulations to have the first 10 flips all be tails. We ran this simulation three times. As a matter of fact, when we actually performed this repeated simulation on a computer for the first time, we found 1010 of those 1,024,000 simulations had their first 10 flips as all tails—very close to what the mathematics predicted.

For each of these 1010 instances in which the first 10 of our 11 flips were all tails, we looked at the next flip, the 11th flip. Surely after 10 tails in a row, our luckless gambler would bet that a heads would almost surely appear in the next flip. Answer: no, and our simulation verified it. As you see in the table, in the first run we saw the 11th flip was tails 495 times and heads 515 times—almost equal. We ran the entire simulation two more times. The second time, the first 10 flips were all tails 1033 times, of which the 11th flip was tails 523 times and heads 510 times—again almost equal. In the third simulation, the first 10 flips were all tails on 955 occasions, of which the 11th flip was tails 491 times and heads 464 times—yet again nearly equal. When it comes to randomness, history does not matter, so heads-up!

	Number of Times 10 Tails	Number of Tails for 11th	Number of Heads for 11th
Trial 1	1010	495	515
Trial 2	1033	523	510
Trial 3	955	491	464

Flipped a coin 11 times and repeated 1,024,000 times. Repeated the whole experiment three times.

From Needle Droppings to an Approximation of π

Do random events ever lead to concrete results? Seems unlikely—after all, they're random. Let's consider the following random experiment. Suppose we have a sheet of lined notebook paper and a needle whose length is equal to the distance between consecutive lines on the paper. We now randomly drop the needle onto

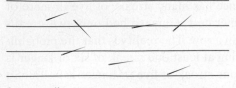

Some needles cross lines—others do not.

the paper. We notice that sometimes the needle crosses one of the lines, and sometimes the needle does not cross any line. What is the probability that the needle will cross a line?

This question was first raised and then answered by the 18th-century French scientist Georges Louis Leclerc Comte de Buffon (if you drop his name on the paper, it will definitely hit a line). The surprising answer is that the probability of the needle hitting a line can be computed and is exactly equal to $2/\pi$. Using this fact, Buffon was able to give estimates for π by, we kid you not, throwing French bread sticks over his shoulders numerous times onto a tiled floor and counting the number of times the bread sticks crossed the lines between the tiles. Although we are told not to play with food, Buffon's food tossing actually gave birth to an entire realm of mathematics now known as *geometric probability*. Hundreds of years after Buffon tossed his bread sticks, atomic scientists discovered that a similar needle-dropping model seems to accurately predict the chances that a neutron produced by the fission of an atomic nucleus would either be stopped or deflected by another nucleus near it—so, even nature appears to drop needles.

Since the probability of the needle hitting the line is equal to $2/\pi$, we can get a good approximation of π by just dropping a needle onto paper many, many times. Here's how: First drop the needle a good number of times and keep track of the number of line hits and the total number of drops. We know that the relative frequency of line hits, which equals the number of line hits divided by the total number of drops, will be approximately the probability of hitting the line. But the probability of hitting the line is $2/\pi$. Therefore, we can solve for π and see that

$$\pi = 2 \times \frac{(\text{total number of drops})}{(\text{the number of line hits})}.$$

In 1864, by making 1100 drops, the English scientist Fox repeated this experiment and used his findings to give an estimate for π of 3.1419. Today we can visit a number of Web sites and watch computers simulate this experiment with thousands of virtual drops. The occurrence of π in the needle-dropping experiment shows that even randomness has a rich and precise structure.

Our next random journey could be an opening scene from the classic 1960s TV series *Mission: Impossible.*

Random Journeys—Mission Impossible?

"Good morning, Mr. Phelps. The road you are standing on runs east and west and goes on forever in both directions. You are holding a penny. Somewhere, although you don't know where, along the side of the road is a small tape recorder that will self-destruct in five seconds after you play the message [thus it has an extremely limited manufacturer's warranty]. Your mission, should you decide to accept it, is to find the tape recorder by walking one block at a time as follows: You flip the penny. If the coin comes up heads, then you walk one block east; if it comes up tails, you walk one block west. You repeat this process indefinitely until you find the recorder. Question: Is this mission possible? As always, if you or any of your I.M. Force is caught or killed, or dies of old age, the secretary will disavow any knowledge of your actions. Good luck, Jim."

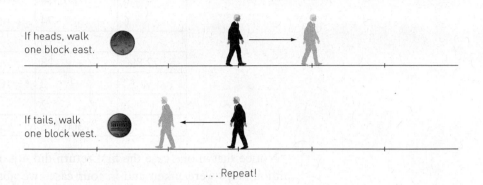

If heads, walk one block east.

If tails, walk one block west.

. . . Repeat!

The preceding walking scenario is an illustration of a *random walk*—a walk whereby the direction we move at any particular stage is selected at random. In this case we are walking one block at a time, and it is equally likely that we walk east or west. For example, suppose our flipping gave a sequence of H H T H H T T T H T H H T T and we started at the marked spot. Draw what our path would look like below.

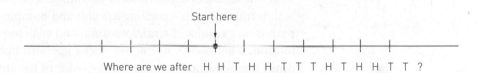

Start here

Where are we after H H T H H T T T H T H H T T ?

Although we have no idea how far away it lies, or even in which direction the poorly made recorder is, the surprising fact is that the probability of finding the recorder in this manner is equal to 1—we will find it with (probabilistic) certainty. Why? If we flipped the coin indefinitely, we should not be shocked to see long, long, long strings of heads. That string would result in a long, long, long journey eastward. This observation illustrates that it is possible and even reasonable for us to migrate far from our starting position (and similarly, eventually return back home). Although this informal observation certainly does not *prove* that the probability is, in fact, equal to 1, it does make the result plausible.

If we walk randomly on a grid in the plane where we move east, west, north, or south at each step, then the probability of finding the hidden recorder remains equal to 1, but we won't try to justify that fact here. In fact, you may not return any too soon. In fact, the following table records 30 simulations of random walks in the plane. In each case, we recorded how many steps were required before the random walk returned to the starting place.

4	36,374,154	28
4	2	2
28	56	>100,000,000
56	80,326	>100,000,000
6	74	>100,000,000
32	16	926
8,072,890	28	118
22	2	>100,000,000
6	2	99,276
4	300,658	2

Notice that in one case the first return did not occur until after more than 36 million steps were taken and in four cases we abandoned the simulation after we had taken 100 million steps without ever returning to our starting point. Nevertheless, with probability 1, we will eventually return to where we start when we are walking in the plane.

Surprisingly, however, if we consider walking randomly in *three* dimensions (we now have wings and can move up and down), then the probability that we will ever return back to where we start is only about 0.699 . . . , and we are not at all certain of finding a hidden tape. Why the change in likelihood? Well, as we add dimensions (and thus new directions to randomly journey), we move in a space with more degrees of freedom. When we have three or more dimensions, it turns out that it is possible for us to get lost in space and never return home.

Random walks actually occur in nature. The path of a liquid or gas molecule is determined by its knocking around and bumping into nearby molecules. This path is an example of a random walk, and such movement is known as *Brownian motion*. The theory of random walks has also been used to study and analyze other phenomena, including the behavior of the stock market.

COINCIDENCES AND RANDOM BEHAVIOR DO occur, often with predictable frequency. A bit of careful thought reveals that coincidences are not as shocking as they may first appear. One of the most famous illustrations of randomness is the scenario of monkeys randomly typing *Hamlet—Hamlet Happens*. Randomness frequently boggles our intuition. We need to learn to expect the unexpected when dealing with random events. Buffon's needle shows how random behavior can be used to estimate numbers, such as π. The theory of random walks is filled with counterintuitive and surprising outcomes and appears in Brownian motion and the behavior of the stock market.

We can apply the principles of probability to understand coincidences and random behavior. The basic definition of probability and how to compute it allows us to gauge more meaningfully how rare or common a seemingly unlikely event really is. We can estimate the probability of no coincidences occurring. This opposite view helps us to understand how likely coincidences really are.

If an experience or an idea seems surprising or vague, often we do not fully understand it. We can understand it more deeply by looking at the opposite point of view or by analyzing it using familiar principles. Often such analyses not only solve the mystery but also lead to deeper insights into related issues.

Life Lessons

Consider all points of view.

•

Apply reason to understand the mysterious or the unknown.

•

Expect the unexpected.

MINDSCAPES ▷ Invitations to Further Thought

In this section, Mindscapes marked **(H)** *have hints for solutions at the back of the book. Mindscapes marked* **(ExH)** *have expanded hints for solutions at the back of the book. Mindscapes marked* **(S)** *have solutions.*

Developing Ideas

1. **Daily deaths.** About 58 million people die every year. About how many die each day, on average?

2. **Wake-up call.** Suppose you wake up each morning and think of an event that has a one-in-a-thousand chance of happening that day. What are the chances the event you thought of in the morning will *not* happen that day? What are the chances you will not experience your event *du jour* two days in a row? What are the chances you will pass an entire year without experiencing one of these coincidences?

3. **More than 12 monkeys.** What does the Infinite Monkey Theorem assert?

4. **Get shorty.** Who was Georges Louis Leclerc Comte de Buffon? What was his role in the history of probability?

5. **Nothing but heads.** If you flip a fair coin three times, what are the chances you will get all heads? If you get all heads and you flip the coin once more, what is the probability that you will get heads on your fourth flip?

Solidifying Ideas

6. **Pick a number.** Pick a number from the following list: 1, 2, 3, 4. Write it down. If each student in a class of 50 were also to pick a number, how many would you guess had selected the same number as you did?

7. **Personal coincidences.** List three coincidences you have experienced in your life.

8. **No way.** It is the last Sunday of spring break and you are flying back to school. You have a connecting flight in Chicago. In the gate area, you see a bunch of friends from school. Are you shocked? Discuss this coincidence.

9. **Enquiring minds.** For a previous year, find the end-of-the-year issue of the *National Enquirer* or a similar fine publication and find all the predictions made by the psychics. How many of them actually happened? Based on your investigation, can psychics predict the future?

10. **Milestones.** Look online for a list of celebrity deaths for two consecutive weeks. How many famous people died? Use your answer to make your own estimate of the number of famous people who die in one year.

11. **Unlucky numbers.** Suppose you randomly picked 1000 people from the telephone book. What would you estimate the probability to be that one of them will die within the next year? Justify your estimate.

12. **A bad block (S).** Suppose 1054 people died in Datasville last year. Why must there be a two-week period during which 40 of those people died?

13. **Coin toss test.** Ask two friends to help you repeat the coin toss test. Ask one to toss a fair coin 100 times and record the sequence of H's and T's. Ask the other to create a "random" sequence of 100 H's and T's without tossing a coin or consulting the first friend. (Have them decide who does which task after you leave the room so that you don't know which friend does which task.) Return to the room and see if you can tell which sequence is which.

14. **Deceptive dice (ExH).** You asked three friends to do an experiment while you were out of the room. Two of them each tossed a fair die 40 times and recorded the results. The third friend generated a sequence of 40 digits from 1 to 6 with an effort to make it look like the random results of tossing an actual die. You return to see the results below. Which sequences do you think came from tossing an actual die? Why?

    a. 2 2 6 3 5 4 4 4 6 2 3 2 5 2 6 6 2 1 3 3 1 4 3 4 1 1 1 5 4 3 2 5 1 3 3 2 6 5 1 2

    b. 5 3 4 1 6 3 4 2 1 6 2 5 3 4 1 6 2 3 3 2 5 4 1 1 6 2 3 5 4 2 5 6 1 4 2 5 6 3 1 3

    c. 5 6 2 5 1 6 2 3 1 6 1 4 4 6 3 4 6 2 3 3 3 3 2 6 6 2 5 1 3 6 5 1 6 4 6 2 4 5 3 6

15. **Murphy's Law.** *If something can go wrong, it will.* Given our discussion of randomness, do you agree or disagree with this law? Using examples, try to place it in the context of our discussion of coincidence.

16. **A striking deal.** Get two decks of ordinary playing cards, give one deck to a friend, and have the person arrange it in any order, either random or planned. Bet the person that you will be able to take the other deck and place it in an order so that, if you both turn over the cards one at a time, at least once both cards will come up exactly the same. Shuffle your deck randomly. Now turn one card from each deck over and continue until there is a match. If there is a match, you win; otherwise, you lose. Play the game six or more times and record the results. Given your data, what would you guess is the probability of your winning?

17. **Drop the needle.** Try Buffon's needle experiment by dropping a needle 50 times and recording your results. Use your numbers to give an approximation for π. Go to the Web and try the computer simulation. Record the results and the associated approximations of π.

18. **IBM again (H).** Suppose we try the IBM stock-prediction scheme of sending half (or as close to half as possible) of the group the "up" message and the other half the "down" message, but this time we start with only 600 people. How many weeks could we go until we are down to just two people seeing a perfect track record?

19. **The dart index.** Take a page of stock quotes from a newspaper. Mount it on a piece of cardboard and throw 10 darts randomly at the page. Record the 10 stocks you hit. Now get a newspaper that is exactly six months old and

look up the prices of those 10 stocks. Suppose you bought 100 shares of each of those 10 stocks six months ago. How would you do if you sold them all today?

20. **Random walks.** Using a piece of graph paper, a penny, and a dime, embark on a random walk on the grid. Flip both coins. If the penny lands heads up, you move one unit to the right; if it lands tails up, you move one unit to the left. If the dime is heads, you move one unit up, and if it is tails, you move one unit down. Mark your trail on the graph paper as you make 50 flips of the coins.

21. **Random guesses (S).** A multiple-choice test has 100 questions; each question has four possible answers from which to choose. Each question is worth one point, and no points are taken off for choosing an incorrect answer. Someone decides to take the test by selecting answers randomly. What is the probability that a person taking this test gets 100%? Suppose now that the person actually reads the questions, is able to eliminate two of the incorrect choices, and then guesses randomly from the other two choices. What is the probability that the person gets 100%?

22. **Random dates.** There is a room filled with exotic people, any one of whom you would be happy to ask out on a date. Suppose that the probability that any particular person agrees to go on a date with you is 0.5. What is the probability that the first person you ask says yes? What is the probability that the first five people you ask say no?

23. **Random phones (H).** Suppose you roll a 10-sided die with sides marked from 0 to 9. If you continued to roll and recorded the outcomes, what do you think the probability would be that at some point you would see seven digits in a row that make up your telephone number? Explain your reasoning.

24. **Four-ever (ExH).** Suppose you roll a die repeatedly, forever. Is it possible that you would roll only 4's? Is it likely? What is the probability of rolling only 4's?

25. **Pick a number, revisited.** In Mindscape 6, did you pick 3? Are you impressed? What is the probability that we correctly guessed your answer?

Creating New Ideas

26. **Good start (H).** Suppose the monkey is typing using only the 26 letter keys. What is the probability that the monkey will type "cat" right off the bat?

27. **Even moves.** Suppose you embark on a random walk on the real number line. Show that, if you return back to your starting point, you must have flipped the coins an even number of times.

28. **Playing the numbers.** Here is a numbers game. You choose a number from 000 to 999 each morning and compare it to the last three digits of the official attendance figures at the nearest racetrack. What is the probability of guessing the correct number at least once if you make a guess each day for three years? (*Hint:* Consider a related example from this section.)

29. **Random results.** Someone looks at a list of 10,000 numbers, each from 1 to 100, generated by a random-number generator and states that the program must not work correctly because there is a string of 17 numbers starting with the 2713th number that reads 1, 2, 3, 4, 5, 6, 7, 8, 9, 10, 11, 12, 13, 14, 15, 16, 17. How do you respond to this conclusion?

30. **Monkey names.** Suppose we have a monkey typing on a word processor that has only 26 keys (only letters, no spaces, numbers, punctuation, and so on). The monkey types randomly for a long, long time. What is more likely to be seen: MICHAELSTARBIRD or EDWARDBURGER? Explain your answer.

31. **The streak.** Suppose you flip a fair coin 10 times on two different occasions. One time you see 10 heads, the other time you see H H T H H H T T H T. Is either one of these outcomes more likely than the other? Which one is random? Explain.

32. **Girl, Girl, . . . (S).** A couple has eight children. Suppose that the probability of having a girl is 0.5. What is the probability of producing eight girls? How does that answer compare with the probability of producing boy, girl, boy, girl, boy, girl, boy, girl? Explain. (*Note:* This does happen.)

What an amazing coincidence. . . . All the daughters have the same outfits!

©AP/Wide World Photos

33. **One mistake is okay.** Suppose we try the IBM stock-prediction scheme with only 128 people. Now, however, we are allowed one mistake. That is, if we send a batch of letters out saying that IBM stock will go down next week, and it actually goes up, we can keep sending letters to this group as long as we never make another mistake. After five weeks, how many people will have seen you make at most one wrong prediction?

34. **Picking and matching.** You and a friend individually and secretly pick a number from 1 2 3 4. What is the probability that you both picked 3? What is the probability that you both picked the same number?

35. **Picking and matching.** You and a friend individually and secretly pick a number from 1 2 3 4. What is the probability that you both picked 3? What is the probability that you both picked the same number? Why is this question here? Think about what this section is about.

36. **Dice are different (ExH).** In Mindscape 14 you probably used ideas similar to those in the analogous text example on coin tosses. Yet how are the two experiments different? What qualities might you look for (or look for the absence of) in a sequence of alleged random die tosses that would not apply in the case of coin tosses? In particular, can you tell which of the sequences below was generated by rolling an actual die and which was created by a person writing down what they think of as random digits (before reading this text)?

   a. 2 6 3 5 4 3 2 1 6 6 5 3 2 4 5 3 6 1 1 4
   b. 4 3 1 1 6 4 4 5 6 5 1 6 5 1 6 3 3 1 3 2

Further Challenges

37. Death row (H). You may have noticed that two pairs of the celebrity deaths we mentioned occurred one day after the other (Brian Keith on June 24 and Jacques Cousteau the next day; Robert Mitchum on July 1 and Jimmy Stewart the next day). Suppose that two celebrities die in one week (Monday through Sunday). What is the probability that they die on the same day? What is the probability that they die one day after the other in that same week?

38. Striking again. Consider the striking deal game described in Mindscape 16. Compute the actual probability that you will win. (*Hint:* It might be easier to first compute the probability that your opponent will win.)

39. Random returns. Suppose we take a random walk on an infinitely long street. Show that, with probability 1, we walk past every point on the street *arbitrarily* often. (*Hint:* Once you land on a particular point, imagine that you are starting a random walk from scratch.)

40. Random natural. Suppose you have a 10-sided die with sides labeled from 0 to 9. You roll it 50 times and record the digits to create a 50-digit natural number. What is the probability that the digit 9 occurs at least once in your random 50-digit number? What do you conclude about the digits of very large random natural numbers?

41. Ace of spades. You randomly shuffle a deck of cards and then look at the first card. If it is the ace of spades, you win; if not, you lose. What is the probability that you will win after 36 tries?

In Your Own Words

42. Personal perspectives. Write a short essay describing the most interesting or surprising discovery you made in exploring the material in this section. If any material seemed puzzling or even unbelievable, address that as well. Explain why you chose the topics you did. Finally, comment on the aesthetics of the mathematics and ideas in this section.

43. With a group of folks. In a small group, discuss the ideas of coincidence and the Infinite Monkey Theorem. After your discussion, write a brief narrative describing these ideas in your own words.

44. Creative writing. Write an imaginative story (it can be humorous, dramatic, whatever you like) that involves or evokes the ideas of this section.

45. Power beyond the mathematics. Provide several real-life issues—ideally, from your own experience—that some of the strategies of thought presented in this section would effectively approach and resolve.

For the Algebra Lover

Here we celebrate the power of algebra as a powerful way of finding unknown quantities by naming them, of expressing infinitely many relationships and connections clearly and succinctly, and of uncovering pattern and structure.

46. Flipping out. You flip a coin eight times and get heads x times and tails y times. Your roommate flips the same coin 41 times and gets 4 times as many heads and 7 times as many tails as you did. Determine exactly how many heads and tails you each flipped.

47. **Random reigns (H).** Your (very egalitarian) school chooses the Homecoming King and Queen by selecting two students at random (gender is irrelevant). If there are 4000 students at your school and 100 have read *The Heart of Mathematics,* what's the probability that both King and Queen have read this book?

48. **Survey says...** A survey in your intro psychology class included a question with four possible answers: A, B, C, or D. Here's how the responses came in: three times as many B's as A's, five more C's than A's, and 13 more D's than A's. If all 108 students in your class answered the question, determine how many chose answer D.

49. **Rock n' roll.** As a project in your *Geology and Art* class, you create a pair of dice out of granite. One day you roll one die 27 times and notice that 1 comes up $(x + 1)$ times, 2 comes up $(x + 2)$ times, 3 comes up $(x + 3)$ times, and the pattern continues. Determine how many times each face of the die appeared. Do you think your homemade die is fair? Explain your answer.

50. **Random pitch.** During a practice game, your baseball coach decides to choose the pitcher at random from all the players on the team, not just the pitching staff. He chooses your starting left fielder, nicknamed Wild Walter, who has a great arm for distance but not much control. Walter throws 80 pitches: x are strikes, $3x$ are balls (caught by the catcher), and $x - 5$ are wild pitches (too wild for the catcher to catch). During this outing on the mound, what's the probability Walter threw a strike?

8.4 DOWN FOR THE COUNT

Systematically Counting All Possible Outcomes

Did you win?

> *No priest or soothsayer that ever lived could hold his own against old probabilities.*
>
> OLIVER WENDELL HOLMES

Let's count. To determine probabilities, we often have to do some serious counting. Since probability is often a fraction of the number of favorable outcomes divided by the total number of all possible outcomes, we must count how many outcomes are involved, and that's not always so easy. Perhaps you are thinking that you learned how to count in kindergarten, so you'll just skip this section. It's true that, when we were children, we learned to count one at a time, and that is a simple task for small collections of objects. But when we count big, complicated collections, such as how many lottery outcomes are possible, we find ourselves perplexed and prone to error.

In principle, counting a collection of objects is easy. All we must do is

- count every object, and
- avoid counting the same object more than once.

These two rules sound so easy and obvious that we might think that counting is a piece of cake, but let's list a few things to count that might convince us that counting is not, in fact, a dessert item.

How many lock combinations are possible in a standard padlock? How many passwords are possible for your e-mail account? How many different poker hands are there? Experienced counters tend to group different counting scenarios into various categories, but, unless we are intending to become professional counters (known as *combinatorists*), sorting counting problems into a taxonomy of types can be a perilous enterprise. The perils center around the possibility of applying an incorrect counting method to a particular situation. Instead, let's explore some principles of analyzing questions on counting so that we can correctly analyze whatever case is at hand.

609

Certainly happy ... but are you surprised?

I've got it!

A Truly Merry Festival

In Florida, gambling is allowed on jai alai games. Jai alai (meaning *merry festival* in the Basque language) is the world's fastest ball game. The players use scoops (called *cestas*) to catch and throw the ball in one motion. During the day, six games are played in a type of round-robin play with eight players contesting each game. One can make a "super 6" bet for $2.00 by guessing who will win each of the six contests. Of course, the chance of anyone winning such a bet is rather small since one has to be correct on all six games. Thus, this type of wager is run like the lottery in that money that is bet is carried over into the jackpot day after day until someone finally wins.

On March 2, 1988, the pool of money at West Palm Beach Jai Alai reached an enormous amount, so one enterprising gambler decided not to gamble. He simply bet on every possible outcome, thus assuring himself of the prize money. Of course, there was the slight danger that some other person would also happen to win that day, in which case he would have had to share his bounty.

Our questions are as follows:

- If each bet costs $2.00, how big would the prize need to be to make it worthwhile to place every possible bet?
- How many bets would we have to make to be absolutely guaranteed of winning?

These questions present us with our first real counting issue. We have to pick the exact winner from each of the six contests. Naturally, we will have to guess all eight players as potential winners of the first game, because any one of them might win. So, for each of those we must choose all eight of the players of the second game. For each of those 64 patterns of potential winners in the first two games, we will have to choose all eight potential winners of the third game. So, there are $8 \times 8 \times 8$ potential sets of winners from the first three games. We see the pattern and conclude that there are 8^6 different possible sets of winners for the six games, for a total of 262,144 possible bets costing a total of $524,288.

If the prize exceeds that amount, then we are sure to come out ahead. In the case of the West Palm Beach game in 1988, the payoff was $988,326 (minus $197,664 that the Internal Revenue Service deducted before issuing the check). Unfortunately, if someone else also wins, the prize must be shared, but fortunately for the Jai Alai bettors, they lucked out. This possibility of having to share the prize is what prevents wealthy people from actually buying every combination in lottery games whenever the prize exceeds the cost of buying every combination. Despite this nagging possibility, this method of making money by covering all the possibilities has been used throughout history. In 1729 the French writer Voltaire used such a method to win the Parisian city lottery, and more recently, in 1992, a group of Australian investors essentially cornered the market on Virginia lottery tickets, winning about $27 million.

Lottery

Let's dream about hitting the jackpot in the lottery and spending the remainder of our lives in the lap of luxury. In a typical lottery, we pay a dollar and choose six different numbers, each from 1 to 50; the order of the six numbers does not matter. If we guess all six, we win the big jackpot. How high would the lottery prize have to get before it would be worth our while to buy a ticket for each possible outcome, thus assuring ourselves the title of "winner"? Or, equivalently, how many collections of six different numbers from 1 to 50 are there?

Choosing six numbers from 50 is far too large a task to think about yet, so let's first examine a simpler task. Remember: When the going gets tough, the smart stop going and instead do something easy.

Rather than choosing six numbers from 50, let's figure out how many ways there are of choosing two numbers from 5. We could first list all pairs of numbers in all possible orders by writing down the pairs whose first number is 1, then 2, then 3, and so on.

It's easy to see how many of these ordered pairs we have, since for each of the five choices of first number there are four choices for second number; therefore there are five times four ordered pairs altogether.

Notice that every unordered pair of numbers—for example, 2,4—occurs twice, once as 2,4 and once as 4,2. So, in our systematic list of ordered pairs, each pair of numbers appears twice instead of once. To get the actual number of unordered pairs, we need to take our 20 ordered pairs and divide by 2 (the number of times each pair that uses the same two numbers appears) to give the number of unordered pairs, namely, 10.

Let's consider one more example before returning to our lottery question. Our goal here is to examine enough "easy" scenarios so that the hard lottery question at hand becomes "easy." Let's count how many unordered selections of three numbers there are taken from the collection of numbers 1, 2, 3, 4, 5. As before,

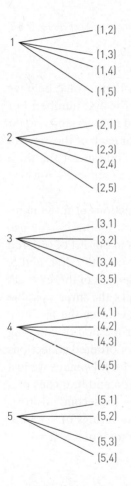

our strategy is first to count how many ways we can select three numbers in which different orderings count as different selections and then see how many times each particular group of three numbers appears on that list.

We know how to list all possible first two numbers since we already did that. But now, for each such pair of numbers, we have three choices for the third number. For example, the ordered pair 4,2 can be extended to three different ordered sets of three numbers, namely, 4,2,1 and 4,2,3 and 4,2,5. Likewise, the ordered pair 2,5 can be extended to 2,5,1 and 2,5,3 and 2,5,4. Although we have not written down every number on the chart,

Just three examples of orderings with a given starting pair.

we can compute how many such ordered triples of numbers there would be if we were to write out the entire chart. There are five choices for first number. For each of those first numbers, there are four choices for second number; thus, so far we have 5×4 choices for the first two ordered numbers. For each of those 5×4 choices of first two numbers, we have three remaining possibilities for the third number. Thus, we have $5 \times 4 \times 3 = 60$ ways of choosing three numbers in which we are viewing different orderings as different choices.

However, our real question was to count how many collections of three numbers there are in a set of five numbers if the order does not matter. So we ask ourselves how often a particular set of three numbers occurs in our list of ordered numbers. So let's just choose three numbers and see. Let's consider 2, 4, and 5. Where are all the places they occur in our list of 60 ordered groups of three? Well, any one of those three numbers could be first; and for each of the three possible first numbers, there are two choices for the second number. Once the first two numbers are chosen, the last number is determined. So our method of counting how many times 2,4,5 occurs (in any order) on our long list of ordered collections of three numbers is really the same method we employed before. Namely, we just look at a tree-like figure that has 2, 4, or 5 as the first position and branches out. Each of the three possibilities for first position has two choices for second position, and then the third position is determined. So the possible orderings of 2,4,5 are 2,4,5 and 2,5,4; then 4,2,5 and 4,5,2; then 5,2,4 and 5,4,2 for a total of $3 \times 2 = 6$.

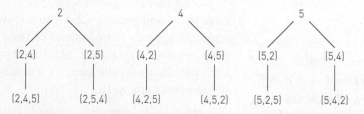

So now we can answer the question of computing how many unordered collections of three numbers can be chosen from the numbers 1 to 5. The answer is to take the number of *ordered* sets of three numbers (which is $5 \times 4 \times 3$) and divide by the number of times the same set of three numbers occurs on that list (which is $3 \times 2 \times 1$). So the number of *unordered* ways to select three numbers from the numbers 1 to 5 is:

$$\frac{5 \times 4 \times 3}{3 \times 2 \times 1} = 10.$$

We can now face the lottery question. We will apply a similar type of analysis to figure out how many different ways we can select six different numbers, each from 1 to 50: We first count all *ordered* collections of six numbers and then divide by how many times each unordered collection was counted.

The number of different ways of selecting six different numbers from 1 to 50 is

$$\frac{50 \times 49 \times 48 \times 47 \times 46 \times 45}{6 \times 5 \times 4 \times 3 \times 2 \times 1},$$

which equals 15,890,700.

Why? We can choose any of the 50 numbers as the first number. For each first choice, we can choose any of the remaining 49 numbers second. So, we have 50×49 ways to choose the first two numbers. Continuing, we have $50 \times 49 \times 48 \times 47 \times 46 \times 45$ ways of choosing six numbers in specific orders.

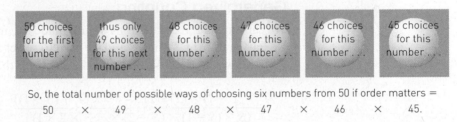

So, the total number of possible ways of choosing six numbers from 50 if order matters =
50 × 49 × 48 × 47 × 46 × 45.

Since we counted the number of ways of selecting six numbers where order matters, we counted

2, 5, 4, 16, 27, 45;

5, 2, 4, 16, 27, 45;

2, 4, 5, 45, 27, 16; and so on,

separately since their orderings are different even though each ordering involves the same collection of six numbers; therefore, we did some major overcounting. In how many ways can we order these six numbers?

There are $6 \times 5 \times 4 \times 3 \times 2 \times 1$ ways to order those six numbers: Any of the six could be the first number, so there are six different possibilities for the first number. Once the first number is determined, any of the remaining five could be the second. Thus, for each choice of first number, there are five different possible choices for the second number. Similarly, there are four different choices for the third number, and so forth. Thus we have $6 \times 5 \times 4 \times 3 \times 2 \times 1$ many different orderings of each group of six numbers. So, since each group of six numbers is counted exactly $6 \times 5 \times 4 \times 3 \times 2 \times 1$ times in our count of $50 \times 49 \times 48 \times 47 \times 46 \times 45$, we simply divide

$50 \times 49 \times 48 \times 47 \times 46 \times 45$ by $6 \times 5 \times 4 \times 3 \times 2 \times 1$

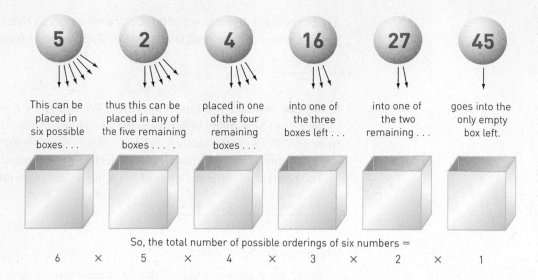

5	2	4	16	27	45
This can be placed in six possible boxes . . .	thus this can be placed in any of the five remaining boxes	placed in one of the four remaining boxes . . .	into one of the three boxes left . . .	into one of the two remaining . . .	goes into the only empty box left.

So, the total number of possible orderings of six numbers =

6　×　5　×　4　×　3　×　2　×　1

to arrive at the number of different six-number groupings selected from 1 to 50.

Generalized Counting

We can extend the calculations we just made to a more general situation. Suppose we have a total of T objects and we wish to select S of them. How many different ways can we do that? We see that the answer is that there are

$$\frac{T \times (T-1) \times (T-2) \times \cdots \times (T-(S-1))}{(S \times (S-1) \times (S-2) \times \cdots \times 2 \times 1)}$$

different ways of selecting S objects from T things where the order doesn't matter.

Calculators and computers have no difficulty counting these unwieldy collections. If from a set of 50 numbers we want to know how many different ways there are of choosing 6 where order does not matter, in real life we simply find the *combination* key on our calculator, enter the numbers 50 and 6, and we have the answer. The trick is to understand thoroughly when we are seeking the number of unordered sets of 6 from a set of 50 and when we really want something else. In this case, the calculator would report 15,890,700; and so if the Lotto jackpot exceeds $15,890,700, we will make money if we buy every single possible combination of six numbers (for $1 each)—assuming that we don't share the prize.

Dealing with Cards

Some of the most challenging counting questions occur in dealing with playing cards. Let's use a standard 52-card deck of playing cards and consider, for example, five-card poker hands. First of all, how many different five-card hands could we be dealt? The answer is the number of ways of picking five things from that group of 52 things. We just figured out how to compute such a count (notice that order does not matter). The answer is (52 × 51 × 50 × 49 × 48)/(5 × 4 × 3 × 2 × 1), which is 2,598,960. Among all those hands, we may now ask how many are of the various poker types, such as four of a kind, flush, straight, full house, and so forth. Many of these hands are not so easy to count—especially if we don't know how to play poker.

To count how many four-of-a-kind hands there are, let's first ask: How many hands have four aces? This question is not too difficult, because after we have been told that there are four aces, there can be only one additional (non-ace) card in the five-card hand. That card could be any one of the remaining 48 cards (52 minus the four aces). So, there are 48 poker hands that have four aces. There are, of course, the same number of poker hands that contain four kings, or four queens, and so on. So, the total number of hands that contain four of a kind is 48×13. If we are dealt a five-card hand, the probability of getting four of a kind is $(48 \times 13)/2,598,960$, or 0.00024—not likely.

Such a hand is rare indeed. Let's now see how it compares to a straight flush. A straight flush is a run of five consecutive cards all of the same suit. For example, 5, 6, 7, 8, 9 all of diamonds and 8, 9, 10, J, Q all of spades are both straight flushes. How many possible straight flushes are there? The answer requires some careful counting. Let's start with one suit, say spades. How many straight flushes are there in spades? Well, we could have the A, 2, 3, 4, 5, or the 2, 3, 4, 5, 6, and so forth until 10, J, Q, K, A. (The ace can be used either as the highest or lowest card.) That is a total of ten straight flushes in spades—the same for hearts, diamonds, and clubs. So the total number of straight flushes possible is 10×4. Therefore, a straight flush is rarer than four of a kind. The probability of being dealt a straight flush is $(10 \times 4)/2,598,960$, or 0.000014—which makes getting four of a kind look pretty easy. Counting the number of hands of each type is the method used to determine which hands beat which other hands. So, the counting we have done here shows why a straight flush beats four of a kind.

"Or" and/or "And"

Suppose we have a die and a coin, and we are going to throw them both. Consider these three questions:

- What is the probability we will roll a 6 on the die **and** flip a heads on the coin?
- What is the probability we will roll a 6 on the die **or** flip a heads on the coin?
- Which of these two outcomes is more likely?

The first two questions are similar and bring up a significant feature of counting. As always, when we want to compute a probability, we must know how many total outcomes are possible, and then we need to know how many of those are the outcomes we desire. In this case, for every one of the six equally likely outcomes from rolling the die, there are two possibilities for the coin. Therefore, there are a total of 12 outcomes altogether.

	1	2	3	4	5	6
H	H, 1	H, 2	H, 3	H, 4	H, 5	H, 6
T	T, 1	T, 2	T, 3	T, 4	T, 5	T, 6

Now let's count how many outcomes have a 6 on the die **and** a heads on the coin. Well, only one. So, the probability of rolling a 6 on the die and flipping a heads on the coin is 1/12. Notice that the probability of rolling a 6, 1/6, multiplied by the probability of flipping a heads, 1/2, yields the answer of 1/12. If we wish to compute the probability that one event happens **and** simultaneously another unrelated event also happens, we need only multiply the individual probabilities together. Why multiply? Because the number of outcomes of the two events is obtained by multiplying the number of outcomes for the first by the number of outcomes for the second.

	1	2	3	4	5	6
H	H, 1	H, 2	H, 3	H, 4	H, 5	**H, 6**
T	T, 1	T, 2	T, 3	T, 4	T, 5	T, 6

Heads **and** 6: 1 out of 12

Probability of Two Events Both Occurring.

*The probability of one thing happening **and** some unrelated thing happening is equal to the **product** of the individual probabilities.*

Now let's count how many of the outcomes have a 6 on the die **or** a heads on the coin. Let's be clear that when we say "6 on the die **or** a heads on the coin" we include the possibility that both a 6 appears and a heads appears. In mathematics, the word "or" always means that either of the options or both of the options happens. Well, we could have a 6 and either a heads or a tails on the coin. So, that's two ways. We could also have a heads on the coin and any number 1, 2, 3, 4, 5, or 6 on the die. That sounds like 6 more, but we must avoid the double-counting of a 6 and a heads. The total number of outcomes that have a 6 **or** a heads is 7. So, the probability of getting a 6 **or** a heads is 7/12.

	1	2	3	4	5	6
H	**H, 1**	**H, 2**	**H, 3**	**H, 4**	**H, 5**	**H, 6**
T	T, 1	T, 2	T, 3	T, 4	T, 5	**T, 6**

When computing the probability of getting something **or** something else, it is often more convenient to compute the probability of the opposite scenario. In the preceding case, for example, we were asked to consider the possibility of getting a 6 or a heads. The alternative is that we do not roll a 6 **and** we do not flip a heads. This question is easier, because we just saw how to find the probability that two such events happen simultaneously: We multiply the individual probabilities. What is the probability of not rolling a 6 on the die? There are five out of the six possibilities that lead to success, so the probability is 5/6. What is the probability of not flipping a heads? We know that it is 1/2. So, the probability of avoiding both a 6 **and** a heads is the product $(5/6) \times (1/2)$, which equals 5/12. We can now deduce that the probability of getting a 6 **or** a heads is $1 - 5/12$, which equals 7/12, as we saw before. So, the opposite of an event where *something happens or something else happens* is that *the first does not happen **and** the second does not happen.* With careful counting and multiplying probabilities, we can now find probabilities of several events happening together. Let's see these ideas in action.

Dicey Issues

Recall that the seminal counting questions posed by our French nobleman concerned the frequency with which a 6 would appear in four rolls of a die. Let's count how many different ways a die can be thrown four times and then count how many of those outcomes contain a 6. These two numbers allow us to compute the probability of throwing a 6 in four rolls of a die. Let's count.

1. Any of the six numbers could come up first.
2. For each of the six possible first numbers, any of the six numbers could come up second.
3. For each of the 36 possible first and second rolls, any of the six numbers could come up third.
4. For each of the $36 \times 6 = 216$ possible first, second, and third rolls, any of the six numbers could come up fourth.

So, altogether there are

$$6^4 = 216 \times 6 = 1296$$

equally likely possible outcomes of rolling a die four times.

How many of those contain a 6? Well, the 6 could be in the first, **or** the second, **or** the third, **or** the fourth place. But beware: We must not double-count the outcomes of rolling two 6's in a row. We might also roll three 6's, and so on, so we have to be careful not to double, triple, and whatever count those outcomes involving 6's. Looking at the counting question in this way is sufficiently complicated and perplexing that it is best to try to think of an alternative approach.

Life Lesson
If the going gets tough, do something else.

In this case, a little thought saves a great deal of toil. Instead of thinking about what's there, let's think about what's not. Often it is best to count what's missing rather than what's present. In this case, we want to know how many of the 1296 outcomes include a 6. Why don't we think about the alternative? Namely, how many outcomes do not contain *any* 6's? That is, let's count how many outcomes involve only the numbers 1 through 5. Well, that is pretty easy once we realize that we just answered that same kind of question. There are five possibilities for the first roll, five for the next, five for the third, and five for the fourth. So, there are $5 \times 5 \times 5 \times 5$ outcomes altogether that involve only the numbers 1 to 5. That is a grand total of 625.

Therefore, of the 1296 possible outcomes of rolling a die four times, 625 do not involve a 6, and so, the rest ($1296 - 625 = 671$) must involve a 6 somewhere. Therefore, the probability of rolling at least one 6 among the four rolls is 671/1296, which equals 0.5177. . . . Since this probability is greater than one-half, a person betting on rolling a 6 in four rolls will tend to win slightly more often than lose.

Dicier Issues

When the French gambler tried his game of rolling a pair of dice 24 times and asked whether there would be a pair of 6's, his winning streak evaporated. He probably reasoned that in 24 rolls of the pair, the first die on average would be a 6 four times. Therefore, the other die would have those four chances to be a 6 as well. This unfortunately incorrect line of reasoning would lead him to think that his winning percentage for getting a pair of 6's in 24 rolls of a pair of dice should be the same as for getting one 6 in 4 rolls. Let's see why his winning streak came to a screeching halt.

Our whole goal is to carefully compute the number of possible outcomes of rolling two dice 24 times and seeing how many equally likely outcomes there are. With any one roll there are 36 possible outcomes. So, let's get rid of the red herring of two dice and instead concentrate on the fact that one of 36 outcomes is possible in each of the 24 repetitions. By the same reasoning as before, there are then a total of 36 to the 24th power outcomes, or about 2.245×10^{37}—which is one heck of a big number. We now need to know how often a particular outcome (a pair of 6's) from the 36 possible outcomes will occur at least once among the 24 trials. Since counting such a thing is difficult, we will instead count the number of outcomes that avoid a pair of 6's. That is to say, in how many ways can each of the 24 rolls result in one of the 35 outcomes other than a pair of 6's? Again, that number is just 35 to the 24th power, which is about 1.142×10^{37}. So, the probability of **not** getting a pair of 6's in 24 rolls is approximately

$$1.142 \times 10^{37}$$

divided by

$$2.245 \times 10^{37}.$$

That fraction equals

$$\frac{1142}{2245},$$

which is

$$0.508 \ldots,$$

and so the probability of actually rolling a pair of 6's is approximately

$$1 - 0.508 \ldots = 0.491 \ldots.$$

Thus, we are slightly more apt **not** to roll a pair of 6's during 24 rolls than we are to roll a pair of 6's, and our French *bon homme*, sadly, is a loser.

From Dice to DNA

With cap pulled down and gloves pulled up, the murderer steals into the enclosed garden and confronts his victims. There is a blood-curdling scream as the murderer slashes the throat of his first victim. In a desperate and vain attempt to save his life,

the second victim wrestles with the murderer, slightly injuring his assailant, but receiving a fatal wound himself. A few drops of the murderer's blood fall from his wound onto the carnage as he flees the gruesome scene.

Soon the police arrive and take blood samples from the scene. Most of the blood belongs to the victims, but a small portion belongs to the murderer. Those drops of blood might prove the culprit's guilt beyond any reasonable doubt, or they might not. Let's investigate.

A strand of DNA, but whose is it?

DNA is the genetic material contained in each of our cells. All the cells of an individual have identical DNA; however, the DNA from one person differs from the DNA of another. DNA is composed of genes, each of which can be present in one of several forms, known as *alleles*. Human DNA has tens of thousands of genes, and each gene has several possible forms in which it might appear. By looking at the distinct alleles of the various genes, we will see how DNA can incriminate or exonerate a defendant.

Suppose we examine 20 genes from a DNA sample, and each of these genes can occur as two different alleles. Let's further suppose that those two alleles are roughly equally present among the human population and that the choices of allele of these 20 genes are independent of one another. In other words, having one form of one gene does not make it more or less likely to have any particular form of any of the other genes. Let's try now to compute the probability that the 20 genes from a random DNA sample would have the same alleles as those in the murderer's DNA.

Each of the 20 genes can be present as one of two alleles. So, the first gene could be in the first or second form. For each of those possibilities, the second gene could be in its first or second form. So, there are four possible patterns for the first two genes. The third gene could also appear as either of two alleles. Thus, we observe that for 20 genes, there will be 2^{20}—which equals 1,048,576—possible patterns for those 20 genes. So, the chance of another person having all 20 of those genes is 1 in 1,048,576.

DNA evidence is potent. How could a lawyer refute such incriminating evidence? The only possible way to refute such evidence is to recognize that the test is only as valid as the accuracy with which it is performed. If samples are switched, if the defendant's blood contaminates the blood from the murder scene, or if the samples are mislabeled, then the reported match may be completely unreliable. Thus, it could be argued that police and lab error, in a sense, wipe out the accuracy of DNA tests. In other words, the more convincing probabilistic analysis would take into account the probability that the laboratory or the police had made errors.

A Look BACK

COUNTING CAN BE HARD. CAREFUL COUNTING can lead to a better understanding of probabilities and perhaps even to breaking the bank at Monte Carlo. In fact, if a casino catches an individual *successfully* counting cards at blackjack, that individual is banned from the casino. However, casinos generally encourage counting, because most people who try counting foul up and lose more money than ever. Counting helps us determine our chances for winning the lottery, being dealt a royal flush in poker, and having the same genetic profile as a murderer.

In all cases, careful counting consists of two challenges: counting every item and not counting anything twice. To deal with difficult counting challenges, we start with simpler versions of the same thing. Instead of choosing six numbers each from 1 to 50 for the state lottery, we start by assuming we are in a very small state where we choose two or three numbers from 1 to 5. From such small examples, we gain experience and see patterns that allow us to deal with the more difficult cases. Sometimes a useful strategy is to count the opposite of what we want. That is, instead of counting how many rolls of a pair of dice result in a 6 or a 3, we can count the number of rolls that result only in other numbers. Then we subtract from the total. Systematic organization is helpful for counting.

Complicated questions are often best approached by thinking hard about simple examples of similar situations. From the simpler situations we can more easily deduce patterns and methods that let us deal with the more complicated settings. Systematically organizing our thinking empowers us to deal with big issues using the same ideas that work for small ones.

Life Lessons

Look at simple cases.

•

Find patterns and methods.

•

Apply the methods systematically to more complex cases.

MINDSCAPES   Invitations to Further Thought

In this section, Mindscapes marked (H) have hints for solutions at the back of the book. Mindscapes marked (ExH) have expanded hints for solutions at the back of the book. Mindscapes marked (S) have solutions.

Developing Ideas

1. **Have a heart.** If you draw a card from a regular deck of 52 cards, what are the chances that your card would be red? A heart? A queen? The queen of hearts?

2. **Have a head.** You flip a coin 10 times. What is the probability that you see at least one head?

3. **Sunny surprise.** Suppose the chances are 1 in 2 that tomorrow will be sunny. Suppose the chances are 1 in 10 that your math teacher will bring donuts to class tomorrow. What are the chances that it will be sunny **and** you get donuts in math class tomorrow?

4. **Elephant ears.** Suppose a quarter of all the elephants in an African wildlife refuge have ear tags. If there are exactly 60 tagged elephants, what is the total number of elephants in the refuge? If you pick an elephant at random, what is the probability it has a tag?

5. **Little deal.** In how many ways can you select three cards from a regular deck of 52 cards? (Note: The order of the cards doesn't matter.)

Solidifying Ideas

6. **The gym lock.** A lock has a disk with 36 numbers written around its edge. The combination to the lock is made up of three numbers (such as 12-33-07 or 19-19-08). How many different possible combinations are there for this lock? What is the probability of randomly guessing the correct one?

7. **The dorm door.** A dormitory has an electronic lock. To unlock the door, students must enter their unique five-digit secret code into the keypad (made up of the digits from 0 to 9, each of which can be used more than once). How many different secret codes are there? Suppose there are 200 students living in the dorm. What is the probability of one of them randomly guessing a code and having the door unlock?

8. **28 cents.** How many different ways can you make 28 cents using current U.S. currency?

9. **82 cents.** How many different ways can you make 82 cents using current U.S. currency?

10. **Number please.** Someone you really wanted to go out on a date with gave you a beeper number. You didn't write it down, so you remember only the area code and the exchange (the first three digits of the number). How many different numbers are there with that area code and exchange? What is the probability that you randomly pick the right number?

11. **Dealing with jack.** Suppose you deal three cards from a regular deck of 52 cards. What is the probability that they will all be jacks?

12. **MA Lotto (H).** To win the jackpot of the Massachusetts lottery game in bygone days, you had to correctly pick the six numbers selected from the numbers 1 through 36. What was the probability of winning the Massachusetts lottery?

13. **NY Lotto (H).** To win the jackpot of the New York lottery game in bygone days, you had to correctly pick the six numbers selected from the numbers 1 through 40. What was the probability of winning the New York lottery?

14. **OR Lotto.** To win the jackpot of the Oregon lottery game in bygone days, you had to correctly pick the six numbers selected from the numbers 1 through 42. What was the probability of winning the Oregon lottery?

15. **Burger King (S).** You take a summer job making hamburgers. The burgers can be made with any of the following: cheese, lettuce, tomato, pickles, onions, mayo, catsup, and mustard. How many different kinds of burgers can you make?

16. **More burgers.** Suppose you are working at a burger place where the burgers can be made with any of the following: cheese, lettuce, tomato, pickles, onions, mayo, catsup, and mustard. You have a picky clientele: They all want the works, but they wish to specify the order of the placement of the items! For example, one person may want burger, cheese, mayo, onion, catsup, pickle, mustard, tomato, lettuce. Someone else may want lettuce, tomato, burger, onion, cheese, mustard, mayo, catsup, pickle. How many different types of burgers-with-the-works are there?

17. **NetFlix.** You have 65 movies on your NetFlix queue. In how many ways can you order them? (*Note:* Don't multiply it out.)

18. **Cineplex (ExH).** Your local Cineplex is showing eight movies. You want to see a different movie on each of Thursday, Friday, and Saturday nights. How many different three-night movie magic experiences can you choose from if the order in which you see the three movies does not matter to you? How many choices if the order does matter?

19. **One die.** You roll a fair die four times. What is the probability that you see at least one 1?

20. **Dressing for success.** You have 5 T-shirts, 10 shorts, and 3 pairs of underwear, and you wear 1 of each. How many different ensembles can you put together? *Bonus:* Assuming all these clothes are clean, how long could you go before doing the laundry?

21. **Band stand.** The Drew Aderburg Band is planning a concert tour of six cities. In how many different orders could the band cover the six cities?

22. **Monday's undies.** You are spending the weekend at a friend's parents' house. You need 3 pairs of underwear and have 10 clean pairs of underwear of different colors. How many different underwear triples can you pull out to impress your friend's folks?

23. **Counting classes.** Your institution will offer 200 courses next semester, each at a different time. You will take four. How many different groups of four courses can you select from?

24. **Cranking tunes.** Your car stereo can be programmed to hold six radio stations of your choice. There are nine stations you really like. How many different ways can you program your six buttons with different collections of your favorite stations? (A different ordering of the same six stations counts as different programming.)

25. **The Great Books.** There are 20 Great Books, from which you know your English prof for next semester will select 10. You are an overachiever and want to figure out how many different combinations of 10 books the prof can pick. How many are there?

Creating New Ideas

26. **Morning variety (S).** You wish to have a different breakfast every morning. Each day you choose exactly three of the following items: eggs, bagel, pancakes, coffee, orange juice. How many days can you go before you must eat a breakfast combination that you have already had?

27. **Blind man's bluff.** You and a friend each pick a card from a regular deck of 52 cards but do not look at them. Both of you then hold your cards up by your foreheads so that you can each see the other's card but you can't

see your own. Your friend is holding a 6. What is the probability that your card is higher?

28. **Crime story (ExH).** Suppose 20 witnesses saw someone commit a crime, and each supplied a piece of information. One witness said the perpetrator was wearing a certain type of shoe. Another said the perpetrator was taller than six feet. A third said the perpetrator had dark hair, and so on. Each piece of information distinguished the perpetrator only from half the people on Earth. The different pieces were independent in the sense that any combination of them was possible. If there are roughly 6.6 billion people on Earth, approximately how many would fit all 20 of the pieces of information?

29. **There's a 4 (H).** Someone you really want to go out with gave you a beeper number. You didn't write it down and remember only the area code, the exchange (the first three digits of the number), and the fact that the last four digits contain at least one 4. How many different such numbers are there? What is the probability of your randomly picking the right number?

30. **Making up the test.** Your math prof says there will be a 15-question test on probability. She also reports that the test will be made up of problems from the Mindscapes section of the chapter as follows: two questions each from Sections 8.2, 8.3, and 8.5, and three questions each from Sections 8.1 and 8.4 (no questions from the In Your Own Words category). How many different possible exams could she make up? What is the probability that this very question is on your exam?

31. **Moving up.** You have a part-time job in a department with 20 other people. Word comes out that 6 of the people from that department will be given a raise. If the 6 people are chosen at random, what is the probability that you will be one of the 6?

32. **Counterfeit bills.** You are given ten $100 bills and told that three of them are counterfeit. You randomly pick three of the bills and burn them. What is the probability that you burned the counterfeit bills?

33. **Car care.** A burglar wishes to break into a car that has a security system. There are five buttons (marked 1 through 5) to disarm the system, and he knows the code is a three-digit number like 253 or 422. How many possible security codes are there? Knowing this number of combinations, the burglar fears that he will enter the wrong number first—thus tripping the alarm (which everyone just ignores). So he writes down his first 20 guesses and then enters his next guess. Does this strategy improve his chances of success? Explain.

34. **Coins count.** On your bureau you have a half dollar, quarter, dime, nickel, and penny. How many different totals can be formed using exactly three coins? How about using four coins? How about using five coins?

35. **Math mania.** There are 90 students enrolled in Math 180. There are three different sections of 30 students, all meeting at the same time. How many different ways can the 90 students be placed into the three sections?

Further Challenges

36. **Party on.** You want to throw a party and can invite only 15 people. You want to invite 3 people from the soccer team (there are 10 people on the tiny team); 4 people from the orchestra (there are 20 people in the orchestra); and 8 people from your math class (there are 30 other students in your class).

Assuming no overlap of the groups, how many different invitation lists could you devise?

37. **No dice (H).** You roll a pair of dice 24 times. What is the probability of seeing at least one total that is 11?

38. **Three angles.** Draw 10 points on a piece of paper with no three points lying on the same straight line. How many different triangles can you make using these points as vertices?

39. **Four parties (ExH).** You want to have a party and you know 10 men and 10 women. Unfortunately, your common room can hold only 10 people. You wish to have enough parties so that each man and woman among your acquaintances would, at some point, meet at a party. How many parties are necessary to accomplish this task? Show that you need no more than four parties.

40. **Making the cut.** In 1988, the ignition keys for Ford Escorts were made out of a blank key with five cuts, each cut made of one of five different depths. How many different key types were there? In 1988, Ford sold roughly 380,000 Escorts. What is the probability that one Escort key will unlock a random Escort? (This story was reported in the April 1989 issue of the *Atlantic Monthly*.)

In Your Own Words

41. **Personal perspectives.** Write a short essay describing the most interesting or surprising discovery you made in exploring the material in this section. If any material seemed puzzling or even unbelievable, address that as well. Explain why you chose the topics you did. Finally, comment on the aesthetics of the mathematics and ideas in this section.

42. **With a group of folks.** In a small group, discuss and work through the details involved in counting the five-card poker hands and the French dice games. After your discussion, write a brief narrative describing the methods in your own words.

43. **Creative writing.** Write an imaginative story (it can be humorous, dramatic, whatever you like) that involves or evokes the ideas of this section.

44. **Power beyond the mathematics.** Provide several real-life issues—ideally, from your own experience—that some of the strategies of thought presented in this section would effectively approach and resolve.

For the Algebra Lover

Here we celebrate the power of algebra as a powerful way of finding unknown quantities by naming them, of expressing infinitely many relationships and connections clearly and succinctly, and of uncovering pattern and structure.

45. **Parental pride.** The number of pictures you have of your loving parents equals T. You want to choose two to put on prominent display in your dorm room. You consider all your choices and discover there are 21 ways to select two pictures (the order of the pictures doesn't matter). How many pictures do you have?

46. **Sister, Sister.** The number of students in a certain sorority equals T. Two of them will be selected as representatives at the next national meeting. There

are 45 ways to select two women (order doesn't matter). How many women are in the sorority?

47. In search of a cubic. This section explains that the number of ways to select S objects from a total of T objects equals

$$\frac{T \times (T-1) \times (T-2) \times \cdots \times (T-(S-1))}{(S \times (S-1) \times (S-2) \times \cdots \times 2 \times 1)}.$$

Now let $S = 3$. Expand and simplify this formula as much as possible so that there are no parentheses.

48. Cancellation! (H) This section explains that the number of ways to select S objects from a total of T objects equals

$$\frac{T \times (T-1) \times (T-2) \times \cdots \times (T-(S-1))}{(S \times (S-1) \times (S-2) \times \cdots \times 2 \times 1)}.$$

The denominator of this formula is the product of a natural number S and all the natural numbers less than S. Mathematicians denote this product $S!$, read "S factorial." So, for example, $3! = 3 \times 2 \times 1 = 6$, $5! = 5 \times 4 \times 3 \times 2 \times 1 = 120$, and $1! = 1$. Verify that $\dfrac{T \times (T-1) \times (T-2) \times \cdots \times (T-(S-1))}{(S \times (S-1) \times (S-2) \times \cdots \times 2 \times 1)} = \dfrac{T!}{(T-S)!}$.

49. Keep on cancellin'! Building on the previous Mindscape, notice that the numerator of the given formula, $T \times (T-1) \times (T-2) \times \ldots \times (T-(S-1))$ counts the number of ways to *order* or *arrange* S objects taken from a total of T objects. That is, there are T choices for the first object, $T-1$ choices for the second object, $T-2$ choices for the third object, and so on, until S objects have been chosen. (Note that there are S factors in the formula, one for each choice.) Verify that $T \times (T-1) \times (T-2) \times \ldots \times (T-(S-1)) = \dfrac{T!}{(T-S)!}$

DRIZZLING, DEFENDING, AND DOCTORING

Probability in Our World and Our Lives

"I need those probabilities *STAT.*"

> *I'm not a fan of facts. You see, the facts can change, but my opinion will never change, no matter what the facts are.*
>
> STEPHEN COLBERT

Probability has important applications in many real-life arenas from rolling dice to diagnosing diseases. Gambling involves random processes, so it is not surprising that probability is a key ingredient in Vegas, but probability also can play a central role in making decisions or making predictions in situations where randomness may not be involved at all. For example, we'll see that when a doctor makes a tricky diagnosis, the best method for guessing the right illness may involve probabilistic reasoning. Also, randomness and probability can help us train our dogs—if we want to teach our dogs to respond to a command, the best method may be to reward them at *random* intervals rather than with any regular pattern. In this way, the dog develops the hope (rather than expectation) that the next reward is just one more good deed away. Should random rewards be involved in child rearing, education, or governmental policies?

Randomness and probability definitely arise in game theory—the study of strategic decision-making. In game theory, often the optimal strategy is one that involves intentionally introducing randomness. Optimal business strategies or sports strategies frequently are probabilistic in nature. Should we run or pass on the football field; should we buy or sell in the trading pit? The best strategic decision-making process may be to flip a coin. But when optimal strategies involve probability, how can we judge whether decisions were made wisely? Even the very best possible strategy might result in some serious losses every once in a while.

Probability is a basic component of clear thinking as we cope with the randomness and uncertainty of our daily lives.

A Chance of Rain . . .

Predicting the future is always fraught with uncertainty. Those zany TV weather people do it every day when they declare a probability of rain tomorrow. We often need to know whether to take our umbrellas and rubbers when we leave home in the morning, so weather prediction has a noticeable impact on our daily lives. We wake up bleary-eyed from an all-too-short slumber, turn on the local news, and hear the prediction: *There is a 30% chance of rain today in the greater metropolitan area*. Now what are we going to do? What does a "30% chance of rain in the greater metropolitan area" even mean? Surprisingly, most people do not know its correct meaning. We'll now cast clear, sunny light on this cloudy issue of the probability of precipitation.

First, we need to know the "official" threshold of how much rain must fall before we can officially declare that "it rained." The answer is 0.01 inch. If we put out a rain gauge and more than 0.01 inch falls into it, then we report that it rained at that spot.

If a forecast of rain were to be given for one single spot, then the meaning of the "30% chance of rain today" prediction is pretty clear. It means that if 100 days had weather conditions nearly identical to those of today, we would expect to see rainfall of at least 0.01 inch at that spot on approximately 30 of those days. Historical weather records would help the weather forecasters make that "spot-on" prediction.

The issue is complicated when we hear that there is a 30% chance of rain in a large region, such as a greater metropolitan area. There are countless different points in the region and some points in the region may get more rain than other points, so we must face and understand these variations.

Let's consider a large region and assume that for every point in half this region there is a 40% probability of rainfall and that for every point in the other half there is a 20% chance of rain. Under these particular conditions, we would declare that the probability of rain for the entire region is the average of 20% and 40%, that is, the chance of rain for the entire region would equal to 30%. The 30% represents an average probability of precipitation weighted by the area. In this case we have [(½ the area) × (40%) + (½ the area) × (20%)] = 30% probability of precipitation for the whole area.

In an extreme case, we might have 30 square miles out of a 100-square-mile region in which the probability of rainfall is 100%, and in the remaining 70 square miles, the probability of rain is 0%. Again the probability of rain for the entire region is equal to the weighted average: (30/100 of the area) × (100%) + (70/100 of the area) × (0%) = 30% probability of precipitation for the whole area. Notice that there is no point in the region in which the probability of rain is actually equal to 30%. We remark that this weighted average foreshadows our discussion of *expected value* in Chapter 9. The 30% is summarizing two features of the situation: (1) It is reflecting the different probabilities of rain at different points in the region, and (2) it is weighing the various probabilities of rain together with the proportion of area that has those probabilities to describe a type of weighted average. In the case in which it will definitely rain on 30 square miles and definitely not rain on the remaining 70 square miles, the 30% probability of precipitation is as good a summary as we can provide for the whole region. But to know whether to pack our umbrella and rubbers, it would be much more useful to know whether our destination is in the area that gets no rain or in the area where it will definitely pour.

Unfortunately, many people believe that a 30% probability of precipitation in a region means that there is a 30% chance that it will rain at *some point* in the region. We now see that this interpretation is incorrect. In fact, a 30% probability of precipitation actually implies that *on average* 30% of the area will receive rain. If conditions were repeated many times where there is a 30% probability of precipitation for any point in the region, and for each repetition the percentage of the area of the region that got wet were recorded, then the average of those percentages would equal 30%.

The definition of probability of precipitation is tricky. An unfortunately incorrect explanation has appeared on the National Weather Service Web site. Their official explanation is, at best, misleading: "Technically, the probability of precipitation (PoP) is defined as the likelihood of occurrence (expressed as a percent) of a measurable amount (0.01 inch or more) of liquid precipitation (or the water equivalent of frozen precipitation) during a specified period of time at any given point in the forecast area. Forecasts are normally issued for 12-hour time periods." The phrase "at any given point in the forecast area" cannot be correct since different points can have different likelihoods, particularly if some points in the forecast area lie in rainy terrains and others in drier areas. Whoops.

The definition *should* read: "Technically, the probability of precipitation (PoP) is defined as the likelihood of occurrence (expressed as a percent) of a measurable amount (0.01 inch or more) of liquid precipitation (or the water equivalent of frozen precipitation) during a specified period of time at a random point in the forecast area. Forecasts are normally issued for 12-hour time periods." Let's hope that someone from the National Weather Service reads this text and will correct the cloudy probability mistake.

Pass the Ball or Run? A Beautiful Mind

Whether we are or are not addicted to football, this sport does provide us with an excellent opportunity to illustrate a strategic decision-making process that employs probability and arises on the playing fields, in the Wall Street trading pits, and in many parts of our lives. *Game theory* is the mathematical area that offers models for strategic decision making. It is heavily used in economics, business, games, sports, war, and other arenas in which strategic decisions must be made. We will consider a somewhat simple illustration of game theory that involves two competing teams, each of which must make one of two choices. The combination of those choices determines how each team fares. Our example presents itself at a moment of high drama on the football field. The fans are screaming because it is third down and long. The offense must decide whether to run or pass; and the defense must decide whether to align itself primarily against the run or the pass. What to do? The clock is ticking. . . .

What should you do if you know nothing about football and care even less about the game? You're not alone (in fact, one or more of the authors feels the same way), so there is no need to punt or even pout. Winning a football game is far more

© Anthia Cumming/iStockphoto

important to overpaid coaches than to most of us. But making decisions in the face of uncertainty is a challenge that each of us faces every day. So looking at the strategy employed on the gridiron is interesting and important, even if you don't care or know much about football.

Game theory utilizes the concept of a *payoff matrix*, a grid of values that describes the payoffs for each combination of choices that the players could make. We introduce the payoff matrix and illustrate its import in the context of a strategic decision during a football game. In football, when the offense faces third down with many yards to go for a first down, the usual options for the offense are to either pass or run. The defending team can align itself to defend more effectively against the pass or defend more effectively against the run.

Below is a possible payoff matrix for this scenario. Each number represents the payoff (in yards from the offensive team's point of view). The defense wants the offense to get as *few* yards as possible, so *lower* numbers are better for the defense.

If the offense chooses the strategy of always passing, the defense will quickly learn to always defend against the pass. The chart shows that this combination gives an expected value of 5 yards for the offense.

	Defend Against Pass	**Defend Against Run**
Pass	5 yards	7 yards
Run	6 yards	1 yards

Pass and Run are the options for the offense; Defend Aganist Pass and Defend Against Run are the options for the defense.

Life Lesson
Be open to new ideas.

If the offense chooses the strategy of always running, the defense will learn to always defend against the run. That combination gives 1 yard for the offense. We might think that those are the only two strategies available for the offense; however, the trick is that the offense does not need to use a pure strategy of always passing or always running. Instead the offense can use the strategy of passing with a certain probability. For example, the offense could flip a coin and run if it comes up heads or pass if it comes up tails. Or instead of a coin, the offense could use some other random device to decide to pass with any chosen probability. Choosing a probabilistic strategy is a great idea, because as we will now see, with a probabilistic strategy, the offense can gain more yards on average than either pure strategy (of always passing or always running) yields.

The probabilistic game theory strategy confirms the intuitive idea that once in a while, at random, doing the nonobvious play is a good idea. Instead of deciding to always pass or to always run, we can decide to pass with some probability p and run with probability $(1 - p)$. So in the huddle, the quarterback could be flipping a coin or consulting a table of random numbers to determine what to call for that play.

Different choices for the probability of passing give the offense a different expectation for yards gained. For example, suppose the offense decides to flip a fair coin and pass when it comes up heads. By using the table, we can estimate how many yards the offense would expect to gain with that strategy. A 0.5 probability of passing means the offense will pass half the time and run half the time. Suppose the defense defends against the pass. Then when the offense passes, the offense will make 5 yards and when the offense runs, they will make 6 yards. So on average, when the defense defends against the pass and the offense passes with probability 0.5, the offense can expect to make

$$0.5 \times 5 + 0.5 \times 6 = 5.5 \text{ yards.}$$

Similarly, suppose the defense defends against the run. Then when the offense passes, the offense will make 7 yards and when the offense runs, they will make 1 yard. So on average when the defense defends against the run and the offense passes with probability 0.5, the offense can expect to make

$$0.5 \times 7 + 0.5 \times 1 = 4 \text{ yards.}$$

Life Lesson

Visualize information or data whenever possible.

We could do the exact same analysis for any choice of passing with probability p by simply putting in p for the first 0.5 and $(1 - p)$ for the second 0.5 in the two previous equations. We can visualize the information using a graph.

The graph below shows how many yards we would expect to gain on average depending on the probability p that we use for passing. There are two lines. The red line indicates how many yards we as the offense will expect to gain, on average, if the defense defends against the pass, and the blue line indicates how many yards we as the offense will expect to gain, on average, if the defense defends against the run. In each case we get a sliding scale for how many yards we would expect to gain depending on the probability p with which we pass.

For each probability p of passing, we can compute how many yards we as the offense would expect to gain on average if the defense defends against the pass (the red line). Specifically, the expected number of yards gained if the offense passes with probability p and defense defends against the pass is $p \times 5 + (1 - p) \times 6$. The graph is just a straight line descending from 6 yards down to 5 yards indicating our expected gain if the defense defends against the pass.

Similarly, if the defense defends against the run, we can display the number of yards we expect to make by the blue line. If we pass with probability 0 (that is, we always run), then we will get only 1 yard on average when the defense defends against the run. While if we pass with probability 1, then we will gain 7 yards when the defense defends against the run. The expected number of yards gained if the offense passes with probability p and the defense defends against the run is $p \times 7 + (1 - p) \times 1$.

The question is: How does the offense select with what probability p to pass in order to maximize the number of yards that it can expect to gain? The answer is that we look at where the red line and the blue line cross.

The red and blue lines cross at one point, specifically, when $p \times 5 + (1 - p) \times 6 = p \times 7 + (1 - p) \times 1$. Applying some basic algebra to solve for p, we can discover that in this example the

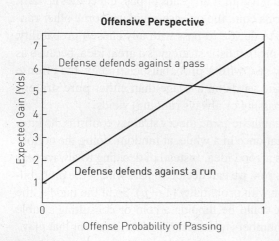

Offensive Perspective

Expected Gain (Yds)

Defense defends against a pass

Defense defends against a run

Offense Probability of Passing

crossing point occurs when our probability of passing p equals 0.71. At that point, we can expect on average to make 5.3 yards regardless of whether the defense defends against the pass or the run. For any other choice of probability of passing, the defense can always defend against the run or always defend against the pass and our expected number of yards will be less.

Thus our conclusion is that the offense should pass 71% of the time, but select the 71% at random. Passing with probability 71% gives an expected value of 5.3 yards gained for the offense, which is a higher value than either of the two pure strategies of always passing or always running.

Likewise, the defense would not want to always defend against the pass or always defend against the run, because then the offense could change its strategy to take advantage of that poor strategy. Using the payoff matrix values again and performing the same analysis for the defense as we did for the offense, we would find that the defense should defend against the pass 86% of the time (again at random).

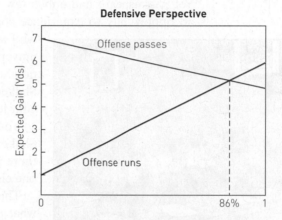

Defensive Perspective

So the next time you're watching a football game and the offense runs on third and long and gets tackled after a one-yard gain, you should not necessarily be angry or frustrated. Your team may, in fact, have been using the best *long-term* strategy, that is, your team may be choosing the play with an appropriate probability. Unfortunately, since the best strategy involves randomness, by definition, the results on each individual outcome will not always be great. But what you do know is that if your team always passes when it's third and long, you should hire a new coach.

This combination of probabilistic strategies for the offense and defense is called a *Nash equilibrium*, that is, a strategy whereby no player can get an advantage by unilaterally changing his or her strategy. It was named for John Nash, who won the Nobel Prize in 1994 for his work on Game theory and who became famous through the book and acclaimed movie *A Beautiful Mind*.

AFP/Getty Images, Inc.

Life Lesson
Consider long-term solutions.

Believe It or Not—The Bayesian Model of Belief

Previously, we considered the experience of repeating trials that had random outcomes and we defined probability as an idea constructed to measure the likelihood of certain outcomes. For example, we said that if we rolled a fair die, then the probability of seeing a 4 was 1/6 because there were six equally likely outcomes from rolling a fair die and 4 was one of the possible six outcomes. This concept of probability is a wonderfully useful way to describe our uncertainty about what will happen when several random outcomes are possible. But another kind of uncertainty occurs in life when we talk about what we believe to be true, but we don't know for certain. Let's look at some examples.

Most scholars believe that Shakespeare wrote *Hamlet*, but a few scholars assert that the plays ascribed to Shakespeare were actually written by someone else. They point out that the Shakespearean plays contain many references to classical knowledge and display a nuanced understanding of sophisticated court behavior. Yet Shakespeare had only a few years of elementary school education and had no opportunity to experience court life. These scholars argue that the person

who wrote the Shakespearean plays must have had a university education and must have lived among the nobility. So what do you think? Do you believe that Shakespeare wrote *Hamlet*? How would you express your level of uncertainty about this question?

You might well say, "I am 98% sure that Shakespeare wrote *Hamlet*, although I admit that there is a small chance, say a 2% chance, that someone else wrote his plays."

This kind of assertion does not mean anything like what our dice-throwing probability meant. Our 98% certainty that Shakespeare wrote *Hamlet* does not mean that if 100 Shakespeares lived, then about 98 of them would have written *Hamlet*. Instead, the 98% figure is expressing a level of belief.

Was it to be or not to be?

We often say things like, "I'm 95% sure I left the keys on the dresser." Or a doctor might tell us, "There is about an 80% chance that you have a cold, but there is a 20% chance that it's more serious."

One of the roles of mathematical thinking is to take everyday experiences and clarify them and, in particular, make them quantitative if possible. So here we'll take a mathematical look at how we can describe the strength of our beliefs about the world. One useful approach is named after Thomas Bayes, a mathematician and Presbyterian minister who lived from 1702 to 1761. The Bayesian analysis that we present below shows us one way to describe beliefs (in our example, what disease you have when you're sick) and how to update those beliefs when we get new evidence.

Calling All Doctors

One day you wake up feeling terrible. You have a fever of 101°, you are coughing, and you've lost your appetite. You feel so lousy that you decide not to go to class but instead to visit the doctor. The doctor looks in your ears and throat and listens to your chest and says, "Right, you're sick. I've seen thousands of patients with exactly

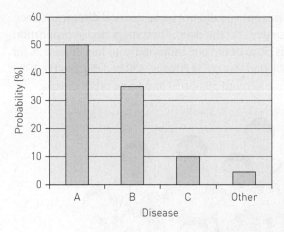

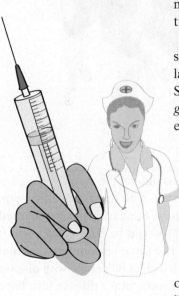

those symptoms. Based on my extensive experience, I'm almost certain that you have one of three diseases A, B, or C and I know for certain that you can't have more than one of them. I'd say you have a 50% chance that it's disease A, a 35% chance that you have disease B, and a 10% chance that you have disease C. There is a 5% chance that it is something else."

This scenario offers us a perfect opportunity to develop a concept of probability to help us quantify our beliefs in an uncertain world. The fact is, in this situation, the truth is only one of four possible states: you have disease A, disease B, disease C, or another disease. So in some sense, there is no probability involved since only one of those states is true. On the other hand, we do not know which of them is true, so it is reasonable to describe our view of the world by ascribing a level of likelihood to each of the four possibilities. Since you definitely have one of those three specified diseases or some other disease, we know that the probabilities must add up to 1. So we could draw a little chart that shows our current state of belief.

The doctor could start treating you for disease A, the most likely disease, but might instead ask you to take a particular blood test that might give some additional information.

You bravely give a considerable amount of your lifeblood to the smiling, blood-sucking lab technician and go home to drink a few quarts of chicken soup. Two days later the doctor calls and says, "Your test indicates that you have *gorp* in your blood. Studies have shown that among people who have disease A, only 10% of them have *gorp*; while 30% of people with disease B have *gorp*; and 80% of people with disease C have *gorp*. Among people who do not have A, B, or C, 10% have *gorp*."

These additional facts give us information about the possible states of the world, so let's see how we can best apply this new information to develop a refined sense of the probabilities that we have disease A, disease B, disease C, or something else. Our goal is to ascribe a probability to each of the four possibilities by giving a numerical value to each possibility so that the sum of those probabilities adds up to 1. So the first question is whether there is a reasonable way to compute those probabilities or whether many different opinions are reasonable.

When we first think about assigning probabilities to the four possible states of the world, we might be inclined to simply ignore our previous sense of the likelihoods of A, B, C, or Other and simply use the evidence related to the *gorp* test. In other words, we might say, well, clearly you have disease C, since *gorp* appears much more frequently among people who have C than among people with either of the other diseases. However, that thinking would be incorrect, because that thinking would be ignoring all the previous evidence about the diseases. Instead, we need to devise a method to give appropriate weight to the previous probabilities of A, B, or C and then let the new evidence modify our opinion. A good way to think about this issue is to imagine a thousand yous.

A Thousand Yous

Suppose you lived in a large town and, fortunately for the town, one thousand residents were clones of you—that's right, there are one thousand yous. Now suppose you all got sick and had identical symptoms, but you did not have identical diseases. In fact, suppose that you had the various diseases that the doctor first suggested in proportion to the likelihoods he assigned. That is, 50% of the one thousand yous had disease A (in other words, 500 had disease A); 35% had disease B (that is, 350 had disease B); 10% had disease C (that is, 100 have disease C); and 5% or 50 had some other disease.

Now let's simply count how many of these thousand people would have *gorp* in their blood. Well, we were told that 10% of people with disease A have *gorp*, so of the 500 yous with disease A, 50 would have *gorp*. Similarly, since we learned that 30% of people with disease B have *gorp*, that would be 30% of 350, which equals 105 people with disease B also have *gorp*. The same reasoning shows that 80% of the 100 yous with disease C have *gorp*, which gives 80 people with disease C and *gorp*. And finally, 10% of the 50 yous with some other disease also have *gorp*, that is, 5 people have some other disease and *gorp*.

So among the 1000 people who initially came to see the doctor with the symptoms you had and then tested positive for *gorp*, we would see the following numbers having the various diseases:

- Disease A: 50
- Disease B: 105
- Disease C: 80
- Other disease: 5

We can record these data in a convenient table called a *two-way table*. The last column records how many yous are presumed to have each disease, A, B, C, or Other, and each row records how many of those have or do not have *gorp*. So reading the column labeled *Gorp* shows how many people whom we assumed to have each disease also have *gorp*.

	Gorp	No *Gorp*	Total
A	50	450	500
B	105	245	350
C	80	20	100
Other	5	45	50
Total	240	760	1000

So a total of 50 + 105 + 80 + 5 = 240 people have all the symptoms that the doctor can now use for evaluating the likelihood of the various diseases. We can simply look to see what fraction of these 240 have the various diseases:

- 50/240 have disease A
- 105/240 have disease B
- 80/240 have disease C
- 5/240 have another disease

We can phrase this information in terms of probability of having the various diseases by saying you have a:

- 21% chance of having disease A
- 44% chance of having disease B
- 33% chance of having disease C
- 2% chance of having another disease

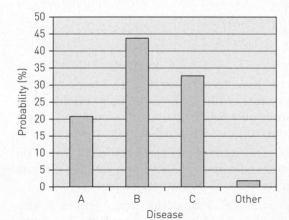

Using this analysis, the best course of treatment might be directed at disease B.

If you watch dramatic doctor programs on TV, the head doctor sometimes asks the assembled underling doctors, "What is the differential diagnosis?" That question is really asking for a Bayesian analysis, such as the one we just performed, of the consequences of some evidence. Given the symptoms, what is the likely disease? What further test will provide evidence that will discriminate among the possible diseases?

Definition of a Closed Mind

If you assign a 0% probability to some possible state of the world, then no amount of evidence will sway your opinion. So this analysis lets us give a mathematical description of a closed mind, namely, someone is completely closed-minded about a possibility if they believe there is literally no possibility that their own beliefs are wrong. Many people on many issues fail to view their world as having the possibility of being wrong. Many wars are fought because of that attitude.

Life Lesson
Keep an open mind.

A Look BACK

PROBABILITY IS APPLIED IN MANY AREAS OF life, from weather prediction to sports strategies to the diagnosis of disease. We daily hear forecasts like, "There is a 30% chance of rain tomorrow in the greater metropolitan area." That prediction is really an average of the likelihoods of rain at all the various points in the region.

In sports or business, the best strategies may involve making decisions using randomness. Most of the time it may be better to do the expected thing, but once in a while, at random, it may be a good idea to do the unexpected.

Often we are not entirely certain about our world. We can use probability and, in particular, the ideas of Bayesian analysis to clarify the strength of our opinions and to learn how best to change our minds when we learn new evidence. In cases of medical conditions, the good use of evidence can improve the accuracy of diagnoses.

Using probability with clarity and understanding can be an important way to hone our perception and application of common sense.

Life Lessons

Understand simple things deeply.

•

Be open to new ideas.

•

Modify your thinking when the evidence points you in a new direction.

MINDSCAPES ▸ Invitations to Further Thought

In this section, Mindscapes marked **(H)** *have hints for solutions at the back of the book. Mindscapes marked* **(ExH)** *have expanded hints for solutions at the back of the book. Mindscapes marked* **(S)** *have solutions.*

Developing Ideas

1. **No pop quizzes.** Your instructor gives "mom quizzes" (she doesn't like pop quizzes). Suppose you expect a score of 100% when you study and a score of 60% when you don't. If you study only half the time, what would you expect your average score to be on all your quizzes?

2. **No easy quizzes (S).** Your instructor starts giving harder quizzes. When you study you expect a score of 80% and a score of 30% when you don't. If you study only half the time, what score would you expect on average?

3. **Silly sickness.** Based on your symptoms, your doctor says, "There's a 40% chance you have meemps, a 30% chance you have moosles, a 20% chance you have chicken pux, and a 10% chance you have some unknown silly disease." Create a bar graph similar to the one on page 635 to illustrate this silly diagnosis.

4. **Making an algebra-down.** Consider the equation $p \times 5 + (1 - p) \times 6 = p \times 7 + (1 - p) \times 1$ from the football scenario given in the section. Solve this equation for p and thus confirm the answer given in the text.

5.   A beautiful mind. Who was John Nash? Give as many similarities and differences as you can between John Nash and Johnny Cash.

Solidifying Ideas

6.   Upon further study. On your instructor's quizzes, you expect a score of 100% when you study and 60% when you don't. Suppose the probability that you study is 0.8. Given these facts, compute the expected average of all your quizzes.

7.   Power pass (ExH). The matrix below gives the yards gained for the offense in each of four football scenarios. If the defense always defends against the pass and the offense passes with probability 0.6, find the number of yards the offense can expect to make on average.

	Defend Against Pass	Defend Against Run
Pass	4	9
Run	6	1

8.   Pass with a *p*. Suppose the defense referred to in the previous Mindscape always defends against the pass and the offense passes with probability p. Given the matrix in the previous Mindscape, write an expression (in terms of p) giving the number of yards the offense can expect to make on average.

9.   Random run. Suppose now in Mindscape 7 the defense always defends against the run and the offense runs with probability 0.4. Find the number of yards the offense can expect to make on average.

10.   Another random run. Suppose now the defense referred to in Mindscape 7 always defends against the run and the offense passes with probability p. Write an expression (in terms of p) giving the number of yards the offense can expect to make on average.

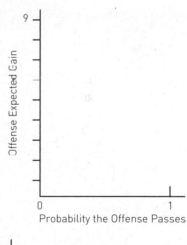

11.   Lining up the defense (S). Using the payoff matrix in Mindscape 7, plot two points on the axes to the right as follows. First, let the probability the offense passes be 0 and suppose the defense defends against a pass. Plot the point showing the gain for the offense. Then, let the probability the offense passes be 1 and suppose the defense defends against a pass. Plot the point showing the gain for the offense. Draw the line through the two points to show the average offensive gain when the defense defends against a pass depending on the probability the offense passes. If you did Mindscape 8, how does your answer there relate to this line?

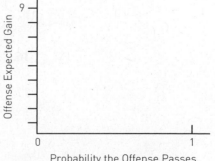

12.   De-lines of defense. Using the payoff matrix in Mindscape 7, plot two points on the axes to the right as follows. First, let the probability the offense passes be 0 and suppose the defense defends against a run. Plot the point showing the gain for the offense. Then, let the probability the offense passes be 1 and suppose the defense defends against a run. Plot the point showing the gain for the offense. Draw the line through the two points to show the average offensive gain when the defense defends against a run depending on the probability the offense passes. If you did Mindscape 10, how does your answer there relate to this line?

13. **Approximate Nash (ExH).** Take the two lines from the previous two Mindscapes and draw them on a single set of axes (as in the graph on page 631). Estimate the coordinates of the point of intersection. What do these coordinates tell you?

14. **Precise Nash.** Use your answers to Mindscapes 8 and 10 to help you find the exact coordinates at which the two lines in Mindscape 12 intersect. (Or find equations of the two lines directly and solve them simultaneously.)

15. **Positive payoff (H).** Below is a payoff matrix for a game involving two players. Player 1 can use either Strategy A or Strategy B; Player 2 can defend against either Strategy A or Strategy B.

	Player 2 Defends Against Strategy A	Player 2 Defends Against Strategy B
Player 1 Strategy A	4	10
Player 1 Strategy B	10	7

Sketch a graph showing two lines on the axes below, one line giving the expected gain for Player 1 if Player 2 defends against Strategy A, the other giving the expected gain for Player 1 if Player 2 defends against strategy B. Label each line appropriately.

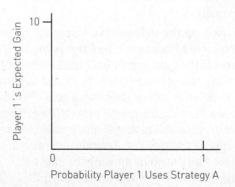

16. **Estimating the equilibrium.** Estimate the coordinates of the point of intersection of the two lines in the previous Mindscape.

17. **Exacting the equilibrium.** Find equations for the two lines in Mindscape 15 and use them to solve for the exact coordinates of the point of intersection. (This point is called the *Nash equilibrium point.*)

18. **Payoffs need not be positive.** Here's a payoff matrix for another game. Notice that some of the payoffs are negative, indicating undesirable states.

	Player 2 Defends Against Strategy A	Player 2 Defends Against Strategy B
Player 1 Strategy A	−5	4
Player 1 Strategy B	2	−3

On the axes below, sketch a graph analogous to that requested in Mindscape 15.

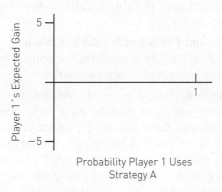

Probability Player 1 Uses Strategy A

19. **Negative equilibrium.** Find equations for the two lines in the previous Mindscape and use them to solve for the Nash equilibrium point, in other words, the intersection of these two lines.

20. **Making negative sense.** In what kind of game or scenario might a negative payoff make sense?

Creating New Ideas

21. **Colin and Tubbes.** When first grader Colin comes home from school each day, his pet hippo Tubbes is waiting to greet him. Because Tubbes is so large and enthusiastic, Colin does not enjoy these blubbery greetings, so he tries to avoid them by switching between the back door and the front door to thwart Tubbes. When Colin succeeds in avoiding Tubbes, he feels a satisfaction "gain" of 5. If Colin enters the front door and finds Tubbes waiting, he feels a satisfaction gain of −5 (in effect, a slobbery loss). If Colin enters the back door and finds Tubbes waiting, his satisfaction loss is only −2 because his mother's presence in the kitchen helps keep Tubbes under control. Create a payoff matrix that illustrates this situation. (Let the top row give Colin's possible payoffs if he uses the front door. Let the first column give the payoffs if Tubbes is waiting at the front door.)

22. **Colin and Tubbes 50–50.** Suppose in Mindscape 21 that Tubbes waits at the front door. If Colin chooses the front door with probability ½, what is his average satisfaction? Suppose Tubbes waits at the back door and

Colin chooses the back door with probability ½. What is Colin's average satisfaction?

23. **Averaging Colin and Tubbes.** Suppose p is the probability that Colin uses the front door in Mindscape 21. Write an expression that gives Colin's satisfaction payoff if he uses the front door with probability p and Tubbes always chooses to wait at the front door.

24. **Colin and Tubbes line up (ExH).** Referring to Mindscape 21, sketch a graph showing two lines on the axes here, one line giving the expected gain for Colin if Tubbes waits at the front door, the other giving the expected gain if Tubbes waits at the back door.

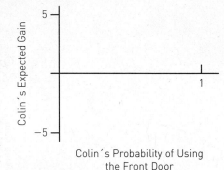

Colin's Probability of Using the Front Door

25. **Colin and Tubbes with a dash of Nash (S).** Find the Nash equilibrium point for the Colin and Tubbes scenario.

26. **Diagnosis.** Based on your symptoms, your doctor says, "There's a 40% chance you have disease A, a 30% chance you have disease B, a 20% chance you have disease C, and a 10% chance you have some unknown disease." Create a bar graph similar to the figure on page 635 to illustrate this diagnosis. Out of 1000 patients with identical symptoms, how many would you expect to have disease A? Disease B? Disease C? None of A, B, or C?

27. **Sploosh test (H).** Given the scenario in the previous Mindscape, one way to enhance the diagnosis is to test the blood for *sploosh*. Studies have shown that among people who have disease A, 20% have *sploosh* in their blood; among people with disease B, 50% have *sploosh*; among people with disease C, 10% have *sploosh*; and among people with neither disease A, B, nor C, 30% have *sploosh*. Suppose there are 1000 people with symptoms identical to yours. How many of those with disease A have *sploosh*? How many of those with disease B have *sploosh*? How many of those with disease C have *sploosh*? How many of those with neither disease A, B, nor C have *sploosh*?

28. **Diagnose the data.** Using the data from the previous Mindscape, create a table similar to the one on page 635. Compute the fraction of those with *sploosh* who also have disease A, disease B, disease C, or some unknown disease. Convert your fractions to percentages. Which disease is most likely for those with *sploosh*?

Further Challenges

29. **My theory ...** You are a paleontologist who finds part of a dinosaur bone. After careful study you determine there's a 50% chance the dinosaur was a Juliasaurus, a 15% chance it was a Noahsaurus, and a 35% chance it is a new dinosaur altogether. Later your graduate student finds a confirmed Maxxasaurus bone at the same site. You know from earlier research that of all confirmed discoveries of Juliasaurus bones, 5% have been found at sites that also contain Maxxasaurus bones. Of all confirmed

discoveries of Noahsaurus bones, 30% have been found at sites containing Maxxasaurus bones. And among all discoveries that are neither Juliasaurus or Noahsaurus, 10% have been found at sites containing Maxxasaurus bones. Imagine you have 1000 identical bone specimens. Complete the table below to help you decide which dinosaur is the most likely source of your original discovery.

	Found with Maxxasaurus	Not Found with Maxxasaurus	TOTAL
Juliasaurus			
Noahsaurus			
Other			
TOTAL			1000

30. **The real (old) story.** What factors affect the reliability of the analysis in the previous Mindscape?

31. **Going stag.** Two Englishmen, Neville and Winston, go hunting. Each may choose to hunt a hare or a stag but must make a choice without knowing the other's decision. A solo hunter will catch a hare easily, but it takes both hunters in cooperation to catch a stag, the more valuable prey. The payoff matrix from Winston's point of view is given below.

	Neville Hunts a Stag	Neville Hunts a Hare
Winston Hunts a Stag	5	0
Winston Hunts a Hare	2	2

Find the Nash equilibrium point for this game. Is this point truly optimal for both hunters?

32. **Selling sweets.** Two snack cake companies, *Snacky Cakes* and *Little Eddie*, compete for the same consumers. If both spend the same amount on advertising, the effect is to cancel each other's message and there is no change in their sales. Yet the extra spending reduces profits (payoffs). Each would like to spend less on advertising, but if one reduces its budget and the other doesn't, then the latter company would see increased sales over and above its additional spending on advertising. Fill in the table below to show a possible payoff matrix from the point of view of Little Eddie snack cakes.

	Snacky Cakes Spends More on Advertising	*Snacky Cakes* Spends Less on Advertising
Little Eddie Spends More on Advertising		
Little Eddie Spends Less on Advertising		

Create a graph illustrating your scenario. Is there a Nash equilibrium point?

33. **Prisoner's dilemma.** Two people are suspected of robbing a bank. They are being interrogated in separate rooms. If both stay silent, they can be convicted of a lesser crime and sentenced to only 6 months. If one confesses (or "defects") and the other does not, the confessor goes free as a reward for cooperating while the other suspect will be sent to prison for 10 years. If both defect, they each go to prison for 5 years. Create a payoff matrix from the point of view of Suspect #1. (Adopt the method from previous Mindscapes.)

In Your Own Words

34. **Personal perspectives.** Write a short essay describing the most interesting or surprising discovery you made in exploring the material in this section. If any material seemed puzzling or even unbelievable, address that as well. Explain why you chose the topics you did. Finally, comment on the aesthetics of the mathematics and ideas in this section.

35. **With a group of folks.** In a small group, discuss the ideas of strategy and game-playing introduced in this section. After your discussion, write a brief narrative describing these ideas in your own words.

36. **Creative writing.** Write an imaginative story (it can be humorous, dramatic, whatever you like) that involves or evokes the ideas of this section.

37. **Power beyond the mathematics.** Provide several real-life issues—ideally, from your own experience—that some of the strategies of thought presented in this section would effectively approach and resolve.

For the Algebra Lover

Here we celebrate the power of algebra as a powerful way of finding unknown quantities by naming them, of expressing infinitely many relationships and connections clearly and succinctly, and of uncovering pattern and structure.

38. **Match point.** Find the coordinates (p, y) where the lines $y = -3 + 10p$ and $y = 2 - 10p$ intersect.

39. **Match lines.** Find an equation for the line passing through the points $(-1, 6)$ and $(2, 3)$ and an equation for the line passing through the points $(-2, 2)$ and $(0, 10)$. Now find the coordinates of the point where these lines intersect.

40. **Unmatched slopes.** Two lines intersect at the point $(6, 2)$. If one line has slope ½ and the other line has slope –3, find an equation for each line.

41. **Plotting percentages.** Of the students in your dorm, 30% are business majors, 15% are engineers, 20% are science majors, 10% are humanities majors, and the rest are social science majors. (There are no double majors.) If there are 600 students in your dorm, how many are social science majors? If you choose a student at random, what's the probability that they are a

science or engineering major? Make a bar graph similar to those in this section to show the probabilities for each type of major.

42. **Itching for** x **(H).** You go to the Health Center with an itchy rash. The nurse says there are four possibilities: poison ivy, poison oak, poison sumac, or something else. Based on your symptoms, she says it's twice as likely to be ivy as oak, 1/3 as likely to be sumac as ivy, and 1/6 as likely to be something else as it is oak. What's the probability you have something else?

CHAPTER NINE

Meaning from Data

9.1 Stumbling Through a Minefield of Data

9.2 Getting Your Data to Shape Up

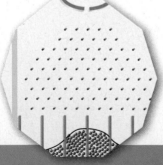

9.3 Looking at Super Models

> *Statistics are like bikinis. What they reveal is suggestive, but what they conceal is vital.*
>

Should young women take cervical cancer (HPV) vaccinations? Evidence for and against such a decision is usually presented in the form of data with statistical conclusions. How good a student will Chris Jones be, if admitted to college? Much of the information we use to make decisions is based on numerical data. Should we buy lottery tickets when the jackpot reaches dramatic heights? Should a coach keep a player on the bench when the player's in a slump? How can we tell if gender discrimination influenced college admissions procedures? Trying to understand the economy, the weather, wars, school systems, grading, risk, the quality of products, and measurements of everything from social trends and marketing to science and sports fundamentally involves data. Most practical aspects of our world involve our ability to make sense of data.

Data are the pixels of information from which our picture of the world can be assembled. The trouble with data is that they come to us with errors and with neither order nor meaning. Data are useless or actually misleading until we learn to draw their secrets from them. The fundamental challenge for statistics is to take a whole bunch of raw numbers, organize them in a meaningful manner, and then employ that newfound structure to interpret that heap of numbers. Statistics provides the conceptual and procedural tools for making sense of data.

Analyzing data correctly is one of the most powerful tools that we have for understanding our world. But it is a two-edged sword. Mark Twain attributed to Benjamin Disraeli perhaps the most famous quip about statistics: "There are three kinds of lies: lies, damned lies, and statistics." But an apt rejoinder is: "While it's easy to lie with statistics, it's easier to lie without them." In this chapter, we will see the two sides of data—their use and their misuse.

Our method for discovering powerful statistical concepts follows strategies we have seen many times before. We consider examples and look for patterns; and as we do so, we will literally see the concepts of statistics emerge in front of our eyes.

Statistics enables us to better understand our world. It transforms a disordered pile of information into a focused picture of measured proportion. It serves as the mathematical foundation of common sense, wisdom, and good judgment. Statistics can be a powerful tool for finding meaningful ways to describe our complex world and for making decisions grounded on clear measures of reality.

9.4 Go Figure

9.5 War, Sports, and Tigers

9.1 STUMBLING THROUGH A MINEFIELD OF DATA

Inspiring Statistical Concepts Through Pitfalls

© Ikon Images/SuperStock

Individuality versus anonymity.

Life Lesson

Always apply common sense.

> *There are three kinds of lies: lies, damned lies, and statistics.*
>
> **MARK TWAIN, ATTRIBUTED TO BENJAMIN DISRAELI**

Fact: *The average American has one testicle and one ovary.*

Aren't statistics great!? We find this perfectly valid statistical summary of the testicular and ovarian endowments of humans funny because we know exactly what is actually going on. We know that the population is roughly equally divided between men and women, and we know some rudimentary facts about the physiology of each sex. Averaging over the whole population doesn't give a meaningful description of reality, because the average does not tell us how these body parts are distributed throughout the population. Statistics can help us see our world with detail and nuance, but it can also lead us astray. Common sense is an essential ingredient in using statistics effectively and accurately.

Our world is afloat in data. Newspapers report statistics on everything: sports, economics, world affairs, elections, and, of course, celebrity meltdowns. Data about the world give us the possibility of grounding our opinions on a firm foundation. But data by themselves are merely lists of lifeless numbers. It is up to us to find methods of digesting the data in order to draw meaning from them. We need to organize, describe, and summarize data before we can apply data effectively to make decisions.

Unfortunately, one of the ugly realities of statistics is that they can mislead us. So in this section we introduce some of the subtle themes of statistical thinking by considering several scenarios in which statistics can mislead, distort, and downright lie. A good strategy for avoiding erroneous conclusions is to understand perilous pitfalls and fallacies and employ practices that will prevent them. Knowing what can go wrong helps us to become sensitive to potential defects in presentations or interpretations of data, and also enables us to be confident in the conclusions drawn from well-constructed studies.

Graphical Distortion

Life Lesson
A picture is worth a thousand words— unless the picture is distorted.

To understand data, we need to organize, describe, and summarize the data. One of the most effective ways to uncover and convey the meaning within data is through good images and graphics. However, one of the most common means to misrepresent statistics is to present the data visually in ways that distort the true picture. Here are some instances in which reality is distorted even though the data presented are accurate.

Terri Schiavo suffered severe brain damage in 1990, which resulted in her living in a persistent vegetative state after that. In 1998, her husband made the decision to have her feeding tube removed, but Terri's parents opposed that decision and years of legal arguments followed. In the spring of 2005, the case became a media sensation and Terri Schiavo became a household name. The final outcome was that the courts ruled that the feeding tube should be removed. A survey was conducted to record how people with different political views reacted to this decision. The graph below displaying the results of the survey appeared on CNN.com.

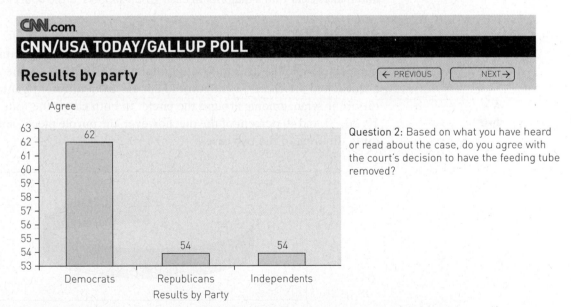

CNN.com
CNN/USA TODAY/GALLUP POLL
Results by party   [← PREVIOUS] [NEXT →]

Agree

Question 2: Based on what you have heard or read about the case, do you agree with the court's decision to have the feeding tube removed?

Results by Party: Democrats 62, Republicans 54, Independents 54

SAMPLE: Interviews conducted by telephone March 18–20, 2005, with | SAMPLING ERROR: +/- 7% pts. 900 adults in the United States. Copyright © 2005 Gallup, Inc. All rights reserved. The content is used with permission; however, Gallup retains all rights of republication.

After CNN.com received complaints that this graph distorted the results, it was replaced by the following graph.

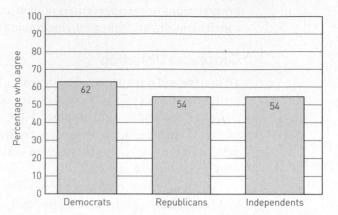

RESULT BY PARTY: CNN/USA TODAY/Gallup Poll
Margin of error: +/- 7%

Question 2: Based on what you have heard or read about the case, do you agree with the court's decision to have the feeding tube removed?

Both graphs report the same information; however, the first graph gives the impression that Democrats vastly differed from Republicans and Independents in their views. The second graph highlights the similarities in the opinions of the three groups and makes clear that a majority in each group supported the court's decision.

Slanted Pies

One of the most common strategies for visually distorting data is to use 3D-perspective graphs. The following graphs both show the same-size sectors of a thickened pie chart, as they would appear from different vantage points and different arrangements around the circle. In both cases, the four pieces contain 10, 20, 30, and 40 percent of the pie; however, the purple piece especially appears quite different in the two cases.

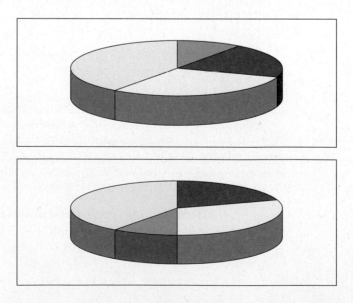

Views of a Tax Cut

Opposing political parties often want to present their cases so as to encourage the public to agree with their positions. During a discussion of a proposed tax cut in the mid-1990s, the Republicans and the Democrats presented the following two pictures of the *same* data about the proposed tax cut. Both graphs are accurate, but perhaps you can see why the two parties chose to present that information in their own particular way.

ON THE OTHER HAND...

Two Views of a Tax Cut

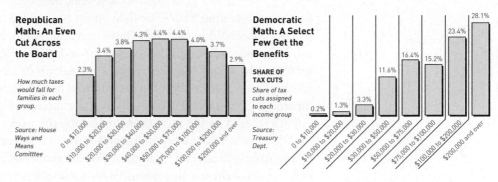

Republican Math: An Even Cut Across the Board

How much taxes would fall for families in each group.

Source: House Ways and Means Comitttee

Democratic Math: A Select Few Get the Benefits

SHARE OF TAX CUTS

Share of tax cuts assigned to each income group

Source: Treasury Dept.

Showing graphical representations of data is extremely helpful and powerful. Of course, we all know that absolute power corrupts absolutely, and in this case the power of graphical images is fraught with the potential for distorting the data or misleading the viewer. The best defenses against being misled are being conscious of the possibility of distortions and discerning, for ourselves, the true meaning of what is depicted in graphs.

> *Life Lesson*
>
> *A graphic tells a story; it is up to us to interpret and determine its meaning and validity.*

Summarizing Data

Collections of data often contain hundreds, thousands, or millions of pieces of information. To make sense of such unwieldy collections of numbers, we often have to summarize them in some useful and sensible manner. Often the summary consists of taking some sort of average. One average is called the mean, which is obtained by adding up all the numbers and dividing by the number of data you have. Unfortunately, this reasonable-sounding summary can sometimes be seriously misleading.

Summaries by definition summarize, which implies that they necessarily leave out information. Sometimes summaries accurately reveal the essential features of reality, but sometimes our summaries have no real meaning at all—once again we must be statistically on our toes and think for ourselves.

School Daze

Suppose we wish to choose a high school for our children. There are many variables involved in such a school choice, but one we might consider is the graduates' prospects of remunerative employment. We are statistically savvy, so we decide to compare the average earnings of graduates of various schools for the year 1997. On our list of schools to consider is a fine school in Seattle, Washington,

Courtesy of Lakeside School

The Lakeside School

called Lakeside School. The Lakeside School has an excellent reputation in many respects, but when we look at the average income of its graduates in 1997, we are so excited that we consider packing up the entire family in the SUV and moving to Seattle immediately. We learn that that year the mean income, including salary and investments, of all the graduates of Lakeside School was more than $2.5 million. In fact, this average includes recent graduates who are still in college and graduates from the 1920s who are now dead. Wow, what a record!

When we read about or hear a fact that is almost unbelievable, it is a good idea to first seek an explanation rather than warming up the gas-guzzling SUV. The data about Lakeside School are readily explained by two household names: Bill Gates and Paul Allen, Microsoft's founders. They are Lakeside School graduates and from 1996 to 1997, Bill Gates's net worth increased by about $18 billion and Paul Allen's by about $7 billion. Lakeside is an excellent school, so it is likely that the typical Lakeside graduate does have an income higher than the country's average income. However, using Gates's and Allen's salaries in the mix when computing the average salary totally misrepresents the facts. Those envy-inducing net worth numbers are the ultimate example of what statisticians call "outliers." For the purposes of computing a meaningful description of the earning potential for Lakeside graduates, we need to use common sense to find a helpful way to present and interpret the data. In future sections we will see several methods of summarizing data in meaningful ways.

> *Ninety-eight percent of all statistics are made up.*
>
> ANONYMOUS

Collecting Data

Before we can organize, describe, or summarize data, and before we can infer estimates of the whole population from information about a part of the population (called a *sample*), we have to gather the data. It may appear at first that collecting data is no big deal. However, a plethora of pitfalls accompanies this seemingly simple process, and it is into these pits that we now peer.

Asking Embarrassing Questions

One of the easiest ways survey data can be misleading and inaccurate is if those surveyed simply lie. Suppose we seek accurate data on an embarrassing or controversial topic. We might expect many people not to answer honestly about the topic if we were to ask them outright, so we would like to employ a method of surveying people that doesn't raise fears that their privacy is being invaded. How can we proceed?

One of the most serious problems facing colleges and universities today is alcohol abuse among students. A possible first step in dealing with this problem is gathering accurate data about alcohol use among college students. Although some students might forthrightly and honestly report their use of alcohol, others would simply lie, and some, for any number of reasons, would exaggerate, thereby skewing the data. Let's suppose the question we want to answer is: What percentage of the student body has been inebriated during the past week? Our goal is to structure the survey in such a manner that we can approximate the number of students who have been drunk without knowing which individual students have been drunk. An easy method might be to ask each student whether he or she had been inebriated and promise to keep the survey anonymous. But people might not believe us, and they might not want us to be able to peek at their answers as they write them. Thus, people still might not feel comfortable answering honestly.

Is a data-collecting method possible whereby, from *public* statements of the group, we are able to deduce what percentage of students have been drunk, and yet no one—including us—knows whether any *specific* individual has or has not been inebriated? Sounds like a pretty tall, frosty order, but the *apparently* impossible is not always *actually* impossible.

If we don't know the behavior of even one student, how can we hope to learn of the behavior of the aggregate? Let's think about this conundrum for a moment. As a first observation, we notice that, if we chose only half the students at random and asked them the question, we would be getting a representative sample of the behavior of the whole group. This observation may at first seem to be of no value, because we would still have the same problem of potentially collecting inaccurate answers from the students we did ask. Let's now add an interesting twist. Suppose we ask only half the group, but no one knows which half we're asking. This strategy is actually a fruitful one, but what in the world does it mean?

Some Anonymity

Let's ask each student to privately flip a coin. If the coin lands heads up, then the student answers "yes" to the question, "Have you been inebriated during this week?" whether or not the student actually has been drunk this week. If the coin lands tails up, then the student answers the question honestly—that is, if the student has been drunk, then the answer is "yes," and if the student hasn't, then the answer is "no." Thus, a student answering "yes" could mean either that the student actually did get drunk or that the student flipped heads but may or may not have gotten drunk.

If the student sees , then student answers "yes." (No matter the truth.)

If the student sees , then student answers honestly.

From the results of that experiment, how can we estimate the actual percentage of students who've been drunk? Well, if no one's been drunk, we would expect 50% of the students to answer "yes" by virtue of their flipping heads. If 40% of the students have been drunk, then we would expect half of them to flip tails and answer "yes." Thus 20% more than half of the students who flipped heads would answer "yes." Thus, if 70% (50% + 20%) answered "yes," we could deduce that about 40% have been drunk.

For example, suppose we try this procedure in a class of 60 students, and 41 answer "yes." We estimate that about 30 people said "yes" because their coin toss resulted in heads, so the remaining 11 people said "yes" because they flipped tails *and* have been drunk. Thus, if 11 people who flipped tails have been drunk, it seems reasonable that about 11 other people flipped heads and have been drunk (the coin-flipping process cut the class randomly into two groups of 30 and 30); so, we would estimate that 22 of the 60 students have been inebriated this week.

Total Anonymity

One defect of the previous coin-flipping data-collection technique is that if a person answers "no," then we *do* know that that person has definitely not been drunk. Think about the different scenarios in which someone would answer "no," and verify that, in each such case we would know for sure that the person has not been drunk. Can we further refine this method so that we do not know *either* way—that is, a person who answers "yes" may or may not have been drunk, *and* a person who answers "no" may or may not have been drunk?

1st coin flip \ 2nd coin flip	H	T
H	Lie (HH)	Tell truth (HT)
T	Tell truth (TH)	Tell truth (TT)

The perhaps surprising answer is that we *can* devise a data-collection method so that we do not know anything about any particular person, regardless of the answers. Suppose, for example, that we ask each person to secretly flip two coins. If the coins both land heads up, then that person reports the *opposite* of the truthful answer. Anyone who flips at least one tail answers truthfully.

To illustrate this process, let's analyze the situation if, in another class of 60 students, 24 people answer "yes," and 36 people answer "no." Notice first that we do not know the behavior of any individual. In fact, we could have the students answer by raising their hands: If someone said "yes," he or she could be telling the truth or not, and if another person answered "no," then he or she could be answering honestly or lying. So, how do we determine the number of students who have been drunk?

If the student sees , then the student lies.

Otherwise, the student answers truthfully.

There are four equally likely coin outcomes: HH, HT, TH, TT. Because these four outcomes are equally likely, the number of students who have been drunk and flipped HH is about the same as the number of students who have been drunk and flipped HT (or TH or TT)—and the same holds for those who've remained sober. Let's suppose that the number of students who've been drunk and who flipped HH is D, and the number of sober students who flipped HH is S. Similarly, for each group of students who flipped HT, TH, and TT, we expect D drunk students and S sober students. So, how many students in the class have been drunk? Answer: $4D$. (D of them flipped HH, another D of them flipped HT, still another D of them flipped TH, and finally, yet another D of them flipped TT.)

® = Number of students who answered "yes" in response to the "I've been drunk" question

D = Number of drunk students for a particular combination of coin flips

S = Number of sober students for a particular combination of coin flips

Estimated Number of Drunk (D) and Sober (S) Students for Each Flip

1st coin flip \ 2nd coin flip	H	T
H	$D\boxed{S}$ (HH)	$\boxed{D}S$ (HT)
T	$\boxed{D}S$ (TH)	$\boxed{D}S$ (TT)

Now, which people answered "yes"? Any student who flipped HH is to answer the *opposite* of the truth, so the S sober students in that group answered "yes." The students who flip at least one tail answer truthfully, so, because there are three such groups (HT, TH, and TT), there are $D + D + D$ students, which is $3D$, who answer "yes" truthfully from those groups. So we see that $3D + S = 24$ (the number who answered "yes").

The rest answered "no." So, $D + 3S = 36$ (the number who answered "no"). Now we have two equations with two unknown values to deal with. We want to find how many students have been drunk, so we need to figure out what D equals. The first equation tells us that

$$S = 24 - 3D;$$

we can substitute this fact into the second equation and get rid of that pesky S:

$$D + 3 \times (24 - 3D) = 36.$$

We now distribute that factor of 3,

$$D + 72 - 9D = 36,$$

which is the same as

$$-8D = 36 - 72 = -36.$$

So, we discover that

$$D = \frac{36}{8} = \frac{9}{2}.$$

How many students in the class have been drunk? Remember that the number of students who have been drunk is $4D$, so we would estimate that $4 \times 9/2 = 18$ students have been drunk this week, and yet we do not know whether any particular student has been drunk or not.

This procedure becomes more accurate and reliable as our pool grows larger, because, as the "Law of Large Numbers" from the last chapter tells us, the percentages of students and flips in the various categories will tend to become closer to the expected numbers when larger numbers of students are involved. By applying this data-collection method, which maintains anonymity, we may be able to overcome the potential pitfall of respondents' answering dishonestly on a survey.

Of course, students as a group may choose to systematically lie by reporting less drinking than is the truth in order to avoid administrative attention to the issue. In that case, the issue is not so much individual anonymity as a group problem. This method would not gather good data if the students as a group decided that they didn't want to divulge how much drinking was going on overall. Obtaining clean data is not an easy task. And dirty data can lead to dirty stats.

Leading and Misleading Questions

Surveys can produce skewed results by phrasing the questions in ways that might bias the answers. Here was a question taken from the New Jersey state Web site in the summer of 2005:

- Do you support Acting Governor Codey's Safe Schools Initiative that will integrate state, county, and local law enforcement measures to ensure that our schools are secure and protected against violence and terrorism? (Yes or No).

How would responses to that question differ from the answers to the following alternative phrasing?

- Do you support Acting Governor Codey's Safe Schools Initiative that would spend millions of dollars that could otherwise be spent to improve education, health care, and economic incentives for the state? (Yes or No).

Both of these questions are leading questions that would strongly bias the responses. When questions are phrased so that they suggest that a particular answer is correct, then the results from the survey may not reflect the true opinions of the people being surveyed.

Even if there is no intent to mislead people, questions can inadvertently elicit incorrect results. For example, suppose we are trying to find out how much people eat and we say on the survey that a "portion" of meat is 4 ounces. We might ask how many portions of meat you ate in a day. If you went to a restaurant where you ate a steak, you might feel that you ate one portion, whereas, using the survey's definition of portion size, your steak may have been two or three portions.

One additional source of inaccurate survey data can result from people responding with their good intentions rather than with reality. For example, if we ask how many people will vote in the next election, we might well find that a far larger percentage of people say, yes they will vote, than actually go to the polls. They are not exactly lying, but they are reflecting the human reality that often our intentions are somewhat more virtuous than our actions.

Life Lesson

Beware of invisible spin that might bias our thinking.

Sample Bias—Polluted Pools

Added to the problem of respondents' lying on surveys or being misled by the questions are the problems associated with the pool itself. We now illustrate the pitfall of a biased sample as one of the potential perils of data collection. Here we see that the answers we get often depend on whom we ask. To understand what opinions people in the population hold, the basic strategy is to select some people and just ask them. Of course the answers we get depend heavily on whom we ask.

The Literary Digest, July 14, 1934

A Statistical Fiasco

Before its demise in 1937, *The Literary Digest* undoubtedly provided its readers with well-digested literature; however, the publication is perhaps most famous for a monumental statistical fiasco caused by sampling bias. As your great-grandparents no doubt recall, the 1936 presidential election was contested between Republican Alfred Landon and incumbent Democrat Franklin Delano Roosevelt. Before the election, *The Literary Digest* conducted a poll. It sent out 10 million surveys and received 2.4 million responses. It predicted that Landon would receive 57% of the popular vote to Roosevelt's mere 43%. This outcome would result in Landon's winning in the electoral college by the substantial margin of 370 votes to Roosevelt's paltry 161. When the election was actually held in November, the electoral college vote was in fact a landslide the other way: Roosevelt 523, Landon 8; Landon won only two states, and Roosevelt received 62% of the popular vote. Some statisticians at *The Literary Digest* may have been encouraged to move on, or at least move to the horoscope department. How can we account for such an incredibly mistaken survey? The answer: a biased sample.

The year 1936 was in the middle of the Great Depression. Most people were struggling financially and had eliminated luxuries and unnecessary expenditures from their budgets. The sample of citizens polled by *The Literary Digest* came from several sources: *Literary Digest* subscribers, telephone directories, and automobile registration records. Most of those surveyed did, in fact, prefer Landon and no doubt voted for him.

The problem was that subscribing to *The Literary Digest* and owning a telephone or a car were luxuries that most people had eliminated from their personal budgets. So the data were gathered from a group of voters who were not representative of the voting population as a whole. *The Literary Digest* was sampling from the wrong population. Instead of taking a random sample among all voters, *The Literary Digest* was taking a sample among relatively wealthy citizens. In addition, only a quarter of the surveys were returned. The surveys that were returned might not be representative even of the people to whom the survey was mailed, because people who voluntarily fill out and return surveys are not necessarily a random cross-section of the population.

George Gallup

George Gallup

As often happens in life, one person's calamity is another person's opportunity. While *The Literary Digest* was hard at work sampling the rich, a young man named George Gallup recognized the serious flaws in *The Literary Digest*'s sampling method. He realized that to get a reasonable picture of the election outcome, the sample that was polled would have to be representative of the entire voting population. So Gallup did his own survey of 50,000 voters who were selected more representatively, and he correctly predicted Roosevelt's victory in the election. He wisely used random sampling methods to avoid bias. Using randomness to obtain a clear picture of reality was a terrific intellectual accomplishment that made statistical analyses valuable and accurate.

Gallup also made another prediction: By taking random samples from the lists that *The Literary Digest*'s poll used, he predicted what *The Literary Digest*'s prediction would be, within one percentage point. *The Literary Digest*'s poll was extremely well known and had correctly predicted the five previous presidential elections, often within a percentage or two of the vote. So Gallup's accurate prediction of the coming *Literary Digest* error gave his organization great credibility. His work in the 1936 election started Gallup on a career that made *Gallup Poll* a household word (or, more accurately, two household words).

Sampling Bias in Everyday Life

The phenomenon of sample bias arises daily in our lives. We all talk and listen to relatively few people, most of whom we know—our family, friends, co-workers, fellow students, and people we see during the day. This collection of people is far from representative of the entire population. Generally, we are surrounded by people having one or two religious or political perspectives. So if we were to use our daily experience as a sampling method for inferring a statistical understanding of the world, we would be way off. We would have the impression that people with other religious views are rare (and perhaps wacky). We would be mystified by election results in which vast numbers of strange people actually disagree with our clearly correct political views. Alas, from a statistical point of view, our own experience is an excellent example of a poor sampling technique.

Another force pushing us away from an accurately balanced view of reality is the daily news as reported in newspapers, Web sites, and on television. All news organizations are designed for the purpose of reporting the *unusual* rather than the *normal*. The ratings of news stations would drop dramatically if they spent 99.999% of their programming time reporting on all the people who did not commit a murder that day, who did not defraud their investors, and who did not attend the Emmys.

The Literary Digest fiasco illustrates the importance of ensuring that the pool from which we gather our data is as clear of bias as possible. To get an unbiased sample, often the best method is to make every effort to select the sample randomly from the whole population. Choosing randomly gives a better chance of having a representative sample but does not guarantee it. One method of increasing the odds that the sample is representative of the variability in the population is to put a large number of randomly selected representatives in the sample pool. But how big does that pool have to be?

Bad, Bad Biltong

© John Peacock/iStockphoto

Slices of tasty spicy beef biltong from South Africa—"It's what's for dinner."

Biltong is a cured, air-dried beef product developed in South Africa and is a staple in many African countries. It is similar to beef jerky and prosciutto. Another popular South African dish is the meat pie. A few years ago, the Onderstepoort Veterinary Institute, located in Onderstepoort, South Africa, made a shocking announcement. It found that three-quarters of the biltong and half of the meat pies tested by the institute in one month contained horsemeat. There is great danger in eating old racehorses: Consumers might be exposed to steroids, antibiotics, and pesticides. This disturbing finding immediately prompted the local meat board to launch a full-scale investigation, which led to big headlines in the local papers: "Meat Pies, Biltong Made with Horse Meat."

It appeared that the Onderstepoort Veterinary Institute made an important discovery that might, in fact, save the lives of many people. The meat board was willing to spend as much money and time as needed to make sure the meat pie manufacturers' dangerous and illegal act was stopped and punished. Sounds like a happy ending, right? But not in the eyes of the meat pie manufacturers. They were downright outraged at the allegations, because they believed that essentially all meat pie manufacturers produced healthy and safe products, free of horsemeat.

A Case of Insufficient Data

When pressed by the meat pie manufacturers, the Onderstepoort food hygiene division manager, Dr. Werner Giesecke, revealed that the institute's investigation consisted of testing *two* meat pies. Are we as outraged and frightened now as we were after first reading about this incident? Well, probably not. The crucial point is that for data and the interpretation of that data to help us answer questions meaningfully, the data must be accurately collected according to a sampling scheme or experimental design that is not biased, and the data set must be large enough to adequately represent the population. The obvious resolution to the biltong controversy was to stop horsing around and use a much larger sample than just two meat pies.

Are We Asking the Right Question?

We have encountered some pitfalls in collecting a sample of data. We want the data to be accurate, unbiased, and sufficient in number to reflect the population's variability. However, perhaps the most fundamental characteristic we desire from our data is that it be pertinent. We want the data to be useful for the purpose of answering some question. Thus, we discover that gathering data is *not* the first step. Before we even begin the statistical portion of a statistical study, we must face two even more fundamental issues—that is:

• What is the question?
• What role will the data play in answering that question?

When we can clearly articulate what we're trying to discover, we can devise studies or experiments whose results will produce data from which to infer the answer. Asking the right question is often the most important step when deciding what action to take. So in order to determine the right question, we should ask ourselves: How will the answer affect our decisions or actions? If the answer to the question won't help us make a decision, then perhaps we are not asking the right question.

Another Round for the Students

Let's revisit our survey of alcohol use among college students. What question did we really seek to answer, and why? Perhaps our real question is: What can we do to reduce the abuse of alcohol in colleges? If you were the dean of students and the president gave you a budget of $100,000 to reduce alcohol abuse among students, then knowing the percentage of students who were drunk this week might not be the most pertinent piece of information. Once the money is allocated for the purpose of reducing alcohol abuse, your goals are different. You now seek to take action, not further confirm the existence of the problem.

In this new scenario, more valuable information might include which students drink excessively, how students decide whether to drink or not, the social pressures involved, and which interventions were most successful with prior abusers. Only after we have decided which questions are actually important and how we will use their answers are we ready to devise a good method to collect pertinent data.

In this case, surveying all students would not be useful. We are really only interested in those students who have abused alcohol and have been successful or unsuccessful at changing their behavior. So we might get a list of past alcohol abusers (for example, police records of alcohol-related incidents, data from other schools, past membership lists of certain fraternities). From this pool, we can then gather data regarding intervention attempts, successes, and failures.

Deciding the right questions to ask is not always an easy task. Sometimes we need help to focus our attention on the relevant issues.

Sweddar—A Delicious Cheese Blend

As we journey through life, occasionally we come up with ideas that seem truly great—to us. Unfortunately, other people don't necessarily share our enthusiasm. Suppose one day when casually placing cheese on a cracker, we think of a great idea. Why not blend our two favorite cheeses, Swiss and Cheddar, to create a new cheese sensation—*sweddar?* Having previously misjudged the popular response to our last great idea, *Termoil* (a franchise combining term life insurance with an oil change), we decide to investigate whether sweddar will swing or lead to more turmoil. At first, we don't quite know which issues are most important. What will the populace find alluring or dubious about sweddar?

One way to home in on the right questions is to discuss the issue with a small number of people, often referred to as a *focus group*. So, we gather a focus group of a dozen people and ask them what their views are on sweddar. Some love the flavor; some respond to the holes in the orange cheese; some remark on its unique texture; some like the Wisconsin-cow-in-lederhosen logo; many comment on its name. What's our next move? Are we ready to cut into the cheese market?

We might be tempted to calculate what fraction of the focus group liked the name or texture, and so on. However, that strategy would be wrong. The right answer is to use the focus group to accumulate a list of issues and concerns arising in the group rather than to do a statistical study of individual participants' reactions. While this small group of people may or may not represent the whole population proportionally, it will likely bring up most of the issues that concern most people in the actual target population. The value of the focus group is to help us discover the correct questions to later ask an appropriately large, random sample of cheese consumers. For example, we might discover the need to ask whether cheese eaters would prefer the name *chiss*.

Jumping to Conclusions

Statistics helps us use the data we do have to make reasonable inferences about aspects of the world that we may not know, for example, the future. Predicting the future is full of potential hazards. Here are some to avoid.

Extending the Line—Extrapolating to Nonsense

When we look at a graph of how some feature of the world has changed over the years, it is natural to extend the graph to guess what will happen in the future. That practice—known as *extrapolation*—sometimes is valid but sometimes leads to ridiculous conclusions. Here is a graph of the world record times for running a mile during the 20th century. The graph illustrates an obvious trend.

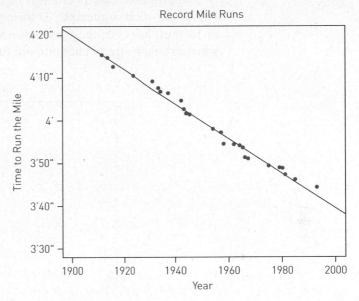

This trend is so clear, it is hard to resist the temptation to extend the graph. If we do extend the graph, we'd see:

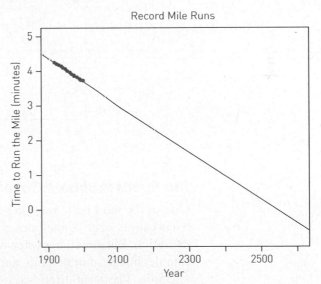

This graph predicts that by the year 2600, runners will be able to complete a mile run before they even start! Wow, that's fast! Even by around 2400, the most talented runners will run the mile in under one minute—now, that's half fast! When we run around the issue of extrapolating trends into the future, there are several serious intellectual hurdles over which we must jump.

A wonderful example of unrealistic extrapolation actually occurred during a brief period from roughly 1997 to 2001 when the U.S. national debt declined. In the early 2000s, we heard politicians discussing how soon the debt would be eliminated and what to do with expected budget surpluses. Unfortunately, the graph took a dramatic upward turn that as of this writing has shown no signs of slowing down any time soon, as shown in the graph below.

Inappropriate extrapolation is one of the most common methods of making bogus statistical arguments. Drawing inferences about the future is valid and important; however, these examples warn us to use common sense and caution when extending trend lines into our future.

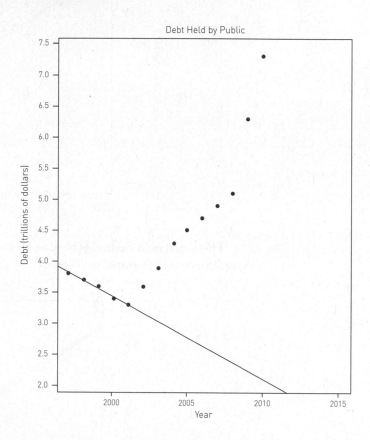

The *Sports Illustrated* Cover Curse

One of the most basic ways in which we understand our world is to think about what causes what. Sometimes we use statistical evidence to support a claim of causation. But sometimes when we make inferences, we drop the ball—we choke, fumble, bungle, double-fault, and foozle. Not good.

Sports Illustrated often features on its cover the athlete or athletes who have recently performed in a notably spectacular way (especially for its swimsuit edition). But one of the legends of appearing on the cover of *Sports Illustrated* is known as the "*Sports Illustrated* cover curse." The "curse" is that athletes who appear on the cover immediately go into a slump. Why is this happenstance actually an explainable and commonsensical reality rather than a force of some evil supernatural sports power? The answer is that athletes appear on the cover of *Sports Illustrated* because they have performed far above their own and everyone's average level of performance. When athletes are at the very top, generally speaking, they will tend to return to their own average, which is lower. So, naturally, they will tend to revert toward their own average and thereby get worse. Curses!

This phenomenon illustrates the concept referred to as "regression to the mean," that is, gathering more data will tend to move toward the overall mean. In baseball, the batting averages of players during the first few weeks of the season will often include a few batters whose batting averages are very high or very low. If you are a betting person, you would do well to bet that those extremely good or bad averages over the course of the season will tend to become more like the average of the whole league.

Quack Medicine—Quack Statistics

Another illustration of this phenomenon of regression to the mean concerns the true observation that quack medicine tends to work. Why do we tend to feel better after taking some miracle elixir, even if the medicine has no actual effect on us whatsoever? There are two reasons. One is the placebo effect—when we think that a remedy will do some good, then it tends to have a good effect because of psychological influences. However, another reason that quack medicines lead to cures is that we take the dose when we are feeling sick. For most illnesses, whether we take medication or not, we later feel better and return to our average health—we regress toward our mean level of health. Since we take the wonder elixir only during those times when we feel worse than usual, then generally we will tend to feel better soon afterwards. Keep the castor oil in your grandmother's medicine cabinet.

Statistics is one of our most powerful and prevalent tools for understanding our world. However, it is up to us to understand how to use statistics wisely by gathering data, describing data, and making inferences from data in a meaningful manner.

A Look BACK

DEALING WITH DATA INVOLVES COLLECTING, describing, and making inferences from the information we have. Our challenge is to gain valuable insight from examining data while not being led astray.

An instructive way to approach statistical issues is to first examine potential pitfalls, because the pitfalls exaggerate and highlight subtle issues of good statistical practice. Many fallacies are possible. Seeing some illustrations of startling errors promotes caution and inspires us to choose appropriate statistical techniques that will lead to valid and useful conclusions.

Looking squarely at mistakes can be a fruitful way to develop insights. The more dramatic and pronounced the error or distortion, the easier it may be to detect the underlying defect. Do not ignore errors—learn from them.

Life Lessons

Beware of invisible spin that might bias our thinking.

•

Don't blithely accept all assertions as fact.

•

Always apply common sense.

•

Learn from mistakes.

MINDSCAPES Invitations to Further Thought

In this section, Mindscapes marked **(H)** *have hints for solutions at the back of the book. Mindscapes marked* **(ExH)** *have expanded hints at the back of the book. Mindscapes marked* **(S)** *have solutions.*

Developing Ideas

1. **Embarrassing data.** Suppose you asked 100 students to answer the question: Have you ever cheated on a test? You used the one-coin method described in this section, and 54 people answered "yes." Estimate how many students in the group have ever cheated.

2. **Hugging both parents (S).** You ask 150 students to answer the question: Do you still hug your parents? You use the one-coin method described in this section, and 83 people answer "yes." Estimate how many students still hug their parents.

3. **Short test.** To save time grading, a professor asks just one question on a test that is graded either right or wrong. Suppose each of the 100 students knows 80% of the answers to potential questions. Approximately how many students will fail the test?

4. **Computer polls.** Suppose dot-com entrepreneurs want to know what percentage of people use the Internet. They send a questionnaire to all e-mail addresses asking whether the recipient uses the Web. What percentage of the respondents is likely to say that they use the Web?

5. **Voluntary grade inflation.** The dean felt that a professor had graded too harshly, so she wrote a questionnaire to all students asking if they would like their grades reevaluated. Of those students who received a grade of A, how many would you think responded to the dean?

Solidifying Ideas

6. **Pornography (ExH).** Suppose you asked 100 students to answer the question: Do you have pornography in your room? You used the one-coin method described in this section, and 98 people answered "yes." Estimate how many students in the group have pornography in their room.

7. **Cartoons.** Suppose 380 students out of 1000 still like to watch cartoons. If you used the one-coin method to sample the 1000 students, approximately how many students would you expect to answer "yes"?

8. **Bias beef (H).** Suppose you have an assignment to estimate what fraction of the students at your college are vegetarians. At the next football game you ask people as they walk into the stands if they are vegetarians. Why might the meaty results of your survey be off?

9. **Drug data (ExH).** You ask 250 students to answer the question: Have you used an illegal drug in the past 72 hours? You use the two-coin method described in this section, and 78 people answer "yes." Estimate how many students in the group have used drugs recently.

10. **Kissing.** You ask 180 students to answer the question: Did you kiss on your most recent first date? You use the two-coin method described in this section, and 50 people answer "yes." Estimate how many students in the group kissed.

11. **Cheating (S).** Suppose 60 students out of 200 cheated during the last year, and you used the two-coin method to survey those 200 students. How many students would you expect to answer "yes" and how many "no"?

12. **Ask them.** In a dining hall or at a gathering of friends (at least 20 or so), ask the participants an embarrassing question and use the two-coin method to extract information. Make sure you clearly explain the instructions and make sure they understand that they will not be revealing embarrassing information! Record the outcomes, estimate the answer to the question, and record the group's reactions.

13. **Dental hygiene.** Suppose you want to know how often people brush their teeth every day, so you take a survey. Explain why the survey results would indicate far more gleaming smiles than actually exist. (In fact, such a survey would indicate about twice as much brushing as actually happens.)

14. **More homework.** In a recent survey of random college students, 50% of them believed that there should be more homework assigned in all classes. What is your guess as to the sample size of this survey? Explain your answer.

15. **Amazing stats.** Think of an appropriate question on a specific topic (politics, fashion, or music, for example) and then ask your friends to answer the question. Try to construct the question so that the statistical result you generate (some sort of average, perhaps) is not accurate due to either an extremely small sample size or a biased sample. Record the question, sample size, and the dubious conclusion. Explain the bias.

Creating New Ideas

16. **PBS.** In the old days, the Nielsen ratings of TV-watching habits would ask people to fill out a card indicating how much TV they watched and which shows they watched. Why did they get different results when they used electronic monitoring devices instead?

17. **9:00 AM versus 9:00 PM (H).** You take a survey of students in the library at 9:00 AM on a Saturday morning and ask: How many hours per week do you study? You ask the same question at 9:00 PM at a wild party. What would you expect to see in your data? Explain the biases.

18. **Internet askew.** Find ads or articles on the Web that show some type of graph (pie chart, bar chart, etc.). Print out or describe your examples. Analyze

each graph for bias or distortion. Explain how you might alter the display of data to reduce or remove possible bias.

19. **Sleazy survey (H).** The mayor of a city seeks to know the hot issue for the next election. He sends a letter to each of the 100,000 households in the city asking whether pornography vendors should be shut down. A majority of the 8000 respondents say to shut down the sleaze. The mayor runs on the clean-up-the-dirt platform and loses badly. What went wrong?

20. **Bread winners?** You want to know in what percentage of households people bake their own bread. Will it make a difference whether you take your survey in the evening or in the afternoon?

Further Challenges

21. **Beemer babies (H).** You want to estimate what percentage of people in your town own a BMW. You call 100 people at random and ask them whether they own a BMW. Even assuming that everyone answers truthfully, why will you still not know accurately what percentage of people own BMWs?

22. **Coffee, tea, or milk?** You're considering whether to start a coffee house near campus. Surveys indicate that a significant percentage of students drink coffee. Is that information what you would need to know to make your decision? Discuss what additional information you would seek before making your decision.

23. **The dog pound.** Otis the puppy grew approximately 11.6 pounds per month between 2.5 and 4 months of age. The graph below shows a line that closely fits the data points. Use the equation of the line given below to predict Otis's weight at 3 months and at 6 months. Explain why you would be less confident in your second estimate.

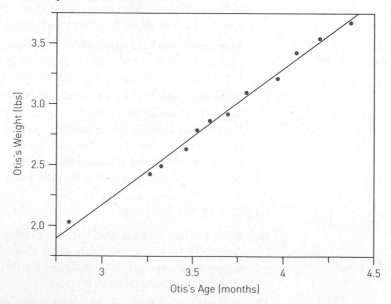

Equation of line

Otis's weight (in lbs) $= -13.4 + 11.6 \times$ Age (in months)

In Your Own Words

24. **Personal perspectives.** Write a short essay describing the most interesting or surprising discovery you made in exploring the material in this section. If any material seemed puzzling or even unbelievable, address that as well. Explain why you chose the topics you did. Finally, comment on the aesthetics of the mathematics and ideas in this section.

25. **With a group of folks.** In a small group, discuss and actively work through the reasoning involved in collecting accurate answers to embarrassing questions. After your discussion, write a brief narrative describing the methods in your own words.

26. **Creative writing.** Write an imaginative story (it can be humorous, dramatic, whatever you like) that involves or evokes the ideas of this section.

27. **Power beyond the mathematics.** Provide several real-life issues—ideally, from your own experience—that some of the strategies of thought presented in this section would effectively approach and resolve.

For the Algebra Lover

Here we celebrate the power of algebra as a powerful way of finding unknown quantities by naming them, of expressing infinitely many relationships and connections clearly and succinctly, and of uncovering pattern and structure.

28. **How's that again?** An ambitious politician in your town describes a recent survey of 100 voters. He claims the survey showed 65 respondents agreed with his policies and only 38% disagreed. What do you think of this politician? Explain.

29. **Smells like team spirit.** A losing basketball team at your school has recently suffered a large slump in attendance at games. Suppose 1200 students attended the first game, 1150 attended the next game, 1100 attended the game after that, and this pattern continues for several more games. Write a formula that matches this pattern, giving the number of students attending a game, y, in terms of the number of games since the first game, x. So when $x = 0$, $y = 1200$, and when $x = 1$, $y = 1150$, and so on. If there are 40 games in the basketball season, how many students will attend the last game, according to your formula. Is this a reasonable prediction? Explain.

30. **Stack me timbers.** Every October, students from the Senior Class at your school stack timbers on the quad for a large bonfire at homecoming. The timbers are stacked in layers. Each year, the seniors try to outdo the previous year's effort by creating a timber stack that is one layer higher than the previous year's stack. Suppose the stack was 12 feet tall in 2010 and that each layer is 12 inches high. Copy the axes shown and plot points showing how high the stack will be for each of the years 2010, 2015, 2020, 2025, and so on through 2050, if you assume that each year a stack is built with one more layer than the previous year. How high (in feet) would the stack be in 2050? Does this seem possible? Does it seem safe?!?

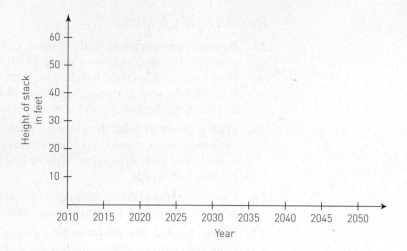

31. **Risky business (H).** In a psychology class, the instructor uses the two-coin method when asking the question, "Have you ever had unprotected sex?" Of the students who have, ¾ will respond truthfully (answering "yes") and ¼ will lie (answering "no"). Let U equal the number of these students who answer "no". Thus the number of those who've had unprotected sex is $4U$. Of the students how have not had unprotected sex, ¾ will respond truthfully (answering "no") and ¼ will lie (answering "yes"). Let N be the number of these students who answer "yes". Thus the total number of students who answer "yes" is $N + 3U$ and the total number of students who answer "no" is $3N + U$.

    Suppose $N + 3U = 32$ and $3N + U = 48$. Solve these two equations in two unknowns. According to the results of the survey, how many students have had unprotected sex?

32. **Risky II.** The psychology instructor from the previous Mindscape asks a second class the same question. This time the results are $N + 3U = 40$ and $3N + U = 64$. According to the results of the survey, how many students in this class have had unprotected sex?

Organizing, Describing, and Summarizing Data

©AP/Wide World Photos

> *You can observe a lot by just watching.*
>
> YOGI BERRA

Suppose you know it all. You know every grade of every student in the history of your school. You know every batting average of every major league baseball player since Doubleday invented the game. You know the Dow Jones average for every day for the last century. You have all the data, but what do all the numbers mean? What can we do to data to entice them to reveal their hidden secrets? Assuming that we have accurate and comprehensive data, how can we cull understanding and insight from large mounds of numbers?

The first three rules of statistics are: Look at a picture, look at a picture, look at a picture. A good visual representation of data would (or should) reveal patterns and relationships. A graph may show important features of the data, such as where the data are centered, whether the data are tightly clustered or spread out, whether the data create a telling form, whether there are unexpected values that do not seem to fit in with the rest. Graphical representations can often be used to tell the story hidden within the data. By organizing data in visually informative ways, we will be led to arithmetical concepts of shape, center, and spread. Choosing meaningful summaries of data, displaying the data artfully, and describing how the data are distributed over their entire range all help us to better understand the otherwise cryptic list of numbers that hold, but may hide, the answers we seek.

Visualizing Data

To attain a global impression of a set of data, the most basic and important step we can take is to simply *look at it*. Visualizing data through graphics and pictures can draw

The Art Archieve at Art Resource

Life Lesson
Whenever possible,
visualize a situation.

our attention to the heart of the story the data tell—in many cases, the picture is far from pretty, but it makes the reality come to life. Here is a graphical rendering of a horrific military campaign that Napoleon undertook about two hundred years ago.

Napoleon

Perhaps the most famous example of conveying data through graphical representation is Charles Joseph Minard's 1861 thematic map of Napoleon's 1812–1813 invasion of and retreat from Russia. This diagram dramatically portrays the disastrous fate of Napoleon's army. The diagram displays a wide variety of information. We see the size of the army (represented by the width of the arrow) dwindle down from a great river to a tiny trickle. We see the locations of the troops during the campaign as well as the corresponding freezing temperatures that contributed so decisively to the decimation of Napoleon's army.

In *The Visual Display of Quantitative Information,* Edward Tufte anoints Minard's diagram as the best statistical graphic ever drawn. Amazingly, this one illustration shows relationships among four varying quantities: date, temperature, size of army, and location. We can all take Tufte's statement as a challenge to use our imaginations to present our ideas and statistical data in visual formats that help the viewer to see significant patterns and relationships.

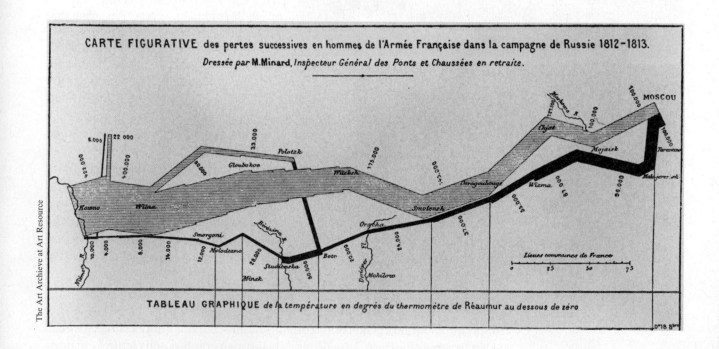

Fortunately, we don't need such artistry to display data in a way that effectively conveys meaning.

Donuts and Pie Charts

Pie charts have culinary appeal as well as being helpful in describing how resources are divided. Here is some donut data to digest:

	Donut Production (in millions of dollars)	Donut Production (in millions of donuts)
Entenmann's	$129.1	41.0
Krispy Kreme	102.5	30.7
Hostess	94.9	32.2
Private Label	80.3	35.7
Dolly Madison	38.6	16.3

Top five donut brands based on data from Information Resources Inc., Chicago, Illinois, in *Snack Food and Wholesale Bakery*, June 2002, p.SI-33.

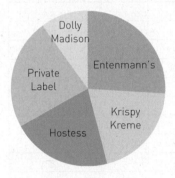

Now look how much more delicious and digestible that same information can be when we use a mouthwatering pie chart that lets us almost taste the data.

This graphic also gives us the opportunity to display possible market share changes in an emphatic way without glazing over anything. Suppose your company had lost ground in the market and you were making a presentation to executives to break the bad news. You could use this graphic to show how the hungry competitors have taken a bite out of your share of donuts.

Bad Grades

In their rawest form, numerical data are simply a collection of numbers—in the same sense that a person is simply a collection of cells. The meaning in the data resides in how those numbers are distributed. The *distribution* of the data refers to the different values the data attain and how many data points attain each value. Understanding how the data are distributed can allow us to answer the questions: Are the data clustered altogether, spread evenly throughout the range of the possibilities, or clumped at several values? How can we describe the distribution of data? Even more basic: What does the phrase "distribution of data" mean precisely? So let's develop some insights by considering an example.

Suppose last semester you took the course "The Open-Minded Self" in the Department of Enlightened Thought and Openness. The whole grade in the course was based on your score on the final, which had 100 points total. You received a C− in this course, which sounded a bit low given your final score of 80. When you approached the instructor about this issue, he curtly dismissed you and said he refused to even consider a grade change on principle alone (so much for the open-mindedness that he preached). You then moved to the chair of the department and made your case. You had the final scores of the entire class of 200 students (they were posted, without names) in r andom order.

62	78	52	63	61	83	69	85	83	51
87	54	87	78	94	97	94	94	60	96
81	87	82	81	69	52	87	63	69	94
51	75	87	87	95	92	91	92	90	91
66	80	58	51	65	77	65	55	94	66
88	72	74	88	90	72	95	96	41	46
86	67	89	85	89	88	62	88	92	98
76	22	86	77	90	73	86	65	88	73
65	89	85	96	60	89	91	89	84	97
90	65	75	85	85	97	85	75	59	86
59	89	80	81	67	59	21	96	88	97
86	85	56	96	94	96	85	68	93	76
94	65	68	74	92	81	97	53	79	74
63	53	92	93	83	56	81	66	88	64
86	78	72	88	93	93	85	92	72	94
90	85	82	57	92	88	97	95	50	86
50	75	94	95	76	64	59	77	88	97
69	90	66	78	68	92	78	98	78	91
95	86	92	95	90	93	66	82	89	56
84	71	95	82	95	84	72	89	84	85

Final scores

Well, if you just plunk down this list of numbers and say, "Look at these scores. I deserve a higher grade," then the reaction of the chair might be similar to that of your esteemed professor.

You Snooze, You Lose

When faced with a seemingly endless ocean of numbers, it is easy to feel dragged down by the undertow of information. Disorderly tables of numbers have a profound effect on most people. Their eyelids start to droop, and they fall into a restful snooze. Of course, many people have trouble sleeping, so escorting people to dreamland may be one of the important uses of data. In fact, sleeping might very well be the appropriate reaction to a seemingly endless list of data, because unstructured data have no meaning. The important step of organization and interpretation is required before data can shake us from our slumber by shedding light on an issue.

The chair, after waking up from her nap and examining the data of the raw scores, would laugh you out of the office, because a list of disordered numbers doesn't mean anything. To make your case, you have to organize your data. You can start by putting the final scores in numerical order.

21	59	66	73	78	85	87	89	92	95
22	59	66	73	79	85	87	89	92	95
41	59	66	74	80	85	87	89	92	95
46	59	66	74	80	85	87	90	93	95
50	60	66	74	81	85	87	90	93	95
50	60	67	75	81	85	88	90	93	96
51	61	67	75	81	85	88	90	93	96
51	62	68	75	81	85	88	90	93	96
51	62	68	75	81	85	88	90	94	96
52	63	68	76	82	85	88	90	94	96
52	63	69	76	82	85	88	91	94	96
53	63	69	76	82	86	88	91	94	97
53	64	69	77	82	86	88	91	94	97
54	64	69	77	83	86	88	91	94	97
55	65	71	77	83	86	88	92	94	97
56	65	72	78	83	86	89	92	94	97
56	65	72	78	84	86	89	92	94	97
56	65	72	78	84	86	89	92	95	97
57	65	72	78	84	86	89	92	95	98
58	65	72	78	84	87	89	92	95	98

Even that simple step helps, but it doesn't help much. At least now you could show that the lowest score was 21% and the highest score was 98%. That's something, but it doesn't give us a very refined sense of whether most of the scores were nearer the high end or the low end. So to see what the data tell us about the scores, it helps to display them visually.

One method for displaying the data is to record each score using what is called a *stem and leaf plot*. We just write down the numbers 1 through 9 in a column with a vertical bar after each digit. In each row, we record the scores that start with that digit in an efficient way. For example, looking at our table of scores, we see that two people scored in the 20's—a 21 and a 22. So in our stem and leaf plot, after the "2|" we write first "1" to record "21" and then a "2" to record "22." So "2|12" means there are two scores in the twenties, specifically, 21 and 22.

1|

2|12

3|

4|16

5|00111123345666789999

6|00122333445555556666677788889999

7|112222233444555566677778888889

8|0011111222233344445555555555566666666677777788888888889999999

9|00000000111122222222223333334444444444555555556666667777777788

This stem and leaf plot gives us some visual sense of where the scores fall while retaining every individual value. Other visual methods show us where the scores fall, but do not record every single value.

An effective way to convey a visual sense of all the data is through the use of a histogram. A histogram is a good method for saying how many people have scores in different ranges. What we do is group similar scores together and record how many scores fall in each interval. In this case, let's use 2-point increments—99 to 100, 97 to 98, and so on—and then for each interval, we just draw a bar whose height represents the number of students whose score was in that range. For example, the height of the highlighted bar on the histogram indicates that approximately 19 people received scores of 85 or 86.

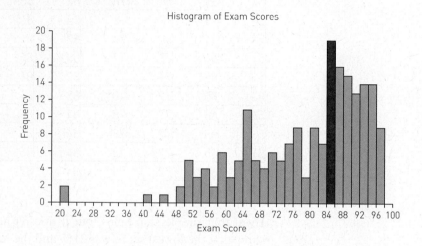

When we take in the panoramic view of all of the bars of this histogram at once, we notice that the bars come together to form a shape, and the shape of the data is a good way of getting some sense of the distribution of the grades, that is, the values of the data—in this case final scores—and how many scores lie in each of the ranges. Thus the shape of the histogram conveys the distribution of the data.

Now, notice the histogram doesn't need to be in 2-point increments. That was just a choice we made. We could have chosen 5-point increments, and we would

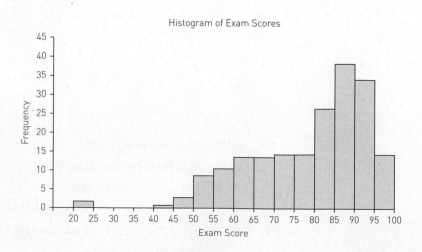

get a different histogram of the same data, a different kind of visual summary of the same data.

Notice that if we took very small increments—for example, if the scores were computed to a tenth of a point and we used just 0.1-point increments—then it may be that our histogram would give us no useful information because perhaps no two people received the exact same final score when computed to the nearest tenth of a point, and therefore, we would just have a collection of equally tall bars (of height 1). Choosing the size of the groupings is an important decision in creating a histogram that gives us a good sense of the shape of the data.

When we show the chair a histogram, we can see where our grade of 80 lies in relation to the rest of the class and we can make a good argument. We can see that many students had far lower scores, so we can argue that C− is too harsh. The relative position of the score of 80 in the whole stream of scores is also clear in the table of ordered scores, but the histogram helps us to see how all the scores are distributed over the whole range.

In creating a histogram, our goal is to create an image that reveals the structure that the data contain, and histograms can be an excellent tool for visually capturing how the data are distributed. Let's keep following Yogi Berra's advice; let's just "watch" some more data and "observe" what we can in order to develop some concepts that will help us to describe what we see.

Emily Cockenlocker's Height

Let's look at an example of a set of data that might come from measuring not the grade of an individual student but the student herself. Let's start with Emily Cockenlocker, a first-year student, and ask her to stand up straight. We use a fancy laser measuring device, which reads 5′2.51″. Then, for practice, we invite our teaching assistant to use the device to measure Ms. Cockenlocker. The fancy laser reads 5′2.37″. This discrepancy confuses us. Which measurement is correct: 5′2.37″ or 5′2.51″? We're not sure, so we carefully measure her again. The silly instrument now reads 5′2.62″. Okay, now we're getting annoyed. Which measurement is her correct height? Since we're not sure, we ask 100 different people to carefully measure Ms. Cockenlocker, and we discover that we have anything but a consensus. On the contrary, we find that the 100 measurements are close to one another, but they are not exactly the same. Measurements of values such as heights are inherently imprecise. The measuring device has limited precision and other variables such as how straight the subject is standing make precision impossible.

In each measurement the laser records a value to a hundredth of an inch. Now let's look at the data graphically. To get a sense of how the measurements are distributed, we could create a histogram of the measurements by grouping them in, say, 0.1-inch height intervals ranging from 5′2.00″–5′2.09″, to 5′2.10″–5′2.19″, and so on until 5′3.00″–5′3.09″. For each of those ranges, we count how many of the measurements fall within each range. Now we create a histogram

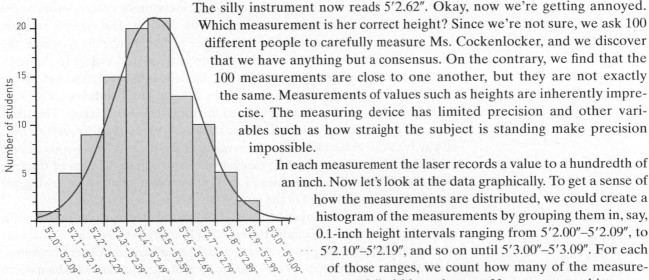

of the data by putting the range intervals along the horizontal axis and the number of measurements within each interval along the vertical axis.

Looking at the 100 individual measurements in this graphical way gives us a pretty good idea about what we should record as Ms. Cockenlocker's height. It is reasonable for us to record her height as the value about which the data seem to be centered. Notice that this value may not be the height recorded by any individual measurement.

This example has brought to our attention a feature of interest regarding a data set, namely, a center value. We have not yet pinned down how to compute this central value, but we have an intuitive idea that the center of a set of data is a value of interest—in this case it gives us our best estimate of Ms. Cockenlocker's height. That central value gives us one way to summarize an entire collection of data.

Measures of Center

Perhaps the most familiar way to summarize a data set is to boil it all down to a single value. A single number that provides an overall sense of the data is referred to as a *measure of center.*

In thinking about measures of center, it is natural to consider an "average," but we caution that the word *average* has different meanings in different contexts. When your grade point average is computed, it means that your grade points for each credit hour are added up and divided by the total number of credit hours. However, when you read in the newspaper that the average house in a city sold for $220,000 last year, that probably means that the sale prices of all the houses were lined up and $220,000 was the middle value in that ordered list of house prices. The price of an average house was probably not found by adding up all the sale prices and then dividing by the number of houses sold. Since grade point "averages" and house price "averages" mean different things, let's be precise about what we mean by *mean* (and *median*).

"Averages": The Mean and the Median

When we use the term *average*—that is, some measure of center—what we're commonly referring to is either the mean or the median. The *mean* is the sum of all the numerical data divided by the number of data points; the *median* is the middle data point when the data are lined up in numerical order. To illustrate these measures of center, let's consider our data about final exam scores above. When we add up the 200 scores and divide by 200, we find that the mean is 79.21. In the chart above we saw the scores lined up from smallest to largest. The 100th value is 84 and the 101st value is 85. Since there is no exact middle value, we take halfway between those two to be the median, in this case 84.5. If we visualize a set of data using a histogram, the median will occur where the area of the bars is equal on either side and the mean occurs at the point at which the histogram would balance if we thought of the bars as weights on a teeter-totter.

Each of these two "averages," the mean and the median, can give useful information. The data set often dictates the appropriate average to use. In the Lakeside School example from the previous section, Bill Gates's enormous income caused the mean to be unrepresentative of the incomes of most graduates. The median would be a far more meaningful piece of information than the mean for assessing

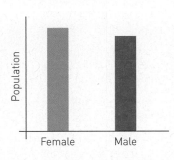

the earning potential of typical graduates. However, if you buy gas every week for 10 weeks and record how much you pay, then the mean amount you spent would probably be more useful, because the mean price times 10, the number of weeks, would correctly tell you the total amount that you spent.

In some cases neither the mean nor the median captures the spirit of the data. For example, in the case of counting the testicles and ovaries of people in the United States, the mean number of testicles and ovaries (one testicle and one ovary per person) is misleading. But so is the median: Because there are slightly more females than males, the median American has no testicles and two ovaries, a summary of the physiology of Americans that might lead sports equipment manufacturers to halt their production of jockstraps.

So the mean and the median are each measures of center that may be useful summaries of the values in a data set; however, they must both be used with common sense. Let's look at two other data sets with an eye toward isolating another characteristic that will help us understand their story.

Batter Up

We used Yogi Berra's quote at the beginning of this section, "You can observe a lot by just watching." Since Yogi was a catcher for the New York Yankees, let's now look at some major league baseball statistics. The batting average of a major league baseball player basically records what percentage of his at-bats resulted in a safe hit. As it happens, the mean of the batting averages over all major league baseball players in the year 1920 and the mean in 2000 were almost identical (.266 and .265). Incidentally, our data consist only of players who recorded at least 80 at-bats so that each datum is a legitimate batting average. Here are histograms of the batting averages for those two years.

Life Lesson

Determine the most appropriate method of distilling a vast body of information down to one digestible nugget.

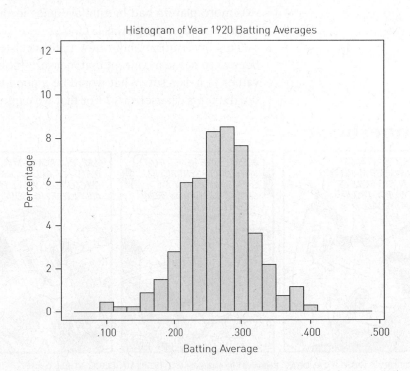

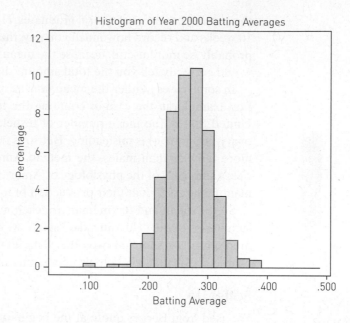

Histogram of Year 2000 Batting Averages

These two histograms are very similar. To make them comparable, we made the height of each bar equal to the percentage of batters whose averages fell in the range. So in both histograms the sum of the heights is 100%; in other words, the total area of the bars is the same in the two histograms.

So what is the difference between these two data sets? Well, the feature that may strike us is that the batting averages in 2000 are more tightly clustered around the central average value compared with the 1920 batting averages. In other words, in 1920 there was more variation among the batters' averages. For example, in 1920, one batter, George Sisler, had a batting average over .400 and two more players had batting averages in the .380s. In 2000, the highest major league batting average was .372.

This observation about how tightly clustered the batting average values are begs us to define a concept that measures how much variation there is among the values in a data set. What would be a good way to measure the extent to which the data in a data set vary? For that we explore the spice of life.

Doonesbury

BY GARRY TRUDEAU

Variation

Life would be dull if nothing ever varied. Recall that variation (or something like that) is the spice of life. How dull it would be if all people looked alike, weighed the same, earned the same, had the same batting average, thought the same, favored the same candidate, watched the same TV shows, and drank the same brand of orange juice. But happily for humanity, life and the world present all manner of variation. Every aspect of our lives or our world that we measure presents us with a bewildering sense of variety—which *truly* is the spice of life.

Measuring Variation

To describe a data set, we saw that it was useful to describe where it is centered, and we saw in a visual representation that some data sets are more spread out and others are more tightly clustered together. We now want to devise methods to *quantitatively* measure how *much* variation there is among the data.

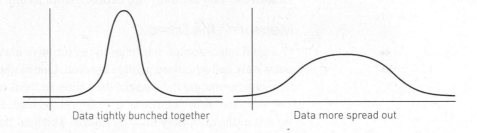

Data tightly bunched together   Data more spread out

Five-Number Summaries

One way to quantitatively describe the spread of the data is to record the range of the entire data set—that is, the minimum value and the maximum value. In the case of our previous list of final exam scores, we see that the minimum is 21% and the maximum is 98%. But that information does not tell us a lot about how the data are distributed between those two extreme values. It does not tell us how many of the grades are near the minimum, how many are near the middle, and how many are near the top. We saw that information visually in the histogram, but it would be useful to measure that spread numerically. One piece of information to add to our developing concept of how the data are arranged over their range is the median value. With the minimum, the median, and the maximum values, we get a sense of whether most of the values are nearer the minimum or the maximum or approximately in the middle. The median may be quite far from halfway between the minimum and the maximum. In the case of the 200 scores, the median value is 84.5, which is much closer to the maximum than to the minimum.

To get a more refined sense of how the data are spread out, we could look at the range of the bottom quarter of the data, the second quarter, the third quarter, and the top quarter of the data. These four intervals give us a quantitative indication of the distribution of the data. There are five data values involved—the minimum value; the value of the datum (called *the first quartile*) below which lie one-quarter of the data points; the median (called *the second quartile*); the value of the datum (called *the third quartile*) below which lie three-quarters of the data points; and the maximum value. These values give us a five-number summary of how the data are distributed over their range.

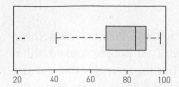

Box and whisker plot for final exam scores data

The five-number summary can be visualized using a *box and whisker plot*, which has the advantage that its name is easy to associate with its looks. The box extends from the first quartile to the third quartile, while the line in the box occurs at the median. The span of the box contains the middle 50% of the data. Then lines extend out the ends of the box and end with hash marks to indicate the minimum and the maximum values that are not outliers. Finally, dots are used to individually point out values that are *potential outliers*. Potential outliers are values that are more than one and one-half times the distance from the first quartile to the third quartile (that is, the size of the box) above the third quartile or below the first quartile. The choice of what criterion to use for designating a value as a potential outlier is rather arbitrary, and alternative guidelines might be used in the context of different data sets. Box and whisker plots are both cuddly and informative. They also have the virtue that the median and the quartile values are *resistant measures*, in the sense that extreme values do not influence the median or quartiles.

A different way to describe how spread out the data are is to give a numerical measure of how far away the data are from a central value.

Measuring the Spread

Our goal is to associate a number to the intuitive idea that some data sets are spread out widely and others are tightly clustered. One method of measuring the spread is to begin by computing the *mean* of the values to focus our attention on a central value. The next step is to determine *how far,* on average, each value is from the mean. If we are in the incredibly unusual case in which all the data in our data set have the same numerical value, then we know that our data set contains no variation (that is, there is no *deviation from the mean*). However, in all realistic scenarios (in which the data are not all equal) we want to know how much the data vary from the mean. For example, if we have computed that the mean height of all students at a university is 5′7″, then we could determine how far any given student's height deviates from that mean—for example, a height of 5′10″ deviates 3 inches from the mean, and a height of 5′2″ deviates 5 inches from the mean. We could then calculate how much, on average, the heights differ from the mean. That measure of average distance away from the mean is a single number that gives us some measure of the spread.

Standard Deviation

In practice, a slightly more delicate (and more complicated) measure of the spread of the data, called the *standard deviation,* is frequently used. The term "standard deviation" reminds us that it is a measure of how far the average data point differs (or deviates) from the mean. The standard deviation is basically the square root of the average of the squared distances from the data to the mean,* but in practice, we

---

*The formal formulaic definition of standard deviation when we have all the values in the population is:

$$\sigma = \sqrt{\frac{(x_1 - \mu)^2 + (x_2 - \mu)^2 + \cdots + (x_{n-1} - \mu)^2 + (x_n - \mu)^2}{n}}$$

where the x_i's are all the values in the data set and μ is the mean. The standard deviation of a sample is defined with $n-1$ in the denominator when we are dealing with a random sample taken from a larger set of data. We will discuss this distinction in Section 9.4 when we talk about inferring facts about the whole population from facts about a sample.

would use a spreadsheet, a calculator, or a statistical computer program to calculate the standard deviation. We'll see later that the standard deviation provides us with important information about the distribution of data in certain bell-shaped distributions. In fact, the reason that the standard deviation is used so much is that the mean and the standard deviation completely determine the shape of the bell-shaped distributions called *normal distributions* that arise commonly in many situations. We will talk more about normal distributions in a later section.

Let's develop a numerical sense of the standard deviation from some of our previous data sets. The standard deviation of the measurements of Ms. Cockenlocker's height (see page 673–674) is approximately 0.2″, which we can see visually gives a reasonable estimate of an average of the distances between the mean and the individual data points.

In the case of the baseball data (see page 675–676) the standard deviation for the 1920 batting averages is .050, while the standard deviation for the 2000 batting averages is .038. As it is intended to indicate, the smaller standard deviation for the year 2000 batting averages indicates that the batting averages are more tightly grouped in 2000 and more spread out in 1920.

The standard deviation for the final exam score data is 14.782. Notice that it is quite large because the very low scores that are outliers strongly affect the standard deviation as well as the mean. Thus these examples illustrate that the greater the spread of data, the larger the standard deviation, and conversely, the more tightly clustered the data are around their mean, the smaller the standard deviation.

A Gallery of Data—Shapely Graphs

So far we have isolated two features of data sets that help us describe them—a measure of center (the mean or median) and a measure of variation or spread (either using a five-number summary or standard deviation). In looking at graphs of data, however, perhaps the most conspicuous differences are the actual shapes of the graphs themselves. Let's look at several graphs with an eye toward identifying characteristics of their curviness.

How can we describe shapes of histograms effectively? In looking at our previous graphs, several characteristics of their shape jump out at us. The measurements of Emily Cockenlocker and the batting averages in both 1920 and 2000 were all essentially symmetrical around a central value. The final exam scores were not symmetrical since there were a few students with very low scores compared to the number with very high scores. We describe such graphs as *skewed to the left*, where the direction of skew indicates the side with the longer tail.

One feature of skewed distributions is that the mean is pulled in the direction of the skew relative to the median. To understand why, think of starting with a symmetric distribution in which the mean and the median are the same. Now move some of the values to the right, as would occur in a distribution that is skewed to the right. The mean will increase while the median will stay the same.

Salaries

Often salaries at a corporation are skewed to the right. As we move up the management chain, the numbers of individuals involved goes down while the salaries may rise considerably. Because the salary distribution is skewed to the right, the mean is higher than the median. The median gives a sense of what the typical employee earns; however, the mean is useful if you want to compute the total payroll of the company by multiplying the mean by the number of employees.

Often the big cheeses at the top give us a chance to revisit the idea of "outliers"—data that are relatively out of the range compared to the other data. We have computed the mean and the standard deviations of these salaries with and without the outlier's very high salary. Notice the significant difference that the one outlier makes in the mean and in the standard deviation. Also notice that the median is not much affected by adding the outlier.

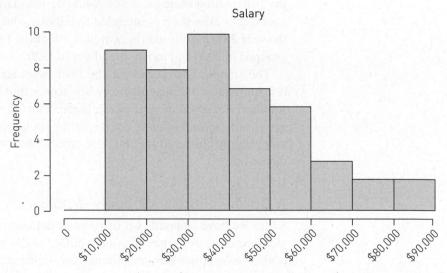

Mean is $44,000: Standard Deviation is $19,000

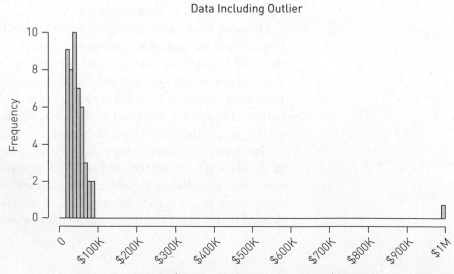

Mean is $64,000: Standard Deviation is $138,000

House Prices

Here is a histogram of house prices in a city. Again, this data set is *skewed* to the right. Again we see the mean pulled in the direction of the skew relative to the median.

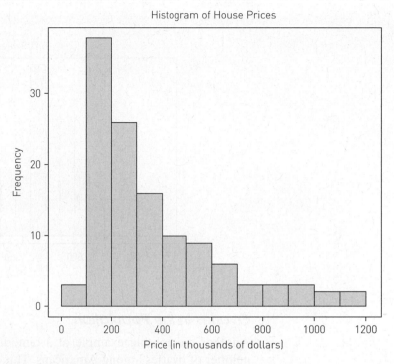

Histogram of House Prices

Bimodal Distributions—Two Populations

Some distributions have two peaks. These are called *bimodal*.

What's the Score?

Woody Allen said that 80% of life is showing up. Showing up helps in math class, too. One semester, a particularly flaky section of a math class had some attendance issues. Half the students attended regularly and absorbed the wit and wisdom of the professor while the other half were slackers whose career paths might have a few potholes up ahead. The final exam scores reflected the students' attendance habits, as shown here.

The exam scores range from a low score of 4% to a high score of 98%, and we notice that the scores seem to cluster around two values—80% and 30%. The shape of the graph tells the story. The scores from students who studied diligently would be distributed around a high score, whereas those from students who partied diligently would hover near a lower mark. Notice that there is variation within each group because we would not expect every prepared student to get everything right nor every unprepared student to get everything wrong. In bimodal distributions neither the mean nor the median score is typical. A bimodal distribution of scores could indicate to an instructor that the class contained two groups of students, each experiencing the class quite differently.

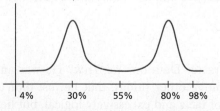

Bimodal distribution with peaks at 80% and 30%

Life Lesson
Find useful representations.

SATs in 7th Grade/11th Grade

Suppose at some school every 7th grader and every 11th grader takes the SAT (the version with a top combined score of 1600). The 7th graders, of course, tend to get much lower scores than the 11th graders. So the collection of all the SAT scores has two peaks and is therefore a bimodal distribution.

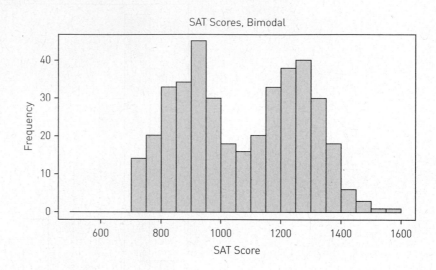

Ovaries in the Population

Another memorable example of a bimodal distribution is the data about the number of ovaries among Americans. This simple graph is clearly bimodal, with one peak at 0 and the other at 2.

The characteristics symmetric, skewed, and bimodal give us a vocabulary for describing shapes of distributions; however, these descriptors do not really pin down the shape with much precision.

Ms. Cockenlocker's Height and Baseball

The measurements of Ms. Cockenlocker and the major league batting averages both were symmetric and had a single peak. We will see many more such bell-shaped curves in the next section.

Overview of Statistics

Do you remember the expressionless character Data on *Star Trek*? He was an android and his name Data suggested, among other things, that he just didn't quite catch on to human life. He was full of facts and excessively logical, but he was confused about how to interpret the facts. Statistics is the study that helps us transform meaningless data into meaningful creative ideas.

The challenge of statistics is to make sense of data. We can break that challenge into two parts: (1) How can we describe and draw meaning from a collection of data when we know all the pertinent values? (2) How can we infer

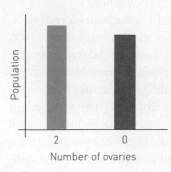

information about the whole population when we know data about only some of the population (a sample)? These two questions form the structural backbone of statistics. In this section we introduced some important concepts and strategies for making sense of data when we know all the data. In future sections we will further explore the shape of data and will investigate questions of how to estimate the characteristics of the whole population from a sample.

DATA CAN HOLD MANY IMPORTANT messages for us, but to find those messages we first must present the data in meaningful ways that make patterns in the data obvious. Looking at data by presenting it in visually clarifying ways is often an excellent first step. Various graphs are useful for seeing the data as a whole. Some people describe this country as a data-driven society, but, in fact, we're not a society driven by data, we're a society driven by the interpretation of data.

One of the most important features about data is the variability it displays. If the data did not show variety, summarizing it would be easy. The different values actually give us a distribution of values, and understanding and describing the variation and distribution of data is one of the goals of statistics. Even in measuring the same object, we find that the measurements vary, and by describing the variation, we are getting at the truth of what is happening. Various means of summarizing data are useful. Among these are the following:

- Graphs and pictures can show the shape of the data.

- One-number measures of the center of the data, such as the mean and median, give us a summary of that data, but they don't help us understand the variation and distribution of the data.

- Five-number summaries of data showing minimum, first quartile, median, third quartile, and maximum data values give us some idea of how much variation there is in the data and how the data are distributed over their entire range.

- The standard deviation gives a numerical measure of the spread of the data.

- Descriptions of shapes of distributions (symmetric, bimodal, skewed) give us a qualitative sense of the shape of the data.

To understand data, we focused on our goal: to describe and display the distribution of the data set. Visualization, representation, and highlighting essential concepts were important themes in the process.

Life Lessons

Search for the most effective means of making your case.

•

Whenever possible, visualize a situation.

•

Find useful representations.

In Mindscapes 13–18, you are asked to analyze the data from a class survey. The survey is given below. Your instructor may ask you to submit your responses.

Class Survey: How Do You Measure Up?

Answer the following twenty-question anonymous survey honestly. DO NOT PUT YOUR NAME ON YOUR RESPONSE SHEET, AND DO NOT ANSWER ANY QUESTION YOU WOULD RATHER NOT ANSWER. For each question, circle the one answer that most closely reflects the correct answer in your case. Give only one answer per question. Your instructor will collect, compile, and return the data from all the students in the class to you for further analysis.

1. What is your gender?

    (1) Male   (2) Female

2. How much do you weigh?

    (1) Less than 90 lbs   (2) 90–119 lbs   (3) 120–149 lbs   (4) 150–179 lbs   (5) 180 lbs or more

3. What is your height?

    (1) Below 4′6″   (2) 4′6″–4′11″   (3) 5′–5′5″   (4) 5′6″–5′11″   (5) 6′ or above

4. What is your favorite part of your own body?

    (1) Hair   (2) Face   (3) Arms   (4) Stomach   (5) Legs

5. On average, how often do you call home?

    (1) Once every two months   (2) Once a month   (3) Twice a month   (4) Once a week   (5) Several times a week

6. How often do you do your laundry?

    (1) Every time you go home   (2) Once a semester   (3) Once a month   (4) Twice a month   (5) Once a week

7. How often do you drink alcohol?

    (1) Never   (2) Once or twice a month   (3) Once or twice a week   (4) Three or four times a week   (5) More than four times a week (seek help)

8. What is the average number of dates you go out on per month?

    (1) At most 1   (2) 2–3   (3) 4–5   (4) 6–7   (5) More than 7

9. When did you see your last movie in a movie theater?

    (1) At least 2 months ago   (2) 1 month ago   (3) 3 weeks ago   (4) 2 weeks ago   (5) Last week or this week

10. How many hours do you surf the Web per day?

    (1) At most 1 hour   (2) 1–2 hours   (3) 2–3 hours   (4) 3–4 hours   (5) More than 4 hours

11. What is the average number of e-mail messages you receive per day?

    (1) At most 4   (2) 5–9   (3) 10–14   (4) 15–19   (5) 20 or more

12. How many hours do you study per week?

    (1) At most 9 hours   (2) 10–19 hours   (3) 20–29 hours   (4) 30–39 hours   (5) 40 or more hours

13. How many courses have you taken in college so far?

    (1) 5 or fewer   (2) 6–10   (3) 11–15   (4) 16–20   (5) More than 20

14. When do you expect to graduate?

    (1) This year   (2) Next year   (3) In 2 years   (4) In 3 years   (5) In 4 years

15. What would you guess you will do after you graduate?

    (1) Pursue graduate studies in the arts and humanities

    (2) Pursue graduate studies in the sciences   (3) Attend a professional school

    (4) Find a job in business   (5) Find a job not in business

16. How many people in this class do you find attractive?

    (1) At most 2   (2) 3–4   (3) 5–6   (4) 7–8   (5) 9 or more

17. Of the choices below, what is your favorite topic in this book?

    (1) Infinity   (2) The fourth dimension   (3) Chaos   (4) Fibonacci numbers   (5) The *Let's Make a Deal* story

For questions 18–20, give a numerical answer. If your instructor gave you a Scantron form so that the answers could be automatically compiled, omit these last three questions unless you were specifically told how to enter the answers on the Scantron form.

18. How many siblings do you have?

19. How many hours do you sleep on average per night?

20. How many hours of athletic activities do you participate in per week?

MINDSCAPES Invitations to Further Thought

In this section, Mindscapes marked (H) have hints for solutions at the back of the book. Mindscapes marked (ExH) have expanded hints at the back of the book. Mindscapes marked (S) have solutions.

Developing Ideas

In 2009, a survey of 23 people who had home access to the Internet asked them how much they paid per month for that access, rounded to the nearest dollar. The following data show each person's response, from smallest monthly bill to largest. Use this data for Mindscapes 1–3.

   $14, $18, $18, $18, $19, $19, $23, $23, $23, $25, $25, $25, $25, $25, $26, $26, $26, $27, $33, $37, $42, $58

1. **Internet costs.** Make a histogram of the data on the cost of Internet access using the intervals 0–4, 5–9, 10–14, and so on. Describe the distribution of values in words.

2. **Internet costs—summaries (S).** Find the mean and median of the data on the cost of Internet access. Which of these gives a better answer to the question "How much did a typical person with Internet access pay for that service in 2009?"

3. **Internet costs—more summaries.** Compute the quartiles of the data on the cost of Internet access and give a five-number summary of the data. Does

the five-number summary give a similar impression of the distribution of the data to that of the histogram?

4. **How variable.** Remember that the standard deviation measures how far a typical score is from the mean. We didn't learn a formula to compute it, and you aren't expected to compute it. However, for the set of exam scores below, if you are told that the standard deviation is either 2.7 or 10.3, which one must it be? Explain why.

$$72, 93, 88, 85, 97, 100, 77$$

5. **Pie or no pie.** Below are two graphical representations of the data from the donut example in the section. Which figure makes it easier to compare the relative values of donut production?

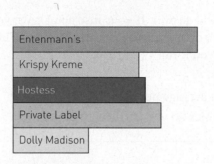 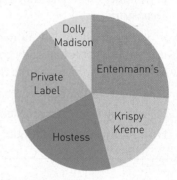

Solidifying Ideas

6. **Stats on steroids.** Barry Bonds and Mark McGwire are two of baseball's home-run kings. Following are data for their season home-run totals. Make histograms of these two sets of data using the scale 0–9, 10–19, 20–29, and so on and compare the distributions. Do this first for all the data, and then just for the data from 1986 to 2001 (McGwire's last year).

Year	'87	'88	'89	'90	'91	'92	'93	'94	'95	'96	'97	'98	'99	'00	'01	'02	'03	'04	'05	'06	'07
McGwire	49	32	33	39	22	42	9	9	39	52	58	70	65	32	29						
Bonds	25	24	19	33	25	34	46	37	33	42	40	37	34	49	73	46	45	45	5	26	28

7. **Who's the best—short summary.** For the home-run data in Mindscape 6, find the five-number summary for McGwire and Bonds. When you compare these five-number summaries, do they show the same patterns that you see in the histograms?

8. **Will tomorrow be different?** At the beginning of this section's Mindscapes, data were given on the cost of home Internet access in 2009. If similar data were collected in 2020, do you think that the distribution of the data would be much different? Why or why not? (*Hint:* Do you think there will be differences in the use and cost of Internet access in 2020?)

9. **How different are they? (S)** The following histogram shows the exam scores for 30 students in a freshman accounting class. Estimate the mean of these scores. Is the standard deviation of these scores likely to be closer to 12 or to 25?

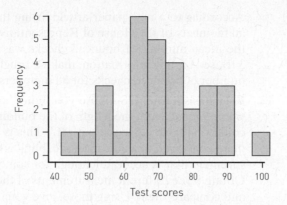

10. **What's normal?** The histograms of three data sets are given below. Classify each data set as having a normal distribution, a skewed distribution, or a bimodal distribution. Estimate the mean for that (those) data set(s) that have a normal distribution.

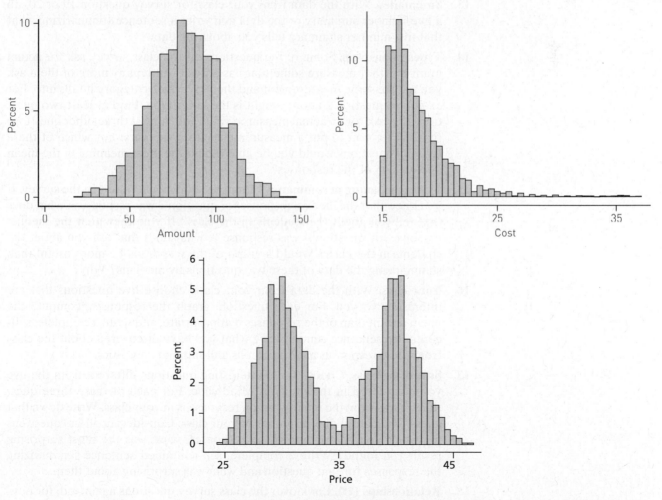

11. **Check the bounce (ExH).** Until 1992, members of Congress could write checks from accounts in an internal bank without incurring penalties for bounced checks (a bounced check is one that causes the account to be overdrawn).

According to a newspaper article listing the number of bounced checks for all members of the House of Representatives over a certain interval of time, the mean number of bounced checks was 47, and the median number was 3. Based on this information, make a rough sketch of the distribution of the number of bounced checks for all members of the House.

12. **Easy measurement.** Suppose you didn't have a long tape measure, but you wanted to measure the length of the building in which your class meets. You could do that by "stepping it off." This is a rather crude measuring instrument compared to a tape measure, but it is often used because it gives a measurement that is accurate enough for some purposes and is easily available. Obtain three different measurements of the same building from this method and compare them. Using those, give your best estimate of the length of the building. How accurate do you think your best estimate is? (You can do it three times yourself or get three different people to do it and use those.)

Mindscapes 13–18 refer to the class survey provided at the opening of these Mindscapes.

13. **Summaries.** With the data from your class for survey question 19 or 20, do a five-number summary of the data and write a sentence summarizing what that five-number summary tells you about the data.

14. **Types of answers.** Some of the questions on the class survey ask for actual numbers (that measure something) as answers and many more of them ask you to take some measurement and then say which category it falls into. For example, question 2 about weight is the latter type. Find at least two questions that ask for an actual measured value and at least three other questions that ask for you to put a measurement into a category. For which of these types of questions would you be able to find the most meaning in the mean and median of the responses?

15. **Is there meaning in summaries?** Compare questions 3 and 4 in the survey. If you knew that the median response for question 3 was response 3, what does that tell you about the students in the class? If you knew that the median response for question 4 was response 3, what does that tell you about the students in the class? Would a graph of the responses be more useful than summarizing the data of these two questions by medians? Why?

16. **Your class.** With the data from your class, choose five questions that are interesting to you. For each question, graph the responses, compute the mean and median of the responses if appropriate, and write a complete, self-contained sentence summarizing what you have discovered about the class from those responses and the graphs and summary statistics.

17. **Surprising class.** Choose three interesting questions different from the five you considered in the previous Mindscape. For each of these three questions, guess what the most common response is in your class. Write down that guess. Then graph the responses of your class. Considering all the questions for which you have graphed the responses, what was the most surprising result you found? Write a complete, self-contained sentence summarizing the responses for that question and what was surprising about them.

18. **Relationships (H).** Look over the class survey questions again and, for now, consider what the answers might have looked like if the variables that put measurements into categories didn't have categories and instead just asked for numerical answers, as in questions 18–20. So, for example, question 2

would have said, "How much do you weigh?" Now think about pairs of the questions. Find three pairs of variables that you think might have a relationship between the answers. For each pair, draw an *x–y* axis, and for each person surveyed, put a dot whose first coordinate is the first variable's value and whose second coordinate is the second variable's value. (This type of graph is called a *scatter plot*, which is discussed in more detail in Section 8.5.) Include the scale on both axes of the graph.

19. **College data.** Using the student survey data given below, create the five scatterplots (see Mindscape 18) for the number of parties per month versus the other five columns. Which pairs are related? For the related pairs, what type of relationship is there?

Data from Student Survey

Student	Hours of Web/Week	Number of E-Mails Received/ Week	Number of Dates/Month	Number of Hours Spent Studying/ Week	Number of Hours Participating in Sports/Week	Number of Parties Attended/ Month
1	16	45	3	25	10	3
2	5	17	7	50	6	1
3	10	26	11	21	15	6
4	13	30	9	25	8	2
5	3	10	7	30	18	4
6	7	19	9	14	5	8
7	9	24	8	35	5	1

20. **Web use.** Using the student survey data given in Mindscape 19, create the five scatterplots (see Mindscape 18) for the number of hours per week spent surfing the Web versus the other five columns. Which pairs are related? For the related pairs, what type of relation is there?

Creating New Ideas

21. **News data.** Find some data in a recent newspaper or Web site. Describe the data and how it was analyzed. If an average was given, was it clear which measure of center was used? Which do you think was used? Why? Compute the other measures of central tendency. What conclusions were deduced from the data? Do you agree with the conclusions? Are there other ways of interpreting the data? Try to give another explanation for the statistics given other than what was mentioned.

22. **Which half?** *Half of the people in the United States are below average.* Is this general statement true or false? Explain.

23. **Grades.** A student received the following quiz grades: 72, 84, 61, 95, 92, 98, 87, 84. Compute the student's mean quiz grade and the median quiz grade. If on the next quiz the student earns an 80, will the mean go up or down? What about the median?

24. **Raising scores (ExH).** A student received the following quiz grades: 90, 83, 80, 72, 78, 63, 79, 90. What is the lowest grade the student can receive on the next quiz that would raise the student's mean grade? Can you generalize your observations?

25. **Mean and median (S).** Give an example of a data set where the mean is greater than the median.

26. **Median and mean.** Give an example of a data set where the median is greater than the mean.

27. **Mean and median.** Using the class survey data, compute the mean and median of the responses to question 1. (If your class didn't do the survey, then answer this for a class with more men than women, say about 25 men and 18 women.) Although the variable isn't even a measurement, in this case, both the mean and median give some useful summary information about the class. One of them tells us whether there are more males or females in the class. The other can be used to easily find the percentage of males in the class. Which is which? (When there are only two categories, the mean and median can be interpreted in this way.)

28. **Mode.** The *mode* of a distribution is another summary statistic sometimes used to estimate the center of a distribution. It is the value that occurs the most often in the data. Look at the data sets in the "Developing Ideas" section and at the data from the class survey. Does every data set have a mode? For which types of questions in these problems is the mode a useful summary?

29. **Some taxing statistics (H).** Politicians love to quote averages that support a desired goal. The table below shows how much taxpayers would save under certain legislation supported by the president in 2007. Calculate the mean savings for all taxpayers. Estimate what percentage of Americans would actually see their savings reduced by an amount at least equal to the mean.

Approximate Percentage of Taxpayers Ranked by Income	Bottom 19%	Next 50%	Next 20%	Next 8%	Top 3%
Mean tax savings	$45	$468	$838	$1454	$7955

(*Note:* These data come from the independent, nonpartisan Tax Policy Center. Their data also show that the top 1% of taxpayers would see an average savings of over $66,000.)

30. **Whoops.** Find an example in a recent newspaper or magazine story where statistics were presented in a misleading manner. Explain why the statistics are misleading.

Further Challenges

31. **Further whoops (H).** A survey of 100 recent college graduates was made to determine their mean salary. The mean salary found was $35,000. It turns out that one of the alums incorrectly answered the survey. He said he earns $29,000 when in fact he earns $42,000. What is the actual mean salary of the 100 graduates?

32. **Is this normal?** Suppose the data in the previous Mindscape had a normal distribution. Is it possible that 45 of the graduates had salaries less than the mean, while the remaining 55 had salaries greater than the mean? Would your answer change if you did not know that the distribution was normal?

33. **Average the grades.** In a math class of 23 men and 25 women, the mean grade on the most recent exam for the women was 89% and for the men was 83%. Is it possible to compute the mean exam grade for the entire class of 48 students? If so, do it; if not, explain why. Is it possible to compute the median exam grade for the entire class? If so, do it; if not, explain why.

34. **Taxes (ExH).** Find an effective graphical representation of the data below on residential taxes that shows the relationship between the tax and the valuation of the home as well as the location of the home.

Tax (dollars)	Valuation (dollars)	Location
1800	110,000	Metropolis
4400	240,000	Metropolis
1500	80,000	Metropolis
2700	150,000	Metropolis
2600	180,000	Metropolis
1400	95,000	Suburbia
2500	140,000	Suburbia
3100	195,000	Suburbia
4300	310,000	Suburbia
1690	125,000	Suburbia
2200	187,000	Suburbia

35. **Doing surveys.** For question 10 on the class survey about surfing the Web, discuss both the advantages and disadvantages of asking for the response as one of five categories instead of asking for a numerical answer. Start by thinking of how you feel answering the survey and then also think about what you are thinking when someone asks you to interpret the results of the survey. Also consider whether your class was given an opportunity to answer questions 18–20 and think about why a class of 120 students might not have been given that opportunity.

In Your Own Words

36. **Personal perspectives.** Write a short essay describing the most interesting or surprising discovery you made in exploring the material in this section. If any material seemed puzzling or even unbelievable, address that as well. Explain why you chose the topics you did. Finally, comment on the aesthetics of the mathematics and ideas in this section.

37. **With a group of folks.** In a small group, discuss and actively work through the reasoning involved in the possible pitfalls of summarizing data sets with the mean or median. After your discussion, write a brief narrative describing the methods in your own words.

38. **Creative writing**. Write an imaginative story (it can be humorous, dramatic, whatever you like) that involves or evokes the ideas of this section.

39. **Power beyond the mathematics**. Provide several real-life issues—ideally, from your own experience—that some of the strategies of thought presented in this section would effectively approach and resolve.

For the Algebra Lover

Here we celebrate the power of algebra as a powerful way of finding unknown quantities by naming them, of expressing infinitely many relationships and connections clearly and succinctly, and of uncovering pattern and structure.

40. **Kids in the hall**. You have three cousins with mean age 11. If the oldest is twice the age of the youngest and two years older than the middle cousin, what are their three ages?

41. **Testing the data**. A math instructor reports the results of an exam as follows: 4 scores equal to x, 6 scores equal to $(x + 2)$, 6 scores equal to $(x + 8)$, 2 scores equal to $(x + 15)$, and 2 scores equal to $(x + 20)$. Find and simplify an expression that equals the mean of all the exam scores. Next, if $x = 70$, find the mean and the median of the exam scores.

42. **One mean mean**. Suppose an instructor reports exam scores exactly the same way as in the previous Mindscape. If you haven't already done so, find and simplify an expression for the mean of all those exam scores. Now suppose the instructor tells you that the mean is 81.5. Find the top score earned on this exam.

43. **Home sweet home**. For a political science project, you review some data that gives housing prices for a town in your home congressional district. Suppose the median home price is $120,000. If the price of the highest-price house is 80% higher than the median, find the price of this house. If the price of the lowest-price house is 45% lower than the median, find the price of this house.

44. **No back up (H)**. You have 120 pieces of data. They are ordered from smallest to largest but unfortunately, your computer loses all the data except the 60th and the 61st values. It turns out that the 60th value is $117 - 8x$ and the 61st is x^2. You're told that the median of the data equals 75. Find the 61st value.

9.3 LOOKING AT SUPER MODELS

Mathematically Described Distributions

Image Source/Getty Images, Inc.

Who looks normal?

What does a set of data look like? Describing all shapes of all distributions of all data sets would be a daunting and impossible task, because any possible shape is the shape of some distribution. Just take a photograph of a glorious sunset over the Rocky Mountains. Then draw horizontal and vertical axes and put little bars that extend up to the heights of the peaks and ridges in the photo—presto, you've just created a histogram!

Happily, many sets of real data have shapes that reflect underlying reality that gives a reason for the shape of the distribution. And that reason often can be captured by a mathematical formula, called a *model*. The shapes of many distributions quite often can be approximated extremely well by certain predictable and describable models. In this section we'll explore various data sets and group them by types that can be usefully approximated by specific mathematical functions. This association of an actual (complicated and jagged) data set with a theoretical (nice and smooth) data set is one of the powerful ways by which statistics brings meaning to data.

Rolling Dice—Uniform Distribution

Suppose we roll a fair die 1000 times and record the number of times it comes up 1, 2, 3, 4, 5, or 6. Here are histograms of four simulations of that experiment.

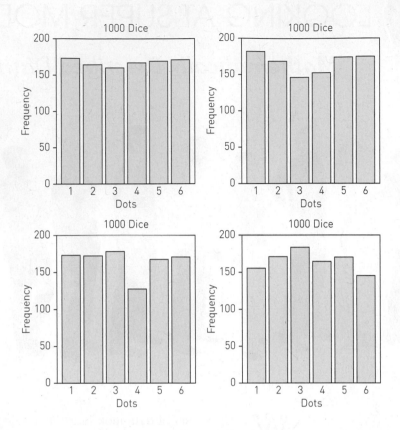

We are not surprised by the data we see, because we expect that each of the six possible values will arise roughly equally. Such a distribution of data is called a *uniform distribution*. Because each value arises in about 1/6 of the 1000 tosses, we expect to see each value about $\frac{1000}{6} = 166\frac{2}{3}$ times. Notice that this value equals the number of rolls, 1000, multiplied by the probability, 1/6, that a particular value arises. We can give a mathematical model for this data by letting the variable x represent a possible value showing on the die and defining the function f to be $F(x) = 166\frac{2}{3}$. This theoretical distribution has a graph that is exactly a straight, horizontal line at a height of $166\frac{2}{3}$ and gives a good approximation to each of our simulations.

Let's highlight a very important observation. That theoretical horizontal line cannot possibly be *exactly* correct for any experiment of throwing a die 1000 times, because no number on the die can possibly come up precisely $166\frac{2}{3}$ times. The power of a theoretical description is that we can compare the real data to the theoretical data. We could instantly see if the real data were far off from flat-lining at $166\frac{2}{3}$. If we saw that the number 6 came up 300 times, for example, then we could be quite sure that we were in possession of a loaded die.

This die-rolling experiment illustrates a powerful strategy for understanding data and our world. A theoretical, mathematically described distribution that is simple and easy to describe can be compared with actual data, which appears more chaotic, to offer us insight.

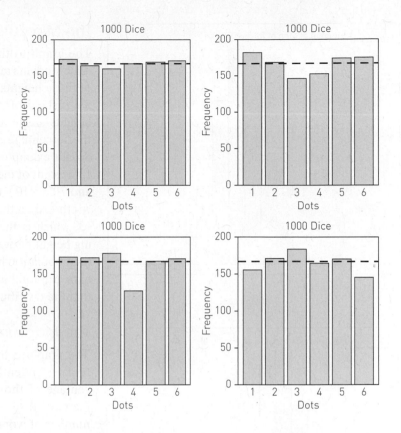

The Beauty of the Bell—Normal Distributions

The most famous shape of distributions is the bell-shaped curve, also called a *normal curve*, a *normal distribution,* or a *Gaussian distribution*. It is symmetrical around its peak and has a specific shape and a specific mathematical formula that describes it. It arises frequently for several reasons, physical and mathematical. Let's look at several examples and then come to understand why we would expect them all to exhibit this characteristic beautiful bell shape.

SATs

The graph here is a histogram of SAT scores for incoming students at a particular college. (They used the version with a top combined score of 1600.) The SAT scores are collected in 40-point increments on the horizontal axis. Each bar has height equal to the number of students who got SAT scores in that range. Notice that the distribution is roughly symmetric about a single peak and it tapers down on each side.

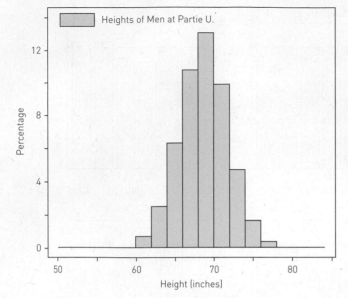

The Measure of Men at Partie U.

The graph to the left records the heights of all the male students at Partie U. by showing the percentage of those heights that fall into each 2-inch increment. So, instead of counting the number of men whose heights are in a given 2-inch range, here we report the percentage of men in that particular height range. So, for example, the highest bar indicates that about 13 percent of men at Partie U. are between 68″ (5′8″) and 70″ (5′10″) tall. If we think of this bar as having width 1, then the area of the bar has numerical value 13, which is the percentage of the population having height between 68″ and 70.″ Because everyone in the population has a height represented in one of the bars, the total area of the bars must equal 1. Notice that the distribution is somewhat bell-shaped.

The Measure of Women, Too

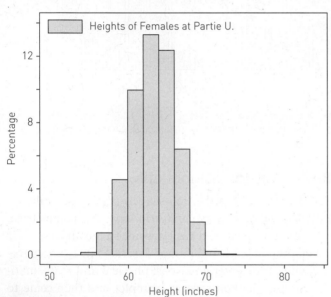

The graph to the left records the heights of all the female students at Partie U. by showing the percentage of those heights that fall into each 2-inch increment. As with the men, instead of counting the number of women whose heights are in each 2-inch range, here we report the percentage of women in that particular height range. The total area of the bars here will be the same as the total area in the men's graph, making comparisons easier.

Notice that the distribution of the heights of females is very similar to the distribution of the heights of men, but the distributions are centered at different values. Men's heights are centered at about 68″ (5′8″) and women's heights are centered at about 63″ (5′3″), but both are bell-shaped.

Random Gas Prices

When we get gas, we know how much it will cost per gallon. But wouldn't it be more fun to let luck play a role in how much we pay? Suppose when you went to the One-Die Random Gas Station, you rolled the die and whatever it came up, that's how much you would pay (in dollars) per gallon. The average price that the station would collect from the customers would be $3.50, but sometimes you could get gas for $1.00 per gallon. Of course, other times you might have to pay $6.00 per gallon. (By the way, the station insists that you buy 10 gallons of gas with each roll of the die.)

We might also imagine the Two-Dice Random Gas Station. It would work the same way, except you would roll two dice and take the average of the two numbers to determine what you pay per gallon. If you rolled two 1's, you would get gas for $1.00 per gallon. If you rolled a 2 and a 6, you would pay $4.00 per gallon.

You can guess the pricing scheme for the Three-Dice Random Gas Station, and the Four-, Five- , and 100-Dice Random Gas Stations. Below are some histograms showing how often you pay various amounts for gas at the different stations. In each graph, the total area under the graph is the same, which we think of as 1 unit of area. The way to interpret the graph is that you take any two dollar values, like $3.00 and $4.00, and you look at the area under the graph above the interval from $3.00 to $4.00. The fraction of the total area is the percentage of throws of the dice whose average lies between those two values. Notice that the more dice we use, the more often we pay close to the $3.50 average price. If you go to the 100-Dice Random Gas Station, you are essentially certain to pay between $3.00 and $4.00 per gallon, because almost all of the area under the graph lies over the $3.00 to $4.00 interval.

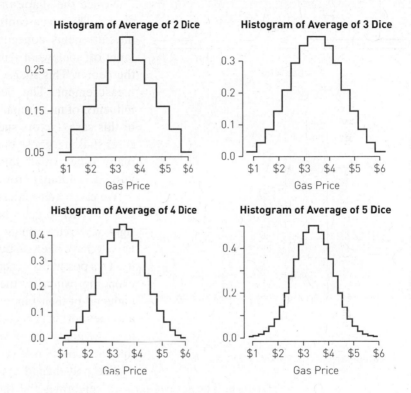

The distributions of how much we can expect to pay form bell shapes. No matter how many dice are used, the total area under each graph is exactly the same (namely 1 unit) and the mean price remains $3.50; however, the variability decreases as more dice are used, meaning that as you use more dice the probability becomes increasingly high that you will pay closer to the $3.50 average price per gallon. Notice how the N-Dice Random Gas Stations have prices whose curves become more bell-shaped as the number of dice increases, as well as becoming taller and thinner.

Why We Should Expect All Those Bells

Let's return to Ms. Cockenlocker's laser measurements of her height. Why do the measurements vary? One of the reasons may be that during the time Ms. Cockenlocker was being measured, her height actually did change several times;

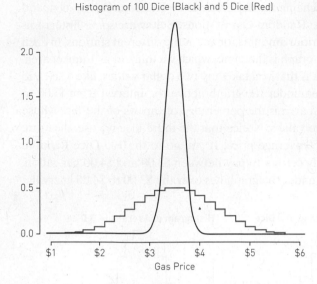

Histogram of 100 Dice (Black) and 5 Dice (Red)

Gas Price

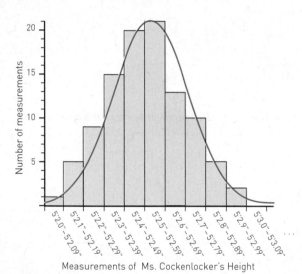

Measurements of Ms. Cockenlocker's Height

that is to say, sometimes she stood taller than others, or her spine may have compressed a bit after lunch, or if she got cold while being measured, the goose bumps on her head may have added a smidgen to her stature. A more fundamental reason is that the measuring process itself is fraught with problems. The measuring device may have been aimed slightly high or low. In other words, measuring precisely isn't possible. (You can confirm this yourself by measuring the width of a couch to the nearest tenth of a millimeter several times—you will typically get an array of slightly different answers.)

Notice the shape of the graph of her height measurements. There is a central point around which most of the measurements concentrate, and then the measurements taper off somewhat symmetrically as they move away from the center. This shape is typical for graphs of this sort of measurement. The bell shape reflects the accumulated influence of many small random and independent factors—in this case, factors such as how tall she stands, the measurer shaking, goose bumps appearing or disappearing, and many other small influences—each individual influence varying randomly around its own central value.

We can better understand the curve's shape by observing that in order to get a height that is, for example, far to the right on the curve, many of the small varying quantities would have to have been on the large side of their range of variability. This possibility occurs very rarely. Most of the time, however, approximately the same number of factors are in the high end of their range as are in the low end of their range, and they sort of cancel each other out. The effect is that most measurements cluster near the middle. The global effect is that the collection of values will result in a bell-shaped curve.

The bell-shaped curve is sometimes called a *Gaussian* curve. The term *Gaussian* celebrates the famous German mathematician Carl Friedrich Gauss, who lived from 1777 to 1855. In 1801, observations were taken of the first asteroid ever observed, the asteroid Ceres. After a few observations, Ceres passed behind the Sun, and the question was how to find it after it emerged. Gauss was able to predict the future position after it passed the Sun by looking at the past measurements and realizing that those observations, like all measurements, include errors.

Gauss fit a curve into the observed data that would minimize the error between the curve and the known observations. That process involved an understanding of the distribution of errors. The bell-shaped distribution was so commonly associated with errors, such as in measurements, that an old name for the Gaussian or normal distribution was the "error distribution."

The term "error distribution" suggests one of the reasons that this symmetric curve arises—that measurement errors often create a set of measurements that are distributed symmetrically on each side of some "correct" value. So, it turned out that, although Gaussian distributions were associated with errors, later, they became associated with the distribution of other sets of data and other measurements in nature.

In the cases of heights of adult males or adult females, again we expect bell-shaped curves, because all kinds of random factors combine to determine a man's or a woman's height. Height is influenced by many genes each acting randomly, as well as the accumulated effects of a child's dietary habits (such as how much protein the student consumed in elementary school). In general, when there are lots of random factors involved, we can expect a histogram of data to create a bell-shaped curve.

So how did the expression "normal distribution" arise? In the 19th century, the Gaussian distribution of measured characteristics was taken almost as a law of nature, as a measurement of what was "normal." So people who had extreme qualities, such as, say, being very tall, began to be thought of as being abnormal—they deviated from the norm. So, maybe this is one reason that some people don't like the normal curve because, in a way, it defined *abnormal.*

Random processes often result in normal distributions. Suppose we perform the experiment of flipping a coin 100 times and record the outcomes. Each experiment will result in a number of heads from 0 to 100. If we were to perform this 100-coin-flipping experiment thousands of times, each time recording the number from 0 to 100 of heads, we would expect to see the distribution of those numbers form a bell-shaped curve centered around 50.

The *N*-Dice Random Gas Stations have prices whose curves become more bell-shaped and thinner as the number of dice increases.

Making the Normal Distribution Physical

A physical realization of this bell phenomenon appears when we use an apparatus in which a ball bounces randomly off a collection of nails and then falls into slots. This apparatus has a name that is useful for throwing around at parties; it is called a *quincunx.* In the quincunx, each ball is dropped from the top in the center. Sometimes the ball will bounce to the right from a nail that it hits, sometimes to the left. Because the ball will tend to bounce left about as many times as it bounces right, balls will tend to end up near the middle. Often there will be a few more right bounces than left bounces, or vice versa, so the ball will end up slightly to the right or left of center. Because each bounce is as likely to be left as right, the distribution of the balls at the bottom will tend to be symmetrical about the center. If we repeat this experiment with lots of balls, we will find that the balls gather in a distribution that is the characteristic bell-shaped normal curve.

Life Lesson
An example or experiment can make an abstract idea concrete.

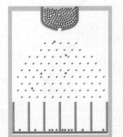 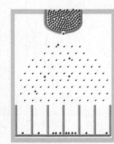 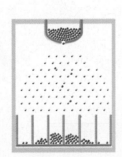 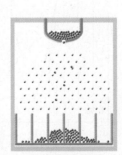 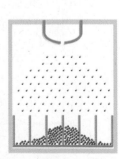

The Bell Curve Normally Fits

The normal curve has a specific mathematical formula that is determined by its mean and its standard deviation. Although it appears a bit complicated, it is easily computed by a calculator or computer. Here is the formula:

$$P(x) = \frac{1}{\sigma\sqrt{2\pi}} e^{-\frac{1}{2}\left(\frac{x-\mu}{\sigma}\right)^2}$$

where μ = mean and σ = standard deviation. One neat feature about this complicated formula is that it contains the two most famous constants in mathematics, π and e. Let's see how well this bell-shaped, normal curve approximates all the examples we have seen before: Emily Cockenlocker's measurements, SAT scores, heights of men at Partie U., heights of women at Partie U., gas prices at the N-Dice Random Gas Stations, and how the balls landed in the quincunx. The normal distributions approximate many sets of data.

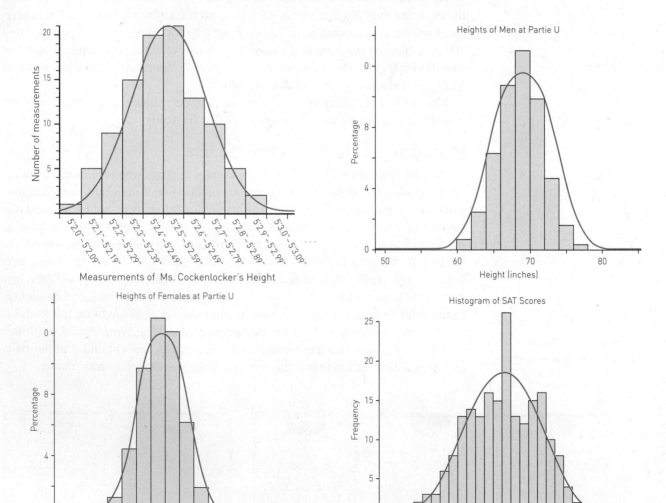

Histogram of 100 Dice (Black) and 5 Dice (Red)

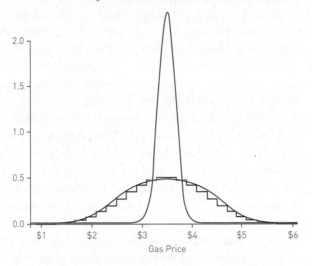

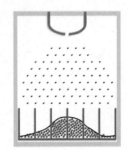

Normal Curves and Standard Deviation

Recall that standard deviation is a measure of deviation from the center of a set of values. In the case of normal distributions, the standard deviation is particularly useful in that it tells us specifically how close the data are to the mean value (the peak). In a normal distribution, about 68% of the data differ from the mean by less than the value of the standard deviation; about 95% of the data differ from the mean by less than two times the standard deviation; and about 99.7% differ from the mean by less than three times the standard deviation. These percentages are valid for normal distributions regardless of whether the distribution has a small standard deviation, making the graph a tall, narrow bell shape, or has a large standard deviation, making the graph wide and squat. The percentages within a certain number of standard deviations from the mean remain the same.

So in a normal distribution, if we know the mean *and* the standard deviation, we know a lot about how much variation there is in the data. The smaller the standard deviation is, the tighter the data is bunched around the center.

We can put together knowledge about the percentage of data in a normal distribution that is within one, two, or three standard deviations from the mean to estimate some facts about heights of females, for example. We saw that the mean height of females at Partie U. is about 5′3″. Suppose the standard deviation of their heights is 3″. Then we can estimate what percentage of

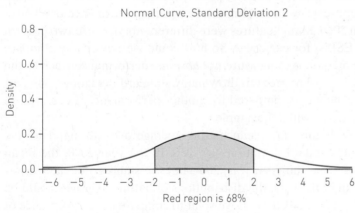

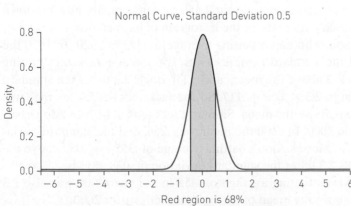

women will be 5′9″ tall or taller. Since 5′9″ is two standard deviations above the mean of 5′3″, our analysis of normal curves indicates that only about 5% of women will be that far away from the mean. Since there are roughly as many shorter as taller, we would expect only about 2.5% of women to be taller than 5′9″.

When we know the mean and standard deviation of a normal distribution, we can estimate what fraction of the values will lie between two values. For example, we saw that in 1920, the mean batting average over all major league players was .265. The standard deviation was .050. So what percentage of the batters probably hit between .315 and .365? Well, that range is between one and two standard deviations above the mean. We expect 32% of the players to have batting averages farther than one standard deviation from the mean, so about 16% of players would have averages farther than one standard deviation above the mean. Similarly, about 2.5% of players would be expected to have batting averages more than two standard deviations above the mean. So we would expect about 13.5% of players in 1920 to bat between .315 and .365.

Understanding the normal distribution gives us an excellent way to estimate how many values will lie within certain ranges in a normal distribution. That information can be crucial for understanding the world more accurately and for making decisions from choosing which sizes of clothes to manufacture to figuring out sport strategies.

Comparing Using Standard Deviation

We can use these insights about normal distributions in arguments about who were the better athletes when the athletes performed in different eras or even in different sports. Suppose we want to compare the sports performance of a batter in 1920 to a batter in 2000. Many features were different about those two eras (in 1920, there was no ESPN, for example). So how could we reasonably compare performances? One method is to measure not *absolute* performance, such as the batting average, but instead to measure how many standard deviations one performance was above the mean compared to another performance. Let's explore this method of comparison with an example.

In 1920, as we saw before, the mean batting average over all major league players was .265. The standard deviation was .050. In the year 2000, the mean was very close to the same thing—it was .266—but the standard deviation was smaller. It was .038. In both years, the distributions of batting averages could be well approximated by a normal curve. If we want to compare the performance of a batter from 1920 to a batter from the year 2000, we can compare how far down the slope of the bell each batter sits on the histogram of his own era.

Look at Joe Jackson, who had a batting average of .382 in 1920. In 1920 the mean was .265, and the standard deviation was .050. So Joe Jackson's batting average of .382 was .117 above the mean, and .117 equals 2.3 times the standard deviation of .050, that is, $2.3 \times .050 \approx .117$. So Joe Jackson's batting average was 2.3 standard deviations above the mean. Similarly, let's look at Moises Alou's batting average of .355 in 2000. In 2000 the mean was .266, and the standard deviation was only 0.038. So Moises Alou's batting average of .355 was .089 above the mean, and .089 equals 2.3 times the standard deviation of .038, that is, $2.3 \times .038 \approx .089$. So Moises Alou's batting average of .355 in the year 2000 was also 2.3 standard deviations above the mean of the batting averages for 2000.

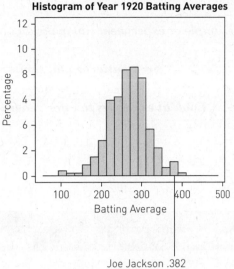

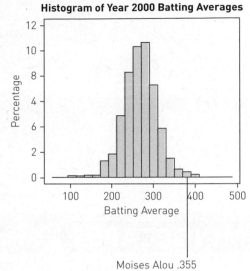

Joe Jackson .382

Moises Alou .355

Each is about 2.3 standard deviations above the mean. Each has *z*-score about 2.3.

This analysis suggests that Joe Jackson's performance relative to his compatriots in 1920 is comparable to Moises Alou's performance of .355 in the year 2000. Both of them are 2.3 standard deviations above the mean. So measuring number of standard deviations above the mean is one way to compare data from different normal distributions.

The number of standard deviations, σ's, that a datum x is from the mean μ is called its *z-score*; therefore, the formula for the *z*-score is:

$$z = \frac{x - \mu}{\sigma}.$$

Notice that the *z*-score will be negative if the datum x is below the mean. Doing a little algebra with this formula, we can solve for x to get:

$$x = \mu + z\sigma.$$

This formula confirms that the datum x is z times the standard deviation σ from the mean μ.

Since Joe Jackson's *z*-score and Moises Alou's *z*-score were both about 2.3, we might view them as having performed equally well in their own eras.

A Look BACK

MANY FEATURES OF our world can be captured with data that are naturally described by some mathematical functions. By understanding the influences that are affecting the data, we can predict which model is likely to create a good approximation of the real data. By looking at the reality, we can sometimes guess the correct family of functions that model the data and then use the data to determine the exact member of that family that fits the data best. That process gives us insight into the world. Also, sometimes we can see data and then look at it to see whether one of our standard model types fits the data well. If it does, then we might well get some insights into what might be the underlying cause of the data we are examining.

Looking for commonalities in examples and then pinning them down with mathematical expressions and more precise definitions is a fruitful strategy for gaining understanding of our world.

Life Lessons

An example or experiment can make an abstract idea concrete.

•

Seek patterns and similarities.

•

Look at examples to infer general principles.

MINDSCAPES   Invitations to Further Thought

In this section, Mindscapes marked (H) have hints for solutions at the back of the book. Mindscapes marked (ExH) have expanded hints at the back of the book. Mindscapes marked (S) have solutions.

Developing Ideas

1. **I.D. please. I**dentify the **D**istributions below as normal or uniform.

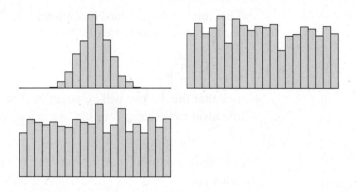

2. **Deviating from the norm (H).** The following three distributions are approximately normal. Which one has the largest standard deviation? Which one has the smallest?

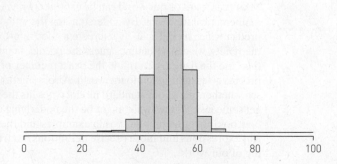

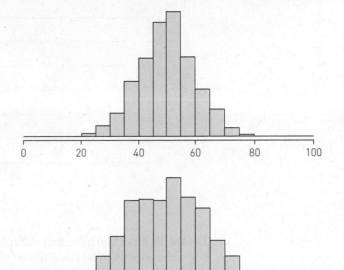

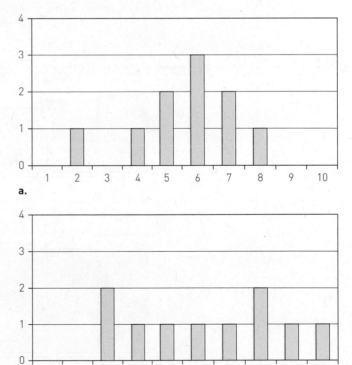

3. **Match quest**. The following three data sets are scores from three quizzes given in a chemistry class. Match each to the distributions below.

Quiz a: 10, 10, 8, 9, 10, 10, 9, 10, 9, 8

Quiz b: 6, 2, 5, 7, 6, 8, 7, 4, 5, 6.

Quiz c: 4, 3, 7, 9, 8, 10, 6, 5, 8, 3.

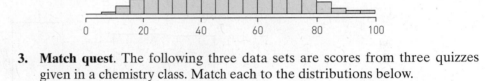

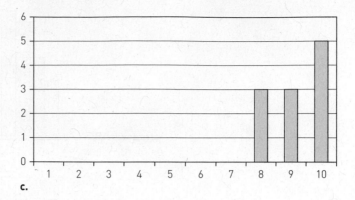

c.

4. **Doubt the dice**. You have four different dice and wonder if each one is fair. You roll each one 100 times and obtain the frequency distributions below and on the next page. Decide which dice are most likely to be fair.

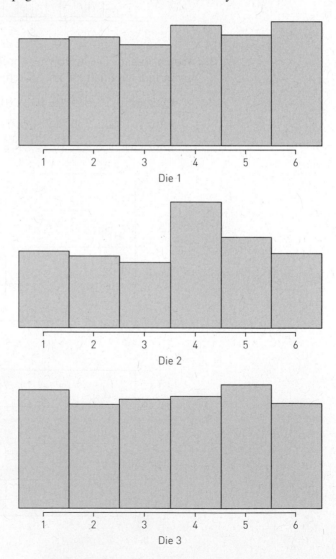

Die 1

Die 2

Die 3

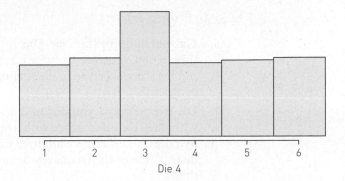

Die 4

5. **More matching**. A class was quizzed on two different occasions. Prior to one of the quizzes, the students stayed up all night watching movies and playing video games. Which of the graphs below most likely shows the distribution of scores for that quiz?

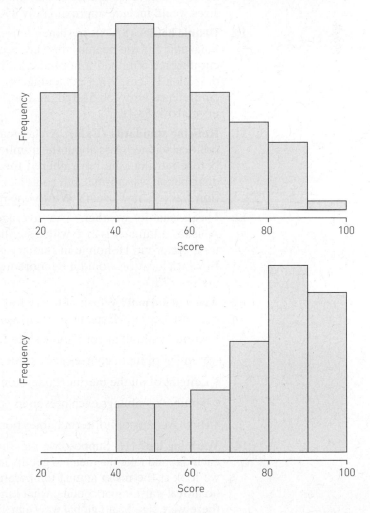

Solidifying Ideas

6. **Gas guzzling.** In the One-Die Random Gas Station scenario described in this section, the average price for a gallon of gas is given as $3.50. Revisit the pricing scenario given in the section and describe how this average price was computed.

7. **Dr. Die.** Suppose you doctor a standard six-sided die by changing the face with three dots into a face with five dots. (So your die now has two faces with five dots, and no face with three dots.) Sketch a histogram of the theoretical distribution of the frequency of each outcome if you were to roll the die 600 times.

8. **Dr. Mean (S).** What is the mean roll you would expect when using the doctored die in the previous Mindscape? How is this different from the mean when rolling a fair die?

9. **A podiatrist in the making.** Suppose you ask the shoe size of each female in your dorm and discover that the data have a normal distribution with a mean of 7. If the standard deviation is 1.5, approximately what percentage of women have shoe sizes in the range from 5½ to 8½? What range of shoe sizes would include approximately 95% of the women?

10. **Beanie babies (S).** All freshmen entering Whatsamatta U. have their heads measured for the beanies they are required to wear. One year the head circumference data had a normal distribution with mean 55 cm and standard deviation 1.7 cm. What percentage of the students that year had a head circumference between 53.3 cm and 56.7 cm? What percentage had circumference above 58.4 cm?

11. **Retiring standards (ExH).** A classmate describes a study discussed in her political science class about the number of years the typical American works before retiring. She remembered the mean value of the data and that the distribution was normal, but couldn't remember whether the standard deviation was 1, 8, or 15 years. Which one must it be? Why?

12. **Unseasonable weather.** The average high temperature in Anchorage, Alaska, in January is 21°F with a standard deviation of 10°. The average high temperature in Honolulu in January is 80°F with a standard deviation of 8°. In which location would it be more unusual to have a day in January with a high of 57°F?

13. **Are you normal?** Which of the following data sets will probably have a normal distribution? Explain your answers briefly.

    • Annual rainfall in your home state for each of the last 100 years.

    • Number of math courses each student in your dorm has taken.

    • Lengths of all the marine iguanas on Darwin Island in the Galapagos.

    • Number of siblings each person in your math class has.

    • Birth weights of full-term babies born in the United States in 2008.

14. **Warming Up?** (H) Suppose we measure the temperature at a given location each day and take the mean to get the mean temperature for the year. Suppose we look at the mean annual temperature for a hundred years. If there were no global warming or cooling, what kind of a distribution would we expect? If there were significant global warming, how would the distribution differ?

15. **Possibly uniform.** Which of the following data sets will probably have a uniform distribution? Explain your answers briefly.
    - Number of times a Web server is accessed per minute.
    - Your body temperature taken every morning during one year.
    - Number of phone calls to a call center per minute.
    - Outcomes of 100 dice rolls.

Creating New Ideas

16. **Under the normally spreading chestnut tree.** A forester measured a sample of trees in a tract of land that is being sold for a lumber harvest. Among 27 trees, she found a mean diameter of 10.4 inches and a standard deviation of 4.7 inches. Suppose her sample gives an accurate representation of the entire tract of land and that the tree diameters follow a normal distribution.
    a. Sketch a graph of the distribution of tree diameters.
    b. What diameter would you expect the central 95% of trees to be?
    c. About what percentage of trees should be less than one inch in diameter?
    d. About what percentage of the trees should be between 10.4 and 19.8 inches in diameter?

17. **Abnormal.** Give two examples of data sets from your experience that would not conform to a normal distribution.

18. **Heavy petting.** You adopt a kitten from the animal shelter. Every month for two years, you weigh your pet and record the data. Will the data be distributed uniformly, normally, or neither? Explain your answer.

19. **Less than normal.** Here are some distributions that look like they might be normal. For those that aren't, point out any graphical features that are not "normal."

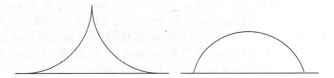

20. **Big digit on campus.** Ask fellow residents of your dorm individually what their favorite digit is (i.e., their favorite number from among 0, 1, 2, 3, 4, 5, 6, 7, 8, 9). Collect at least 50 responses. If you create a graph, would your data conform more to a uniform distribution or to a normal distribution? Explain. Just using your graph, estimate the mean and median of your data set. Compute the actual values and compare with your estimates.

Further Challenges

21. **Dr. Die revisited.** Recall the doctored die from Mindscape 7 (a standard six-sided die altered to have two faces with five dots, and no face with three dots). Suppose you have two such dice. Roll them both and record the sum. Sketch the distribution you would expect if you did this many times.

22. **Gas guzzling (ExH).** Consider the Two-Dice Random Gas Station described in the text. The price you pay for a gallon of gas is equal to the average of the numbers showing after rolling two dice. What are the chances that you will pay only $1 per gallon of gas? What are the chances that you will pay $6 per gallon? Use your answers to the two previous questions to determine the chances that you will pay something more than $1 but less than $6 per gallon.

23. **The 68–95–99.7 rule.** Referring back to a discussion in this section, what do you think this "rule" means? To what kind of distribution does it apply?

In Your Own Words

24. **Personal perspectives.** Write a short essay describing the most interesting or surprising discovery you made in exploring the material in this section. If any material seemed puzzling or even unbelievable, address that as well. Explain why you chose the topics you did. Finally, comment on the aesthetics of the mathematics and ideas in this section.

25. **With a group of folks.** In a small group, discuss and actively work through the reasons that normal distributions arise so frequently. After your discussion, write a brief narrative describing the reasons in your own words.

26. **Creative writing.** Write an imaginative story (it can be humorous, dramatic, whatever you like) that involves or evokes the ideas of this section.

27. **Power beyond the mathematics.** Provide several real-life issues—ideally, from your own experience—that some of the strategies of thought presented in this section would effectively approach and resolve.

For the Algebra Lover

Here we celebrate the power of algebra as a powerful way of finding unknown quantities by naming them, of expressing infinitely many relationships and connections clearly and succinctly, and of uncovering pattern and structure.

28. **Heads up.** The marching band is getting new uniforms. Measurements taken include the head circumference of all band members. Suppose these data have a normal distribution and that 68% of all head measurements lie between 54.1 cm and 57.9 cm. Use this information to write two equations that relate the mean μ and standard deviation σ for the head measurements. Now solve for μ and σ.

29. **Knees up.** Building on the previous Mindscape, suppose an in-seam measurement was also taken for each band member. Suppose these data have a normal distribution and that 99.7% of all in-seam measurements lie between 28 inches and 36 inches. Use this information to write two equations that relate the mean μ and standard deviation σ for the in-seam measurements. Now solve for μ and σ.

30. **Mother Brown (H).** The most popular pizza place near campus is Mother Brown's, which is famous for the amount of cheese put on the pizzas. Your school orders 200 pizzas for a big party. Assume the amount of cheese per pizza is normally distributed. If the average (mean) amount of cheese per pizza is 12 ounces, and only 5 pizzas have more than 13.6 ounces of cheese on them, what's the standard deviation for this cheesy data?

31. **Scoring some Z's.** Refer to the formula for the z-score of a datum x on page 703. Use this equation to solve for x in terms of z, μ, and σ. Use the same equation to solve for σ in terms of z, x, and μ.

32. **Catchin' some Z's.** You happen to know the z-score for two values in your normal data set. The datum $x = 7$ has z-score -2 and the datum $x = 14.5$ has z-score 1. Use this information and the formula for the z-score on page 703. to determine the mean μ and standard deviation σ for your data set.

GO FIGURE

Making Inferences from Data

Birch Forest (1905), Gustave Klimt. Seeing the forest through the trees.

> *The statistics on sanity are that one out of every four Americans is suffering from some form of mental illness. Think of your three best friends. If they're okay, then it's you.*
>
> RITA MAE BROWN

Suppose we want to know how people will vote in an upcoming election, so we take a poll asking random voters how they will vote. Even if we assume that voters will answer our survey honestly, the problem remains that we cannot ask every single voter. If 100 million people will vote in the election, then we're not going to contact each and every one of them (we *do* have a life!). Instead, suppose we ask only 1200 of them how they will vote (okay, we don't have *much* of a life!). We know how the 1200 people will vote, but how confident can we be that their voting habits reflect the preferences of all of the 100 million voters? We need to develop strategies for answering two questions: How closely does our sample represent the whole population? And how confident can we be that our estimate is that close?

Let's now move from running candidates to runny noses. Suppose we are deciding whether to take a new remedy called Honk-Stop with their clever bumper sticker ad campaign, "Don't Honk if You Take Honk-Stop." An experiment was done involving 200 people with runny noses. One hundred received Honk-Stop and 100 received a placebo. After three days, 60 of those who received Honk-Stop stopped honking and 40 of those who got the placebo also were cured. Should we be confident that the drug Honk-Stop works?

Using a data sample to make an inference about the world arises not only in polling and medicine, but also in marketing, in scientific studies, in the social sciences, and many other areas. Randomness and probability are important tools that enable us to know how confident we can be in our conclusions. Since those inferences can literally become matters of life or death when the technique is applied to issues involving, for example, the efficacy of medical procedures, we see that statistical inference is one mathematical area that can literally change (and even save) our lives.

The Ideas Behind Statistical Inference

There are at least three types of situations in which statistical inference arises.

Setting 1: There exists a fixed collection of data, but we only know a sample of it. Our goal is to infer the data of the entire population from analyzing that sample. Examples include:

- We test a little blood for cholesterol when we want to know how much cholesterol is in our entire bloodstream.

- We want to know the distribution of heights of male adults in the United States, so we select a thousand at random and see how they measure up.

- We want to know how the 100 million people will vote in an upcoming election, so we take a poll of 1200 future voters.

Concepts involved:

- The mean, shape, and spread of a random sample should resemble the mean, shape, and spread of the whole population.

- More data gives a higher probability that the mean of a random sample will be close to the mean of the entire population. Similarly, larger random samples will tend to reflect other features of the whole data set such as population proportion and standard deviation.

- We can give a probabilistic measure of how confident we are that the mean of the random sample is within a certain distance of the mean of the whole population, and similarly for population proportion and standard deviation.

Setting 2: Some fact about reality is unknown, and so we employ statistical analyses to help us determine what is most likely true. Examples include:

- *The Federalist Papers* are a collection of essays written in 1787–1788 mostly by Alexander Hamilton and James Madison, but published anonymously. The authorship of most of these essays was known, but an historical mystery remains as to who actually wrote about a dozen of these 83 essays. Statistical evidence about word usage sheds light on the Hamilton versus Madison conundrum.

- You pick up a bunch of playing cards from the floor. Instead of carefully looking at each card to see if you are playing with a full deck, you randomly select one card at a time, replace it, randomly select again, and repeat. After a few thousand selections, you can be reasonably confident as to which cards you have and which are missing.

- When an individual is on trial for a crime, the jury's task is to determine whether or not the defendant is guilty of committing the crime. Only if it is *unreasonable* that the evidence presented would exist assuming that the defendant was innocent, does the jury find the defendant guilty.

Concepts involved:

- If the world is a certain way, we expect certain data to reflect that reality. For example, if our deck of playing cards is complete, in the long run we expect to choose each card with approximately the same frequency as any other card.

- In the criminal justice system we are instructed to presume that the defendant is innocent. Only if the evidence would be extremely unlikely under that assumption do we find the defendant guilty.

Setting 3: Reality contains some probabilistic feature and we use a random sample to determine what the chances are. Examples include:

- We might test a medicine to determine what fraction of the time it is effective.
- If you spin a penny on its edge and then let it fall, is there a 50% probability that it will land heads up? We can interpret evidence to decide.

Concepts involved:

- We apply a method that mirrors the criminal justice system. We presume a hypothesis (like spun pennies have a 50% probability of landing heads up) and test whether the data we find would be unlikely if the hypothesis were true.
- Measuring the strength of the evidence leads us to the idea of statistical significance, which—as we'll soon see—is significantly different from ordinary significance.

In the previous sections of this chapter, we developed a number of different concepts that allow us to organize, describe, and summarize a collection of data. Those strategies help us to understand a set of data when we know *all* the data. But often, in reality, when we are faced with the unknown, we don't know it all. Instead, we want to use data from a sample of the population to infer what is true about the whole population.

Is Blood Thicker Than Stats?

Given the option, we would rather *not* have a heart attack. Scientists have shown that large amounts of cholesterol in our bloodstream may increase the chances that our arteries get clogged up, which, in turn, can result in a much-dreaded heart attack. In order to take action against this bleak possibility, we periodically have our blood tested.

Our bloodstream has a certain amount of cholesterol within it. That amount is what we want to know, but we may be a bit squeamish about lying down on a cot while the nurse sucks out every single corpuscle and carefully measures the total cholesterol count. Such a cholesterol test would make a possible heart attack an attractive alternative. Instead of draining us until our skin becomes raisin-like, the nurse merely takes a small random sample of our blood. The sample is tested for cholesterol. By seeing what amount of cholesterol there is in the sample, we confidently infer that we know how much cholesterol we have altogether.

We are really relying on the idea that the cholesterol is well mixed within our bloodstream. We are assuming that the cholesterol is so well distributed that we can expect every sample of our blood to contain the same proportion of cholesterol as any other sample. But is this assumption correct? Might one sample of blood just happen to contain a higher amount of cholesterol than is generally present in our bloodstream altogether? It *is* possible; however, the sample is sufficiently large that we can be quite confident that the sample's level of cholesterol is representative of the whole bloodstream. Notice that cholesterol is checked with a sample taken with a syringe rather than a pinprick to obtain just a drop. A single drop of blood is a smaller sample, and that amount of blood might have an unrepresentative proportion of cholesterol. Happily we don't need to drain the entire body, but we do need to take a large enough sample to be *confident* that the cholesterol in that sample reflects the cholesterol within the whole body. So lesson one is that your sample must be large enough. But how large is that?

Measuring Up

In the previous sections, we developed a number of different concepts that allow us to organize, describe, and summarize a collection of data. The shape, center, and spread of the data enabled us to bring the picture of a set of data into focus. Let's see how we can infer the shape, center, and spread of the entire population from the shape, center, and spread of just a sample. We illustrate this technique by trying to infer the mean height of a whole population of adult men by just measuring the heights of a sample.

We will use a database of heights of 100,000 men, which we will think of as our entire (unknown) male population. We'd like to get a picture of that whole population by selecting a few good men. That is, we'd like to choose a sample of a few men whose mean height just happens to be the mean height of the whole population. The problem is, of course, that we don't know what the mean height of the entire population is before we start, and so we don't know which men to select.

So, the question is, "How can we find those few good men without knowing what the population looks like in advance?" This sounds like a mission impossible. How can we possibly find a sample whose mean is close to the mean of the whole population when we don't know what the mean of the population is?

The answer is: Just pick people randomly. Let's try it now.

Suppose we randomly choose one adult male from our total population of 100,000. Well, that person's height may be totally unrepresentative. He might be extremely short. He may be extremely tall. But, as a matter of fact, only a small fraction of the population is extremely short or tall, so the chances are good that a randomly selected male will be somewhat near average height. Of course, "chances are good" and "somewhat near" are pretty vague terms. Let's see how we can improve our odds and make our measures more precise.

We performed several simulations using our data set. We selected a sample of 100 men, at random, from our data set and took the mean of those 100. The first sample of 100 had a mean of 68.61 inches. Then we chose another 100 at random. The mean of that sample of 100 was 68.57. We performed this process 10 times and the chart on the next page shows the means of each sample. The actual mean of the whole collection is 68.68 inches. Notice how close the means of the samples are to the mean of the population. Among our 10 experiments, the largest sample mean was 69.11 inches, which was 0.43 inches above the actual mean, and the smallest mean was 68.37 inches, which was 0.31 inches below the actual population mean. So, when we studied random samples of size 100 from our population, the mean of each sample just by chance alone was very close to the mean of the whole population. It was possible to have chosen all 100 in a sample to be far taller than average or all 100 to be far shorter, but randomness makes such selections extremely unlikely.

The Mean of Each of Ten Random Samples of 100 Heights of Adult Males (Inches)	
68.61	
68.57	
69.11	
68.85	
68.77	Actual Population Mean was 68.68.
68.37	
68.77	
68.67	
68.91	
68.69	

Fortunately, random chance tends to produce results that are close to the actual correct value. That is, luck is on our side. If you download a data set of heights from the Web, you can attempt this experiment for yourself. You can randomly select some number of data points from the population of heights. Notice that if you randomly select a larger sample, your sample mean will tend to become increasingly close to the actual population mean.

What we can infer from these experiments is that if we randomly select 100 adult males from a population of 100,000, the mean of that group of 100 is likely to be quite close to the actual mean height of the entire male population. In fact, as we will see later, if we select 100 adult males randomly from an even larger population such as the whole United States, the mean of the sample is likely to be quite close to the mean of the whole population. And if we choose a random sample of 1000, we can expect our answer to be even closer to the actual value.

In fact, from our knowledge of the world, we know that heights of men are basically distributed in a bell-shaped curve centered around the mean value. So we can also use our sample data to infer the standard deviation of the entire population, that is, how spread out the heights are. That inference then would give us a good description of the distribution of heights for all adult males. Again, the larger the random sample size, the more confidence we can have that the sample's mean is close to the population's mean and that the sample will have a bell shape to it that will approximate the shape of the entire population's heights.

Pollsters and Confidence Intervals

During election cycles, we frequently hear and read about the results of polls. A poll typically asks voters whom they intend to vote for. A news headline might trumpet: "Poll shows that Arnold Schwarzenegger will receive 46% of the vote with a ±3% margin of error."

What does that statement mean? It is reporting the outcome of a sampling survey that found that among a random sample of voters, 46% of them said they will vote for Schwarzenegger, a.k.a., the Terminator. (For the purpose of this discussion, let's assume that all the voters answered honestly and will not change their minds.)

But what does the $\pm 3\%$ margin of error mean? Here's the scoop: When the votes are actually counted, Arnold will get a certain percentage of the votes. At the time we collect the data from the sample (before the election), however, the *exact* percentage of the vote that Arnold will get is unknown—it might be 47%, it might be 43%, it might be 55%. In all of those cases, there is some chance that 46% of the voters randomly surveyed before the election state that they will vote for Arnold.

Perhaps 43% of all voters actually will vote for Arnold and we just happened to choose a sample of voters that included a higher proportion of Schwarzenegger supporters than were representative of the whole voting population. Or perhaps 55% of the voters actually will vote for Arnold, but just by luck only 46% of our sample will vote for him. Because we are not asking every single voter, we cannot guarantee that our sample will have the same percentage of Schwarzenegger voters as there are Schwarzenegger voters altogether. So the margin of error of 3% indicates that the sample size was sufficiently large that the actual percentage of Schwarzenegger voters is likely to be within the range 43–49%.

Of course, "likely" is not a very quantitative adverb. To get a better idea of the poll's accuracy, we need more information. For example, the pollster might state that, "We are 95% confident that Arnold will receive 46% of the whole vote with a $\pm 3\%$ margin of error." This statement means that for a random sample of this size, the probability of our sample proportion being within $\pm 3\%$ of the true population proportion is 95%. Having observed the sample proportion of 46%, we say that we are "95% confident" that the true proportion is between 43% and 49% (but we can't know for sure if it is or it isn't until we see the election results). Our confidence was justified because there was only a 5% chance that we would select a random sample of that size whose proportion of Schwarzenegger supporters differed by more than 3% from the true value. The range of 43%–49% is called a *confidence interval*.

If we were to sample a larger number of voters, we would have more information, and we could report that improved knowledge in one of two ways. We could report that, "We are 95% confident that Schwarzenegger will receive 46% of the vote with a $\pm 1\%$ margin of error." In this case, we have chosen to shorten the confidence interval while leaving the confidence level at 95%. Alternatively, we might prefer to conclude something like this: "We are 99% confident that Schwarzenegger will get 46% of the vote with a $\pm 3\%$ margin of error."

Sample Size, Chicken Soup: When Is Enough Enough?

We can have a great deal of confidence that the opinions of 1200 people chosen randomly from a population of a 10,000 will reflect the opinions of the entire population, but would 1200 people be enough if we are trying to determine the preferences of a million or even 100 million people? We might think that to determine the opinions of a huge population, such as 100 million, we would have to take a much larger sample. Surprisingly, the fact is that what determines our level of confidence in the information we get is the number of people in the sample, while the total number of people in the population from which the sample is drawn is basically irrelevant. Here we will try to make this surprising assertion less surprising.

The sample size is more important than the sample's percentage of the overall population. Let's illustrate this idea with chicken soup. Suppose we're preparing two pots of chicken soup, one for the family in a 4-quart pot and one for the community soup kitchen in a 40-quart pot. We wish to determine whether each pot has enough salt, so we stir the soup well and then taste a spoonful from each pot. Do we need to take a larger spoonful from the larger pot than from the smaller pot? No, because as long as we stirred the pots well before we took the spoonful, we wouldn't expect there to be a lot of variation in the saltiness in the various parts of the pot, so one spoonful is adequate in either case. In other words, to get a meaningful result, our sample—in this case, the spoonful of soup we're tasting—must simply be large enough to reflect the variability in the population; it need not contain a certain proportion of the population.

In trying to estimate the percentage of the population who favor a certain candidate, there is a heuristic that lets us estimate the size of the margin of error given the size of the sample. Specifically, for 95% confidence, a sample of size n will have a margin of error of approximately $1/\sqrt{n}$. (We won't explain here why this estimate is valid.) Notice that the size of the population does not enter into this heuristic. So if we take a random sample of 1200 voters from any sized population, we can be 95% confident that the true proportion of the population for a given candidate will lie within $\pm 1 / \sqrt{1200} \approx \pm\frac{1}{34.64} \approx \pm 0.029$ of the sample proportion. So a random sample of 1200 will give us a margin of error of less than 3% at the 95% level of confidence.

The Federalist Papers

Life Lesson
Apply quantitative reasoning in every arena.

Sometimes statistical reasoning can contribute decisively to arguments in matters that don't appear to have any statistical component to them at all. Here's an example in which statistics gave important insight into an historical conundrum.

In 1787 and 1788, Alexander Hamilton, James Madison, and John Jay wrote *The Federalist Papers*, a collect of 85 essays advocating ratification of the United States Constitution. The essays were all published anonymously under the pseudonym Publius. Although they were published anonymously, the authorship of most of the essays was known. Hamilton was known to have written 51 of the essays; Jay, 5; Madison, 14; and Madison and Hamilton co-authored 3. However, people disagreed about who wrote the remaining 12 *Federalist Papers*, although they agreed that they were written either by Hamilton or by Madison.

Historians had disputed this question for well over a century, but in 1964, statistical reasoning entered the debate. Statisticians Frederick Mosteller and David Wallace published a book called *Inference and Disputed Authorship: The Federalist* in which they analyzed the question by looking at the frequency with which various words appeared in the disputed essays.

In speaking or writing some words are more or less interchangeable. For example, "on" and "upon" are often interchangeable. Here's an excerpt from one of the disputed *Federalist* essays: "The advantage of biennial elections would secure to them every degree of liberty which might depend on a due connection between their representatives and themselves." Changing "on" to "upon" would not change the meaning of the sentence. It would then read, "The advantage of biennial elections would secure to them every degree of liberty which might depend upon a due connection between their representatives and themselves."

Mosteller and Wallace focused on such word choices in which Hamilton and Madison had a tendency to use the two more or less interchangeable options but with different frequencies. Hamilton and Madison had written extensively, so each one had established patterns of word usage. Hamilton was fonder of the word "upon" than was Madison. Looking at the writing of Hamilton and Madison, Mosteller and Wallace found that Hamilton used "upon" much more often than Madison did. Hamilton averaged six uses of "upon" per 1000 words, whereas Madison used "upon" less than one time per 1000 words.

By examining the usage of "on" versus "upon" and several other words, Mosteller and Wallace found that word choices in the disputed essays corresponded much more strongly with Madison's usage than with Hamilton's. This evidence strongly supports the conclusion that the dozen disputed *Federalist Papers* were all written by Madison.

Fifty-Three Card Pick-Up

Suppose you start with a few identical decks of playing cards and after playing poker with "the gang" from your dorm, by the end of the night the different decks get all mixed together into one huge pile of cards. You ask your friends to separate the cards into their respective decks. Someone hands you a pack of cards with a rubber band wrapped around it, but you can't really be certain whether it is a complete deck or if it might have some cards missing or some extraneous cards from another deck. Of course, the natural way to determine whether it is a complete deck is to carefully sort the cards and see; however, there is another method that illustrates the power of random sampling.

We could simply shuffle the deck and select a card at random, record what it is, return the card to the deck, shuffle again, select a random card, record it, and repeat the process. Suppose we repeated the process 4000 times. Our data would give us very persuasive evidence about missing cards or repeated cards because we know what to expect if we had a complete deck of 52 cards: We would expect each card to appear the same number of times. We simulated this process on a computer and here is a histogram of how many times the different spades appeared.

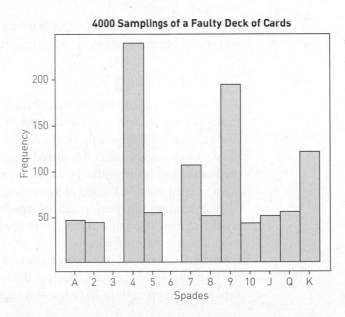

4000 Samplings of a Faulty Deck of Cards

This histogram of how many times we selected each spade card suggests several features of our collection of cards. Since we never selected a 3 of spades or a 6 of spades, it is likely that neither of these cards appears in our deck. If they were in the deck, the probability is rather small that we would select the 3 of spades or the 6 of spades on any single random draw, but there is essentially zero probability that we would not select them on any of the 4000 draws. We expect our sample to reflect reality. Notice that we selected the 4 of spades about four times as frequently as the ace, 2, 5, 8, 10, jack, or queen. That is evidence that there may be four 4 of spades in our deck. Likewise, there appear to be an extra 7, an extra king, and perhaps three times as many 9's as 5's. From random sampling, we obtain very strong evidence about the reality of our deck. The logic of this statistical inference is that the sample we obtained would be far more likely to occur from a deck with no 3 or 6, with one extra 7 and king, two extra 9's, and three extra 4's than it would from a normal deck of cards. In fact, that is the situation from which this random simulation was created. By analyzing probabilities, we can quantify how unusual it would be to get the outcome we got if the deck were in fact a regular deck. It would be possible to randomly select cards from a complete deck and find the data reported in the histogram; however, the probability of never selecting a card that is present in the deck would be extremely small.

Do the Crime, Do the Time

Suppose you are arrested for robbing a jewelry store. You are put on trial and evidence is presented. One witness testifies that he saw a car leaving the scene that was the color of your car. Another testifies that the culprit's hair color was the same as yours. Another says your height is the same as the burglar's, and so on. After the evidence has been presented, the judge will instruct the jury that you, the defendant, are to be presumed innocent. The members of the jury are to weigh the evidence under the assumption that you are innocent. Only if it is unreasonable that the evidence would have occurred under the assumption of innocence are they to declare you guilty.

The pattern of criminal justice is used extensively in statistical inference, as we will see in our description of hypothesis testing below.

The Lady Tasting Tea

A famous story involving the statistician Ronald Fisher and a lady tasting tea illustrates important ideas of statistical inference and experimental design. In the 1920s, Fisher attended a dinner party at which one of the ladies claimed that if she were handed a cup of tea with milk stirred in, she could usually tell by the taste whether the hot tea had been poured into the cold milk and then stirred or the cold milk had been poured into the hot tea and then stirred.

Fisher reportedly performed an experiment to test the lady's claim. In another room, he took two cups. In one he poured a measured amount of cold milk first, then the hot tea, and stirred. In the other, he poured the hot tea first, then the cold milk, and stirred. He carried the two cups to the lady and asked her to taste the beverages and then say which of the two she thought was the tea-first cup. He repeated this procedure 10 times—on each occasion presenting the lady with two cups and soliciting the lady's answer. This experiment allowed him to make a reasoned conclusion about whether or not the lady could actually tell a difference.

We are skeptical that there really is a difference in the taste of the two beverages. So we might adopt the predisposition that the lady really is just guessing. Of course, we understand from the nature of guessing that there is a chance that she could simply guess and get quite a few correct. In fact, by luck alone, she might get all 10 correct. So we might ask ourselves this question about evidence: How many of the 10 pairs of cups would the lady have to get right for us to feel confident that she can actually detect a difference in taste between a tea-first and milk-first beverage, that is, that her selections are not just a matter of random guessing?

Here is one way for us to think about what we can infer from various possible outcomes of this experiment. At one extreme, if she got all 10 correct, we might say, "Well, certainly the lady can identify the difference." But there is some chance that she could simply guess and just accidentally get them all correct. So let's be specific about how likely it is that she simply guessed and got them right.

Let's make the hypothesis that she was just guessing. Then what is the probability that she would get them all right by random luck alone? The answer is easy to compute. She has two choices for the first pair, two choices for the second pair, two for the third, etc. So altogether there are $2^{10} = 1024$ different ways she could make her 10 selections. Only one of those is correct every time. So her probability of getting them all correct if it she were just guessing would be 1 out of 1024. So she would be extremely unlikely to get them all right by luck alone. That low probability of her guessing them all correctly by accident would be strong evidence that she didn't just randomly guess. If she got them all right, that is strong evidence that she has some ability to detect the difference.

We could ask what would be the probability of getting 9 or more right by luck alone. To get 9 correct she could incorrectly identify the first cup, or the second, or the third, and so on to the tenth, or she could get them all correct. So there are 11 out of the 1024 possible ways to guess with the 10 pairs that would include at least 9 correct. So the probability of getting at least 9 correct by luck alone is 11/1024, which is a little more than 1%. So there is nearly a 99% chance that she would miss at least two if she really could tell no difference and instead she was selecting by chance alone. So if she got 9 or more correct, we could be quite confident that she could tell a difference between the milk-first and tea-first beverages.

The evidence becomes less compelling if we allow up to two mistakes. We must ask ourselves the question: What is the probability that a person doing this experiment would appear discerning enough to get 8, 9, or 10 correct but really just be guessing randomly? It would be a mistake to ask for the probability of guessing exactly 8 correct, because in doing the experiment, getting 9 or 10 correct would be considered even better. So when we start the experiment and we are wondering how persuaded we should be by an outcome, we want to evaluate the probability that random guessing would have a result better than some specified level of accuracy—in this case, no more than two misses. There are 56 out of the 1024 possible ways to guess with the 10 pairs that would include at least 8 correct. So the probability of getting at least 8 correct by luck alone is 56/1024, which is about 5.5%. So there is nearly a 94.5% chance that if she randomly selected the cups for each of the 10 pairs, she would miss more than 2 of the cups, but 5.5% of

Life Lesson
Craft your questions clearly and precisely.

Life Lesson
Make it quantitative.

the time, she would correctly guess 8 or more by chance alone. That means that more than 1 out of 20 times, guessing by chance alone she would get 8 or more right, so that evidence is less persuasive.

We don't want to perform these computations after the results are in. Instead, we adopt a basic strategy of classical statistical inference. We decide *in advance* what level of confidence we will accept as persuasive. Often people decide that if an occurrence would happen less than 5% of the time by chance alone, then they will declare the evidence statistically significant. If we choose to use that level of persuasiveness, then in the case of the lady tasting tea, if we selected the 5% level of significance, we would say that her correctly identifying 9 or more cups would be statistically significant, since that would happen only about 1% of the time by luck alone; however, getting 8 correct would not be persuasive evidence that she could tell the difference. We would argue that since she would guess 8 or more correctly more than 1 out 20 times (about 5.5%) by luck alone, we would not find the evidence persuasive that she really could tell the difference.

Fisher became famous for his work among other things on design of experiments, which means collecting data in such a way that we can make reasoned inferences from the data we gather. He told the story of the lady tasting tea in one of his books but did not divulge how many cups she got right. Someone else at the party claimed she got them all correct.

This story illustrates how we can use statistics to base our understanding of the world on evidence.

Throwing Up Statistical Figures

What cures the blues? According to a *Washington Post* article about a study of treatments for depression that compared St. John's wort, an herbal remedy, with the prescription drug Zoloft, "St. John's Wort fully cured 24 percent of the depressed people who received it, and Zoloft cured 25 percent—but the placebo fully cured 32 percent." No wonder we're depressed—we're not taking enough sugar pills.

Which treatment would you pick?

Many decisions we make are ultimately based on statistics. We want to know whether a certain medicine has a dangerous side effect and whether taking vitamin C helps us live longer. We want to know when to hold 'em, know when to fold 'em, know when to walk away, and so on, at the poker table.

One way to address these questions is to assert a hypothesis for what we believe to be the case and then test our hypothesis by doing experiments, collecting data, and computing the likelihood that the data we *actually* found would fall into the range that they did if the hypothesis *were* correct. If the data do not correspond to the hypothesis, then we can conclude that the hypothesis is likely to be false. For example, suppose we've developed a new medicine and we assert that there is only a 10% chance that it will make people vomit. To investigate whether this assertion is true, the FDA might give the medicine to a large group of people chosen randomly. If 80% of them vomit after taking the medicine, the FDA might strongly doubt our assertion that the medicine has only a 10% chance of making people toss their cookies. Of course, there remains a slim chance that the group of people we chose randomly just happened to have unusually sensitive stomachs even though the medicine actually only sickens 10% of the entire population.

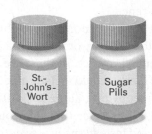

Which would you pick?

But given the collected data, we can be very confident that our 10% hypothesis is not valid.

Spinning Pennies

How will it land?

To gain a better understanding of the process of hypothesis testing, let's consider an experiment that you can do for yourselves, namely spinning pennies. Imagine that you take a penny, place it on its edge on a table, and while resting your left index finger on the top of the penny, you flick the side of the coin or otherwise put the penny into motion like a spinning top. As the penny slows down, its balance on its spinning edge gets more and more tenuous, and finally it falls exhausted onto the table. It lands either heads up or tails up. If we do this experiment again and again, will we see more heads, more tails, or about the same number of each?

It is typical with hypothesis testing that we state a hypothesis (often called the *null* hypothesis) that we wish to refute, rather than a hypothesis that we wish to support. In the earlier analogy of a defendant at a trial, the defendant is on trial because someone (the prosecutor) thinks the person is guilty. Then the hypothesis of innocence is tested against the evidence.

Perhaps surprisingly, it turns out that from our experience we (the authors) suspect that a spinning penny does not land heads up and tails up with equal probabilities. So we want to refute the hypothesis that heads and tails are equally probable. So we make the hypothesis that the probability that a spun penny will land heads up is 0.5. Our strategy is to gather data and to determine whether the data we find would be unlikely under the assumption that heads and tails are equally likely. Let's first state our null hypothesis clearly and then devise some experiments to test it:

Null Hypothesis: There is a 0.5 probability that a spun penny will land heads up.

Life Lesson

Make a hypothesis and then determine its validity by measuring the evidence.

One-Spin Experiment

We spin a penny once. It comes up heads. Can we conclude anything from this experiment beyond the profound insight that a penny has at least one head? This experiment sheds no light on the veracity of our hypothesis because our hypothesis is based on probabilities. And probabilities reflect a measure of what happens if pennies are spun over and over and over again, not just once. In fact, saying that the probability that a spun penny will land heads up is 0.5 means that if a penny is spun millions of times or billions of times or trillions of times, then the fraction of times the penny lands heads up will get increasingly close to half of the number of times the penny is spun. You may recall that this fact about the relationship between outcomes of an experiment and probability is referred to as the Law of Large Numbers. So this experiment can neither support nor refute our hypothesis. Larger numbers of spins give better evidence.

Twelve-Spin Experiment

Suppose we spin a penny 12 times, and it lands heads up four times. If our only data are from this experiment, do we have enough information to draw a conclusion? How strong is this evidence against the hypothesis that each spin has a 0.5 probability of landing heads up? This question is difficult to answer because even if the probability of the penny landing heads up is exactly 0.5 for each spin, it certainly *could* happen that we could get only four heads in 12 spins. The question is: Does this experimental outcome provide strong evidence that our hypothesis is wrong?

First we must realize that this question is connected with probability. There is some chance that a 0.5-probability event will occur only four times in 12 tries. But how likely is that?

To answer that question, we must think carefully about the meaning of probability. Recall that to compute a probability, we must think about equally likely outcomes of an experiment. Let's think very carefully about our current situation of spinning pennies in order to understand what the equally likely occurrences are.

Our experiment is to spin a penny 12 times. Our hypothesis is that each spin might land heads up or tails up with equal probability. In a particular 12-spin experiment, spin 1 might land heads, spin 2 tails, spin 3 tails, spin 4 heads, and so on, resulting in some specific sequence like HTTHHHHTHHTHT. This particular sequence has 7 heads. That outcome is one possibility. Any other list of heads or tails for each spin comprises another equally likely outcome. The collection of all possible outcomes of heads and tails for each spin is the collection of equally likely possibilities. There are two outcomes (heads or tails) for spin 1, two for spin 2, and so on. So, there are a total of $2^{12} = 4096$ possible outcomes—each one as likely to occur as any other if the hypothesis that the probability of heads is 0.5 is correct. Here is a chart summarizing all the possibilities:

Number of Heads	0	1	2	3	4	5	6	7	8	9	10	11	12	Total
Frequency	1	12	66	220	495	792	924	792	495	220	66	12	1	4096

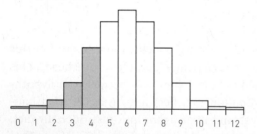

- Number of spins = 12
- Probability of one spin yielding heads = 0.5
- Number of equally likely possible outcomes: 212 = 4096
- This graph shows the distribution of the number of heads, that is, how many of those 212 = 4096 outcomes had 0 heads, 1 head, 2 heads, and so on.

The shaded histogram gives a clear view of these data. If the probability of a single spin of the coin coming up heads is really 0.5, then this graph accurately tells us how often in 12 spins we are likely to see 0, 1, 2, 3, . . . , or 12 heads among the 12 spins.

Now let's note where the result of getting 4 heads is on the graph. Notice that 4 heads is a little off to the left side of the center. If the probability for each spin were 0.5, then most of the time when we perform 12 trials, we would see 5, 6, or 7 heads. Getting only 4 heads is not *extremely* unusual, but it does look a little atypical, that is, a little out of the center pack.

We can be specific and ask ourselves how unusual it is to get a number of heads in 12 spins that differs from the center as much as or more than 4 differs from the center. Among the 4096 equally likely possible outcomes with 12 spins, 2508 of those outcomes have 5, 6, or 7 heads. Therefore, missing those most common values happens 1588/4096 or about 3/8 of the time. In other words, if landing heads up really happened with probability 0.5, then the probability of spinning a penny 12 times and getting 0, 1, 2, 3, 4, or 8, 9, 10, 11, or 12 heads is 1588/4096 ≈ 0.38. This amount of deviation from exactly half heads is not too unusual, and, therefore, getting 4 heads out of 12 spins certainly does not provide extremely persuasive evidence against our hypothesis that the probability for each spin is 0.5. If we find that even with larger numbers of spins only about a third of the outcomes are heads, then the evidence against the 50–50 hypothesis becomes much more compelling. Let's see.

Hundred-Spin Experiment

Suppose we spin a penny 100 times. It comes up heads 35 times. What can we conclude? Since only 35 heads appeared in 100 spins, how strong is that evidence that our 0.5 hypothesis is wrong?

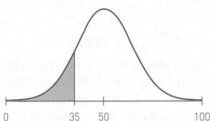

Again, to answer that question we must describe the equally likely outcomes and see where our results fit into the distribution of possible outcomes. The shaded graph here shows the possible outcomes: If the probability of a single coin spin coming up heads is really 0.5, then this graph accurately tells us how often in 100 spins we are likely to see 0, 1, 2, 3, . . . , or 100 heads among the 100 spins. Notice that the graph has the characteristic bell shape of a normal distribution.

When we see where the 35-heads outcome fits into all possible outcomes, we find compelling evidence that our hypothesis is false. The 35-head outcome is in the thin part of the graph. Therefore, deviating from 50 heads by 15 or more (such as 35 heads does) would be *very* rare for 100 spins if there really were a 0.5 chance of getting heads with each spin. The probability of getting 35 or fewer or 65 or more heads out of 100 spins would be so unusual, if there really were a 0.5 probability of getting heads on each spin, that we find such evidence very compelling that the hypothesis is wrong. Let's do one more experiment to pin down this idea.

Thousand-Spin Experiment

We spin a penny 1000 times. It comes up heads 329 times. What can we conclude? How confidently can we reject our hypothesis given that the number of heads that appeared differed that much from 500 heads?

As always, to answer that question we must describe the equally likely outcomes and see where the 329-heads result fits into the distribution of possible outcomes.

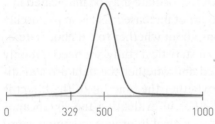

- Number of spins = 1000
- Probability of one spin yielding heads = 0.5
- Number of equally likely possible outcomes: 2^{1000}
- This graph shows the distribution of the number of heads, that is, how many of those 2^{1000} outcomes had 0 heads, 1 head, 2 heads, and so on.

The graph shown here gives the usual view of the 2^{1000} possible outcomes. If the probability of a single coin spin coming up heads is really 0.5, then this graph accurately tells us how often in 1000 spins we are likely to see 0, 1, 2, 3, . . . , up to 1000 heads among the 1000 spins. The graph displays an even thinner bell-shaped curve.

Statistical Significance

If there really were a 0.5 chance of getting heads with each spin, our getting a number of heads in 1000 spins that is as far off from the center as 329 is from 500 would be extremely rare. So our data—getting only 329 heads in 1000 spins—give us very strong evidence that the hypothesis—that there is a 0.5 probability of a spun penny landing heads up—must be false. The probability, computed assuming the null hypothesis is true, of obtaining a result as extreme or more extreme than the result in our sample is called the *p-value*. So the lower the *p*-value is, the stronger is the evidence against the null hypothesis. The probability of deviating from the mean of 500 heads by as much as or more than 329 does is less than 0.001, so the *p*-value is less than 1/1000. When an outcome would occur very rarely by randomness alone, we say that the outcome is *statistically significant*.

In the 12-spin experiment, getting four heads was not statistically significant evidence against the hypothesis, since deviating from the center by that much or more would not be unusual by chance alone.

A result becomes more statistically significant when the chance of the outcome happening by accident becomes more and more remote, in other words, when the *p*-value gets increasingly small. The data in our 1000-spin experiment allow us with great confidence to reject the hypothesis that a spun penny has a 0.5 probability of landing heads up.

Statistical Significance versus Practical Significance

Suppose we are trying to decide whether eating chocolate leads to heart attacks. Perhaps we would take as our null hypothesis that eating chocolate has no effect on the probability of a heart attack. Since this study is of such importance to so many people, suppose we get a large grant and hundreds of thousands of randomly selected people are involved in the study. With large numbers of people involved, even small effects will become statistically significant. So perhaps people who eat chocolate have a tiny tendency to have more heart attacks. Then the study might report that eating chocolate causes a greater probability of heart attacks and that the result is statistically significant. However, if the differences in heart attack rates between the chocolate eaters and the chocolate abstainers are very tiny, then the distinction may have no practical significance in making a decision. The study may be statistically significant in asserting that chocolate increases the rate of heart attacks, meaning that chocolate may be implicated in a greater risk of heart attacks, and yet the amount of increased risk is so minuscule that it should have no effect on your decisions about whether to eat chocolate.

When deciding how we should *respond* to a statistical study, we need not only evaluate whether the study was well designed and whether the data are statistically significant, but we must also evaluate whether the amount of the effect is sufficient that we should care about it. Whether in evaluating the significance of statistical significance or any other use of statistical inference, common sense is always an important ingredient—in the final analysis, we must always think through all the evidence before we make our next move.

A Look BACK

OFTEN WE LEARN some data by taking a poll or doing some experiments. We thereby collect some data about the sample we are examining, but we really want to draw larger conclusions. We look at some of our blood and want to know how all of it is. We ask some voters how they will vote and we want to know who will win the coming election. We test a medicine on a few people and we want to know how effective it will be for every sick person.

A basic component of our strategy is to select our samples at random so that we can use probabilistic analysis to determine how confidently we can make an interpretation. By choosing 1200 voters randomly to take a poll, we can be 95% confident that the actual percentage of voters for a given candidate will be within 3% of the percentage in the sample who are for the candidate. Larger samples lead to more precise estimates and more confidence in our inferences. The concept of statistical significance captures a notion of the strength of the evidence against a particular null hypothesis, but a result being statistically significant does not necessarily mean it is of any practical importance.

Common sense and probability calculations allow us to gain important insights into the implications of statistical studies, which help us describe and understand our world.

MINDSCAPES   Invitations to Further Thought

In this section, Mindscapes marked (H) *have hints for solutions at the back of the book. Mindscapes marked* (ExH) *have expanded hints at the back of the book. Mindscapes marked* (S) *have solutions.*

Developing Ideas

1. **Sophomore survey.** You want to survey all students at your school about their study habits. Which of the following methods are more likely to result in a random sample? Why?

   • Go to the library on Saturday afternoon and ask all the students you find there.

   • Stand at the entrance to the only dining hall on campus throughout the dinner hour and interview every 20th student.

   • Go to the fitness center on Friday afternoon and ask all the students you find there.

   • E-mail a survey to all students and see who replies.

2. **Interval confidence.** Which of the following best describes the meaning of "95% confidence interval"?

   • Ninety-five out of every 100 people will believe your results.

   • Your sample includes 95% of the population about which you are gathering data.

   • This is a numerical interval generated by a procedure that, 95 times out of 100, will produce an interval that contains the true value for the whole population.

3. **Voting bodies.** Calculate the percentage of the population that each of the following samples represents. Put the three percentages in increasing order.

   • 20 voters from a population of 1000

   • 265 voters from a population of 265,000

   • 1400 voters from a population of 14,000,000

4. **Play the percentages.** If $n = 100$, what is the value of $1/\sqrt{n}$ as a percent? What if $n = 10,000$?

5. **Play the percentages backwards (S).** If $1/\sqrt{n} = 0.05$ (equivalent to 5%), then what is the value of n? What if $1/\sqrt{n} = 0.02$?

Solidifying Ideas

6. **Election up for grabs.** In the upcoming election between two close candidates, a poll reports that Jones leads with a 51% preference with a margin of error of ±3% at the 95% confidence level. How confident are you that Jones will win?

7. **U.S. samples (S).** Suppose you can be quite confident of a survey result with a sample size of 500 for the whole United States. Missouri has about 2% of the population. Can you have the same confidence with the result of that survey for Missouri if you sample 10 people from Missouri? Why or why not?

8. **Bigger confidence?** Suppose a pollster reports that a brand is preferred by 33% of the people with a ±2% margin of error at a 95% confidence level. For the same data set, would a 99% confidence interval be wider or narrower? Why?

9. **Overlapping confidence.** Suppose two surveys are taken to determine what percentage of the population find green catsup appealing. One survey reports that 45% of the population likes green catsup with a ±12% margin of error at a 95% confidence level. The second, bigger survey reports that 41% of the population likes green catsup with a ±1% margin of error at a 95% confidence level. Is the percentage of people in the whole population who like green catsup more likely to be closer to 45% or 41%? Why?

10. **Does size matter?** (ExH) You want to gather data about voters in Wyoming and California. California has approximately 14,000,000 registered voters. Wyoming has approximately 265,000. Which of the following samples will give you more reliable results: 265 voters from Wyoming or 1400 voters from California? (Assume both samples are chosen at random.)

11. **More voting bodies (H).** You want to gather data about voter preferences. Which of the following samples will give you more reliable results: 1200 voters surveyed from a population of 120,000 or 1200 voters chosen from a population of 200,000? (Assume both samples are chosen at random.)

12. **National poll.** Suppose you do polling in each of the 50 states and construct a 95% confidence interval for your results in each state. In how many states would you expect that state's interval not to contain the actual voter preference in that state?

13. **Dizzy Lincoln.** The text describes a penny-spinning experiment. The graph below shows the theoretical distribution of the number of heads occurring for 1000 spins. Suppose you repeat the experiment and obtain heads 412 times. Locate 400 on the horizontal axis below. Does this look like an extremely rare occurrence? What if you obtained 550 heads? What about 650 heads?

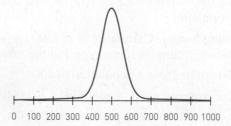

0   100 200 300 400 500 600 700 800 900 1000

14. **Dizzy friends.** Get several friends together, gather some pennies, and spin away until you have completed 1000 spins. Record the number of times a penny lands heads up. Use the graph from the previous Mindscape to decide

whether your results are consistent with a 0.5 probability that each spun penny will land heads up.

15. **Hire a consultant.** A consulting firm analyzed records and discovered a statistically significant difference in the mean value of contracts obtained in 2003 and 2004. The mean values differed by $93. Given that the typical contract for this firm was for more than $100,000, is this difference of any practical importance? Explain.

16. **House of cards.** Your friend hands you a deck of 52 cards along with the histogram below. It shows the results for hearts if a card is chosen at random from the deck, its rank recorded (A, 2, 3, . . . , 10, J, Q, K), the card replaced, and then the process repeated 4000 times. How do you know the heart suit in the deck is incomplete? Which heart cards are duplicated in the deck? How many duplicates do you think there are?

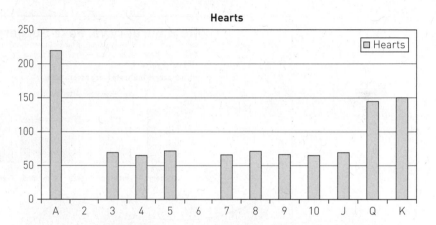

17. **Rig the deck.** To mimic the experiment from the previous Mindscape on a more manageable scale, doctor a deck of cards by removing half the hearts, half the diamonds, and half the clubs. Then draw one card at random, record its suit, and return the card to the deck. (This is much easier with a friend.) Repeat the experiment A LOT. See if you get data that reflects the disproportionate number of spades.

18. **A fish story (ExH).** You are a biologist sampling the population of fish in Razorback Lake. You estimate the total fish population to be about 10,000. You catch and return 1500 fish (all at once) and obtain the following data. What percentage of the fish you caught were bluegills? How many bluegills do you think there are in the lake altogether? What assumptions are you making about the fish in the lake?

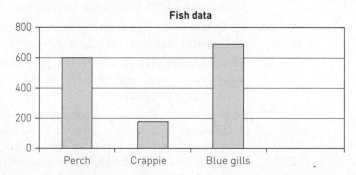

19. **Political gamble.** The mayor of a small city has proposed that a gambling casino be built in the downtown area, arguing that the construction project, resulting jobs, and tax revenue will boost the local economy. A total of 183 residents attend a public hearing on the proposal and a show of hands finds only 31 in favor of the casino. What does this vote tell the city council about public support for the project?

20. **The power of placebo.** A study of pain relief following arthroscopic knee surgery used two control groups to compare to a surgery group. One of the control groups had their knees opened and "lavaged," where the inside of the knee is flushed out, but no scraping or other surgical manipulation is performed. The other control group had their knees opened and then closed again, with no treatment at all. The graphs below display the average pain levels reported for the three groups. Which, if any, treatment appears to be effective?

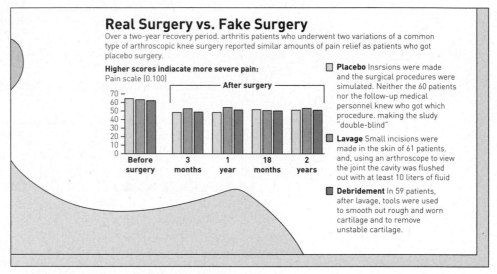

Source: *Philadelphia Inquirer*, 2002, from *New England Journal of Medicine*.

Creating New Ideas

21. **Marge Innoverra.** You test 100 light bulbs randomly out of a large shipment and find 6 of them to be defective. What margin of error would you need if you want a 95% confidence interval? What is the 95% confidence interval in this case?

22. **How many math students does it take . . . (S).** How many light bulbs should you test to have a 95% confidence interval with a margin of error equal to 2%?

23. **. . . to screw in a lightbulb?** How many light bulbs should you test to have a 95% confidence interval with a margin of error equal to 1%?

24. **Snack sample (H).** A survey asking students whether or not they snacked most days reports a margin of error of 5%. About how many students do you think were surveyed?

25. **In the balance.** Balance a penny on end and then pound on the table so that it falls over. Repeat this 25 times, recording the percentage of times your coin lands heads up (you can also balance several pennies at a time and do fewer trials). Find a 95% confidence interval using the $1/\sqrt{n}$ formula for

margin of error. Does your data suggest that the probability of heads is different from 0.5 for your penny in this experiment?

Further Challenges

26. **Stop the study (ExH).** A 2006 NIH study of circumcision and HIV-AIDS in Kenya was stopped early because it became so evident that circumcisions reduced the probability of contracting AIDS. Uncircumcised men without AIDS were randomly divided into two groups. Both received safe sex education and condoms, and one group had circumcisions performed by a physician in a hospital. After one year, 22 of the 1,393 circumcised men had AIDS, compared to 47 of the 1391 uncircumcised men.

    Make a table and compute the rate of AIDS in the two groups. Explain what it means that this difference is statistically significant. Do you think it is also practically significant?

27. **The sniff test.** In 2004 the *British Medical Journal* reported on a study showing that dogs can be trained to smell cancer. In one test, six dogs each performed nine trials for a total of 54 trials. On each trial, the dog would sniff seven bowls of urine, one of which was from a bladder-cancer patient. The dogs were trained to sit down by the bowl they thought smelled cancerous. Overall, the dogs were correct on 22 of 54 trials (about 41% of the time).

    a. If the dogs were guessing on each trial, on what proportion of trials would you expect them to be correct? (This is the probability of success by randomness alone, denoted p.)

    b. There is a more accurate formula for the 95% confidence interval margin of error than the $1/\sqrt{n}$ formula. Using p as defined in part (a), the margin of error is $1.96 \times \sqrt{p \times (1-p)/n}$. Calculate this margin of error using the value of p you found in part (a).

    c. If each dog has probability p of successfully identifying the correct bowl, use the result from part (b) to construct a 95% confidence interval for p.

    d. Is there evidence the dogs were doing better than chance?

28. **Roving report.** In 2006 there was a scandal involving the firing of 9 U.S. attorneys. In fact, 12 U.S. attorneys had been on a list for replacement, and another 3 resigned under pressure without being fired. Attorney General Alberto Gonzales said, in 2007, "I would never, ever make a change in a United States attorney for political reasons...." However, in 2006 White House advisor Karl Rove mentioned 12 states as being especially important in the 2008 presidential election. Six of the 12 attorneys on the list were from these states.

    Overall there were 93 U.S. attorneys in 2006, of whom 12 were replaced. Seventeen of the 93 attorneys were in the states mentioned by Rove; and 6 of these were replaced. The remaining 76 U.S. attorneys in the other 38 states were not mentioned by Rove, and 6 of these were replaced.

    a. Compute the proportion of attorneys from states mentioned by Rove who resigned or were fired. Compute the analogous proportion from states not mentioned by Rove.

    b. Does the difference in the two proportions from part (a) seem unusually large, or might it have happened due to chance alone? If the difference seems too large to be due to chance alone, what does that suggest about the decision making with regard to replacing U.S. attorneys?

In Your Own Words

29. Personal perspectives. Write a short essay describing the most interesting or surprising discovery you made in exploring the material in this section. If any material seemed puzzling or even unbelievable, address that as well. Explain why you chose the topics you did. Finally, comment on the aesthetics of the mathematics and ideas in this section.

30. With a group of folks. In a small group, discuss and actively work through the reasoning involved in hypothesis testing. After your discussion, write a brief narrative describing the methods in your own words.

31. Creative writing. Write an imaginative story (it can be humorous, dramatic, whatever you like) that involves or evokes the ideas of this section.

32. Power beyond the mathematics. Provide several real-life issues—ideally, from your own experience—that some of the strategies of thought presented in this section would effectively approach and resolve.

For the Algebra Lover

Here we celebrate the power of algebra as a powerful way of finding unknown quantities by naming them, of expressing infinitely many relationships and connections clearly and succinctly, and of uncovering pattern and structure.

33. Polling roundup (H). A polling organization wants to survey voters in Montana, Idaho, Utah, and Wyoming. The total population of these four states is about 6 million. The population of Montana is about 1 million, Idaho is about 1.6 million, Utah is about 2.8 million, and Wyoming is about 0.6 million. If the pollsters plan to survey 1200 people altogether, about how many should they survey from each state if they want the proportion of people surveyed from each state to equal the proportion of that state's population of the whole 6 million?

34. Rodent roundup. Your psychology class is planning a study involving mice and rats. There are two experiments in the study. For the first experiment, each mouse will complete 2 trials and each rat will complete 3 trials. For the second experiment, each mouse will complete 3 trials and each rat will complete 5 trials. The goal is to do 90 trials for the first experiment and 145 trials for the second. Let x equal the number of mice you need and y equal the number of rats you need. Create an equation for each of the two experiments that relates x and y to the total number of desired trials. Then solve these two equations for x and y to determine the number of mice and rats you need for your study.

35. Even more confident. Mindscape 27 gives a more accurate formula (than $1/\sqrt{n}$) for the margin of error corresponding to a sample size of n trials as follows. If p equals the probability of success by randomness alone on each trial, then the margin of error for a 95% confidence interval is $1.96 \times \sqrt{p \times (1-p)/n}$. Suppose $p = 0.4$ and $n = 200$. Find the more accurate value for the margin of error associated with a 95% confidence interval.

36. ESP. Your psychology class does some experiments to test for extrasensory perception. For a single trial, a subject must guess, er…, sense, which of five cards has been chosen without seeing those cards. Suppose your class conducts a total of $n = 80$ trials. Use the formula in the previous Mindscape to compute the margin of error associated with a 95% confidence interval. (*Hint*: with five cards, what is the probability p of a correct answer if a subject guesses at random?)

37. **Trying trials.** Your roommate is supervising an experiment where each trial consists of rolling a ten-sided die and guessing what will come up. (What's the probability of success?) After n trials, she uses the formula given in Mindscape 35 to compute the margin of error associated with a 95% confidence interval. She tells you the value is 4.8% (or 0.048). Use this value and the formula from Mindscape 35 to determine the number of trials, n.

WAR, SPORTS, AND TIGERS

Statistics Throughout Our Lives

Handle with care.

> *More people die in hospitals than in bars, so if you're really sick, tell the ambulance driver to take you to the nearest bar.*
>
> **ANONYMOUS**

At a very fundamental level, our understanding of the world comes from taking information, processing it, and applying what we find to inform our beliefs about ourselves and our world. One of the most profound developments during the last century has been the extent to which statistical knowledge and statistical methods have become pervasive in how we know what we know. Science, social science, business, and even sports describe their domains in terms of statistical findings and inferences.

One of the basic concepts that we use to describe reality is the principle of cause and effect. It is so basic an idea that we take it for granted. But when we look deeper, we soon realize that cause and effect can lead us to a minefield of difficulties. The evidence for cause and effect often comes to us in the form of statistical data. People who successfully complete more years of schooling tend to have higher earnings. Does the schooling make people more able to earn more money? Or could it be that the individuals who excel in school would naturally be better at performing other tasks as well and so would earn more even without the extra schooling? People in prison have a higher rate of illiteracy than people who do not commit crimes. Does illiteracy cause people to become criminals? Or do people who are less intelligent or who attend poor schools or have a harder time learning to read also have a harder time learning to be productive citizens?

Statistics can help us describe the extent to which two attributes of a person (years in school and income, for example) are connected and move together. Statistics can point out potential instances of cause and effect and it can identify unjust outcomes. For example, we can deduce statistically that the lottery that determined those who would be drafted during the Vietnam War was unfair. It turned out that men whose birthdays occurred in the last half of the year were more apt to be drafted than those with earlier birthdays, because the balls used to randomly select the dates were not thoroughly mixed.

Statistical methods empower us to deduce the principles that underlie many facets of our world. Mendel's experiments with inherited characteristics of peas led to our modern understanding of genetics. Statistical methods helped the Allied forces estimate the number of tanks in the German army during World War II. Statistical methods can allow us to count the number of tigers in a jungle. And statistical methods can reveal whether acceptances to a university program are discriminatory.

Statistical methods are fundamental to essentially every area of the modern world. The subtle logic of how to use statistical data to gain understanding of the world requires us to be clear-minded and realize that, in a fundamental way, statistics is a quantitative extension of common sense.

> *Life Lesson*
>
> *Whenever possible, create an experiment and study the outcomes.*

Courtesy of Edward Burger

The Birth of Genetics

Examining data and drawing conclusions from it can have profound consequences. One of the most significant instances of drawing conclusions from data occurred in the 1860s when an Augustinian monk named Gregor Mendel performed some experiments concerning inherited characteristics of peas from one generation to the next. Mendel's work led to our modern understanding of genetics and heredity. His analysis is a powerful illustration of probability and statistics at work.

One of Mendel's experiments was to cross-pollinate two varieties of peas—those with green pods and those with yellow pods. Some green-podded peas, when crossed with yellow-podded peas, will have only green-podded offspring, and some will have both green-podded and yellow-podded offspring. He began his experiments with green-podded peas whose offspring had only green pods and with yellow-podded peas that always produced only yellow-podded offspring. Mendel's experiment was to cross-pollinate the pure green-podded peas with the yellow-podded peas, which we refer to as the first generation. He discovered that their offspring (the second generation) had only green pods. However, when he cross-pollinated the second-generation plants with one another, approximately one-quarter of their offspring (the third-generation plants) had yellow pods. From this basic result, and with some appreciation for randomness and probability, he was able to deduce important basic features of genetics.

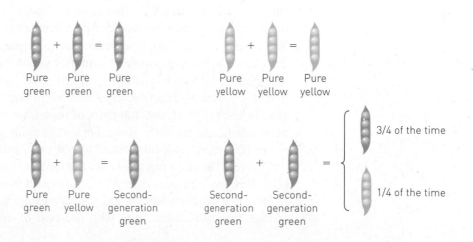

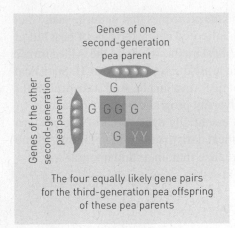

Genes of one second-generation pea parent

Genes of the other second-generation pea parent

G   Y

G	G G	G
Y	Y G	Y Y

The four equally likely gene pairs for the third-generation pea offspring of these pea parents

He realized that some of the genetic material from the yellow-podded peas must be in the second-generation offspring, even though the pods themselves were green, because in the third generation the yellow-podded variety reappeared. He also noticed that in all generations the color of the pods was never a blend of yellow and green. From these observations he hypothesized that offspring of the cross-pollinated plants contained genes from each parent and that the gene for green pods must be *dominant*—that is, if the gene for greenness is present, the pod will be green regardless of what the other gene is.

Finally, he realized the significance of the fact that one-quarter of the third-generation peas were yellow-podded. The second-generation plants had two parents—one yellow-podded, one green-podded—so Mendel assumed that the second-generation plants must have one gene for greenness and one gene for yellowness. If one gene from a yellow-green plant is randomly chosen and combined with a randomly chosen gene from another yellow-green plant, what would we expect to see?

By comparing expectations with reality we can confirm or refute Mendel's hypotheses. Let's count the possibilities. From the first-parent pea, we would receive either a green gene or a yellow gene. For each of those possibilities, we would inherit either a green gene or a yellow gene from the other parent. Therefore, four possibilities occur: green-green, yellow-green, green-yellow, or yellow-yellow. In the first three of those four cases, the pods of the offspring will be green. In the fourth and last case, the pods will be yellow.

This simple example of counting the possible outcomes and counting how many of these result in yellow-yellow genes gave powerful evidence for the important theory of how genetic information inherited from parents manifests itself in their offspring. Mendel's seminal work was fundamental to our understanding of genetics.

Cooking the Books Rather than the Peas

After Mendel's work became famous, statisticians carefully analyzed Mendel's data and discovered that the data fit the model *better* than random variation would predict that it should. This pea-culiar finding raises some eyebrows and some interesting questions. Apparently Mendel or some of his assistants had the mathematical model of genetic inheritance in mind and knew what data would support such a model. With that foreknowledge in mind, the data magically came out a little better than we would expect from carefully reported real experiments. We actually would expect experiments to display a wider range of values that were close to but not *exactly* one-quarter yellow pods.

To gain a better understanding of this issue, let's think about what would happen if we were to flip two coins and count how many heads-heads results we saw. In each trial of, say, 100 pairs of tosses, we would expect about 25 pairs to be heads-heads; however, if we flipped the coins many times (that is, performed *many* 100-pair trials), our results would vary, and occasionally they would vary quite a bit. Mendel's reported data are more consistently very close to one-quarter yellow-podded peas than we would expect from randomness alone.

In other words, when we know what results we're seeking, it is very easy to "fudge the data" a bit, even with the best of intentions (and even if one is a monk). In Mendel's case, plants that did not support his hypothesis were probably not counted as "fully" as those that did. For example, some plants may have been prematurely pronounced dead a little before they breathed their last pea-breath.

A famous statistician, Ronald Fisher, pointed out another indication that Mendel's data were not quite what one would expect from observations. This twist on the tale illustrates an interesting relationship between mathematics and expectation.

As we saw, Mendel performed experiments in which he crossbred peas with green-yellow genes and grouped the offspring into three groups according to the genes they have: green-green, yellow-yellow, and green-yellow. The genetic model and probability predict that we should expect: ¼ green-green, ¼ yellow-yellow, and ½ green-yellow offspring. But how could Mendel tell whether a green-podded plant had green-green or green-yellow genes since both such plants have green pods? He couldn't look directly at the genes. The only way to tell was to crossbreed the plant with itself. If any of the incestuous offspring were yellow, then the plant definitely has green-yellow genes, since only plants with yellow-yellow genes have yellow pods. Mendel crossbred the plant with itself 10 times. If all 10 offspring were green, he declared that the plant had green-green genes.

But let's contemplate this scenario a bit further. There is a ¼ chance on any self-crossbreeding of a green-yellow plant that the offspring will be yellow-podded. So the probability of 10 experiments never coming out yellow is $(3/4)^{10}$, which is about 5.6%. So Mendel should have *wrongly* classified roughly 5.6% of the green-yellow plants as green-green. Since we expect this misclassification using Mendel's method of determining his classification of green-green plants, we actually should expect an outcome chart that is not actually ¼ green-green, ¼ yellow-yellow, and ½ green-yellow offspring. Because of the expected misclassifications, we should expect 5.6% more of the offspring that are actually green-yellow to be wrongly classified as green-green. So if in reality the distribution were ¼ green-green, ¼ yellow-yellow, and ½ green-yellow offspring, then we would expect the report to mistakenly classify 5.6% of the green-yellow offspring (which we are assuming are ½ of the plants) as green-green, with corresponding reductions in the reported green-yellow numbers. So the percentage of plants that we would expect to be classified as green-green would actually be 27.8%, which is the 25% that are actually green-green plus 5.6% of ½, or 2.8%, which are actually green-yellow but misclassified as green-green. Likewise, only 47.2% would be classified as green-yellow, since 2.8% of those that are actually green-yellow would be mistakenly classified as green-green. And 25% would be correctly classified as yellow-yellow, since only plants with yellow-yellow genes have yellow pods.

The data that Mendel reported, however, were much closer to the values that would be expected if the classification scheme were perfect. This fact about his data shows again that the data were somewhat modified to correspond with expectations rather than strictly reporting what was actually seen.

Scientists, businesses, governmental agencies, and individuals all may have strong incentives to report what they hope to find rather than merely the facts. The natural bias toward good results is one reason that important experiments should be and are duplicated to confirm them. Underreporting negative results remains an ongoing challenge in statistical studies. Careful guidelines describe appropriate practice in scientific studies, and suspiciously good data invite special scrutiny.

Life Lesson
If it's too good to be true, probably someone is stretching the truth.

Life Lesson

Correlation does not imply causality.

Relationships versus Cause and Effect

Post Hoc, Ergo Propter Hoc (After This, Therefore Because of This)

One of the goals of understanding anything is to find what causes what. So when we observe that two quantities vary in a related manner, it is natural to wonder if one is the cause of the other. But beware: Deducing that one thing *causes* something else just because the two vary in related ways is a rich source of potential error. Even if we can accurately predict one thing from another, we can't conclude that the one causes the other. The fallacy of thinking that one thing causes another just because it follows the other is a famous logical flaw that has a fancy Latin name, namely, *post hoc, ergo propter hoc* (after this, therefore because of this). Here are some examples:

- While writing our section on randomness, five famous people died. Therefore, writing a section on randomness is deadly because it results in the deaths of five famous celebrities.

- Ninety-four percent of all CEOs have personal assistants. Therefore, if you want to be a corporate leader, hire an assistant pronto.

- People who fly in first-class seats are wealthy. Therefore, if we buy first-class tickets, we will be rich.

- Almost two-thirds of the inmates in prison have divorced parents. Thus, divorce leads children to criminal behavior.

- Overall, people who wear bigger shoes know more (because babies have such small feet). Therefore, the next time you're walking into an important examination, forget studying and just wear size-16 sneakers.

Twin Studies: Randomness versus Causality

Identical twins Jim and Jim, reunited

Jim Lewis and Jim Springer were identical twins who were separated at birth and brought up without any contact with each other. Thirty-nine years later, on February 9, 1979, they met for the first time. Researchers subjected the two to elaborate tests and reported that the two men were remarkably similar in many ways. Both were named James. Both lived on a block with only one house on it in towns in Ohio. Both had a white bench around a tree in their yards. Both liked Miller Lite beer. Both first married women named Linda, and both remarried women named Betty. Lewis had a son named James Alan, and Springer had a son named James Allen. Both, at one time, owned dogs named Toy. Both had workshops in their basements. Both drove Chevrolets. Both chain-smoked Salems. Both liked stock-car racing but disliked baseball. Both vacationed at the same three-block-long beach on the Florida Gulf coast, and both were guests on the *Tonight Show Starring Johnny Carson* (well, okay, they appeared together after all the coincidences were revealed). Given all this evidence, it must surely be the case that there is a strong genetic component to personality, right?

What additional information would we need to know about the study to be confident that it supports the theory that personality is significantly influenced by genetics? The answer is that we would have to compare the frequency of coincidence in the twin study with the frequency of coincidence in studies of random pairs of

> *"Genetics may not be the main reason that identical twins raised apart seem to share so many tastes and habits,"* said Richard Rose, a professor of medical genetics at Indiana University. *"You're comparing individuals who grew up in the same epoch, whether they're related or not. If you asked strangers born on the same day about their political views, food preferences, athletic heroes, clothing choices, you'd find lots of similarities. It has nothing to do with genetics."*
>
> K.C. COLE, *LOS ANGELES TIMES*, 4 JANUARY 1995

people to make sure that the twin study's frequency was in fact greater than the frequency with random pairs of people.

Genetics appears to influence personality and tastes; however, we cannot evaluate the significance of a list of coincidences without additional information. In fact, some of the coincidences listed in the twin study cannot be more than coincidence. For example, their both being named Jim has nothing to do with their being twins—they didn't name themselves. That both chose wives with the same name is interesting, but what pertinence does it have to the question of genetic influence on personality? Is the study suggesting that the preference for names is genetically based? It's a good thing one of the twins was not raised in Japan—it may be hard to find a Linda there. More pertinent might be statements about the personality types of the Lindas. How many questions were asked of these twins? If they were asked thousands of questions, our sense of amazement at finding a list of surprising coincidences would drop dramatically. If we ask any two people thousands of questions about their preferences, life experiences, and physical surroundings, we could easily pick out many examples of "amazing" coincidences.

To evaluate the significance of the twin studies, we must compare the results of the twins' answers with results from similar tests administered to random people or adoptive families in similar socioeconomic conditions. The observational study must be designed so that the data extracted enable us to draw meaningful conclusions about the question of nature versus nurture, genetics versus experience. Perhaps the studies were designed properly; however, reading newspaper accounts of such studies does not give us enough information to know how much of a child's future personality is determined at birth and how much is determined by imaginative or repressive child-rearing techniques.

Measuring Relationships—Correlation

Statistics cannot, in itself, tell us whether one thing causes another; however, we can develop meaningful ways to display and measure the extent to which two attributes of the same individual are related to one another throughout the population. Let's illustrate this idea with several examples.

Let's examine the SAT score and grade point average (GPA) for all students in a college. We can represent the relationship between SATs and GPAs graphically. Each student is represented by a dot on the graph below, using the person's SAT score to determine how far to the right to place the dot and using the person's GPA to determine how far up to place the dot. This graph is called a *scatterplot*. In this example, we see that the two quantities, SAT score and GPA, appear to be somewhat related to one another, in the sense that higher SATs tend to have higher GPAs, but there are many exceptions. We can draw a line on the graph that tracks the associated data as much as possible. Then we can give a quantitative measure of how well the data move together by measuring how close the data points in the scatterplot are to that straight line. In this way, we can give a quantitative value, called the *correlation*, to the extent to which a linear relationship exists between the SATs and GPAs.

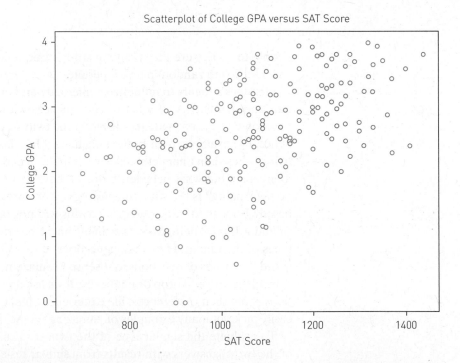

The correlation is 1 if the data lie *exactly* on a straight line that slopes up to the right and −1 if the data lie *exactly* on a straight line that slopes down to the right. If the data are tightly clustered around the line that best approximates the direction of the data, then the correlation gets increasingly close to 1 if it's sloping upward to the right, or close to −1 if the points are close to a line that's sloping downward to the right. A correlation of 0 means that there is no linear relationship between the attributes—the scatterplot may resemble a random cloud

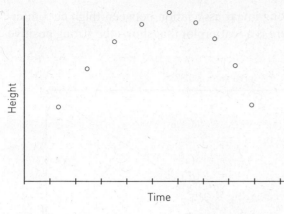

of dots, or the association between the attributes may exist but not be linear. For example, if we throw a ball in the air and measure time versus height, the two values have a clear association; however, the correlation is 0. Looking at a scatterplot before blindly computing the correlation is a good strategy for finding some possible nonlinear association even though the correlation is 0. As always, clear thinking and common sense are important parts of good use of statistics.

A statistical calculator or a spreadsheet program or a statistical package on a computer can calculate the actual value of the correlation for a given scatterplot. Below we offer some visual examples to give a sense of how the correlation measures the extent to which two quantities are moving together by being associated in a linear fashion.

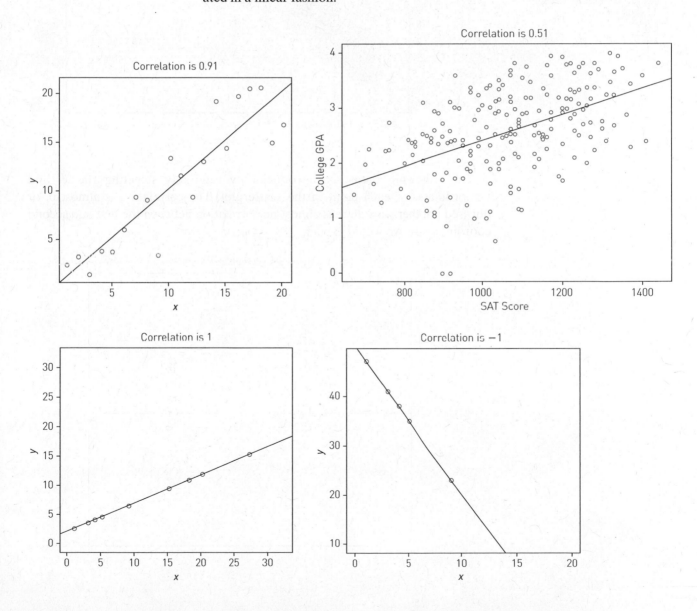

It turns out that there is a strong linear association between thigh circumference in men and their weight. Here is a scatterplot that shows the strong positive correlation of 0.87.

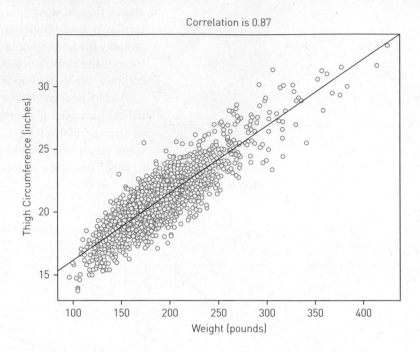

The following data were produced by randomly selecting the *x*- and *y*-coordinates of each point in the scatterplot. The correlation is almost 0, or phrased another way, there is almost no correlation between the first and second coordinates, as we would expect.

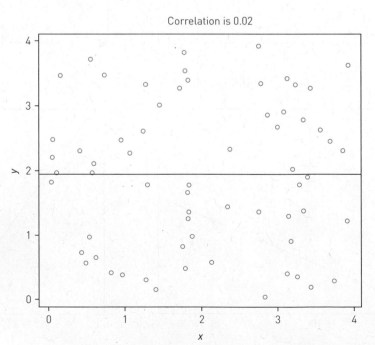

Scatterplots allow us to visualize relationships. In cases where two varying quantities, like SAT score and rank in high school class, are associated with a third value, the college GPA, we can use a 3D scatterplot. Here are three different views of the same 3D scatterplot that reveal relationships when you view them with your *Heart of Mathematics* 3D glasses.

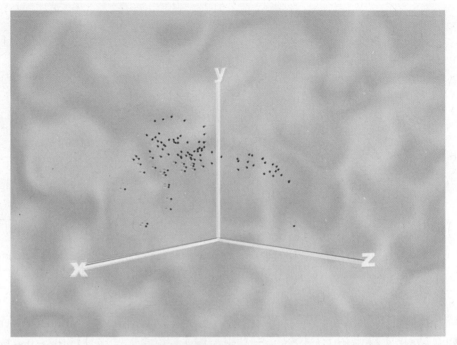

Looking at complicated data sets from different points of view sometimes leads to unexpected correlations—put on your 3D glasses and correlate away!

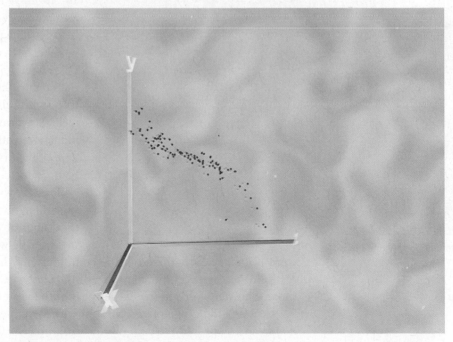

A different view of the data.

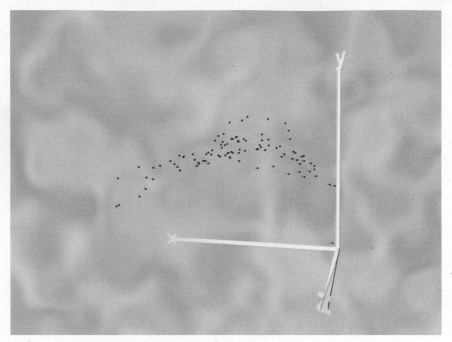

Another different view of the same data set. Connections and correlations become visible.

A lottery nobody wanted to win!

Vietnam War Draft

During the height of the Vietnam War a lottery was conducted to determine which 18- to 26-year-old men would be drafted into military service. A large glass container held 366 blue plastic balls representing every possible birth date. The drawing was held on December 1, 1969, and young men around the country huddled around television sets to watch their fate be decided by random luck. The drawing determined the order of induction, with men who were born on the date of the first ball selected, September 14, being inducted first, and so on.

This drawing literally determined who was put in danger of death through war for thousands of young men. It turned out that the drawing was not fair. The birthday balls were not sufficiently mixed in the container and the effect was that men whose birthdays occurred in the later part of the year were more apt to be drafted than those who were born in the earlier part of the year. Correlation proved it.

Although the dots appear randomly located, the best line fit shows a definite downward slope and the computed correlation is −0.23. Perhaps a more convincing method for seeing the trend is to make a box plot for each month where the

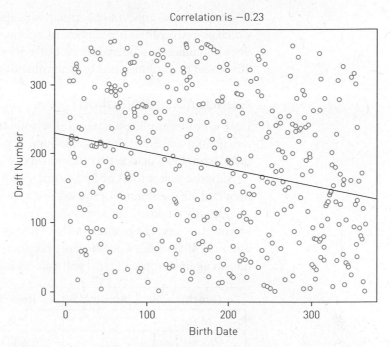

Correlation is −0.23

draft numbers associated with the days of that month are plotted. In that representation, we see more clearly that the later months have lottery numbers, shown by the boxes, that are lower than for the earlier months, meaning that the later months' draft numbers were lower than the numbers for the earlier months.

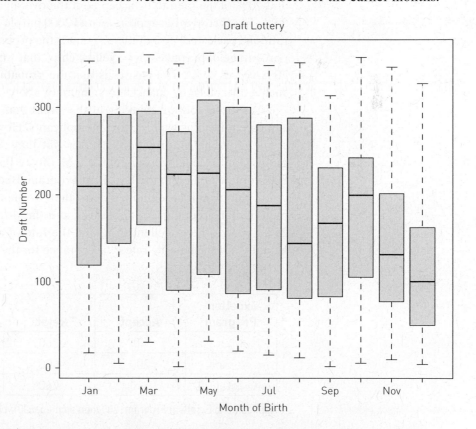

Draft Lottery

The statistical concept of correlation helps us identify and quantify the extent to which paired qualities of members of a population are varying together. Correlation does not indicate cause and effect, but it does indicate an association. Statistical evidence can be used in courts to determine discrimination, but it can be tricky.

Gender Discrimination?

Let's consider the legal issue of discrimination on the basis of gender in acceptances to a university. You be the jury. If you were presented with the following facts, would you decide that the university is practicing discrimination?

Suppose that a total of 2000 students apply to this particular university: 1000 men and 1000 women. For the sake of argument, assume that all of the applicants are equally qualified for admission. The prosecutor presents the following chart to you, the jury. He points out that 70% of the men who applied were accepted and only 40% of the women were accepted. Is this an example of blatant discrimination against women? How will you decide?

Overall	Accept	Reject	Total	Accept Rate
Men	700	300	1000	70%
Women	400	600	1000	40%
Total	1100	900	2000	55%

We might ask how probable it would be for only 40% of women to be accepted and 70% of the men to be accepted if the overall acceptance rate, as it is here, is 55%. In other words, suppose we had 2000 purple and yellow poker chips and we randomly selected 55% of them. What is the probability that the difference in the percentage of purple chips to yellow chips that we select would be as extreme as 70% versus 40%? The answer is that the probability is extremely small to have such a difference by chance alone, namely, about 0.0000000000000002.

As a juror, you are thinking at this point that the admissions committee has some genuine bias against women applicants. However, the defense now reveals some additional information. It turns out that the university had two subprograms—an Excellent Program and a Mediocre Program. Two hundred men and 800 women applied to the Excellent Program. Because it is a more selective program, the overall acceptance rate to the Excellent Program was only 24%. The men had an acceptance rate of 20%, and the women had an acceptance rate of 25%. So, for the Excellent Program, the rate of acceptance for the women was actually higher than the rate of acceptance for the men.

Excellent Program	Accept	Reject	Total	Accept Rate
Men	40	160	200	20%
Women	200	600	800	25%
Total	240	760	1000	24%

Applicants to Excellent Program: 200 men applied; 800 women applied

Now let's look at the Mediocre Program. In the Mediocre Program, 800 men applied and 200 women applied. From these groups, 82.5% of the men were accepted and a whopping 100% of the women who applied to the Mediocre Program were admitted.

Mediocre Program	Accept	Reject	Total	Accept Rate
Men	660	140	800	82.5%
Women	200	0	200	100%
Total	860	140	1000	86%

Applicants to Mediocre Program: 800 men applied; 200 women applied

So overall, across both subprograms, the acceptance rate of the men was 70%, and the acceptance rate of the women was 40%. But in each of the subprograms the acceptance rate for the women was *higher* than the acceptance rate for the men. This surprising conundrum arises from the fact that more women than men applied to the more selective program. The number of women applying to the more selective program is an example of a *lurking variable*—a feature of the situation that may be pertinent but is not originally displayed. Lurking variables can account for surprising data and can be difficult to ferret out. In this case, as a jury member, you can now be outraged that the poor male students are being discriminated against. Our first impression of the data appeared to indicate one conclusion, but a second look at the data appeared to imply something entirely different.

The phenomenon in which the overall distribution behaves one way whereas each of the subparts behaves another is called *Simpson's Paradox*.

Bart Simpson—a different kind of paradox.

Which Hospital?

Suppose we need to make a decision about which hospital we should enter for a serious operation. There are two hospitals and we have data about each. In Hospital Heaven, of the last 1000 patients who entered the hospital with this medical issue, 71% survived, and at Hospital Hope, only 54% survived. We might say it certainly seems clear, from these data, that Hospital Heaven is the preferred hospital—let's check in there.

	Hospital Heaven	Hospital Hope	Survivors from Hospital Heaven	Survivors from Hospital Hope
Total Patients	1000	1000	710 (71%)	540 (54%)

Heaven seems better than Hope.

But, before putting on one of those revealing backless hospital gowns, let's look at the data broken down into subgroups. There are three different types of patients (in terms of their condition before they entered the hospital): patients may enter in fair condition, serious condition, or critical condition.

Suppose that in Hospital Heaven, most of the patients are in fair condition, which is the best of the three conditions. More precisely, 700 enter in fair condition; 200 are in serious condition; only 100 enter who arrive in critical condition. Hospital Hope has a different profile of patients upon entry; namely, there are only 100 who enter in the rather benign fair condition; 200 are in serious condition; and 700 enter in critical condition.

Next we acquire data on all three of these subpopulations. We discover that for those who enter in fair condition, in Hospital Heaven, 600—that is, 86%—survive; whereas, in Hospital Hope, of those among the 100 who enter in fair condition, 90—or 90%—survive. Of those in serious condition—200 in each hospital—only 100 from Hospital Heaven survive, for a 50% survival rate, and 150 from Hospital Hope survive, for a 75% recovery rate. Then, for those who enter in critical condition, of the 100 who enter Hospital Heaven, only 10 survive. That is, only 10% of the critical-condition patients who enter Hospital Heaven survive, whereas an impressive 43% from Hospital Hope survive.

The overall percentages agree with what we first were told: Seventy-one percent survive from Hospital Heaven and 54% survive at Hospital Hope. The absolute numbers were not changed, but by studying the subcategories, we discover that, in fact, Hospital Hope actually has a higher success rate in each of the subpopulations. This reality suggests that Hospital Hope is actually the hospital that we should consider for facing this serious condition—anyway, that's the hope.

	Hospital Heaven	Hospital Hope	Survivors from Hospital Heaven	Survivors from Hospital Hope
Fair condition	700	100	600 (86%)	90 (90%)
Serious condition	200	200	100 (50%)	150 (75%)
Critical condition	100	700	10 (10%)	300 (43%)
Total	1000	1000	710 (71%)	540 (54%)

Hope seems better than Heaven.

Hope versus Heaven presented us with another example of Simpson's Paradox. In this case, the number of critically ill patients was a lurking variable.

Counting Tigers

Statistics often helps us count in our world, which is extremely useful since counting the number of criminals, for example, can be a pretty frightening task. We have encountered some pretty scary counting, but perhaps none as dangerous as counting tigers in the wild. Suppose someone forced us to go to the wilderness and count the tigers. How would we begin? We might ask the tigers to line up

neatly for us to count, but a better method would be to think about probability and statistics.

Before facing the tigers, let's think about a related, less threatening counting conundrum. Suppose we have a large jar of jawbreakers and we want to know approximately how many there are. We do not have time to count them all, so how can we estimate the number? Even this question seems too hard to chew on and, to some, perhaps too threatening, so let's make it easier still. Suppose we are told that in the jar of jawbreakers, exactly 50 of them are green and the green ones are thoroughly mixed in with the others. Can we use that information to help us count the total number?

Let's reach in and start taking the jawbreakers out of the jar. Suppose we take out 40 jawbreakers, look at them, and see that 20 are green. What can we now conclude about the total number of jawbreakers in the jar? There are 50 green jawbreakers in the jar. When we randomly selected 40, half were green. If that proportion holds up, then roughly half of the jawbreakers in the entire jar will be green. Since there are 50 green jawbreakers, there must be a total of roughly 100 jawbreakers in the jar.

Now back to the jungle. Remember that we must count the tigers. Unfortunately, we do not know that 50 are green. In fact, none are green—they all are striped. So what will we do? We'll create a situation we can understand. We now understand how to estimate an entire population if we know how many are green. So let's see whether we can make some number of tigers green. They don't start green, so we'll have to capture them first. Suppose we capture some number of tigers, say 50, and mark each one by putting a tag on its ear (probably green) indicating that it is one of the captured tigers. Next we let them all go, wait a few months, then recapture some number of tigers, and see what proportion of them have tags. Suppose we capture 30 tigers and 10 have tags. Assuming that we have picked a random collection of tigers, we can guess that the tagged tigers represent about a third of the total tiger population. Since there are 50 tagged tigers, we deduce that there must be approximately 150 total tigers in the population.

Of course, there are several assumptions in this process that need to be considered. For example, are we really capturing a random set of the tigers? Perhaps we are capturing only the dumb ones. In which case, all we can conclude is that the population of tigers dumb enough to be captured is about 150—an important piece of information nonetheless.

A variation of the capture-recapture method is used to evaluate the quality of census data. Random blocks are selected and census workers attempt to get complete data on who lives in that block. These counts are compared to the people whom the census counted in those blocks to estimate what proportion of people in different regions the census missed.

Counting Tanks

An important counting conundrum arose in World War II that may be even more frightening than counting tigers. The Allied forces (the Good Guys) were combating the powerful Nazi army. Among the weapons in the Nazi army were Mark V tanks. It was crucial to estimate the number of these Mark V tanks, and in helping to do so statistical methods became true war heroes. Here's the story.

1943: The Nazi army enters Tunisia. They were retreating, pursued by the victorious 8th Army

During World War II, when a Nazi tank was captured, analysts noticed that the tanks had serial numbers and it appeared that the serial numbers were consecutive, starting with 1 and increasing as each new tank was built. So the challenge of estimating the number of tanks in the whole army could be viewed as a statistics question. We know some information about a sample of the population, namely the serial numbers of the captured tanks, and we wish to infer information about the whole population, namely how many tanks there are altogether.

So let's assume that there are an unknown number N of Mark V tanks. Let's also assume that numbers of the captured tanks are a random selection from the numbers 1 to N. We want to decide what would be our best guess of the total number of tanks.

Let's consider several possible *estimators*, that is, methods or strategies for calculating an estimate, and then decide on one. Here's a specific example. Suppose we've captured tanks whose numbers are {68, 35, 38, 107, 52}. What estimate for the number of total tanks would we make? The first two methods are based on the idea that the center of the sample should be close to the center of the entire population. The center (mean or median) of all the tank numbers {1, 2, . . . , N} is $(N + 1)/2$, so doubling our estimate of the center and subtracting 1 should give us an estimate of N, because $2 \times (N + 1)/2 - 1 = N$.

Method 1: Take the mean of the sample of five numbers, getting 60. Double this result and subtract 1—giving an estimate of 119.

Method 2: Take the median of the sample, getting 52. Double this and subtract 1—giving an estimate of 103.

The problem with these methods is that either one might produce an estimate that is less than the number on one of the tanks we actually captured, which would be definitely too small a guess for the total number of tanks N. So these intuitively appealing methods have a drawback.

We could avoid this drawback by just taking the average of the largest and smallest number among those we captured and estimating double that average.

This method is certain to at least give us a value larger than the biggest number that we actually captured:

Method 3: Add the largest and smallest numbers that we've captured (107 + 35), take their average (71), double it, and subtract 1—giving an estimate of 141.

Another method for avoiding the drawback of guessing a number that is less than a number actually captured is to simply guess that we have captured the last tank produced:

Method 4: Let the estimate be the value of the largest serial number captured—giving an estimate of 107.

Of course, this method is silly, because we would not expect to capture that particular tank, the highest-numbered tank. We want our estimate to be somewhat bigger than the biggest serial number we have captured.

We can think of a more refined method that is superior to all of the preceding:

Method 5: Suppose we captured k tanks. We take the maximum serial number we captured, multiply it by $(k + 1)/k$ and subtract 1. Notice that we have to subtract 1 to make the method work in the case where we have captured all N tanks. In that case, we would have captured N tanks and the largest numbered tank we captured would be N, so our estimate would be $N \times ((N + 1)/N) - 1 = N$. In our example of capturing 5 tanks, we would estimate the total number of tanks to be: $(5 + 1)/5 \times$ maximum of $\{68, 35, 38, 107, 52\} - 1 = 6/5 \times 107 - 1 = 127.4$.

How can we decide which estimator to use? One property that we might want a method to have is that if we perform the method many times and take the mean of the estimates produced, then the mean of the estimates would equal the true number of tanks. Such an estimator is called an *unbiased estimator*.

Methods 1, 2, and 3 are unbiased estimators, but we saw that Methods 1 and 2 had other drawbacks, namely, sometimes producing estimates that are definitely too low. Method 5 is also an unbiased estimator. In addition, we want an estimator that produces estimates that tend not to vary too much. Method 5 has this virtue as well. We have done some simulations to demonstrate the use of Method 5 where we suppose the actual number of tanks in the German army is 3000 and we have randomly captured 100 of them. We performed the simulation of selecting 100 numbers from 1 to 3000 randomly, then we took the maximum number selected, multiplied it by 101/100, and subtracted 1. We did that simulation 1000 times, as though we were fighting World War II 1000 times. Here is a histogram of the results of those simulations of our estimates.

The minimum estimate this method produced was 2844; the maximum estimate was 3030; and the median of the estimates was 3011.

1000 Estimates: Sample Size 100, Number of Tanks 3000

The mean of the estimates, as we expected from the fact that this method is an unbiased estimator, was extremely close to the actual value of 3000, namely, the mean was 3002.

Another desirable quality we want in an estimator is to have as small a variation as possible because that would mean that not only are the estimates close to the correct answer, but they don't vary too much from each other. So, we actually are looking for an estimator that is not only an unbiased estimator—meaning that on average, it gives us the right answer—but also has small variation from the mean. It turns out that our estimator of taking $(k + 1)$ divided by k, times the maximum, minus 1, is an unbiased estimator that has the *smallest* variance of estimators possible.

Here is what actually happened, according to some sources. The Allies feared that the Nazis were producing up to 1400 tanks per month during the critical period 1940 to 1942. By using the methods above, statisticians estimated the rate of production at 246 tanks per month. After the war, it was ascertained that the actual German tank production during that period averaged 245 per month. This real example shows that statistical reasoning can have profound applications in the world.

A Look BACK

STATISTICAL INSIGHTS CAN BE inferred from either experiments or observational studies. In an experiment, we deliberately select the values of some features of the situation that we believe cause a specific result, while in observational studies we draw conclusions from the data we observe. A good experiment assigns treatments at random in order to balance any potential lurking variables over the various treatments. A good observational study attempts to record and adjust for any potential lurking variables, but generally, results from observational studies do not make as strong a case for causation as do results from a carefully conducted experiment.

We have just scratched the surface of the role of statistical reasoning in the real world. Statistics can be applied to business and economic realms as well as determining which foods or medicines are healthful. Every business or school or household has a myriad of occasions in which statistics plays an important role. Pick up a random issue of *The New York Times* and look for the many places in which statistics figures prominently in an article. When done correctly, statistics can really be "All the news that's fit to print!"

In making personal and business decisions and in forming our opinions about political or social issues, clear statistical reasoning can play a prominent role.

MINDSCAPES   Invitations to Further Thought

In this section, Mindscapes marked **(H)** *have hints for solutions at the back of the book. Mindscapes marked* **(ExH)** *have expanded hints at the back of the book. Mindscapes marked* **(S)** *have solutions.*

Developing Ideas

1. **Mendel's snapdragons.** Another of Mendel's experiments involved red-flowered and white-flowered snapdragons. Through crossbreeding experiments he concluded that each flower had two genes. The red flowers had two red genes while the white had two white genes. We will call those parent plants the first generation. When two flowers bred, the offspring received one gene from each parent. If the offspring inherited one red and one white gene, unlike the case of the peapod color, in the flower the genes combined to produce a pink flower. Suppose a red-flowered snapdragon breeds with a white-flowered snapdragon. What are the possible gene outcomes for the second generation of this crossbreeding?

2. **Telephone/soda twins.** Suppose your school has 1000 students. Why can you be pretty sure that some pair of them will have telephone numbers that end in the same two digits and that they will like the same brand of soda?

3. **Correlation comparison.** Below are three scatterplots. Which set(s) of data show a positive correlation? A negative correlation? No correlation?

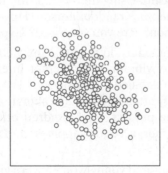 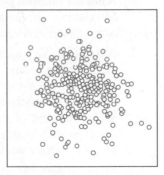 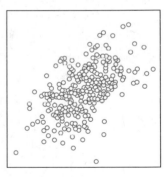

a.   b.   c.

4. **Percent practice.** Convert the fractions below to percentages.

12/400, 16/100, 2/100, 28/400, 1/5.

5. **Family dinner.** Recent studies have reported on the value of families having dinner together. Children in families who regularly have dinner together are less likely to drink or do drugs, among other things. Do you think this is evidence of a causal relationship between family dinners and child behavior? What other factors might be influencing this association?

Solidifying Ideas

6. **More Mendel (S).** Given Mendel's snapdragon experiment from Mindscape 1, suppose that two second-generation snapdragons breed. Construct a two-by-two table showing all the possible gene pairs for the third generation of snapdragons. If it is equally likely to inherit a red or white gene from the second-generation parent, what is the probability that a third-generation plant has a white flower? A red flower? A pink flower? What proportions of the colors would you expect to see in a bed of such flowers?

7. **Oedipus red.** Given Mendel's work described in Mindscape 1, suppose that a second-generation plant breeds with a red-parent plant. What is the probability of the offspring having a white flower? A red flower? A pink flower?

8. **Oedipus white.** Given Mendel's work described in Mindscape 1, suppose that a second-generation plant breeds with a white-parent plant. What is the probability of the offspring having a white flower? A red flower? A pink flower? What proportions of the various colors of flowers would you expect to see in a bed of such flowers?

9. **Pure white (ExH).** Given Mendel's work described in Mindscape 1, suppose that a second-generation plant breeds with a third-generation plant. Is it possible for these hybrids to produce an offspring with a pure white flower? If so, explain how. If not, explain why not.

10. **Random person.** Pick a person at random (in your dorm, your class, or a dining hall) whom you do not know well. Find seven unusual common features you share with that person (your mothers were both born in San Francisco, for example). Devise a test of 10 questions that would make these observations seem remarkable.

11. **Another random person.** Using the 10-question test created in the previous Mindscape, survey another random person. How many answers do the two of you have in common? Are you surprised? Explain why or why not.

12. **Astrology.** An experiment is done to test the value of astrology for patients at a nursing home. A vibrant experimenter interviews the patients extensively and then casts detailed horoscopes of 100 elderly patients. They are observed immediately afterward and an estimate is made of how happy they are at the end of the process. One hundred other patients who were not interviewed are also measured for happiness and are found to be less happy. Does astrology work?

13. **Stressful diet.** One hundred young women were given a diet survey and were classified as having either a healthy diet or an unhealthy diet. Each was then given a questionnaire designed to assess her degree of stress. It was found that those in the healthy diet group tended to have significantly lower levels of stress than those in the unhealthy diet group. Is this good evidence that a healthy diet causes lower stress levels?

14. **Abstinence evidence?** President George W. Bush emphasized "abstinence-only" sex education during his two terms beginning in 2000 and 2004. Teen pregnancy rates rose by 3% between 2005 and 2006, the first annual increase since 1991. Is this evidence for a causal relationship between the Bush policies and teen pregnancy rates?

15. **Matching correlations.** Below are some scatterplots showing the "best-fit" line in each case. Match each graph to its correlation value from the following set of values: 0.5, −1, 0.9, −0.7, 0, 0.2.

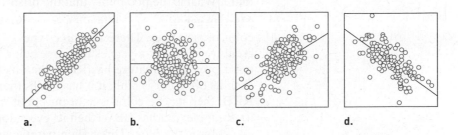

a.   b.   c.   d.

16. **Fast data (H).** If you plot the winning Olympic men's 1500-meter time against the year the Olympics was held, the correlation since WWII is about −0.7. What does this mean?

17. **College town.** You go to school in a college town. You know that there are 2000 students enrolled in the school, but you don't know the population of the town (without students). You walk up and down the main streets of the town, stop people, and ask them if they are students or not. You ask 100 people, and 60 of them say they are students. Estimate the nonstudent population of the town.

18. **Car count (S).** You wonder how many cars there are in your area. You call up the Honda dealership and ask how many Hondas they've sold. They report that they've sold a total of 10,000 cars. You then take a lawn chair and camp out on the most traveled roadway in town and count cars. You count 800 cars and note that 250 of them are Hondas. Estimate the number of cars in your area. Do you believe you would have gotten more, less, or the same accuracy if you had called the Lexus dealership instead of the Honda? Explain.

For Mindscapes 19–22, say whether or not each is an example of *post hoc, ergo propter hoc.* **Explain your reasoning.**

19. **Lucky charms.** *I was doing lousy on the soccer field. Then my girlfriend gave me a necklace to wear. Ever since then, I've blocked every shot to our goal. That necklace brought me good luck—if I keep wearing it, we'll keep winning.*

20. **Read my lips.** *Shortly after a Republican tax bill bringing tax breaks to the wealthy becomes law, the economy moves into a recession. The Democrats campaign that this new law was the cause of the country's economic crisis and thus want to repeal it.*

21. **Penny luck.** *Right before going in to buy my lottery ticket, I saw a penny on the ground. Usually I don't bother to pick up pennies, but this time I did. That night I won $5000! I'm always going to pick up any pennies I see on the ground from now on.*

22. **The lick.** *I was visiting some friends and their dog, Golden, licked my hand. Two days later I felt achy and had a mild fever. I must have caught some bug from that darn dog lick.*

Creating New Ideas

23. **Mendel genealogy (H).** We return once again to Mendel's snapdragon experiment from Mindscape 1. Suppose two second-generation crossbred plants mate. What is the possibility that the offspring will be white-flowered?

24. **Martian genetics.** A certain alien species has three genes that determine the color of their eyes. There are three types of genes: red, white, and blue. If a creature has all three genes the same color, then the creature will have that color eyes. If a creature has RRW, then its eye color is red; if it has RWW, then its eye color is pink; if it has RWB, then its eye color is purple; if it has RRB, then its eye color is light purple; if it has RBB, then its eye color is dark purple; if it has BBW, then its eye color is blue; if it has BWW, then its eye color is sky blue. Three alien parents are required to produce one offspring. Each parent contributes one gene to the new alien, and this gene is randomly selected. Suppose an RRR and a WWW and a BBB all mate and produce an offspring. What are the possible eye colors of this second-generation offspring, and for each of those colors, what is the probability that the offspring will have that eye color?

25. **More Martians.** Given the scenario from the previous Mindscape, suppose that a second-generation alien with genes RWB mates with a WWW and another WWW. What are the possible eye colors of their offspring, and for each of those colors, what is the probability that the offspring will have that eye color?

26. **Politics as usual.** During the presidential campaign of 1968, it was reported that Richard Nixon made the following statement about his opponent, Hubert Humphrey: "Hubert Humphrey defends the policies under which we have seen crime rising 10 times as fast as the population. If you want your president to continue a do-nothing policy toward crime, vote for Humphrey. Hubert Humphrey sat on his hands and watched the United States become a nation where 50% of the American women are frightened to walk the streets at night." Is this a logical reasoning fallacy? If so, what kind of fallacy is it, and what is the fallacy?

27. **Going postal.** Your company will sign a contract with one of two delivery services. Below are data showing each service's on-time record during a trial period.

Service	Type of Delivery	Number of Deliveries	Number of Late Packages
RedEx	Regular	400	12
	Overnight	100	16
United Package Service	Regular	100	2
	Overnight	400	28

a. For each service, compute the percentage of late deliveries overall.

b. Compare on-time percentages for the two services in each of the two types of service.

c. Which service should your company choose? (Might there be mitigating factors?)

28. **Fast paradox (ExH).** The correlation between year and winning Olympic 1500-meter time since WWII is about −0.7 for men and −0.2 for women. But if you combine men's and women's times, the overall correlation between year and winning time is +0.1. The men's times go back to 1948, but the women's 1500-meter race only goes back to 1972. How does this fact help to explain the apparent paradox?

29. **Up and down.** In the past year, the grade point average of women at a college went up. The grade point average of men at the college also went up. The grade point average of all students at the college went down. Is this possible? Explain.

Further Challenges

30. **Modified Mendel.** We again return to Mendel's snapdragon experiment from Mindscape 1. Suppose that in the crossbreeding of second-generation pink flowers, the probability that a pink snapdragon passes on the white gene is 1/3 and the probability it passes on the red gene is 2/3. What would be the possible outcomes for the third-generation offspring if two second-generation flowers bred? What would be the probabilities of the various colors? What color would you expect to see?

31. **Going extinct.** A paleontologist has gathered data on fossils for a certain species of brachiopods. This species is now extinct and the scientist wants to estimate when this extinction occurred. A total of nine fossils have been discovered and dated as shown in the table below. (These data are estimated from the work of Y.G. Jin et al, "Patterns of Marine Mass Extinction," *Science*, 289 [2000]: 432–436.)

Fossil	Date of Fossil (Years Ago in Millions)	Time After Earliest Fossil Find in Millions (Date −253.7)
1	253.7	0
2	254.69	0.99
3	256	2.3
4	256.15	2.45
5	256.4	2.7
6	256.72	3.02
7	256.75	3.05
8	256.82	3.12
9	256.88	3.18

Using each of the five methods described in the text to count tanks, compute five estimates of the date around which this species went extinct. (Use the data in the far right column of the table, then convert your answers to a date millions of years ago.) Which estimate do you think is the best? Why?

In Your Own Words

32. Personal perspectives. Write a short essay describing the most interesting or surprising discovery you made in exploring the material in this section. If any material seemed puzzling or even unbelievable, address that as well. Explain why you chose the topics you did. Finally, comment on the aesthetics of the mathematics and ideas in this section.

33. With a group of folks. In a small group, discuss and actively work through the reasoning involved in estimating the number of tanks from the captured tanks' serial numbers. After your discussion, write a brief narrative describing the methods in your own words.

34. Creative writing. Write an imaginative story (it can be humorous, dramatic, whatever you like) that involves or evokes the ideas of this section.

35. Power beyond the mathematics. Provide several real-life issues—ideally, from your own experience—that some of the strategies of thought presented in this section would effectively approach and resolve.

For the Algebra Lover

Here we celebrate the power of algebra as a powerful way of finding unknown quantities by naming them, of expressing infinitely many relationships and connections clearly and succinctly, and of uncovering pattern and structure.

36. Diners delight. There are five dining halls on your campus. One night, 10% of the students eat at the first dining hall, 20% eat at the second, 40% eat at the third, 15% eat at the fourth, and the remaining students eat at the fifth dining hall. If there are 2000 students at your school, how many students ate at the fifth dining hall?

37. The candy man. Your math instructor brings a large, opaque jar to class one day, claiming it is filled with red, blue, and green candies. He says that there are 200 candies in the jar, all mixed up. Each of the twenty students in the class takes one piece of candy from the jar at random. Let x equal the number of selected red candies, let y equal the number of selected blue candies, and z equal the number of selected green candies. Give an expression in terms of x that gives your best estimate of the number of red candies in the jar. Do the same for blue candies (in terms of y) and green candies (in terms of z).

38. The candy jars. Your math instructor brings two opaque jars of candies to class one day. All the candies are red mints or green mints. The first jar contains 15 candies; the second jar contains 40 candies. Suppose the second jar contains twice as many red mints as the first jar, and three times as many green mints as the first jar. Determine how many mints of each color are in the two jars.

39. Correlated chorus (H). The chorus has data on the attendance at their concerts over the past ten years. In 2001, attendance was 200. In 2002, attendance was 220. In 2003, attendance was 240. If the plot of all ten data points shows a correlation of +1, what was the attendance at the chorus concert in 2010?

40. Correlated Britney. The Britney S. Fan Club has data on attendance at their weekly meetings over the last semester. In Week 1, 40 students attended. In Week 2, 37 students attended. In Week 3, 34 students attended. If the plot of all the data points shows a correlation of -1, what was the attendance in Week 12? Write an equation that gives the number of students attending, y, as a function of the number of the week of the semester, x. (So x will take on the values 1, 2, 3, …, up to 12.)

CHAPTER TEN

Deciding Wisely

Applications of Rigorous Thinking

10.1 Great Expectations

10.2 Risk

10.3 Money Matters

> *Few people think more than two or three times a year. I have made an international reputation for myself by thinking once or twice a week.*
>
> GEORGE BERNARD SHAW

Life is one decision after another. We make decisions every waking moment, from huge to tiny: whom to date, whom to marry, whether to eat dessert tonight, what to think about, what socks to wear, whom to vote for, whether to use correct grammar, what college to attend, what investments to make, what car to buy, what medicines to take, how much insurance to buy, whether to hold 'em or fold 'em, whether to buy a lottery ticket or study. These decisions can alter a moment or an entire life.

Each realm of decision making contains its own surprises, and sometimes the surprises are discouraging. For example, consider the seemingly simple task of determining the candidate most preferred by voters. Sounds simple enough—just count the ballots. Surprise: We'll see that the whole notion of "most preferred candidate" is essentially meaningless. On the encouraging side, we'll see that we can allocate scarce resources in ways that leave everyone satisfied. In fact, we will show that a highly desirable cake can be cut into three pieces so that three greedy and hungry claimants definitely all feel that they have the best piece. We can productively view insurance decisions as games of chance not much different from roulette. And money matters present issues that range from chaotic to inevitable. When we throw in the risks of life and death involved in issues of medicine and health, the world of decisions takes on an immediacy and importance that encourages us to give math a chance.

With all these decisions to be made—public and private, large and small—improving our chances of making good decisions is crucial. The strategies of insight that have worked over and over in mathematics are equally potent when applied to making decisions. These strategies include understanding simple cases deeply, isolating essential elements from a complex situation, and describing a mathematical model that captures salient features of the decision situation. We now embark on an excursion into the world of thoughtful decision making.

10.4 Peril at the Polls

10.5 Cutting Cake for Greedy People

GREAT EXPECTATIONS

Deciding How to Weigh the Unknown Future

The Card-Players by Pieter Bruegel the younger (Flemish, ca. 1564–1638).

> *Chance favors only the prepared mind.*
>
> **LOUIS PASTEUR**

Often in life we must make decisions whose wisdom or lack of wisdom becomes clear only in the future—buying stock versus investing in bonds, buying term life insurance versus investing in mutual funds, buying this book versus investing the money in lottery tickets. Perhaps after a year has passed, we can measure how sound our decisions were, but at the time we make these decisions, how can we assess their wisdom? Looking into the future, we must determine, as best we can, the value of the various possible actions we may elect to take, even with the understanding that, in the end, they are all gambles.

We have a vague idea of what we seek: a measure of the value of future actions. Our strategy for developing a concept of future value is to begin with concrete examples. By focusing on situations we know, we can extract essential features and then apply them to settings that are not so clear.

Roulette for Scarlett

The best illustrations of the value of future events emerge in the smoky confines of casinos. In gambling, we can analyze the consequences of our decisions in dollars and cents. We might visualize a scene from a James Bond film: handsome tuxedo-clad men and beautiful, elegantly dressed women in Monte Carlo or Saint Moritz. But almost all will leave the casino a little less well-dressed than when they arrived, or at least lighter in the pocketbook. Let's head to an American-style roulette table and see how we fare.

The roulette wheel has 38 slots, which are numbered 1–36 plus 0 and 00, in random order. Eighteen of the numbers 1 to 36 are red, and 18 are black; the 0 and 00 are green. On a play of roulette, the roulette wheel spins gracefully around and the ball spins swiftly in the reverse direction. As the ball slows down, it meets the wheel in an abrupt collision. After several moments of bouncing randomly about, the

ball finally settles into one of the 38 slots. As players, we bet on where the ball will come to rest. Many bets are allowed, including betting on one particular number or block of numbers, betting that the number will be even or odd, or betting that the number will be red or black. If we bet on red, and a red number comes up, we double our money. Now, doubling our money sounds promising, so let's learn vicariously how wise this investment truly is.

Gambling is addictive for some people, and one striking, though apocryphal, example of compulsive gambling will illustrate the theme of assessing future values of our actions. Ms. Scarlett was attracted to both roulette and red (not Rhett). One day she arrived at the casino, sat down at the roulette table, bought 3800 one-dollar chips, and proceeded to bet $1 on red for each of the next 3800 spins of the wheel. As luck would have it, every number on the roulette wheel came up exactly 100 times. How did Ms. Scarlett do? (Remember, the response "Frankly, my dear, I don't give a damn" has already been used.)

Since 18 numbers are red, during these spins red numbers came up 1800 times, black numbers came up 1800 times, and the green numbers (0 and 00) came up 200 times. On each of the 1800 occasions that red came up, she won a dollar and retrieved her bet. On the other 2000 spins she lost her dollar bet. Thus she ended up with $3600. Poor Ms. Scarlett lost $200, and while she may still have the shirt on her back, she will probably leave clad only in that shirt and her green curtains. On the plus side, she may have learned a valuable lesson about the future value of her bets. She realized that over the course of her 3800 bets, she lost a total of $200—and that money is gone with the wind. So her average loss was $\frac{200}{3800} = \$0.052 \ldots$ per bet.

Average Value

This average value per bet is a measure of the expected value of each bet on red, on average. We arrived at this value by following the repeated exploits of our compulsive, and now red-faced, Ms. Scarlett. However, we can understand this value a bit better by looking at it from a different point of view. Each time Ms. Scarlett plunked down her dollar, there was some chance she would win and some chance she would lose. Specifically, the probability of getting a red number is 18/38, whereas the probability of getting something else is 20/38. So, another way of computing the expected value of her red bet is to note that 18/38 of the time she will win a dollar, whereas the other 20/38 of the time she will lose a dollar. So her expected value is

$$\left(\$1 \times \frac{18}{38}\right) + \left(-\$1 \times \frac{20}{38}\right) = -\$\frac{2}{38},$$

the same value we found before.

The negative sign indicates that by betting on red, we are losing on average (we're in the red). Thus, we see that our expected value of betting a dollar on red is $-\$\frac{2}{38}$. So, if we played and played this game again and again, overall we would see an average net loss of 5.26¢ per game. Should we play? No way: We lose more than a nickel each time, on average. However, there is a positive side to this negative outcome.

Expected Value

The important positive side to our story is that it led us to an idea of how to determine the value of playing a game of chance. The *expected value* is the average net gain or loss that we would expect per game if we played the game many times. Consequently, the expected value is the sum of the products of the value of each possible outcome multiplied by the probability of that outcome. In other words, to find the expected value of playing a game, we list all the possible outcomes of the game and the positive or negative value of each possible outcome. We then compute the probability of each outcome and multiply the value of each outcome by its probability and add up all these products. This sum is the value, on average, we can expect to get from playing the game if we play over and over again. If the expected value is positive, then that means, on average, we would profit; if the expected value is negative, then, on average, we would experience a loss.

Computing Expected Value.

To compute the expected value, we multiply the value of each outcome with its probability of occurring and then add up all those products.

Not Everything Is Fair Game

To further illustrate the meaning and usefulness of expected value, let's consider a different game. This game costs a dollar to play. After paying the dollar, we roll a die. If we roll a 1, then we get our dollar back (the game is a draw). If we roll a 2 or a 3, then we win and are given $3. If we roll a 4, a 5, or a 6, then we lose. What is the expected value of this game? Let's try to compute it by carefully using the definition of expected value as a sum of products of values and probabilities.

The probability of rolling a 1 is 1/6, and the value of this event is $0, since we neither made nor lost money (we paid a dollar to play and got it back). The probability of rolling either a 2 or a 3 is 2/6, which equals 1/3, and, for this event, the value is $2 ($3 winning − $1 initial bet). The probability of rolling a 4, a 5, or a 6 is

3/6, which is 1/2, and the value of any of these outcomes is $-\$1$. So, the expected value of this game for us is

$$(\$1.00 - \$1.00) \times \frac{1}{6} + (\$3.00 - \$1.00) \times \frac{1}{3} + (\$0.00 - \$1.00) \times \frac{1}{2}$$

$$= (\$0.00) \times \frac{1}{6} + (\$2.00) \times \frac{1}{3} + (-\$1.00) \times \frac{1}{2}$$

$$= \frac{\$2}{3} - \frac{\$1}{2} = \frac{\$1}{6} = \$0.1666\ldots$$

Therefore, the expected value is 16.7¢, which means that, if we play this game over and over, we would expect to gain an average profit of about 17¢ per game. If we were inclined to gamble, this game is one we might consider playing.

A game is called a *fair game* if the expected value equals zero. That is, if the game is played again and again, we would expect all players to break even. Suppose, in the spirit of sportsmanship, we wish to make the previous die-rolling game fair. Given the rules and payoffs, how much should we pay to play the game for the game to be fair? Should we pay more or less for the game to make it fair? Answer these questions before reading on.

To make the game fair, we have to pay an amount to play that would make the expected value equal to 0. Let's call the price we have to pay to play P. Then we can write out the expected value, just as before, but now replacing the $1.00 with P dollars to obtain:

$$(\$1.00 - \$P) \times \frac{1}{6} + (\$3.00 - \$P) \times \frac{1}{3} + (\$0.00 - \$P) \times \frac{1}{2}$$

$$= \frac{\$1}{6} - \frac{\$P}{6} + \$1 - \frac{\$P}{3} - \frac{\$P}{2} = \frac{\$7}{6} - \$P.$$

So, the expected value if you pay P dollars is $\$7/6 - \$P$. We can check this answer by substituting \$1.00 for P and see if we get the answer we found before. Verify this check for yourself. Now to make the game fair, we must have the expected value equal to zero. So, we set the expected value equal to zero and solve for the price P:

$$\$\frac{7}{6} - \$P = 0,$$

so that means

$$\$P = \$\frac{7}{6} = \$1.1666\ldots$$

Therefore, we should be charged around \$1.17 to have the game be a fair game. It does make sense that we should have to pay an additional amount equal to our previous expected winnings to have the game be fair.

Life Insurance

Games of chance include more realms than just gambling. Insurance policies (life, home, auto, Chihuahua) are powerful but hidden gaming prospects whose expected values are used to determine the price of a policy. Let's examine term life insurance from the point of view of the insurance company. The insurance company keeps elaborate mortality tables that might state that roughly 1% of 60-year-old men will die within the year. The insurance company may then choose to sell a one-year term life policy to a 60-year-old man for $1000 that would pay $50,000 if the man dies. Since on average 99% of the policyholders will live the whole year and only 1% of the policyholders will die and cash in on their investment, the expected value of this policy to the insurance company is:

$$(\$1000 \times 0.99) + [(\$1000 - \$50,000) \times 0.01] = \$500.$$

That is, the insurance company would expect to make $500 profit on average for such a policy. Of course, this computation is overly simplistic, since we have not computed the expenses incurred by the insurance company. Nevertheless, this example does illustrate how computing the expected value is a central consideration for insurance companies in weighing premiums, coverage, and risk.

Should You Buy Lottery Tickets or This Book?

We are now about to analyze a treacherous issue that few authors wish to even discuss: the expected value of buying and reading their books. Specifically, we are going to attempt to answer the following question. Suppose this book costs $130. Is it better to use the $130 to buy this book and read it or to spend the $130 on lottery tickets in the hopes of hitting the big one?

We have already seen in Chapter 8 that the probability of winning the lottery is $1/15,890,700 = 0.00000006292 \ldots$ (pretty slim chance). Let's assume that we play only when the payoff is $6,000,000, and let's assume we are the only ticket purchaser, so we don't worry about sharing the bounty if we hit it big. The cost of each ticket is $1, so the expected value of purchasing a lottery ticket is

$$\left[(\$6,000,000 - \$1) \times \frac{1}{15,890,700}\right] + \left[(\$0 - \$1) \times \frac{15,890,699}{15,890,700}\right] = -\$0.62 \ldots.$$

Therefore, the expected value of buying 130 lottery tickets is roughly $130 \times (-\$0.62) = -\$80.60$, so we would expect the value to be $-\$80.60$. Not a great payoff—you lose more than $80. Now consider this book.

As this book is only in its fourth edition, we must make estimates as to the effect of the book on readers' lives. It is difficult to estimate the personal satisfaction of discovering some of the most beautiful and deep ideas of humankind. Thus, we will limit ourselves to cold, hard cash. We believe that the ideas in this book can empower readers to successfully understand and creatively resolve problems, cut through complicated issues, discover the heart of the matter, ask the right questions, and generally think more effectively. These activities are extremely valuable in the workplace, so, armed with such skills, those who've read the book will be promoted faster, move up quicker, and reap the benefits of merit raises and bonuses.

We now make, in our minds, *extremely* conservative estimates as to the value of carefully reading this book. We believe that, with probability 0.5, the ideas honed from this book would result in at least an additional $40-per-year increase in salary each year. We believe that, with probability 0.4, this book would result in an additional $20-per-year increase in salary each year. Finally, although we honestly do not believe it possible, with probability 0.1, we will say that the reader sees no additional increase in salary resulting from this book.

Let's consider the case of a 25-year-old person who will work for the next 40 years before retiring. An increase of $40 per year in salary for 40 years would result in a $40 increase in salary the first year, a total of $80 increased salary the second year, a total of $120 increased salary the third year, and so on. These increases over 40 years would result in total extra income of $32,800. A yearly extra increase of $20 per year in salary for 40 years would result in total extra income of $16,400. Therefore, the expected value of buying this book for $130 (and carefully reading it) would be equal to

$$[(32,800 - \$130) \times 0.5] + [(\$16,400 - \$130) \times 0.4] \\ + [(\$0 + \$130) \times 0.1] = \$22,874.$$

Thus, we believe a conservative figure for the value of buying this book is $22,874. Plainly, the purchase of this book is a financially sound investment.

Perhaps comparing the expected value of the lottery versus this book is a bit ridiculous, but the idea of expected value is not. Measuring the expected future value of events that have not yet occurred is an important way of assessing prospects for the future. Certainly the strategy of giving more weight to more probable future events and less weight to less probable future events seems eminently reasonable. But the future is a tricky realm, and sometimes reasonable ways to analyze it run contrary to other commonsense ideas about cause and effect, as we will now see.

Newcomb's Paradox

Throughout history, paradoxes have played important roles in stimulating new ideas. A paradox presents a situation that has two possible interpretations or resolutions. Each view appears irrefutable, and yet the views are diametrically opposed to each other. The philosopher William Newcomb proposed a paradox that to this very day has not been fully resolved. Many have found this paradox intriguing to wrestle with and debate. Perhaps you will find, as we have, that people who discuss this paradox are nearly evenly divided—usually believing that anyone who thinks the opposite way is completely wacko. Such firm convictions make for interesting exchanges and may reveal important insights. Exploring difficult questions often exposes misperceptions. Good luck in resolving this intriguing puzzle.

 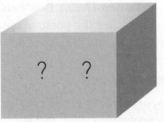

Here is the premise of Newcomb's Paradox. A famous psychologist is showing off her predictive powers by giving you the chance to become rich. You will be led into a room in which stands a table holding two boxes, the Thousand-Dollar Box and the Zero-or-Million-Dollar Box. You must now choose one of two possible courses of action. You may choose to take the contents of both boxes or you may take the contents of the Zero-or-Million-Dollar Box only.

The Thousand-Dollar Box has clear glass sides and visibly contains $1000—*cash*. The other box, the Zero-or-Million-Dollar Box, is opaque. When you are led into the room, you will not know the contents of the Zero-or-Million-Dollar Box, but the experimenter informs you in advance that in the Zero-or-Million-Dollar Box there is nothing at all or there is $1,000,000. The experimenter further informs you of the criterion under which she placed $0 or $1,000,000 in the Zero-or-Million-Dollar Box. The criterion is that, if she predicted that you will take the contents of the Zero-or-Million-Dollar Box only, then she placed $1,000,000 in the Zero-or-Million-Dollar Box. However, if she predicted that you will take the contents of both boxes, then she placed $0 in the Zero-or-Million-Dollar Box.

Let us assume that the psychologist is exceptionally good at making her predictions. Specifically, let's suppose that she correctly predicts an individual's decision 90% of the time, meaning that, of those who choose to take the Zero-or-Million-Dollar Box only, she correctly predicts their behavior 90% of the time, and, of those who choose to take both boxes, she correctly predicts their behavior 90% of the time. You are led into the room and the door is closed. You are all alone and the two boxes are sealed and affixed to the table. What do you do? We now invite you to pause and ponder this conundrum.

When we enter the room, we are faced with two persuasive arguments, leading to two different conclusions. Here are the two arguments.

Argument 1—Choose Both Boxes

The $1,000,000 is either in the Zero-or-Million-Dollar Box or it is not. What we do after we have entered the room does not influence the existence or absence of the $1,000,000. Therefore, we may as well take the contents of both boxes. In either case (that is, if the Zero-or-Million-Dollar Box contains $0 or contains $1,000,000), by taking both boxes we will be $1000 ahead.

Argument 2—Choose only the Zero-or-Million-Dollar Box

If the psychologist predicts that we will take both boxes, she will have put $0 in the Zero-or-Million-Dollar Box. Consequently, perhaps it is better to choose only the Zero-or-Million-Dollar Box in the hopes that our decision to take the contents of only the Zero-or-Million-Dollar Box will somehow have been apparent to the psychologist when she made her prediction. If she is in fact a good predictor, then we take the Zero-or-Million-Dollar Box only and hope she predicted our actions correctly. We can formalize the persuasiveness of choosing just the Zero-or-Million-Dollar Box by computing the expected value of each possibility.

Quantifying Expected Values

Recall that the psychologist correctly predicts an individual's decision 90% of the time. That is, 0.9 is the probability that she is accurate in her predictions. So, the expected value of taking the Zero-or-Million-Dollar Box only is

$$(0.9 \times \$1,000,000) + (0.1 \times \$0) = \$900,000.$$

However, the expected value of taking both boxes is

$$0.9 \times (\$1000 + \$0) + 0.1 \times (\$1000 + \$1,000,000) = \$101,000.$$

A Paradoxical Situation

Thus, we have a paradoxical situation. One plausible analysis (either the $1,000,000 is there or it is not) leads to one conclusion (take both boxes), whereas another plausible analysis (the expected value of choosing only the Zero-or-Million-Dollar Box is $900,000, while the expected value of taking both boxes is $101,000) leads to a different conclusion (take the Zero-or-Million-Dollar Box only). However, the expected value is the value we would expect on average if we repeated the experiment many, many times. Here we get to play only once. But, suppose we were in line with 100 people in front of us who all do the experiment. The psychologist correctly predicts 90 actions and misjudges a mere 10. Now it's our turn. What would we do?

Posing Variations

What would *you* do? How can we think about this paradox? If a question is hard, a helpful strategy is to pose a variation on the question. For example, we might see how our choices change as we alter the amounts in the two boxes. Does our concept of free will play a role here? Suppose all our actions are predetermined and the psychologist can detect the preexisting conditions for our eventual choices. Or suppose before we are led into the room, we see the experimenter flip a coin and make her decision on the basis of the outcome of that coin toss. Suppose we are taken into the room without ever meeting or even seeing the experimenter. Would that alter our decision?

Many other ideas will come to mind as we discuss this paradox. There is no general consensus about whether Newcomb's Paradox is completely understood. Perhaps this paradox underscores the significance of the expected value perspective. After thinking about the alternatives, get a wealthy friend to try the experiment on you. Good luck and choose carefully.

THE NOTION OF EXPECTED value allows us to determine how much on average we are likely to benefit or lose from a future event. Decisions about insurance and pension plans, horse racing and the lottery, and political policy matters are usefully informed by an expected value point of view. Life decisions we make that run contrary to the wisdom of their expected value are taken at our peril. Occasionally people do win the lottery, but, if we consistently make decisions whose expected value is negative, over the long run we are likely to lose. Likewise, if we weigh the gains against the losses in the proportionate fashion embodied in the determination of expected value, we will tend to benefit in like proportion. We will discover great value in expected value: It is a powerful idea that can help guide us through our uncertain future.

Our strategy for developing a practical way to measure the value of future possibilities is to start with simple, concrete examples. We envision that uncertain situations are repeated many times to get a proportional sense of the possible results. These thought experiments give us experience with measuring the value of the unknown future, rather than just guessing. By finding a method for determining the average value in a special case, we can use that idea to formulate a definition for *expected value*. Armed with that definition, we can apply the concept to analyze different types of situations.

We gain experience by looking repeatedly at clear and simple cases. Then we can extract specific methods and ideas. As the ideas crystallize, we can form clear definitions and develop broadly applicable methods. These new notions and methods can then be applied to new settings, which, in turn, lead to new insights.

MINDSCAPES ⟩ Invitations to Further Thought

In this section, Mindscapes marked **(H)** *have hints for solutions at the back of the book. Mindscapes marked* **(ExH)** *have expanded hints at the back of the book. Mindscapes marked* **(S)** *have solutions.*

Developing Ideas

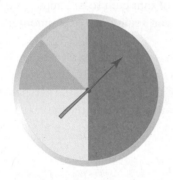

1. **What do you expect?** How do you compute an expected value?

2. **The average bite.** Your little sister loves visits from the Tooth Fairy. Suppose half the time the Tooth Fairy leaves $1 and half the time she leaves $0.50. What is the expected value of the Tooth Fairy's payoff?

3. **A tooth for a tooth?** Suppose your cousin's Tooth Fairy leaves $1 one-third of the time, $0.50 one-third of the time, and $0.80 one-third of the time. What's the expected value of her Tooth Fairy payoff?

4. **Spinning wheel.** Suppose the spinner shown is balanced, so the arrow is equally likely to land in any direction. What's the probability you spin red? What's the probability you spin blue or green?

5. **Fair game.** What does it mean for a game to be fair?

Solidifying Ideas

6. **Cross on the green (S).** A standard roulette wheel has 38 numbered slots for a small ball to land in: 36 are marked from 1 to 36, with half of those black and half red; two green slots are numbered 0 and 00. An allowable bet is to bet on either red or black. This bet is an even-money bet, which means if you win you receive twice what you bet. Many people think that betting black or red is a fair game. What is the expected value of betting $1000 on red?

7. **In the red.** Given the bet from Mindscape 6, what should the payoff be if the game is to be a fair game?

8. **Free Lotto.** For several years in Massachusetts, the lottery commission would mail residents coupons for free Lotto tickets. To win the jackpot in Massachusetts, you have to correctly guess all six numbers drawn from a pool of 36. What is the expected value of the free Lotto ticket if the jackpot is $8,000,000 and there is no splitting of the prize?

9. **Bank value.** What is the expected value of keeping $100 in a bank for one year if the bank pays 3% interest, compounded annually?

10. **Newcomb your neighbor.** Explain Newcomb's Paradox to some friends and ask them what they would do and why. Explain the expected value argument. Record your friends' reactions.

11. **Value of money.** In Newcomb's Paradox, first suppose that you have no money and have not eaten in two days; next suppose your net worth is $800,000. How would these different scenarios affect your decision?

12. **Die roll.** What is the expected value of each of the outcomes (1, 2, 3, 4, 5, or 6) of rolling a fair die?

13. **Dice roll (ExH).** What is the expected value of each of the outcomes (2, 3, 4, 5, . . . , 12) of rolling two fair dice?

14. **Fair is foul.** Someone has a weighted coin that lands heads up with probability 2/3 and tails up with probability 1/3. If the coin comes up heads, you pay $1; if the coin comes up tails, you receive $1.50. What is the expected value of this game? Would you play? Why or why not?

15. **Foul is fair (S).** Someone has a weighted coin that lands heads up with probability 2/3 and tails up with probability 1/3. You pay $5 to flip the coin. If the coin comes up heads, you lose and receive nothing. For this game to be a fair game, how much would you have to receive if you flip tails?

16. **Cycle cycle (H).** You live in an area where the probability that one's bicycle is stolen is 0.2. You care deeply for your $700 road bike. What is a fair price to pay to insure your bike against theft over the life of your bike? (*Postscript*: On Friday, June 13, 1997, an uninsured mountain bike belonging to one of the authors was stolen out of the garage of the other author. What are the chances?!)

17. **What's your pleasure?** You have three options for the evening. (1) You could watch some sitcoms on TV that you are certain to enjoy and that will provide you with a relative pleasure rating of 4. (2) You could go to a movie that is supposed to be good. You would enjoy the movie with probability 0.5; if you enjoy it, it will provide you with a relative pleasure rating of 11, but if you don't enjoy it, it will provide a negative rating of −2. (3) You could go on a date. With probability 0.3 you will experience a pleasure rating of 21, and with probability 0.7 the date will provide a negative rating of −2. What are the expected values of pleasure for these individual activities? List the activities in order of expected pleasure. What would you do?

18. **Roulette expectation.** A standard roulette wheel has 38 numbered spaces for a small ball to land in: 36 are marked from 1 to 36, half black and half red; 0 and 00 are green. If you bet $100 on a particular number and the ball lands on that number, you are paid a whopping $3600. What is the expected value of betting $100 on red 9?

19. **Fair wheeling.** You are at the roulette table and bet $100 on red 9. What payoff should you receive to make the game fair? (See Mindscape 18.)

20. **High rolling (H).** Here is a die game you play against a casino. You roll a fair die. If you roll 1, then the house pays you $25. If you roll 2, the house pays you $5. If you roll 3, you win nothing. If you roll a 4 or a 5, you must pay the house $10, and if you roll a 6, you must pay the house $15. What is the expected value of this game?

21. **Fair rolling.** Suppose you are considering the game described in Mindscape 20. How much would you have to pay, or be paid, to make the game a fair game?

22. Spinning wheel. You pay $5, pick one of the four spinners below, and spin that spinner. Each spinner is balanced, so the spinner is equally likely to land in any direction. Wherever the spinner lands, you receive that much cash. Which spinner would you pick? Justify your answer with an analysis of the expected values of the various spinners.

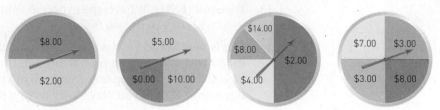

It costs ONLY $5.00 to play . . . pick a spinner, spin, and win!!

23. Dice (ExH). You place a bet and then roll two fair dice. If you roll a sum of 7 or 11, you receive your bet back (you break even). If you roll a sum of 2, 3, or 12, then you lose your bet. If you roll anything else, you receive half of the sum you rolled in dollars. How much should you bet to make this a fair game?

24. Uncoverable bases. Show by a specific example how it is possible to lose at the lottery even if the jackpot exceeds the cost of purchasing every possible ticket.

25. Under the cap. A national soda company runs a promotional contest. Under the cap of one in a million bottles there is a message saying that you won $1,000,000. Suppose that the two-liter bottle costs $2. What is the expected value of buying a bottle of this soda (assuming, of course, that the soda itself is worthless)?

Creating New Ideas

26. Two coins in a fountain. You pay $1 for two coins to toss in a fountain and see how they land. If you see two tails, then you receive $3; otherwise you lose. What is the expected value of this game? Is there one single possible outcome whereby you would actually gain or lose the exact amount computed for the expected value? If not, then why is the expected value *expected*?

27. Three coins in a fountain. You pay $5 for three coins to toss in a fountain and see how they land. If you see no heads, then you receive $20. If you see exactly one head, then you receive $5 (the game is a draw), and if you see at least two heads, then you lose. What is the expected value of this game? Is there one single possible outcome whereby you would actually gain or lose the exact amount computed for the expected value? If not, why do we call the expected value, the *expected* value?

28. Insure (S). You own a $9000 car and a $850 mountain bike. The probability that your car will be stolen next year is 0.02, but the probability that your bike will be snatched is 0.1. An insurance company offers you theft insurance for your car for $200 and insurance for your bike for $75. What is the expected value of the car insurance? What is the expected value of the bike insurance? (This question pains one of the authors . . . see Mindscape 16.)

29. **Get a job (H).** You search for a job. Three companies are interested in you, and you will receive at most one offer. The first company has a job open with a salary of $21,000, and the probability of getting an offer is 0.5. The second company has a job open paying $32,000, and the probability of getting an offer is 0.3. The last company has a job paying $45,000, and the probability of getting an offer is 0.1. There is a 0.1 probability of not landing any job. What is the expected income you will make next year?

30. **Take this job and . . .** Given the employment scenario described in the previous Mindscape, suppose that the companies do not consult one another about offers, so you may receive more than one offer. Suppose the first company calls you on the phone, offers you the job, and says you have to accept or reject it right there on the spot. You have not heard from either of the other companies yet. Do you accept or reject the offer? What if the call were from the second company? What if it were the third company? Justify your answers using expected value.

31. **Book value.** Refer back to our analysis of the expected value of reading this book (pages 766–767). Suppose you took the $130 and put it in a bank account guaranteed to pay 10% interest compounded annually for 40 years. Which is the better investment: depositing the money in a bank or buying and reading this book? (*Hint*: You can use the formula $A = P \times (1 + r)^n$, where A is the amount you will have by investing P dollars with an interest rate of r compounded annually for n years. So, in this case, $P = 130$, $r = 0.10$, and $n = 40$. You may need a calculator.)

32. **In search of . . .** A group of deep-sea divers approaches you with a proposition. They are 60% certain that they know where an ancient shipwreck is; they are also 50% certain that there is a treasure worth about $2,000,000; and finally they are 70% certain that they will be able to get to it. They want you to invest $200,000 in this expedition. If they find the buried treasure, you receive $1,000,000. What is the expected value of this investment?

33. **Solid gold.** There is a 50% chance that the price of gold will go up $25 an ounce; a 20% chance that it will remain the same; and a 30% chance that it will drop to $40 an ounce. What is the expected value of a purchase of gold? Given your answer, would you invest in gold at this time if gold cost $375 per ounce?

34. **Four out of five.** In Newcomb's Paradox, suppose that the psychologist predicts the choice of the subjects correctly four out of five times. What is the expected value of selecting both boxes? What is the expected value of selecting just the Zero-or-Million-Dollar Box? What would you do?

35. **Chevalier de Méré.** Suppose that the Chevalier de Méré bets 1000 euros that a 6 will appear at least once in four rolls of a fair die. If he wins, he receives 2000 euros. What is the expected value for the Chevalier de Méré? (*Hint*: Return to Section 8.4, page 618.)

Further Challenges

36. **The St. Petersburg paradox.** Here is an interesting game: You pay a certain amount of money to play. Then you flip a fair coin. If you see tails, you flip again, and the game continues until you see a head, which ends the game.

If you see heads on the first flip, you receive $2. If you see heads on the second flip, you receive $4. If you see heads on the third flip, you get $8, and so forth—the payoff is doubled every time. What is the expected payoff of this game? How much would you pay to play this game? Suppose you pay $1000 to play. What is the probability that you would make money? Why is this game a paradoxical situation given the expected value?

37. **Coin or god.** In Newcomb's Paradox, first suppose that the psychologist just flips a coin to determine whether to place the million dollars in the box. What is the expected value of selecting both boxes? What is the expected value of selecting just the Zero-or-Million-Dollar box? Suppose, instead, that the experiment is run by an all-knowing godlike being. What would you do?

38. **An investment.** You wish to invest $1000, and you have two choices. One is a sure thing, and you will make a 5% profit. The other is a riskier venture. If the venture pays off, you will make a 25% profit; otherwise, you lose your $1000. What is the minimum required probability of this riskier venture paying off in order for the expected value to exceed the value of the first investment?

39. **Pap test (H).** Assume that the insurance value of a life is $1,200,000. Suppose Pap smear tests will save one life in 3000. A Pap smear costs about $30. What is the expected value of a Pap smear?

40. **Martingales.** A game is played with a fair coin. You bet some amount of money on heads. If you flip heads, then you receive even money. (You get your bet back, plus an extra amount equal to the bet.) If you flip tails, then you lose your bet. The *doubling strategy* is one where you continue to double your bet after each loss until you win. So, for example, if you first bet $1 and lose, then the next time you would bet $2. If you lose again, you would then bet $4, and then $8, and so on. How much would you earn if you used this strategy and lost seven times in a row before finally winning? How much money would you need to play eight times?

In Your Own Words

41. **Personal perspectives.** Write a short essay describing the most interesting or surprising discovery you made in exploring the material in this section. If any material seemed puzzling or even unbelievable, address that as well. Explain why you chose the topics you did. Finally, comment on the aesthetics of the mathematics and ideas in this section.

42. **With a group of folks.** In a small group, discuss and actively work through the reasoning involved in Newcomb's Paradox. After your discussion, write a brief narrative describing the proper Newcomb choice in your own words.

43. **Creative writing.** Write an imaginative story (it can be humorous, dramatic, whatever you like) that involves or evokes the ideas of this section.

44. **Power beyond the mathematics.** Provide several real-life issues—ideally, from your own experience—that some of the strategies of thought presented in this section would effectively approach and resolve.

For the Algebra Lover

Here we celebrate the power of algebra as a powerful way of finding unknown quantities by naming them, of expressing infinitely many relationships and connections clearly and succinctly, and of uncovering pattern and structure.

45. **Spin to win.** To play a certain carnival game, you spin a wheel to obtain one of the nine equally likely outcomes. Three of the outcomes have a payoff of x^2 dollars and the remaining six have a payoff of x dollars. Write an expression that gives the expected value of the payoff for this game in terms of x. If the expected value of the payoff is $5, determine the two actual payoffs.

46. **Spinner winner.** To play a different carnival game, you spin a wheel to obtain one of the 30 equally likely outcomes. Five of the outcomes have a payoff of x^2 dollars and the remaining 25 have a payoff of x dollars. Write an expression that gives the expected value of the payoff of this game in terms of x. If the expected value of the payoff is $6, determine the actual values of the payoffs.

47. **Insurance wagering (H).** From the point of view of an insurance company, the expected value of an insurance policy is the profit the company would make on average from such a policy. Suppose an insurance company sells a computer-theft policy. The cost they charge for the policy is $$C$ per year, the probability a computer will be stolen in a given year is p, and the value of a typical computer is $$V$. Write an expression in terms of C, p, and V that gives the profit the company will make on average from such a policy. Simplify your expression as much as possible.

48. **Probable cause.** Continuing the scenario from the previous Mindscape, suppose the cost of the computer theft insurance policy is $80, the value of a typical computer is $1000, and the profit (expected value for the company) of each policy on average is $40. What is the probability that a computer will be stolen in a given year?

49. **The bicycle thief.** Some entrepreneurial classmates start a business selling insurance for bicycle theft at your school. The data they get from the office of public safety suggests that the probability a bicycle is stolen is about 0.05. By surveying friends, they find the typical bike is worth about $500. If they want to make an average profit of $20 per policy, how much should they charge for the insurance policy? (Use your work from Mindscape 47.)

10.2 RISK

Deciding Personal and Public Policy

Sea of Galilee (1633), Rembrandt Harmens van Rijn. *How risky is a three hour tour? . . . A three hour tour.*

> *Nothing in life is certain except death and taxes.*
>
> BENJAMIN FRANKLIN

Every move we make, every step we take (or don't take) involves risk. The question is not whether to take risks—we can't avoid them. Disaster is always a possible consequence of a decision, but how likely is it? It's possible that we will be struck by lightning if we walk outside in a thunderstorm. But it is also possible that an earthquake will split open the Earth and our house will plunge into a deadly crevasse while we sit inside, avoiding the danger of lightning. The question is how to measure risk so that we can make decisions about actions or inactions with our eyes open.

When we fly on an airplane, what are the chances that it will crash? When we take a medical test, how worried should we be that we'll get an unpleasant result? When we invest our life savings in the next big dot-com boom, should we acquire a shopping cart and old trench coat to prepare for the future? How often should women have Pap smears? Should we avoid cheeseburgers? Should we bother to study? To make good decisions, we need to measure the risks and the rewards.

Measuring risks and rewards involves counting, randomness, and data, but most importantly it involves clear thinking. The first challenge is simply to *try* to measure the risk. That is, just taking the trouble to estimate the likelihood of various possible results of our actions will lead us to insights that facilitate our decision-making processes. Describing clearly what could happen is often a big step toward illuminating risk. So let's stroll down a risky lane lined with decisions—life or death, riches or poverty, health or sickness. Which will we choose?

Goal

When facing issues, we want to take steps to help us make informed decisions. We do this by measuring the risk: enumerating the potential consequences, measuring the probability of different outcomes, seeing how changing the decision alters the probabilities, and, finally, drawing conclusions by looking at those different likelihoods.

Public Policy

As a society we make many decisions that affect our health and safety. What standards of safety should the law require? What health-care practices should we institute nationally? We can't afford to make every car as safe as we know how to make it, to treat every illness with the best medical care we know how to give. So how do we decide what to do and what not to do? One method is to look at some data and describe the likely consequences of various decisions. Here we will look at several public-policy issues: HIV/AIDS testing, airplane safety, terrorism, Pap smears, poverty, and education. Our goal is to develop some techniques for evaluating the wisdom of various approaches to making public-policy decisions.

HIV/AIDS Testing

Analyzing data is important in coping with the world of death and disease. AIDS (Acquired Immune Deficiency Syndrome) has been the most lethal disease for young people world-wide for the last 30 years. A tremendous research effort has resulted in an understanding of some of the mechanisms by which the disease works, and clinical research has given some hope for an eventual cure; however, for decades AIDS was viewed as inevitably fatal and, as such, remains one of the most feared diseases in the world. AIDS is the active disease that results from a person's being infected with HIV (Human Immunodeficiency Virus). People carrying HIV will develop AIDS after some interval of time has passed—sometimes 10 to 15 years.

In 2001, approximately 42 million people worldwide were living with HIV/AIDS, of whom nearly half, approximately 19.2 million, were women. In 2001, approximately 3.1 million people died as a result of AIDS. Approximately 500,000 people are living with HIV/AIDS in the United States. These figures are painful to contemplate.

A person who is HIV-positive can transmit the disease to others through sexual contact or contact with blood. If we knew who is HIV-positive, we could potentially curb the spread of this dreaded disease. Scientists have worked hard to develop reliable tests for HIV. One such test, the ELISA blood-screening test, is quite accurate. The test is 95% accurate in that it tests positive for 95% of people who actually have the disease, and it tests negative for 99% of people who do not. The 5% error in detecting HIV in carriers may be due to low concentrations of the virus or antibodies to the virus among some people. The 1% false-positives error may or may not be caused by mistakes in running the test. Some false positives might be due to some special but unknown idiosyncrasies in the body chemistries among that 1% of the population.

Infected people should be warned to avoid the kind of contact that would spread the disease. But falsely telling an uninfected person that he or she is infected has tremendous costs: It may frighten the person half to death and the person may have to change his or her lifestyle unnecessarily. A fundamental policy question is whether we should undertake universal testing for HIV/AIDS. Should every man, woman, and child in the country be tested?

Let us suppose for a moment that the policy of universal testing has been instituted. You have had your test and soon thereafter you receive the dreaded result: *We regret to inform you that your test was positive.* When faced with this devastating news, you need to control your instinct to panic and force yourself to think clearly and rationally.

Given the accuracy of the test and the data about what percentage of the population is HIV-positive, what is the probability that you are HIV-positive? At first glance, this question seems to have been answered in the description of the reliability of the test. The test gives a positive result to an uninfected person in only 1% of the cases. It is 99% accurate, so you might conclude that you are 99% certain to be HIV-positive—a most frightening prospect.

But let's think a bit more deeply about this case by actually looking at the numbers. Let's write down a few of the facts we know:

1. The population of the United States is about 315,000,000.
2. About 1,000,000 people in the United States are HIV-positive.
3. Of the 1,000,000 who have the virus, the test will come out positive 95% of the time; that is, in 950,000 cases.
4. There are 314,000,000 people in the United States who do not have the virus (that is, 315,000,000 − 1,000,000).
5. Of the 314,000,000 people who do not have the virus, the test will come out falsely positive 1% of the time, which equals 3,140,000 cases.
6. So the total number of people receiving a positive test result is:
   950,000 + 3,140,000 = 4,090,000.
7. Of the 4,090,000 who get positive test results, only 950,000 actually have the disease. Therefore, if you get a positive test result, your chance of actually having the virus is 950,000/4,090,000, which is less than a 24% chance.

How can we reconcile this apparent paradox? On the one hand, the test is 99% accurate. On the other hand, among people who receive a positive result, the chance of having the disease is less than 24%.

The Hidden Test

This result is surprising at first or second glance. We need to understand this apparent paradox so that instead of appearing counterintuitive, it becomes natural and reasonable. If truth seems strange, we do not understand it well enough.

How can we change our view of a result from surprising to obvious? Mathematics shows us a way, and this book explores several such situations. The ideas of infinity, irrational numbers, the fourth dimension, and distortions of objects all required human beings to shift their perspectives. One way to retrain our intuition is to look at issues from different points of view and at examples that illustrate some feature in a clear way.

One way to look at the HIV/AIDS testing situation differently is to recognize that the scenario really presents *two* tests for the virus, but only one was called a test. The hidden "test" is the statistic that 1,000,000 out of 315,000,000 people in the United States are HIV-positive. This implicitly gives us a test: Take a person at random. The probability that person is not HIV-positive is 314,000,000/315,000,000, which equals 0.997. It doesn't seem like much of a test, but we have to admit, it is pretty accurate. If we take a person at random, look at him or her and say, "You are not HIV-positive," we will be correct 997 times out of 1000. That's a pretty impressive success rate. By comparison, the test explicitly mentioned in the scenario states that the blood test is accurate only 99 times out of 100. So the blood test is actually far *inferior* to the hidden "test" of just looking at the proportion of people in the population who are HIV-positive.

Now let's change the presentation of the situation slightly and see if we get a different sense of the significance of the outcome. Instead of saying that 1,000,000 out of 314,000,000 people in the United States are HIV-positive, suppose we say that you will be taking two different tests: One test, called *PoorerTest*, is accurate 99% of the time, and the other test, called *BetterTest*, is accurate 99.7% of the time. Suppose that according to *PoorerTest* you have the virus and according to *BetterTest* you do not. Of course, if both tests told you that you were healthy, you would be even more confident; however, as you have taken two tests in this example, the result of the more accurate test is more important than the result of the less accurate test. That disparity in the accuracy of the two tests is what accounts for the rather low probability that you actually have the virus.

The Big Picture

Here is yet another alternative view of the HIV test issue: There are roughly 950,000 people in the United States who have HIV and get a positive test result. Roughly 4,000,000 people receive positive test results. So, only about 1 in 4 positive test results can be correct.

The surprising part of the HIV testing scenario was the mismatch between the accuracy of the test (99%) and the confidence with which we could assert that a person has the virus if the test is positive (23%). We strove to understand this example better by rephrasing it. Although this result might now be a bit more intuitive and make more sense, let's be certain to understand that the 99%-accurate test's positive results make a large difference in predicting the *probability* that someone has the virus. Specifically, it changes the probability that a person is HIV-positive from roughly 3 in 1000 to roughly 1 in 4.

In practice, the paradoxical situation in HIV testing does occur and has resulted in methods of screening for diseases that use a preliminary, inexpensive test to winnow the population and then a more refined test that is more conclusive on the subpopulation who receive positive test results. This powerful example shows how understanding data and probability can have a significant effect on our thinking about policy decisions. To further illustrate how data and probability impact public policy issues, we now consider another serious topic: travel safety.

The Friendly Skies

No one wants to be in an airplane crash, and, happily, our chances of being in one are extremely low. Each day U.S. commercial airlines log approximately 1.7 billion passenger miles, and yet accidents are so rare that we remember big crashes for years. U.S. commercial airplane crashes are so infrequent that during the last several years, on average, there was only about 1 accidental fatality per 3.4 billion passenger miles. An individual would have to fly 500 miles every day for over

18,000 years between crashes. That's a long wait, and it doesn't even include waiting for luggage.

Nevertheless, as safe as planes are, when one does crash, people naturally turn their attention to air safety, and there is always room for improvement. Here we consider *accidents* rather than *terrorism*—a painful reality that we'll discuss later. Thus, here we will not consider the lives lost on September 11, 2001, when four planes crashed due to terrorism, but we will consider the November 12, 2001, crash of American Airlines flight 587 in Queens, New York, where over 260 people were killed in a tragic accident.

Should We Support Increased Air Traffic Safety?

Air safety could be improved, but at a price, and that price is real money. We could probably make the airlines 10 times as safe as they are now, but ticket prices would have to increase to pay for the safety improvements. We are responsible citizens and must decide whether regulations should require the airlines to institute more precautions, and we are naturally interested in the consequences of various possibilities. If air safety were increased next year as described above, what would be the effect on the loss of life? Let's take the trouble to figure out the quantitative consequences instead of relying entirely on a qualitative impression about safety.

From the facts given, we can estimate how many commercial airline fatalities are expected per year. So we can estimate how many lives would be expected to be saved if airline fatalities were only one-tenth as great.

Let's do the arithmetic and see how many people could be expected to live or die if we improved airline safety as described. First let's note that because U.S. airlines fly about 1.7 billion passenger miles each day, and there is an airline fatality about every 3.4 billion passenger miles, then, on average (the mean), there is one death every two days. Of course, what really happens is that many people are killed all at once instead of at a steady one-death-per-two-days rate. However, one fatality every two days means that the estimated number of fatalities expected due to U.S. commercial airline accidents with current safety practices is roughly 183 per year. Now, suppose that air travel became 10 times safer. Then, instead of 183 people dying, only about 18 commercial air fatalities would be expected each year, a savings of 165 lives. This safety proposal looks pretty darn good.

Consider Unintended Consequences

We must remember that this improved safety will cost money and the cost of airplane tickets will go up. When ticket prices go up, some people will drive instead of fly. Let's suppose that if airlines made their planes 10 times safer and increased the cost of tickets to pay for that safety, then 10% of travelers who now fly would choose to drive instead. Sadly, fatal automobile accidents are much more common than airplane crashes. In fact, our chances of dying in an automobile accident are about 34 times greater per mile than in a commercial airplane accident. What will be the result of this extra driving?

If 10% of the people who now fly drove instead, there would be an extra 170 million miles of driving each day. Driving is 34 times more dangerous per mile than commercial flying. The fatality rate for flying is 1 death per 3.4 billion passenger miles; the fatality rate for driving is 34 deaths per 3.4 billion miles, or 1 death per 100 million miles. With people driving 170 million more passenger miles per day (instead of flying), on average there would be 1.7 more automobile deaths per day. Therefore, there would be an additional 620 people killed in automobile accidents each year. Uh-oh.

The consequences of our increased air safety are somewhat counterproductive. It is true that 165 fewer people would die in airplane crashes, but 620 more people would die in car crashes. Unfortunately, that means that approximately 455 additional people would be killed if airline safety were improved as outlined. Maybe air traffic safety is not such a great idea after all. In fact, maybe more lives would be saved by making airlines even less safe if that would make airline tickets cheaper, thus getting more people to fly instead of drive.

This scenario illustrates an important lesson. Even simple arithmetic combined with a rough analysis of data can lead to striking conclusions. In this case, we see that a side effect of increased safety measures—namely, higher prices for airline tickets—could well result in the opposite of what we intended. When we make decisions, anticipating all the likely consequences of our actions is extremely useful in trying to visualize the whole picture.

Life Lesson

The moral: Beware of unintended consequences.

Terrorism

Terrorism has become a dominant theme in our society, particularly in the media. The painful events of September 11, 2001, and their unfolding aftermath weigh heavily on our hearts and minds. How can we measure the risks involved with living in a world marked by terrorism? Should we be vaccinated for smallpox? Perhaps the risks associated with the vaccine itself are greater than the risk of smallpox. What is the risk of smallpox? What about anthrax—the toxin that was mailed to senators? Perhaps we should not open our mail—or at least avoid opening our bills.

Suppose we are considering a trip to Israel but are concerned about terrorism in that country. For the last 20 years we have heard about violence and terrorism in Israel. Should we avoid this country? Can we estimate the risk of our encountering terrorism in Israel? Let's look at the data: Over the past four years, the average population of Israel has been approximately 6,500,000. During this time, 288 people a year on average were killed through terrorist acts in Israel. Thus, the probability that any particular person in Israel would be a victim of such an attack is roughly 288/6,500,000 = 0.000044, which represents about a 1 in 22,500 chance if we were to stay a full year.

For comparison, let's consider the likelihood of dying in an automobile accident in the United States. More than 40,000 people are killed in the United States each year in automobile accidents. The population of the United States is approximately 280,000,000. So our chance of dying in a car crash is approximately 42,000/280,000,000, which is a little more than a 1 in 7000 chance. That probability is more than three times the probability of dying as a result of a terrorist act in

Which would you rather be in?

Israel. So we might actually be safer vacationing in Israel than driving our Isuzu in the States.

Of course, if we drove in Israel, we would add the risk of dying in a car crash there to the chance of dying in a terrorist incident. The extra risk would be something to consider. However, by looking at the data we can obtain a rough estimate of how significant the danger of terrorism in Israel is—roughly one-third the danger of driving in the States, a danger we accept every day without much anxiety.

We include this example because it clearly illustrates how distorted our sense of a particular danger can be when we get our information from news media that naturally focus on rare and tragic events. When we make decisions, looking at a rough estimate of the data can at least help us put various choices in perspective. Some choices are more risky than we might intuitively imagine, and some are less so. Intelligent decisions balance the reality of the data—as best as we understand them—with an intuitive sense of the danger that may reflect features of the situation not captured in the available data. Risk analysis can be a valuable aid in making decisions.

The Price of Life

No society can afford to provide each citizen with the best possible medical care money can buy, because the whole society does not have enough money to pay for it. We are put in the difficult position of having to make life-or-death decisions about how to spend our limited health-care dollars. These decisions involve many ethical, philosophical, political, and medical components. Making informed decisions requires having a clear-sighted view of the options.

One basic issue we're called on to decide as a society is how much of our limited resources we should spend on preventive medicine and screening. Vaccines and tests for cancer and other diseases are among the most common tools available for preventive medicine. How can we assess their worth? One method of measuring the relative value of one investment in health care over another is to measure or estimate the cost per life saved. This approach, placing an economic value on life, may seem cold and heartless, but the practical reality is that dollars spent in one arena may save far more lives than dollars spent in another.

Pap Smears

Let's begin by analyzing the costs and benefits of one of the common cancer-screening tests for women—the Pap smear, a test that detects cervical cancer. The goal of this test is to detect the cancerous growth early enough so that treatment is effective. It is estimated that each Pap smear, taken every two years as recommended for most women, has a 1 in 3000 chance of saving the life of the woman. If we gave the test only to women who are most at risk for cervical cancer, the chance of saving a life might increase. The cost of a Pap smear is approximately $30 per test. So the total cost per life saved is approximately $30 \times 3000 = \$90,000$.

As a society we must decide whether or not we can afford that expense. A reasonable way to address this issue is to compare the lives saved per dollar spent for Pap smears with the lives saved per dollar spent for some other medical procedure, safety improvement, or other means of improving life expectancy. To make the comparison, let's look at some other deadly and preventable maladies.

Asbestos

During the last 30 years, the media have frequently highlighted the dangers of asbestos. Many buildings have had asbestos removed to avoid the increased number of lung cancer deaths that inhaling asbestos can cause. But the cost per life saved is more than $100,000,000. This cost is enormous and far exceeds the cost per life saved of other potential expenditures for health and safety. So, for example, if as a society we had made different social decisions and instead of paying for asbestos removal had invested in more Pap smears, thousands more women would be alive today who died of cervical cancer for lack of screening than people whose lives were saved due to asbestos removal.

Costs per Life Saved

To place the previous examples in perspective, we list some costs per life saved for other initiatives our society might consider in the table.

Initiative	Cost per life saved
Provide smoke alarms in every U.S. home	$120,000
Provide improved traffic signs	$31,000
Offer blood tests to detect colon or rectal cancer	$20,000

LIFE = $???

While the government has not enacted any of these initiatives, it *has* supported the following measures:

Initiative	Cost per life saved
Environmental Protection Agency [EPA] safety measures for coal miners	$1,000,000
EPA removal of abnormally high levels of radium in water	$5,000,000
Nuclear Regulatory Commission [NRC] regulations on emissions of radioactive iodine	$100,000,000

Trade-offs have real and deadly consequences when we make decisions about investing our society's money in reducing risk. It is estimated that if the money currently spent in the United States on life-saving interventions were reallocated to more cost-effective options, perhaps 60,000 people each year would live who now die. Estimating the effectiveness of regulations on a cost-per-life-saved basis can be a helpful tool in making wise societal decisions.

Because health and longevity correlate with income, well-intentioned but costly safety measures can have the opposite of the desired effect. Not only could the money have been spent on more cost-effective life-saving measures, but even if the money had not been spent on safety at all but simply spent to raise the standard of living of the poor, for example, one could argue that more lives would have been saved simply by increasing people's wealth. (Of course, these good intentions may have unintended consequences as well.)

Loss of Life Expectancy

Death removes all the potential remaining years of life from an individual. By estimating the average shortening of life due to various life circumstances and choices, we can compare the relative seriousness of different risks. This measure of risk is known as *loss of life expectancy*. We calculate this value simplistically as follows:

> The loss of life expectancy due to an action = (the additional years of life the individual is expected to live) × (the chance of death due to the action under consideration).

From our earlier discussion, we may recognize this value as an expected value. As an illustration, suppose that statistics indicate that an 18-year-old in the United States will live on average for another 60 years. If that individual decides to take an action that has a 2% chance of immediate death, then the loss of life expectancy due to this action equals $60 \times 0.02 = 1.2$ years. If thousands of 18-year-olds took that particular risk, 2% of them would die, causing the average length of life for that group to be shortened by 1.2 years.

Poverty

Perhaps the largest contributor to a shortened life expectancy is poverty. In the United States, people living in poverty live an average of nearly 10 years less than those who don't. In developing countries, the disparity in life expectancy between those who live in poverty and those who don't is much larger. In Afghanistan, for example, a person living in poverty has a life expectancy of 30 years less than someone who is not.

Education

We all know intuitively that practicing risky behavior, such as smoking, obesity, or driving while intoxicated, significantly lowers the expected length of our lives. But one of the riskiest choices a young person can make in terms of shortening life expectancy is to drop out of school. The life expectancy of a person who does not complete elementary school is 2.2 years less than a person with a college degree. So in some real sense, not studying literally kills.

When we hear such statistics, we are naturally drawn to ask what specifically causes the earlier deaths of less-educated people or people in poverty. In the United States, poverty does not usually result in death by starvation. But poverty and a lack of education often lead people to make, on average, poorer decisions about personal health care than those who are more educated or are not living in poverty. It should also be noted that those in poverty often have less access to health care. Obesity, smoking, irresponsible drinking, drug use, and violent behavior are all more prevalent among less-educated people. One of the goals of education is to enable people to develop the skills necessary for weighing the consequences of various life choices.

Because poverty and lack of education are such strong indicators of earlier death, programs that promote education and economic development may actually be the best investment we can make in health care, even though these programs do not directly fund medicine, doctors, or hospitals.

Fuzzy Math, Focused Thinking

Numbers often lose their meaning when they're flung at us at high velocity. Let's illustrate a possible misinterpretation of numbers by drawing some dubious conclusions from data. Shown here is a chart of various factors and their associated loss of life expectancy.

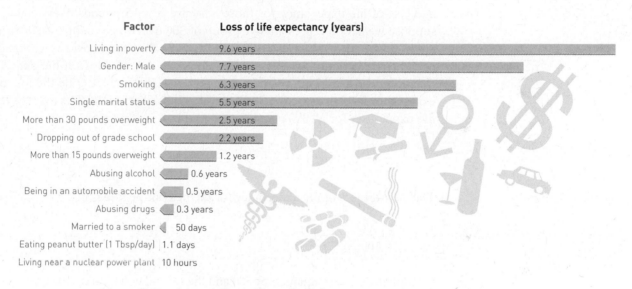

Factor	Loss of life expectancy (years)
Living in poverty	9.6 years
Gender: Male	7.7 years
Smoking	6.3 years
Single marital status	5.5 years
More than 30 pounds overweight	2.5 years
Dropping out of grade school	2.2 years
More than 15 pounds overweight	1.2 years
Abusing alcohol	0.6 years
Being in an automobile accident	0.5 years
Abusing drugs	0.3 years
Married to a smoker	50 days
Eating peanut butter (1 Tbsp/day)	1.1 days
Living near a nuclear power plant	10 hours

Tick . . . tick . . . tick . . . Is "the end" near?

Because U.S. life expectancy is about 78 years, it might appear that the average 45-year-old poor, single, male smoker who dropped out of grade school, is 45 pounds overweight, and eats peanut butter is already dead. The dubious logic leading to this erroneous conclusion is that he loses 9.6 years of life expectancy for poverty, 7.7 for being male, 6.3 for smoking, 5.5 for being unmarried, 2.5 for being 30 pounds overweight, another 1.2 for being 15 pounds overweight, 2.2 for dropping out of grade school, plus 1 day for eating peanut butter. Because he has lost 35 years from his fleeting dream of 78 years of expected life, statistically speaking he's been dead for 2 years.

Something seems awry. To better understand the problem, let's revisit the meaning of loss of life expectancy. For example, what is meant when we say that the loss of life expectancy for living in poverty is 9.6 years? We mean that if we examine the ages of people who have died, and compare those who lived in poverty with those who did not, then we would find that those who lived in poverty were, on average, 9.6 years younger when they died than those living high on the hog.

When we try to put several risk factors together, logical difficulties arise. Just adding up the loss-of-life-expectancy numbers in order to determine the risk of multiple factors does not work. Let's see why.

Fat and Fatter

Suppose we are 30 pounds overweight. Then we are also 15 pounds overweight, of course. On our chart it says that a person who is 15 pounds or more overweight has a loss of life expectancy of 1.2 years, and a 30-pound excess gives a loss of life expectancy of 2.5 years. So if we are just 15 pounds overweight, what is our loss of life expectancy?

The problem is that the 30-pounds-overweight people are increasing the average loss of life expectancy for those who are just 15 pounds overweight. Let's suppose we have a representative group of 300 overweight people—200 of them are at least 15 pounds overweight but less than 30 pounds overweight, and 100 of them are 30 pounds or more overweight. Let P be the loss of life expectancy for those who are 15 but not 30 pounds overweight, and let Q be the loss of life expectancy for those who are 30 or more pounds overweight. The average loss of life expectancy for all 300 people is given to be 1.2 years. So

$$\frac{200P + 100Q}{300} = 1.2.$$

The average loss of life expectancy for the heavier people is just

$$\frac{100Q}{100} = 2.5.$$

Multiplying the first equation by 300 and the second equation by 100 gives

$$200P + 100Q = 360$$
$$100Q = 250.$$

Subtracting the two equations gives

$$200P = 110.$$

Therefore,

$$P = 0.55.$$

So the loss of life expectancy of people who are 15 pounds overweight but not 30 pounds overweight is actually only 0.55 year, whereas the people 30 pounds or more overweight have the much greater loss of life expectancy of 2.5 years. Interpreting data always requires us to think carefully about what the numbers mean. This analysis suggests the wisdom of indulging only in light desserts.

Personal Risk

As individuals, we don't have the authority to alter government policies or priorities. But in a democracy, we are each responsible to be informed citizens and to vote and lobby for wise choices. Understanding how to evaluate risk makes us better citizens. But for issues of personal choice, the effect of understanding risk is much more immediate.

In our personal lives we can choose which risks we'll avoid and which we'll take. Weight, exercise, sex, drugs, smoking, and investments—these are all arenas within which we make choices. Of course, it is easy to say that choosing the less-risky alternatives in each category is the best thing we can do, but in reality we will not be perfect. We can attempt to get some quantitative sense of the consequences of various personal life choices so that we can choose which ones are worth the risk. Not every smoker dies young, nor does every obese person. Choosing to quit smoking or lose those excess pounds does not guarantee anything. We can say for certain that more smokers than nonsmokers will get lung cancer, but we can't say that a specific individual, Naomi Nicotine, will get lung cancer if she continues to smoke. We know that a person who is 30 pounds overweight lives 2.5 years less, on average, than someone who is not, but we cannot say that "Fast Food" Filbert, who weighs 100 pounds too much, will himself die years sooner than he would if he were slimmer. For one thing, the smoker and the weighty one could be run over by a drunk driver before they have a chance to die of lung cancer or clogged arteries.

Making better decisions about life circumstances typically alters the probability of life success, but it's not a guarantee. However, we know from our study of expected value in the previous section that people who act in opposition to the expected value analysis do so at their peril. The strategy most likely to result in longer life and greater success is to analyze the likelihood of various outcomes and take actions that tend to lower the probability of calamity.

Changing Risk

The examples given above in the loss-of-life-expectancy list of personal risk factors give us some quantitative information that is helpful for our decision making. Those statistical averages indicate the relative riskiness of various activities. When we make life choices, we can look at these data and get a sense of which behaviors are most significant. All the above examples are quite significant in terms of lessening life expectancy. Airplane crashes, on the other hand, are negligible risk factors. So if we want to live a long life, giving up smoking is far more significant than giving up flying.

The Thrill of It All—Risk as Arousal

Risk is not always a bad thing. Many people crave the adrenaline rush that comes, for example, from hanging by their fingertips from a steep rock face,

©Scott Hailstone/iStockphoto

gambling in a high-stakes game in Las Vegas, or indulging in unsafe sex. This desire may stem from the need for a chemical change in the brain linked with the neurotransmitter serotonin. According to this line of reasoning, the compulsive gambler gets the same rush whether he wins or loses—it's the risky behavior acting on the brain's chemistry that produces the "high."

Sometimes the actual risk is not as high as our intuition tells us it is. Rock climbing, for example, is not nearly as dangerous as it appears. Sure, that climber is dangling off a thin ledge hundreds of feet from the ground below, but, in fact, very few people die from this endeavor (climbers can, however, get pretty dinged up and sometimes break bones). The point is that from the vantage point of the ledge with our fingertips clenching on for dear digits, we perceive this activity as riskier than it actually is—thus we get the rush. Many people misjudge risk, whether it is the not-so-risky act of rock climbing or the riskier act of indulging in unsafe sex. The point remains that some people crave risky behavior as a means of arousal. The prudent arousal seeker can improve her chances of living to risk another day by choosing activities that *appear* riskier than they actually are.

A Look BACK

EVERY DECISION WE MAKE is obtained at some cost that can be viewed as risk. How much risk we can tolerate is a function of the particular activity, our individual tolerance for risk, and our value systems about what is important to us (skydiving versus never leaving the library). While many events are random and outcomes are always uncertain, we are not just drifting down the river of life. In fact, we have a paddle and can steer and alter our course. Will we hit those rocks? Perhaps we will. But our actions and decisions can alter the likelihood of unfortunate and unforeseen obstacles. Coolly assessing the likelihood and severity of the various possible consequences of a decision gives us a huge advantage in our attempts to make life-enhancing choices.

If we can quantify an issue or examine data, great; however, in many cases of personal decision making, we don't have complete information. Nevertheless, we can think qualitatively and carefully and make decisions that will tend to lessen the probabilities of disaster. Whether taking a quantitative or qualitative approach to a question, elements common to most good decision making are clear thinking and careful weighing of potential consequences.

Life Lessons

Make it quantitative.

•

Beware of unintended consequences.

MINDSCAPES   Invitations to Further Thought

*In this section, Mindscape marked (**H**) have hints for solutions at the back of the book. Mindscapes marked (**ExH**) have expanded hints at the back of the book. Mindscapes marked (**S**) have solutions.*

Developing Ideas

1. **Remarkably risky.** List two activities that are more risky than they may first appear.

2. **Surprisingly safe.** List two activities that are less risky than they may first appear.

3. **Infectious numbers (H).** Suppose a disease is expected to infect 2% of those living in the United States. If the U.S. population is 280,000,000, how many people could we expect to become infected with this disease?

4. **SARS scars (S).** Suppose a new vaccine that prevents the SARS (Severe Acute Respiratory Syndrome) virus is discovered. Each injection costs $10 and is expected to save 1 in 15,000,000 lives. What is the total cost per life saved for this vaccine? Would you support a government-sponsored program to vaccinate each citizen? Would your response change if a huge epidemic of SARS broke out and the injections would save 1 in 1000 lives? Explain why.

5. **A hairy pot.** At a certain famous school of witchcraft and wizardry, drinking a certain potion from a fur-covered pot will cause instant death (and, even worse, expulsion from the school) 7% of the time. What is the loss of life expectancy associated with taking a swig from the old hairy "potter" for a 12-year-old student who would otherwise expect to live to the age of 78?

Solidifying Ideas

6. **Blonde, bleached blonde (H).** You have high standards with respect to truth in advertising, particularly when it comes to hair color. One day at the laundromat, you meet an attractive blonde stranger named Chris and wonder if you should pursue a relationship. Unfortunately, you have the nagging belief that Chris's golden locks may have been the result of peroxide—presenting the specter of a dark (haired) future. However, you also know several facts about the incidence of dyed hair and about your ability to detect fraudulent follicles. You know that 90% of blonde people in the world are naturally blonde. You have done a personal survey and learned that you are 80% accurate in your ability to correctly categorize fake hair color as fake and real hair color as real. What is the probability that Chris's hair is fair and that your bleached beliefs were incorrect? Given these facts, should you pursue your relationship with Chris?

7. **Blonde again (S).** Given the scenario in Mindscape 6, now suppose that 70% of blonde people are naturally blonde and that you are 85% accurate in your ability to correctly categorize fake hair color as fake and real hair color as real. What is the probability that Chris's hair is fair and that your bleached beliefs were incorrect? Given these facts, should you pursue your relationship with Chris?

8. **Bleached again.** Given the scenario in Mindscape 6, now suppose that 80% of blonde people are naturally blonde and that you are 50% accurate in your ability to correctly categorize fake hair color as fake and real hair color as real. What is the probability that Chris's hair is fair and that your bleached beliefs were incorrect? Given these facts, should you pursue your relationship with Chris?

9. **Safety first.** Suppose a particular car is widely believed to be the safest car made. You might expect people driving this car to be less severely injured in accidents, but why would you expect that model of car to be involved in fewer accidents per million miles than a sports car?

10. **Scholarship winner (ExH).** You apply for a national scholarship along with 100,000 other students; 200 scholarships will be awarded. You call to inquire whether you are one of the winners. The absent-minded professor in charge of the program reports that the letters have just been sent out, but no list of all the winners is available; however, the professor recalls that you were selected. The professor correctly recalls information 90% of the time. Assuming that the scholarships are awarded randomly (not a completely ridiculous assumption), what is the probability that you won one?

11. **Less safe (ExH).** Given the scenario in our air safety discussion earlier in this section, now suppose that if planes were made only 5 times safer, then airplane ticket prices would rise less than before, and thus only 1% of travelers who now fly would choose to drive instead. Assuming all the other data still hold from our discussion, what is the net result on lives lost if we make the planes safer?

12. **Aw, nuts!** Suppose that the loss of life expectancy for eating at least 1 tablespoon of peanut butter a day is 1.1 days, whereas the loss of life expectancy for eating at least 2 tablespoons of peanut butter a day is 3 days. In a group of 100 people where 75 eat at least 1 tablespoon of peanut butter a day and the remaining 25 people eat at least 2 tablespoons a day, what is the loss of life expectancy in this group for eating 1 to 2 tablespoons of peanut butter a day?

13. **Don't cell! (H)** Suppose you are a U.S. senator and a lobbyist approaches you to report the fantastic news that the group she represents has just discovered a device that will block reception for cell phones in automobiles and thus make driving much safer. The device would cost a mere $100 per vehicle, and your lobbyist wants you to offer a bill that would subsidize the expense. What additional information would you need in order to determine the cost per life saved of this gizmo?

14. **Risk to order.** Select two risky activities you undertake and analyze, as best you can, their risk level, using one of the measures described in this section. Explain your analysis. Might your findings change your behavior?

15. **Buy low and cell high (H).** The microwaves produced at cell-phone antennas could result in brain tumors, cancer, and, even worse, hair loss. Suppose that 1 in 100,000 cell-phone users will die due to cell-phone use. The government will only fund safety features for which the cost per life saved is $12,500. How much is the government willing to spend on increasing the safety of each cell phone?

Creating New Ideas

16. **Taxi blues (H).** An eyewitness observes a hit-and-run taxicab accident in a city in which 85% of the cabs are green and 15% are blue. The witness is 100% certain that the cab was blue. Given all this information, how likely is it that the cab actually was blue?

17. **More taxi blues (S).** An eyewitness observes a hit-and-run taxicab accident in a city in which 85% of the cabs are green and 15% are blue. The witness is 80% sure that the cab was blue. Given all this information, how likely is it that the cab actually was blue?

18. **Few blues.** An eyewitness observes a hit-and-run taxicab accident in a city in which 95% of the cabs are green and 5% are blue. The witness is 80% sure that the cab was blue. Given all this information, how likely is it that the cab actually was blue?

19. **More safety.** Given the scenario of our earlier air safety discussion, suppose that planes could be made 10 times safer with government support, so that airplane ticket prices would rise less and thus only 5% of travelers who now fly would choose to drive instead. Assuming that all the other data still hold from our discussion, what is the net result on lives lost if we make planes safer?

20. **Reduced safety.** Given the scenario of our air safety discussion, suppose that the FAA reduced many of its safety requirements, and air travel safety dropped by 5% (that is, 5% more people would be killed in air travel). However, the airlines passed this savings on to the consumers, and lower ticket prices resulted in a 15% increase in air travel and a likely reduction in travel by car. Assuming all the other data still hold from our discussion, what is the net result on lives lost if we make the planes less safe?

Further Challenges

21. **HIV tests.** Recall that, in the United States, approximately 1 person in 300 is estimated to be HIV-positive. Suppose a person decides to take two independent tests. Test *A* determines whether *or not* a person is infected with HIV with 95% accuracy, whereas test *B* is 99% accurate. Suppose this person learns that both tests came back positive—that is, both predicted that the person was infected with HIV. What is the probability that this person is actually infected?

22. **More HIV tests.** Given the tests described in the previous Mindscape, assume now that test *A* came back positive and test *B* came back negative. What is the probability that this person is actually infected?

In Your Own Words

23. **Personal perspectives.** Write a short essay describing the most interesting or surprising discovery you made in exploring the material in this section. If any material seemed puzzling or even unbelievable, address that as well. Explain why you chose the topics you did. Finally, comment on the aesthetics of the mathematics and ideas in this section.

24. **With a group of folks.** In a small group, discuss and actively work through the reasoning involved in HIV/AIDS testing and false positives. After your discussion, write a brief narrative describing the reasoning in your own words.

For the Algebra Lover

Here we celebrate the power of algebra as a powerful way of finding unknown quantities by naming them, of expressing infinitely many relationships and connections clearly and succinctly, and of uncovering pattern and structure.

25. **Super sale.** The bookstore is having a super sale on sweatshirts emblazoned with the school mascot, Mr. Woof. The sweatshirts were on sale last week for 30% off the original price. Now they've been marked down another 30% *off the sale price.* If the original price was $50, what's the super sale price?

26. **Wisk risk (H).** You always sort your laundry into dark clothes and light clothes. Each time, there is a 20% chance that a red shirt gets in with your light load, turning all your white underwear pink. If you do laundry five times a semester, what are the chances that your tighty whities are still white for final exams?

27. **Bag for life.** An insurance company estimates that an air bag in a car reduces driver deaths at a rate of 1.4 deaths for every 100 million miles driven. Suppose the average car is driven 100,000 miles. Calculate the probability that an airbag in such a car would save a life. If that airbag cost $400, calculate the cost per life saved.

28. **Mooving sale.** Plush toy versions of your college mascot, the purple cow, are very popular items sold at the bookstore. The manager decides to order some cows in a size 50% larger than the original. They are so popular with alumni that she orders cows in an even larger size, two-and-a-half times larger than the original. If the three sizes are designated original, large, and extra large, how much larger (as a percentage) is the extra large cow than the large cow?

29. **Reweighing life expectancy.** An example in this section (page 786) recalculates the loss of life expectancy for people more than 15 pounds overweight by taking into account the effect that even heavier people (those more than 30 pounds overweight) have on the loss of life expectancy. The text example considers the case of 300 people, 200 of whom are at least 15 pounds overweight but less than 30 pounds overweight, and 100 of whom are at least 30 pounds overweight. Replicate this example assuming you again have 300 people; however, this time only 100 are at least 15 pounds overweight but less than 30 pounds overweight, and the remaining 200 people are at least 30 pounds overweight. For this new sample, what's the loss of life expectancy for those who are at least 15 pounds overweight but less than 30 pounds overweight? Explain the difference between your example and the example from the text.

MONEY MATTERS

Deciding Between Faring Well and Welfare

© Antagain/iStockphoto

> *Lack of money is the root of all evil.*
>

Money, with its tendency to dribble in and drain out, looms large (or, unfortunately more often, small) in our daily lives. Money is an area where lessons from *The Heart of Mathematics* definitely generate high returns. Nearly all issues about money and its growth and decline involve repeating processes—in particular, most are variations on the straightforward example of money growing through the repeated accumulation of interest. We described this process in Section 7.5, recalled below.

Money

Suppose we deposit $1000 in a savings account that pays 5% interest compounded annually. After one year, our account will have a balance of $1000 plus 5% of $1000, which equals $1.05 \times \$1000$ or $1050. If we leave the money in the account for another year, we will accrue 5% interest on the new balance. So after two years we will have $1.05 \times \$1050 = \$1102.50$. If we leave our money in the account yet another year, we will earn 5% on our new total, so after three years we will have $1.05 \times \$1102.50 = \$1157.63$. The pattern is clear—not exciting, but clear. Take what we have at the beginning of the year, multiply that number by 1.05, and that gives us the amount of money we will end up with after one more year. We can just sit back and watch our money grow. Simple enough.

Who Wants to Be a Millionaire?

One lesson we have learned in this book is that the appropriate representation of resolving an issue is helpful in describing and reasoning about it. In the case of money matters, a good way to keep track of the reasoning is first to think through the ideas in words and then abbreviate those sentences using symbols that are easier

Before   After

Life Lesson

Find Patterns.

to keep track of. Also, when dealing with money issues, computations that measure what is happening can bring the abstract reasoning down to Earth. Using the symbols (that is, algebra) and doing computations are valuable tools that will help us plan our financial futures.

For example, the pattern of growth described in the previous scenario shows that the amount of money in the account after n years can be expressed as $1.05^n \times \$1000$. This happy example shows an easy way for us all to become millionaires. We could simply invest $1000 in a bank account that pays 5% interest compounded annually, live 142 more years, and realize that our account would then have $1.05^{142} \times \$1000$. That amount equals \$1,020,658.53. We are millionaires! Now we can go out and have some real fun, or at least as much fun as possible considering that we will have been dead for more than 50 years.

We do need to point out that this enormous amount of money is a bit deceiving. Suppose that during the same 142 years, while we are watching our money grow, inflation is growing at 5% per year. Then the refrigerator that we can buy today for about $1000 would cost about \$1,000,000, so maybe we shouldn't be too excited about being millionaires after all. Of course, on the bright side, we will be able to sit around the old folks' home and say, "When I was a young whippersnapper, that there hamburg' only cost a buck. Now the Golden Arches charges $1000. 'Happy Meal,' my patootie. What's this world coming to?"

Compounding More Frequently

Instead of compounding interest once each year, savings accounts often compound the interest more frequently. Suppose a savings account compounds the interest each quarter, that is, every three months. Let's see what effect this has on the growth of the account. As usual, we are really just doing an iterative, repeating process. What is the process in this case?

Well, let's begin by depositing $1000 into an account paying an annual interest rate of 5%, but compounded quarterly. In the first quarter of the year, the account will have earned one-quarter of 5% in interest. That is, the account will have $1000 + \$1000 \,(0.05/4)$, or

$1000 + \$1000 \times 0.0125$ or 1000×1.0125, which equals $1012.50.

During the next quarter, that entire balance of $1012.50 will earn another $0.05/4$—in other words, 1.25%—interest. So after two quarters (half of a year), the account will have

$1012.50 \times 1.0125 = \$1000 \times 1.0125^2 = \$1025.16$.

After three quarters, it will have

$1025.16 \times 1.0125 = \$1000 \times 1.0125^3 = \$1037.97$.

And after the final quarter of the year it will have

$$\$1037.97 \times 1.0125 = \$1000 \times 1.0125^4 = \$1050.95.$$

After a year, the quarterly compounding has added an additional 95¢ to our swelling pocketbooks over and above what we would have earned with just annual compounding.

The annual percentage rate (APR) equivalent to this 5% compounded quarterly is an effective rate of 5.095%, meaning that earning 5.095% interest compounded *annually* will give the same rate of return as 5% compounded *quarterly*.

The net effect of quarterly compounding is that we will become millionaires sooner than in our previous example. Instead of waiting 142 years to become a millionaire, we will find that after merely 140 years our $1000 compounded quarterly at 5% has grown to $1000 \times 1.0125^{4(140)}$, which equals $1,050,069.05. Let's party like it's 2099! Well, you are probably thinking that your party days will be pretty much behind you even at the ripe age of 140, but this is a step in the right direction.

A Compounding Pattern

At this point, we can see the pattern of how to compute compound interest. If we deposit $1000 into an account that pays 5% compounded quarterly, we will have $\$1000 \times (1 + 0.05/4)^4 = \$1050.95$ at the end of the year. Following this pattern, we see that if we deposit $1000 into an account that pays 5% compounded monthly, then we will have $\$1000 \times (1 + 0.05/12)^{12}$ at the end of the year. Similarly, we see that if we deposit $1000 into an account that pays 5% compounded daily, then we will have $\$1000 \times (1 + 0.05/365)^{365}$ at the end of the year. More frequent compounding makes some difference, but nothing dramatic. In this example, in which our 5% interest account is compounded every day, our $1000 will grow to be $\$1000 \times (1 + 0.05/365)^{365}$ at the end of the year, which equals $1051.27—not that much bigger than the $1050 we would have with no compounding during the year.

So the annual percentage rate equivalent to 5% compounded daily is 5.127%, meaning that earning 5.127% interest compounded annually will give the same rate of return as 5% compounded daily.

Of course, if we decided to let our $1000 ride at 5% compounded daily for 139 years, then there would be $365 \times 139 = 50,735$ intervals for which interest will be paid. Thus, we would have $\$1000 \times (1 + 0.05/365)^{50,735}$, which equals $1,042,653.33. So, with daily compounding, our $1000 initial investment would make us millionaires a whopping three years earlier than if we had just annual compounding.

Our compounding pattern can be summed up in a simple observation: If we start with a certain amount of money (let's call it our INITIAL AMOUNT), we have an interval of time between compounding, we earn a certain amount of interest during that interval (let's call that our INTEREST PER INTERVAL), and we hold the money for a certain number of such intervals of time (let's call that number our NUMBER OF INTERVALS), then at the end of those intervals of time we will have:

FINAL AMOUNT = INITIAL AMOUNT
$\times (1 +$ INTEREST PER INTERVAL$)^{\text{NUMBER OF INTERVALS}}$.

To illustrate the applicability of the previous relationship, let's consider a general banking situation. Let's suppose a bank offers an annual interest rate of r% compounded t times a year—so the INTEREST PER INTERVAL would be equal to $\frac{r/100}{t}$. If we kept our INITIAL AMOUNT in that account for y years, then our NUMBER OF INTERVALS would be $(t)(y)$, and at the end of that time we would have a grand total of

$$\text{FINAL AMOUNT} = (\text{INITIAL AMOUNT}) \times \left(1 + \frac{r}{100t}\right)^{ty}$$

in our account.

Life Lesson
Summarize insights.

2 Versus 3—The Difference that One Percentage Point Makes

Many people think that 2 and 3 are pretty much the same. But there is a world of difference between them, especially if we are investing over the long haul. According to some authorities, Adam and Eve were born in 4004 BCE. Suppose at that time Adam invested 1¢ in an account that paid 2% interest compounded annually, while Eve, who was more financially savvy, invested 1¢ in an account that paid 3% interest compounded annually. How much would each of their accounts have after 1000 years? Using our trusty Excel worksheet or computing $(0.01)(1.02)^{1000}$, we find that Adam's account balance would be

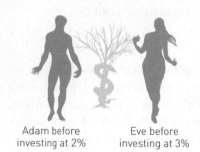

Adam before investing at 2%   Eve before investing at 3%

$3,982,646.52.

Not bad for good old Adam. On the other hand, Eve's balance, $0.01(1.03)^{1000}$, would be approximately

$68,742,402,310.

Adam after 1,000 years   Eve after 1,000 years

Adam is a millionaire, while Eve has over 68 billion dollars—putting her on par with Bill Gates. So Eve's account has more than $17,000 for every dollar that Adam's has. Eve considers opening up the very first business school.

You may be curious to know how much Adam and Eve would have in their respective accounts today. Here are their current balances, about 6000 years after investing their 1¢ at 2% and 3%, respectively. Adam would have about

$44,939,800,000,000,000,000,000,000,000,000,000,000,000,000,

whereas Eve would have roughly

$12,600,000,000,000,000,000,000,000,000,000,000,000,000,000,000,
000,000,000,000,000,000.

Not bad for a 1¢ investment. To put these enormous numbers in perspective, for every dollar Adam's account earned, Eve's would have earned about $28,037,5 00,000,000,000,000,000,000. In fact, Eve would have more dollars than there are atoms in the universe. You go, Eve!

Lottery

The lottery is society's way of taxing people who are bad at math. Nevertheless, when the jackpot has risen so high that buying a lottery ticket is a prudent investment, we are faced with a pressing issue of financial wisdom. If we win, should we take the cash-now option, or the multiyear payout? In December 2002, Jack Whittaker of West Virginia won the $315 million Powerball jackpot. He could have received the entire $315 million in equal yearly installments over 29 years; however, instead, he selected to receive one lump sum of a mere $170 million. Was that decision a wise one? Let's compute how much money each scenario will produce over 29 years if both sums are put in a savings account earning 5% per year, compounded annually.

The 29-year annual payment method would be worth $676,953,595 in 29 years. On the other hand, putting $170 million in a savings account earning 5% per year would yield $699,743,051 in 29 years. So if Jack can earn 5% interest annually, he made a wise decision. Of course, these calculations do not include many confounding factors, such as taxes, inflation, and long-lost friends and relatives who will no doubt come out of the woodwork and want their cut.

Saving for College

Birth is one of the miracles of life. Frequently, this supreme moment takes place in a delivery room or taxicab. The crying newborn has just emerged into the world. Mother and father are awestruck. The doctor or cab driver says, "Congratulations, it's a girl." And the banker says, "You'll need $100,000 in 18 years to pay for this kid's college education." How bankers get into delivery rooms or cabs we do not know, but they make good sense.

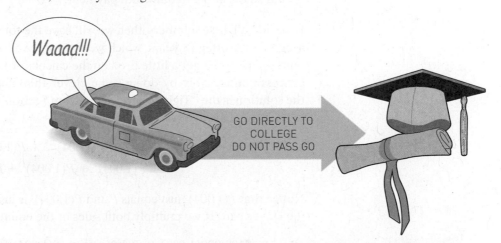

Fortunately, the banker has the answer and comes armed with many forms for us to sign. Before we arrive at the recovery room, we have agreed to pay a fixed amount of money each month into an account earning 5% annual interest compounded monthly so that in 18 years young Cabriel (Cabby for short) will have the $100,000 necessary to arrive onto the college party scene. The big question, of course, is:

How Much Must We Pay Each Month?

Luckily, we know exactly how to think about this question because we know how to calculate how much our first payment will have grown to in 18 years (that is, $18 \times 12 = 216$ months); we know how much our second payment will have grown to in the 17 years and 11 months that it will be accumulating interest; we know how much our third payment will have grown to in the 17 years and 10 months that it will be accumulating interest, and so forth. All we need to do is add them up. So, if our monthly payment is represented by the letter P, then:

18 years from now, our 1st payment will have grown to $P\left(1 + \dfrac{0.05}{12}\right)^{216}$;

18 years from now, our 2nd payment will have grown to $P\left(1 + \dfrac{0.05}{12}\right)^{215}$;

18 years from now, our 3rd payment will have grown to $P\left(1 + \dfrac{0.05}{12}\right)^{214}$;

18 years from now, our 4th payment will have grown to $P\left(1 + \dfrac{0.05}{12}\right)^{213}$;

$$\vdots$$

and so on, down to our 216th payment.

If we add all these numbers, then we will have the total amount of money we have accumulated after 18 years, which we want to be $100,000. You might think that our fingers would get a little tired on the calculator keys, but we can simplify the process considerably by doing some clever thinking. Actually, we already saw the solution in the "Tight Weave" story from Section 1.3, "Fun and Games." Let's look at the sum we are trying to compute:

$$\$100,000 = P(1.004)^{216} + P(1.004)^{215} + P(1.004)^{214}$$
$$+ \cdots + P(1.004)^2 + P(1.004)^1 + P(1.004)^0.$$

Notice that $P(1.004)^0$ just equals P and $P(1.004)^1$ is just $P(1.004)$. Now here comes the clever part. If we multiply both sides of the equation by 1.004, we get:

$$\$100,000(1.004) = P(1.004)^{217} + P(1.004)^{216} + P(1.004)^{215}$$
$$+ \cdots + P(1.004)^2 + P(1.004).$$

The clever idea is now to line up like terms from these two equations and subtract:

$$\$100,000(1.004) = P(1.004)^{217} + P(1.004)^{216} + P(1.004)^{215}$$
$$+ \cdots + P(1.004)^2 + P(1.004)$$
$$\$100,000 = P(1.004)^{216} + P(1.004)^{215}$$
$$+ \cdots + P(1.004)^2 + P(1.004) + P$$

After subtracting, the result is:

$$\$100,000(1.004) - \$100,000 = P(1.004)^{217} - P.$$

Notice how almost all the terms on the right side of the equals signs were subtracted and just disappeared. If we factor out the $100,000 on the left side and the P on the right side, we see that:

$$\$100,000(1.004 - 1) = P(1.004^{217} - 1).$$

Dividing both sides by the number $1.004^{217} - 1$ gives us

$$\frac{\$100,000(1.004 - 1)}{1.004^{217} - 1} = P.$$

If we plug those numbers into a calculator, then we see that our monthly payment must be $290.27.

Paying each month into a savings account is called an *annuity*. Children causing big expenses is *life*, and the power of mathematical thinking: *priceless*.

Draining Our Lifeblood

The previous examples all involved pouring our money into an account and watching it grow, but in real life we frequently watch our money disappear. Often money happily pours in when we borrow and, sadly, disappears when we have to pay it back. Unfortunately, paying the money back is part of the deal of a loan or mortgage, as loans are often called. This process is deadly.

Morticians bury dead people. *Mortuaries* are where dead people are praised before they are buried. *Mortality* means death is in our future. There's a pattern here. *Mort* means *death*, and words with mort in them are often deadly. Among the deadliest are *mortgage* and *amortize*. Mortgage comes from the Latin meaning a "death pledge." To amortize a debt means to "kill the debt." It all sounds bleak, but we must all face the grim reaper, also known as the mortgage broker.

To understand loans and mortgages, interest, and payments, we adopt the effective thinking strategy that we have encountered so frequently in this book, namely, we start with simple cases, understand them deeply, and then apply the ideas we have learned to more complicated situations.

A One-Payment Loan

To understand how loans and mortgages work, let's visit the simplest possible place to borrow money: the One-Day-a-Year Savings and Loan Company. This

_Life Lesson_____

Understand simple cases.

unique S&L is open just one day a year, February 2, Groundhog Day. We walk into the One-Day-a-Year S&L on February 2 and ask to borrow $1000 for one year. The interest rate is 12%. We sign the papers and walk out with their (now our) money. Next Groundhog Day, we return to the One-Day-a-Year S&L and pay back the $1000 plus 12% interest, which is $120, for a total of $1120. We had the $1000 for a year, the One-Day-a-Year Savings and Loan Company worked only one day and earned $120, and everybody is happy as a hog (a groundhog, that is).

A Two-Payment Loan

The previous example was very simple, but it gets us started in the right direction. Let's now consider the next easiest loan, which is offered by the Twice-a-Year Savings and Loan Company. This S&L is open an incredible two days during the year, February 2 (Groundhog Day) and August 2 (the day in 1909 that the first Lincoln penny was minted, when the S&L's slogan was "A penny for your loans"): We go to the Twice-a-Year S&L on February 2 and borrow $1000 for one year. The interest rate is 12%. Here are the terms of the loan: We must pay back the money in two *equal* installments, one on August 2 and the final payment on the next Groundhog Day. The big question is:

How Much Is Each Payment?

Our first guess might be $500. This would be fine, except for the interest. Because one year's interest on $1000 at 12% is $120, we might guess that we should pay $60 of interest in each payment, that is, we would pay $560 on August 2 and $560 on February 2. Unfortunately, this reasonable-sounding guess is not correct. It is true that on August 2, we will owe $60 in interest on the $1000 we borrowed on February 2. So if we pay $560 on August 2, we will have paid all the interest that we owe up to that point and we will have paid $500 of the $1000 back. But now we would have only $500 left of the original $1000 that we borrowed. So during the next half-year, we would pay interest on that $500; we would owe only $30 in interest for the second half of the year. Our final payment would be $530. Well, those payments are not equal, and the terms called for equal payments. So how can we figure out exactly how much to pay?

We will have to resort to a bit of thought and keep track of it with a little algebra. To begin, we'll let P represent the payment that we need to make. On August 2, we will owe $1000 $(1 + \frac{0.12}{2}) =$ $1000 (1.06) and we will pay P dollars, which will leave an outstanding balance of $1000(1.06) - P$.

On February 2, when our next payment is due, we will owe that amount, $1000(1.06) - P$, plus 12%/2 = 6% interest on it.

So we will owe $[\$1000(1.06) - P](1.06)$. But that amount must equal our first payment, so we have an equation:

$$[\$1000(1.06) - P](1.06) = P \text{ or,}$$

$$\$1000(1.06)^2 - P(1.06) = P \text{ or,}$$

$$\$1000(1.06)^2 = P(1.06) + P \text{ or,}$$

$$\$1000(1.06)^2 = P(1.06 + 1) = P(2.06) \text{ or,}$$

$$P = \frac{\$1000(1.06)^2}{2.06}.$$

Three-Times-a-Year Savings and Loan Company

"Why hog all your money when you can gamble on us for peanuts?"

You can bank on them!

Using a calculator to evaluate P, we see that our payments on August 2 and on February 2 must be $545.44.

Digging into the Mortgage Pattern

We are beginning to see a pattern, but let's consider one more example before we figure out the mortgage pattern in general. So, you've guessed it, we will now visit the Three-Times-a-Year Savings and Loan Company. This S&L is open three days during the year, February 2 (Groundhog Day), June 2 (the date New Jersey legalized gambling in Atlantic City in 1977), and October 2 (when the comic strip "Peanuts" first appeared in 1950). This S&L's slogan is: "Why hog all your money when you can gamble on us for peanuts?" Despite their slogan, we enter the Three-Times-a-Year S&L on February 2 and borrow $1000 for one year at an interest rate of 12% under the following terms: We must pay back the money in three *equal* installments, one on June 2, one on October 2, and the final payment on the next Groundhog Day. So once again:

How Much Is Each Payment?

Again, we will resort to some thought and one equation. If we call the payment we need to make P, here is what will happen:

On June 2, we will owe $1000(1+0.12/3) = \$1000(1.04)$ and we will pay P dollars, leaving a balance of $\$1000(1.04) - P$ that we still owe.

On October 2, we will owe $[\$1000(1.04) - P](1.04)$ and we will pay P dollars, leaving a balance of $[\$1000(1.04) - P](1.04) - P$ that we still owe.

On February 2, we will owe that balance plus $\frac{12\%}{3} = 4\%$ interest on it, so we will owe

$$([\$1000(1.04) - P](1.04) - P)(1.04).$$

But that amount must equal our last payment! Because all the payments are equal to P, we have:

$([\$1000(1.04) - P](1.04) - P)(1.04) = P$ or, eliminating the brackets

$(\$1000(1.04)^2 - P(1.04) - P)(1.04) = P$ or, getting rid of a pair of parentheses,

$\$1000(1.04)^3 - P(1.04)^2 - P(1.04) = P$ or

$\$1000(1.04)^3 = P(1.04)^2 + P(1.04) + P = P(1.04^2 + 1.04 + 1).$

So,

$$P = \frac{\$1000(1.04)^3}{1.04^2 + 1.04 + 1}$$

Using a calculator, we find that our payments on June 2, October 2, and on February 2 must be $360.35. But better than that, perhaps we can see the pattern now.

Life Lesson

Follow through in pursuing patterns.

The Real World Savings and Loan

Suppose we buy a house and borrow $100,000 at 9% interest for a 30-year loan (the mortgage). The mortgage is to be repaid with *equal* monthly payments. How much will our monthly payment be?

The number of payments will be 30 years times 12 months per year, which equals 360 payments. The interest per month will be the 9% annual interest rate divided by the 12 months in a year, that is, $0.09/12 = 0.0075$. If we call our payment P, we can skip down to our last equation from the preceding example and, following the pattern that we learned there, we see that:

$$\$100,000(1.0075)^{360} = P(1.0075)^{359} + P(1.0075)^{358} + P(1.0075)^{357}$$
$$+ \cdots + P(1.0075)^3 + P(1.0075)^2 + P(1.0075) + P.$$

To actually solve for P, we use the subtraction trick that we saw previously. Namely, we first multiply both sides of the equation by 1.0075 to get:

$$\$100,000(1.0075)^{361} = P(1.0075)^{360} + P(1.0075)^{359} + P(1.0075)^{358}$$
$$+ \cdots + P(1.0075)^3 + P(1.0075)^2 + P(1.0075).$$

Subtracting, we get:.

$$\$100,000\,(1.0075)^{361} - \$100,000\,(1.0075)^{360} = P\,(1.0075)^{360} - P.$$

Factoring out $\$100,000(1.0075)^{360}$ on the left side of the equals sign and P on the right side gives:

$$\$100,000(1.0075)^{360}(1.0075 - 1) = P[(1.0075)^{360} - 1] \text{ or}$$

$$\frac{\$100,000(1.0075)^{360}(1.0075 - 1)}{[(1.0075)^{360} - 1]} = P.$$

Plugging those numbers into a calculator reveals that the monthly payment will be $804.62.

Double Death—Pass the Cigarettes and Pass the Wallet

Smoking is harmful to many parts of one's body, especially the pocketbook. Let's consider Pat and Chris, two college students who seek fame and fortune. Pat is a chain-smoker, and Chris is a chain-saver. Each spends $5 per day on their respective habits. Pat invests $5 daily in cigarettes, while Chris invests $5 daily into a savings account that earns 5% per year, compounded daily. With appropriate lung substitutes for Pat, both Pat and Chris attend their 50th reunion and compare life experiences. Pat, of course, has no savings, but does have attractive yellow teeth to complement two blackened lungs. Chris has a savings account.

How Much Is In Chris's Savings Account?

To find the answer, we proceed this time as a person does in real life. That is, we simply plug the numbers into the appropriate Excel function (in this case it is called FV, for Future Value—*not* Filtered Virginia Slims) and learn that Chris's account contains $408,084.90.

WOW! Chris buys a pearly white Mercedes to match his pearly white smile, takes friends on trips around the world, and buys a cute mountain cabin in Aspen.

Stocking Up on Other Investments

Of course, we all know that there are many other investment opportunities besides savings accounts. There are stocks, bonds, mutual funds, real estate, gold, and even a guy who'll be willing to sell you the Brooklyn Bridge. All of these potential money-makers can be compared and measured by considering the rate of return on the investment.

NYSE TICKER: HoM

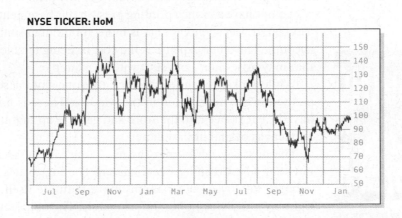

As an illustration, suppose that on January 1, 2003, we purchased 100 shares of HoM stock at $50 per share for a total investment of $5000. In September 2003, HoM sent us a dividend check for $185, which we cashed and stored under the mattress. In January 2004, HoM had a 2-for-1 stock split, which meant that instead of 100 shares, we then held 200 shares. In September 2004, we received a dividend check for $97.50. On January 1, 2005, we sold all 200 shares at the current trading price of $26.15 per share. So how did we fare? Well, from the sale of the shares we received 200 × $26.15 = $5230. But we also received two dividend checks totaling $282.50. Thus, after two years, our $5000 turned into $5512.50. So, what was our average rate of return?

Rather than answering this question directly, let's look at it from a different angle. Suppose we deposited our $5000 in a savings account that paid interest annually and two years later our investment had grown to $5512.50. What interest rate was paid on this account? Using the formula we found in the previous examples, we see that if the account paid $r\%$ (that is, $r/100$ or $0.01r$) interest compounded annually, then after two years we would have $\$5000(1 + 0.01r)^2$, which we know is equal to $5512.50. Thus, if we set these amounts equal to each other we have

$$\$5000(1 + 0.01r)^2 = \$5512.50.$$

If we divide both sides by $5000, then we have

$$(1 + 0.01r)^2 = 1.1025,$$

which, after taking the square root of both sides, becomes

$$1 + 0.01r = 1.05.$$

Solving for r, we see that $r = 5$, that is, we were earning 5% interest. Finding the average rate of return allows us to compare different investment possibilities and see which ones yield the highest returns. Of course some investments are riskier than others, but purchasing HoM is never a bad investment—by the way, HoM, of course, stands for *The Heart of Mathematics*.

Opportunity Costs

After we spend all our money, we don't have any left to spend, unless, of course, we are the government, in which case we simply print some more. But assuming that we do not have a money printing press in our basement, there is an extra expense associated with spending that is more subtle than simply depleting the old bank account. The other expense is the cost of being unable to embrace opportunities that may arise later. Let's start with an example.

tonite only GRINDGORERECORDS presents
CRUSHING CRANIUMS
w/ DEERHOOF & ERASE ERRATA
@cacaphoney 1515 Ave D doors at 10 show at 11

While walking by the electronics store one day, Pat sees a portable DVD player painted bright neon green in the window. There is no doubt about it; Pat must have it. Pat rushes into the store and gathers up that treasure (and a few DVDs) and leaves behind his last dime.

The next day, while enjoying his *Porky's XI* DVD, a friend runs in to inform Pat that The Crushing Craniums are coming to town for a once-in-a-lifetime concert. Pat loves the Craniums more than life itself. In fact, Pat would give a right arm for a ticket, but unfortunately, Ticketmaster takes only cash. What to do? Pat has spent every penny for the DVD player. So the Craniums must crush without Pat.

However, on the bright side, Pat has learned a valuable lesson about *opportunity costs*. Having spent his money on the DVD player, he was no longer in a position to buy concert tickets. In other words, the cost of the DVD player also entailed the cost of *not being able* to embrace other opportunities that might arise.

Let's take a more monetary look at opportunity costs. Suppose we have $10,000 to invest. We could buy a 30-year CD (certificate of deposit) that pays 2% interest annually. But let's suppose it is impossible to withdraw the money until the CD matures. We are making an investment decision today, but we don't know whether the economy might change next year and interest rates might increase to 3%. We saw the difference in return in a 3% versus a 2% investment (remember Adam and Eve). So having our money available to take advantage of potentially higher rates is valuable. How valuable? What is the lost opportunity cost of investing now? Let's consider the options.

To make the question specific, let's suppose there is a 10% chance that the interest rates will rise to 3% in any given year and a 90% chance they will stay at 2%. Let's also assume that we can buy a CD only once a year on January 1 and that we can buy a CD that expires 30 years from *today*. In other words, next year's option will be to buy a 29-year CD; the year after that, a 28-year CD, and so on.

For simplicity, we will suppose that there are no other possibilities and that our money earns nothing until we invest it in a CD. The question is, what is the best strategy for investing?

We can determine this strategy by thinking through the possibilities. Obviously, as soon as we have the option to buy the 3% CD, we should jump on it. The question is how long to wait for that chance. Should we buy the 2% CD now? Should we wait one year and hope that the interest rate is 3% next year, but buy at 2% anyway if it isn't? Should we wait two years and then buy even if it does not reach 3%? Three years? Four years? Five? Ten? Fifteen? Twenty-eight? How could we possibly decide what is best? The answer, as always, is to think things through.

Let's begin by determining the expected value of wait times. For example, if we are considering the strategy of hoping that 3% comes up by the fifth year and then giving up and buying anyway, what value would we assign to that strategy?

Again, we need to break down the possibilities and see how each one plays out, that is, how much money we would end up with 30 years from now.

If 3% comes up next year, we would earn 3% for 29 years.

If 3% comes up in two years, we would earn 3% for 28 years.

If 3% comes up in three years, we would earn 3% for 27 years.

If 3% comes up in four years, we would earn 3% for 26 years.

If 3% comes up in five years, we would earn 3% for 25 years.

If 3% does not come up in five years, we would buy the 2% CD and earn only 2% for 25 years.

The expected value of this decision is the probability of each possibility times the payoff for each. Let's consider the probability of 3% coming up for the first time in various years. The probability of 3% arising in the first year is given to be 0.1. The probability of 3% not arising in the first year is 0.9. The probability of 3% not arising in the first year and then arising in the second year is the product of the probabilities, namely, 0.9×0.1. Likewise, the probability of 3% not arising in the first two years, but then arising in the third year, is $0.9^2 \times 0.1$. This reasoning lets us conclude that the expected return on the wait-five-years strategy equals

$$\$10,000[0.1(1.03)^{29} + 0.1(0.9)(1.03)^{28} + 0.1(0.9^2)(1.03)^{27}$$
$$+ 0.1(0.9^3)(1.03)^{26} + 0.1(0.9^4)(1.03)^{25} + 0.9^5(1.02)^{25}].$$

We need to compute the corresponding expected values for each of the possible wait times and then see which expected value is largest. Such a computation can be programmed in a spreadsheet or on a calculator or computer. The table here shows the values of what we would expect to have after 30 years with each of the potential wait times, from waiting 0 years to waiting 28 years (at which time we may as well take the 2% return for the last year rather than getting nothing).

Years to Wait Before Buying the CD at 2%	Expected Value at the End of 30 years	Years to Wait Before Buying the CD at 2%	Expected Value at the End of 30 years
0	$18,114	16	$18,960
1	$18,339	17	$18,945
2	$18,518	18	$18,929
3	$18,658	19	$18,912
4	$18,766	20	$18,896
5	$18,848	21	$18,881
6	$18,909	22	$18,865
7	$18,952	23	$18,851
8	$18,981	24	$18,837
9	$18,999	25	$18,824
10	$19,008	26	$18,811
11	$19,010	27	$18,800
12	$19,006	28	$18,789
13	$18,999	29	$18,779
14	$18,988	30	$18,769
15	$18,975		

From this table of expected values, we see that our best strategy is to wait up to the 11th year for the 3% rate, but if the 3% rate does not materialize by then, to go ahead and buy the CD at 2%. Notice that using this strategy of waiting (and hoping) for a more favorable rate for 11 years will, on average, yield almost $900 more than if we were to invest immediately in the 2% CD.

The lessons here are twofold: (1) tying up assets has a definite cost in lost opportunities in the future, and (2) a more global second lesson, we can make better financial and life decisions if we calmly and clearly consider the alternatives and weigh one against the other. Making decisions based on this kind of methodical analysis is particularly potent in dealing with money matters.

A Look BACK

REPEAT, REPEAT, REPEAT. Money matters are frequently the result of repeating simple processes over and over. To understand the general pattern, we start with very simple cases and then gradually increase the number of time periods involved or increase some other complicating factor. When dealing with money matters in the real world, we are often confronted with many technical terms that refer to a blizzard of different types of mortgages and loans, annuities and saving accounts. Only professionals in the field can keep them all straight. What we can do is to see how they all result from repeating patterns that we can understand by looking first at simple scenarios.

Life Lessons

Understand simple things deeply.

•

Look for patterns.

MINDSCAPES   Invitations to Further Thought

*In this section, mindscapes marked (**H**) have hints for solutions at the back of the book. Mindscapes marked (**ExH**) have expanded hints at the back of the book. Mindscapes marked (**S**) have solutions.*

Developing Ideas

1. **Simple interest (H).** Suppose you deposit $500 into a savings account that earns 2% simple interest per year (*simple interest* means that there is no compounding). After one year, you close out the account and withdraw all the money. How much money would you have? How would you react if the bank charged you an $11 processing fee for that transaction?

2. **Less simple interest.** Suppose that at the beginning of party animal Sam Student's college career, he deposits $500 into a savings account that earns 3% interest per year, compounded annually. Exactly 10 years later, when Sam graduates, he closes the account. How much money does the bank give good old Sam?

3. **The power of powers (H).** In this section we discovered that the power of compounding arises from raising quantities to large powers. Which number is larger: 2^3 or 3^2? 4^5 or 5^4? In general, if n and m are large numbers and n exceeds m, which would you guess is larger, m^n or n^m? If you like money, would you rather have $\$2^{100}$ or $\$100^2$?

4. **Crafty compounding.** Two thousand years ago, a noble Arabian king wished to reward his minister of finance. Although the modest minister resisted any reward, the king finally put his princely foot down and insisted. Impishly the minister declared that he would be content with the following token: "Let us take a checkerboard. On the first square I would be most grateful if you would place one piece of gold. Then on the next square twice as much as before, thus placing two pieces, and on each subsequent square, placing twice as many pieces of gold as in the previous square. I would be most content with all the gold that is on the board once your majesty has finished." This request sounded extremely reasonable, and the king readily agreed. Given that there are 64 squares on a checkerboard, roughly how many pieces of gold did the king have to give this "modest" minister of finance? Why did the king have him executed?

5. **Keg costs.** List some of the opportunity costs involved in taking part in a Thursday night keg party.

Solidifying Ideas

6. **You can bank on us (or them) (S).** You wish to invest $1000 for a year, and you have two investment options: Happy Bank will pay you 3% compounded annually, while Glee Bank offers 2.5% compounded quarterly. To maximize your income, which bank would earn your business and give you a real smile? Justify your answer by analyzing both options.

7. **The Kennedy compound.** You wish to invest $1000 for five years, and you have two investment options: Bobby Kennedy Bank offers 3% compounded

annually, while Jack Kennedy Bank offers 3% compounded daily. Clearly it's all the way with JFK. Here we ask not what your bank can do for you. Instead, we ask you to analyze both options and determine how much more money you would have with Jack than with Bobby.

8. **Three times a lady.** The Three-Times-a-Year Savings and Loan Company has a Ladies' Nite on June 2. On that evening, women can borrow at a lower rate. Your Aunt Edna borrows $2000 to buy her favorite collegiate relative (that's you) a fancy plasma TV. She takes out a one-year loan at an interest rate of 7%. How much are her three equal payments?

9. **Baker kneads dough (ExH).** Your favorite baker, Adrian, from Adrian's House of Fattening Finger Foods, needs a new oven. If Adrian borrows $10,000 from The Bread Bank at 6% annual interest for five years, what would Adrian's equal monthly payments be?

10. **I want my ATV!** You want to purchase a cool, yellow all-terrain vehicle. With all the bells and whistles, the cost of the vehicle is $34,000 (well, actually there are no whistles, but a bell does go off if you forget to turn off the lights). You make a down payment of $5000 and borrow the rest at 4% annual interest for 10 years. What are your equal monthly payments?

11. **Lottery loot later?** You have a big problem: You've just scratched off an instant lottery ticket and discovered that you hit the jackpot and won $50,000! The lottery folks inform you that you will receive five annual payments of $10,000. Sounds good. As this is soaking in, a very genteel bookie named Jimmy "The Wallet" Banana offers to buy your winning ticket for $45,000 in a lump sum. If in either case, you were to deposit your winnings into a savings account paying 3% compounded annually, how much money would you have by the fifth year (at the moment the lottery folks pay off the last $10,000) in each scenario? Is the Banana making you an offer you can't afford to refuse?

12. **Open sesame (S).** Bert and Ernie each open a savings account on the same day with an initial deposit of $1. Bert's account pays 5% interest compounded daily, while Ernie's account pays 6% interest compounded daily. What is the balance in each of their accounts after 50 years? What if they each initially deposited $10,000?

13. **Jelly-filled investments (H).** Suppose you purchase $2000 worth of stock in Krispy Kreme donuts. Three years later you sell your shares for $2300. If instead you had invested that money in a three-year CD paying 5% interest compounded monthly, would your investment have been as sweet? Don't glaze over the details: Justify your answer.

14. **Taking stock.** Suppose that a stock transaction yields a profit of $155. If the initial investment of $2000 was made three years ago, then what was the average annual rate of return of the investment; that is, what rate on a savings account compounded annually would produce the same outcome as the stock scenario?

15. **Making your pocketbook stocky.** Suppose that a stock transaction yields a profit of $248. If the initial investment of $3500 was made two years ago, what was the average annual rate of return on the investment; that is, what rate on a savings account compounded annually would produce the same outcome as this stock scenario?

Creating New Ideas

16. **Money-tree house.** You decide you wish to build your dream home in the woods. You must borrow $250,000. One bank offers you a 30-year loan at a rate of 7.5%. Another bank offers you a 15-year loan at a rate of only 7%. Given that each bank would expect monthly installments of equal payments, how much would the monthly payments be for each loan? How much would you end up paying (in total after the loan is completely repaid) with each of the two offers?

17. **Future value (S).** What is the future value of $3000 invested in a six-year CD that earns 4.5% interest compounded quarterly?

18. **Present value (ExH).** On the first day of your college career you decide that you want to make one deposit into a mutual fund so that exactly four years later you have enough to purchase a car costing $20,000. Given that the account will generate a yield of 5.7% interest compounded monthly, what should your initial deposit be in order to have that big pot of money at graduation?

19. **Double or nothing (H).** You decide you wish to double your wealth. So you take all your savings—$1500—and deposit it into a savings account generating 3% interest compounded daily. How long would you have to wait in order for your money to double? Would it take longer to double your money if you started with $5000? Explain your reasoning.

20. **Triple or nothing.** You decide you wish to triple your wealth. You take all your savings—$3500—and deposit it into a savings account generating 2.5% interest compounded weekly. How long would you have to wait in order for your money to triple? Would it take less time to triple your money if you started with $500? Explain your reasoning.

Further Challenges

21. **Power versus product (S).** In this section we derived the formula:

$$\text{FINAL AMOUNT} = (\text{INITIAL AMOUNT})$$
$$\times (1 + \text{INTEREST PER INTERVAL})^{(\text{NUMBER OF INTERVALS})}.$$

Carefully explain why the "NUMBER OF INTERVALS" is an exponent rather than a multiplier such as:

$$(\text{INITIAL AMOUNT}) \times (1 + \text{INTEREST PER INTERVAL})$$
$$\times (\text{NUMBER OF INTERVALS}).$$

22. **Double vision.** Suppose we have $P and we invest it in an account that pays a rate of r% compounded t times a year. Produce a formula for the number of years we would have to wait in order for our money to double. Does the formula depend on P? Does the answer surprise you?

In Your Own Words

23. **With a group of folks.** In a small group, discuss and actively work through the ideas about opportunity costs. After your discussion, write a brief narrative describing this concept in your own words.

24. **Creative writing.** Write an imaginative story (it can be humorous, dramatic, whatever you like) that involves or evokes the ideas of this section.

For the Algebra Lover

Here we celebrate the power of algebra as a powerful way of finding unknown quantities by naming them, of expressing infinitely many relationships and connections clearly and succinctly, and of uncovering pattern and structure.

25. **Adding up the bucks (H).** You have a job every summer. Most of your earnings go to help pay your college expenses, but you save some money each year. Suppose you start at age 16, and each summer for six years on August 31, you put $500 into a savings account that pays 5% interest on August 30 of the following year. How much money do you have in your savings account on August 30 when you're 17? How much do you have the next day after your annual deposit? How much is there in your account on August 30 when you're 18? When you're 19? When you're 20? When you're 21?

26. **Fiddling for dollars.** As presented in the section on page 796, if you invest P dollars in a savings account paying r percent interest compounded t times per year, the amount of money you have at the end of y years is equal to

$$P \times \left(1 + \frac{r}{100t}\right)^{ty}$$ dollars. If the interest is compounded once a year, then $t = 1$

and the formula becomes $P \times \left(1 + \frac{r}{100t}\right)^{ty}$. Suppose you buy a violin for

$5000 and sell it four years later for $8000. What was the average annual rate of return on this investment?

27. **Facebank.** Your roommates are developing some social networking software they expect to go big some day. They ask you to invest some money, promising you an annual return of 10%. If you believe their hype, how much of your money (or really, your parents' money) should you invest if you want to have

$50,000 after ten years? (*Hint:* Use the formula $P \times \left(1 + \frac{r}{100}\right)^{y}$ described in

the previous Mindscape.)

28. **Boatload o' cash.** At age 12 you dream of sailing around the world some day. You know it will take big bucks, but you are lucky enough to have a wealthy aunt who adores you. She promises to invest $100,000 in an account that pays an interest rate of r% compounded monthly. She will give you all the money in this account in ten years if you can tell her what interest rate

will double her money in that time. (*Hint:* Use the formula $P \times \left(1 + \frac{r}{100t}\right)^{ty}$

described in Mindscape 26. What is the value of P? Of t? Of y?)

29. **Houseload o' cash.** You want to buy a house by age 30. Right after college, at age 22, you decide to save money every month towards a down payment of $20,000. If you can invest your money at 3% compounded annually, how much should you invest each month to reach your goal in 8 years?

10.4 PERIL AT THE POLLS

Deciding Who Actually Wins an Election

Harry Truman, after winning the 1948 presidential election. Election outcomes may surprise even the winner—there's more to voting than meets the eye.

> *. . . Democracy is the worst form of government except all those others that have been tried from time to time.*
>
> **WINSTON CHURCHILL**

Everyone has an opinion. So what do we do when a group of people must make a decision? In a democracy, for example, citizens must choose which laws to enact, which people to lead them, and which programs to fund. Families, couples, schools, and companies all must make decisions: *What DVD should we rent? Which restaurant should we go to? How do we rank students? Whom do we hire? Who gets the promotion?*

Every issue of social choice involves taking into account many different opinions and somehow generating one collective decision. Different situations may require different methods for making the group decision. For example, when a family decides to buy a new car, the person doing the driving or signing the loan documents may have more say about the make and model than the person bawling in the car seat. In a committee meeting, the chairperson may set the agenda, but the committee members may determine whether a given proposal is approved or not. When a company hires a new employee, all of those involved in the process may have different priorities, values, and roles in the organization, and those opinions are somehow combined to reach a final decision.

Certainly, one of the most common methods of decision making is voting. Voting is a simple matter when only two people are running for office. Voters cast their ballots for their preferred candidate, and the candidate who receives more votes wins. But as soon as a third candidate appears in the pool of possibilities, the voting waters become mighty muddy.

In this section we will consider two counterintuitive realities in the election process. First, an election's outcome may have less to do with the voters' preferences

811

for the candidates than the actual voting method employed—that is, the method by which the voters' preferences are combined to determine the winner. Second, every voting method is seriously flawed and, in some sense, the only consistent voting method is a dictatorship. Thus, we see that there may be more to Churchill's quote than meets the eye.

An Election Conundrum

Several years ago Gwyn was running for vice president of her sixth-grade class against two other candidates—Chad and Courtney. Gwyn prepared a persuasive speech ("Hey, vote for me!"), delivered it well, and even had buttons that read: "Help Gwyn Wyn." The sixth-grade class, roughly half boys and half girls, liked all three candidates. The election took place, and the candidate with the most votes was declared the winner. Without any further knowledge of the election issues, why do we know for certain that Gwyn lost the election?

beats

beats

But... beats and

To solve this riddle, we must return to sixth-grade thinking—not such a far trip for some of us. Sixth graders hold various opinions on many hot political questions, such as the availability of chocolate milk at the cafeteria. But there is one issue on which all sixth graders agree: In general, boys vote for boys and girls vote for girls. Since the girl vote was split between Gwyn and Courtney, Chad easily walked away with the election. Was that outcome fair? Not necessarily. If there had been slightly more girls than boys in the class, then either girl would have defeated Chad in a head-to-head contest.

While this elementary illustration may not seem crucial to the national welfare, clearly, electing the president of the United States is no time for recess. In the 2000 presidential election, for example, George W. Bush won despite the fact that more people voted for Al Gore and even though Gore would have beaten either Bush or Ralph Nader, the third major candidate in the race, in a head-to-head contest. The electoral college system used for electing presidents in the United States allows such outcomes. However, as we will soon see, every method of selecting one winner among three or more candidates is doomed to have serious defects. One of the most surprising facts about making choices in society, *Arrow's Election Disaster Theorem*, asserts that every voting method will contain fundamental flaws. In other words, when it comes to elections, voters can't win.

Our strategy for understanding voting issues and their paradoxes involves the same methods of analysis we've used throughout this book to develop many great mathematical ideas. We will clarify what we want to know, look at simple cases, and then apply our insights from these simple cases to the more complex situations.

Simple Voting Methods

What's all the fuss about voting? Doesn't voting simply mean that people cast their ballots, the ballots are counted, and whoever gets the most votes wins? Well, not exactly. When there are three or more candidates, many methods can be used to choose the winner. Let's take a look at three methods.

Plurality Voting

In plurality voting, each voter votes for one person, and the candidate with the most votes wins. In the sixth-grade class election, for example, Chad got the most votes, and therefore won the election, because the girls split their votes between Courtney and Gwyn.

Vote-for-Two

In vote-for-two voting, each voter must vote for two different candidates and the candidate with the most votes wins. No problem. This method of voting overcomes the problem that arose in the vote-splitting situation in the sixth-grade class election. In that scenario, for example, if each student voted for two candidates, all the girls would vote for both girls, of course, but each boy would have to cast his second vote for one girl or the other. Since no girl would vote for Chad, he would lose using this vote-for-two method of voting. If most of the boys cast their second votes for Courtney, she would win.

Borda Count

In the Borda count method (named after Jean-Charles de Borda, 1733–1799), each voter ranks all the candidates: 1, 2, 3, and so on. The highest ranking is 1. The rankings are then tallied for each candidate, and the candidate with the *lowest* total wins. Suppose in our sixth-grade election that Gwyn is more popular among the girls than Courtney. So, most girls' votes would give 1 to Gwyn, 2 to Courtney, and 3 to Chad. In this case, using the Borda count method, Gwyn might win the election. Fine, so now Gwyn is finally elected.

Although each of these methods appears reasonable, we see that these different voting methods lead to dramatically different outcomes. In fact, in the case of the sixth-grade election scenario, we see that the three different voting methods result in three different winners. The voting *method* rather than the voters' *preferences* actually determines the victor.

The Method Counts

The sixth-grade election scenario is a plausible situation in which three reasonable voting methods could be used to determine a winner. While the voters' preferences do not change, we consider different methods by which those preferences are combined to determine the winner. The candidate who ultimately wins depends on the voting method used, not on the students' preferences. Using one method, Chad wins. Using another method, Courtney wins. And using a third method, Gwyn wins.

To quantify this seemingly paradoxical sixth-grade situation, let's consider a chart that shows how each student ranked the candidates. Using this information, we can analyze the election outcomes under the three voting methods.

	6 boys prefer	4 boys prefer	8 girls prefer	4 girls prefer
First choice	Chad	Chad	Gwyn	Courtney
Second choice	Courtney	Gwyn	Courtney	Gwyn
Third choice	Gwyn	Courtney	Chad	Chad

Notice that the chart shows that there are 22 students in the sixth-grade class altogether—10 boys and 12 girls. All 10 boys prefer Chad over either girl. Six of the boys rank Courtney as second choice and Gwyn as third choice. The other columns similarly show how other groups of students rank the three candidates.

Using the chart, we can see how each student will vote using each voting method.

Plurality Voting

Under plurality voting, each student votes for one candidate, namely, his or her first choice. Notice that 10 students—all the boys—rank Chad first, 8 rank Gwyn first, and 4 rank Courtney first. So Chad will get 10 votes, Gwyn will get 8 votes, and Courtney will get 4 votes. So under the plurality voting method, Chad will win the election. Congratulations, Chad!

Vote-for-Two Method

Under the vote-for-two method, all students vote for their top two choices. So Chad gets 10 votes again (since he doesn't pick up any girl votes), while Courtney now gets 18 votes (all 12 girl votes plus 6 votes from boys who ranked Courtney second), and Gwyn gets 16 votes (the 12 girl votes plus 4 boy votes). So Courtney wins under the vote-for-two method. Congratulations, Courtney!

Borda Count Method

Finally, if we use weighted votes, that is, the Borda count method, Chad gets 46 points (10 number-1 votes plus 12 number-3 votes), while Courtney gets 44 points (4 number-1 votes, 14 number-2 votes, and 4 number-3 votes), and Gwyn gets 42 points (8 number-1 votes, 8 number-2 votes, and 6 number-3 votes). In the Borda count method the lowest total wins, so using this method, Gwyn wins. Congratulations, Gwyn!

Therefore, we discover that the same students, with their same rankings of the candidates, elect three different winners depending on which voting method is used. Given the preferences of the voters, who do *you* think should win? *Hint*: There is no right answer. Perhaps Gwyn should win since she would beat Courtney 12 to 10 and Chad 12 to 10 in head-to-head contests. But this opinion is certainly debatable.

This sojourn into sixth-grade voting may be troubling, but Donald Saari, a mathematician who has studied voting methods, has proved that the situation can get far worse. Indeed, he makes the following challenge in his book *Chaotic Elections! A Mathematician Looks at Voting* (American Mathematical Society, 2001):

> For a price, I will come to your organization just prior to your next important election. You tell me who you want to win. I will talk with the voters to determine their preferences over the candidates. Then, I will design a "democratic voting method" which involves all candidates. In the election, the specified candidate will win.

Professor Saari claims that such a feat is usually easy in practice (at least for him). You will attempt (and succeed at) this feat in some of the Mindscapes at the end of this section. These examples all underscore that the method used to determine the aggregate choice given the voters' rankings of the candidates can greatly alter the outcome.

So far we have looked at three voting methods. We'll now consider a few others, but we will see that all voting methods have serious problems. In fact, later we will prove that *every* possible voting method has serious drawbacks.

The Runoff

Runoffs are familiar fixtures of several voting methods when three or more candidates are running. One runoff method asks voters in the first round of voting to vote for their top candidate and then pits the two highest vote-getters against each other in a runoff election. Holding a runoff appears to be a good method for selecting a winner; unfortunately, the runoff method has a serious flaw, namely, getting more votes may paradoxically actually be bad for a candidate. Let's illustrate this potential defect by adding a few more candidates to our sixth-grade election and seeing what happens when we try a runoff method to determine the winner.

Suppose now that two boys (Chad and Cliff) and three girls (Courtney, Gwyn, and Sarah) are all candidates. As before, all the sixth-grade boys vote for boys and girls vote for girls.

Scenario 1

Suppose the boy votes are almost evenly split between Chad and Cliff, and the girl votes are split three ways so that no girl receives as many votes as either boy. In this case, the runoff is between Chad and Cliff. It's a cliffhanger, but Chad hangs on to win in the runoff.

Scenario 2

Suppose we have the same candidates, but this time Chad campaigns more vigorously and gets even more of the boy votes in the first round of voting compared with what he received in Scenario 1. Now look what happens: Chad gets more votes, and therefore Cliff gets fewer, so a girl, perhaps Sarah, now becomes Chad's runoff opponent. Because the class has more girls in it than boys, Sarah wins the runoff and becomes vice president. Getting more votes actually led to Chad's political downfall.

This example illustrates that the runoff method fails to satisfy a basic and reasonable feature that a voting method might have, namely:

The Better Is Better Principle

If a voter changes his or her vote to now favor a given candidate, then that change should not lead to a worse result for that candidate.

Throwing Out Losers

Another plausible method of selecting a winner from several candidates is to eliminate candidates one by one until only one remains. That is, we could ask voters to indicate their *least favorite* candidate. The candidate who is the least

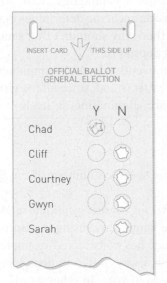

A hanging chad

favorite of most voters would be eliminated from the ballot. The voters would then repeat the process; that is, they would indicate who among the remaining candidates is their least favorite. That next most unpopular candidate would be removed and the process would continue until only one candidate remained.

Scenario 1

Suppose we hold our sixth-grade election with five candidates: Chad, Cliff, Courtney, Gwyn, and Sarah. Suppose the girls split their negative votes between Chad and Cliff, but all the boys vote against Sarah even though Sarah is the first choice of all the girls. In this case, Sarah is eliminated in the first round, and after several more eliminations, someone else, perhaps Chad, wins.

Scenario 2

Suppose in the same election, Cliff senses defeat in his future and decides not to run at all. In this case, all the girls vote against Chad in their first round, so Chad is eliminated and a girl is elected.

This method has at least two unfortunate possible effects. One is that in the first round, voters could eliminate the candidate who might defeat every other candidate in a head-to-head election. Scenario 1 illustrates this possibility, since Sarah could beat anyone in a head-to-head contest, and yet she is eliminated in the first round.

The second problem arises from an apparently reasonable feature of a voting method, namely, that eliminating one or more of the *losing* candidates should not change who wins an election. In other words, if our voting scheme determines that some candidate is the best of, say, five candidates, then why shouldn't that same candidate be elected if one of the losing candidates doesn't run?

In the Throwing-Out-Losers scenario, if Cliff, an eventual loser, is eliminated from the race, then Chad would be eliminated in the first round, since all the girls would vote him out. So Sarah would win. Therefore, eliminating an irrelevant candidate alters the outcome.

Voting for More Than One Candidate

Under this voting method, each voter votes for some specified number of candidates, and the candidate with the most votes wins.

As we saw in the sixth-grade election, each voter voting for one candidate made Chad the winner, voting for two made Courtney the winner, and using the Borda count method made Gwyn the winner. Donald Saari has constructed examples using any number of candidates where the outcome is always different depending on how many candidates for whom voters are allowed to vote. In other words, if you have four candidates, then the same voters with the same preference rankings of the candidates will elect one candidate if the voters just vote for their first choice, a different candidate if they vote for two candidates, yet another candidate if they vote for three, and still another if they use the Borda count method.

Approval Voting

We could let voters vote for as many candidates as they find acceptable and then declare the winner to be the candidate who receives the most votes. In an expanded sixth-grade election with six candidates—Chad, Cliff, Courtney, Sarah, Hannah, and Gwyn—each student could choose to vote for one to six candidates. Suppose the boys all vote for both boys, but the girls just each vote for their favorite three girls instead of voting for all four girls. In this case, the boys could tie for the win even though either of the boys would lose in a head-to-head contest with any girl. In other words, the outcome of the election is determined not by the preference ordering of the student voters but by how many candidates each chose to vote for.

Bullying

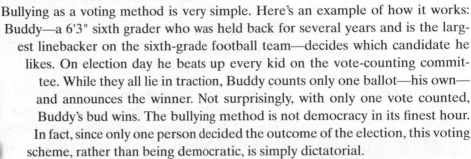

Bullying as a voting method is very simple. Here's an example of how it works: Buddy—a 6'3" sixth grader who was held back for several years and is the largest linebacker on the sixth-grade football team—decides which candidate he likes. On election day he beats up every kid on the vote-counting committee. While they all lie in traction, Buddy counts only one ballot—his own—and announces the winner. Not surprisingly, with only one vote counted, Buddy's bud wins. The bullying method is not democracy in its finest hour. In fact, since only one person decided the outcome of the election, this voting scheme, rather than being democratic, is simply dictatorial.

So far, we have seen a wide array of possible voting methods, each of which exhibits one or more serious defects. But there are yet more difficulties about making choices, and these involve how we rank choices.

Transitivity

Ranking the preferences of a group of people is not a simple matter. It sounds reasonable that if a voter prefers A over B and prefers B over C, then that voter would prefer A over C. When our preferences are consistent in this manner, we say that they are *transitive*.

No dominating die

But not all orderings are transitive, as the Cool Dice from your kit show. Story 8, "Rolling Around in Vegas," in Chapter 1 described these nontransitive dice. Those strange dice have different numbers on their faces—one has a 3 on each face, one has four 2's and two 6's, one has three 1's and three 5's, and the fourth has four 4's and two blanks. These dice demonstrate that we can have four dice such that if we roll two dice against each other and see which die comes up with the bigger number, then die A beats die B two-thirds of the time, die B beats die C two-thirds of the time, die C beats die D two-thirds of the time, but then, amazingly, die D beats die A two-thirds of the time. In other words, the dice form a circle of dominance, so it is impossible to say that one of them is the best. If we think of the dice as candidates in an election, we would not be able to choose any candidate as clearly the best, because there is always another candidate who can beat that candidate. Can you think of three political candidates where you would personally prefer A to B and B to C but C to A?

Condorcet's Paradox

Even if we are in the satisfying situation where every individual voter's ranking is transitive, the *cumulative* ranking of the group as a whole may not be transitive—that is, the ranking may have a circle of preferences. One of the early investigators of this puzzling possibility was the French mathematician and philosopher

Marie Jean Antoine Nicolas Caritat, Marquis de Condorcet (1743–1794), which is why it's called Condorcet's Paradox.

We illustrate this paradox by considering the preferences of 15 voters, each of whom ranks three candidates: A, B, and C. Five voters rank them A, B, C; five rank them C, A, B; and five rank them B, C, A, as shown in the chart. Who should win the election?

	5 voters' ranking	5 voters' ranking	5 voters' ranking
First place	A	C	B
Second place	B	A	C
Third place	C	B	A

Most voters (10, or two-thirds) prefer A over B. Likewise, most prefer B over C. Unfortunately, most voters also rank C over A! Hmmm . . . We have to face the circular conundrum that in a head-to-head contest, A would beat B, B would beat C, and C would beat A. So this circular pattern of preferences can arise even when each individual voter has a definite preference order that is not circular.

Arrow's Election Disaster Theorem—We Just Can't Win

Every election system we've considered suffers from at least one glaring defect. We would love to find a voting method that is truly equitable, but sadly, we will never find one. We base our gloomy assertion on the most famous result in the theory of social choice and voting, namely, *Arrow's Impossibility Theorem*, which might better be called *Arrow's Election Disaster Theorem*. Kenneth Arrow discovered this surprising insight in 1952; we illustrate his ideas here within the context of the high-stakes drama of preseason football polls.

Bettmann/©Corbis

Kenneth Arrow

Party On!

The characters in the following drama are named to remind us of the political setting in which voting paradoxes often arise. As we all know, the two principal political parties in the United States are the Democratic Party and the Republican Party. The symbol of the Democratic Party is the donkey. According to the Web site of the Democratic National Committee, "When Andrew Jackson ran for president in 1828, his opponents tried to label him a 'jackass' for his populist views and his slogan, 'Let the people rule.' Jackson, however, picked up on their name calling and turned it to his own advantage by using the donkey on his campaign posters." Thus, the donkey, another name for the animal called an ass and whose males are called jackasses, became the Democratic Party's symbol. The Republican Party's official symbol is the elephant. This pachyderm appears most memorably in literature in the classic story Dumbo the Flying Elephant. So as the following story unfolds, the names selected will remind us that, although the saga concerns football polls, the underlying message lies at the heart of our political process.

Slings, Arrows, and Outrageous Football

Jack Hammer and Dom Bo were two of the most respected analysts of the football world. Jack had Mr. Spock-like donkey ears and brayed frequently. Dom on the other hand was known for his weighty wisdom—or at least his weight.

Jack and Dom were scheduled to make a guest appearance on the Antilles Sports Show (A.S.S.) and offer their preseason football rankings for the Little-3

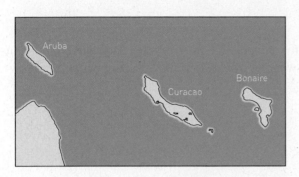

Conference consisting of UA (University of Aruba), UB (University of Bonaire), and UC (University of Curacao). The show airs live from Washington, D.C., where votes are supposed to count. The host of A.S.S., Artie Arrow, had to be prepared to take all predictions under the sun that Jack and Dom might bring and combine them to make an A.S.S.-endorsed ranking using the collective wisdom of the predictors. Artie began to think about how to combine his guests' rankings under every contingency. He wanted his method of deriving the cumulative ranking to be reasonable, so he decided it must satisfy two principles:

1. *Go along with consensus*. If Jack and Dom both ranked one team over another, then the cumulative ranking should rank those two teams in that order as well.

2. *Ignore the irrelevant*. The relative ranking of one team over another should depend only on Jack's and Dom's relative rankings of those two teams, not on how they ranked the third team.

With these principles in mind, Arrow sat down to devise his scheme for combining Jack's and Dom's potential rankings. Because Arrow wanted to be prepared for any contingency, he needed to specify the cumulative ranking that would result from any ranking Jack and Dom might give. Of course, Jack and Dom would probably not always agree, so in some cases he would just have to make an arbitrary decision.

He first considered the simplest scenario in which Jack and Dom might not totally agree. He decided to consider what to do if Jack ranked the teams UA, UB, UC, and Dom ranked the teams UB, UA, UC, as shown in the chart.

Jack's ranking	UA	UB	UC
Dom's ranking	UB	UA	UC

Jack	Dom	A.S.S.
UA	UB	UA
UB	UA	UB
UC	UC	UC

To determine what the cumulative ranking would be, Arrow arbitrarily decided that in this case, he would go along with Jack instead of Dom—that is, the A.S.S. ranking in this case would be UA, UB, UC.

The Surprise

After making that one concession to Jack, there is no choice about how to produce the cumulative rankings. As we will now discover, that tiny concession to Jack forces Arrow to concede *all* decisions to Jack. And doing so is the only way to keep the two principles, *go along with consensus* and *ignore the irrelevant*.

In other words, after just one concession to Jack, there is no method of producing the cumulative ranking that respects the two principles except the method of making Jack the dictator and ignoring Dom altogether. That one concession to Jack is the leading edge of the wedge. Give Jack an inch, and he'll take 100 yards. Let's see why all other decisions are forced on Arrow by those two reasonable-sounding principles.

Another Decision

Since Arrow had to be ready for any possible rankings that Jack and Dom might bring, he decided to consider what he would do if they came up with the following rankings:

Jack	UB	UA	UC
Dom	UC	UB	UA

We might think that Arrow could choose to go along with Dom this time just to be fair and have the official ranking be UC, UB, UA. At this moment, there is no obvious reason why such a choice would be bad. But, in fact, his doing so would be inconsistent with our two principles and, rather than dance around the issue, let's consider the two-step reason why.

Step 1: Remember that the only decision Arrow has made so far is to go along with Jack's opinion about UA and UB whenever Jack ranks UA over UB. What would Arrow do if Jack ranked the teams UA, UB, UC, and Dom ranked them UB, UC, UA?

Jack	UA	UB	UC
Dom	UB	UC	UA

Well, we know that UA would have to be above UB, since by *ignoring the irrelevant* UC, we would have the same relative rankings we had in the very first case where Arrow went along with Jack. But also both Jack and Dom agreed that UC was below UB, so the A.S.S. ranking must be UA, UB, UC—again in agreement with Jack. In order to abide by the principle of *ignoring the irrelevant*, Arrow has

Jack	Dom	A.S.S.
UA	UB	UA
UB	UC	UB
UC	UA	UC

no choice but to make the cumulative ranking agree with Jack again.

Notice that this cumulative ranking put UA above UC even though Dom had put UC above UA—Arrow was forced into this pick.

Step 2: Now let's look at the case Arrow tried to consider before, as shown in the chart here.

Notice that Arrow can't go along with Dom after all, because if he did and ranked UC as number 1, then, since UB is not the winner, we could ignore UB and see what remains. If we ignore UB, Jack's and Dom's rankings of UA and UC are precisely what we considered in Step 1. But in that case we were forced to go along with Jack

Jack	UA	UB	UC
Dom	UB	UC	UA

and rank UA above UC. So to preserve the principle of *ignoring the irrelevant*, Arrow must rank UA above UC in this case also. Finally, since both Jack and Dom put UB above UA, then Arrow must make the cumulative ranking UB, UA, UC. So we see that again Arrow is forced in this case to let Jack's preferences prevail.

Jack Dictates

Once Jack is given the authority on a single case, Arrow is then forced by the two principles to go along with Jack on more and more cases. In fact, the only consistent thing to do is to make Jack the dictator of the whole show. Arrow must simply consider rankings in which the two principles force the decision, and then those decisions force even more decisions, and so on, and so on. Some of the Mindscapes at the end of this section ask you to set up sequences of forced rankings that demonstrate how every decision is now forced to simply agree with Jack's ranking. On the bright side, we can now draw the inescapable conclusion: Jack, A.S.S. endorsed.

One Concession—Dictatorship

Suppose we enlarged the ranking panel to include many panelists, but for some reason Arrow still gave Jack the authority of UA over UB, even when all the other panelists voted the other way. That one concession to Jack would essentially make Jack the dictator over the whole panel, because the analysis from the previous example applies in this case as well. If we ever let an individual panel member dictate the result of one pair of teams in the face of the opposite opinion of all the other panelists, then that one panelist would have to dictate the whole show, even if there are many panelists.

Of course, why would we be so foolish as to let one panelist's pick be the official pick when all the other panelists disagree? It's not that we're foolish; it's just that we have no choice—any voting method that respects a few simple principles, such as those that our show-host Arrow tried to keep, is forced to concede all authority to one person, who then becomes the absolute dictator.

Arrow's Election Disaster Theorem

What we have really established with this football-ranking scenario is a special case of one of the most important theories of voting and social choice:

Arrow's Impossibility Theorem.

When there are three or more alternatives; it is impossible to devise any voting scheme (that is, a rule for taking the voters' transitively ranked orderings of the candidates and producing one cumulative ranking for the whole group) other than a dictatorship that will order the alternatives or select the winner in a manner that satisfies the following principles:

1. *Go along with consensus—that is, if it's unanimous, then we have a winner.*
2. *Ignore the irrelevant—that is, throwing out a loser will not change the election.*
3. *Better is better—that is, changing our vote to favor a candidate will not result in that candidate's ranking dropping.*

Of course, a dictatorship is not what we have in mind when we think about democracy, so Arrow's Election Disaster Theorem really implies that every non-dictatorial voting scheme must fail to satisfy one of our three reasonable principles. This result is discouraging, but it certainly shows why controversies about elections and football rankings will always be with us.

The implications of this result are unsettling. It means that we can never realize a view of democracy that we would hope to, because every voting scheme that is not a dictatorship will have a major defect. Thomas Jefferson would not be amused.

Arrow's Election Disaster Theorem is proved with the same style of argument that we employed to deduce that the A.S.S.–endorsed football poll ranking must simply copy one panelist's ranking. More voters (or panelists) or more teams or candidates merely make the argument more complicated, but the strategy remains that we find ourselves step by step being forced into a corner where we must conclude that there is a dictator.

A Look BACK

DEVISING A FLAWLESS VOTING scheme when there are three or more candidates is impossible. Arrow's Theorem and Condorcet's Paradox prove that in the game of voting, we just can't win.

We now realize that the method we use to elect a candidate often has more influence on the outcome of an election than the preferences of the voters.

We learned a great deal by carefully examining simple cases of voting in the sixth grade. There we saw the need to consider several voting methods, because just choosing the candidate with the most votes is not always reasonable. We then discovered that all other alternatives also have problems. We investigated the reasoning behind this negative result by looking at how two sports analysts might produce a preseason ranking of a three-team football league. These simple cases contain the fundamental reasons that any voting scheme involving any number of voters and any number of candidates more than two must have severe defects.

On the bright side, we saw that our strategies of analysis served us well in looking at this messy voting morass. Simple cases looked at carefully revealed the essential reality of the issue.

Life Lessons

Understand simple things deeply.

•

**Look at specific examples
to find general principles.**

MINDSCAPES Invitations to Further Thought

*In this section, Mindscapes marked (**H**) have hints for solutions at the back of the book. Mindscapes marked (**ExH**) have expanded hints at the back of the book. Mindscapes marked (**S**) have solutions.*

Developing Ideas

1. **Landslide Lyndon.** The two candidates in the 1948 U.S. Senate race in Texas were then-Congressman Lyndon Johnson and then-Governor Coke Stevenson. After the statewide election, early indications were that Johnson lost. Nearly a week after the election, it was discovered that at the very last minute, 203 people in Alice, Texas, Precinct 13, voted in the exact order their names appeared on the tax rolls. Incredibly, 202 of those newly found votes were for Johnson. Search the Internet and find out who won that statewide election and by how many votes. Explain why the later-President Johnson earned the nickname "Landslide Lyndon."

2. **Electoral college.** Briefly outline a voting scheme used in the United States where an election between two candidates can be held and the winner is not the person having the most popular votes. Give a specific example where this paradoxical situation occurred.

3. **Voting for voting.** What are some differences between plurality voting, approval voting, and the Borda count?

4. **Voting for sport.** Given an example (ideally from life, but, if not, then fictional) of Condorcet's Paradox within a sports-ranking context.

5. **The point of the arrow (S).** What does Arrow's Impossibility Theorem assert?

Solidifying Ideas

6. **Dictating an election through a dictator.** Suppose that an election among five candidates yields the following outcome:

Candidate:	Jack	Mary	Owen	Adam	Ellen
Votes received:	100	100	100	100	101

Is Ellen really the winner, given that roughly 80% of the population did not vote for her? Offer a suggestion as to what to do next and then describe a problem with your proposed method.

7. **Pro- or Con-dorcet? (S)** Consider the following voter rankings for an election of a group's favorite carbonated beverage (Austin Arctic; Boston Breeze; and Cleveland Club):

	3 voters rank	3 voters rank	3 voters rank
First place	A	C	B
Second place	B	A	A
Third place	C	B	C

Is this scenario an example of Condorcet's Paradox? Explain your answer.

8. **Where is Dr. Pepper? (S)** Given the voting data from Mindscape 7, which soda would win if the group were to use the Borda count method? Explain your answer.

9. **Approval drinking (H).** Returning to the voting scenario from Mindscape 7, it turns out that the voters in the first column find both *A* and *B* acceptable;

the voters in the second column find only C acceptable; and the voters in the third column cannot drink anything but B. Given this additional information, if approval voting is employed in this situation, which soda bubbles to a victory?

Mindscapes 10 through 15 are based on the following musical scenario.

The editors of the five top rap e-zines vote on the best rapper of the year (winning the "Wrapper Award," which is made of aluminum foil). The candidates for this year's Wrapper rapper are some of the best musical artists around: *A*cid Burn Baby Burn, *B*illie Hooker, *C*ool KK, and *D*octor DoDo. The chart below shows the vote tally. Suppose that each voter considered his or her first and second choices acceptable winners and the last two as unacceptable for this prestigious honor.

	Voter 1	Voter 2	Voter 3	Voter 4	Voter 5
1st choice	A	B	C	D	A
2nd choice	B	C	B	B	D
3rd choice	C	D	D	C	C
4th choice	D	A	A	A	B

We now wonder who should win this year's Wrapper Award.

10. **Long live Acid Burn!** You are such a diehard fan of Acid Burn Baby Burn that some call you an "Acid-head." Which voting scheme discussed in this section could you employ in order to have A win the Wrapper Award?

11. **The Hooker rules! (H)** You are such a diehard fan of Billie Hooker that you are going to make it to her concert by hook or by crook. Which voting scheme discussed in this section could you employ in order to have *B* win the Wrapper?

12. **A Cool win, oh-KK?** You are such a diehard fan of Cool KK that the only breakfast you would even consider is an awesome bowl of Special K. Which voting scheme discussed in this section could you employ in order to have *C* win the Wrapper?

13. **Gotta love DoDo!** You are such a diehard fan of Doctor DoDo that you've decided to enroll in medical school. Which voting scheme discussed in this section could you employ in order to have *D* win the Wrapper?

14. **The Hooker scandal.** News breaks that the ultra-hip Billie Hooker was caught singing show tunes at a SoHo club. She is immediately disqualified from the Wrapper. If *B* drops out but the relative voting orders remain (for example, now Voter 1 votes for *A*, then *C*, then *D* as her third choice), then which artist would win using plurality voting?

15. **Borda without Billie.** As in the previous Mindscape, where *B* is out of the running, what is the outcome if you now use the Borda count method?

Creating New Ideas

16. **What's it all about, Ralphie?** Many people believe that, had Ralph Nader dropped out of the 2000 U.S. presidential race, Al Gore would have won the election. Which voting principle discussed in this section best describes this situation?

17. **Two, too (ExH).** Given an election between just two candidates, is it possible that the Borda count method and the plurality voting method give two different outcomes? If so, illustrate why with an example; if not, explain your reasoning.

18. **Two, too II (ExH).** Given an election between just two candidates, is it possible that the Borda count method and the approval voting method give two different outcomes? If so, illustrate why with an example; if not, explain your reasoning.

19. **Instant runoffs.** One way to avoid the lengthy process of runoff elections between the two top candidates is a method called *instant runoff voting*. In this voting scheme, voters not only vote for their first overall choice but consider each possible pair of candidates, and for each pair, vote for the candidate they prefer out of the two. San Francisco adopted this voting procedure in 2002 for their municipal elections. If there were four candidates for mayor of San Francisco, then how many votes would each voter be asked to cast? (Don't forget that the voter first votes for his or her overall first choice.)

20. **Run runoff.** Given the method of instant runoff voting from Mindscape 19, is it possible that the runoff could result in a need for a runoff? If so, illustrate your answer with an example; if not, explain why not.

Further Challenges

21. **Coin coupling.** For this challenge, you will need five pennies, five nickels, and five dimes. The goal is to fill each cell of the following "bank" with a pair of two different coins paired in all six ways the coins can be coupled, that is, penny-over-nickel, nickel-over-penny, penny-over-dime, dime-over-penny, nickel-over-dime, and dime-over-nickel.

The Bank

Cell 1	Cell 2	Cell 3	Cell 4	Cell 5	Cell 6
Penny					
Nickel					

The rules: You start with the penny-over-nickel pair already shown in the bank. You also start with a "Moveable Coins" column and a penny, a nickel, and a dime ready to arrange in it, but don't put them in yet. To get another pair of coins into the bank, look at a pair of coins you already have in the bank. Arrange the three coins in the Moveable Coins column by starting with the ordering of the pair from your bank and then putting the third coin above or below the pair. (Don't take the coins from your bank. Once a pair is in the bank, it stays there. Just arrange the corresponding moveable coins appropriately.) Since penny-over-nickel is in your bank, you could create the column as shown, penny-over-nickel-over-dime (or, alternately you could create dime-over-penny-over-nickel for example), in the Moveable Coins column:

Moveable Coins	Penny	Nickel	Dime

Now you are allowed to put the top-over-bottom pair that is in your moveable column, in this case, penny-over-dime, in another cell in your bank. Your challenge: Can you get all six possible pairs in your bank?

22. **From money-mating to cupid's arrow.** Explain how the coin coupling challenge in Mindscape 21 connects with the argument behind Arrow's Impossibility Theorem.

In Your Own Words

23. **With a group of folks.** In a small group, discuss and actively work through the implications of Arrow's Impossibility Theorem and its justification. After your discussion, write a brief narrative describing the ideas of this theorem in your own words.

24. **Creative writing.** Write an imaginative story (it could be humorous, dramatic, whatever you like) that involves or evokes the ideas of this section.

For the Algebra Lover

Here we celebrate the power of algebra as a powerful way of finding unknown quantities by naming them, of expressing infinitely many relationships and connections clearly and succinctly, and of uncovering pattern and structure.

25. **Vote night.** There are four candidates running for Student Government. All students vote for one candidate. Carl gets 30% of the votes, Emmy gets 25%, George gets 20%, and Sonya gets the rest. If there were 1200 votes cast, how many votes did Sonya get?

26. **Wroof recount.** The election in the previous Mindscape is held again due to accusations of voter fraud (one of the candidate's dogs ate some ballots). This time, Carl got twice as many as George and 2/3 as many as Emmy. Sonya got 157 more than Carl. If there were 1200 votes cast, who won this time?

27. **Biggest loser?** Who was the biggest loser in the election in the previous Mindscape? That is, who received the fewest votes?

28. **The *X*-act winner.** Your school's math club has 73 members. In a recent election for club president, Ada received x^2 votes and Bernhard received $7x - 5$ votes. If all 73 members voted, who won the election?

29. **Borda rules.** Candidates A, B, and C are running for an election in which the votes are counted by the Borda method. Suppose the table here shows the percentages of 1st place votes, 2nd place votes, and 3rd place votes for each candidate. Who wins the election? (*Hint*: suppose there are 100 voters. Tally the Borda count total for each candidate. Who has the lowest total?)

	A	B	C
First choice	60%	20%	40%
Second choice	20%	40%	40%
Third choice	40%	40%	20%

CUTTING CAKE FOR GREEDY PEOPLE

Deciding How to Slice Up Scarce Resources

Can everyone get their fair share?

One of humanity's biggest challenges is dividing scarce resources among competing people. This process can cause tremendous strife—envy, greed, and justice can all come into play. From one point of view, most of the world's problems arise from the difficulty of allocating scarce resources fairly. What one person views as reasonable and equitable, another may regard as exploitative and grossly biased. When large societal groups carry on these disputes, the results can be anything from a baseball strike to a world war.

Perhaps surprisingly, mathematical thinking can contribute substantial insight to this most human of problems. On our journey through mathematical ideas, we have learned methods to understand difficult questions. We've seen the power of stating clearly what we want to know, identifying essential ingredients of the issue, and understanding simple cases deeply. Surely one of the premiere puzzles for humanity is the fundamental challenge of fair division. We'll find that our strategies of thinking are effective in exploring the contentious question of how to divide scarce resources. To explore this question we'll use an appetizing model: cake.

Let Us Cut Cake

We present this section in terms of cutting a cake because cakes are good battlegrounds for divisive allocations. Cakes also have a variety of features that are important in understanding a basic tenet of fair allocation, that is, different people have different opinions about what is valuable. Some people like icing considerably more than the inside portion of the cake. They may be willing to take a smaller piece of cake if it has a lot of icing on it (a corner piece, for example). Where the balance of value lies is a matter of individual taste. If everyone had the same concept of value, the problem of dividing things would be much simpler. However,

827

we will soon see that individual differences in people's preferences actually may allow us to give everyone *more* than they think they rightfully deserve.

The Two Steps of Allocation

One of the themes of this section is realizing that disputes about allocating resources (such as a cake) often can and should be divided into two steps: (1) agreeing on what proportion of the resource each party deserves and then (2) making a division that realizes that agreement. In this discussion of cake-cutting, we concentrate on the second step.

In some cases the decision about what proportion of the resource each party gets is decided in advance. For example, wills may declare that an estate is to be divided equally among the heirs. Divorce settlements—which often lead to splitting headaches—fall into this category as well. However, the question of what allocation of the resources accomplishes that particular division often represents a considerable obstacle. The principles and methods developed here for cutting a cake can be applied directly or used to provide a perspective to help resolve problems of resource allocation.

One of the principles that will emerge from our discussion is that negotiation among the parties is not necessarily a correct or desirable procedure in determining a good allocation. After the decision has been made regarding what proportion of the cake each person deserves, it is not desirable to have the people enter into disputes about their idiosyncratic value systems. Instead, as we will see, we can be successful simply by asking all individuals to privately convey their preferences regarding various hypothetical divisions of the cake, and then have an outside party find from that information a division that is satisfactory to everyone.

The Setting for the Cutting

To focus our discussion, let's assume that we are dividing a cake among people who all agree that they have equal claim to their proportional share of the cake. And let's assume that each person wants a part of the cake that maximizes its value in his or her eyes and that no part of the cake has negative value to anyone; in other words, no one is on a diet, and there are no Brussels sprouts on the cake that a person might wish to (logically) avoid. So more is better, but different parts of the cake may be of greater relative value to one person than to another. For example, if it is a birthday cake, then perhaps the birthday honoree may value the part of the cake containing her name written in script, while another person may value the portion of the cake that contains that beautiful sugar-loaded candy rose.

We will also assume throughout this section that the cake does not contain any indivisible object. For example, if we are thinking of a wedding cake, the little statues of the bride and groom on top may present a problem because, as much as the bride might want, we cannot cut the groom in half. We will work only with cakes where all parts can be divided.

Throughout this book we have seen the power of starting with simple cases to build insights, and we apply that strategy here.

The One Person Case

Suppose we have a cake that we want to divide among one person, say, Alice. This case is not too hard. Alice looks at the cake carefully; assesses its various qualities; weighs the relative value of icing versus size; eyes the candy rose with due appreciation; and, after deep contemplation, takes the whole cake.

Two People—One Cake

Adam and Becky seek an amicable method for dividing a cake in two. In the "I-cut, you-choose" method, Adam cuts the cake into two parts that are equally valuable in his opinion, and then Becky is allowed to choose either piece. Becky is happy because she's given the first choice; so the piece she chooses represents at least half the value of the whole cake in her eyes. Adam is happy because he cut the cake in such a manner that either piece was equally valuable to him.

Is it always *possible* for Adam to cut the cake in such a manner that he would be equally satisfied with either piece? Suppose that Adam holds the knife over the cake and starts with the knife to the left of the cake and slowly moves the knife to the right until finally the entire cake is to the left of the knife. Now if Adam lowers the knife and cuts the cake at any point, then there would be two pieces: the left piece and the right piece.

Notice that at the start, with the knife to the left of the cake, if Adam were to cut it right there, he would actually miss the cake. The entire cake is to the right of the knife. Thus, with this "cut," the left piece is nonexistent and the right piece is the entire cake. Obviously, in this case, Adam would prefer the right piece. Now suppose that he cuts the cake with the knife in the far right position.

Then the entire cake is the left piece. Thus, in this case, Adam would select the left piece. Because he moves the knife continuously from left to right and at the start he would prefer the right piece and at the end he would prefer the left, then there must exist a knife placement somewhere in between such that he would be equally content with either piece.

Life Lesson

Clarify the question.

Cake-Cutting Question.

Given a cake and three people, is there a method of cutting the cake equitably?

The first question is, "What is the question?" Clarifying a question is frequently *the* most important step in resolving the issue. So let's look at the question again carefully.

The difficult word is "equitably." Does equitable mean "fair" in the sense that each person gets a piece that he or she views as having at least one-third of the value of the whole cake? That is a reasonable definition; however, we must face an unfortunate fact: Human beings are typically dissatisfied with just getting their fair share. We want the biggest piece; greed prevails. What we really want is for no one else to get a better piece than we're getting. This greedy attitude leads to a more precise version of the cake-cutting question.

Greedy Division Question.

Given a cake and three people, is there a method for cutting the cake into three pieces so that each person gets the piece that he or she believes has the greatest value? In other words, can the cake be divided into three pieces so that, of the resulting slices, everyone gets their favorite piece?

In the Greedy Division Question, no participant would covet another person's piece. Remember that each of the three people has a different concept of the value of the various cake parts. Why should it be possible to divide the cake into three pieces in such a way that each person's first choice out of the three pieces is different from the first choices of the other two people? We might think that their preferences could be irreconcilable. Perhaps there is no envy-free method, you say? Read on.

A Knife-Moving Method

Let's start with a cake and three people. Suppose someone takes a knife and, starting from the left side of the cake, slowly moves it across the cake. The three potential cake eaters watch intently, probably drooling. As soon as any one of the three believes the knife has reached the one-third mark in her valuation scheme, she yells, "Stop!" The cutter cuts the cake and the yeller gets the left piece. Since the other two people did not yet yell, they each believe that the remaining piece is worth at least two-thirds in their value systems. So they could employ the I-cut, you-choose method to divide the remaining piece or, alternatively, they could let the knife resume its motion, yelling "Stop!" when either of them believes that half the value of the remainder is reached.

A Challenge—Fair but Not Greedy

Does the preceding knife-moving method always work? When might it *not* satisfy the conditions of the Greedy Division Question? Remember that the Greedy Division Question asks whether it's possible for everyone to get his or her first choice after the entire cake is cut and everyone examines all the pieces.

In fact, the knife-moving method always gives a fair division, but not necessarily an envy-free division of the cake. The failure of this cutting scheme further reinforces our skepticism about finding an envy-free division of a cake for three people. To make progress, we examine the issue in a slightly different manner.

The Point of a Division

Cakes come in all shapes and sizes, and we can imagine many different potential ways to cut those cakes. We'll first consider a triangular cake and delve into other-shaped cakes later.

Given a point in a triangular cake, we can cut along the three straight lines from that given point to the vertices of the cake.

These lines divide the cake into three pieces, a North piece, an East piece, and a West piece. Thus, every point in a triangular cake corresponds to a division of the cake into three pieces by cutting from that point to the vertices of the cake.

Suppose Alice, Becky, and Claire wish to share the cake. With the cutting scheme just described, every point on the triangular cake presents each person with a decision, that is, if the cake were cut from that point to the three corners,

All fair, but #1 thinks #2's is better.

Life Lesson
Choosing a convenient representation of an issue often allows us to see new possibilities.

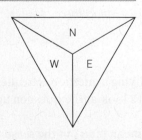

Life Lesson
Representing questions in different ways is often a valuable step toward finding solutions.

which of the three resulting pieces, the North, the East, or the West piece, would be her first choice? To satisfy the conditions of the Greedy Division Question, we seek a point in the triangular cake such that Alice prefers one piece, Becky prefers another, and Claire prefers the third. For example, perhaps at that point Alice prefers the West piece, Becky prefers the East piece, and Claire prefers the North piece. If we cut the cake from that point to the corners of the cake, everyone would get her first-choice piece.

The big question remains: Is there always such a point of universal satisfaction? Notice how we have changed the question from cutting cakes to finding a point in a triangle that possesses certain properties.

What's Your Preference?

Although it may appear at first counterintuitive, our method for finding an envy-free division of the cake will avoid all negotiations and even discussions among the parties. Instead, we privately ask each person a long list of hypothetical questions. Namely, for each point in the triangular cake, we ask each person to tell what her preference would be (the North, East, or West piece) if the cake were to be cut from that point to the vertices of the cake. We then analyze all that information from each of the three people and deduce that there must be some point in the triangular cake where all three people gave three different preferences—one had declared that she would want the East piece if the cake were cut from that point to the vertices of the triangle; the second person said that she would want the West piece if the cake were cut from there; and the third person said that she would want the North piece if the cake were cut from there.

But even from here, we can hear your complaints about the impracticality of this method. You are no doubt saying to yourself, "Great, there are infinitely many points on the cake. So Alice, Becky, and Claire have to make infinitely many decisions about which piece they would each prefer from each point in the triangular cake. So this whole method is totally impractical, and this whole section is a useless crock."

Easy now. Let's try to overcome this infinite obstacle. Suppose we put ourselves in Becky's position. We are given a triangular cake and we are asked to label every point East, West, or North depending on our answer to the question, "Which piece of the cake would you choose if the cake were divided from that point by cutting to the three vertices?"

Preference Diagrams

Faced with labeling trillions and trillions of points (in fact, *infinitely* many), we will not get out our microscope. Instead, we now realize that the points that we want to label East, for example, are not scattered randomly around the triangle. Instead, those points form a region of the triangle on the upper-left side. To understand why, first note that at the upper-left corner of the cake, everyone would definitely pick East. Why? Well, if we "cut" from the upper-left vertex, there would be no northern or western piece at all—only an East piece, that is, the entire cake. Thus, Becky would definitely pick the East piece at that corner. In fact, for any point very near the upper-left vertex of the triangle, she will again choose East. The East piece is dramatically larger than the North or West slivers when the cake

If cut from here, Becky tells us that she prefers the North piece (perhaps N contains extra icing).

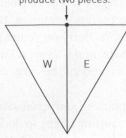

If cut from here, we only produce two pieces.

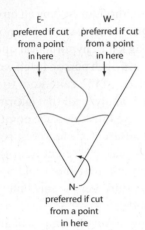

E-preferred if cut from a point in here

W-preferred if cut from a point in here

N-preferred if cut from a point in here

Becky's preference diagram

is cut from any point near that upper-left vertex of the cake to the three vertices of the cake. Likewise, any point in the lower region of the triangular cake will be labeled North and any point near the upper right will be labeled West.

Becky does not have to look at billions of points individually in those particular areas.

Similarly, as Becky considers points on the upper edge of the cake that runs horizontally from left to right, we are certain that she would never pick the North piece, since any cut from a point on that side produces only two pieces: an East and a West piece.

If the cake is cut from the upper-left vertex, we know Becky will prefer the East piece and if the cake is cut from the upper-right vertex, we know she picks West. Therefore, somewhere along the top edge, her preferences will change from East to West.

At the *exact* point of her change in preference, Becky prefers each of the two equally—they are both her first choice (we have a tie). As Becky considers points as she moves down from the top, her preference will eventually change to North.

To label all points in the triangle, all Becky really needs to show is where her preferences change. Where are the boundaries between the regions where she prefers East, West, and North? Roughly speaking, her East-West boundary, her East-North boundary, and her West-North boundary will create a possibly wavy Mercedes-Benz symbol as illustrated. Such a Mercedes-Benz symbol creates her *preference diagram* because it conveys the information about what piece she would choose—the North, East, or West—for every possible division point in the triangle.

So if the cake were to be cut from any point within Becky's "E-preferred" region, then we would know that she would select the East piece. If the cake were cut from any point within Becky's "W-preferred" region, then we know that she would find the West piece the most appealing. Similarly for the "N-preferred" region.

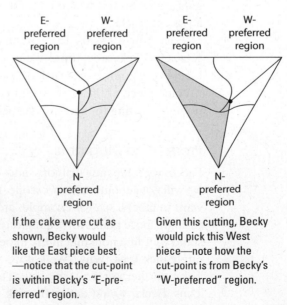

If the cake were cut as shown, Becky would like the East piece best—notice that the cut-point is within Becky's "E-preferred" region.

Given this cutting, Becky would pick this West piece—note how the cut-point is from Becky's "W-preferred" region.

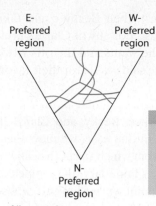

E-Preferred region   W-Preferred region

N-Preferred region

Alice = red
Becky = green
Claire = blue

Superimpose the Preference Diagrams

Once Alice, Becky, and Claire each have privately drawn their personal preference diagrams on three identical pictures of the cake, the three diagrams can be superimposed.

> **The Greedy Division Theorem.**
> *Suppose three preference diagrams are superimposed. Then there will be a point where the three people have indicated that they all prefer different pieces.*

In order to answer the Greedy Division Question affirmatively, we must show that there will be some point or points at which Alice, Becky, and Claire have each made a different choice, that is, if the cake were cut from that point, everyone would have a different favorite piece. In fact, such a point can *always* be found.

Proof of the Greedy Division Theorem

Before proceeding with our argument, we would like to make an impassioned plea. What makes the argument ahead challenging is the delicate analysis of the preference diagrams and their boundaries. Thus, we urge you to carefully study the figures together with the prose. Move slowly, think, and stop often to draw your own diagrams in order to help further your understanding. The argument below is definitely *not* a "quick read."

Let's take Alice's preference diagram and superimpose Becky's preference diagram on it. We will first deal with the possibility that the branch point of Becky's preference diagram lies exactly on a boundary of Alice's diagram. Suppose, for example, that Becky's branch point lies on Alice's East-West boundary curve.

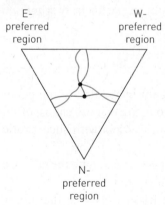

Alice = red
Becky = green

E-preferred region   W-preferred region

N-preferred region

Then that point would be a suitable place from which to cut. We give Claire the piece she most prefers out of the three. Alice takes East or West, whichever is left, because she believes those are equally valuable, and Becky takes the remaining piece since she thinks all three are equally valuable when cut from that point.

In most cases, of course, Becky's branch point will not lie exactly on a boundary of Alice's diagram. Let's suppose that Becky's branch point lies in Alice's West-preferred region, for example.

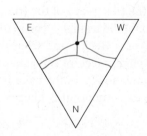

E   W

N

Then Becky's North-East boundary must go from there to some point on the left side of the triangle. So Becky's North-East boundary must somewhere cross Alice's West-preferred border, which consists of a West-East boundary and a West-North boundary. Suppose Becky's North-East boundary crosses Alice's West-East boundary. Then that point is a suitable place from which to cut the cake. At that point Claire has some preference, say North. Then Claire could have North, Becky could take East, and Alice could take West. If Claire preferred East, then Becky could take North, and Alice could take West.

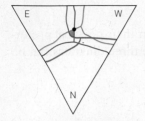

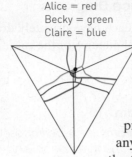

Alice = red
Becky = green
Claire = blue

Finally, if Claire preferred West, then Becky could take North, and Alice could take East. For any of Claire's preferences, she can have what she wants and the other two can take a piece that they have said is a tie for their favorite piece.

Notice that usually we will find near that branch point a whole region of points where Alice, Becky, and Claire all prefer different pieces of the cake. In that case, we may select any point in that region as the place from which to cut the cake to the three corners. Only if one of Claire's boundary lines coincides exactly with Alice's or Becky's near that point will we have to choose the point exactly. Therefore our argument is now complete.

Recapitulation

Notice that Alice, Becky, and Claire *never conferred* with one another during the entire cake-cutting process. Instead, they independently conveyed to an outside party their hypothetical preferences for which piece they would prefer if the cake were cut to the vertices from each point in the cake. Those decisions were all made privately. Later we proved that no matter what preference decisions each made, there must be a point from which we can cut the cake to the three vertices such that each of them can have their first choice. So this method of allocating resources emphasizes the power and value of separating the participating people.

How important is the triangular shape? Well, now that we have learned something about triangular cakes, let's explore variations of it, particularly when we have a cake of a different shape.

Non-Triangular Cakes and Pies

At real bakeries, occasionally we see a cake that is not triangular. Some peculiarly shaped cakes are rectangular or even round (go figure). Some cakes, often angel food cakes, have holes in the middle. How can we deal with such exotic cakes?

The easiest method for dealing with non-triangular cakes is to take the cake we have and put it inside a big triangle as illustrated.

Now we can use the method for the triangle. Notice that every point in the triangle, whether it's in the cake or not, represents a division of the cake—that is, the part of the cake that is in the North piece of the triangle, the part of the cake that is in the East piece of the triangle, and the part of the cake that is in the West piece of the triangle. The method above will give a point in the triangle where Alice, Becky, and Claire will have different views about their preferences for which piece of the cake they want. Notice that the central cutting point may or may not be in the cake itself. In particular, if we consider an angel food cake, the division point is likely to be in the hole.

From a Piece of Cake to the Heart of Texas

One time long ago, there was a wealthy Texan who owned all of Texas. He was rich and tall (in fact, he only shopped at rich and tall men's clothing stores). When he died he left Texas to his three children, Alice-Bob, Becky-Bob, and Claire-Bob, saying they should divide Texas among themselves. Having read this book (well, actually just an early draft of the book, since this was quite a while ago), they simply drew a map of Texas inside a triangle, made hypothetical choices for each potential cutting point in the triangle, and found a point from which to cut that made every one of the three happy because they all got their favorite piece of the Lone Star State. This example demonstrates that our method of dividing resources is not only useful for cakes, but for land or other resources as well.

Four or More People

What happens if we want to divide a cake among four people? Faced with this new problem, we'll first try to solve it by using a variation of the method we used for dividing a cake among three people.

We might guess that a way to proceed would be to consider a square cake. However, this great idea does not seem to work. What parts of the cake-cutting method that we learned above fail to work if we try using a square cake and four people? In particular, can you draw four preference diagrams in which no point divides the cake such that each of the four people prefers a different piece? Although this sensible attempt does not work, we do not give up.

Just because one idea fails, we don't stop. There are often several different ways of generalizing an idea. If one fails, we'll try another. Here we seek to take a technique that worked successfully with three people and generalize it to the case with four people. Going from a triangle to a square seems reasonable, but does not work. It turns out to be more effective to go from a triangle to a tetrahedron. We invite you to discover how to adopt the previous cake-cutting strategy to the case of four people dividing a tetrahedral cake. Notice how each point in a tetrahedron corresponds to dividing the tetrahedron into four sub-tetrahedra by drawing triangles from the point out to the vertices of the big tetrahedron. This generalization to four people requires us to visualize tetrahedra in three dimensions, which is a difficult task.

Splitting the Rent

This idea of cake-cutting can be applied to other divisions as well, for example, dividing the rent of an apartment among three roommates. Suppose three people share a three-bedroom apartment. The total rent is $1000 per month. But the three bedrooms are not equally attractive. One is bigger than the others, one has a view, and one is very small. It would not be fair to make all three people pay the same rent. The person who gets the small room should pay less, and people might prefer the room with the view. But how much should each person pay?

The first step is to clarify the question. The three bedrooms will be called the Big Room, the Tiny Room, and the Room with a View, and the three renters will be Alice, Becky, and Claire. The total rent paid must equal $1000. The question is, "How much should the occupant of each room pay?" As an example, one possible division of the rent would be for Becky to occupy the Big Room and pay $450, for Alice to stay in the Tiny Room and pay $150, and for Claire to take the Room with a View and pay $400.

The question is, "Can the rent and rooms be divided in such a way that no one would want to change rooms if they had to pay the rent associated with the other room?" In other words, we want to divide the rent as above so that Becky does not say to herself, "Well, I like the Big Room, but at $150, I would prefer to take the Tiny Room and save $300 per month." We want all three people to feel that they would prefer their room at its price rather than wanting to switch to either of the other rooms and paying its rent.

Let's make the assumption that any renter would prefer a free room if the other two rooms charge rent, and that for any division of the rent, each renter would like some room.

Often the way to solve one question is to realize that it is similar to another question that we do know how to answer. In this case, we may feel that the rental harmony question is similar to the cake-cutting question. Can we answer it in a similar way?

Let's start by seeing if we can represent all possible ways to divide the rent. Again, a triangle does the trick. Each vertex represents the place where the total rent is paid by the occupant of one room, respectively, the Big Room, the Tiny Room, and the Room with a View. Each other point of the triangle is labeled with three dollar values that add up to $1000 where the first number is the rent for the person in the Big Room, the second number is the rent paid by the person in the Tiny Room, and the third number is the rent paid by the person in the Room with a View.

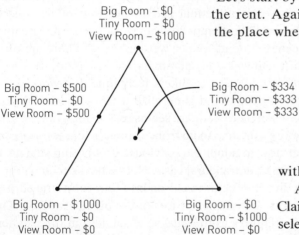

Big Room – $0
Tiny Room – $0
View Room – $1000

Big Room – $500
Tiny Room – $0
View Room – $500

Big Room – $334
Tiny Room – $333
View Room – $333

Big Room – $1000
Tiny Room – $0
View Room – $0

Big Room – $0
Tiny Room – $1000
View Room – $0

At each point in the labeled triangle, Alice, Becky, and Claire each has an opinion about which room she would select if the prices were divided as listed. The three renters make their decisions independently. Each person can record her preferences by making a preference diagram as in the proof of The Greedy Division Theorem. In this case, however, all the boundary lines between preferences end up going to the three vertices of the triangles, because everyone will prefer the same room on each side since each such point gives a room away for free.

Now the three diagrams are superimposed and the same arguments similar to those used in cake-cutting demonstrate that there is at least one point where the three renters prefer three different rooms.

Finding a way to represent the situation with renters allows us to see that this situation is very similar to the cake-cutting issue that we already dealt with. So the same method of solution solves a different question as well.

WE CAN DIVIDE a cake among three people so they all get their favorite piece with regard to that division. Surprisingly, the method does not involve negotiation or psychological issues. Instead it is an unexpected application of mathematical reasoning.

Our strategy is to represent the question in a convenient way. We associate each point of a triangular cake with a division of the cake into three pieces by cutting to the corners. This correspondence of points in the triangle with divisions of the cake lets each person express his or her hypothetical preferences for each possible division point in the triangle. We then superimpose the three diagrams that record each person's preferences from each potential division point. A geometric insight shows that at some point, the three people must have different first choices. By transforming a question about cake into a question about drawing wavy Mercedes-Benz symbols in a triangle, we can find a solution.

The rent-splitting question shows that three renters can divide the rent for an apartment with unequal rooms in such a way that each renter is happy to pay his or her rent rather than switch to another room with its rent. Representing this new question in an effective way allows us to see that at its root the question has the same essential solution as the cake-cutting question. Exploiting successful methods in different settings is an important way to take advantage of insights.

Questions can be formulated in many different ways, so we can sometimes choose the arena in which to fight our battles. Part of successful thinking is to extract the essential ingredients from a complicated situation. That way, we can avoid distractions and get to the heart of the matter. Often the hearts of quite different matters are similar and can be conquered with the same insight.

Life Lessons

Clarify the question.

•

Abstract the essence.

•

Choose a convenient representation of an issue.

MINDSCAPES Invitations to Further Thought

*In this section, Mindscapes marked (**H**) have hints for solutions at the back of the book. Mindscapes marked (**ExH**) have expanded hints at the back of the book. Mindscapes marked (**S**) have solutions.*

Developing Ideas

1. **You-cut, you-lose.** In the I-cut, you-choose method, would you rather be the cutter or the chooser? Which role would give you a piece of cake that may be more valuable in your opinion?

2. **Understanding icing (S).** Suppose a person who had not read this section declares that cutting cake fairly for three people is simple—just weigh the cake very carefully and give each person a piece that weighs one-third of the total. What idea is that person missing?

3. **Liquid gold.** Suppose you and your two brothers are dividing a mound of solid gold. You could use the cake-cutting methods to get a fair division, but

instead you melt it down and give each sibling an equal weight of gold. Why does this method work for gold, but not for cake?

4. **East means West.** Suppose you have a triangular cake and you find the point in the middle such that if you cut from that point to the three vertices of the triangular cake, you like all three pieces equally. Now suppose that you cut the cake from a point on the line from that equal-liking point to the east vertex (that is, upper-right vertex). If the cake were cut from that new point, would you prefer the West piece, the East piece, or the North piece? Why?

5. **Two-bedroom bliss (H).** Suppose you and a roommate are renting a two-bedroom apartment that costs $1000 per month. The bedrooms, the Big Bedroom and the Small Bedroom, are not equally desirable, but they are each OK. So you and your roommate decide that you will pay unequal rents depending on who takes the Big Bedroom and who takes the Small Bedroom. Why is there a rent so that you would be equally happy to live in the Big Bedroom for that rent or live in the Small Bedroom for $1000 minus that rent? Why can you find prices for living in the apartment where both you and your roommate are happy to pay the price you pay for the room you each get?

Solidifying Ideas

6. **Your preference.** Suppose the accompanying figure is your preference diagram. Several points in the triangle are highlighted. For each point complete the sentence, "If the cake were cut into three pieces from this point by straight lines to the vertices, I would prefer. . ."

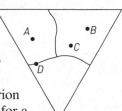

7. **Bulk.** Suppose for you, bigger is better, so your only criterion is volume. What will your preference diagram look like for a triangular cake?

8. **Don't move that knife.** Give a specific scenario to show that the three-person moving-knife method does not always give everyone their favorite piece.

9. **On the edge (H).** Why are there only two viable choices for your preferred piece if the point from which the cake is cut is a point on an edge of the triangular cake? Mark a triangle with the different possible choices on each edge. What must happen at the vertices?

10. **Just do it.** Get three people together and have them share something. Have them fill out preference diagrams and actually proceed with the cake-cutting method. Afterwards, explain to them why it had to work. Write up their preferences, the cutting solution, and their reactions to why such a cutting is always possible.

11. **The real world.** Give three real-world examples where this cake-cutting method might be useful for resolving disputes beyond sharing cakes.

12. **Same tastes (H).** If you are dividing a cake among three people using the cake-cutting method from this section, what would happen if all three people have the exact same preference diagram? Where would you cut the cake? Why?

13. **Crossing the line.** In each triangle shown on the next page, each edge is marked with a red and a blue dot. In each picture, connect the same colored dots by a possibly distorted, inverted "Y." What is the smallest number of times they must intersect?

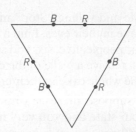 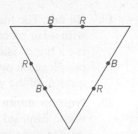

14. **Cutting up Mass (S).** You, Joan, and John want to share Massachusetts. Suppose that you like Williamstown, Joan likes Springfield, and John likes Boston. Make the preference diagrams and superimpose them to find a place from which to cut.

15. **Where to cut (H).** The accompanying figure pictures three overlaid preference diagrams. Where should the cake be cut so each person is happy with the part of the cake they get and why?

Creating New Ideas

16. **Land preference (ExH).** Suppose you are preparing to divide a triangular piece of land to share with two other people. You assess the value of each small triangle of land and mark it as pictured on the next page. A triangle with no number in it has value 0. Approximately what would your preference diagram look like?

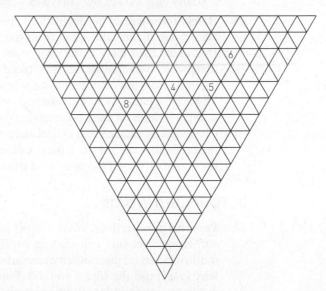

17. **Uneven pair (S).** Suppose two people want to divide a cake, but because one person has paid more, both agree that this person should receive two-thirds of the cake while the other should get only one-third. How could they cut the cake?

18. **Diversity pays.** Explain why having differences of opinion about the value of various parts of the cake may make it easier (than if everyone had identical preferences) to find a division where everyone feels they are getting more than anyone else.

19. **Be fair.** The moving-knife and yelling "Stop" method leaves all three people with a fair piece of the cake in their eyes. Find a moving-knife method for giving a fair division for four people, five, six, and so on. That is, if there are four people, each person must receive a piece of the cake that is worth at least one-quarter of the value of the whole cake in their eyes.

20. **Nuclear dump (ExH).** Suppose there is a nuclear waste disposal site in your state and you very much do not want to own that property, so it has negative value in your eyes. Given this aversion, explain why the accompanying preference diagram is reasonable.

Further Challenges

21. **Disarming (H).** Two nuclear superpowers decide to disarm and agree to reduce their arsenals by 3% per year. Each superpower lists the weapons in its arsenal and gives its own assessment of the value of each of its weapons as a percentage of its total arsenal. For example, Power A might declare that a particular submarine is worth 0.6% of its arsenal. The superpowers switch lists. Each superpower can require the other power to destroy any collection of weapons that add up to 3% in the eyes of the other power. Why would each superpower feel they are getting the better deal?

22. **Cupcakes.** Suppose you had 100 different cupcakes that you wanted to divide among three people. Why is it possible to divide them into three portions such that everyone gets his or her favorite of the three, and at most three cupcakes actually need to be cut? (Hint: Consider first a large triangle and scatter the cupcakes as though they were points.) Extra challenge: Show that actually only two of the cupcakes need to be cut.

23. **Barely consistent.** It is possible for Chris to prefer the West piece if the cake is cut from the point marked W and the East piece if the cake is cut from the point marked E, even though the West piece is actually larger when cut from the point E than it is when cut from the point W. Suppose each part of the cake has a value measured by the number written on the accompanying figure. What value could Chris view the shaded part of the cake as having in order to make the preference of East when the cake is cut from E consistent with the choice of West if the cake is cut from W?

In Your Own Words

24. **Personal perspectives.** Write a short essay describing the most interesting or surprising discovery you made in exploring the material in this section. If any material seemed puzzling or even unbelievable, address that as well. Explain why you chose the topics you did. Finally, comment on the aesthetics of the mathematics and ideas in this section.

25. **With a group of folks.** In a small group, discuss and actively work through the argument showing that we are always able to cut a cake with three people so that each person gets their favorite piece with regard to that cutting. Include a careful discussion of the preference diagrams. After your discussion, write a brief narrative describing the ideas in your own words.

26. **Creative writing.** Write an imaginative story (it can be humorous, dramatic, whatever you like) that involves or evokes the ideas of this section. Be as

creative as possible, attempt to capture the ideas and essence of the section, and avoid technical details as much as possible.

27. **Power beyond the mathematics.** Provide several real-life issues—ideally, from your own experience—that some of the strategies of thought presented in this section would effectively approach and resolve.

For the Algebra Lover

Here we celebrate the power of algebra as a powerful way of finding unknown quantities by naming them, of expressing infinitely many relationships and connections clearly and succinctly, and of uncovering pattern and structure.

28. **Your X.** You and your ex-roommate happen to share a birthday. Your friends bring one cake to the party. Because you and your ex-roommate both love cake, and because all your friends are on low-carb diets, you split the cake between just the two of you. You get a piece that has volume $4x - 10$ and your ex gets a piece with volume $2x + 20$. If the volume of the entire cake is 58 cubic inches, who got the piece with the larger volume?

29. **Musical X's.** You play the violin in a chamber trio that includes a pianist and a cellist. After one very hot summer rehearsal, you all decide to split a large ice cream cake. You get $x^2 - 20$ cubic inches, the pianist gets $2x + 15$ cubic inches, and the cellist gets $3x + 9$ cubic inches. If the total volume of the cake was 88 cubic inches, who got the piece with the largest volume?

30. **Cake plot.** Imagine a cake in the shape of a rectangle that is placed on the xy-plane with corners at the points A-$(0, 0)$, B-$(0, 3)$, C-$(5, 3)$, and D-$(5, 0)$. You plan to cut the cake along a line from point A to point C. Find an equation for this line. What if you wanted to cut the cake along the line from B to D? Find an equation for this line.

31. **Cake trisection.** Imagine a cake in the shape of a triangle that is placed on the xy-plane with corners at the points A-$(0, 0)$, B-$(3, 6)$, and C-$(6, 0)$. You plan to cut the cake into three equal pieces that meet at the central point of the cake D-$(3, 3)$. Find an equation for the cutting line from A to D, for the cutting line from B to D, and for the cutting line from C to D.

32. **Roommate wrangling.** You and a friend rent a two-bedroom apartment together. The rent is $1000 per month. One bedroom is nicer than the other, so you want to adjust the rent fairly. The table below shows the maximum rent each of you is willing to pay for each bedroom.

	Max. rent for small bedroom	Max. rent for large bedroom
You	$400	$600
Roommate	$300	$700

Before deciding on actual rents, is it clear who should get the small bedroom and who should get the large bedroom? Why or why not? To determine a possible way to split the rent, plot the points (300, 700), (400, 700), (300, 600), and (400, 700) on a pair of xy-axes. (What do the x and y coordinates represent?) Use two line segments to connect the points in an X shape. Estimate the coordinates of the point where the line segments intersect. Do these coordinates seem reasonable as rents for the two bedrooms? Which person gets which room?

Farewell

As you read these final pages, we wonder what you will take away from your journey. After you close this book for the last time, we hope that some parts of it will remain open in your mind. Perhaps a few specific concepts will become part of your permanent collection of ideas. Perhaps your sense of mathematics has changed. Maybe mathematics will no longer seem quite so dry; maybe it seems richer now. We hope the life lessons of this book expanded your repertoire of strategies and modes of thought. These lessons are simple, yet profound. We hope that they help you strengthen your confidence to face challenging life issues and conquer them. Mathematics is an empowering force. If you can grapple with infinity and the fourth dimension, what can't you do? We hope that in 20 years you will still be contemplating some of the ideas you encountered here.

We now invite you to look back over your experience. Perhaps some concepts were challenging and appeared always just out of reach; some you struggled with before making them your own; and some are so solid in your mind that they are now a part of you. In any case, we hope that you will recall all of them with great satisfaction.

It all began, as it should, with fun and games. Our mission was to pose some entertaining and thought-provoking conundrums that would introduce certain modes of logical thought and problem solving. Simultaneously, we foreshadowed ideas to come.

What do you now think of when you contemplate the notion of *number*? Perhaps you think about the power of counting, as in the Pigeonhole principle, or you recall intriguing patterns of natural numbers, such as the Fibonacci numbers or the prime numbers. Perhaps your mind drifts to mod clock arithmetic, and you remember that arithmetic serves our technological world in unexpected ways, from error-correcting digits to public key cryptography. Perhaps you now see and appreciate subtle distinctions between the rational and the more mysterious and less understood irrational numbers.

There are a handful of mathematical arguments that are so artful that they may be viewed as elegant. Three of these arguments are from the study of numbers: There are infinitely many prime numbers; the square root of 2 is irrational; and 1 equals 0.99999 . . . . We hope you made these proofs your own and that you will share them with others.

What could be more empowering than understanding the nuances of infinity? That some infinities are actually larger than others is a profound, counterintuitive idea that has shaken the foundations of many people's beliefs. Certainly, when you think of the infinite, part of what we hope you will remember is the power of a simple and clear definition—that of a one-to-one correspondence—and how carefully following its consequences opened a new world of insight.

> . . . the primary question was not What do we know, but How do we know it.
>
> ARISTOTLE

> Oh moment, one and infinite!
>
> ROBERT BROWNING

> Wherever there is number, there is beauty.
>
> PROCLUS

We hope that the Pythagorean Theorem and its jigsaw-puzzle proof are yours for life. Once the aesthetically appealing Golden Rectangle becomes part of your consciousness, you will see it often—in architecture, art, and snails. The jumbled pinwheel pattern with the same symmetry of scale as a regular checkerboard pattern, despite its chaotic appearance, is an enticing illustration of the beauty and the nuance of the aperiodic.

Can you sit in a room and see in your mind's eye the outline of its dual octahedron? Maybe when you see a soccer ball, you will recognize the wedding of a Platonic solid and its dual. We hope you have developed a sense of the geometry of the fourth dimension—where being inside a box may not be an obstacle. Indeed, building some intuition about the fourth dimension is an outstanding example of the power of analogy and abstraction.

The amorphous universe of topology thrives on the unexpected. Whether you are removing a rubber vest without removing your outer coat, untying a knotted ring, or dissecting Möbius bands, we hope you discover important lessons in visualization and the wonderful consequences of bending and twisting space.

Topology also has unexpected consequences in real life—including the life of the twisted DNA molecule. Perhaps you now understand why at any instant there must be two diametrically opposite places on Earth where the temperatures are identical. We believe it is exciting to discover abstract mathematical issues that lead to interesting new facts about the real world. We certainly hope you share our appreciation for the wonderful interplay between the physical and the abstract.

Fractal pictures are literally infinitely intricate. They all arise from repeating a simple process infinitely often, and they seem to capture the complexity of nature. Repetition is at the heart of both chaos and fractals. As ordered as we wish our universe to be, in reality, chaos reigns.

We hope that you now see that amazing coincidences are nearly certain to happen to you. After one thinks about chance and randomness, surprises appear in a whole new light. Randomness is a powerful force; it permits us to collect answers to embarrassing questions that people would normally never answer truthfully. We hope you have developed a feel for expectation and that you will look at lotteries and other games of chance with a critical eye.

You've seen that we must be cautious with statistics and not read more into an average than we rightfully should—remember what the average American has. We hope that you have a better sense of the significance of false positives in medical tests. Perhaps you are more sensitive to the possibilities of unintended consequences. If improved airline safety raises prices and therefore makes more people drive, maybe safer airplanes would lead to more accidental deaths—but not in the air.

As we approached the end of this *Farewell*, we considered copying it in a smaller font that would, of course, contain the entire *Farewell* in an even smaller font, which would, of course, contain another copy of the whole *Farewell*, and so on. That fractal font farce would be fun, but we resisted the temptation.

Nevertheless, we do end with a paradox, because we hope that this end is also a beginning. One of the great features about mathematics is that it has an endless frontier. The farther you travel, the more you see over the emerging

> *Mighty is geometry; joined with art, resistless.*
>
> **EURIPIDES**

> *The true spirit of delight . . . is to be found in mathematics as surely as in poetry.*
>
> **BERTRAND RUSSELL**

> *God has put a secret art into the forces of Nature so as to enable it to fashion itself out of chaos into a perfect world system.*
>
> **IMMANUEL KANT**

> *Chance, too, which seems to rush along with slack reins, is bridled and governed by law.*
>
> **BOETHIUS**

horizon. The more you discover, the more you understand what you've already seen. How many more ideas are there for you to explore and enjoy? How long is your life?

LESSONS FOR LIFE

1. *Just do it.*

2. *Make mistakes and fail, but never give up.*

3. *Keep an open mind.*

4. *Explore the consequences of new ideas.*

5. *Seek the essential.*

6. *Understand the issue.*

7. *Understand simple things deeply.*

8. *Break a difficult problem into easier ones.*

9. *Examine issues from several points of view.*

10. *Look for patterns and similarities.*

Acknowledgments

Fourth Edition

First and foremost, we wish to thank the many students and teachers who have used previous editions and shared with us their comments, reactions, questions, projects, and ideas. This text is all about using mathematics as a metaphor for effective thinking, and thus your comments are most meaningful and gratifying to us. We also wish to thank those colleagues who carefully reviewed portions of this new edition and offered insightful comments and suggestions. Those reviewers are:

- Libby Arnesen, Thomas Nelson Community College
- Sandra Becker, Eastern Michigan University
- Kiran Bhutani, Catholic University
- Stephen Gendler, Clarion University
- Kerin Keys, City College of San Francisco
- Erika King, Hobart and William Smith Colleges
- Roger Knobel, University of Texas Pan American
- Oscar Levin, University of Northern Colorado
- Margaret Michener, University of Nebraska Kearney
- Cindia Stewart, Shenandoah University
- Cornelia Van Cott, University of San Francisco

A special thanks goes to Erika King, who went through the new graph theory chapter node by node, and also De Cook, who checked both the text and the solutions manual for accuracy. Of course, any rogue vertex or stray error is the sole fault of the authors.

We wish to acknowledge the wonderful work and deep dedication of the John Wiley & Sons "The Heart of Mathematics family." Their support and contributions continue to improve and refine this work. In particular, we would like to thank: Laurie Rosatone (vice president and publisher), Jennifer Brady (project editor), Anne Scanlan-Rohrer (developmental editor), Thomas Kulesa (product designer), Melissa Edwards (editorial operations manager), Beth Pearson (associate content editor), Christine Kushner (director of marketing), Ken Santor (senior production editor), Sheena Goldstein (photo editor), James O'Shea and Maureen Eide (designers).

Finally, we wish to thank our colleague, collaborator, and friend Dr. Deborah J. Bergstrand, from Swarthmore College, for her continued contributions to this work. Beyond her leadership and tireless work on the Instructor Resource material, she offered her considerable graph theory expertise in the new chapter and helped craft the algebra-enriched Mindscapes. Thank you, Deb, for your tremendous loyalty to this effort.

Third Edition

As we complete this third edition of *The Heart of Mathematics: An invitation to effective thinking*, we wish to express our deep appreciation to all those students and faculty who have used our text. In particular, we wish to thank all the students and colleagues who have written to us over the years with their reactions, comments, and generous words. We wrote and designed this book to be read and we are truly heartened to discover that both students and instructors alike have found using the book to be a pleasurable and meaningful experience.

While this third edition includes a number of significant changes—including an entire new chapter—perhaps the most dramatic is our change in publishers. With this new edition we begin what we believe will be a long and productive collaboration with John Wiley & Sons. We are delighted to be associated with such a distinguished publisher and to have the enthusiasm and support of the entire Wiley family. Members of the executive team Peter Wiley (the current Chairman of the Board), Brad Wiley (past Chairman), Bonnie Lieberman, MJ O'Leary, and Joe Heider have been extremely welcoming and encouraging. Our wonderful publisher Laurie Rosatone has led the *Heart of Mathematics* project at Wiley from the beginning. We wish to acknowledge and thank all those at Wiley who have made this third edition a reality. From editorial, to design, to publishing, to marketing, to sales we have been fortunate to work with a most talented and creative team of individuals. In particular we would like to thank Jennifer Brady, Anne Scanlan-Rohrer, Sarah Davis, Heather Johnson, Jeof Vita, James O'Shea, Patricia Nelson and the many others who have worked on this third edition of *The Heart of Mathematics*. We are excited to work with the entire Wiley "family" in bringing mathematics to life for real students around the world.

This edition contains a new chapter on statistics, which was greatly enriched by the outstanding contributions of our colleague Philip Everson from Swarthmore College. His insights, suggestions, and reactions were tremendously helpful and we wish to sincerely thank him for all his excellent work. Deborah Bergstrand, also from Swarthmore College, continues to be an invaluable collaborator especially on the *Instructor's Resources* that accompanies this text. We wish to thank Deb for her outstanding contributions, her dedication to this project, and her loyal friendship.

We also wish to thank Jodi Cotten at Westchester Community College and Tina Carter at Buffalo State College who provided student and instructor resources available online through the book's website www.heartofmath.com and through WileyPlus and also our friends at MATTI Associates for the enhanced web-based applets. These individuals' enthusiasm and hard work have been an inspiration to us. We also wish to thank the following faculty members who took the time to review various versions of the third edition and share their insights and suggestions:

Linda Barton
Ball State University

Deborah Bergstrand
Swarthmore College

Terence Blows
Northern Arizona University

Valleri Bond
Austin Community College

Christina Carter
Buffalo State College

Hei-chi Chan
University of Illinois, Springfield

Tim Chartier
Davidson College

Jodi Cotten
Westchester Community College

Anthony Cutler
Suffolk University

Candace Dance
Onandaga Community College

Sloan Despeaux
Western Carolina University

Steven Dunbar
University of Nebraska, Lincoln

Chuck Dunn
Linfield College

Philip Everson
Swarthmore College

Mark Farris
Midwestern State University

Russell Goodman
Central College

Richard Grassl
University of Northern Colorado

Sarah Greenwald
Appalachian State University

Roger Hammons
Morehead State University

Kenneth Hoover
California State University
Stanislaus

Tim Hsu
San Jose State University

Tarcia Jones
Austin Community College

Sue Hurley
Siena College

Michael Keynes
American University

Kerin Keys
City College of San Francisco

Firooz Khosraviyani
Texas A&M University

Donald Krug
Northern Kentucky University

Judy Lalani
Central New Mexico Community
College

Andrew Long
University of Northern Kentucky

Vincent Maltese
Monroe County Community College

Bernard McDonald
George Washington University

Colm Mulcahy
Spelman College

Cornelius Nelan
Quinnipiac University

Amber Puha
California State University
San Marcos

Harold Reiter
University of North Carolina,
Charlotte

Kathleen Shannon
Salisbury University

Barbara Shipman
The University of Texas,
Arlington

Deirdre Smith
University of Arizona

Mary Jane Sterling
Bradley University

Tatiana Tatarinova
Loyola Marymount University

Sarah Triana Stovall
Stephen F. Austin State University

Amy Szczepanski
University of Tennessee, Knoxville

Ron Taylor
Berry College

Agnes Tuska
California State University, Fresno

Roger Vogeler
Central Connecticut State University

Elizabeth Walters
Loyola College, Maryland

Deborah Wood
University of Arizona

We know how busy faculty are and thus we truly are humbled by the amount of time and energy these individuals have given of themselves to help in the crafting of this new edition. In this spirit, we wish to also acknowledge and thank all our colleagues who have shared their stories, comments, or suggestions with us regarding their classroom experiences with *The Heart of Mathematics*. These faculty members include:

Helmer Aslaksen
National University of Singapore

Marion W. Athearn
Lincoln School

Adrienne Behrmann
Eastside Preparatory School

Ann M. Bledsoe
Columbia College

Russell Blyth
Saint Louis University

Martin Cohen
Brooklyn College

Anthony Cutler
Suffolk University

Scott Eberle
The University of Texas at Austin

Moira Fearncombe
The Illinois Institute of Art

Alfinio Flores
Arizona State University

Gary Gordon
Lafayette College

Glen Granzow
Salish Kootenai College

Raymond N. Greenwell
Hofstra University

Eloise Hamann
San Jose State University

Jason Howald
SUNY Potsdam

Art Israel
Eastern Connecticut State University

Dan Kalman
American University

Edward C. Keppelmann
University of Nevada at Reno

Mark Krusemeyer
Carleton College

Andre Kundgen
California State University San Marcos

Laurel Langford
University of Wisconsin at River Falls

Tom LoFaro
Gustavus Adolphus College

Lew Ludwig
Denison University

Vinnie Maltese
Monroe County Community College

Attila Mate
Brooklyn College

Colm Mulcahy
Spelman College

Stephen New
University of Waterloo

Jeff Parent
Oakland Community College

Gary Peterson
University of Idaho

Stephen Robinson
Wake Forest University

Peter Ross
Santa Clara University

James Schaefer
Rhode Island College

Melinda S. Schulteis
Concordia University

Rich Stankewitz
Ball State University

Ron Taylor
Berry College

Marc Zucker
Marymount Manhattan College

Finally, as our primary goal is to touch the life of the mind of real students, we wish to express our gratitude to the following faculty who generously surveyed their students regarding their reactions to *The Heart of Mathematics* in the spring 2009 term:

Christina Carter and her colleagues
Buffalo State College

Kerin Keys
City College of San Francisco

Robert Whitton
Davidson College

Second Edition

First and foremost we wish to thank all the students and instructors who have read *The Heart of Mathematics: An invitation to effective thinking* and shared their thoughts and reactions with us. Their comments informed our thinking on the second edition and also confirmed our vision that students appreciate a text that is lively and offers great mathematical ideas together with methods of thinking and analysis that transcend mathematics.

Thank you to those who have provided comments through e-mails, focus groups, and reviews of the revision proposal and the new material.

Vivian Anderson
SUNY Oswego

William Bloch
Wheaton College

James Brown
Northern Essex Community College

Michael J. Caulfield
Gannon University

Beth Chance
California Polytechnic State University, San Luis Obispo

Janis Cimperman
St. Cloud State University

Michael Clapp
California State University Fullerton

Michael Colvin
California Polytechnic State University, San Luis Obispo

Jodi Cotten
Westchester Community College

Joyce Blair Crowell
Belmont University

Candice Dance
Onondaga Community College

Lily Eidswick
University of Montana

Sandra Fillebrown
St. Joseph's University

Thomas Foley
St. Joseph's University

Steve Gendler
Clarion University of Pennsylvania

Sarah Greenwald
Appalachian State University

Raymond Greenwell
Hofstra University

Linda D. Henderson
The University of Texas at Austin

David Kennedy
Glenville State University

Diane Laison
Temple University

Nicole Lang
North Hennepin
Community College

Lew Ludwig
Denison University

Teresa Magnus
Rivier College

Liz McMahon
Lafayette College

Jackie Miller
The Ohio State University

Ken Millett University of
California Santa Barbara

David Phillips
Georgia State University

Phil Pickering
Genesee Community College

Vadim Ponomarenko
Trinity University

Agnes Rash
St. Joseph's University

Altha Rodin
The University of
Texas at Austin

Allan Rossman
California Polytechnic State
University, San Luis Obispo

Thomas Rousseau
Siena College

Alan Russell
Elon University

Melinda Schulteis
Concordia University Irvine

J. Sriskandarajah Madison
Area Technical College Truax

Sarah Triana Stovall
Stephen F. Austin
State University

Ron Taylor Berry College
Ileana Vasu Holyoke
Community College

Michael Veatch
Gordon College

Nancy Wyshinski
Trinity College

Sarah Ziesler
Dominican University

Two users of the first edition surveyed their students and shared those comments with us. We appreciate the feedback from the students of Lowell Abrams, George Washington University, and Karl Schaffer, De Anza College.

We are grateful for the comments that came from a survey commissioned by our publisher, Key College Publishing. Those who responded to the survey include the following.

Peter Blaskiewicz
McLennan Community College

Mark Bollman Albion College
David DeCoste
St. Francis Xavier University

Sloan Despeaux
Western Carolina University

Johnny Duke
Shorter College

Stephanie Fitchett
Florida Atlantic University

Frank Garcia
North Seattle Community College

Scott Garten
Northwest Missouri State University

Cathy Gorini
Maharishi University of
Management

Dorothea Grimm
Cardinal Stritch University

Roger E. Haglund
Concordia College

Kira Hamman
Hood College

David Hartz
College of St. Benedict

Jeralynne Hawthorne
Warner Pacific College

David W. Henderson
Cornell University

Vernon M. Kays
Richland Community
College

Frederick Lane
Palm Beach Community
College

Myrtle Lewin
Agnes Scott College

Andy Long
Northern Kentucky University

Carol Marinas
Barry University

Richard Mason
Indian Hills Community College

Julia Massey
University of West Alabama

Susann Mathews
Wright State University

Cynthia McGinnis Okaloosa-
Walton Community College

Sr. Jeanne Moenk
Notre Dame College

Bette Nelson
Alvin Community College

John Noonan
Mount Vernon Nazarene University

Mary L. Platt
Salem State University

K. Brooks Reid
California State University,
San Marcos

Melissa Reeves
East Texas Baptist
University

Richard S. Rempel
Bethel College

Theresa Riel
Pima Community
College

Kathy V. Rodgers
University of Southern
Indiana

Jan Roy
Montcalm Community
College

Robin Ruffato
Ball State University

Peter Sandberg
Judson College

Cameron Sawyer
Southwestern University

Harry Sedinger
St. Bonaventure University

Judith Silver
Marshall University

Rick Silvey
University of Saint Mary

Larry Smith
Snow College

Brian Snyder
Lake Superior State University

Jamie Thomas
University of Wisconsin,
Manitowoc

Gideon Weinstein
Montclair State University

Julia Wilson
SUNY College at Fredonia

Fred Worth
Henderson State University

Connie Yarema Abilene
Christian University

Li Zhou Polk Community
College

Many people were involved in the creation of the supplements for this second edition. Deborah Bergstrand, Swarthmore College, helped us expand the instructor resource materials and made those materials more robust. Jodi Cotten, Westchester Community College, provided suggestions for the Web site and evaluated all of the electronic materials that accompany this text. She also provided content for many of the electronic supplements. Joshua Laison, Colorado College; Michael Veatch, Gordon College; Teresa Magnus, Rivier College; and Joyce Blair Crowell, Belmont University, gave us their test questions for the Test Bank. Bob Heal, Larry Cannon, and Joel Duffin of MATTI Associates developed the applets for the Interactive Explorations. Kevin Briley and Summit Interactive seamlessly filmed and produced the Instructor Videos and provided a platform for the Instructor CD-ROM.

We wish to express our gratitude to several people who have made significant contributions to our work on the second edition. Mary Parker provided us with a wealth of information, advice, and guidance on the new statistics sections. Stephanie Nichols provided solutions for the Instructor Resources and Adjunct Guide for the Mindscapes in the new sections. Joni Harlan read and line edited each chapter of the text, providing insight and sound suggestions all along the way.

We would like to thank Ingrid Mount, our project manager at Elm Street Publishing Services; and at Wiley publishing: our production project manager Beth Masse, our designer Marilyn Perry, and our art editor Jason Luz. We especially wish to thank Mike Simpson, Richard Bonacci, Allyndreth Cassidy, and Nigel Fenton for their many years of support, encouragement, and enormous effort on behalf of *The Heart of Mathematics.*

Finally, we wish to express our deep appreciation for the creativity and imagination of our first publisher, Jerry Lyons. Jerry was the first person to truly understand our vision, and through his unbridled enthusiasm and insight he was able to turn seemingly impossible ideas into reality. This book is a testament to his outstanding talents. We were very saddened to have lost our friend Jerry Lyons and we continue to miss him.

First Edition

One theme of this book is that ideas arise from many sources, and certainly many sources were important in the creation of this book. Here we thank those who contributed to *The Heart of Mathematics: An invitation to effective thinking.* We began writing this book in February 1995. Since then, hundreds of people have helped to inspire us and shape our thinking. First and foremost we would like to thank all our students who used various draft versions of this text (many without figures). Our students' input had an enormous impact on what we wrote and how we wrote it, but more fundamentally, they gave and continue to give us the motivation to make mathematical thought a pleasurable part of their lives.

We hope this textbook helps teachers bring mathematics to life for students. We are therefore especially grateful to the many professors of mathematics who reviewed drafts of the manuscript. Our reviewers included:

Thomas Banchoff
Brown University

John Emert
Ball State University

Phillip Johnson
Appalachian State University

Darrell Kent
Washington State University

Earl Fife
Calvin College

Rich France
Millersville University

Kay Gura
Ramapo College of NJ

Edwin Herman
St. Thomas University

Fred Hoffman
Florida Atlantic University

George Kertz
University of Toledo

Joe Malkevitch
CUNY, York College

John Orr
University of Nebraska

Edward Thome
Murray State University

Stan Wagon
Macalester College

We would especially like to thank Earl Fife and Joe Malkevitch, who read several versions of this text. Their detailed comments were of great value. We also wish to thank Mary L. Platt from Salem State College; Bret A. Simon and Angela Vanlandingham, from the University of Colorado, who used a manuscript version of the text in their classes; and Stephen A. Kenton, from Eastern Connecticut State University, who used a draft of the infinity chapter in his course *Infinity: Math and Philosophy*. Their comments and their students' reactions were extremely valuable.

The Editorial Director at Springer-Verlag, Jeremiah Lyons, was a constant source of enthusiasm, inspiration, and insight. From the very beginning, Jerry supported our vision of an enticing book that would be truly enjoyable for real students to read, and he remained confident of its potential. He enabled us to turn our more unconventional ideas into reality. Throughout this process he has demonstrated a great sense of humor, and it is our hope that one day he will win the Superfecta at Saratoga. It has been a privilege to work with Jerry Lyons.

Our developmental editor, Jeanne Woodward, helped to focus us on our meta-message of effective thinking. Jeanne inspired us to reexamine the format and organization of the sections. Her comments and those of John Bergez resulted in a much more readable text. Eric Gerde accomplished the monstrous task of answering each and every Mindscape question. Larry Cannon, Robert Heal, and Richard Wellman from Utah State University allowed us to tap their creative talents for the CD-ROM component of the kit. Daniel Symmes and Jim Carbonetti created the wonderful 3D art. The production of a complicated project such as this text requires many editors, artists, designers, and others. Their imagination and talent helped bring this book to life. We thank everyone for their creativity, hard work, and dedication.

Finally, we would like to thank the Educational Advancement Foundation for their support. The EAF fosters methods of teaching that promote independent thinking and student creativity.

We close with our individual acknowledgments.

EDWARD B. BURGER: The two greatest influences on my life have been my parents, Florence and Sandor Burger. They have been a constant source of love, support, joy, and inspiration.

My interest in this course started in 1988. I was intrigued by a course offered at Austin Community College entitled *Mathematics: Its Spirit and Use*. I want to thank Stephen Rodi from ACC for giving me the opportunity to teach liberal arts students. Also from those early years, I wish to thank Jeffrey Vaaler, my Ph.D. advisor, who has been and continues to be one of the greatest teachers, mentors, and friends I've ever had.

In 1993, I developed a similar course at Williams College. My colleagues and students at Williams have always been a great source of inspiration, and I thank them all. Two who deserve special recognition are Stephen Fix, from the English Department, and my student Eric Cohen, both of whom provided me with many valuable specific comments.

For me, the work on this book began and ended at the University of Colorado at Boulder, where I spent sabbatical leaves in 1995 and 1999. I wish to thank the Department of Mathematics and the Center for Number Theory for all their interest, encouragement, and support in my work.

The greatest aspect of this project for me was my association with Michael Starbird. Mike is a sheer joy and inspiration to be around. He has the wonderful ability to turn work into play and play into innovative ideas. Far beyond this project, Mike has taught me life lessons too numerous to mention. My respect and admiration for Mike are immense, and I am most fortunate and grateful to have him as a role model, mentor, and close friend. My collaboration with Michael Starbird has been one of the most valuable experiences of my life.

MICHAEL STARBIRD: I am incredibly lucky to have a family whose support and encouragement have inspired me throughout my life. My parents gave me curiosity and independence. My wife, Roberta, and children, Talley and Bryn, are the joyful foundation for all I do. I want to thank all my family from the bottom of my heart.

I began teaching a mathematics course for liberal arts students because Betty Sue Flowers, the Director of the Plan II Honors Program at UT, needed a new mathematics course to enrich the technical side of the program. That request opened the door to a decade-long project that includes this book. Paul Woodruff, the subsequent Director of the Plan II Honors Program, continued to support the course and encouraged me to continue its development. Friends are often most significant for what they ask rather than what they give. I thank my friends for what they asked.

I thank all my friends and colleagues from the Department of Mathematics and across UT who have been uniformly supportive of the idea of making mathematics fun. They encouraged me by showing a real interest in the mathematical ideas that are at the heart of this book.

Finally, I want to thank my coauthor. Certainly the single greatest joy from this whole project was working with Ed. His imagination is boundless. His energy is endless. His insight and humor are uncontainable. He has a clear vision of the whole project and can see every aspect of what needs to be done. He does global thinking and attends to details. But best of all, the collaboration we created brought out the strengths of each of us. It is an honor and a privilege for me to learn from and work with my friend and coauthor, Edward Burger.

Hints, Expanded Hints, and Solutions

H = Hint; ExH = Expanded Hint;
S = Solution

CHAPTER 2: NUMBER CONTEMPLATION

2.1: Counting

6. (ExH) One Life Lesson we can apply here is: *Visualization and experimentation often lead to surprising and even counterintuitive results.* You want to estimate the weight of the million dollars to see if the challenge is at all reasonable. Find a large paperback book and estimate how many one-dollar bills will fit on one page. Now check the number of pages in the book and estimate the weight of the whole book. Can you use these estimates to get a rough idea of how much one million dollars in single bills will weigh? Remember that you don't need an exact answer, just a general sense—that is, the *order of magnitude*—of how heavy the money would be. We hope you are surprised by this weighty question.

8. (H) Don't worry about the total number of pieces on the board. Just estimate how many pieces are on the 64th square.

11. (S) Let's first estimate the thickness of an ordinary piece of paper by noting that packages of 200 sheets of paper are more than 1/2-inch thick. So, a single piece of paper is at least 1/400-inch thick. Now, after one folding, the paper is twice the original thickness. After two foldings, the paper is $4 = 2^2$ times as thick. After 50 foldings, the paper will be 2^{50} times as thick. The resulting paper is then at least $2^{50}/400$-inches thick. (That's more than 2.8×10^{12} inches, which is more than 40 million miles!)

14. (ExH) One Life Lesson we can apply here is: *Consider all the possibilities.* When you're trying to find one matched pair, what are all the possibilities if you pull out only two socks? What are the pairing possibilities when you pull out a third sock? When you want two matched pairs, you certainly need four socks, but what's the worst that could happen if you pull out only four socks? What happens when you pull out a fifth? When you want a pair of black socks, what's the worst that could happen as you pull out socks one by one?

17. (H) How many numbers less than 10,000 have no 3 in them? How many choices are there for each of the four digits of such a number?

20. (S) Within the next 100 years, virtually all of the 6.2 billion people currently populating Earth will die. If fewer than 50 million people died each year, then, at the end of 100 years, only 5 billion people would have died, which means well over 1 billion people would live at least 100 years. This contradiction shows that at some point more than 50 million people will die in a year. Alternatively, note that the average number of people who will die each year is $6,200,000,000/100 = 62$ million. Since this is the average, there must be at least some year in which more than 62 million people will die (otherwise the average would be *lower*).

24. (H) Ramanujan tells you that 4^3 and x^3 have a special relationship to 2261. Express that relationship as an equation. Can you use this equation to find x?

2.2: Numerical Patterns in Nature

7. (S)

n	1	2	3	4	5	6	...
$(F_n)^2$	1	1	4	9	25	64	...
$(F_{n+1})^2$	1	4	9	25	64	169	...
sum	2	5	13	34	89	233	...
	F_3	F_5	F_7	F_9	F_{11}	F_{13}	...

Note that we are getting all the odd Fibonacci numbers. This leads to the formula, $(F_n)^2 + (F_{n+1})^2 = F_{2n+1}$.

9. (ExH) One Life Lesson to apply here is: *Devising a good representation of a problem is frequently the biggest step toward finding a solution.* Creating a table to keep track of everything is extremely helpful here. The table below has been partially filled in. Why is the number of new babies at Month 4 equal to 1? Why is the number of mature pairs at Month 5 equal to 2? Fill in the rest of the table and extend it through Month 10. Add up the total number of rabbit pairs in each month. How is the total number of pairs in Month 6 related to the numbers in previous months? Do you see a general pattern?

Time in months	Start	1	2	3	4	5	6	7
Number of mature pairs	0	0	1	1	1	2	3	?
Number of pairs of new babies	1	0	0	1	1	1	2	?
Number of pairs of older babies	0	1	0	0	1	1	?	?
Total number of rabbit pairs	1	1	1	2	?	?	?	?

15. (S) If you look closely, you'll notice that the pieces don't line up exactly. The triangle with sides 3 and 8 appears to be similar to the big "triangle" with sides 5 and 13. If this were true, then the corresponding ratios would be equal. But 8/3 isn't 13/5. Since these are ratios of consecutive Fibonacci numbers, the ratios are close, and that's why this is a convincing trick!

16. (H) Start with the largest Fibonacci number smaller than the given number and work your way backward. $52 = 34 + 18$. $18 = 13 + 5$, so $52 = 34 + 13 + 5$.

27. (S) Mindscape II.8 showed us that $F_{2n} = (F_{n+1})^2 - (F_{n-1})^2$, which can be factored as the product of $(F_{n+1} - F_{n-1}) \times (F_{n+1} + F_{n-1})$. This means that besides 2, none of the even-indexed Fibonacci numbers are prime!

34. (ExH) One Life Lesson we can apply here is: *Break a difficult problem into easier ones by considering simple examples.* Before trying to come up with an argument that works in general, look at small examples. We're told that N is a natural number that is *not* a Fibonacci number, so let's try $N = 4$. What's the largest Fibonacci number that does not exceed 4? That's your value of F. Now what does $4 - F$ equal in this case? Is it less than the Fibonacci number that comes right before F? What if $4 - F$ were equal to the previous Fibonacci number? Then how would N compare to the Fibonacci that comes after F?

Do you have an argument that will work for a general N? Try it for $N = 6$ and look for a recurring circumstance that can be generalized.

35. (H) Start with a natural number N. If N is a Fibonacci number, then we're done; if it isn't, then grab the largest Fibonacci number smaller than N, and call it F. Now repeat with $N - F$.

40. (H) Rewrite the fraction under the square root as $F_{n+2}/F_n = (F_{n+2}/F_{n+1}) \times (F_{n+1}/F_n)$.

46. (H) Note that $1 + \dfrac{1}{x} = \dfrac{x+1}{x}$. (Why?)

49. (H) To expand $\left(\dfrac{1-\sqrt{5}}{2}\right)^2$ you'll have to FOIL, but before you start, notice that $\left(\dfrac{1-\sqrt{5}}{2}\right) = \dfrac{1}{2}\left(1 - \sqrt{5}\right)$

This means that $\left(\dfrac{1-\sqrt{5}}{2}\right)^2 = \left(\dfrac{1}{2}\right)^2 \left(1 - \sqrt{5}\right)^2 = \dfrac{1}{4}\left(1 - \sqrt{5}\right)^2.$

In other words—factor out the ½ before you FOIL!

2.3: Prime Cuts of Numbers

11. (ExH) One Life Lesson to apply here is: *Ground your understanding in the specific.* It's easy to answer this question with "Yes." What's interesting is providing a concise and convincing proof. Rather than offering a possibly rambling argument, can you precisely describe a specific infinite collection of natural numbers that are definitely not prime?

15. (H) One of the two primes in the sum will have to be 2.

16. (S) The harder question is, Are any of these prime? We can describe each element in the list by its number of digits. If it has an even number of digits, then the number is divisible by 11. If it has $3, 6, 9, \ldots$ digits, then the number is divisible by 3. By computer search, the first three primes in the sequence have 19, 23, and 317 digits.

20. (H) Try to factor the $n = 16$ and $n = 17$ cases. (Factor the $n = 40$ and $n = 41$ cases for the bonus.)

24. (S) Write the original number as $X = 13A + 7$. Adding 22 yields $X + 22 = 13A + 7 + 22 = 13A + 29 = 13A + 13 \times 2 + 3 = 13 \times (A + 2) + 3$. So 13 goes into our new number $(A + 2)$ times with a remainder of 3. Alternatively, just add $7 + 22$ and find its remainder when divided by 13.

26. (H) Write the two numbers, X and Y, in the following way: $X = 57A + r$, and $Y = 57B + r$. The difference is $(X - Y) = 57A - 57B = 57 \times (A - B)$.

35. (S) By exhaustively looking at differences between successive primes you will find that the first string of six nonprimes appears between 89 and 97. But it's more interesting to consider the hint. Let $M = 2 \times 3 \times 4 \times 5 \times 6 \times 7$. Both M and $M + 2$ are divisible by 2, so $M + 2$ isn't prime. Similarly, $M + 3$ is divisible by 3, $M + 4$ is divisible by 4, and so on up to $M + 7$, which is divisible by 7.

37. (ExH) One Life Lesson to apply here is: *Simple observations often have deep consequences.* Write down the first 10 or 12 natural numbers. What do you notice about every third number? How many of these numbers are prime? Does this help you answer the question? Do you have to be a little careful with the triple 1, 2, 3?

40. (H) Suppose there were only finitely many primes, $p_1, p_2, p_3, \ldots, p_L$. Consider $N = p_1 p_2 p_3 \ldots p_L + 1$. Think about the prime factors of N.

49. (H) Your answer will require some kind of root sign.

2.4: Crazy Clocks and Checking Out Bars

4. (ExH) One Life Lesson to apply here is: *Understanding a specific case well is a major step toward discovering a general principle.* Dividing 7 by 3 gives a remainder of 1. So 7 mod 3 equals 1. Dividing 7^2 (= 49) by 3 also gives a remainder of 1, so 7^2 mod 3 equals 1. When you square the value of 7 mod 3, what do you get? So how are $[7 \bmod 3]^2$ and 7^2 mod 3 related? How does this help you reduce 7^{1000} mod 3?

8. (H) Express each number as a simpler number mod 12.

9. (S) Note that $(3 \times 0) + (1 \times 7) + (3 \times 1) + (1 \times 7) + (3 \times 3) + (1 \times 4) + (3 \times 0) + (1 \times 0) + (3 \times 0) + (1 \times 2) + (3 \times 1) + (1 \times 8) = 43$. Since the sum is not evenly divisible by 10, it's not a correct UPC. The corresponding sums for the next two codes are 40 and 42, respectively. So the second code is the correct one.

13. (H) If the covered digit were D, then the sum would be $55 + 3D$. Now find a value for D that will make the sum divisible by 10.

19. (S) There are three unknown digits, the 9, the 1, and the 7. Since each digit could be one of two different numbers, we have eight possible combinations in all to try: 903068823517, 903068823511, 903068823577, 903068823571, 403068823517, 403068823511, 403068823577, 403068823571. Of all these numbers, only 903068823577 is a valid code. This is your best guess.

22. (ExH) One Life Lesson to apply here is: *Just do it.* Let D represent the unknown check digit. Then the bank error-checking system described in the section gives the expression $7 \times 6 + 3 \times 2 + 9 \times 9 + 7 \times 1 + 3 \times 0 + 9 \times 0 + 7 \times 2 + 3 \times 7 + 9 \times D$, which equals $171 + 9D$. The check digit D must have a value that makes $171 + 9D = 0 \pmod{10}$; that is, the number $171 + 9D$ must be a multiple of 10. What value of D makes $171 + 9D$ a multiple of 10?

26. (S) $129 = (9 \times 13) + 12$, so 12 is the remainder when 129 is divided by 13. We can also say $129 = 12 \pmod{13}$. You would spin around 9 times and then move the clock hand ahead 12 hours more.

33. (H) Let D represent the check digit, and compute the resulting sums. Now find the values of D between 0 and 9 that will make the sum divisible by 11.

37. (H) Look at numbers of the form $x\,y\,0000000\,z$ and $y\,x\,0000000\,z$ whose sum is $3x + y + z \pmod{10}$, and $3y + x + z \pmod{10}$, respectively. When does $3x + y = 3y + x \pmod{10}$?

49. (H) Your formula will be linear: $y = mx + b$. Try to find m and b so that when $x = 1$, $y = 10$.

2.5: Public Secret Codes and How to Become a Spy

8. (ExH) One Life Lesson to apply here is: *Seek the essential.* Notice that this challenge asks only that you encode the message "2" and then *explain how* to decode it. Is there a place in the text that encapsulates this process for the sender and receiver? (Look for a table with two silhouettes.) Remember that you don't actually have to decode the received message, just explain the method in this case.

9. (H) Encrypting a message involves raising it to the public power 7 mod the public number 143. Decoding an encrypted message involves raising the encrypted message to the secret power 103 mod 143.

11. (S) Note first that $m = (3 - 1) \times (5 - 1) = 8$. Since e must be relatively prime to m, we need only consider the values $e = 1, 3, 5,$ and 7. For each possible value of e, find d and y that satisfy $de - 8y = 1$. For example, for $e = 1$, fill in the following blanks: ____ $\times 1 - 8 \times$ ____ $= 1$. Since $1 \times 1 - 8 \times 0 = 1$, $(e = 1, d = 1)$ is a pair. Similarly, since $3 \times 3 - 8 \times 1 = 1$, $5 \times 5 - 8 \times 3 = 1$, and $7 \times 7 - 8 \times 6 = 1$, $(e = 3, d = 3)$, $(e = 5, d = 5)$, and $(e = 7, d = 7)$ are possible pairs.

15. (H) Remember that decoding an encrypted message involves raising the encrypted message to the secret power.

16. (S) The hint asks us to recall that $5^6 \equiv 1 \pmod{7}$. This means that $(5^6)^k \equiv 1^k \equiv 1 \pmod{7}$ for any integer k. In particular, since $600 = 6 \times 100$, it is convenient to choose $k = 100$, giving us $5^{600} \equiv (5^{6 \times 100}) \equiv (5^5)^{100} \equiv 1^{100} \equiv 1 \pmod{7}$. Similarly, since $1{,}000{,}000 = 10 \times 100{,}000$, $8^{1{,}000{,}000} \equiv 1 \pmod{11}$.

17. (ExH) One Life Lesson to apply here is: *Carefully understand and analyze the facts at hand.* Consider how Fermat's Little Theorem applies here: Because 7 is prime and 5 is a number that does not have 7 as a factor, we have $5^6 \equiv 1 \pmod{7}$. Now, given what you learned in Mindspace I.4 from the previous section, you know that 5 raised to any multiple of 6 is equivalent to 1 (mod 7). In particular, $5^{60} \equiv 1 \pmod{7}$ and $5^{600} \equiv 1 \pmod{7}$. What does $5^{600} \times 5^{60}$ equal? Does this help you analyze $5^{668} = 5^{600} \times 5^{60} \times 5^8$?

2.6: The Irrational Side of Numbers

8. (ExH) One Life Lesson to apply here is: *Use properties you know and follow assumptions to their logical conclusions.* If you can write a given number as a ratio of two integers, then it's rational, so the first number, 4/9, is clearly rational. What about 1.75? Can you write this number as a ratio with denominator 100? A similar idea works for the last number, 3.14159. For the third number, start by recalling that $\sqrt{20} = \sqrt{4}\sqrt{5} = 2\sqrt{5}$. For the fourth number, suppose it were rational. Then you would have $\dfrac{\sqrt{2}}{14} = \dfrac{a}{b}$ for some integers *a* **and** *b*. What happens when you multiply both sides of this equation by 14? Do you contradict a fact you know about $\sqrt{2}$? What does this mean about your assumption that $\frac{\sqrt{2}}{14}$ is rational?

10. (H) Imitate the proof of the irrationality of $\sqrt{2}$, except replace the notions *even* and *odd* with *divisible by 5* and *not divisible by 5*.

15. (S) We need a modification of the proof in Mindscape II.10. Assume $\sqrt{10} = c/d$ with c and d having no common factors. Squaring gives $c^2 = 10d^2$. Since the right-hand side is divisible by 5, the left-hand side is also divisible by 5, so that c is divisible by 5. (If c weren't divisible by 5, then c^2 wouldn't be divisible by 5 either.) We can replace c with $5n$ to get $25n^2 = 10d^2$, or $5n^2 = 2d^2$. We want to show that 5 divides d. Imagine writing out the prime factorization for the left and right sides of the equation. On the left we have all the prime factors of n (listed twice since we are squaring n) and 5. On the right we have 2 and all the prime factors of d (listed twice). Since the two sides are equal, we call on the uniqueness of prime factorizations to argue that the list of primes for both numbers is the same. Since the prime 5 appears on the left side, it must also appear on the right side. And since it can only come from the prime factorization of d, we must have that 5 is a prime factor of d. So d is divisible by 5, and we have our contradiction.

17. (H) Assume that E is rational—that is, $E = n/m$, so that $12^{n/m} = 7$. Raise both sides to the mth power; $12^n = 7^m$ and look at the prime factorizations of each side.

20. (ExH) One Life Lesson to apply here is: *Experimentation is an effective means of resolving difficult issues.* One way to show that E is a rational number is to find a rational value for E that satisfies the equation. So you want to find integers n and m so that $8^{n/m} = 4$. Follow the Hint for Mindscape 17 to get $8^n = 4^m$. Now do some experimenting with small natural numbers to find values of n and m that work.

When contemplating the second challenge of the Mindscape, think about why the equation $8^n = 4^m$ does not yield a contradiction, whereas the corresponding equations in Mindscapes II.17, II.18, and II.19 do.

25. (S) Assume $\sqrt{(2/3)} = a/b$ (with no common factors). Squaring gives $2b^2 = 3a^2$. At this point it doesn't matter whether you choose 2 or 3, but you must stick with it! (We'll choose 3.) The right side is divisible by 3, and so $2b^2$ is divisible by 3. Since 3 is prime and 2 isn't divisible by 3, b^2 must be divisible by 3. Again, since 3 is prime we conclude that b is divisible by 3. Writing $b = 3n$ and substituting it into the equation gives $18n^2 = 3a^2$, or $6n^2 = a^2$. Using the same reasoning, we conclude that 3 divides a. So 3 divides both a and b, contradicting the fact that a and b had no

common factors. We conclude that our original assumption was wrong, and therefore $\sqrt{(2/3)}$ is irrational.

30. (H) Use the $\sqrt{2}$-is-irrational proof as a template, but begin by *cubing* both sides instead of squaring both sides.

33. (S) Not always. The not-so-satisfying counterexample: π and $(-\pi)$ are both irrationals, yet their sum is zero, which is rational. The numbers 1.01001000100001 ... and 0.10110111011110... are both irrational because their decimal expansion doesn't repeat. Their sum is 1.11111111111111 ... $= 10/9$, a rational number. Keep in mind that sometimes the sum is irrational, for example, $\pi + \pi = 2\pi$.

34. (H) Remember that you can multiply an irrational number by itself.

40. (H) Assume that they are both rational—that is, $(a + b) = m/n$ and $(a - b) = r/s$, and solve for b.

47. (H) Consider the case that x is rational. What kind of number would $7x^3 - 19x^2 + 10x$ be? Now solve for the $\sqrt{2}$ term.

2.7: Get Real

8. (ExH) One Life Lesson to apply here is: *Seek the essential.* The requested number lies to the right of 12.0345691 on the number line and to the left of 12.0345692. Thus its decimal expansion must begin 12.0345691 .... What are the possibilities for the next decimal digit? If you choose 0, can you stop there?

10. (H) Take an irrational number that you know and stick its nonrepeating decimals after 5.70.

14. (S) Use long division (or a calculator!). 7 goes into 60 eight times with a remainder of 4; bring down the zero. 7 goes into 40 five times with a remainder of 5; and so on. 6/7 = 0.8571428571 42857142857142857143 ....

21. (S) Call our elusive number E, so that $E = 43.12121212...$ Since there are two digits in our repeating segment, multiply E by 100 to shift the decimals by two digits: $100E = 4312.121212 ....$ Subtracting gives:

$$100E = 4312.12121212 ...$$
$$-E = 43.12121212 ...$$
$$99E = 4269.$$

Thus we see that $E = 4269/99$.

24. (ExH) One Life Lesson to apply here is: *Look for patterns and similarities that allow for the use of previous knowledge.* Try the approach used to show that 0.999 ... = 1 by letting $N = 71.23999 ....$ Multiply N by 10 to obtain $10N$.

Then compute $10N - N$. Is the result a number with a decimal expansion that terminates? Can you write that number as a fraction? Now set it equal to $9N$ and take one more step to obtain N as a fraction.

25. (H) This number has arbitrarily long sequences of zeros. Why does that imply that there is no repeating sequence and that the number is irrational?

28. (H) $(10^n + 1)$ has $n - 1$ adjacent zeros. So, the decimal expansion contains arbitrarily long sequences of zeros. Could this number have a repeating sequence of length N?

31. (S) Using the same argument used in Mindscape II.25, let's show that $y = 0.21221222122221 \ldots$ is also irrational. Suppose there were a repeating sequence of length N. If the repeating sequence were all 2's, then we'd end up with a rational number. If the repeating sequence were not all 2's, then eventually we would see a non-2 digit after every N digits. But this isn't the case.

34. (ExH) One Life Lesson to apply here is: *Devising a good representation of an issue is frequently the biggest step toward finding a solution.* First let $L = 1/100,000,000,000$ just for convenience. Now, try drawing a schematic of the first few steps of this challenge. Given the difficulty of actually drawing a line segment with length only L, it's OK to exaggerate your length. (Remember, you're the line segment!) Below, you are shown in red.

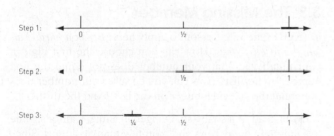

What will the schematics look like as you repeatedly cut the distance from your center in half? Will you (as the line segment) eventually contain 0? How is this different from Mindscape III.33?

37. (H) If the two real numbers were farther than 1 apart, then there would be an integer between them. Can you find a sequence of rationals that are evenly spaced over the whole line (like the integers) where the spacing is less than the difference between the two given reals?

49. (H) Add the third equation to the first equation to get $5a + 2c = 29$. Then add the third equation to the second equation to get $3a + 4c = 37$. Now you have two equations in two unknowns. (What happens when you subtract the second new equation from twice the first new equation?)

CHAPTER 3: INFINITY

3.1: Beyond Numbers

6. (H) Weighing two jars of pennies against each other indicates which jar has more pennies without giving any clue as to the number of pennies in each jar. The process of passing out tests to students quickly tells you whether you have more tests, more students, or exactly the same number of tests as students.

7. (S) If the correspondence were not one-to-one, we would have one of the following situations:

a. A symbol representing no company: No one would buy stock in a company that didn't exist. Conversely, if a company has no symbol, then no one will ever purchase its stock.

b. A symbol representing two different companies: This poses a problem if you want to invest in one of the companies but not in the other.

c. A company having no symbol: It's in the company's best interest to have an exchange symbol, because it makes the buying and selling at stocks easier.

13. (H) Does each resident of the United States have a Social Security number? Are they different?

15. (ExH) One Life Lesson to apply here is: *Be open to new ways of looking at a situation.* Put aside laundry for a moment and think about playing poker. Is it easier to tell how many chips you have when they are in a big pile or when they are in little, equal-sized stacks? If two stacks are the same height, does this give a one-to-one correspondence between the chips in one stack and the chips in the other?

Now empty out the bag of quarters onto your desk. What size stack would be useful here?

17. (S) The number of possible Social Security numbers is 10^9, but the population is only about 250 million. If a one-to-one correspondence existed, then there would be the same number of elements in each set.

20. (H) Given the information in the question, there may be more rooms than students, more students than rooms, or the same number of students and rooms. Why? Ask your roommate.

3.2: Comparing the Infinite

9. (ExH) One Life Lesson to apply here is: *Simple observations often have deep consequences.* To describe an explicit one-to-one correspondence between two collections, it can be very useful just to begin listing the members in an orderly fashion, if possible. If you do this carefully for **N** and 6**N**, a one-to-one correspondence may appear quite naturally.

N: 1 2 3 4 ...
6N: ? ? ? ? ...

In this case, it's also possible to describe a one-to-one correspondence using a formula. If n designates an arbitrary natural number, to what member of the collection 6N might n correspond?

11. (H) Make a chart with two columns: one containing the list of natural numbers and the other containing the elements of *TIM*. The correspondence can't be given by a nice formula.

16. (H) What if the hotel manager instructs all guests to move one room down the hall?

18. (ExH) One Life Lesson to apply here is: *Just because a specific attempt failed does not mean that the task at hand is impossible.* The challenge for the beleaguered night manager is to move each of the current guests to a specific new room in such a way that infinitely many rooms are made available for the new guests. He can't just ask everyone to move down "infinitely many rooms" because this would put everyone out on the street! Can you think of an infinite collection of just some of the rooms that has the same cardinality as the entire hotel? Look at Mindscape 7 for an idea. Is there an orderly way for all the current guests to move to this collection of rooms? (Keep in mind that current guests can all move to their new rooms at the same time.)

25. (S) If the sets had the same cardinality, then there would exist a correspondence that matched every element from one set with every element of the other, something the question states specifically cannot happen. Therefore, the sets have different cardinalities, and the set that intuitively seems larger in fact is.

26. (H) First list all fractions with 1 in the denominator. Then list all new fractions that can be obtained by using 2 for the denominator. Continue increasing the denominator and listing in order all the new fractions that can be created.

31. (ExH) One Life Lesson to apply here is: *By doing, we often discover valuable insights.* You clearly have infinitely many peanuts! Is there a one-to-one correspondence between the nuts and the natural numbers? In other words, as we progress down the list of natural numbers: 1, 2, 3, 4, and so on, can we also progress through our piles of peanuts in an orderly fashion, being sure that no peanut is left out? Count the peanuts starting with the first pile, then the second pile, etc. As you count them, are you establishing a one-to-one correspondence? Will every peanut appear somewhere on your list if you count long enough?

32. (S) Let's suppose that the streets are laid out as in the figure below. Now pick an intersection, label it 1, and start spiraling outward. As we move outward, label each intersection with the next natural number. Since we will eventually "hit" any single

intersection, we've provided a valid one-to-one correspondence between intersections and natural numbers.

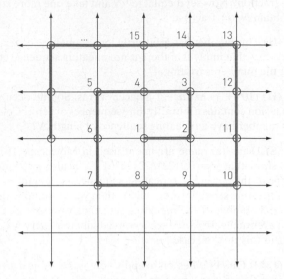

36. (H) Is ball number 1 in the barrel? No, it was gone after the first step. How about number 2? Number 1294959?

47. (H) Once you have a formula that gives y when you plug in x, use this formula to help you find another formula that gives x when you plug in y.

3.3: The Missing Member

4. (ExH) One Life Lesson to apply here is: *Break a hard challenge into easy ones.* How can you choose the first digit in your number so that your number does not equal the first number on her list? How can you be sure your number does not equal the second number on the list? And the third?

7. (S) Instead of a six-by-six Dodgeball game, Cantor's game has infinitely many rows and infinitely many columns. Since Cantor starts with a listing of the reals, it is as if Player One has filled out the entire Dodgeball chart before Player Two begins. Finally, instead of filling a table with X's and O's, Cantor creates his new number using two digits (say, 2's and 4's) and places them in such a way that the new number differs from the nth number on the list in at least the nth decimal place.

13. (ExH) One Life Lesson to apply here is: *Look for patterns and similarities.* Can we relate the endless run of colored dots to endless decimal expansions of real numbers and thus adopt Cantor's diagonalization to this new challenge? Assume that the set of all possible colorings has the *same* cardinality as the natural numbers. You can now use Cantor's diagonalization approach to show that this assumption is faulty—that is, for any claimed one-to-one correspondence between the natural numbers and the set of all colorings, you can find a coloring that is not included.

Create a sample of how a claimed one-to-one correspondence would look:

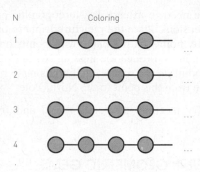

N   Coloring

1   ...

2   ...

3   ...

4   ...

Now describe how you would create a coloring that you know cannot appear on this list. Can you now present your argument for an arbitrary claimed correspondence?

15. (H) Make a list of decimal numbers in which the first number begins with a 7 after the decimal point. Now write down the remaining digits of that first number (just use 2's and 4's), and then complete the list so that our first number turns out to be the same as *M*.

16. (H) Modify Cantor's diagonalization argument. Note the connection between this Mindscape and both Coloring revisited (13) and A penny for their thoughts (14).

17. (S) This procedure does work; an example follows:

0.24434322234 . . .
0.43223253242 . . .
0.23424432133 . . .
0.23435442432 . . .

. . .

$M = 0.22442222 . . .$

Note that each pair of digits in *M* is determined by a single number on the list. Also, *M* is different from each number on the list. For example, it differs from the third number in the third pair, the fourth number in the fourth pair, and so on.

19. (H) Consider the Real Number Life Insurance Company.

27. (H) What point on the graph do you get when you let $x = 2$? What about $x = -2$? So, using your graph, how many values of *x* correspond to the value $y = 4$?

3.4: Travels Toward the Stratosphere of Infinities

6. (ExH) One Life Lesson to apply here is: *Break a hard challenge into easier ones.* There are four family members, so use the Subset Count result in the section to find how many groupings are possible. (Be clear whether you want to count the empty grouping—a rather pathetic dinner!) To list all the groupings of the set {Mom, Dad, Bro, Sis}, for example, be systematic. List all the singleton groupings first: {Mom}, {Dad}, {Bro}, {Sis}, then all pairs, {Mom, Dad}, {Mom, Bro}, etc.

7. (H) Each agenda item either appears or does not appear on the final agenda. The number of different agendas is equal to the number of different ways that eight named boxes could be checked with either Y's (yes's) or N's (no's).

8. (S)

@ and ! are in *S*,

{!, #, %} and {#} are in $\mathcal{P}(S)$,

{{!}, {@}, {#}, {%}, {&}} and {{@}, {$, !}}

are in $\times (\times (S))$

{{{@, !}}, {{$}}} and {{{@}}, {{#, $}}, {{!},

{%, &}}} and {{{!}}} are in $\mathcal{P}(\mathcal{P}(\mathcal{P}(S)))$

Note that the set {@, {#}} isn't in any of the sets!

11. (ExH) One Life Lesson to take away from this challenge is: *Look at simple things deeply.* You want to find a subset of {1, 2} that is not on the list; call it your Mystery Set. Look at the first row in the table, which shows that the element 1 is paired with the subset {2}. Is 1 an element of {2}? No, so put 1 in your Mystery Set. (If 1 *were* in its paired subset, then you would *not* put 1 in your Mystery Set.) Now apply analogous reasoning to the second row in the table. What do you end up with as your Mystery Set? Is it on the list in the right column of the table?

14. (H) Is @ in {@, !, $}? Yes, so leave it out. Is & in { }? No, so put it in, and so on.

17. (S) This question is equivalent to "Is there a largest cardinality?" If there were a largest set, how would it compare in size to its power set? Consider the following proof-by-contradiction argument. Assume that there is a largest set, *S*. Cantor's theorem states that the power set of *S* is even larger, contradicting the assumption that *S* is the largest set. So our assumption is wrong; there is no largest set.

19. (H) Classify all men as either self-shavers or non-self-shavers. If the barber belongs to the self-shavers group, then would he shave himself? If he belongs to the non-self-shavers group, would he shave himself? Similarly, if we classify all sets as either self-included or non-self-included, could *No Way* be a self-included set? A non-self-included set?

28. (H) If the set *S* has 3 members, then the power set of *S* has $2^3 = 8$ members. Now, how many members are in the power set of a set with 8 members?

3.5: Straightening Up the Circle

4. (H) The red line consists of all the points (x, y) that satisfy the equation $y = -x + 3$. The line L consists of all the points (x, y) that satisfy the equation $y = 2$. The point of intersection must satisfy both equations.

6. (H) Place the circles inside one another, and pair points by drawing rays emanating from the center of the smaller circle.

9. (ExH) One Life Lesson to apply here is: *Ground your understanding in examples.* Start with the point a on the real line. Following the stereographic projection method described in the section, you can find the point on the rolled-up line segment that corresponds to a by drawing a line from a to the point at the "North-Pole hole." Where does this line intersect the rolled-up line segment? Can you do something similar for b, c, d, and e?

11. (H) Un-shuffle the digits of the given number, and then recall that we decided to express all numbers, whenever possible, with trailing 9's instead of trailing 0's.

15. (S) Write the digits of **p** as 0.1 2 0001 0001 0001 0001 0001 0001 . . .

$x = 0.1$ 0001   0001   0001   0001 . . .
$y = 0.2$ 0001   0001   0001   0001 . . .

p gets paired with the point: (0.1000100010001 . . . , 0.2000100010001 . . .).

17. (H) Cut the two letters up as in the figures below. Note that both figures are made up of a single solid line segment (including the two endpoints) and one half-open line segment, where one of the endpoints is missing.

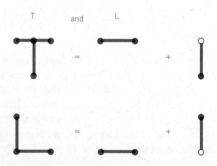

19. (ExH) One Life Lesson to apply here: *Consider alternatives.* Asking the basic question, "What if . . ." is a powerful way to develop new insights and new ideas. For the point at the North Pole to correspond to a point on the real line, those two points would have to determine a line that does not intersect the circle at any other point. Can you describe a line that passes through the North Pole but no other point on the circle? Does this line intersect our real line?

To show that the cardinality of a circle is the same as that of the interval (0,1], first note that the interval (0, 1] can be rolled up to form a circle with no points missing or overlapping. Now look at Mindscape 6.

20. (S) We need to extend the stereographic projection to three dimensions. Place the punctured sphere on the plane, and identify points on the plane with points on the sphere by drawing lines through the missing North Pole. Given an arbitrary point on the sphere, we find its mate by constructing the line from this point to the North Pole.

30. (H) Try plotting the points $(0, 1)$ and $(6, 4)$ in the xy-plane.

CHAPTER 4: GEOMETRIC GEMS

4.1: Pythagoras and His Hypotenuse

9. (S) "Directly above her" means that the triangle formed by the student, the kite, and the spectator is a right triangle. If the height is A, then the Pythagorean Theorem says that $A^2 + 90^2 = 150^2$. (If you make the 3-4-5 right triangle 30 times bigger, you get a 90-120-150 right triangle, and that makes solving for H easier!) The kite is 120 feet high.

10. (H) The tip of the mast, the base of the mast, and the stern of the sailboat form a right triangle, and so the height (L) satisfies the equation $L^2 + 50^2 = 130^2$.

12. (ExH) One Life Lesson to apply here is: *Understand the issue.* If an isosceles right triangle has legs each of length 3 and hypotenuse of length c, then you know $3^2 + 3^2 = c^2$, which you can solve to find c. (The last step involves taking a square root.) But what does the scarecrow actually claim? Read his assertion carefully and then write what he's saying using the values from your sample triangle. Pay careful attention to the words "square roots." Is he really stating the Pythagorean Theorem? Should the wizard rescind the diploma?

14. (H) In a right triangle, the lengths of the sides would satisfy $a^2 + b^2 = c^2$. Do the numbers given satisfy that relationship? If $a^2 + b^2 < c^2$, is the angle opposite c larger or smaller than 90°?

15. (H) Remember to convert miles into feet, and notice that the base of each right triangle in the figure is only half a mile long.

19. (ExH) One Life Lesson to apply here is: *Apply ideas widely.* Draw a triangle using the diameters of the three pizzas. Does your triangle look like a right triangle? What does this tell you about the two pizza choices? Remember that the area of a circle is πr^2, where r is the radius. How is the radius of a circle related to the diameter?

20. (S) The longest side has length $r^2 + s^2$, so we need to check the following equation: $(2rs)^2 + (r^2 - s^2)^2 = (r^2 + s^2)^2$. Expanding gives $(4r^2s^2) + (r^4 - 2r^2s^2 + s^4) = (r^4 + 2r^2s^2 + s^4)$, which is true. So yes, there are infinitely many integer-valued Pythagorean triplets.

28. (H) Use the Pythagorean Theorem to find the value of *x*. Then use it again to find the length of the unknown leg of the larger sail.

4.2: A View of an Art Gallery

9. (H) There are numerous correct triangulations. Here is one.

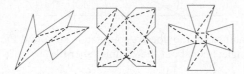

11. (ExH) One Life Lesson to apply here is: *Follow through on the consequences of an action or idea.* To begin, you may pick any pair of adjacent vertices and assign them distinct colors of choice, but after that, the colors of the rest are determined. Here's a start to a coloring for the leftmost gallery.

Remembering that each triangle must have vertices of three different colors, what color must vertex 3 be? Now what color must vertex 4 be? Finish the coloring.

15. (S) See the zig-zag museum in the figure.

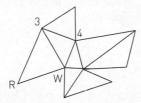

17. (H) Look in the section!

18. (ExH) One Life Lesson to apply here is: *Be open to new ways of looking at a situation.* Try some examples. (See Mindscape II.13.) If the given natural number is 6, then 1/3 of 6 is 2. Write 6 as the sum of three natural numbers in several different ways. Does each sum have a term that is 2 or greater?

This example does not lend itself to a general argument, however. Try looking at the situation from the other side. Is it possible to have three natural numbers with sum of 6 so that each of the three numbers is *less than* 2? That is, if $a < 2$, $b < 2$, and $c < 2$, then what can you say about $a + b + c$?

To generalize, suppose you have a given natural number *n*, and three other natural numbers *a*, *b*, and *c*, with $a < n/3$, $b < n/3$, and $c < n/3$. What can you say about $a + b + c$?

19. (S) Here are some possible guard placements.

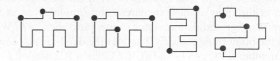

There are several ways to do this. The museums that look like 3's need at least three guards; the museum that resembles the number 2 needs only two guards!

20. (H) Modify the comb-shaped museum from Mindscape III.17.

4.3: The Sexiest Rectangle

5. (H) For each equation, begin by clearing the denominators using your preferred method ("cross-multiplying" both sides by each denominator). With equation (*b*), for example, multiplying both sides by $x - 4$ yields $(x/3)(x - 4) = 2$. Multiplying both sides by 3 yields $x(x - 4) = 6$. Now expand and simplify to get $x^2 - 4x - 6 = 0$. Use the Quadratic Formula to solve for *x*.

9. (H) Suppose the first rectangle had side lengths *b* and *h*, with *b* the larger. What are the base and height of the folded rectangle? What is the ratio of base to height for each rectangle?

11. (H) If our original rectangle had lengths 2 and $(1 + \sqrt{5})$, then our new rectangles would have dimensions 2×1 and $2 \times \sqrt{5}$. Why? Are either of these Golden Rectangles?

12. (H) If the original lengths are 1 and $(1 + \sqrt{5})/2$, then the new side will have length $(3 + \sqrt{5})/2$. Simplify the ratio $(3 + \sqrt{5})/(1 + \sqrt{5})$ by multiplying both top and bottom by $(1 - \sqrt{5})$.

14. (S) Consecutive Fibonacci numbers (1, 1, 2, 3, 5, 8, 13, . . .) are wonderful approximators of the Golden Ratio. In an 10×10 grid, an 8×5 rectangle is your best bet. You can convince yourself by computing the 100 possible ratios and seeing which is closest to $(1 + \sqrt{5})/2$. Surprisingly, it doesn't help to consider nonhorizontal right triangles! (Write a small computer program to check this . . . but don't do it by hand!)

16. (ExH) One Life Lesson to apply here is: *Apply ideas widely.*

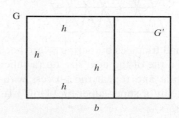

A sketch is extremely helpful here. What are the dimensions of the rectangle G'? What is the area of G? What is the area of G'? Look at the ratio of these two areas and simplify. Now divide the numerator and denominator of this ratio each by h. What does the numerator equal? What does the denominator equal? (Remember that both G and G' are Golden Rectangles!)

18. (S) Let the starting rectangle be two units wide and one unit high.

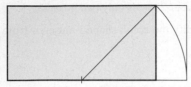

The new longer rectangle has length $(1 + \sqrt{2})$, width 1, for a corresponding ratio of $\sqrt{2} + 1 = 2.141 \dots$. The smaller rectangle has width $(\sqrt{2} - 1)$ and a ratio of $1/(\sqrt{2} - 1) = 2.141 \dots$. It has the same proportions! To see that these two expressions represent the same number, multiply both numerator and denominator by $(\sqrt{2} + 1)$.

19. (H) There are two similar triangles in the drawing, both having a leg that corresponds to a long side of a Golden Rectangle. Use the relationships between sides of Golden Rectangles and properties of similar triangles.

31. (H) To expand $\left(\dfrac{1 - \sqrt{5}}{2}\right)^2$ you'll have to FOIL, but before you start, notice that $\left(\dfrac{1 - \sqrt{5}}{2}\right) = \dfrac{1}{2}\left(1 - \sqrt{5}\right)$. This means that $\left(\dfrac{1 - \sqrt{5}}{2}\right)^2 = \left(\dfrac{1}{2}\right)^2\left(1 - \sqrt{5}\right)^2 = \dfrac{1}{4}\left(1 - \sqrt{5}\right)^2$. In other words—factor out the ½ before you FOIL!

4.4: Soothing Symmetry and Spinning Pinwheels

5. (ExH) One Life Lesson to apply here is: *Look for patterns and similarities.* First, draw your own versions of each tiling using more tiles, and designate a tile near the middle (M). So the second tiling might look like this:

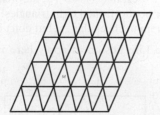

How many small triangles bordering M are needed to create, along with M, a tile of the next size up (a super-tile)? Shade in this super-title and repeat the process with these bigger triangles to create a super-super-tile. How many small tiles does it include?

9. (H) Fix an interior triangle in the center of the floor of a large warehouse and argue that, eventually, the whole warehouse floor will be covered.

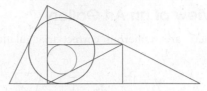

The 5-unit super-tile doesn't completely surround the original triangle; how about the 25-unit super-tile?

13. (S) Look at the 5-unit super-tiles containing the tile marked with a star and the one marked with a 1. Because they are in different positions in those super-tiles, none of the surrounding tiles will line up.

14. (H) Yes, but it won't look anything like the pinwheel construction—it will look regular.

15. (H) The starting tile can be in any of four relative locations in the 4-unit super-tile.

16. (S) Consider:

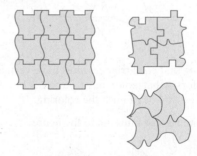

The nine tiles represent the basic tiling pattern. The top and bottom edges are identical, and so are the left and right edges. More complicated patterns with one or two tiles can be made by grouping four of them together (see figure) to make one standard tile.

17. (ExH) One Life Lesson to apply here: *Don't be afraid to experiment, especially when outcomes are uncertain.* Some of these tiles can be used to tile the plane easily, such as the rectangle. In a simple tiling with these rectangles, how many come together at a point? What is the measure, in degrees, at one corner of the rectangle? What is the sum of the degrees of the rectangles that come together at a point of your simple tiling? (You should get 360°.) Now look at the hexagonal tile. Inspired by the honeycombs of bees, can you fit hexagonal tiles at a point so that the angles sum to 360°? How many tiles fit? (Do you know the angle measurement of one angle in a regular hexagon? Try drawing diagonals to divide the hexagon into equilateral triangles, and recall that such a triangle has angles of 60°.) Now look at the octagon. How does its angle measurement compare to that of the hexagon?

20. (S) To see that the order matters, it is easier to look at the end results of the modified square pattern of Mindscape 16. Flipping diagonally and rotating by 90 is equivalent to flipping about a horizontal line. Reversing the steps is equivalent to flipping about a vertical line.

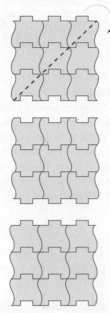

29. (H) Make a schematic that shows your first super tile surrounded by super tiles. How many super tiles are in your schematic? Now, how many tiles are in one super tile?

4.5: The Platonic Solids Turn Amorous

8. (H) Visit our Web site **www.heartofmath.com** to find java applets for displaying these stereographic projections.

12. (ExH) One Life Lesson to apply here is: *Look at ordinary things in extraordinary ways.* Which is easiest to count: vertices, edges, or faces? You probably answered faces, so start there. Use a piece of chalk to mark off faces as you count them. How many pentagons are there? How many hexagons? Next count vertices, but be clever. Look at the soccer ball. Does each vertex lie on exactly one pentagon? You know how many pentagons you have, so does this suggest a quick way to count the vertices? Finally, to count the edges, first notice that each vertex has exactly 3 edges coming out of it. So 3 times the number of vertices should count the edges, right? Not quite! How many times does each edge get counted in that product? Can you revise your answer?

Now go back and answer the question posed in the Mindscape!

14. (S) Cubes, tetrahedrons, and dodecahedrons yield triangles when their vertices are sliced. Octagons yield squares, and icosahedrons yield pentagons. The number of sides of the boundary corresponds to the number of faces that meet at a vertex.

16. (H) Try to find a plane that intersects all six sides of the cube! Can you find other planes that intersect 3, 4, and 5 sides?

17. (S) The edges of the octagon connect the centers of the square faces. Each edge is the hypotenuse of a right triangle with legs of length 1/2. Therefore, by the Pythagorean Theorem, the edges of the octagon have length $\sqrt{2}/2$.

20. (ExH) One Life Lesson to apply here is: *Don't be afraid to experiment, especially when outcomes are uncertain.* This is a great opportunity to get out your scissors and tape to build something. The key here is to think outside the box (or cube). Cut a bunch of congruent triangles—equilateral triangles work best, but they don't have to be exact. Start taping them together so that edges and vertices match up, but don't worry about how many triangles come together at a corner. It's OK if your solid is floppy and asymmetrical, or if its surface is concave at some vertices. In fact, see what happens if you have 7 triangles meet at a vertex!

4.6: The Shape of Reality?

9–11. (H) If you don't have a globe, then visit **www.heartof math.com**.

13. (S) The triangle formed has angles 76°, 88°, and 63°, for a total of 227°.

21. (S) The figure shows the optimal unfolding of the box. The minimum distance (using the Pythagorean Theorem) is

$$D = \sqrt{(2.5^2 + 1.5^2)} = 2.19 \ldots.$$

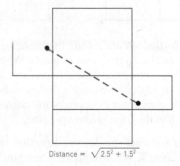

Distance = $\sqrt{2.5^2 + 1.5^2}$

23. (H) On the plane, a triangle's angles add to 180°; on the sphere, they add to more than 180°; on a hyperbolic surface, they add to less than 180°.

26. (H) Every triangle divides the sphere into two triangular regions that cover the entire sphere. We can use either of those two triangles—one may have a much bigger sum of its angles.

27. (S) If all three great circles intersect at a point you get *no* triangles, but, otherwise, you will get eight triangles whose total sum is $(8 \times 180) + 720 = 2160$ (the sum of the angles of eight planar triangles plus an excess of 720°). The simplest example is three perpendicular circles that cut the sphere into eight equilateral triangles, all of whose angles are 90°.

30. (ExH) One Life Lesson to apply here is: *Ground your understanding in examples.* Start with a familiar sphere: a globe, and a familiar line: the equator. Pick the point that marks the location of New York City. Notice that it is not on the equator. Draw (if you actually have a globe) or imagine (if you don't) a great circle through that point. Is it possible that your new great circle does not intersect the equator? Draw (or imagine) a circle through your point that does *not* intersect the equator. Could this circle be a great circle? (Remember that a great circle on a globe is a circle with length equal to the equator.)

Now imagine a general sphere and an arbitrary great circle on it. Does your analysis above apply in this more general situation?

32. (ExH) One Life Lesson to apply here is: *Keep an open mind and be willing to embrace new ideas that first appear counterintuitive.* The figure on the left shows the cube with points and lines as described. (The points are enlarged for clarity.) The figure on the right shows just one "triangle." Note that the sides of the triangle are bent, yet we can still consider these strange-looking shapes to be triangles.

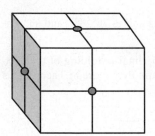 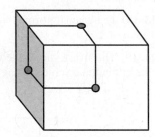

36–39. (H) When we create a cone by removing a thin sliver, the cone is almost like a flat piece of paper. When we remove a bigger sliver, the cone becomes more pointed, like a wizard's hat. The sums of angles of triangles that go around the cone point will differ more from the sums of the angles of a triangle on the plane when the cone is more pointed.

47. (H) If you subtract the second equation from the first, you get $2w - 2h = 2$. What happens when you add this new equation to the third equation, $2w + 2h = 14$?

4.7: The Fourth Dimension

7. (S) Object 1: Either a sphere or two cones—a diamond spun about its axis. Object 2: A tetrahedron. Object 3: A vaselike object—a capped-off cylinder with a bulge in the middle. Object 4: A doughnut.

11. (H) Take a small portion of one of the rings and push it into the fourth dimension.

14. (H) It's an artistic feat to make a 2-dimensional drawing of a 6-dimensional cube. Here are 4D and 5D edge drawings: A 3D cube has $4 + 4 + 4 = 12$ edges. A 4D cube has $12 +$

$12 + 8 = 32$ edges. A 5D cube has $32 + 32 + 16 = 80$ edges, and a 6D cube has $80 + 80 + 32 = 192$ edges.

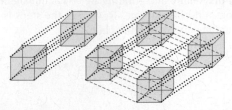

15. (ExH) One Life Lesson to apply here is: *Ground your understanding in the familiar.* Look at the 3-dimensional cube below. Recall that we are measuring distance between two vertices as the minimum number of edges needed to create a path from one of the vertices to the other. So two vertices on an edge are distance 1 apart. The two vertices shown on the cube below are distance 2 apart, because we have to traverse two edges to get from one to the other. Can you find two vertices distance 3 apart? What about 4?

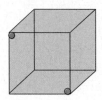

Now think about a 4-dimensional cube. How far apart could two vertices be? Think about the possible locations for the two vertices. What if both are in the same 3-dimensional "face"? What did you discover above? How many edges are needed to traverse from one 3-dimensional cube to the other?

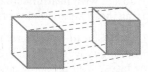

Now see if you can generalize your answers to dimensions higher than 4.

16. (ExH) One Life Lesson to apply here is: *Seek reasons for patterns.* From what we know of 1-, 2-, and 3-dimensional triangles, the first three rows of the table are easy to fill in.

Triangle's dimension	Number of vertices	Number of edges	Number of 2D faces	Number of 3D faces
1	2	1	0	0
2	3	3	1	0
3	4	6	4	1
4				
5				
n (general)				

Do you notice any patterns? Think about starting with a 2-dimensional triangle (solid lines in the following figure) and then adding a vertex and edges (dashed lines below) to create a 3-dimensional triangle.

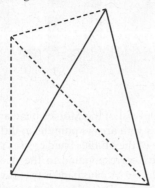

How many new edges do you create? Why? How many new 2-dimensional faces do you create? Why? What happens when you add a fifth vertex and a new edge to each of the four old vertices?

18. (S) Glue together faces that have the same label in the figure below.

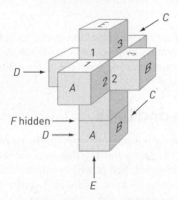

CHAPTER 5: CONTORTIONS OF SPACE

5.1: Rubber Sheet Geometry

6. (S) Removing the two intersection points breaks the theta curve in three pieces, while removing *any* two points on the circle always leaves exactly two pieces. Suppose there were a deformation that turned the theta curve into a circle. Mark the two intersection points on the theta curve in red and follow the deformation process. At any stage, the removal of the two red points should break the distorted theta curve in three pieces; yet when we get to the final stage, the circle, we find that the object falls into just two pieces. This contradiction shows that our assumption is wrong: The theta curve is not equivalent to the circle.

7. (H) Imagine each letter made out of rubber-bandlike material. Any letter with a loop cannot be distorted into a

letter without a loop. What other characteristics distinguish one type of letter from another?

11. (H) Take a Gumbylike coffee mug and flatten out the part that holds coffee so that it looks like a flat disk connected to a ring (the handle).

13. (S) Yes. First rotate your hand 180° by bringing the glass under your armpit. Roll your shoulder forward and continue rotating your hand (while straightening out your arm) until you've rotated the palm of your hand 360°. You might feel close to dislocation, but don't stop here. Swing your arm in front of you and bend your elbow so that the glass is just above your head (540°). Continue swinging your arm above your head until the glass is back in its original position. In fact, you've rotated the glass 720° with your arm intact!

27. (H) Grab the four corners of the large squarish hole and pull them toward the center.

32. (ExH) One Life Lesson to apply here is: *Visualization and experimentation often lead to surprising and even counterintuitive results.* Look at the punctured inner tube. Imagine deforming the hole so that it stretches beyond the width of the upper half of the tube. What do you get? Alternatively, look at the other object. Imagine a drawstring attached to the outer perimeter of the flat rubber disk. Now tighten the drawstring so that the disk begins to close up below the tube. What is the result?

33. (S) Fold the sheet into a cylinder by gluing two opposite sides together. Now bend the tube around so that the top circle is right up against what was formerly the bottom circle, and glue these circles together.

36. (ExH) One Life Lesson to apply here is: *Be open to new ways of looking at a situation; consider the counterintuitive.* Remember that the knot is thickened and the tube is solid, so you can make a model out of clay. Arrange your model to look like the middle picture in the text and focus on the part where the tube meets the knot. Notice that the top of this figure looks like a **T**.

You can slide the left half of the **T** down until it joins the bottom of the tube in front of the other segment of knot on the left side of the tube. Now slide the right half of the **T** down the right side of the tube. Slide it *underneath* the other segment attached to the right side of the base of the tube, then slide it back up to the top of the tube.

Can you perform one more "segment slide" to complete the challenge?

38. (H) Consider stretching out the knot so that you can pull parts of it around the sphere.

47. (H) Given the goal of this Mindscape, if you discover that the solutions to the equation are $n = 1, 2,$ and 5, then the three kinds of bagels being served are 1-holed, 2-holed, and 5-holed bagels. (After all, it is funny bagel day!) We could call a bagel with no holes a roll. If there are any rolls, the value $n = 0$ would have to satisfy the given equation. Are there any rolls?

5.2: The Band that Wouldn't Stop Playing

10. (H) The original strip has only one edge. Cutting along the center line will not cut through that edge. So, the result can have only one piece.

11. (S) With the left and right ends identified appropriately, a rectangular strip of paper of length L represents a Möbius band whose boundary edge has length $2L$. The centerline of the Möbius band still has length L, though. Cutting along this centerline produces a longer two-sided strip whose edges have lengths $2L$, but the centerline of this new strip is $2L$ as well.

15. (H) Try constructing two such Möbius bands by layering one strip on top of another before you make your twist. Use different-colored paper.

21. (ExH) One Life Lesson to apply here is: *Powerful and important discoveries can come from play.* Note that to pull the drawstring completely together, there cannot be material from the Möbius band sticking out. Now, suppose the answer to the question is "Yes." What would you hold in your hand once the drawstring is pulled? Would it resemble a pouch? What qualities does it have? Are those qualities consistent with a Möbius band? If you are confused, construct a model out of cloth with a drawstring and give it a try!

22. (H) The two edges were not connected. How about the sides?

23. (ExH) One Life Lesson to apply here is: *Consider the counterintuitive.* Use the schematic given for Mindscape II.24. Imagine that you (the ladybug) are on the "outside" of the bottle just above the level of the cider and your gentleman friend is "inside" on the other side of the glass from your position. Move down the "outside" of the bottle. Which way can you move once you reach the "bottom"? Can you reach your friend?

25. (S) Compare with Mindscape 13 in Section 5.1. Glue the left and right sides to make a cylinder. Now bend the cylinder into a **U** shape, pass one end through the side of the cylinder

(in the fourth dimension this can be done without cutting into the surface), and then join the two ends.

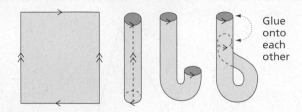

Glue onto each other

27. (S) The half twist of the Möbius band arises from identifying two edges with arrows pointing in different directions. The centerline of the Möbius band crosses this identification edge one time—corresponding to the one half twist. The centerline of the cut Möbius band is twice as long and crosses the identification band *two* times in the same direction.

31. (H) Start drawing lengthwise anywhere in the strip and continue until you get back to where you started.

37. (H) In fact, you can draw a map on the Möbius band with six countries all bordering one another (showing that you need at least six colors to color this map). See **www.heartofmath.com** for a discussion of how many colors are needed for maps on other surfaces.

48. (H) Feel free to use guess-and-check to determine your answer. It may also help to make a sketch of a large rectangle and try some cutting configurations. (Remember that y is divisible by 3.)

5.3: Knots and Links

9. (ExH) One Life Lesson to apply here is: *Just do it.* Get a bunch of toothpicks and try this out. Use small balls of clay (or chewing gum, in a pinch) to attach sticks at corners. Remember that your knot exists in 3 dimensions; that is, it doesn't lie completely flat in the plane. So some of your sticks must poke out at funny angles.

11. (H) Look at one end of the figure (either one). Can you reduce the number of crossings by shifting or twisting part of the Slinky? Can you repeat this process?

14. (S) No, it isn't a knot. It only seems like a knot when you hold the ends. As long as you don't cut the rope, an unknot will remain an unknot, because you can always reverse your steps!

19. (H) Switching any crossing in either figure will work.

23. (H) How can you remove the one crossing?

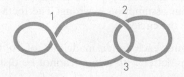

24. (S) Yes. However, if the two loops are painted red and blue, the number of red-blue crossings will always be even.

27. (S) Pick any point and start moving along the knot. Flip crossings when necessary so that it is as if you are constantly traveling under crossings. When you visit a crossing for the second time, do not flip it. This crossing is part of a loop that lies under the rest of the knot. (Imagine shrinking this part of the loop so that all the corresponding crossings vanish. This leaves you with a smaller problem that you can simplify in the same manner.) Continue the process.

32. (H) Yes! Try to draw a curve on the torus that goes three times around the long way while it goes twice around the short way.

37. (H) After you have made your rubber band set of Borromean rings, move them around and lay the ensemble on a table so they are roughly in a line with the left one round, the right one round and completely separate from the left one, and the middle one somewhat jumbled. Look for a way to put in more loops.

38. (ExH) One Life Lesson to apply here is: *Often a clever idea can be more potent than conventional wisdom.* Notice that you can link one loop within another loop as illustrated.

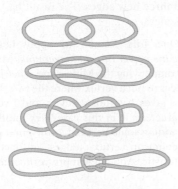

Can you apply this idea, using more loops, to create an escape route? (Assume you have lots of loops!)

46. (H) As illustrated in Mindscape 20, a trefoil knot has three crossings and a figure-eight knot has four.

5.4: Fixed Points, Hot Loops, and Rainy Days

8. (H) For each figure, place a compass on the green circle that continually points to the corresponding point on the yellow curve. Now move the compass clockwise around the green circle (one time) and notice how many times the compass hand spins around (and in what direction). The winding number for the first figure is 1 (clockwise).

10. (S) Yes, for exactly the same reason given in the text as to why there exist two antipodal points having the same temperature. Let Diff(p) equal the difference between the

pressure at a point p and the pressure at p's antipodal cousin. If A and B are antipodal points, then $\text{Diff}(A) = -\text{Diff}(B)$. Since the difference (Diff) varies continuously from a positive number to a negative number, there exists a point C (between A and B) where $\text{Diff}(C)$ is 0. Therefore, C and its antipodal point have the same pressure.

12. (H) Draw a circle with a 5-foot diameter somewhere in your room. Must some pair of opposite points on that circle have the same temperature?

15. (H) Look in the section for a way to distort (or in this case just "spin") the red disk that leaves no fixed points. This challenge is one demonstration of why all the conditions in the Brouwer Fixed Point Theorem are essential.

16. (H) Yes. Measure how far each red point is moved left or right. The right end is either fixed or moves left. The left end is either fixed or moves right. If the left end moves right and the right end moves left, why must there be a point between them that doesn't move at all? The Brouwer Fixed Point Theorem is true for line segments as well as disks.

18. (ExH) One Life Lesson to apply here is: *Just because a specific attempt fails does not mean that the task at hand is impossible.* The spinning idea from the hint in Mindscape II.15 doesn't work in a simple way here, but there's another idea that does. Think about a distortion that shrinks the red disk. What if all the points on the red disk were moved toward the puncture? Can you make this precise?

20. (S) You drove 140 miles in 2 hours, for an average speed of 70 miles per hour. At some point, you must have been traveling at least 70 miles per hour (otherwise your average speed would be lower). So, you will get a ticket for driving at least 5 m.p.h. over the speed limit sometime while you were on the tollway.

23. (ExH) Several Life Lessons apply here: *Seek the essential. Ignore the superfluous. Devising a good representation of a problem is frequently the biggest step toward finding a solution.* The backtracking moments during the hike are red herrings. All that matters is that the hiker traveled up and then down the mountain during identical time intervals on the two days. Imagine plotting a graph showing time on the horizontal axis and altitude on the vertical axis. If you plot the hiker's altitude going up the mountain from 8 a.m. to 5 p.m. and then plot his altitude coming down the mountain over that same portion of the horizontal axis, what will you see?

CHAPTER 6: MODELING OUR WORLD THROUGH GRAPHS

6.1: Circuit Training

9. (S) It is not possible to add loop-d-loops to the given drive to obtain a trip that traverses each road exactly once and returns to the entrance. The given drive begins and ends at the entrance, but there is only one additional road that starts at the entrance. Any additional trip that traverses this road would not have an unused road on which to return to the entrance to complete a loop.

10. (H) When you list all the edges, does each vertex occur an even number of times? How many edges do you need to add to make the answer to the previous question "Yes"?

16. (ExH) One life lesson to apply here is: *Seek the essential.* If you want to model this scenario with a graph, here's one way to designate your vertices:

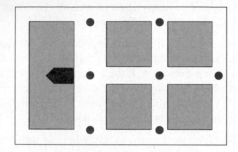

Now label the vertices, list the edges, and analyze.

17. (ExH) One Life lesson to apply here is: *When confronted with a new situation, apply what you already know.* The expanded hint for Mindscape 16 suggests a way to begin modeling this scenario with a graph. You will need to insert additional edges in your graph to reflect the need for the letter carrier to walk along each side of the street. In particular, if an edge is flanked by two gray blocks, then traversing this edge will allow mail delivery on only one side of the street.

26. (S) The center graph has one vertex of degree 5, one of degree 4, two of degree 3, and three of degree 1. (Remember that a loop—an edge from a vertex back to itself—contributes twice to the degree of that vertex.) There are nine edges in the graph and the sum of all the vertex degrees is 18.

30. (H) Imagine that you model this scenario with a graph. Let the teams be the vertices with an edge between two vertices denoting a game between those teams. Now think about what you learned in the previous mindscape.

36. (H) The equation $(1/2)n(n - 1) = 45$ simplifies to $n(n - 1) = 90$, which then becomes $n^2 - n - 90 = 0$. Now factor.

6.2: Feeling Edgy?

9. (H) Either both ends of the new edge are attached to existing vertices or one end of the new edge is attached to an existing vertex and the other end is a new vertex.

12. (S) Connected graphs on the plane satisfy the formula $V - E + F = 2$, where F represents the number of pieces of the plane after removing the edges. If the plane isn't separated into different pieces, we have $F = 1$, so $V - E = 1$.

14. (H) Substituting $E = 151$ into the formula $V - E + F = 2$ gives $V + F = 153$. Could $F = 153$?

19. (S) Three faces, two edges, and two vertices. $V - E + F = 3$, not the expected number 2. The Euler Characteristic Theorem makes statements about connected graphs on the sphere, and the graph described in this Mindscape is not connected.

21. (S) The gem has 13 facets (faces): the top facet (which has six edges); the six bevel-facets adjacent to the top (three we can see and three we can't); and the six triangular facets at the bottom (three we can see and three we can't). There are 13 vertices: six on the top facet, six more below those, and the bottom vertex. There are 24 edges: six around the top facet and six at the tops of the triangular facets, then six dividing the triangles and six dividing the bevel-facets. Thus we have $V - E + F = 13 - 24 + 13 = 2$.

22. (H) The leftmost solid is a truncated tetrahedron. You start with a tetrahedron, which has 4 vertices, 4 faces, and 6 edges. By truncating (cutting off) each vertex as pictured, you replace each vertex with three new vertices, adding a new face and three new edges. The result has 12 vertices, 8 faces, and 18 edges.

24. (ExH) One Life Lesson to apply here is: *Never underestimate the power of simple counting.* Make copies of the figure accompanying the Mindscape. The simplest graph you can create is to put a vertex where the two existing curves intersect. This gives 1 vertex, 2 edges, and 1 region. What's the Euler characteristic in this case? Try another graph on the torus by adding some vertices to the curves given in the figure. Now count up vertices, edges, and faces to find the Euler characteristic. Add a few more vertices and edges and count again.

27. (S) Let V_1, E_1, F_1 represent the vertices, edges, and faces of the first piece and V_2, E_2, F_2 represent the analogous components of the second piece. The total number of vertices is $V = V_1 + V_2$; edges: $E = E_1 + E_2$; faces $F = F_1 + F_2 - 1$, where the -1 counteracts the double counting of the outside face. In all we have $V - E + F = (V_1 + V_2) - (E_1 + E_2) + (F_1 + F_2 - 1) = (V_1 - E_1 + F_1) + (V_2 - E_2 + F_2) - 1 = 2 + 2 - 1 = 3$. The simplest example of such a graph is two dots: $V - E + F = 2 - 0 + 1 = 3$.

30. (H) When pasting the two rectangular faces, we identify four vertices and four edges, and lose two faces entirely. Remember this when you count vertices, edges, and faces of the glued object.

48. (H) None of the figures enclose any regions, so altogether they leave the plane undivided. So there is one region. Each figure has 7 vertices and 6 edges.

6.3: Plane Old Graphs

2. (S) The graph is planar. Here's one way to redraw it so that none of the edges cross.

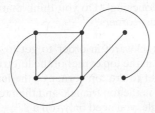

3. (H) Remember the outer region. (So there are a total of four regions here.)

11. (S) It is possible. Put the new vertex inside the triangle region. Then you can draw an edge from the new vertex to each of the old vertices without crossing any edges.

22. (ExH) A subdivision of K_5 in the given graph will have 5 vertices, each of degree 4. Which vertices in the given graph have degree at least 4? Any vertex of degree less than 4 will end up as a freckle in your version of K_5. Remember also that you can chose to ignore any edge that isn't helping you. Just don't include it in your construction of K_5.

24. (H) Vertex #2 will be yellow, and when you get to it, vertex #4 will be red.

28. (ExH) Remember that you want to find the smallest number of colors you need so that you can assign a color to each vertex in such a way that vertices joined by an edge get different colors. For graph (a), try a greedy approach. Pick a vertex and color it. Then go to that vertex's neighbors and color them a different color. Introduce a new color only if you need to. Reuse an old color whenever you can. How many colors have you used when you're done? What about graph (b)? Does the same approach work?

38. (H) Remember that a planar graph with V vertices, E edges, and no loops or multiple edges must have V and E satisfying $E < 3V - 6$.

6.4: Networking

4. (S) Here's one route: Beijing –> Seoul –> Tokyo –> Shanghai –> Bangkok –> Mumbai –> Katmandu –> Beijing. The only other possibility is to reverse the order of this route.

11. (H) Remember that a spanning tree must contain all the vertices in the graph and it must be connected. That is, it must be possible to travel from any vertex in the graph to any other vertex in the graph using only edges in the spanning tree.

13. (H) The cheapest edges all have cost 1. You can choose them in any order and will not create a cycle. What about the cost 2 edges?

22. (S) The graph has at least one Hamiltonian circuit: AFBEHCGDA.

24. (ExH) This question is easier to answer if you have a cube-shaped object to experiment with. (A boutique tissue box works well, or any box. Just pretend it's a cube.) To look for a possible Hamiltonian circuit, pick a corner (vertex) on the top of the box (cube). Can you visit the other three corners by following the edges? When you get to the fourth corner, can you follow an edge down to the bottom of the box?

30. (ExH) How do you create larger trees from smaller trees?? To keep the graph connected, you need to add a new edge for every new vertex you add. Look at the relationship between the number of vertices and number of edges in the smallest trees you drew. Does that relationship change as your trees grow?

39. (H) A Hamiltonian circuit has exactly the same of number of edges as vertices.

CHAPTER 7: FRACTALS AND CHAOS

7.1: Images

2. (S) There are a variety of answers. The largest such sub-figure has side length half the size of the original picture. Note that the original picture can be subdivided into three of these sub-figures. Also find sub-figures that are 1/4, 1/8, 1/16, . . . as large. (Any power of 1/2 will do.)

10. (H) You could start with a border around your picture. Then you could draw four reduced copies of the border, making a picture of four rectangles. In each of those four, draw four smaller rectangles. Continue forever.

7.2: The Infinitely Detailed Beauty of Fractals

13. (S) The edge of every triangle at each stage is eventually replaced with a scaled version of the standard Cantor Set. In fact, each point of the Sierpinski Dust can be associated with a pair of numbers in the Cantor Set on the base of the triangle. Specifically, each point of the Sierpinski Dust above the x-axis is the apex of an equilateral triangle whose other two vertices are points of the Cantor Set on the base of the original triangle.

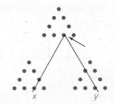

14. (ExH) One Life Lesson to apply here is: *Seek the essential; look for patterns.* Looking at the figures you can easily see that at the first stage there is 1 segment with no bends, at the second stage there are 4 segments and 3 bends, and at the third stage there are 16 segments and 15 bends. Do you see a pattern? You may want to gather your data in a table to make the pattern more evident:

Stage	Number of segments	Number of bends
1	4	0
2	4	3
3	16	15
4		
n		

At each stage, the number of segments increases by a factor of 4. Can you express the number of segments at stage *n* as a power of 4 in terms of *n*? How is the number of bends related to the number of segments?

15. (H) Your third stage will look like 64 dots (each dot representing a tiny copy of the original picture) arranged in four groups of four groups of four.

16. (S) The third stage looks like a flock of birds flying in V formation. After infinitely many stages, the burgers vanish, leaving instead three copies of the entire picture at different scales.

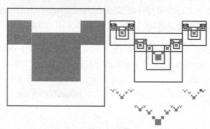

19. (H) The original Z is 24 cm long, so each of the small Z's at the second stage is 1/6 as long, or 4 cm. At the third stage, each smaller Z would be 1/6 as long again. How many small Z's are there at the second stage? How many will there be at the third stage?

21. (ExH) One Life Lesson to apply here is: *Simple repeated processes can lead to complex and interesting outcomes.* To make your drawing as accurate as possible, note that the five line segments in the figure all have the same length. This should hold true with each subsequent iteration. The next iteration would look like this after you have finished modifying three of the five segments:

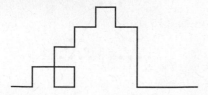

Notice how some of the "bumped-out squares" actually touch corners. Finish this iteration. When you complete the next iteration, be sure you modify each and every line segment in the figure from the previous stage. (There will be lots of corners touching!) Do you think that two edges will ever touch, or just corners?

24. (H) Work backward; in order to get to triangle *b*, you must have been in the top corner of the left triangle and then rolled a 3. To get to the top corner of the left triangle, you must have been in the top triangle and then rolled a 1. To get to the top triangle, you need only roll a 2.

25. (H) *Moving 2/3 the way toward zero* is equivalent to *dividing your current position by 3.* You can move all the points in the Cantor Set at once by reducing the set by a factor of 3 and lining it up with the origin. Notice that the Cantor Set neatly lines up with itself, because the left and right thirds are each 1/3 replicas of the entire set. Moving toward 1 is essentially the same.

27. (S) Like the Sierpinski Dust question above, the edges at each stage will be replaced with Cantor Sets after infinitely many iterations. The coordinates of each point in the Cantor Square have the form (x, y) where both x and y are members of the Cantor Set.

38. (H) Start with any picture. Reduce it by 50%, make six copies, and arrange the centerlines of the six pictures to form a hexagon. Experiment with rotating and/or slightly shifting the positions of the six reduced copies.

49. (H) At Stage 1, the segments will have length 1/3. At Stage 2, the segments will have length 1/3 of that, which is 1/9 or $(1/3)^2$.

7.3: Between Dimensions

7. (H) Four copies of the rectangle can make a rectangle twice as large. So $S = 2$ (scale), and $N = 4$ (number of copies). $2^d = 4$ implies $d = 2$, which agrees with our expectation.

10. (S) The eight squares surrounding the center are a scaled version of the larger square. Therefore, eight copies make a square three times as large. $S^d = N$ becomes $3^d = 8$, implying a fractal dimension of $d = \ln(8)/\ln(3) = 1.89\ldots$.

12. (H) Two copies of the Cantor Set can be arranged to make a Cantor Set that is three times as large.

14. (ExH) One Life Lesson to apply here is: *Look for patterns and keep an open mind.* At the third stage, each of the 4 line segments of length 1/3 is replaced by 4 shorter segments, each 1/3 of the length of the piece they are replacing, or length $1/3(1/3) = 1/9$. There are 16 of them, so their total length is 16/9. Put your data in a table to help reveal the pattern.

Stage	Number of segments	Length of a segment	Total length of curve
1	1	1	1
2	4	1/3	4/3
3	16	1/9	16/9
4			
n			

Can you guess a formula for the length of the curve at stage n? What happens to your formula as n gets larger and larger without bound? What do you think the length of the final Koch Curve equals?

15. (H) If 4 copies reduced by 1/3 can be used to make the original, how many copies of the original do you need to make a version 3 times as large?

16. (S) The collage-making instructions imply that three copies of this object (Sierpinski Dust) can be arranged to make a new object three times larger. $3^d = 3$ implies that $d = 1$. We have a totally disconnected set of points whose fractal dimension is 1.

17. (ExH) One Life Lesson to apply here is: *Follow through on the consequences of an action or idea.* Recall that the Menger Sponge was introduced in Section 7.2. Each stage of the construction of a Menger Sponge requires 20 reduced versions of the previous stage. So to scale up a Menger Sponge would also require 20 copies. What should the scaling factor be? If you imagine a Menger Sponge larger than the one shown in the figures in Section 7.2, how much wider would it be?

20. (H) Find a way to replace each segment with five segments that are each a fourth as large.

29. (H) For the rightmost expression, remember the rules for a negative exponent: $a^{-b} = \dfrac{1}{a^b}$ and $a^c = \dfrac{1}{a^{-c}}$.

7.4: The Mysterious Art of Imaginary Fractals

7. (ExH) One Life Lesson to apply here is: *Use familiar tools to help understand the unknown.* To add complex numbers we group like terms to simplify. In particular, we group the i terms together and the real number terms together. So, for example, $(1/2 + 3i) + (5/2 - i) = 1/2 + 5/2 + 3i - i = 3 + 2i$. Multiplying complex numbers is similar, expect we need to remember that $i^2 = -1$. So, for example, to compute $(3 - 2i)(3 + 2i)$, first we "FOIL" to obtain $(3)(3) + 6i - 6i + (-2)(2)(i^2)$. Then because $i^2 = -1$, we get $9 + (-4)(-1) = 9 + 4 = 13$.

10. (H) $(1 + i)$: angle 45, length $\sqrt{2}$. $(1 + i)^2 = (2i)$: angle 90, length 2.

12. (S) $0 \to i \to -1 + i \to -i \to$ (periodic with period 2).

18. (S) $2 + 3i \to -4.73 + 12i \to -121.4 - 113.5i \to 1841.0 + 27552.9i \to$ (blows up!)

29–30. (H) Those numbers corresponding to connected Julia Sets will lie *inside* the Mandelbrot Set; otherwise, the point lies outside the Mandlebrot Set.

31–34. (S) If the Julia Set is connected, then zero will lie inside the Julia Set.

36. (H) Consider the geometric method of squaring and adding complex numbers. If z_1 has length greater than 2, then z_1^2 has length greater than 4, and $z_1^2 + a$ has length greater than 3. How long must the next iterate be?

37. (ExH) One Life Lesson to apply here is: *Follow through on the consequences of an action or idea.* Imagine what would have to happen for a Julia Set to be unbounded: It would have to contain points corresponding to complex numbers that are arbitrarily large. In fact, for each large number we might find in the Julia Set, there would have to be an even larger one further out. But could this happen, given how we define Julia Sets? What happens when we square a very large number? It will get even larger! Now the squaring process for a Julia Set also requires adding a fixed number each time we square. Could adding that fixed value "rein in" all these very large numbers? Even if this works for some large numbers, what about the ever larger numbers in our collection of values?

45. (H) For the middle equation, notice that $A = 2$, $B = -6$, and $C = 5$. So you'll find that $\sqrt{B^2 - 4AC} = \sqrt{-4} = \sqrt{4}\sqrt{-1} = 2i$.

7.5: The Dynamics of Change

8. (H) After one year you have $(1.05)(\$1000) = \$1050$. This money earns 5% interest for the next year, so after two years you will have $(1.05)(\$1050) = \$1102.50$.

9. (S) Let's estimate first: The rule of thumb from the previous Mindscape says that the population will double roughly every 70 years. So, in 700 years the population will double 10 times. $2^{10} = 2 \times 2 \times 2 \times 2 \times 2 \times 2 \times 2 \times 2 \times 2 \times 2 = 1024$, or about 1000. So, in 700 years our population will grow by about a factor of 1000, to about 5.7×10^{12}. In another 700 years, we get another factor of 1000 for a total of 5.7×10^{15}, which already exceeds the total square yardage! So, in far less than 1400 years, the population growth will decline. The precise solution to $(5.7 \times 10^9) 1.01^y = (2.5 \times 10^{15})$ is $y = \log(2.5 \times 10^{15}/5.7 \times 10^9)/\log(1.01) = 1305.6\ldots$ years.

10. (H) Don't focus on the total number of grains of rice on the checkerboard. Instead, just estimate the total number of grains on the last square! Since each square has twice the number of grains as the previous one, the last square will have 2^{64} grains of rice. That's more than 1.8×10^{19} grains of rice! (For comparison, Earth has a volume of only 10^{21} cubic millimeters.)

12. (S)

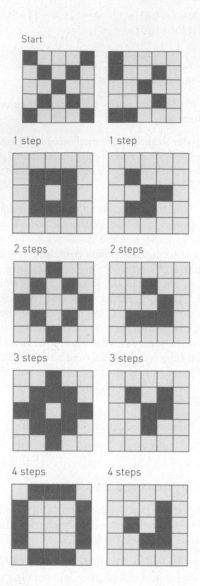

23. (H) Combining a figure having period 2 with one having period 3 is one way to generate a pattern that repeats only after six generations!

29. (ExH) One Life Lesson to apply here is: *Often simple observations have deep consequences.* Draw some figures and record your data in a table to reveal the pattern. Here are the first two stages, and a start on the third.

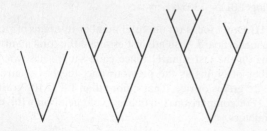

Stage	Number of segments' endpoints
1	2
2	4
3	
4	
n	

30. (ExH) One Life Lesson to apply here is: *It requires time and effort to make ideas and strategies our own.* The cobweb plots in these figures record the results of a repeated process. You start with an x value (on the x-axis) and apply a function $f(x)$. Take the answer you get and plug it back into the function to get a new value, and keep repeating. The figure in the Mindscape in the upper left shows a cobweb plot with a starting value of $x = 0.2$ and a function $y = 4x(1 - x)$. The graph of this function is a parabola, shown in dark blue in the figure.

The path starts at $x = 0.2$. Applying the function yields $y = 4(0.2)(1 - 0.2) + 0.64$, so the path moves straight up from the point $x = 0.2$ on the x-axis to the point $(0.2, 0.64)$ on the parabola. We let this y value be our new x value, so the path moves horizontally to the point $(0.64, 0.64)$ on the line $y = x$ (the red line in the figure). Applying the function with this new x value yields $y = 4(0.64)(1 - 0.64) = 0.9216$, so the path moves straight up from the point $(0.64, 0.64)$ on the line $y = x$ to the point $(0.64, 0.9216)$ on the parabola. This process keeps repeating, with the path alternating between vertical and horizontal motions between the parabola and the line $y = x$. Notice that sometimes the path must move down or to the left in order to reach the next target point. Look at the other two figures with the same parabola to see cobweb plots with different starting values.

Now try creating your own cobweb plot using the figure on the lower right and the function $f(x) = 2.7x(1 - x)$.

31. (S) Periodic. The x-coordinate oscillates between 0.79 and 0.52.

34. (H) Let x represent the distance from the left end. For $x < 1/2$, $x \to 2x$; for $x > 1/2$, $x \to 2 - 2x$. To find points having period 2, assume x first moves right (so $x \to 2x$), and then that result moves back left (so $2x \to 2 - 2(2x)$). Now solve the equation $x = 2 - 2(2x)$.

36. (H) Traditional cobweb iteration takes the form "go vertically to the function and horizontally to the diagonal; repeat." Reverse this to find the intervals of points that are moved into the interval $(1/3, 2/3)$.

7.6: Predetermined Chaos

6. (ExH) One Life Lesson to apply here is: *Just do it. Make the effort to get experience.* Your cobweb plot will alternate between the diagonal and the inverted **V**. You start on the diagonal and always move vertically to the **V**. In this example, your first move will take you straight up to a point just to the left of the top of the **V**. The next move goes to the diagonal, so you move horizontally to the right. (It's OK to cross over the **V**.) Next you move to the **V**, so you move vertically down.

9. (H) Work backward from one of the two fixed points (0 and 2/3). For example, what goes to 0? 0 and 1. What goes to 1? 1/2. What goes to 1/2? 1/4 and 3/4. Therefore, both 1/4 and 3/4 are two starting points that will get fixed at 0 after three iterations. For the fixed point 2/3 we have: 1/3 goes to 2/3; both 1/6 and 5/6 go to 1/3; both 1/12 and 11/12 go to 1/6; and both 5/12 and 7/12 go to 5/6. So 1/12, 11/12, 5/12, and 7/12 all get fixed at 2/3 after three iterations.

11. (S) The diagonal line intersects the inverted **V** at two points: 0 and 2/3. The equation of the right line is $y = 2 - 2x$, and the equation of the diagonal is $y = x$. Solving this system gives $x = 2/3$.

12. (S) $x = 0.4$. This is the cobweb plot of Mindscape II.6. Aside from guessing and trial and error, you can find the solution via algebra. Let's assume that x goes up to the left curve ($y = 2x$), over, and then down to the right curve ($y = 2 - 2x$). So, $x \to 2x \to 2 - 2(2x)$. Solving the equation $2 - 2(2x) = x$ gives the answer.

15. (H) Where does the point 0.1 go? Moving horizontally, to what point on the diagonal does this correspond? What about all the points less than 0.1?

17. (H) 0 stays at 0 while $0.01 \to 0.02 \to 0.04$, and so on. How many steps before the whole interval is covered?

30. (S) The iterates are 0.48, 0.8077225673, 0.5025834321, 0.8089953967, 0.5000431939, 0.8090169884 . . . , which get closer and closer to the repeated values starting with $0.5 \to 0.8090169945 \ldots \to 0.5 \to \ldots$.

32. (II) Try about a dozen iterates and see how every fourth value compares.

36. (ExH) One Life Lesson to apply here is: *Carefully understand and analyze the facts at hand.* In the first iteration the equator will be rotated 30 degrees and then shifted up toward the North Pole. What happens to these points in the second iteration? What does the Mindscape say about which points are fixed by this transformation? Will any other points be fixed?

39. (H) We can get a precise answer by numerically solving the 8th degree equation, $F(F(F(x))) = x$, where $F(x) = 4x(1 - x)$. There are eight solutions corresponding to the two fixed points, and two 3-periodic cycles.

48. (H) When you write x/y as a ratio and convert it to a fraction, you might get something like 1.6. This means that x is 1.6 times larger than y. As a percentage, this means the value of x is 160% that of y. You could also say that x is 60% larger than y.

CHAPTER 8: TAMING UNCERTAINTY

8.1: Chance Surprises

11. (S) You've seen one black card, and there are 25 other black cards in the remaining deck of 51. So, the probability of seeing another black card is $25/51 = 0.490 \ldots$. You should find that roughly half of the black-first pairs had a second black card.

12. (H) There are four ways to pick two cards: BB, BR, RB, RR. Since we are concerned only with pairs containing a black card, we have only three allowable outcomes: BB, BR, and RB. What fraction of these cases has two black cards?

8.2: Predicting the Future in an Uncertain World

12. (S) Answer: 2/9. There are six ways to roll a 7 (1 and 6, 2 and 5, 3 and 4, 4 and 3, 5 and 2, 6 and 1) and two ways to roll an 11 (5 and 6, 6 and 5). Since there are 36 equally likely ordered rolls, the chances of rolling 7 or 11 are 8/36, or 2/9.

14. (ExH) One Life Lesson to apply here is: *Understand the issue.* The questions posed here might seem difficult until we put them in context. For the first question, observe that of the given numbers, only 3, 6, and 9 are multiples of 3. For the second question, note that from our list, only the numbers 2, 3, 5, and 7 are prime. (Caution: recall that we do *not* consider 1 to be a prime number.) By continuing this thought process, each probability can be determined.

15. (H) How many equally likely outcomes are there? How many equally likely outcomes contain a head? How many equally likely outcomes include seeing Lincoln?

20. (H) You know that 62% of the pennies landed heads up. So 38% landed tails up. If the total number of pennies equals T, then 38% of T equals 1900. Create an equation and solve for T. Then convert to dollars.

24. (S) Answer: 0.1087 . . . . The probability that the second person doesn't match the first is 365/366. Now two birthdays are used up, so the chance that the third doesn't match either of the first two is 364/366. The probability that the first doesn't match AND the second doesn't match is $(365/366) \times (364/366)$. Continuing in this way, the probability that the first 40 people don't match is $(365/366) \times (364/366) \times \cdots \times (327/366) = 0.1087681902 \ldots$.

26. (S) Probabilities are easiest to calculate when we can start with a list of equally likely outcomes. The simplest way to do this is to view the coins as different and order the outcomes. With this in mind, there are four equally likely outcomes that contain at least two heads: (T, H, H), (H, T, H), (H, H, T), and (H, H, H). So, the chance of three heads is ¼ provided the flipper is not lying.

29. (ExH) One Life Lesson to apply here is: *Deduce general principles and apply them to more complex settings.* When the original numbers are chosen, the resulting fraction will fall into one of our categories: even/even, even/odd, odd/even, or odd/odd. We know that half the numbers from 1 to 1,000,000 are even and half are odd, so each of these categories is equally likely. Which of these categories have the property that such a fraction will definitely *not* reduce to a fraction of the form odd/odd? Are there fractions in the remaining categories that you know will *not* reduce to the form odd/odd? Does this help answer the question?

30. (H) To help your analysis, assume that each side of each card is a different shade of the colors given—for example, (red and hot pink), (blue and deep blue), and (ruby-red and lagoon-blue). You are shown a side of a card with one of the red hues. It is equally likely to be the hot-pink side, the red side, or the ruby-red side.

34. (S) Matching at least one number is the opposite of not matching any numbers, an event whose probability is easier to determine. The chance that the first number isn't a match with 2 or 9 is 8/10, whereas the chance that the second number also isn't a match is 7/9 because, of the nine remaining balls, only seven are different from 2 and 9. The probability of neither matching is $(8/10) \times (7/9) = 56/90 = 28/45$. So, the probability that you have at least one match is $1 - 28/45 = 17/45$. There are 10 possible balls selected first and nine selected second for a total of 10×9 total ways of selecting two balls in order. Of those 90 possibilities only 2-9 or 9-2 will match both numbers. So, your probability of matching both numbers is $2/90 = 1/45$.

36. (ExH) One Life Lesson to apply here is: *Analyze simple things deeply.* For some pairings of two of the cool dice, it's fairly easy to see that one die has a greater than 0.5 probability of beating the other die. For example, look at the 444400 die versus the all-3's die. What are the chances you roll a 4 with the first die? Now look at the all-3's die versus the 662222 die. What are the chances you roll a 2 with the second die?

Some other pairings are a little trickier, but you can always display in a 6×6 grid the possible outcomes when two dice are rolled. For example, when rolling the 662222 die against the 555111 die, the possible outcomes are listed below. Which die has more than a 0.5 probability of beating the other?

	6	6	2	2	2	2
5	5 vs 6	5 vs 6	5 vs 2	5 vs 2	5 vs 2	5 vs 2
5	5 vs 6	5 vs 6	5 vs 2	5 vs 2	5 vs 2	5 vs 2
5	5 vs 6	5 vs 6	5 vs 2	5 vs 2	5 vs 2	5 vs 2
1	1 vs 6	1 vs 6	1 vs 2	1 vs 2	1 vs 2	1 vs 2
1	1 vs 6	1 vs 6	1 vs 2	1 vs 2	1 vs 2	1 vs 2
1	1 vs 6	1 vs 6	1 vs 2	1 vs 2	1 vs 2	1 vs 2

40. (H) Think of the coins as different—penny, nickel, dime. Saying that two will be the same is different from saying the penny and dime are the same. Always think about the eight equally likely outcomes.

49. (H) Let x equal the number of 5's Ed rolled and y equal the number of 3's Ed rolled. Create an equation showing the total value of Ed's roll of the dice. Now do the same for Mike. (You can use the same x and y. Why?)

8.3: Random Thoughts

12. (S) Suppose only 39 people died during every two-week period. This accounts for only $39 \times 27 = 1053$ deaths. So, 39 deaths per two weeks are not enough.

14. (ExH) One Life Lesson to apply here is: *Expect the unexpected.* Many people think that "random" means the same as "thoroughly mixed up." Thus they mistakenly presume that runs of repeated numbers are rare when rolling a die. But remember that the die has no memory of the previous rolls, and so each time it is rolled, each outcome is as likely as every other outcome. (You might also look for a similar example in the text involving coin tosses.)

18. (H) After the first week, 300 people see a perfect record; second week, 150; third week, 75. Continue the pattern.

21. (S) With a 1/4 chance of correctly answering each question, the chance of a perfect score is $(1/4)^{100}$, or approximately 10^{-60}. With educated guessing, the chances go up to $(1/2)^{100}$, or roughly 10^{-30}, still unfortunately small.

23. (H) Forever is a *long* time. The probability of rolling any seven-digit number right off the bat is $(1/10)^7$. How many chances do you have?

24. (ExH) One Life Lesson to apply here is: *Apply reason to understand the mysterious or the unknown.* With a fair die, the probability of rolling one 4 is $1/6 = 0.1666\ldots$. The probability of rolling two 4's is $\frac{1}{6} \times \frac{1}{6} = \frac{1}{36} = 0.02777\ldots$.

Compute the probability of rolling ten 4's in a row. Each time you want to roll another 4, your probability gets multiplied by 1/6. What happens to the result as you multiply by 1/6 over and over?

26. (H) The chance of typing "c" first is 1/26. The chance of then typing an "a" is also 1/26. What is the chance of doing both those things?

32. (S) The probability of producing any particular sequence is $1/2^8 = 1/256 = 0.004\ldots$.

36. (ExH) One Life Lesson to apply here is: *To better understand a possibility, consider the opposite possibility.* In fact, this Life Lesson calls our attention to one big difference between the results of repeatedly tossing a coin and repeatedly rolling a die: with a coin, the opposite of heads is tails, but with a die, "the opposite of 6" is not a single outcome. With a sequence of coin tosses, it's usually easy to see if the heads and tails are balanced. How easy is it to see the analogous balance with a sequence of die rolls? If you roll a die only 20 times, are you very confident that you will see each number appear about the same number of times?

37. (H) List all the pairs of days on which Celebrity 1 and Celebrity 2 could die—for example, Mon-Mon, Wed-Mon, Mon-Wed, ...

47. (H) If you were selecting only one student out of 4000, say, for Homecoming King, then there's a 100/4000 = 1/40 = 0.025 chance you would pick someone who had read *The Heart of Mathematics*. Now how many choices do you have for Queen? How many students are left? How many of them have read *The Heart of Mathematics*?

8.4: Down for the Count

12. (H) How many different ways can you choose six numbers from 36? There are 36 ways to pick the first number, 35 ways to pick the second number, and so on, so there are $36 \times 35 \times 34 \times 33 \times 32 \times 31$ ways to pick six numbers in an *ordered* way. But we have each combination several times; for example, (123456), (263415), (354162), and so forth are all the same combination. How many different times have you counted each set of six numbers?

13. (H) This is identical to the previous Mindscape, but we'll answer it using a different method. Imagine that we are watching the lottery balls come out one by one. The chance that the first lottery number matches one of your six numbers is 6/40. The chance that the second lottery number matches one of the remaining five numbers is 5/39 (since there are only 39 balls left in the bin). What is the probability that all six events happen?

15. (S) There are eight questions in all: Do I want cheese? Do I want lettuce? and so on, and each answer has two possibilities (yes or no). There are $2 \times 2 \times 2 \times 2 \times 2 \times 2 \times 2 \times 2$ different ways to answer all the questions corresponding to $2^8 = 256$ different possible hamburgers.

18. (ExH) One Life Lesson to apply here is: *Look for patterns and find methods.* The first question asks you to count the number of ways to choose an *unordered* collection of 3 movies from the 8 possibilities. The second question asks for the number of *ordered* collections. Which question is easier to answer? Look in the text for some examples to guide your work.

26. (S) The question is really, How many different ways can you choose three things out of five? First imagine that order counts. There are five ways to pick the first item, four ways to pick the second, and three ways to pick the third. There are $5 \times 4 \times 3$ different ways to pick three items (where [eggs, bagel, coffee] represents a different choice than [bagel, coffee, eggs]). We overcounted each combo by exactly the number of orderings of three things. There are $3 \times 2 \times 1$ orderings of a three-item menu, and so the total number of distinct combinations is $(5 \times 4 \times 3)/(3 \times 2 \times 1) = 10$.

28. (ExH) One Life Lesson to apply here is: *Look at simple cases.* Before tackling the given question, consider simpler versions: Approximately how many of the Earth's 6.6 billion people would fit one characteristic? (What's half of 6.6 billion?) What about two characteristics? (What's one-quarter of 6.6 billion?) What about three?

29. (H) The number fits the form $(4xxx)$ $(x4xx)$ $(xx4x)$ or $(xxx4)$, where each x represents one of nine possible numbers (not a 4). How many total possibilities like that are there?

37. (H) Either you roll an 11 at least once, or you never roll it, and the odds of the second case are easier to compute. Of the 36 equally likely ordered outcomes for a pair of dice, only two total 11.

39. (ExH) One Life Lesson to apply here is: *Look at simple cases.* What if you organized only one party and invited 5 of your 10 male friends and 5 of your 10 female friends? The next weekend you invited the same women but then invited the other 5 of your male friends. How many more parties do you need?

48. (H) Notice, for example, that

$$\frac{7 \times 6 \times 5}{3 \times 2 \times 1} = \frac{7 \times 6 \times 5 \times (4 \times 3 \times 2 \times 1)}{(3 \times 2 \times 1) \times (4 \times 3 \times 2 \times 1)} = \frac{7!}{3!4!}.$$

8.5: Drizzling, Defending, and Doctoring

2. (S) Half the time you expect a score of 80% and half the time a 30%. Your average expected score will thus be the average of 80% and 30%, which is (80% + 30%)/2 = 55%.

7. (ExH) One Life Lesson to apply here is: *Apply methods systematically to more complex cases.* Here you can use the same reasoning used in the example in the text. Because you are told that the defense always defends against the pass, which column of values from the table will you use? If the offense passes with probability 0.6, what's the probability they will run? So six-tenths of the time they will pass and the rest of the time they will run. On average (if you imagine many plays), how many yards will they make?

11. (S) Your expression from Mindscape 8 is an equation for the line above.

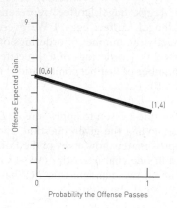

13. (ExH) One Life Lesson to apply here is: *Apply general principles to specific circumstances.* Sketch the two lines carefully to get the best estimate for the point of intersection. Does the horizontal coordinate look to be about $\frac{1}{2}$? This gives gives a value of p, the probability that the offense passes. About how many yards on average does the offense expect to make using $p = \frac{1}{2}$? If they chose to pass with a different probability, does your graph suggest it is possible for the defense to reduce the yards gained by the offense?

15. (H) To find two points on the line giving the expected gain for Player 1 if Player 2 defends against Strategy A, you'll use the values in the first column of the table. One point goes on the vertical axis, with first coordinate 0. What strategy is Player 1 using then? The other point has first coordinate 1. What strategy is Player 1 using then?

24. (ExH) One Life Lesson to apply here is: *Apply methods systematically.* The points in the figure below should help you sketch your lines. Where did these points come from? What Colin and Tubbes scenario is represented by each point?

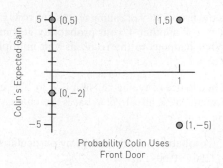

25. (S) The two lines from Mindscape 24 have equations $y = 7p - 2$ and $y = -10p + 5$. The Nash equilibrium point is the point where these two lines intersect, which is found by solving the two equations simultaneously. Subtract the second equation from the first to obtain $0 = 17p - 7$, which yields $p = 7/17$. Substituting back into either of the two original equations yields $y = 15/17$. Thus the Nash equilibrium point is (7/17, 15/17).

27. (H) From Mindscape 26, of the 1000 people with your symptoms you expect 400 (40% of 1000) to have disease A. Of these, 20% have sploosh. What's 20% of 400?

42. (H) Let p be the probability that you have poison oak. Now write the probability that you have each of the other ailments in terms of p. Then the sum of these four probabilities should equal 1.

CHAPTER 9: MEANING FROM DATA

9.1: Stumbling Through a Minefield of Data

2. (S) Answer: 16. Only the nonhuggers will contribute to the no answers in the survey. If there are X nonhuggers, then we'll see roughly $0.5X$ no answers. Since 83 answered yes, there are 67 no answers, and solving $0.5X = 67$ gives $X = 2 \times 67 = 134$. This leaves an estimated 16 people that still hug both parents.

6. (ExH) One life lesson to apply here is: *Even the uncertain can be quantified.* In the one-coin method, each respondent flips a fair coin, then answers "yes" if the coin shows heads, and answers truthfully if the coin shows tails. Out of the 100 students questioned, how many would you expect to answer "yes" just because they flipped heads? So of the 69 students who answered "yes," how many did so because they flipped tails and have porn in their rooms? Of those students with porn in their rooms, how many would flip tails? How many would flip heads? So how many are there all together?

8. (H) Would you expect the same fraction of vegetarians to attend a football game as meat eaters?

9. (ExH) One Life Lesson to apply here is: *Choose a convenient representation of a situation.* In the two-coin method, each respondent flips two fair coins, then answers truthfully if at least one tail is showing, and lies if both coins show heads. Following the example in the text, suppose the number of drug users who flipped HH is U. Then the number of users who flipped HT, TH, or TT is also U. But those who flipped HH will lie, so only the remaining $3U$ users will answer "yes." If we let N denote the number of nonusers who flip each of HH, HT, TH, or TT, how many will answer "yes"? Does this give you an expression equal to 78 (the number of "yes" answers)? The remaining students answer "no." Does this give you an expression equal to $250 - 78 = 172$? Can you use these to solve for U? How does this give you the total estimate for the number of drug users?

11. (S) If 60 students cheated, we would expect that one-fourth, or 15 students, would flip HH, HT, TH, and TT, respectively. That leaves 140 students, of the 200 surveyed, who didn't cheat. We would expect one-fourth, or 35 of them, to flip HH, HT, TH, and TT, respectively. We know that the 35 students who didn't cheat and flipped HH will answer "yes." The only other students who will answer "yes" are the students who did cheat and flipped HT, TH, and TT. This is 3×15 or 45 students. Therefore we would expect $35 + 45$, or 80 students to answer "yes" and the other 120 students to answer "no" ($15 + 35 \times 3$).

17. (H) Where would you expect to find more studious students, at the library at 9:00 a.m. on Saturday or at a wild party?

19. (H) Who would respond the most to the mayor's question and strongly urge the mayor to stop pornography? Would the people who did not feel strongly about the mayor's issue respond as often to the mayor's question? Incidentally, the circumstances of the question actually happened.

21. (H) Only a small fraction of people own BMWs. So when you survey only 100 people, what might happen?

31. (H) Focus on solving the two equations for N and U. Then note that the Mindscape states that the number of students who've had unprotected sex is $4U$.

9.2: Getting Your Data to Shape Up

2. (S) The median of the data on the cost of Internet access is 25, and the mean is 26.14, or 26. Both of these numbers are close together; however, more than half of the people paid $26 or less, approximately 77%; the mean is higher than the median because the high data points pulled the mean higher. Therefore, the median would give a better answer to the question "How much did a typical person with Internet access pay for that access in 2001?"

9. (S) The mean score is approximately 70%. The standard deviation is more likely to be closer to 12, because about half of the scores are within 10 of 70 and the other half are further than 10 but less than 30 away. Therefore, it seems more likely that the standard deviation would be closer to 12 than 25.

11. (ExH) One Life Lesson to apply here is: *Understand basic definitions and search for logical explanations.* There are 435 members of the House of Representatives, so how many data points (values) are in the data set? The median data value is the middle value when all the data are lined up in numerical order. With a median value of 3 in these data, how many values are equal to 3 or less? With all these very small values, what else must be true about the data to result in a mean value as large as 47?

18. (H) Answers will vary. Some examples of possibly correlated variables include weight and height or hours surfing the Internet and number of emails per week.

24. (ExH) One Life Lesson to apply here is: *Look at specific examples to reveal a pattern.* Compute the mean quiz grade. Add to your list one new quiz grade that is lower than that mean and compute a new mean. Now go back to the original list and add to it one new quiz grade that is higher than the original mean and compute a new mean. What do these two experiments demonstrate? Can you answer the question? Do a few more experiments if you need to. Can you generalize?

25. (S) The following is an example of a data set where the mean is greater than the median.

50,000	75,000	95,000	125,000
135,000	150,000	175,000	200,000
1,500,000			

This data set has a mean of 278,333 and a median of 135,000. In this data set, most of the data are somewhat close together and there is one outlier that is much higher than the other values.

31. (H) Write down the formula for the mean, namely, the sum of the 100 salaries divided by 100. What would happen to the sum if one of the salary figures increased by $13,000, which is the difference between $29,000 and $42,000? What happens when the new sum is divided by 100?

34. (ExH) One Life Lesson to apply here is: *Find useful representations of information.* It's reasonable to display the numerical data given here using a scatterplot (see Mindscape IV.18) with, say, the home valuation on the horizontal axis and the tax on the vertical axis. But then each dot in the graph shows only two of the three "values" given for each home. Is there a way to distinguish dots for homes in Metropolis from those in Suburbia?

44. (H) The median value of a set of data is the "middle" value. In this case, you have an even number of data values, so the median will be the average of the two "middle" values.

9.3: Looking at Super Models

2. (S) Remember that standard deviation is a measure of how far the data deviates from the mean.

8. (S) The doctored die has six faces with values 1, 2, 4, 5, 5, 6. The mean roll is therefore $(1 + 2 + 4 + 5 + 5 + 6)/6 = 23/6 = 3.8333\ldots$ For an ordinary die, the mean roll is $(1 + 2 + 3 + 4 + 5 + 6)/6 = 21/6 = 3.5$. So the mean roll for the doctored die is larger than that for the ordinary die, which is what we would expect.

10. (S) For data with a normal distribution, about 68% of the values differ from the mean by less than one standard deviation. The normally distributed head measurements have mean 55 cm and standard deviation 1.7 cm, so heads within one standard deviation of the mean will measure between $55 - 1.7 = 53.3$ and $55 + 1.7 = 56.7$ cm. Thus approximately 68% of the freshmen have head circumference between 53.3 and 56.7 cm. A head measuring more than 58.4 cm is more than 3.4 cm, or two standard deviations, above the mean. For the second question, recall that approximately 95% of the values in a normal distribution are within two standard deviations, so only 5% lie above or below those limits. Thus in this case, roughly 2.5% will be more than 58.4 cm.

11. (ExH) One Life Lesson to apply here is: *Look at examples to infer general principles.* Recall that with a normal distribution, 95% of the data lie within two standard deviations of the mean. If you guess a value for the mean, say 30 years, which value makes more sense for the standard deviation? A standard deviation of 1 year would imply that 95% of all Americans retire after working between 28 and 32 years (2 years more or less than the mean value of 30 years). Does this seem reasonable? A standard deviation of 15 would imply that 95% retire after working between 0 and 60 years. What does this imply about the remaining 5%? Do these comparisons change very much if you try a different mean value?

14. (H) If there were no global warming or cooling, how does the number of annual temperatures that are above the mean of the distribution compare with those that are below? What about in the case of global climate change?

22. (ExH) One Life Lesson to apply here is: *To understand a situation, first understand the basics.* To pay only $1 per gallon, the average showing on your two dice must be 1. To find the probability that this happens, you must find the number of outcomes that give this result and divide by the total number of outcomes. How many outcomes of rolling two dice give an average of 1? How many total outcomes are there? (Think about the fact that there are 6 possibilities for each die.) Can you use a similar analysis to find the probability that your two dice show an average of 6? To answer the last question, think about how to compute the probability that something does *not* happen.

30. (H) This Mindscape will require the use of one of the three key percentages associated with a normal data set: 68%, 95%, or 99.7%. But five pizzas equal only 2.5% of the total order of 200 pizzas. Hummm. Remember that these five pizzas all have *more* cheese than the mean.

9.4: Go Figure

5. (S) Given $\frac{1}{\sqrt{n}} = 0.05$, we multiply both sides by $\sqrt{n}$ to obtain $1 = 0.05\sqrt{n}$, Then dividing both sides by 0.05 yields $20 = \sqrt{n}$, so we have $n = (20)^2 = 400$. Similarly, if $\frac{1}{\sqrt{n}} = 0.02$, then we find that $n = 2500$.

7. (S) Even though 2% of 500 does equal 10 (because $0.02 \times 500 = 10$), 10 is too small a sample size to represent the data.

10. (ExH) One Life Lesson to apply here is *Focus on the essential.* A random sample of 265 in Wyoming represents about 1 out of every 1000 voters. A random sample of 1400 in California represents about 1 out of every 10,000 voters. Does this alone mean that the Wyoming sample is more reliable? Of what importance is the actual size of the sample? Does the sample size tell you something about your margin of error? If this question makes you feel a bit ill, have some chicken soup (see the text).

11. (H) What matters more here: the proportion of the data sampled or the actual sample size?

18. (ExH) One Life Lesson to apply here is: *Seeing just part of a picture might allow you to see the entire thing.* If 50% of the caught fish were bluegills, then 50% of the 10,000 estimated fish in the lake, or about 5,000, would be bluegills. The graph shows 700 of the 1500 fish caught were bluegills. How do you turn this into a percentage? Regarding the assumptions we're making about the fish, could you answer the Mindscape's questions so readily if bluegills were easier to catch than other fish? What if most bluegills tend to populate the side of the lake opposite from where you caught the fish in your sample? What other assumptions could affect your conclusions?

22. (S) For a 95% confidence interval, a sample size of n will have a margin of error of $\frac{1}{\sqrt{n}}$. So you need to find the value n that makes $\frac{1}{\sqrt{n}} = 0.02$. Multiplying both sides by $\sqrt{n}$ and dividing both sides by 0.02 yields $50 = \sqrt{n}$. Thus you need to test $n = (50)^2 = 2500$ light bulbs.

24. (H) If n is your sample size, how does $\frac{1}{\sqrt{n}}$ relate to the margin of error?

26. (ExH) One Life Lesson to apply here is: *Take the step from a qualitative impression of a situation to a quantitative understanding of it.* Fill in the following table and calculate for both groups the rate of AIDS as a decimal number. Do you think the results could happen by coincidence or chance?

	# with AIDS	Total #	Infection rate $= \dfrac{\text{# with AIDS}}{\text{Total}}$
Without circumcision	47	1391	_____
With circumcision	22	1393	_____

33. (H) The one million people in Montana represent 1/6 of the six million total population in the area. So of the 1200 people polled, 1/6, or 200, should be from Montana.

9.5: War, Sports, and Tigers

6. (S)

	Dad gives white gene	Dad gives red gene
Mom gives red gene	(red, white) → pink	(red, red) → red
Mom gives white gene	(white, white) → white	(white, red) → pink

The third-generation flower is red with probability 1/4, white with probability 1/4, and pink with probability 1/2. You should see twice as many pink flowers as either red or white flowers.

9. (ExH) One Life Lesson to apply here is: *Consider all possibilities.* In order to produce a white flower, a plant must have WW genes. Thus each parent plant must have at least one W gene. In the first generation of the given scenario, one plant has RR genes and the other has WW genes. What are the possible gene pairs in the second generation? What about in the third generation? Could you find one second-generation parent and one third-generation parent with the desired genes?

16. (H) What does a correlation of −1 mean? What about −0.9? What about −0.7? As the years go by, what does a negative correlation tell you about the winning times?

18. (S) We'll assume all cars sold at the dealership stay in the area. 250/800 = 0.3125, so 31% of all the cars in the area are Hondas. If C represents the total number of cars, then $0.31 \times C = 10{,}000$. Dividing by 0.31 to solve for C yields $C = 10{,}000/0.31 = 1$ roughly 32,300 cars.

23. (H) To help visualize the situation, you could think of starting with 64 plants: 32 red, 32 pink. You could pair them in the correct proportions: eight red-red pairs (accounting for 16 red plants), 16 red-pink pairs, and eight pink-pink pairs. What colors result in what proportion?

28. (ExH) One Life Lesson to apply here is: *To explain an apparent contradiction between two claims, look for differences in how they were formulated.* Imagine plotting the times for men and women on the same graph, with the year along the horizontal axis and the time on the vertical axis. For the Olympic years from 1948 through 1968, only the men's times appear. Then from 1972 on, both the men's and women's times appear. Will the women's times be faster or slower than the men's? If you try to fit a line to the data points, will it slope downward (negative correlation) or upward (positive correlation)?

39. (H) Having a correlation of +1 implies there is perfect positive correlation between the years and the concert attendance. Thus, if you plotted the data on a pair of axes, year vs. attendance, the points would lie on a straight line with positive slope.

CHAPTER 10: DECIDING WISELY

10.1: Great Expectations

6. (S) The two green slots give the house an edge. The chance of winning on red is 18/38, so the expected value is $2000(18/38) = \$947$.

13. (ExH) One Life Lesson to apply here is: *Even the uncertain can be quantified.* You need to know the probability for each outcome. If you think of rolling two dice, one red and one green, then there are actually thirty-six possible outcomes, but many show the same sum. For example, the sum 7 can occur in six different ways: red 1 and green 6, red 2 and green 5, red 3 and green 4, red 4 and green 3, red 5 and green 2, red 6 and green 1. Thus the probability of rolling a 7 is 6/36 = 1/6. On the other hand, the probability of rolling a 2 is just 1/36 because a sum of 2 can occur in only one way: red 1 and green 1. (See Section 8.2.) A similar analysis for the remaining outcomes in the collection {2, 3, 4, 5, 6, 7, 8, 9, 10, 11} can be used to complete the table of probabilities below:

OUTCOME	2	3 4 5 6 7	8 9 10 11 12
PROBABILITY	1/36	1/6	

Once you know all the probabilities, computing the expected value is straightforward.

15. (S) Suppose you get D dollars if you flip tails. The value of the game is then $(-5) \times (2/3) + (D - 5) \times (1/3) = D/3 - 5$. Since the value of a fair game is zero, set $D/3 - 5 = 0$, and solve for D. So, $D = \$15$.

16. (H) The best way to approach insurance questions is from the point of view of the insurance company. What is their expected value from this deal? If the policy costs P, then 80% of the time they will make P and 20% of the time they will make $P - \$700$. (Here, making negative money means losing money!)

20. (H) The expected value (a.k.a. average return) is the sum of the individual probabilities of each roll multiplied by the corresponding payback.

23. (ExH) One Life Lesson to apply here is: *When possible, measure rather than guess.* Suppose you bet P dollars to play this game. To compute your expected value, you need the probability for each possible outcome. (Look at the Extended Hint for Mindscape 13 for guidance on computing the probabilities you need.) It may help to organize your data in a table. The payoff for each outcome is the amount you receive back minus your bet of P.

Once you have filled in the table, compute the expected value. What is a fair bet for this game?

27. (H) When tossing three coins, there are 8 possible outcomes: Each coin can show heads or tails, for a total of $2 \times 2 \times 2 = 8$ outcomes. How many of these show exactly one head? (If the coins were red, blue, and green, then one way to show a single head would be: red head, blue tails, and green tails. For more on these probabilities, see Section 8.2.) Regarding the last question posed, recall that the expected value represents the average outcome if you were to repeat the process indefinitely.

OUTCOME	2	3	4	5	6 7 8 9	10	11	12
Probability	1/36				1/6			
Payoff	$-\$P$	$-\$P$	$\$2-\$P$	$\$2.50-\$P$	$\$0$	$\$5-\$P$	$\$0$	$-\$P$

28. (S) Look at this game through the eyes of the insurance company. The value of the car insurance to the company is $0.02 \times (\$200 - \$9000) + 0.98 \times (\$200) = \$20$. The expected value of the bike insurance is $0.1 \times (\$75 - \$850) + 0.9 \times (\$75) = -\$10$. The company makes enough money on the car insurance to cover its losses on the bike insurance.

29. (H) Consider an equivalent question: Suppose you knew 100 people who were in a similar situation. After all the offers were accepted, what would their average salary be?

39. (H) View these questions from the point of view of an insurance company that has to shell out $1.2 million for the death of each customer. Is it worth it for the company to pay for all the women to have a Pap smear? How much does it cost the company to pay for 3000 Pap smears compared to paying a death claim? What is the expected value to the insurance company of paying for a Pap smear?

47. (H) If a computer is stolen, the insurance company owes the policy-holder V dollars. But the policy-holder already paid the company the cost of the insurance, which is C dollars. So the company's payout is actually $C - V$ dollars. According to the Mindscape, this payout occurs with probability p. What's the probability that a theft does not occur? How much does the company earn (or pay out) in this case?

10.2: Risk

3. (H) Two out of each 100 people will be infected.

4. (S) In order to determine the total cost per life saved for the vaccine, you need to multiply the cost per injection by the number of shots you would have to give to save just one life. If the injection would save 1 in 15,000,000 lives, the cost per life would be $150,000,000. In this case it seems unlikely that people would support this action because of the huge cost per life. However, if the injections would save 1 in 1000 lives and the cost per life were $10,000, it would be much more reasonable to support a program to vaccinate each citizen.

6. (H) Suppose there are B blondes in the world (fake or real). How many will you suspect of being fake? You'll mistake 20% of the 90% that are real blondes, and you'll correctly identify 80% of the 10% that are fake. Of those, how many are actually fakes?

7. (S) Answer: 29%. Use the reasoning in the previous Mindscape. There are B blondes. You'll correctly identify 85% of the bleached blondes $(0.30B)$ and mistake 15% of the real blondes $(0.70B)$. If you judged every blonde, you would be suspicious of $[0.85 \times (0.30B) + 0.15 \times (0.70B) = 0.36B]$ 36% of all the blondes. Since you correctly suspected only $0.255B$ people, the probability of your suspicion being correct is $0.255B/0.36B = 0.708\ldots$, or 71%. So suspecting that Chris's hair color is fake, the probability that Chris is a true blonde is 29%. Keep waiting.

10. (ExH) One Life Lesson to apply here is: *Make it quantitative.* Suppose all the applicants call the absent-minded professor. The probability you are an actual winner equals the fraction of actual winners among all those students who are *told* they are winners. The professor recalls information correctly 90% of the time, so how many of the 200 actual winners are told they are winners? And how many of the 99,800 actual losers are told they are winners?

11. (ExH) One Life Lesson to apply here is: *When possible, measure rather than guess.* With 1% of travelers who normally fly choosing to drive instead, their flying miles would become driving miles. This means an increase of 1% of 1.7 billion, or 17 million highway miles. How many additional highway deaths would result?

13. (H) In order to determine the cost per life saved, multiply the cost per device by the number of vehicles that would need to have the device installed in order to save one life.

15. (H) In order to determine how much the government is willing to spend on increasing the safety of each cell phone, you can set up the following equation:

$12,500 = C \times 100,000,$

where C is the cost the government is willing to spend on increasing the safety of each cell phone.

16. (H) If the witness is 100% sure, how relevant are the other percentages?

17. (S) Answer: 41%. Let's interpret the statement, "The witness is 80% sure . . ." as, "With 80% accuracy, the witness sees a blue cab as blue and a green cab as green." Suppose that the witness were to look at all the cabs in the city—say, a total of C cabs. Of the $0.85C$ green cabs, she'll identify 20% of them as blue, and of the $0.15C$ blue cabs, she'll identify 80% correctly. She'll claim that $0.20 \times 0.85C + 0.80 \times 0.15C$ cabs are blue, when in fact only $0.80 \times 0.15C$ of those cabs were blue. Her chances of correctly identifying a blue cab in this city are $(0.80 \times 0.15C)/(0.20 \times 0.85C + 0.80 \times 0.15C) = 12/29 = 0.41 \ldots = 41\%$.

26. (H) Each time you do laundry, there is an 80% chance (0.8) your whites stay white. If you do laundry a second time, that same chance applies, so the chances that both white loads stay white is $(0.8)(0.8)$, which equals 0.64.

10.3: Money Matters

1. (H) The total amount you will have will be the amount you start with plus the amount of interest you will have earned.

3. (H) To see why you would much rather have 2^{100} than 100^2, just multiply out 100^2 and then start multiplying out 2^{100} until you find that you would have more money than there is on Earth.

6. (S) If you invest your money at Happy Bank you will receive $1000 \times (1 + 0.03)$, or $1030, after a year. If you invest at Glee Bank you will receive $1000 \times (1 + 0.025/4)^4$, or $1025.24, after one year. So, although Glee Bank compounds interest quarterly while Happy Bank compounds interest only once over the year, you will end up with more money from Happy Bank after one year. Therefore, Happy Bank should earn your business.

9. (ExH) One Life Lesson to apply here is: *Apply ideas widely.* Adrian is borrowing $10,000 at 6% annual interest for 5 years. Assuming that the interest is compounded monthly, you can use the formula in the text developed for finding a monthly mortgage payment P. Here there will be $5 \times 12 = 60$ equal payments. The interest rate each month is 4% divided by 12, which equals $0.04/12 = 0.0033$.

12. (S) After 50 years Bert will have $(1.00) \times (1 + 0.05/365)^{(365 \times 50)}$, or $12.18, whereas Ernie will have $(1.00) \times (1 + 0.06/365)^{(365 \times 50)}$, or $20.08. If their initial deposits were $10,000, the only thing that would change in each formula is the (1.00) to $(10,000)$. All this does is multiply the final amount you found earlier by 10,000, leaving Bert with $121,180 and Ernie with $200,080. Ernie always was the clever one.

13. (H) You're investing $2000 at 5% compounded monthly for 3 years. Is there a formula in the text for your *FINAL AMOUNT* after 3 years?

17. (S) To determine the future value of the $3000, you need to calculate the following: $3000 \times (1 + 0.045/4)^{(6 \times 4)}$, which will give you $3923.97.

18. (ExH) One Life Lesson to apply here is: *Understand simple things deeply.* You want to invest some money at 5.7% interest compounded monthly for 4 years. You want your *FINAL AMOUNT* to be $20,000. Is there a formula in the text to help you determine the *INITIAL AMOUNT* you need to invest?

19. (H) You want to know what value of t would make the following equation true.

$3000 = 1500 \times (1 + 0.03/365)^{(365 \times t)}$

Why would it take the same amount of time to double your money if you started with $5000, or any other amount?

21. (S) When you determine the amount of money that is made when interest is compounded, notice that the first time you earn interest the initial investment is multiplied by 1 plus the interest per interval. When this amount is used and the interest is compounded again, you multiply by 1 plus the interest per interval again. Do several steps of this process to see that raising to powers is involved.

25. (H) You put $500 in the account on August 31 at the end of the first summer (you're 16). The following summer, on August 30, the bank pays 5% interest on your money, so you now have $500(1.05) = 525 in the account. How much do you have the next day when you make you annual deposit?

10.4: Peril at the Polls

5. (S) Arrow's Impossibility Theorem asserts that every nondictatorial voting scheme must fail to satisfy one of the following reasonable principles: go along with consensus, ignore the irrelevant, or better is better.

7. (S) This situation is not an example of Condorcet's Paradox, because 6 of the 9 prefer A over B, 6 out of 9 prefer B over C, and 6 out of 9 prefer A over C. Therefore, this is an example of transitive ranking.

8. (S) Given the voting data from Mindscape 7, using the Borda count method A would receive 15 votes (3 from the first 3 voters, 6 from the second group of voters, and 6 from the third group of voters). B would receive 18 votes (6 from the first 3 voters, 9 from the second group of voters, and 3 from the last group of voters). C would receive

21 votes (9 from the first 3 voters, 3 from the second group of voters, and 9 from the third group of voters). This would leave you with Austin Arctic as the winner because A had the smallest total number of votes.

9. (H) Given that the voters in the first column approve of A and B, they are likely to cast their votes for both in the approval voting method. The voters in the other columns will only vote for one.

10–15. (H) The chart allows you to compute how many votes each candidate receives using various voting strategies. For example, in plurality voting, take each band and see how many people rank it as number 1.

11. (H) Does plurality voting work? What about the Borda count method? (In each case, justify your answer.)

17. (ExH) One Life Lesson to apply here is: *Choose a convenient representation of an issue.* Suppose candidates A and B are running. Let a be the number of voters who choose B as their first choice. If $a > b$, then candidate A wins by plurality voting. In this case, who wins by Borda count? Note that the voters who ranked A first must have ranked B second and vice versa, so we have the table below.

Number of votes	a	b
First choice	A	B
Second choice	B	A

Compute the Borda count for each candidate and compare. (Remember that you're considering the case $a > b$.)

18. (ExH) One Life Lesson to apply here is: *Look at specific examples to guide general principles.* Try some examples using candidates A and B. Suppose you have only 5 voters. What are some of the ways their votes could be divided between A and B? Choose a specific scenario and compute the Borda counts for each candidate. Now consider some approval voting scenarios. Keep in mind that some voters might approve of only one candidate, while other voters might approve of both. Try different voting outcomes and see if any help you answer the question posed in the Mindscape.

10.5: Cutting Cake for Greedy People

2. (S) This person is missing the idea that some people prefer different parts of the cake based on features other than size. Some are icing fanatics and are willing to take a smaller piece in order to maximize the amount of icing they get. Some people prefer the writing; some prefer the candy rose. Therefore, dividing the cake based on size alone would not necessarily satisfy everyone.

5. (H) Consider the following:

| Big Bedroom—FREE | B.B.—$500 | B.B.—$1000 |
| Small Bedroom—$1000 | S.B.—$500 | S.B.—FREE |

At the far left of the line your preference is for the Big Bedroom because it is free. At the far right of the line you prefer the Small Bedroom. Is there a point on the line as you move from left to right at which your preference changes from the Big Bedroom to the Small Bedroom? At this point you would equally prefer either the Big Bedroom for the given price or the Small Bedroom for $1000 minus that price. Now do the same analysis for your roommate and superimpose the diagrams.

9. (H) How many pieces are produced when you cut from a point on the edge to the three vertices of the triangle?

12. (H) Look at the branch point in the middle of the preference diagram. What does this point signify?

14. (S)

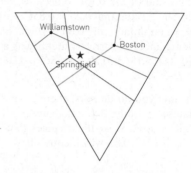

One possible set of preference diagrams for you, Joan, and John are drawn above. Your preference diagram is shown in red, Joan's is shown in blue, and John's is shown in green. All three preference diagrams have their branch points on the preferred city for each person. If you cut at the star, John will pick the East region (since the star lies in his East preference region). Joan will pick the West region, and you will pick North. Therefore, all three will have a region that contains their city of preference.

15. (H) Remember you are looking for a single point where, if the cake is cut from there to the three vertices, then one person prefers the North piece, one prefers the West piece, and one prefers the East piece.

16. (ExH) One Life Lesson to apply here is: *Abstract the essence.* Only four tracts of land have value here, with a total value of $8 + 4 + 5 + 6 = 23$. Presumably, fractions of these valuable parcels have value equal to some portion of their whole value. You want to draw your preference diagram so that your preferred section (North, West, or East) has value at least one-third of 23, or 7 and 2/3. To achieve this balance, your branch point should correspond to a cut that gives three pieces of equal value. Thus the tracts valued 5, 6, and 8 must lie in different pieces when the total parcel is cut from the branch point. What about the tract valued 4? So in which tract does your branch point lie? If you move to a point slightly south of your branch point and cut the parcel, which portion do you prefer? If you move slightly east of your branch point and cut the parcel, which portion do your prefer? Continue experimenting to determine the boundaries in your preference diagram.

17. (S) In order to divide the cake into two-thirds and one-third, have each person draw a preference diagram as though the cake were being divided between three people. You are guaranteed to get one intersection point that lies on a preference border for each person. At this point the first person will prefer two pieces equally and the second will also prefer two pieces equally. However, one of these preferred pieces must be the same for both people and one will be different. Therefore, give the person who gets one-third the piece they prefer that is NOT one of the two preferred by the second person. Then give the second person the two pieces that they prefer. Therefore, the first person believes that their piece is equally valuable to one of the second person's pieces and more valuable than their other piece, so they believe they got more than one-third the worth of the cake. The second person got the two pieces that they found more valuable than the third piece, so they believe they got more than two-thirds the worth of the cake. Everybody wins.

20. (ExH) One Life Lesson to apply here is: *Ground your understanding in examples.* The preference region in the upper left is marked East, indicating that you prefer the East piece for cuts drawn from the region. Pick a point in the upper left preference region and draw the cut lines to the vertices. What do you notice about the East piece? Why might you prefer that piece even though it contains the waste dump? Can you do a similar analysis for the region in the upper right? What about the other two preference regions in the top of the triangle? Why do they alternate?

21. (H) Suppose you and someone else each secretly mark the value of a pile of goods so that each of your totals is $100. If you find an item that you evaluated at $2 and the other person evaluated at $1, then why would you be happy to buy it at the other person's evaluation of $1?

Index